MICROBIOLOGY

MICROBIOLOGY

Cynthia Friend Norton

Department of Biology
University of Maine at Augusta

Addison-Wesley
Publishing Company

Reading, Massachusetts
Menlo Park, California
London
Amsterdam
Don Mills, Ontario
Sydney

This book is in the Addison-Wesley Series in the Life Sciences

Sponsoring Editor: James H. Funston
Production Editor: William J. Yskamp
Designer: Catherine L. Dorin
Illustrator: Oxford Illustrators, Ltd.
Cover Design: Ann Scrimgeour Rose
Cover Photograph: Dr. Hans Paerl

The cover photograph shows a string of freshwater cyanobacteria with attached aquatic bacteria. The cyanobacteria, which are producers, are manufacturing useful nutrients that the bacteria are absorbing and consuming.

Library of Congress Cataloging in Publication Data

Norton, Cynthia F   1940-
  Microbiology.

  Bibliography: p.
  Includes index.
  1. Microbiology.   I. Title.
QR41.2.N67   576   80-23350
ISBN 0-201-05304-7

This book is affectionately dedicated to five fine teachers, each of whom has provided a major inspiration to me in choosing the direction of my professional and personal growth.

John B. Friend, M.A. English
My first teacher

Edwin R. Frude, B.A.
Head of the Science Department
Arms Academy

Cecelia M. Kenyon, Ph.D. D.Lit.
Professor of Government
Smith College

Galen E. Jones, Ph.D.
Professor of Microbiology
University of New Hampshire

Jerome M. Eisenstat, Ph.D.
Professor, Department of Human Genetics
Yale University Medical School

Preface

A new microbiology test is justified by new subject matter and new student needs and interests. This text focuses on changing viewpoints of the relationship between human beings and microorganisms, and incorporates current and relevant materials to interest and challenge the student.

The presentation is organized in four parts. In Part I, "Fundamentals of Microbiology," the structure, classification, growth patterns, metabolism, and genetics of procaryotic and eucaryotic microorganisms and viruses are presented. Woven into this section, and continued throughout the book, is an emphasis on the ecological interaction of the microbial forms with each other and higher organisms. This part features a comprehensive chemistry review in Chapter 2.

Part II, "Host–Parasite Interactions," develops an understanding of the factors involved in microbial virulence and how the multiple levels of human defense react to counter this virulence. The interactions between parasitic microbes and the individual host are examined first; then the text considers the impact that these microbes have on populations. A significant chapter is devoted to the dysfunctional aspects of immunity. The point is made that immune dysfunctions are aberrations of the normal immune mechanism, occurring at the wrong place, the wrong time, too forcefully, or not forcefully enough. The emphasis again is on the ecological interactions of species.

Part III, "Infectious Diseases," presents the major infectious diseases of human beings. The anatomical approach has been chosen here, for several reasons. It allows students to relate the information to knowledge they previously acquired of anatomy and physiology. It also promotes a continuing understanding of the ecological aspects of the host–parasite relationship. The disease is understood as a disorder within a microenvironment, caused by an abnormal relationship between microorganisms and the human tissues. Each major human pathogen is given a thorough discussion that features the pathogen's cultural characteristics, pathogenic potentials, and other key information. Included in this section is a major chapter on oral microbiology that introduces students to the microbial diseases most commonly to afflict human beings, dental caries and periodontal disease.

Part IV, "Control of Microorganisms," brings together a group of topics all of which have the theme of controlling the unwanted aspects of microbial activities. These are presented in sufficient detail to give students solid working knowledge. The concluding chapter on hospital infection control is, I believe, the most complete and useful material available in an introductory textbook.

In choosing the level at which this book was to be written, I made the assumption that most students would have some basic biology and chemistry background, but that the knowledge might not be too extensive nor too recently acquired. Thus the first chapters of the book, in addition to introducing microbiological concepts, are also designed to introduced and reinforce basic biochemistry and cellular biology. In addition to the systematic chemistry review in Chapter 2, I review cell structure and function in Chapter 3. I have always attempted to write in a clear, direct style and to define carefully all new scientific terms as they are introduced. The liberal use of figures and tables provides students with several routes of access to the material.

We have attempted to design this book so that it will be useful for a variety of different types of student population. The choice of topics places strong emphasis on the health-related aspects of microbiology. This focus is especially appropriate for students in health-related programs because it tells them exactly what they want and need to know. However, it also is a valid approach for the student with general interests because the area of the infectious diseases is one that deeply interests and motivates all students. Please note also that there is much material, placed strategically in the introductory section, that stresses the beneficial features of microbiology.

The text has been designed to maximize its flexibility and usefulness. I have included a cross-referencing system that enables students and teacher to pull together all the information presented on the subject. Thus when capsules are discussed as a structural feature in Chapter 3, readers are referred to discussions of the capsules' role in the pathogenesis of pneumonia in Chapter 19. The instructor will find that this feature makes it easy to change the order in which chapters are presented. For example, the chapter on disinfection and sterilization may be used at any time during the course even though it is located in the concluding part of the book. Certain chapters may be skipped if not needed.

Each chapter contains special-interest Boxes. In the early chapters, the Boxes are selected to introduce the connections between theoretical material and the real world as the student views it. For example, a discussion of the structural features of the bacterial cell in Chapter 3 is made relevant by a Box entitled "Why Do Antibiotics Work?" Boxes help not only to increase student interest and motivation but also introduce significant and valuable material.

Several features have been included to make this a useful reference book students will want to keep. The figures and tables contain a wealth of specific information. Each has been designed to fulfill a teaching function. In addition, we have made a great effort to find new electron micrographs that will show clearly and pleasingly what the microbial world looks like. The book provides a particularly extensive and useful glossary that includes a guide to pronunciation.

This book, in its final form, is an outgrowth of my own educational experiences as student, as a researcher, and as a professor. I have served my turn as a student in several institutions, and I have taught students possessing a wide range of capabilities and interests.

Today's students differ from the microbiology student of the fairly recent past. Contemporary students are, on the average, older and more committed, with strong vocational and family interests. Because of their interests and deeper commitments, today's students are wonderful persons to introduce to microbiology. I have tried to present microbiology as the fascinating interdisciplinary science I believe it to be.

ACKNOWLEDGMENTS

I gratefully acknowledge the vital assistance of the many people who have made this book possible. These include my reviewers.

Philip Achey
Department of Microbiology and
   Cell Science
University of Florida

Frank L. Binder
Associate Professor and
   Program Director
Medical and Cytotechnology
Marshall University

R. Blakemore
Department of Microbiology
University of New Hampshire

Joan Handley
Department of Microbiology
University of Kansas

Professor Francis M. Maxin
Head of Biology Department
Allegheny Campus
Community College of
   Allegheny County

Diane Gay Michaels
San Diego Mesa College

M. Charline Mims
San Antonio College

Richard C. Tilton
University of Connecticut
   Health Center

Many teachers and researchers in microbiology have generously sent me photographs to be used. These kind people are too numerous to mention individually, but their names are to be found in the photograph acknowledgments, which appear after the glossary.

Last, I acknowledge the patience and support of my husband, John W. Norton, and of my son Jack, now aged five, who has heard all too many times, "Not now, dear. Mommy has to work on her darned old book again."

East Vassalboro, Maine
January 1981

C. F. N.

Contents

xi

10
The Study of Viruses 283

11
Animal Viruses 305

PART II   HOST–PARASITE INTERACTIONS

12
The Host–Parasite Relationship 345

13
The Immune Response 372

14
Dysfunctional Immunity 408

20

Gastrointestinal Tract Infections 575

Anatomy and physiology 575   Normal flora of the GI tract 580
Bacterial infections 586   Viral infections 597   Protozoan infections 601   Food intoxication 606

21

Urinary and Reproductive Infections 613

Anatomy and physiology of the urinary tract 613   Anatomy and physiology of the reproductive tracts 616   Normal flora 618   Urinary tract infection 621   Genital tract infection 624   Infections of the prenatal period 636   Infections of the neonatal period 641
Postpartum infections of the mother 647

22

Nervous System Infections 650

Anatomy and physiology 650   Bacterial infections 654   Fungal infections 659   Viral infections 659   Protozoan infections 668

23

Wound Infections 670

Pathology of wounds 670   Flora of wounds 671   Accidental trauma 677   Burns 681   Surgical wounds 684

24

Circulatory and Lymphoreticular Infections 687

Anatomy and physiology 687   Bacterial infections of the circulatory system 693   Viral infections of the circulatory system 703
Bacterial lymphoreticular infections 704   Viral lymphoreticular infections 706   Protozoan infections 708

PART IV CONTROL OF MICROORGANISMS

MICROBIOLOGY

I
FUNDAMENTALS
OF MICROBIOLOGY

Amphora coffeaeformis,
a Diatom

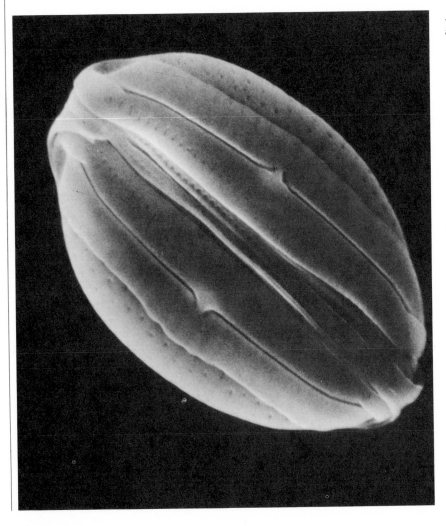

1

Introduction
to Microbiology

The world of microbiology is a world of very minute organisms (Fig. 1.1). This world is a biological universe human beings have begun to explore only recently. Microorganisms were first seen by human eyes barely 300 years ago, and serious study began perhaps 200 years ago. Yet not only have we come to understand these

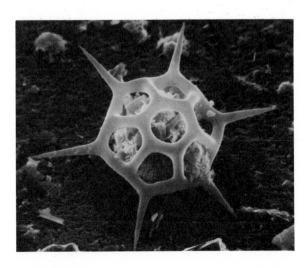

1.1
Distephanus speculum A single-celled organism may be both complex and beautiful. This is the silicate test or outer skeleton of a marine photosynthetic form. In life, the cell body of the organism was found inside this casing. Now the cytoplasm is gone and in its place is deep ocean sediment.

"new" biological kingdoms but our studies have given us tremendous new insights into the world of larger plants and animals.

Microorganisms—those living things that can be seen only with microscopes—carry out the same functions that other life forms carry out, but in more simplified ways. Because these processes are simpler, they are easier to study in depth. Their study has helped scientists to formulate a working definition of life that can be applied to all organisms, large and small.

LIFE

Characteristics of Living Organisms

We can observe five major characteristics when we study any life form. We may use these characteristics to describe life, to decide whether some object is indeed alive, or to increase our appreciation of the life process.

Cell walls, cell membrane; Chapter 3, pp. 77–80

Characteristic 1   Each life form is surrounded by boundary layers that protect, that communicate with the environment, and that may confer shape and rigidity. Boundary layers make it easy to tell where the organism ends and the environment begins. When the boundary layer is destroyed, the organism is weakened or dies.

All cells maintain a stable internal environment that may be quite different from the external environment. Multicellular organisms have specialized surface tissues called epithelium.

Characteristic 2   A living organism is capable of capturing, utilizing, and storing various forms of energy. In the life process, chemical entities are constantly joined together and taken apart in intricate ways. The term **process** implies change taking place. It is a general rule of the physical world that whenever change occurs, **energy** will be involved. Therefore, we can reasonably say that the life process is one of orderly use of energy. Life reverses the trend toward disorder. The earth's biosphere provides many sources of available energy. Each organism's place in the biosphere depends on which energy source it utilizes.

Energy capture may involve obvious processes, for example, the intake of food. The capture of **photons,** units of radiant energy, by tiny, photosynthetic algal cells, is a much less apparent process.

Active transport; Chapter 8, pp. 226–229

Energy utilization is a controlled release of the captured energy, followed by its orderly reuse. Energy must be available to drive the processes of **biosynthesis,** or molecular growth; **active transport** of materials into the organism; and **movement** of the organism (Fig. 1.2). Energy utilization is part of the cell's metabolism.

Energy storage is accomplished by making energy-rich chemical substances and depositing them in suitable locations for future use. Photosynthetic organisms frequently store starch, and nonphotosynthetic organisms may store fats and oils.

Biosynthesis; Chapter 9

Characteristic 3   Only living organisms biosynthesize large molecules. **Inorganic** molecules (those without carbon or containing carbon joined only with oxygen) are rarely larger than ten or so atoms in size. Living organisms put together **organic**

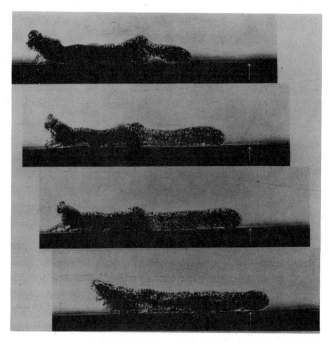

1.2
Movement *Amoeba proteus*, which moves very slowly, does so by extending a pseudopodium. Then the cell's cytoplasm streams forward into the leading portion. With time-lapse photography, at intervals of 20 seconds, the progress is distinctly visible.

(carbon-containing) molecules. Some of these organic molecules, such as proteins, may contain thousands of individual atoms linked together. Nucleic acids are similarly composed of tens of millions of atoms. These **macromolecules** are notable not just for their impressive size but also for the uniqueness of their structures. The change of one atom may alter the function of the whole gigantic molecule, and render it useless. It is no surprise, then, that the assembly of macromolecules is not left to chance, but takes place under very sophisticated cellular controls.

Characteristic 4 All living organisms contain, express, and replicate their own genetic information. The genetic information of each organism is a "blueprint" for growth and reproduction. The molecular structure of DNA is the information source. A nucleic acid blueprint is a "design manual" that directs the manufacture and assembly of the molecules that make up the organism's structure. It is also an "operation manual" giving the organism instructions for specific jobs to be carried out and the rate at which they are to be performed.

DNA; Chapter 2, pp. 55–56

Characteristic 5 All living organisms react to their environments, adapting to changing conditions as they occur. No organism can operate successfully on an inflexible set of blueprints because all natural environments are in continuous change. All organisms possess mechanisms through which they can make internal adjustments, shifting gears in their metabolism. This ability to change in response to environmental pressures is called **adaptation.** Microorganisms respond to differing food sources or quantities of foods, to the presence or absence of signaling molecules such as hormones, and to contact with inhibitory or toxic substances. Movement allows

Motility; Chapter 3, pp. 86–88

them a partial ability to choose their environment or to escape from environments to which they cannot adapt.

Each species is successful only in those environments to which it can adapt. One of the outstanding characteristics of the microbial world is that microorganisms as a group are very versatile and can flourish in a great variety of environments. Microorganisms are widely distributed in all global habitats; some microbial species will be successful in almost any niche, no matter how hostile.

The Origin of Life on Earth

Living organisms are the outstanding feature of the small planet we call the earth. As of now, our limited knowledge suggests that the earth may be unique in having developed life.

The prebiological era   Our galaxy, the Milky Way, may have been ten billion years old when a portion of the galactic cloud of gases and dust particles started to condense causing intense heat. Most of this condensing cloud coalesced into the sun, which began to give off radiant energy as a result of the thermonuclear reactions within. Smaller aggregations became the planets—so hot that their materials were in a liquid form. Heat was progressively lost to space, however, and roughly four billion years ago, some material on the earth cooled sufficiently to solidify and form a crust of rock. There are no signs of living things in these rocks. Some later rocks, believed to be three and a half billion years old, may contain fossilized cells that resemble present-day cyanobacteria. At some time prior to the formation of these rocks, living organisms appeared on the earth.

The early earth surface was very hot, bombarded with continuous lightning discharges, and exposed to intense solar radiation. The atmosphere was very thin. It was composed largely of hydrogen gas (the most plentiful element) and other elements in combination with hydrogen (Table 1.1). There was no free oxygen. Certain

TABLE 1.1
Comparison of the earth's atmosphere four billion years ago with the present atmosphere

MAJOR COMPONENTS OF PRIMORDIAL ATMOSPHERE		MAJOR COMPONENTS OF MODERN ATMOSPHERE	
H_2		N_2	78.09
N_2		O_2	20.94
NH_3		Inert gases	0.93
CO		CO_2	0.03
CH_4	Proportions	CH_4	trace
H_2O	not known	H_2	trace
Cl_2			
H_2S			

In the primordial atmosphere the prominent features were the absence of free oxygen and the presence of large amounts of hydrogen compounds. The modern atmosphere is rich in oxygen but has only tiny traces of hydrogen and its compounds.

Box 1.1

Are These Fossil Microorganisms?

The fossil algae shown in these micrographs may, to the untrained eye, look like undistinguished blobs. How can convincing information as to their age and origin be obtained?

The source of these fossils is the Swartkoppie chert formation in Swaziland in Africa. Using several methods, scientists have determined these rocks to be 3.4 billion years old. Chemically, their carbon isotope pattern and organic carbon content are similar to those of younger rocks created during periods in which biological activity is known to have occurred. The fossil algae have a regular appearance and size (average diameter 2.5 μm), similar not only to present-day algae but also to less ancient algal fossils. Fully 25 percent of all fossils observed were in an arrangement strongly suggestive of binary fission; that is, they were lying in pairs. This arrangement may mean not only that these were cells capable of reproducing themselves but that their environmental conditions were highly suitable for growth. There is possible evidence that the fossils have been subject to postmortem microbial degradation that may indicate the presence of heterotrophic microorganisms in the same environment.

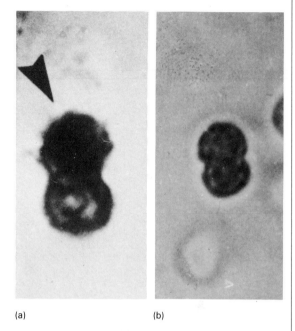

(a)　　　　　　　(b)

Fossil Algae　In (a), the electron microscope has been used to reveal fossil cells from rock that is about 3.4 billion years old. Notice the double forms, suggestive of ongoing cell division. Then compare with (b), a micrograph of *Aphanocapsa*, a current algal species.

There remains an element of doubt concerning whether these markings in the rocks actually **are** algal fossils. They may simply be random, inorganic, crystalline forms.

organic compounds that present-day life utilizes for growth were abiologically synthesized and accumulated to form a "hot soup." The molecules of these organic compounds tended to enlarge and aggregate progressively, and may have given rise to the first self-replicating structure, the cell.

The biological era　The organisms comprising the first life form, whatever it was, proliferated quickly. They were **heterotrophic,** which means that they derived their energy from breaking down organic chemicals. The hot soup was their source of food. They were also **anaerobic,** surviving in an atmosphere devoid of oxygen.

These organisms were followed by, and partially displaced by, a new group of organisms that had developed the capacity for photosynthesis. These self-feeders, or

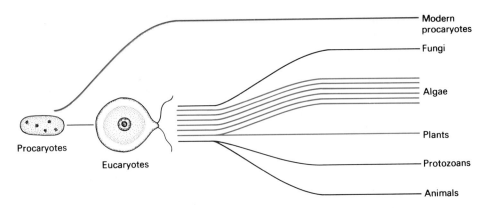

Procaryotes

Eucaryotes

Modern procaryotes

Fungi

Algae

Plants

Protozoans

Animals

1.3

Microbial Evolution   The microorganisms were the first life forms to appear on earth. Although the first events in their evolutionary diversification are hard to deduce because they occurred so long ago, this is one suggested microbial family tree. An ancestral procaryote line (or perhaps several) gave rise to the modern procaryotes. Eucaryotes then probably arose from the procaryotes. One heterotrophic group gave rise to the fungi, and another to the protozoa. A number of algal groups developed, probably separately, from autotrophic procaryotes.

autotrophs, gave off free oxygen as a by-product of splitting water molecules. The atmosphere became **aerobic,** or oxidizing, and was lethal to many of the organisms comprising the first life form, the heterotrophs. Those heterotrophs that survived either found environments in which they were not exposed to air or else developed mechanisms to utilize oxygen productively.

The first heterotrophs (whose descendants are modern bacteria) and autotrophs (now represented by, for example, the cyanobacteria) had very simple cellular architectures. These organisms are called **procaryotes.** A cell structure more complex than that developed by the procaryotes arose later. This cell structure is called **eucaryotic** and is found in the dominant multicellular forms such as human beings. The eucaryotic cell type may have arisen independently or from the procaryotic type (Fig. 1.3). Eucaryotic cells benefit from increased specialization and the capacity for sexual reproduction that promotes rapid genetic change. With these advantages, eucaryotic organisms evolved rapidly, first into multicellular aggregates (the first life forms that were microscopic in size) and later into the higher plants and animals.

Procaryotic cell;
Chapter 3, pp. 74–91

Eucaryotic cell;
Chapter 3, pp. 91–97

ORGANIZATION OF LIVING THINGS

The Cell Theory

The origin of living things, that is, the moment at which life could be said to have begun, coincided with the development of the first self-sufficient **cell.** The science of biology is based on the **cell theory,** first stated in the 1830s. Cell theory states that all living matter is made up of independently functioning units called cells. All living organisms thus far studied have been shown to be cellular in nature. The viruses are not cellular; it is debatable whether they should be called "living" or not. Let us examine some of the important types of cellular structure and organization.

Procaryotes and eucaryotes   The earliest forms of life were cells of the simple procaryotic type. These are characterized by the simplicity of their genetic structures, usually a single, ring-shaped molecule of DNA not separated from the rest of the cell contents by a membrane barrier. The cells also lack internal membrane-bound compartments. Procaryotes are represented today by the bacteria and cyanobacteria.

The much more complex eucaryotic cell, which evolved later than the procaryotic cell, takes its name from the fact that it has what biologists think of as a "true" nucleus. Its DNA is organized into rod-shaped chromosomes, in association with proteins. The chromosomes, usually in pairs (one from each parent), are surrounded by a nuclear membrane that partly isolates them from the rest of the cell. In addition, eucaryotes have a number of **organelles.** Organelles are subcellular units surrounded by unit membranes. Each organelle is specialized to carry out a particular function. The structural differences between the two cell types will be explored in detail in Chapter 3.

Unicellular contrasted with multicellular organisms   Many species of procaryotes and eucaryotes are **unicellular.** They carry out their life processes as single cells without more than casual physical contact with other cells. These unicellular species include most bacteria, many algae, and almost all protozoa.

Organisms that are made up of many cells linked together in a regular arrangement are **multicellular.** There may be varying degrees of cooperation and interdependence among the cells. Organisms large enough to be seen with the unaided eye are multicellular.

Aggregates and tissues   The multicellular organism may be made up of a number of cells all of which are similar in appearance and function. These cells are more or less **undifferentiated.** The giant seaweed *Macrocystis* frequently grows to be many feet long, yet because it lacks cellular specialization, it is an aggregate rather than a multicellular organism. Even when reproductive cells are formed in this type of organism, they are of the same type as their parent cells. Cooperation among cells is seen, but it is minimal.

Algae; Chapter 5, pp. 130–136

At a higher level, organisms possess **differentiated** cells. This means that the generalized type of cell found in the embryo state has divided and, during development, certain of its daughter cells have undergone changes in structure and function. They have developed distinctive shapes and arrangements. They also have acquired special efficiency at one particular task while losing their capacity to do others. Blocks of differentiated cells are called **tissues.**

Tissues make up **organs**—anatomical structures in which groups of different tissues cooperate to carry out complex processes.

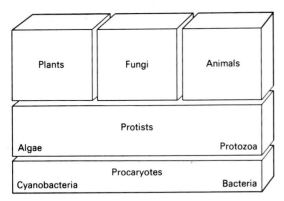

1.4
The Five Kingdoms   The most commonly accepted taxonomic scheme in current use divides the living world into five kingdoms. While dealing with microorganisms, we will be concerned with members of the procaryote, protist, and fungus kingdoms. However, one of our main concerns will be the effects of the microbial forms on higher plants and animals. Thus microbiology cannot limit its field of interest solely to any one box in a classification scheme.

Biological kingdoms   Classification of living things was easy when only the two macroscopic groups of organisms—plants and animals—were known. Once bacteria, fungi, algae, and protozoa were discovered, the problem of classifying them into appropriate kingdoms became more difficult. This book will use the five-kingdom system that, although not strictly accurate, is the most widely accepted at present (Fig. 1.4).

Five-kingdom system; Chapter 4, pp. 103

Levels of Organization

Businesses and institutions draw organizational charts that show the levels of responsibility of each employee. A similar chart has been drawn for the organization of living things (Fig. 1.5). This chart is very informative, because it shows us that al-

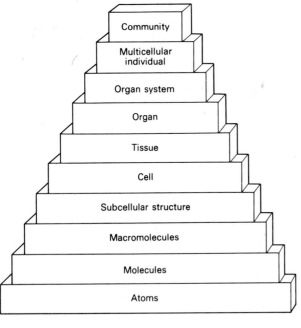

1.5
Levels of Organization   When we consider a multicellular organism, such as ourselves, we can view our organization scheme as a pyramid. The fundamental unit of life, the cell, can exist only because of the systematic organization of nonliving biochemical entities. Cells in turn can be further organized into more and more complex levels. A unicellular organism, of course, does not have tissues and organs. It does, however, always form part of the highest level of biological organization, the community.

though the primary unit of life is the cell, there are levels of organization above and below it. In fact, in the various levels, the cell comes in the middle. In order to completely understand a cell's function, we must first be aware of the nature of the **atoms** that compose it. Next, we must study the forces that hold those atoms together in **molecules.** Finally, we must study the molecules themselves. The assembly of **macromolecules** is the next step up toward a living cell. All cells possess **subcellular structures,** although only eucaryotes have the highly specialized ones we call organelles. Subcellular structures have a prescribed macromolecular composition. The essential chemical nature of the structure allows it to function in a particular way. A corporation president knows that the success of the corporation is based on the skilled work of those employees on the basic levels. By analogy, the success of an organism is based on its subcellular levels of biological organization.

Implications of an Organism's Size

Microorganisms are very small, and sheer smallness has some very important effects on an organism's biological potential. Considering these effects will reveal why microorganisms are different from macroorganisms.

Surface to volume ratio   The surface area of an organism (Fig. 1.6), expressed in square centimeters (cm^2), or other squared units, represents the area available to the organism in which to make contact with its environment, to obtain food, and to get rid of wastes. The organism's volume, expressed in cubic centimeters (cm^3), determines the internal contents that need nourishment. The smaller an organism is, the larger the ratio of surface to volume. In more practical terms, this means there

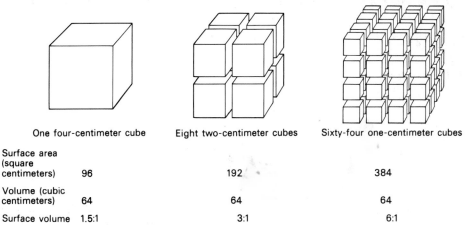

	One four-centimeter cube	Eight two-centimeter cubes	Sixty-four one-centimeter cubes
Surface area (square centimeters)	96	192	384
Volume (cubic centimeters)	64	64	64
Surface volume	1.5:1	3:1	6:1

1.6
Surface to Volume Ratios   If a cube of 64 cubic centimeters volume is divided into progressively smaller cubes, the total volume remains unchanged. However, the total surface area increases rapidly. To see how this principle applies to cellular life, we need only understand that a small cell has less cytoplasmic volume to "feed" by means of gathering nutrients across a proportionally greater membrane surface. A smaller cell can then grow more rapidly than a larger one, even when both are grown under the best possible conditions. It will have a higher metabolic and reproductive rate.

exists a larger amount of food-transporting potential to supply a smaller food demand. Procaryotic cells are usually much smaller than eucaryotic cells.

Growth rate   With a large surface–volume ratio, the cell or organism is capable of a rapid growth rate. Rapid growth is a prerequisite for rapid reproduction. Some of the smallest procaryotic cells under good conditions divide (double their number) every 20 minutes. By contrast, human epithelial cells, existing within a large eucaryotic multicellular organism, replicate themselves about every ten hours. It will become clear that rapid growth is the microorganisms' major advantage in the competition for survival with more highly evolved, specialized organisms. Their small size and simplicity allow them to take advantage of any opportunity for growth by rapidly increasing in numbers.

Growth kinetics;
Chapter 7, pp.
206–207

THE DEVELOPMENT OF MICROBIOLOGY

A brief history of microbiology will quickly show how its development as a science has influenced the growth of biology in general. An historical review can also provide a sense of the usefulness of the study of microbiology (Table 1.2).

Study of the natural world began a long time ago in several parts of the world, some as divergent from one another as China and Greece. However, the fall of the Roman Empire and the rise of Christianity ushered in a long period in which scientific study in the West was in eclipse. The physical sciences—physics and astronomy—were the first to revive from the "Dark Ages." Then alchemy gradually developed into the science of chemistry.

There remained a general belief that living things operated under rules different from those that governed nonliving things. The possession of life seemed to place

TABLE 1.2
Some key developments in the history of microscopy

1665	Robert Hooke publishes first drawing of cells from cork.
1684	Anton van Leeuwenhoek publishes drawings of bacteria made with a handheld single-lens microscope at approximately 270 × magnification.
1877	Robert Koch begins to develop methods for staining bacteria using aniline dyes.
1878	Ernst Abbe describes oil immersion lens. He also constructed a superior light condenser.
1884	Hans Christian Joachim Gram describes useful differential staining technique.
1911	Oskar Heimstadt invents fluorescence microscope.
1932	Max Knoll and Ernst Ruska describe electron microscope.
1935	Frits Zernike describes phase-contrast microscope.
1938	M.V. von Ardenne builds first true-scanning electron microscope.
1979	V.E. Cosslett and co-workers achieve images of individual atoms with high-voltage electron microscope.

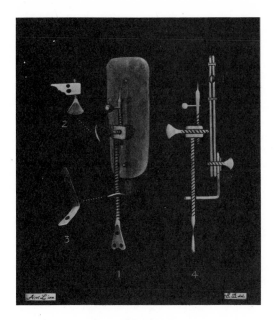

1.7
Leeuwenhoek's Microscope   This tiny, handheld device contains a single, very carefully ground lens. The fluid specimen was placed in a tiny drop on the top of the screw point, which could be raised or lowered. Then the viewer gazed through the lens at the droplet in a strong light. With this device Leeuwenhoek observed most of the major groups of microorganisms. He made careful drawings, which he sent to colleagues at the British Royal Society.

living organisms in the realm of the spirits and made them therefore incomprehensible by the **empirical** (based on observation) means of science. Many natural processes such as fermentation, decay, and disease fell under the same umbrella of mystery. Some startling discovery was required to start changing the opinions of the masses.

Origins of Microscopy

For quite a while, curious individuals had been looking at various forms of matter, living and nonliving, through fairly crude magnifying lenses. In 1674, a persistent and inquisitive Dutch merchant named Anton van Leeuwenhoek produced a very much improved lens in a handheld microscope (Fig. 1.7). He started examining all sorts of materials—hay infusions, tooth scrapings, blood, semen—and reported to the British Royal Society his observation that every environment he sampled seemed to be filled with various microscopic creatures in unbelievable numbers. He was the first person to see microorganisms. His publications contributed to a change in attitudes, a new curiosity, a new willingness to study and search for an understanding of the scientific basis of life. Biology began to take its place among the other natural sciences. Systematic improvements in microscopic equipment and preparation methods continue today.

Microscopic techniques; Chapter 3, p. 64–74

The Spontaneous Generation Controversy

Up until about a hundred years ago, both learned and unlearned believed that mice, insects, maggots, and microorganisms arose spontaneously. Spontaneous generation means the creation of living things from nonliving matter, living things with no

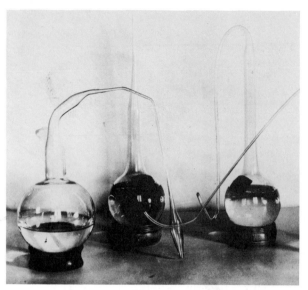

1.8
Pasteur's Flasks In order to put an end to the spontaneous generation theory, it was necessary to devise an experimental setup that effectively excluded random microbial contamination. However, the conditions would also have to be such that no conceivable adverse change in the medium or the air supply was occurring that might "prevent" spontaneous generation from happening. Pasteur's flasks provided the ideal conditions. It could be shown that the medium within was capable of supporting microbial growth, but remained sterile as long as airborne bacteria were prevented from entering by the curve in the neck.

parents. The maggots that so often infested meat in the days before refrigeration were assumed to be spontaneously generated. Francesco Redi (1665) showed that if the fresh meat was covered in such a way that flies could not reach it to lay eggs, the maggots that hatch from those eggs did not manage to be "spontaneously generated." The controversy raged on, but the proponents shifted their attention to the mysterious microbes that Leeuwenhoek had just revealed. A battle raged over the issue of whether microorganisms could arise *de novo* in broths or infusions, or whether they entered from the outside. A duel of wits and scientific reputations began that engaged the talents and vanities of many for at least a century.

For technical reasons, spontaneous generation turned out to be difficult to disprove. In order to disprove spontaneous generation, it was necessary to first take a suitable infusion and sterilize it to remove any microbes present. Because microbes may be very resistant to killing, this is not as simple as it sounds. Second, once sterility was achieved, it became necessary to keep microbes from entering, also a difficult problem. Many pages could be consumed in describing the moves in what became a great intellectual chess game for European scientists. In any case, the last word was uttered by one of the giants of all scientific history, Louis Pasteur. The French researcher designed an ingenious piece of glassware to contain a nutrient medium (Fig. 1.8). This container could be effectively sterilized, yet remained open to the air. Pasteur demonstrated on numerous occasions that such sterile swan-necked flasks remained clear and unchanged in any surroundings as long as the long spout remained intact. Some of Pasteur's original sterile preparations still exist—still sterile. It is now accepted that in the present circumstances life develops only from preexisting life. Remember, however, that most present-day scientists think that spontaneous generation did occur under **other** conditions about four billion years ago.

The Germ Theory of Fermentation

Leeuwenhoek's discovery influenced the direction of the spontaneous generation dispute. The resolution of this dispute similarly influenced the direction of research on such vital processes as fermentation, decay, and disease. Although people had used and enjoyed such products of microbial fermentation as yogurt, cheeses, leavened breads, and alcoholic beverages for countless generations, the processes involved in their production remained unclear.

There were two schools of thought. The **nonvitalists** felt that the changes occurred by reason of chemical agents called ferments and that the microorganisms observed in the material were irrelevant. The **vitalists** felt that the microorganisms were the cause of the change. Schwann, in 1837, clearly demonstrated the role of yeasts in fermentation. Later, it was shown that the type of change depended on the type of organism. Pasteur showed that when undesirable changes, such as the "sickening" of wine, occurred, the cause was an unwanted microbial species that had been introduced. He pioneered the technique of **pasteurization,** or partial sterilization by heat, to deal with the problem. This technique has been further developed and is now used widely for ensuring the safety of milk as well as for quality control in alcoholic beverages of all types.

From Pasteur's work came not only the understanding that specific microbes cause specific chemical changes but also an understanding that fermentation processes typically occur when microbes grow in the absence of air. The discovery of **anaerobic** growth was very surprising at the time. Research shortly revealed that many types of microorganisms not only grew without oxygen but even found it toxic.

1.9

Antibiotic Production Many millions of doses of antibiotic drugs are manufactured and sold annually to control and cure microbial infections. Here the drug mefoxin is being dispensed into vials.

Fermentation pathways; Chapter 8, pp. 236–237

Chemists started studying the fermentation processes, too. They were able to develop methods for microbial production of industrially important chemicals, such as acetone and alcohols, on a large-scale commercial basis. This was the first use of microorganisms in technology.

Enzymes; Chapter 8, pp. 212–221

Development of fermentation science also showed that microbes, indeed all cells, metabolize by using biological catalysts, or **enzymes.** The term "enzyme" was first attached to substances that catalyzed the production of alcohol by yeast. Hundreds of enzymes with other functions have now been isolated in pure form and studied. Some, such as the enzymes used in laundry products, have even found their way into commerce.

The Germ Theory of Disease

Just as the germ theory of fermentation provided the breeding ground for an understanding of microbial physiology and the industrial applications of microbiology, so the germ theory of disease (Table 1.3) was the precipitating force for much of modern medicine and public health protection. The term **germ** was originally applied to microbes indiscriminately because it means "seed." As the study of microbes has advanced and microbes have been more adequately described, the term has given way to more specific terms such as **bacterium** or **virus.**

Disease transmission; Chapter 15, pp. 440–450

The European tradition up until the nineteenth century considered the causes of all sorts of disease to be supernatural. Plagues were believed to be punishments for sin. In 1546, an observant Italian, Frascatoro, described the modes through which diseases could be transmitted. These, he said, were by direct physical contact with

TABLE 1.3
The development of the germ theory of disease

1530	Gerolomo Frascatoro, writing a poem about syphilis, described its venereal origin, suggesting that it is a transmissible disease.
1762	Marc Antony von Plenciz made the theoretical proposal that disease was caused by living organisms and each disease had its own causative agent.
1861	Ignaz Philipp Semmelweis, concluding that puerperal fever was infectious and spread by contaminated hands and instruments, published his recommendations.
1865	Joseph Lord Lister, convinced of the microbial origin of wound infection, introduced the practice of treating wounds with antiseptics, and antiseptic surgery.
1873	William Budd, noting that typhoid outbreaks were caused by contaminated milk or water, proposed the existence of a transmissible agent several years before the bacterium was isolated.
1876	Robert Koch isolated the anthrax bacillus in pure culture and started experiments to establish its infectious nature.
1884	Robert Koch published a formal statement of his postulates.
1886	Adolf Mayer recognized the first virus—tobacco mosaic virus—and started the process of understanding the role of viral agents in disease.

TABLE 1.4
The golden age of bacteriology

DATE	DISEASE	WORKERS
1876	Anthrax	Koch
1879	Gonorrhea	Neisser
1880	Malaria	Laverans
1880–1884	Typhoid fever	Eberth, Gaffky
1882	Tuberculosis	Koch
1883	Cholera	Koch
1883, 1884	Diphtheria	Klebs, Loeffler
1889	Tetanus	Kitasato
1894	Plague	Kitasato, Yersin
1905	Syphilis	Schaudinn and Hoffman
1906	Whooping cough	Bordet, Gengou

Over a remarkably short period of time the causative agents of a number of microbiological diseases were identified and studied. In most cases, effective vaccination and control measures became available soon after.

an infected person, by handling of contaminated materials, and by contact with infected air. However, his work had no immediate effect on untrained and superstitious medical practitioners. As various individuals demonstrated the contagious nature of disease, a gradual process of enlightenment followed.

An important conceptual step occurred when it was established that a specific microbe caused a specific disease. The first convincing demonstration of this principle was carried out by the German scientist Robert Koch. He was studying anthrax, a disease that was, at that time, a terrible source of economic loss to European livestock breeders. Koch, using technical advances such as the petri dish and agar-based solid media developed by his co-workers, developed the **pure culture** technique. He could isolate one organism from a mixture and maintain it so that its unique properties could be assessed.

Anthrax; Chapter 17, p. 506

Pure cultures; Chapter 7, pp. 188–194

In 1876, Koch cultured the anthrax organism from sick animals. He then saw that it multiplied in pure culture. Later he showed that the cultures, reintroduced into a healthy animal, caused clinical anthrax. After the disease developed, the identical organism could be reisolated from this second generation of cases. This sequence of experimental steps became known as Koch's Postulates. It is the standard procedure for showing the causal relationship between a specific organism and the disease it causes. Other European scientists adopted Koch's methods and, in rapid sequence, identified the causative agents of tuberculosis, diphtheria, tetanus, and other major killers (Table 1.4).

TABLE 1.5
The introduction of vaccination

DISEASE	IMMUNIZING MATERIAL USED	DATE OF FIRST CLINICAL USE	WORKERS
Small pox	Serum from cowpox lesion	1796	Jenner
Anthrax	Attenuated bacteria	1881	Pasteur
Rabies	Infected brain tissue	1885	Pasteur
Tetanus	Toxoid	1920	Glenny, Ramon
Diphtheria	Toxoid	1923	Ramon
Polio	Killed virus	1954	Salk
Measles	Live virus	1963	Enders

Increasing knowledge of disease organisms and the human system has made possible the development of practical vaccinations for many microbial diseases. These dates indicate the beginning of effective control. In many cases, significant improvements have been made in the original vaccine since it was first developed.

The Development of Preventive Medicine

When causes of a disease are known, cure and prevention of that disease become possible. In most cases, those who identified disease agents were also the pioneers who proposed methods for their control. The methods have developed along two major lines. **Immunization** involves treatment to render the individual capable of resisting the disease. **Sanitation** involves processing foods, water, and sewage in ways that prevent them from spreading disease in the population. Both methods are preventive in nature.

Smallpox; Chapter 17, p. 520

Building on the observation that the survivors of the contagious disease smallpox had permanent immunity to the disease, the ancient Chinese developed variolation, a technique of controlled exposure to the agent in a mild form to cause a mild immunizing disease. This process occasionally backfired, causing severe disease and death. It was introduced into England by the courageous traveler, Lady Mary Wortley Montagu, who had observed the process in her journeys as wife of the British ambassador to the Ottoman Empire. A refinement became possible when Edward Jenner observed that recovery from a similar but much milder disease, cowpox, also produced immunity. He developed a technique called **vaccination** in which he took the infectious agent from a milkmaid, introduced it into healthy individuals, and showed them that at the cost of having a small sore at the injection site, they could be protected for life from disfiguring, often lethal, smallpox. The cow thus has

Vaccination; Chapter 22

given its Latin name, *vacca*, to all medical procedures in which an agent of disease in some mild or altered form is used to immunize. We now routinely administer vaccinations for measles, polio, and various other diseases (Fig. 1.19).

In the field of immunity, Pasteur again was the leader. He devised procedures for **attenuating** infectious agents, which means weakening them so that they lose their disease-causing ability and may be safely introduced to stimulate immunity. This was a necessary advance because in most cases (smallpox being an exception) **only** the disease agent itself, or its products, will induce the body to produce immunity. Pasteur developed ways to immunize against anthrax and rabies. His methods and the methods of others soon resulted in vaccination procedures for diphtheria, tetanus, and other infections (Table 1.5).

Microbiology in the Twentieth Century

Three great areas of research have dominated the growth of microbiology in this century. The first of these is the accumulation of knowledge about microbial physiology—how the microbial cell—and by comparison, higher cells—metabolize, grow, and control their vital processes. From this area has come our knowledge of DNA, RNA, and protein synthesis and its regulation, the enzymatic pathways of cellular metabolism, and much of what we know about active transport and biological membranes. An offshoot of such research with vast medical consequences has been the discovery and use of **antibiotics.** These invaluable therapeutic aids such as penicillin, streptomycin, and tetracycline are effective because they block microbial metabolism in a specific way.

Antimicrobial drugs; Chapter 26

A second research area developed in the study of viruses. Because viruses are difficult to grow, even difficult to **see,** their nature was not at all clear until the middle of this century when cell culture methods and electron microscopy became available. In fact, some critical details of how viruses infect, interact with, and change the cells they parasitize remain to be discovered. Despite our remaining areas of igno-

Viruses; Chapters 10 and 11

Box 1.2

A Privileged Lady's Health

It is very easy for us to be unaware of the effect of disease on lives in the past. We know there were plagues, we know many children died in infancy, but our history books deal with the lives of kings and high-born ladies. How did these elite people fare?

Astounding knowledge comes to us from the recent excavation of a 2,100-year-old Chinese tomb. It contained the remarkably preserved body of a noblewoman, probably in her fifties, who was in all likelihood a consort of the emperor. She lived in a peaceful era in total luxury. Yet, examination revealed that she had suffered a poorly set fracture of the right arm. X-rays showed scars from tuberculosis in her lungs. She had gallstones that may have caused excruciating attacks. She probably walked with a cane. Her intestines contained not one but three kinds of parasites—pinworms, whipworms, and schistosomes. She was overweight, had arteriosclerosis, and probably died of a heart attack. Her life had clearly brought her much suffering. We can only guess at the horrors of a peasant woman's life.

If the great lady lived today under the most modest circumstances, most of her pain would be prevented or relieved. Among her ailments, only the problems of obesity and heart disease remain widespread in our country.

rance, major advances have been made in diagnosis and prevention of communicable viral diseases.

Microbial ecology;
Chapter 6, pp.
164–165
The third area of research that has been developed in this century has been environmental microbiology. We tend to think of the microbes mainly in connection with disease. This is a false perspective. Only a few species cause disease. The rest play essential roles in the functioning of the biosphere. As decomposers, they recycle crucial nutrients and remove noxious material pollutants, both natural and synthetic. Research into the biogeochemical activities of the microorganisms of soil and water is a fundamental part of the interdisciplinary science of ecology.

An important step in the developmental of environmental microbiology was the recognition that the nitrogen-fixing microorganisms—those which convert nitrogen gas into the combined forms that plants use for growth—are essential for global productivity. Other microorganisms play roles in chemical reactions involving sulfur, iron, manganese, and other minerals.

MICROBIOLOGY IN TODAY'S WORLD

Microorganisms as Research Tools

Microorganisms are ideal research tools for a number of reasons. It is a relatively simple matter to maintain tens of millions of them in a very small space. The methods are technically easy and, compared with the cost of maintaining equal numbers of mice, dogs, or geraniums, quite cheap. Microbial populations are genetically controllable, and one member varies little from another. This certainly is not true of a group of human beings. In addition, large numbers of experimental organisms simplify the task of showing that an experimental result is statistically significant. Not to be forgotten is the fact that there are few restrictions on manipulating, smashing up, or killing these organisms. They are not endangered species! Only a few types of research, deemed to have potential risk to human beings, take place under strict regulation.

As a result, a very large number of today's biological researchers, although they might consider themselves biochemists, geneticists, or pharmacologists first, are also microbiologists. An understanding of this interdisciplinary science can be expected to help you respond to new developments in a wide range of related fields.

Microbiology in Industry

Several industries utilize microbiological techniques, either because the industries need to control unwanted microbial growth or because they depend on the activities of microorganisms to manufacture their products.

Microbial metabolism is used constructively in the production of leavened bread and other bakery products, fermented dairy products, alcoholic beverages, aged meats and sausages, and sauerkraut and other pickled foods. It is also harnessed to make commercial acetic and propionic acid, solvents such as acetone, starting materials for synthetic hormones, enzymes, insecticides, antibiotics, and vitamins. Microorganisms grow rapidly on simple organic materials or waste products and can then be harvested as a cheap nutritious food supplement for human beings or animals.

In the food-processing industry, unplanned microbial growth causes spoilage during storage, cooking, or preserving processes. It causes losses of edible fresh foods in markets and at home. Much research is directed at minimizing these losses. The cosmetic industry develops products such as deodorants and mouthwashes that promise to reduce microbial growth on the body; cosmetics such as hand lotion are in turn often degraded by microbial growth. Microbial degradation causes losses in the lumber and paper industries. It rots fabric; corrodes glass, leather, and metal; and causes delicate electronic instruments to stop working. Drinking-water and waste-water treatment technology is concerned with the use of microorganisms both to purify water and to remove infectious pathogens. Microbiologists may have many different jobs and be members of widely divergent groups—from spacecraft sterilization teams to groups developing improved, disposable diapers.

Changing Public Health Patterns

Living in this time and place, we are the recipients of a gift no other human beings have ever had. We can expect to live for seventy-plus years, long enough to give us time to do and try more things than were ever possible before. This added time will also be relatively free of the crippling and disability that ruined so many lives in other eras. No longer do we see the smallpox-disfigured face, the wasted limb of the polio victim, the blindness of the child infected at birth with gonorrhea. We also have the benefit of readily available relief from pain—from the bacterial sore throat, the abscessed tooth, the bladder infection—pain our predecessors simply put up with.

A comparison of the leading causes of death at the turn of the century and in more recent times shows very clearly the route we have come (Table 1.6). In 1900, the average life expectancy was estimated to be 47.3 years for all races. Most of the leading causes of mortality were infectious diseases. Infant mortality was very high. More than 10 percent of the children did not survive their first year. Now by contrast, the combined life expectancy for all races is 72.0 years. Infant mortality is down to less than 1.7 percent. Life expectancy in the United States continues to rise slowly.

The leading causes of death now are not infectious. They are primarily degenerative problems of older people, problems for which there are no "vaccines," no easy answers. The rapid rise in life expectancy in the first half of this century resulted almost entirely from the discovery and application of sanitation, vaccination, and antibiotics. We are now left with the more difficult and intractable health problems to solve. Many of these, paradoxically, leave more persons in more hospitals for longer periods. While hospitalized, seriously ill, they are highly susceptible to infection. Infectious disease in the United States still accounts for about 29 million hospital occupancy days per year, about 10 percent of the total days the population spends in hospital. The direct cost in 1977 was $4.8 billion and is significantly higher now.

Hospital infections; Chapter 28

TABLE 1.6
A comparison of the ten leading causes of death in 1900 and 1967

	1900				1967		
RANK	CAUSE OF DEATH	DEATHS PER 100,000 PERSONS	PERCENTAGE OF ALL DEATHS	RANK	CAUSE OF DEATH	DEATHS PER 100,000 PERSONS	PERCENTAGE OF ALL DEATHS
1	Pneumonia/ influenza	202	11.8	1	Heart disease	365	39.0
2	Tuberculosis	194	11.3	2	Cancer	157	16.8
3	Diarrhea/ enteritis	143	8.3	3	Cerebral hemorrhage	102	10.9
4	Heart disease	137	8.0	4	Accidents	57	6.1
5	Cerebral hemorrhage	107	6.2	5	Pneumonia/ influenza	29	3.1
6	Nephritis	89	5.2	6	Diseases of early infancy	24	2.6
7	Accident	72	4.2	7	General arteriosclerosis	19	2.0
8	Cancer	64	3.7	8	Diabetes mellitus	18	1.9
9	Diphtheria	40	2.3	9	Other circulatory diseases	15	1.6
10	Meningitis	34	2.0	10	Emphysema and related diseases	15	1.6

Active public health measures in the areas of sanitation and immunization contributed to surprising changes in the causes of death in the United States. Despite the fact that infectious disease is no longer a major cause of mortality, it still has a major impact on daily life in terms of days of work lost, doctors' office visits, and days in hospital.

Avenues for Future Progress

In what expanding areas can microbiology hope to make significant contributions? Much remains to be discovered about viruses. Antibiotics as we know them do not help in viral diseases and there is urgent need for therapeutic drugs to help the patient recover from viral infections. Some of these drugs are in the testing stage, and it seems reasonable to assume that effective antiviral drug therapy is just over the horizon.

Cancer viruses; Chapter 11, pp. 328–338

Viruses are strongly implicated as causative agents in certain types of cancer as well as in a number of slow-developing, degenerative diseases such as multiple sclerosis. These are precisely the types of diseases that resist modern medical controls. Some successes in virus research may lead to the control or cure of these diseases and to yet another reshuffling of the list of the "ten most wanted" killers. We must remember that there is no indication that any scientist will ever succeed in making us immortal. There will always be such a list.

Our knowledge of the immune system seems at this time just large enough so that it is very difficult to see the forest for the trees. Many serious health problems—allergy, graft rejection, autoimmunity, perhaps also susceptibility to cancer—are caused by misdirected or inadequate functioning of the immune system. We can only offer treatment directed at the symptoms. We cannot cure, at this time. As progress continues in sorting out immunity's many components, realistically it should become possible to prevent or manage all of these conditions.

Microbiology, then, has come a long way. It led the way as the biological sciences emerged in the nineteenth century. Its techniques provided the means for the major biological discoveries of the twentieth century. We can confidently expect that it will continue to be pivotal in the efforts to solve health problems and further improve the human condition in the twenty-first century.

SUMMARY

1. All biological systems share many common characteristics. We will divide these into five areas: (1) remaining separate from, but communicating with, the exterior environment; (2) capturing, utilizing, and storing energy in various forms; (3) manufacturing the large complex macromolecules found only in the biological world; (4) maintaining, using, and replicating an internal source of genetic information; and (5) reacting to changes in the environment by movement and adaptation.

2. To explain the appearance of life forms on the earth and not, so far as we presently know, anywhere else in the solar system, a complex set of hypotheses explaining the origin of life has been developed. Inferring the physiochemical conditions of the surface of the cooling planet earth four billion years ago, it is proposed that inputs of energy first caused the formation of small, and then more complex, organic molecules. Patterns of interaction among these molecules led to the formation of aggregations that could grow and divide. At first, such growth was unregulated or random. Systems eventually developed to regulate the processes. At some point, the aggregates developed all the traits necessary to be defined as living forms—cells. This is proposed to have taken place sometime between four billion years ago—the age of the oldest rocks—and 3.5 billion years ago—the age of the oldest proposed fossils.

3. The primitive life form underwent evolution from heterotrophy to autotrophy. The development of photosynthesis led to the accumulation of atmospheric oxygen and thus to a conversion of most organisms from anaerobic to aerobic metabolism. Simple cell forms, the procaryotes, may have given rise to more complex cells, the eucaryotes. These, in turn, developed into multicellular organisms.

4. All living things are composed of cellular units. Several levels of organization underlie the complete organism. To thoroughly understand cells, we must analyze atomic and molecular structure and interactions.

5. Size of the organism places limits on its function. Particularly in the smallest single cells, the procaryotic bacteria, we see very large surface–volume ratios that contribute to extremely rapid growth rates.

6. Microbiology as a science is interdisciplinary. Its methods and discoveries depend on and contribute to studies in many other areas, including biochemistry, cell biology, genetics, medicine, pharmacology, public health, and ecology.

7. At the beginning of the nineteenth century, scientists began to take an active interest in living things. This interest was fueled by the discoveries of Leeuwenhoek, Pasteur, Koch, and many others. In particular, the discovery of bacteriologically pure culture techniques laid the groundwork for studies of the role of microorganisms in fermentation, disease, and natural environmental changes. Bacteria as research tools made possible many of the discoveries of the twentieth century.

8. Health problems such as cancer, immunological malfunction, and degenerative disease are now the leading causes of death. Pollution, waste disposal, and the scarcity of certain commodities present environmental problems. Food shortages and population growth are worldwide threats that will continue to worsen. Microbiology as a science is in a pivotal position to contribute toward solving all of these problems.

Study Topics

1. Compare how the five characteristics of life are manifested in a higher plant (such as a tree), in a bird, and in yourself.

2. Compare environmental conditions as they have been hypothesized to have existed on the earth four billion years ago and today.

3. Where, would you think, could an anaerobic heterotroph survive in today's world?

4. Give some examples of aerobic autotrophs encountered in daily life.

5. Human bodies contain eucaryotic cells in tissue. Consider the implications in terms of the growth rate of our cells.

6. Find ten examples of microbiological applications in your daily life.

7. For a modern person, what are the direct consequences of Louis Pasteur's research?

Bibliography

General biology

A good, up-to-date general biology text will be useful frequently. Many are available, and the three listed have shown themselves to be particularly useful.

Curtis, Helena, 1979, *Biology* (3rd ed.), New York: Worth.

Kimball, John W., 1979, *Biology* (4th ed.), Reading, Mass.: Addison-Wesley.

Wolfe, Stephen L., 1977, *Biology: The Foundations*, Belmont, Calif.: Wadsworth.

Evolution

Fox, S. W., and K. Dose, 1972, *Molecular Evolution and the Origin of Life*, San Francisco, W. H. Freeman.

Oparin, A. I., 1957, *The Origin of Life on the Earth* (3rd ed.), New York: Academic Press.

History of microbiology

Brock, Thomas (ed.), 1961, *Milestones in Microbiology*, Englewood Cliffs, N.J.: Prentice-Hall.

DeKruif, Paul, 1926, *Microbe Hunters*, New York: Harcourt Brace.

LeChevalier, H. A., and M. Solotorovsky, 1974, *Three Centuries of Microbiology*, New York: Dover.

2

The Basic Chemistry
of Life

All living organisms are made up of nonliving chemicals. These chemicals, when removed from the cells they compose, can frequently execute a single function of the cell in the test tube (*in vitro*). But no chemical is **living.** Scientists have not yet put together a living organism. Only the intricate, precise mixture of thousands of chemicals we call a cell is living.

In this chapter we will study chemicals, from the simple to the more complex. We will also survey the forces that cause chemicals to interact in predictable ways. We will see that the laws that govern chemical interactions are those of **thermodynamics** or energy exchange.

ATOMIC AND MOLECULAR STRUCTURE

Atomic Structure

The atom is the smallest unit of matter that participates in chemical reactions. It cannot be subdivided by chemical means, although treatment of the atom with various strong physical forces will cause it to split into a variety of **subatomic** particles (Table 2.1).

An atom is made up of a certain number of three types of particles—protons, neutrons, and electrons. The type of an atom is determined by the number of pro-

TABLE 2.1
The common subatomic particles

PARTICLE	MASS	CHARGE
Proton	1 AMU	1 +
Neutron	1 AMU	None
Electron	0.005 AMU	1 −

The chemist views the atom in terms of these three common subunits. The physicist, however, using atomic fission or fusion as research tools, has arrived at a much more detailed picture of atomic structure.

The "heavy" proton and neutron are the two most prominent members of the class of subatomic particles called **hadrons.** There are several hundred other hadrons, observed only under special conditions. Hadrons, in turn, are composite structures built up of smaller particles called **quarks,** held together by a force called **charm.**

The electron is the most common of the **leptons,** or light particles. These are truly elementary, and cannot apparently be divided further. Each has a corresponding antiparticle, which is opposite in charge and direction of movement.

tons it contains. This number coincides with the atom's **atomic number.** An **element** is a substance in which all of the atoms have the same atomic number. All the known elements may be arranged by increasing atomic number in tabular form—the **periodic table.**

The nucleus   An atom is structurally arranged in two parts. It has a central, dense, massive, electrically positive **nucleus.** This is surrounded by an **electron field,** non-dense and electrically negative.

Within the nucleus, neutrons are associated with protons and held together by a strong nuclear force still not completely understood. The number of neutrons is usually close to, but not necessarily the same as, the number of protons.

Isotopes   Elements may also occur in slightly varying forms called **isotopes.** Isotopes of an element have fewer or more neutrons than the common element has. This causes isotopes to have different **atomic weights.** Atomic weight is equal to the sum of the numbers of protons and neutrons, each of which has one AMU (atomic mass unit).

The importance of isotopes in biology is considerable (Table 2.2). For most atoms, there is a certain ratio of protons to neutrons that is most stable. Any isotope that has a different number of neutrons is unstable and will tend to **decay,** losing subatomic particles to attain a stable state. These changes in atomic nuclei release radiant energy referred to as **radioactivity.** Even small amounts of radioactivity can be detected by sensitive instruments. Because of their detectability, radioactive materials can be used as probes, as in brain scanning, to reveal anatomic abnormalities. In biochemistry radioactive atoms can be traced as they pass through complex metabolic pathways. Isotopes can also be used to determine the age of fossils and geological formations. However, large amounts of radiation are destructive to many of the complex biochemicals of the cell and thus to life itself.

Radiation sterilization; Chapter 25, p. 732

TABLE 2.2
Isotopes of particular importance in biology

ELEMENT	ISOTOPE	HALF-LIFE	IMPORTANCE IN BIOLOGY
Hydrogen	H^3	12.3 yr	As radioactive tracker used in studies of nucleic acids, for radioautography, etc.
Carbon	C^{14}	5,700 yr	As mechanism for dating organic fossils; as tracker used in all types of cellular research
Phosphorus	P^{32}	14.2 days	As label for nucleic acid research
Potassium	K^{40}	10^{-9} yr	In research on ion transport
Cobalt	Co^{60}	5.17 yr	To deliver lethal radiation to tumor cells in cancer therapy
Technetium	Tc^{97}	10^6 yr	Used as a radioactive probe in diagnostic scanning of brain and other tissues
Iodine	I^{131}	8.05 days	Tests for activity of thyroid gland; label for radio-immunoassay of hormones and other antigens
Uranium	U^{235}	$7.1 \cdot 10^6$ yr	Used in nuclear reactors and bombs

Isotopes have been widely adapted to biological research and practical applications. They may be used in primary research, in diagnosis of disease, and in treatment of cancer. Each isotope emits a particular type of radiation that can be detected with great accuracy by electronic equipment.

Electron shells   The electrons of an atom are equal in number to its protons. They are arranged around the nucleus in **shells** according to their energy level. Within the shell, the electrons are arranged in complex **orbitals.** An orbital is described as the volume of space in which the electron will be found 90 percent of the time. For our purposes, it is necessary only to understand that the chemical behavior of an element is determined by the number of electrons in its outermost shell. The outermost shell will always contain from one to eight electrons. Eight is a stable configuration. Atoms, in combining, seek to complete the outermost shell with eight electrons. The innermost shell, however, reaches maximum capacity and stable state with two electrons. The number of electrons found in an element's outer shell can be illustrated by dots (Fig. 2.1). Inner shells are not shown, because electrons in that shell do not normally participate in chemical reactions.

2.1
The Carbon Atom   (a) Carbon has six electrons. They are shown here in terms of their place in the electron shells, in a two-dimensional representation (b) A more accurate picture is obtained by showing the three-dimensional nature of the electron orbitals. The angles of four covalent bonds are shown by the arrows.   (a)

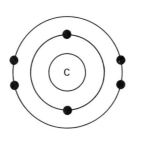

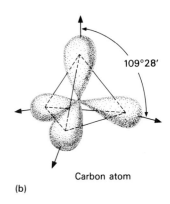

109°28′

Carbon atom

(b)

Valence   It is useful to know the number of outer shell electrons for each element. This is because the element's combinations with others will be governed by the requirement to complete its outer shell. If an element, such as oxygen, has six outer-shell electrons it will tend to acquire two more, to make eight. It then has a combining requirement, or **valence** of two. If an element, such as sodium, has an outer shell with only one electron, it will tend to lose the single electron by some means. Thus it has a valence of one. The valences of certain elements important to life appear in Table 2.3.

TABLE 2.3
Chemical elements in biological systems

ELEMENT	SYMBOL	ATOMIC NUMBER	VALENCE
Hydrogen	H	1	1
Carbon	C	6	4
Nitrogen	N	7	2, 3 ,5
Oxygen	O	8	2
Sodium	Na	11	1
Magnesium	Mg	12	2
Phosphorus	P	15	3, 5
Sulfur	S	16	2 ,4,6
Chlorine	Cl	17	1
Potassium	K	19	1
Calcium	Ca	20	2
Iron	Fe	26	2,3
Iodine	I	53	1

When an element has multiple valences, the number underlined is the most common in biological molecules.

Molecules

Atoms tend to combine with other atoms, and in nature few atoms are found free. Most atoms are tied up in more or less stable combinations called molecules. Combinations form when the energy required to make them is available, and persist for varying periods depending on their stability. A **compound** is composed of a single type of molecule. Let us consider the compound sulfuric acid. It has an **empirical formula** (H_2SO_4) that tells us that the compound contains certain atoms in fixed proportions. Larger or more complex molecules will also have **structural formulas.** These show the precise bonding relationships of the atoms. The compound's total weight is derived by adding the atomic weights of the constituents. (Thus the weight of H_2SO_4 is 98 AMU.) The giant biological molecules may contain millions of atoms. Their weights are usually expressed in **daltons,** an alternative term for AMU.

Molecules are held together by interactions called **chemical bonds.** In order for a bond to form, it must satisfy the mutual valences of the atoms involved. Also the energy to form the bond must be available.

Strong Bonds

We will discuss first the formation of three types of strong bonds—ionic, covalent, and polar covalent. Before discussing ionic bonds, we must first define ions and see how ions come to exist. Ions are derived from atoms. Sodium is an atom with a strong tendency to give up an electron to attain a stable outer shell. Chlorine needs only one electron to complete its outer shell. When these two come in contact, a reaction occurs in which the single valence electron of the sodium atom is transferred to the shell of the chlorine atom (Fig. 2.2).

$$Na + Cl \rightarrow Na^+ + Cl^-.$$
reactants   products

The **equation** for the reaction shows that both **reactants** are electrically neutral and that both **products** are electrically charged. An **ion,** then, is an atom which has become electrically charged due to gain or loss of electrons.

2.2
Ionization The unpaired outer electron of a sodium atom is pulled away to fill the single space in the incomplete outer shell of a chlorine atom. As sodium loses an electron, it becomes oxidized and is converted into a positive ion or cation. As chlorine gains an electron, it becomes reduced and is converted into a negative ion, or anion.

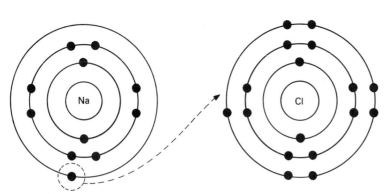

Redox reactions   The equation above illustrates a process in which electrons are transferred. This is called an **oxidation–reduction**, or **redox reaction.** Let us dissect this reaction into two parts. **Oxidation** is the loss of electrons by a substance, as shown in the following half-reaction,

$$Na \rightarrow Na^+ + 1e^-.$$

Reduction is the gain of electrons by some substance, as follows.

$$Cl + 1e^- \rightarrow Cl^-.$$

When the two half-reactions are added together, and repeating substituents removed, the result is as originally given.

$$NA + Cl \rightarrow Na^+ + Cl^-.$$

Note that oxidation and reduction occur simultaneously; as sodium is oxidized, chlorine is reduced. Complex redox reactions form the basis of cellular metabolism.

Coupled reactions; Chapter 8, p. 222

Ionic bond formation   Ions are charged, and opposite charges attract. When two oppositely charged ions make contact, they remain together as an ionic compound, held together by an ionic bond.

$$Na^+ + Cl^- \rightarrow \qquad NaCl.$$
$$\text{common table salt}$$

Molecules of ionic compounds are usually found in regular crystals in the solid state. When subjected to forces that neutralize their attraction for each other, they separate into free ions.

Covalent bond formation   Atoms that less readily form ions react in a different fashion. They share electrons among each other to fill shells, forming covalent bonds. Two hydrogen atoms may react because each needs one electron to complete its shell. When a covalent bond is formed, both hydrogen atoms acquire the use of two electrons. These revolve around both atoms with formation of H_2, a **diatomic** molecule. Thus a covalent bond is the sharing of a pair of electrons.

Two pairs may be shared to yield the diatomic gas O_2.

$$2O \longrightarrow O_2$$

$$O + O \longrightarrow O=O$$

One atom may share with atoms of a different element.

$$C + O_2 \longrightarrow CO_2$$

$$C + 2O \longrightarrow O=C=O$$

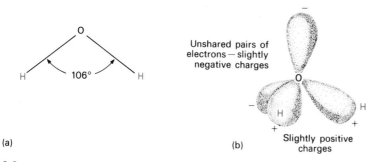

Unshared pairs of electrons — slightly negative charges

Slightly positive charges

(a)   (b)

2.3
The Water Molecule   (a) The shared electrons are indicated by lines.
(b) The electron orbitals are shown in a space-filling simulation.
Water molecules have a four-cornered pattern of charges, so each
water molecule can hydrogen bond to four other water molecules or
other polar molecules.

Single covalent bonds in a structural formula are represented then by single lines. Double lines, showing **double bonds**, signify the sharing of two pairs of electrons.

When covalent bonds are formed, the paired electrons are used by both. They may be held by the participants with varied degrees of force, however. If the bonded atoms are of roughly equal **electronegativity**, or electron-binding strength, the electron pair will be equidistant between the atoms. Such a bond has a uniform distribution of electrical charges and it is called **nonpolar**.

When atoms of unequal electronegativity bond, the electron pair will be displaced toward one partner. This is the case in water (Fig. 2.3).

Each of the atoms has weak electrical charges, smaller than ionic charges, but measurable. These are caused by the attraction of the electron pairs toward the oxygen atom. Such covalent bonds are **polar**. Polar bonds create molecules or portions of molecules with electrically charged surfaces.

If the attractions of the two atoms or ions forming a bond are accurately measured, it becomes clear that there is a continuous gradient from nonpolar covalent bonds through bonds of increasing degrees of polarity to the completely noncovalent ionic bond. In fact, the two partners in a strongly polar bond may on occasion separate, giving rise to free ions.

Three-dimensional models   Atoms do not necessarily combine in straight lines or at right angles. Each atom forms its bonds at specific **bond angles**. Bonds also have fixed **bond lengths**. From experimentally derived knowledge of the lengths and angles of the bonds, the chemist can construct three-dimensional models of molecules with considerable accuracy, even though the molecules are far too small to study with modern microscopy.

Covalent and ionic bonds are strong bonds. These are the bonds that maintain the primary structure of molecules. When you look at an empirical formula such as H_2SO_4 or $C_6H_{12}O_6$, recall that this type of bond is what holds the atoms together.

TABLE 2.4
Some covalent bonds

BOND	RELATIVE POLARITY	BOND ENERGY Kcal/mol
H—H	Nonpolar	104.2
O=O		33.2
≡C—C≡		83.1
=C=C=		147.0
≡C—H	Weakly polar	98.8
≡C—N=		69.7
=C=N—		147.0
=N—H	Polar	93.4
≡C—O—		84.0
=C=O		171.0
—O—H		110.6

Bond energies are the measured amount of energy necessary to break the bond. The exact polarity of a given bond can be calculated. The calculated value is of limited use, however. This is because the polarity of one bond in a molecule may be further increased or decreased by interaction with adjacent bonds.

Weak Bonds

Bonds with little individual bond energy are called weak bonds. They are readily formed and broken without using much energy. Their individual effect is slight, but their cumulative effect is considerable.

Hydrogen bonds The **hydrogen bond** is a weak electrical interaction. It involves the positively charged atom at one end of a polar covalent bond, pulled toward a negatively charged atom at one end of another polar covalent bond. The participating poles can be in two different molecules of the same compound, two chemically different molecules, or in adjacent regions of the same molecule.

Hydrogen bonds form only between molecules that are partly or completely polar. All polar molecules are normally hydrogen-bonded to their polar neighbors.

Van der Waals force There is a second important weak force which attracts parts of molecules together called the **van der Waals force** (Fig. 2.4). When any two atoms approach each other, there is a critical distance at which they are maximally attracted. Each molecule can interact intimately only with another molecule whose atoms are arranged in a complementary way for a good van der Waals "fit."

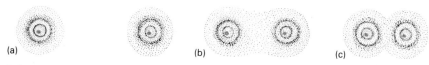

2.4

Van der Waals Forces Every atom has a characteristic radius, the distance from its center to its outermost electrons. (a) When two atoms are at a distance, there is little attraction. (b) At a specific distance, which can be determined for each different pair of atoms, there is a strong attraction of each positively charged nucleus to the other atom's negatively charged electron cloud. This creates a binding force called the van der Waals Force. (c) When the two atoms are too close, the repulsive force of the two positive nuclei pushes them apart.

Self-assembly; Chapter
2, pp. 58–59
The cumulative effect of the two types of weak bonds is that they provide a means for the aggregation of molecules of similar nature. Hydrogen bonds join polar molecules, and van der Waals forces, nonpolar ones. The two opposite types of molecules either repel each other or are indifferent toward each other.

THE KINETICS OF CHEMICAL REACTIONS

The Kinetic Theory

All particles, including atoms and molecules, are in constant motion. The rate of the random motion increases with temperature. Increased heat energy increases the kinetic energy of the moving particles, and these increases cause state changes from solid to liquid and from liquid to gas. At absolute zero ($-273°C$) all motion stops.

The Direction of Reactions

When a reaction, expressed by the simple equation,

$$A + B \rightleftharpoons C + D,$$

is under study we may ask two basic questions. First, we are told by the arrows that the reaction is reversible, at least theoretically. Yet in ordinary circumstances the reaction will be going mainly in one direction, and we need to know in which.

The answer depends on two crucial variables. The first of these is the **concentration** of the four substances. Reaction requires collision. In general, any reaction will go in the direction that converts concentrated substances to ones present in lesser amounts. If other influences are discounted, the reaction will establish an equilibrium in which the number of conversions from left to right is exactly equal to the number from right to left (Fig. 2.5).

In real chemical reactions, equilibria are rarely established at what might be called dead center. Just as the locations of a heavy child and a lighter one success-

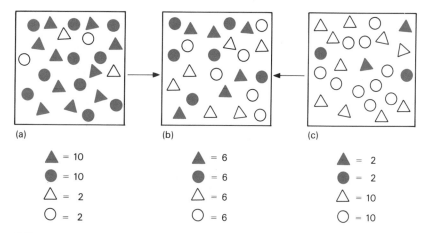

(a)   (b)   (c)

▲ = 10	▲ = 6	▲ = 2
● = 10	● = 6	● = 2
△ = 2	△ = 6	△ = 10
○ = 2	○ = 6	○ = 10

2.5
Establishing an Equilibrium Imagine a reversible reaction,

$$▲ + ● \rightleftharpoons △ + ○$$

in three different boxes containing different proportions of the four materials. Assume that there are no forces other than this that favor one direction or the other. At the start, substances ▲ and ● predominate in (a), and chances of a reaction between them are greater than chances for a reaction between △ and ○. In (c), the situation is exactly opposite. Neither, however, remains that way, because the more common substances will be used up faster. Both extremes will in time come to resemble (b), which does not show net change during the time period.

fully balancing a seesaw will not be equidistant from the center, the balance in many chemical reactions will be very much off-center. As equations are conventionally written, the left-to-right orientation of the equation indicates the favored direction for the reaction.

Energy is the second variable affecting the reaction path, and it sets the equilibrium point. When a reaction takes place, bonds are broken and new bonds are made. Because bonds are a source of chemical potential energy, energy may be either given off or required to make the process go.

$$A + B \rightleftharpoons C + D + \text{energy}.$$

When this reaction runs from left to right as written, it releases energy. This is called an **exergonic** reaction. The energy may be given off in a perceptible form, as heat and light are given off when a candle's wax combines with oxygen. In such a reaction, the bond energy of the reactants was greater than that of the products. The excess energy was given off.

If the reaction runs from right to left, however, it **requires** energy—it is **endergonic.** Photosynthetic manufacture of sugars requires visible light, for example. In every case in which a reversible reaction exists, one direction is ender-

Photosynthesis; Chapter 8, pp. 239–243

gonic—energetically "uphill"—and the other is downhill. Unless energy is specifically supplied, the reaction will run predominantly downhill. Only a few molecules will acquire sufficient kinetic energy from their environment to make the uphill reaction possible. The equilibrium will lie far to the right. Should a source of energy be applied suddenly, however, the equilibrium would quickly shift in the other direction.

The Rate of Reactions

Within the metabolizing cell, reaction conditions are strongly affected by the presence of catalytic enzymes and by the varying concentrations of substances. These may have striking effects on the **rate** of the reaction. Rate measurements have two parts. It is necessary to determine how much of the desired change has occurred (A and B being converted to C and D) over a given time. This could be expressed as (decrease A/min). The everyday term, miles per hour (miles/hr) is another rate expression. If reaction depends on collision, then anything which increases the likelihood of collision increases rate. Because heat energy increases the kinetic energy of molecules so that they move more rapidly, they collide more frequently and rate increases. Addition of energy preferentially favors the rate of the energy-requiring reaction.

When the concentration of the reactants is increased, it improves the chance of collisions leading to the forward reaction. The rate of the reverse reaction can be decreased if one of the products is constantly removed. In many enzyme-catalyzed metabolic reactions, the molecular products are consumed in other reactions as soon as they are made, shifting the equilibrium farther toward production of the desired substance.

Enzyme function; Chapter 8, pp. 217–221.

Enzymes increase the rates of reactions by facilitating combinations that would normally take place at a slower rate. Thus they can cause a rapid reaction rate among compounds present in quite low concentrations, at the moderate temperatures found within living things. Enzymes cannot cause a reaction to take place that would not take place at a slow rate in their absence.

COMPOUNDS IMPORTANT TO LIFE

Inorganic Molecules

Water   Water is the single most important compound on this planet for the origin and present existence of life. Its molecular structure explains its curious and indispensable properties. Each water molecule has points of electrical charge at which it may form hydrogen bonds. In the pure form the compound exists in a complex lattice structure. Each molecule is bonded to four others. Multiple hydrogen bonds have a significant cumulative effect. Thus water molecules have an extraordinary tendency to stick together. This accounts for the high **surface tension** of water—its capacity to bead up on a surface such as a window pane. Water also moves by **capillary action,** being drawn through narrow tubes or vessels simply by the unwillingness of one water molecule to let go of another.

TABLE 2.5
The physical properties of three small molecules

NAME	FORMULA	MOLECULAR WEIGHT	MELTING POINT °C	BOILING POINT °C
Methane	CH_4	16	− 182.6	− 161.4
Ammonia	NH_3	17	− 77.7	− 33.4
Water	H_2O	18	0	100

The strong polarity of the water molecule is responsible for its being a liquid at room temperature (20°C) at which the other compounds are gases.

The conversion of solid to liquid or liquid to gas requires that a compound absorb sufficient kinetic energy so that individual molecules can move more rapidly. More rapid motion disrupts the stable relations among the molecules and causes a change of state. For each compound, the quantity of energy to be absorbed in order to cause a change of state is a constant directly related to the strength of the intermolecular attractions (weak bonds again). It is logical to predict that water, with very strong intermolecular bonding, would require an exceptionally large amount of energy input to melt from its solid form or to boil from its liquid form. Table 2.5 shows that by comparison with closely related compounds, water needs to absorb much more energy and to be raised to a higher temperature before it will turn into a gas, or will boil. Therefore, whereas CH_4 and NH_3 are gases at normal temperatures, water is a liquid. As a liquid, water serves to bathe, nourish, and transport cells. There is no available substitute for water on our planet!

Because water can absorb large amounts of heat without boiling, and give up large amounts of heat without freezing, the terrestrial water masses such as oceans and atmospheric water prevent the extremes of temperature experienced by other waterless planets. Venus, for example, has a daily surface temperature fluctuation of from + 530°C to well below zero. On the earth, the daily range rarely exceeds 15°C.

Water is frequently, if incorrectly, called the **universal solvent**. Most of the molecules important to life dissolve in it. Others, however, do not, and it is important to ask why. First, let us define our terms. When one substance (the **solute**) is mixed with a larger amount of another (the **solvent**) and the two mingle completely so that their molecules are evenly dispersed, a **solution** has been formed. In order for solution to occur, the two substances must be able to form compatible patterns of weak bonds with each other. Water can form hydrogen bonds with any type of molecule, such as NaCl, that is ionic, or that contains many polar covalent bonds, such as the sugars. These types of compounds dissolve readily. Molecules containing almost entirely nonpolar bonds, such as fats and oils, do not dissolve or even mix with water since they have no points of weak bonding in common.

Membrane structure; Chapter 3, pp. 82–84

Some compounds, on dissolving in water, release certain types of biologically active ions. Water causes ionic compounds to **dissociate**

$$NaCl \rightarrow Na^+ + Cl^-,$$

with each free ion becoming surrounded by oriented water molecules.

Water can also cause partial dissociation of polar covalent bonds. For an example, let us use the water molecule itself. About one in every ten million water molecules at any given moment is dissociated to yield a hydrogen ion (H^+) and a hydroxyl ion (OH^-):

$$H_2O \rightleftharpoons H^+ + OH^-.$$

Pure water thus has a predictable concentration of hydrogen ions per liter. In any solution the concentration of H^+ can be measured and expressed logarithmically as a **pH** value. The pH value of pure water is 7. Any solution with a higher concentration of H^+ than pure water is an acid, and its pH value will be less than 7. A base or alkali is a solution with fewer H^+ ions than pure water. It will have a pH value higher than 7.

We can define an acid compound, then, as one that dissolves in water, dissociates, and in so doing increases the H^+ ion concentration of the water. An acid, simply put, is a **hydrogen ion donor. Strong acids** are usually inorganic acids such as hydrochloric acid (HCl), that are completely dissociated. All of their hydrogen ions are free and contributing to the acidity of the solution.

$$HCl \rightarrow H^+ + Cl^-.$$

Weak acids, usually organic compounds, are those that, like water, are only partially dissociated. Even though there is a large amount of the compound present, only a tiny amount of hydrogen ion is free and contributing to the acidity.

acetic acid   acetate ion   hydrogen ion

Note that dissociation is reversible. Should some portion of any one of the three chemicals in the reaction above be removed, then the equilibrium would restore the original concentration.

We can define a base as a substance that dissociates in water releasing ions that are **hydrogen ion acceptors.** These ions remove H^+ and raise the pH. Strong inorganic bases such as sodium hydroxide (NaOH) release OH^-. When the hydrogen ion (H^+) meets the hydroxyl ion (OH^-) a reaction occurs that forms water.

$$H^+ + OH^- \rightarrow H_2O$$

This is called **neutralization.** It reduces the number of H^+ ions available. Weak organic bases frequently contain an **amino group** ($-NH_2$) that can accept hydrogen ion.

Mixtures of ions that can act to regulate or maintain pH are called **buffers.** They contain both hydrogen ion donors and hydrogen ion acceptors.

TABLE 2.6
Some pH values of liquids and foods

FLUID OR FOOD	pH	CONDITIONS
Gastric secretions	0.9	Empty stomach, rises when food added
Carbonated beverage	2.5–3.5	Depends on components, degree of carbonation
Wine	3.0–3.5	Type of grape, ripeness
Citrus fruit juice	2.6–4.4	Ripeness, species
Urine	4.8–7.5	Body allows wide range of pH, contributes to internal pH balance
Milk	6.6–6.9	May become very much more acid as microbial souring occurs
Blood	7.4	Strictly maintained at this value
Intestinal secretions	7.0–8.0	Contribute to neutralization of stomach contents
Distilled water	7.0	In sealed vessel; if CO_2 from air becomes dissolved in it, pH goes down
Fresh drinking water	6.0–7.0	Most sources slightly acid
Sea water	8.0–8.3	Lower pH values found near shore, in river mouths, etc., because of mixture with fresh water

Acidic environments are much more common than basic environments in nature. Many microbial species tolerate relatively acid pH values and grow well.

Water is of crucial importance for life because it is a highly efficient solvent, because it has a high degree of cohesiveness, because it absorbs and stores large amounts of heat energy, and because it dissociates ionic and polar covalent compounds giving rise to acidic and basic solutions. When we begin to discuss metabolic processes, it will also become clear that water supplies almost all of the hydrogen atoms used in organic compounds.

Carbon dioxide   Carbon dioxide is the biological starting material for the construction of larger, energy-rich, organic molecules. It is found in the atmosphere in small but adequate amounts. It is also dissolved in fresh and salt water. **Carbon fixation** (conversion of CO_2 to organic carbon) is the first step in ecological cycles. Carbon dioxide later is regenerated as a waste product of metabolism and is returned to the surroundings.

Carbon fixation; Chapter 6, pp. 164–165

Nitrogen sources   There are several compounds of nitrogen, an element essential for life, available in the inorganic environment. Most common is the diatomic gas, N_2, that makes up almost 80 percent of the atmosphere. This substance can be pro-

Nitrogen cycle; Chapter 6, pp. 165–167

Box 2.1

A Bacterium That Carries a Compass

The spiral bacterium MS-1 moves through the bog waters of its natural habitat, a cedar swamp, by means of the two flagella at its ends. However, being a primitive organism, it lacks conventional sense organs to find its way. It is anaerobic, so its survival will be favored if it migrates downward, toward the heavy, oxygen-free sediments. How does it distinguish up from down?

Each cell contains a series of regular, cubical inclusions aligned with its long axis. These are crystals of magnetite (Fe_3O_4), a major component of iron ore. When subjected to a magnetic field, the magnetobacterium rotates to align with the field, like a compass needle. Killed bacteria, no longer swimming independently, line up "head to tail." When cells are grown without much iron, there is no intracellular crystal formation, and the cells lose the magnetic response.

The earth's magnetic field has a large vertical component. By orienting themselves magnetically to the geomagnetic forces, the magnetobacteria may be able to selectively move to-

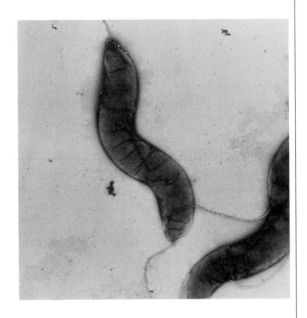

wards their habitat. These discoveries also raise questions about the possible role of magnetic crystals in the abdomens of honeybees and the skulls of homing pigeons.

ductively used by very few organisms, and most plants rely on nitrate ion (NO_3^-) or ammonium ion (NH_4^+). Many heterotrophs can use only organic nitrogen forms.

Minerals   The term **mineral**, from a nutritional point of view, refers to an inorganic substance needed in trace amounts in the diet. Some important minerals are iron, calcium, and magnesium. There are many others, and quantitative requirements for species other than *Homo sapiens* have been worked out only partly. It is worth noting that exposure to unneeded trace metals, or minerals in excessive concentrations, may be harmful to cells. Iodine- and mercury-containing preparations have long been used to inhibit microbial growth.

Disinfectants; Chapter 25

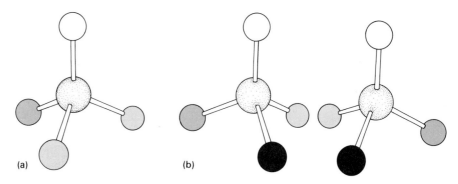

2.6
Asymmetry of Organic Molecules   (a) When the central carbon atom is bonded to
three different functional groups, a plane of symmetry exists to divide the molecule
into two identical halves. (b) When four different groups are bonded to one carbon,
there is no symmetry. Then the four groups can be oriented in two different
patterns. They resemble each other like mirror images or a right and left hand.
Asymmetry is seen in a number of biochemicals, especially sugars and amino acids.
One form (either D or L) always predominates in nature. Organisms that synthesize
the material attach the groups in one preferred orientation. Only "left-handed" or L-
amino acids are incorporated into proteins. Thus the α-helix always twists counterclockwise. However, some D-amino acids occur in the peptide linkages within the
bacterial cell wall substance called peptidoglycan.

Organic Compounds Necessary to Life

Organic chemistry is the chemistry of carbon compounds. Virtually every chemical
compound found in a living cell is organic or carbon-containing. Life on the earth is
possible because of the availability on this planet of adequate amounts of carbon.

Carbon atoms normally form four covalent bonds. If these are all single bonds,
they form at equal angles, and the molecule has a three-dimensional configuration
(Fig. 2.6). Carbon is also noted for its ability to bond to other carbon atoms forming
chains. The carbon-to-carbon bond is strong, stable, and very rich in energy.
Double bonds between carbons are flat and even richer in energy.

The simplest type of organic compound, the *hydrocarbon*, contains just carbon
and hydrogen. Hydrocarbons are found in fossilized plant and animal remains—our
dwindling fossil fuels—petroleum, natural gas, and coal. The characteristic molecular structure of a hydrocarbon is a chain of carbon atoms, with attached hydrogen
atoms. Such molecules may have the chain closed into a ring or may include mixtures of single and double bonds.

Just as carbon atoms may be assembled to form chains, so too can simple organic subunits or building blocks be hitched together. This process, called **polymerization**, is the primary means by which macromolecules are put together. The theory
is simple. Small molecules called **monomers** are linked together in a linear fashion to

Group	Name	Characteristics	Biochemical sources
R—OH	Hydroxyl group	Polar, forms hydrogen bonds. Compounds with this group may be called *alcohols*.	Sugars, alcohols Amino acids
$R-\overset{\displaystyle O}{\underset{\displaystyle OH}{C}}$ (carbon double-bonded to O, single to OH)	Carboxyl group	Because of intense electron pull towards the double bond, the hydrogen of the hydroxyl may ionize: $$-\overset{O}{C}\diagdown_{O^-} + H^+$$ This group is thus a hydrogen ion donor or acid; compounds with this group may be called carboxylic acids.	Fatty acids Amino acids
$R-\overset{\displaystyle R}{C}=O$	Carbonyl group or keto group	Strongly polar, highly reactive group. Compounds with this group may be called *ketones*.	Sugars, polypeptides, triglycerides, steroids
$R-\overset{\displaystyle H}{C}=O$	Aldehyde group	When the carbonyl group has a hydrogen associated it is even more reactive. Compounds with this group may be called aldehydes.	Sugars Aldehydes
$R-\overset{\displaystyle H}{\underset{\displaystyle H}{C}}-H$	Methyl group	Nonpolar grouping	Hydrocarbons Fatty acids Amino acids
$R-N\diagdown^{H}_{H}$	Amino group	Polar, a potential hydrogen ion acceptor or base, becoming the positively charged ammonium group. $$R-\overset{H}{\underset{H}{N}}-H^+$$ Compounds with this group may be called *amines*.	Amino acids Adenine Nucleic acids Amino sugars Chlorine Nucleotides, nucleic acids
$R-\overset{\displaystyle O}{\underset{\displaystyle OH}{P}}-OH$	Phosphate	May ionize once or twice to yield hydrogen ions; a strong acid. $$R-\overset{O}{\underset{O^-}{P}}-O^-$$ Charges may be neutralized by cations (Na^+, K^+, Mg^{++}) bonds to carbon or other phosphates have large energy content.	Phospholipids
R—SH	Sulfhydryl	May form high-energy linkage to carbonyl group. $$\begin{matrix} R \\ | \\ C=O \\ | \\ S \\ | \\ R \end{matrix}$$ May form disulfide bridge to another sulfhydryl by dehydrogenation. $$R-S-S-R \xrightarrow{\quad} 2H$$	Cysteine Methionine Proteins Coenzyme A

2.7
Biologically Important Functional Groups

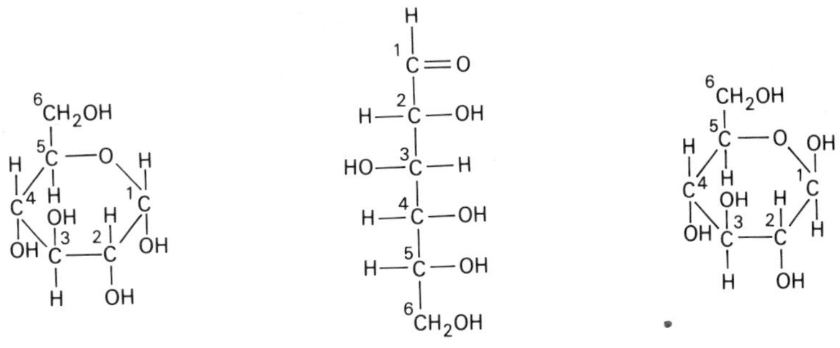

Alpha-glucose
(a)

Glucose straight-chain form
(b)

Beta-glucose
(c)

2.8

Glucose in Various Forms   There are several ways in which the atoms of glucose can interact. Since they are all roughly equivalent in energy content, a glucose solution will contain some molecules in each arrangement. (a) alpha-glucose, a ring-shaped structure, is the lowest-energy configuration and thus predominates at moderate temperatures. It is used in the assembly of starch and glycogen. (b) The straight chain form increases proportionately as the temperature is raised. (c) Beta-glucose may form when the straight-chain structure reenters the ring form. It is the building block for synthesis of cellulose. All three forms are readily interconverted.

form **polymers** just as boxcars are hitched up to form a train. We shall see, in looking at the major classes of biological molecules, that the separate units can be identified and may have particular uses by themselves. Usually, however, they are found assembled into larger groupings in which they contribute to highly specialized activities. Herein lies one of the basic differences between the inorganic and the organic worlds. Inorganic molecules may interact to form crystals. However, organic molecules can be assembled to form unique macromolecules.

The carbohydrates   There are four major groups of biochemicals—carbohydrates, lipids, proteins, and nucleic acids. The carbohydrates contain carbon and water, characteristically in the ratio $(CH_2O)_n$. Small amounts of other elements are less typically present.

   All carbohydrates are composed of simple building blocks, the **sugars** (Fig. 2.8). Simple sugars or **monosaccharides** consist of a chain of some number of carbon atoms. **Pentoses** are simple sugars with five carbons and **hexoses** are those with six. For each carbon in the chain, the sugar will have two hydrogens and one oxygen. In water, sugar structures usually are found in the ring form that is the lowest energy level of bonding. Sugars, having numerous alcohol side groups, are strongly polar in nature, and therefore readily soluble in water. Monosaccharides may also occasionally contain carboxyl groups, amino groups, or other unusual features (Table 2.7).

TABLE 2.7
Common and not-so-common sugars

GROUP	NAME	COMMENTS
Hexose	Glucose	Found in almost all organisms
	Fructose	Keto sugar, found in fruits, honey
	Galactose	Found in milk in disaccharide lactose
	Mannose	From walnut
Pentose	Ribose	In ribonucleic acid
	Deoxyribose	In deoxyribonucleic acid
	Xylose	From various wood products
	Arabinose	From pine gums
Alcohol sugars	Inositol	Cyclical 6-C compound; human B-vitamin
	Mannitol	From seaweed polysaccharides
	Xylitol	Reduced form of xylose; used in "sugarless" chewing gum
	Sorbitol	From various berries, also used in "sugarless" chewing gum
	Dulcitol	Reduced form of galactose
	Erythritol	4-C sugar alcohol from algae
Amino sugars	Glucosamine	Major component of chitin
	Galactosamine	Major component of cartilage
	N-acetyl-glucosamine	Bacterial cell wall polymer
	N-acetyl-muramic acid	9-C, bacterial cell wall polymer
	Neuraminic acid	9-C, important component of mucoproteins
Substituted sugars	Salicin	Glucose-salicyl alcohol from poplar bark
	Rhamnose	Methylated glucose, from poison sumac
	Ascorbic acid	Vitamin C, 6C ketosugar acid
Disaccharides	Sucrose	Glucose plus fructose; cane syrup
	Maltose	Glucose plus glucose; starch digestion
	Lactose	Glucose plus galactose; milk
	Cellobiose	Glucose plus glucose, from cellulose
	Trehalose	Found in insect hemolymph
Trisaccharide	Raffinose	From sugar beets and eucalyptus

Many of these sugars are used in diagnostic microbiology. Each microorganism has the ability to ferment only certain sugars, or none at all. By determining which of these an unknown organism attacks, it is often possible to identify the organism.

Much carbohydrate material is found in stable **disaccharides**, units of two sugars held together by covalent bonds. The formation of a disaccharide from two monosaccharides takes place by an enzymatically catalyzed reaction called **conden-**

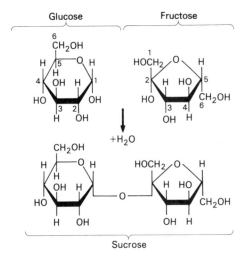

2.9
The Formation of Sucrose To join a molecule of glucose to one of fructrose, the specific enzyme must first remove a hydrogen from one and hydroxyl from the other, combining these to form water. Then the exposed carbon and oxygen can be bonded and a disaccharide results. This type of reaction is called a condensation because it removes water. It is the process used whenever building blocks (monomers) are linked together to form larger groups or polymers.

sation (Fig. 2.9). Such reactions, when taking place within a cell, normally require a source of chemical energy. These reactions involve formation of an energy-storing bond.

Biosynthesis; Chapter 9, pp. 247–248

This linkage process can also be reversed when the cell needs monosaccharides. The reaction separating two covalently linked monomers, by the introduction of a water molecule, is called **hydrolysis**. It is usually catalyzed by a different enzyme than was used for the condensation

$$\text{maltose} + H_2O \xrightarrow{\text{enzyme}} \text{glucose} + \text{glucose}.$$

Actively metabolizing cells move sugar units freely from the single to the polymeric form and back again.

Many cells produce and store large quantities of energy-rich carbohydrates. Sugars are converted into **polysaccharides**—large, macromolecular complexes. These are more stable than the smaller units. Also, their huge size hinders them from dissolving in water. They can be stored in cells or tissues. Other polysaccharides have considerable rigidity and, like cellulose, are used for structural support. The different properties of polysaccharides depend on the exact position of the bond that holds the sugars together. Cellulose has a type of linkage that defies the hydrolytic digestive enzymes of human beings. It is completely indigestible and is therefore not useful as a human food energy source.

Cellulose digestion by microorganisms; Chapter 6, p. 176

The role of the carbohydrates in nature is varied. Carbohydrates are quantitatively the most common biochemical entities. Most heterotrophic organisms derive the major portion of their energy requirement from carbohydrates. Carbohydrates also are the major structural and support materials used by microorganisms and plants. Cellulose alone makes up over 50 percent of the dry weight of the world's biomass.

The lipids A large, diverse, group of organic chemicals sharing certain properties is called the lipids. These compounds are made up largely of carbon and hydrogen

TABLE 2.8
Important, naturally occurring polysaccharides

NAME OF POLYMER	MONOMERIC UNITS	PHYSICAL CHARACTERISTICS	SOURCES, USES
Amylose	Glucose	Straight chain, flexible	Found in starch. Energy storage, bacteria and plants, food source for most organisms
Amylopectin	Glucose	Branched chain, flexible	
Cellulose	Glucose	Straight chain, rigid, strong, insoluble	Structural material plant cell walls, unable to be digested by most heterotrophs. Wood, paper, cotton
Dextran	Glucose	Cross-linked straight chain	Formed by bacteria from sucrose. Important in dental caries as basis of plaque. Used as plasma extender.
Glycogen	Glucose	Extensively branched	Energy reserve for bacteria, animal tissue
Pectin	Galactose Galacturonic acid Arabinose	Jelly-like, water soluble	Intercellular adhesive in plant tissue, used for jellies, applesauce preparation, paints.
Inulin	Fructose Glucose	Straight chains	Replaces starch in some plants. Useful in identifying some bacteria.
Chitin	N-acetyl-gluosamine	Unbranched, very rigid	Cell wall in some fungi, exoskeleton of crustaceans, insects
Sodium alginate	Mannuronic acid	Sodium salt, gel-like, highly water absorbant	From cell wall of brown algae. Used in food preparation for stabilizer, thickener.
Agar	Galactose, sulfate	Gel-like mixture, water absorbant	From red algae. Used in bacteriological media, foods.
Carrageenan	Galactose, sulfate	Gel-like mixture, forms emulsion in water	From red algae. Used in dairy products as stabilizer.

These materials are polymers of five- and six-carbon sugars. The monomers are sometimes altered by addition of alcohol, acid, or amino groups. These polysaccharides may fulfill either nutritive or structural functions. Many have important uses in food processing, medicine, or industry.

and contain only small amounts of oxygen, and in some cases traces of nitrogen and phosphorus. Because they are composed largely of the nonpolar H—H and C—C bonds, lipids are nonpolar or **hydrophobic** and will not mix with water. There are five groups, each with a unique biological function.

Triglycerides (neutral fats) (Fig. 2.10) are most familiar in daily living as fats and oils. They are composed of three **fatty acid** molecules linked to one molecule of a three-carbon alcohol called **glycerol**. Fatty acids are hydrocarbon-like substances, even-numbered chains of carbon more or less **saturated** with hydrogen. A fatty acid

2.10
Neutral Fats   A triglyceride or neutral fat is formed when three fatty acids are joined to glycerol backbone. Three condensation reactions are required to assemble the fat. Stearic acid is a saturated fatty acid. Oleic acid is unsaturated, and linoleic acid is polyunsaturated. Fatty acids of more than one type may be linked up to form a "mixed" molecule.

is saturated when it has the maximum number of hydrogen atoms per carbon atom. When carbon atoms are linked by double bonds, however, each can hold only one hydrogen atom, and the fatty acid is unsaturated. If two or more double bonds occur in a single fatty acid, then it is **polyunsaturated.** Triglycerides are usually called fats if they are solids at room temperature and oils if they are liquids at room temperature. Most natural fats and oils are mixtures of triglycerides. They are the biological result of the random linking together of a glycerol molecule with any three fatty acids available to be linked. If within the mixture a predominance of the fatty acids used are unsaturated (as they usually are in plant cells), the substance will be a mixture of the sort called an oil. If the fatty acids are mostly saturated (as they are in most animal tissues), then the mixture will be a fat. Fats and oils alike are used by the organism as an energy storage mechanism. Gram for gram, lipids contain more calories than any other type of biological chemical. They also, like polysaccharides, are readily deposited in special areas of the cell or tissues.

Waxes are structures similar to the triglycerides in that they contain alcohols and fatty acids. However, the alcohol in waxes is usually a larger molecule, up to 20 carbon atoms in length, esterified to a single fatty acid. All waxes are highly viscous pastes or solids at ambient temperatures. They function as protective waterproofing and lubricating coatings on exterior surfaces.

Phospholipids (Fig. 2.11) are found in biological membranes. Their function in membranes is to separate the aqueous contents of the cell from the aqueous surrounding environment. They are able to do this because they have a structure that contains both highly polar and nonpolar regions. Phospholipids differ from triglycerides only in that the third position on the glycerol carbon chain is occupied by a highly polar phosphorus-containing unit rather than a nonpolar fatty acid. Such molecules, when placed in water, will orient themselves in such a way as to place all nonpolar portions in contact only with nonpolar neighbors, to which they are joined

2.11
The Phospholipids This molecule is a lecithin, one of a group of phospholipids containing choline as the polar group. They are widespread in cell membranes and are concentrated in human nervous tissue, especially the myelin sheaths that insulate the axons of nerve cells. The polar phosphate-bearing head is hydrophilic, and can be wetted by water. The nonpolar tails are hydrophobic, and orient themselves away from the aqueous environment.

Box 2.2

Unusual Lipids in Microbial Outer Layers

Modified chromatography or electrophoresis techniques can be used to separate and identify the lipids of microbial cell walls or cell membranes. Often, unexpected materials or lipid distributions are found. These may correlate with the organism's biological properties.

The *Mycoplasma* genus of bacteria is unusual in that it lacks a cell wall. Its cell membrane contains sterols, like an animal cell, making up to 36 percent of its total substance. The mycoplasmas are animal parasites; they live in a uniquely close association with the surface of the tissue they feed off. In fact, under the microscope it often looks as if the membrane of parasite and host have fused. Fusion is possible largely because of the biochemical similarity between the two.

Mycobacterium tuberculosis, which causes TB, is unusually resistant to destruction by drying and chemicals. This allows it to persist and be spread by drafts of air after a person with the active disease has coughed it out. The resistance is conveyed by a high concentration of lipids called mycolic acids in its cell wall. These lipids also cause its characteristic acid-fast staining trait.

Legionella pneumophila, which causes Legionnaire's disease, among its other odd traits possesses large amounts of branched-chain fatty acids. Recently, identical types and amounts of these substances were demonstrated in an unclassified rickettsia-like organism that had been isolated in 1947. This demonstrates that *Legionella* is not a "new" pathogen, but has been around a long time.

Halobacterium is an **extreme halophile**—it grows in saturated salt solutions. Its membrane phospholipids do not contain fatty acids, but long chain terpenes, linked to glycerol by an uncommon and highly stable linkage. These biochemical oddities are directly related to the organisms' ability to tolerate the high salt concentrations and alkaline pH of the salt lakes in which they live. These conditions would turn ordinary membrane lipids into soap!

by van der Waals forces. All polar, phosphate-containing portions will be oriented toward the water molecules, with which they will form hydrogen bonds. These arrangements result in the formation of **lipid bilayers.** All real biological membranes seem to be sophisticated variations on this simple plan.

Membrane structure; Figure 3.14

Steroids are lipid molecules with a four-ring structure as shown in Fig. 2.12. If they contain alcohol groups in certain positions, they are called **sterols.** The best-known sterol is cholesterol, a substance used by the human body in digestion. With other sterols it is an important component of cell membranes in yeast, protozoans,

2.12
Other Classes of Lipids   The four-ringed structure is characteristic of the steroids. Cholesterol, which like all sterols has an alcohol group at position 3, is an important cell membrane constituent. Cortisone is a steroid hormone, produced by the adrenal cortex in human beings. It is widely used as a drug. Its beneficial effects are the control of inflammation and hypersensitivity responses, but it has many undesirable side effects, such as suppression of immunity. B-carotene, and vitamin A, which is derived from it, are terpenes. Their alternating double and single bonds allow them to capture electrons in visible light. B-carotene directs captured electrons to photosynthetic pathways, while vitamin A directs them to produce a visual stimulus in the retinal cells of the eye.

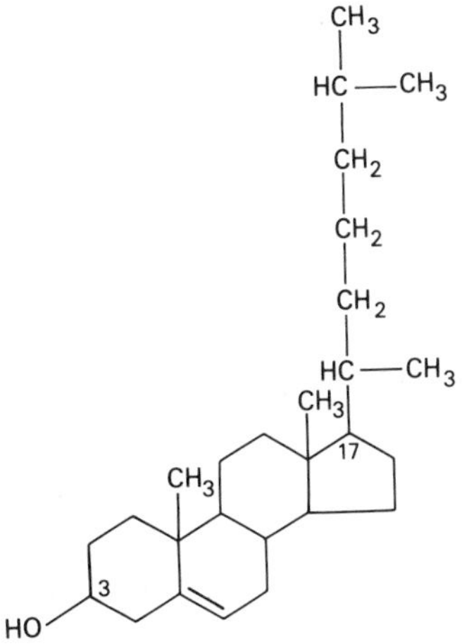

(a)   Cholesterol

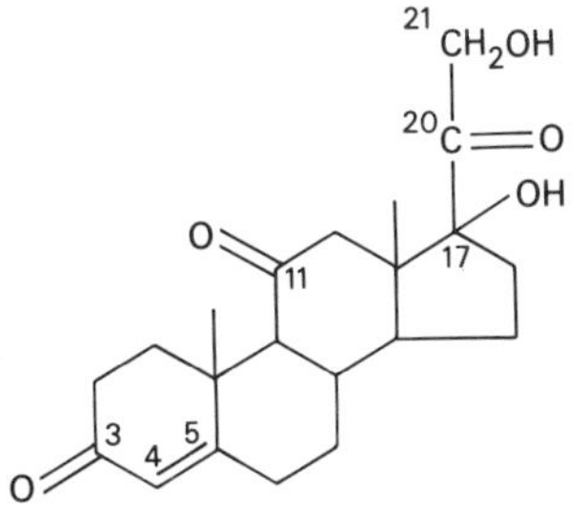

(b)   Cortisone

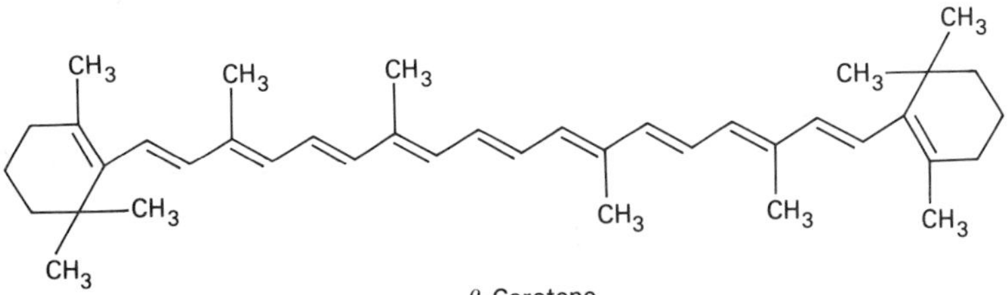

(c) β-Carotene

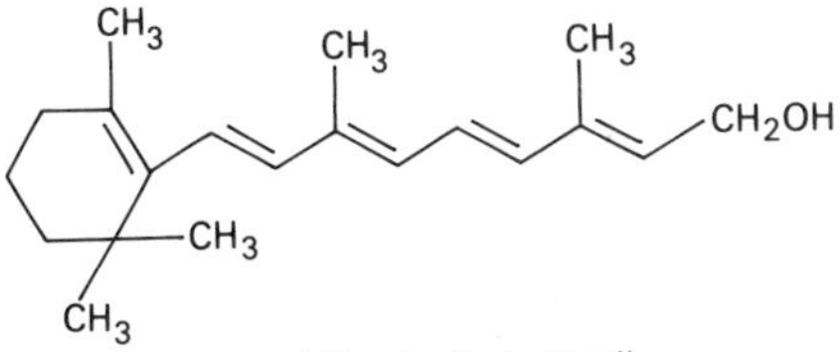

(d)   Vitamin A$_1$ (retinol)

and animal cells. Sterols are found in only one group of bacteria, the *Mycoplasma*. The steroids are important animal hormones regulating growth, development, and reproduction.

Terpenes are compounds assembled from five-carbon groups. They have double bonds regularly alternating with single bonds, which gives them the capacity to transmit electrons. Terpenes are represented by the side-chains of the chlorophyll molecule, carotene pigments, and vitamins A, E, and K.

Chlorophyll; Chapter 8, pp. 239–241

Proteins   The assembly or synthesis of proteins is the single most important task to which a cell devotes its energy resources. A complex set of biochemical machinery is devoted to this job.

Proteins are polymers of small subunits, the amino acids. They differ from polymers such as the polysaccharides in that they are assembled from 20 different subunits rather than one or two. Another difference is that the crucial thing that distinguishes one protein from another is not the nature of the bond holding the units together but rather the sequence of the units. Because they are assembled from many different units, the number of unique sequences that can be formed is infinite. Each unique sequence has its own characteristic biochemical activity.

Amino acids   Amino acids are the monomers from which proteins are assembled. Their basic structure includes an amino group, which functions as a base in water, and a carboxyl group, which functions as an acid. Because the molecule is both acidic and basic, it is called an **amphoteric** molecule. Because it can both add and withdraw hydrogen ions from solution, it tends to maintain pH at or around neutrality and thus functions as a buffer. Each amino acid has a side-chain, an organic radical that gives it particular characteristics. Certain side-chains contain another carboxyl group, making the amino acid an acidic one. Similarly, extra amino groups render the compound more basic. Furthermore, the central portion of each amino acid is quite polar because of the amino and carboxyl groups, but this polarity can be either increased or decreased by the character of the side-chains. Some of the nonpolar amino acids such as tyrosine have very limited water solubility. Two amino acids contain sulfur, and one of these, cysteine, is particularly important because its **sulfhydryl** group (-SH) can become covalently linked to other sulfhydryl groups.

In addition to the 20 amino acids shown in Fig. 2.13, which are those that are genetically incorporated into proteins, there are in excess of one hundred other, mostly uncommon, amino acids found in specialized structures. We will encounter some when discussing bacterial cell walls.

Cell wall; Chapter 3, pp. 76–82

2.13

The Amino Acids   Twenty amino acids, in the left-handed L form, are assembled into proteins under direct genetic coding. All share a common backbone, and vary in their side-chains (shaded). Those amino acids with polar side-chains are readily water soluble. The groups designated aromatic and nonpolar are not.

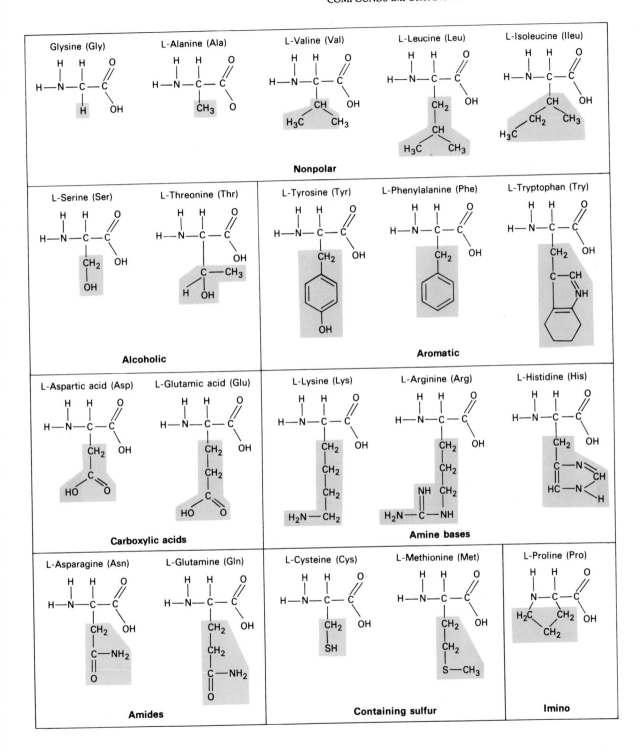

Glysine (Gly) L-Alanine (Ala) L-Valine (Val) L-Leucine (Leu) L-Isoleucine (Ileu)

Nonpolar

L-Serine (Ser) L-Threonine (Thr) L-Tyrosine (Tyr) L-Phenylalanine (Phe) L-Tryptophan (Try)

Alcoholic   **Aromatic**

L-Aspartic acid (Asp) L-Glutamic acid (Glu) L-Lysine (Lys) L-Arginine (Arg) L-Histidine (His)

Carboxylic acids   **Amine bases**

L-Asparagine (Asn) L-Glutamine (Gln) L-Cysteine (Cys) L-Methionine (Met) L-Proline (Pro)

Amides   **Containing sulfur**   **Imino**

Asp   Cys   Pro   Val   Gly

2.14
Peptide Formation   The amino acid glycine is being added to a peptide composed of four amino acids, asp-cys-pro-val. Peptide bonds (shaded) are formed by condensation. Elongation of peptides occurs at the carboxyl end of the chain. Note that when proline is peptide-bonded, it is the only amino acid that does not retain a hydrogen on its nitrogen atom. For this reason, it is unable to form hydrogen-bonded secondary structures such as the alpha helix.

Primary structure of proteins   *In vivo* assembly of proteins requires the presence of an elaborate genetic mechanism, to be discussed in detail in Chapter 9. Limiting ourselves here to the chemistry of protein synthesis, let us look at the basic polymerization step that links amino acids together. The carboxyl terminal of the first amino acid undergoes a condensation reaction with the amino terminal of the second amino acid. Water is withdrawn and a single covalent bond is formed (Fig 2.14). This type of bond is frequently referred to by the special name of **peptide bond**. As a protein is synthesized, the chain will continue to grow by addition of each amino acid in turn to the carboxyl group of the one before it until the sequence is complete. A protein's primary structure, then, is its covalently bonded sequence of amino acids. This first level of structure is very stable and is not usually altered or disrupted by anything short of enzymatic hydrolysis.

Protein synthesis; Chapter 9, pp. 258–260

Secondary structure—the alpha helix   The primary structure is neither flat nor unreactive. It has a strong tendency to undergo natural rearrangements in solution to reach more stable and lower-energy conformations. One of these is the alpha or left-handed **helix**. Bond angles in the protein are such that the chain twists on itself. The twisting brings into proximity oppositely charged polar regions on every fourth amino acid. Stabilizing hydrogen bonds lock the twisted amino acid chain into a regular helix. Most proteins in an aqueous environment will spontaneously form the alpha helix wherever possible. The helix cannot form in areas of a sequence in which the amino acid proline is found, however. Its presence will cause a kink in the regular arrangement. Certain proteins of a structural nature tend to stabilize in another pattern—highly regular hydrogen-bonded sheets or fibers called **beta structures**.

Secondary structures are interactions within portions of the primary structure through complementary hydrogen bonding. They are dependent on the sequence of amino acids laid down during assembly of the primary structure. Because secondary structures represent lower energy levels, input of energy, for example when a pro-

Temperature ranges for growth; Chapter 7, pp. 194–195

tein is heated, may disorder the regular bonding pattern. When this occurs, the protein frequently will not reorder itself but will be left in aggregates of random coils. Since this results in a loss of biological function, the protein is said to be **denatured** when this occurs.

Tertiary structure An alpha helical secondary structure is, as we have seen, a sequence of helical regions with bends between. The bent coil tends to form weak bonds with other portions of the coil. Weak bonds cause the folding of the molecule into a roughly spherical or **globular** form. Note that a globular protein has a surface that is far from smooth. Depending on the folds, the surface has a number of variously sized, shaped, and charged indentations and projections. Its **surface configuration** clearly depends on the way in which the helix is formed, and that depends on the amino acid sequence. Thus for each unique primary structure, a unique tertiary structure will result (Fig. 2.15).

Tertiary structure is loosely stabilized by a combination of hydrogen bonds and van der Waals forces. It is more strongly maintained by the formation of covalent **disulfide** bonds between cysteine molecules in different parts of the sequence. Much of the tertiary structure is held sufficiently loosely so that it can undergo reversible changes. That is, in the course of performing its function, the protein can change its surface conformation temporarily, then return to its original form. These **allosteric** changes seem to be crucial to an understanding of how certain enzymes and transport proteins work.

Feedback inhibition; Chapter 9, pp. 219–221

2.15
Quaternary Protein Structure Hemoglobin, the oxygen-carrying protein, resembles cytochrome C. It is a quaternary complex of four units, each of which is in turn a polypeptide-heme complex. The characteristic spacing of the four heme groups is essential if the protein is to have the correct affinity for oxygen.

Quaternary structure   Certain proteins in their functional state consist of complexes of several **polypeptide chains**, another name for a protein or portion of one. For example, hemoglobin is made up of four chains—two alpha chains and two beta chains (Fig. 2.15). RNA polymerase, the enzyme that assembles RNA, is made up of six different polypeptide subunits. The quaternary structure is held together, like the tertiary structure, by disulfide bonds, hydrogen bonds, and van der Waals forces. Because of the folding patterns, most globular proteins have many polar amino acids on the surface, and therefore have **hydrophilic** surfaces, whereas their foldings and inner surfaces may be strongly hydrophobic. This difference from area to area allows the protein molecule to interact with other cell components by a variety of means.

Proteins have two important functions in living material. Firstly, they are **enzymatic** or catalytic. Globular proteins can hold and interact with reactants to increase tremendously the probability that they will react with each other. This speeds up a process. For an enzyme to function, it must have precisely the surface configuration needed to interact closely, but not too tightly with its particular **substrates** or reactants.

Other proteins form parts of cell structures, such as membranes, ribosomes, or flagella. They confer elasticity rather than the rigidity found when polysaccharides are used. **Structural** proteins such as collagen are usually immobilized, when incorporated in the structure, by complex interactions with neighboring molecules.

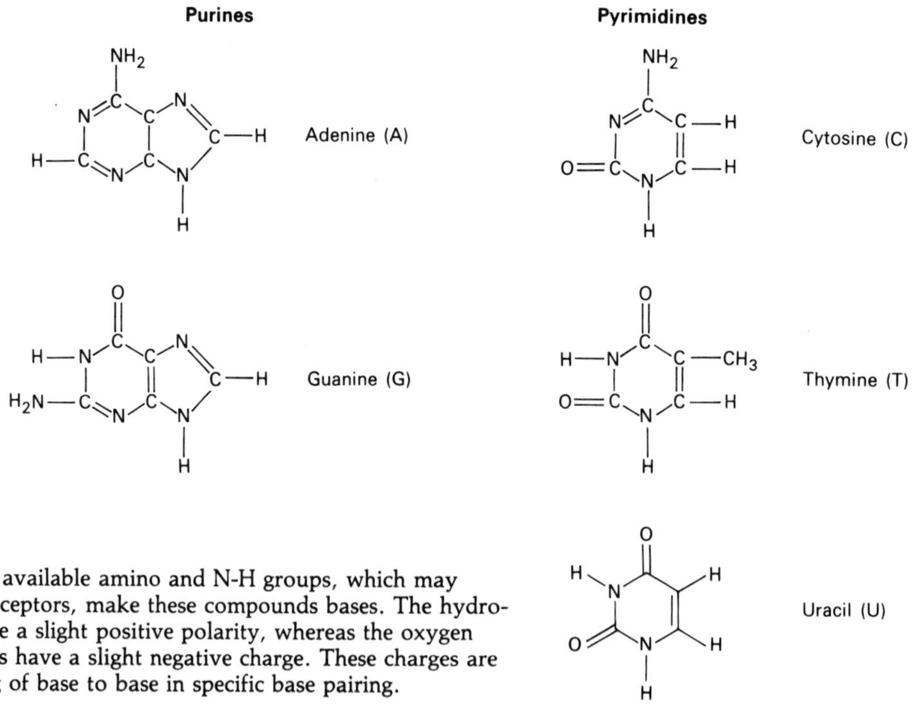

Purines

Adenine (A)

Guanine (G)

Pyrimidines

Cytosine (C)

Thymine (T)

Uracil (U)

2.16
Nitrogenous Bases   The available amino and N-H groups, which may serve as hydrogen ion acceptors, make these compounds bases. The hydrogens of these groups have a slight positive polarity, whereas the oxygen atoms of carboxyl groups have a slight negative charge. These charges are the basis for the bonding of base to base in specific base pairing.

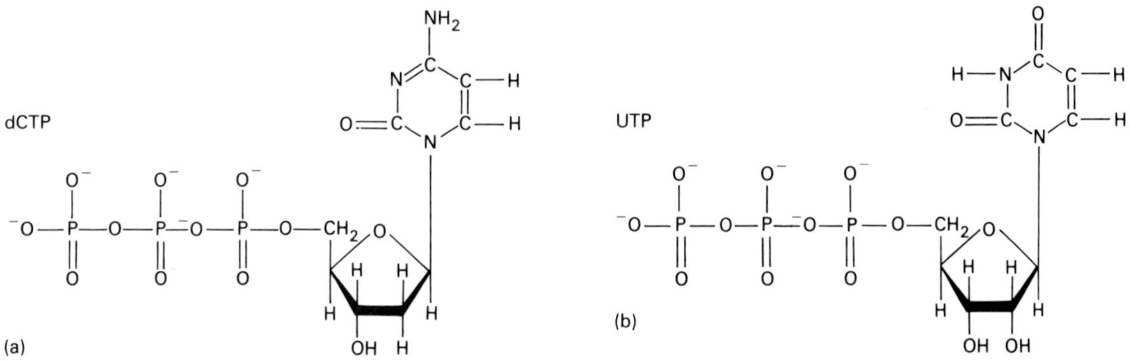

dCTP

UTP

(a)

(b)

2.17
Representative Nucleotides (a) Deoxyribocytidine triphosphate (dCTP) normally is found in solution in the anionic form, strongly charged. It may be linked with other deoxyribonucleotides in the triphosphate form to synthesize DNA. (b) Uridine triphosphate (UTP) contains ribose; it is a ribonucleotide and is incorporated into RNA.

The Nucleic Acid

Nucleic acids fulfill some of the cell's most complex functions but they fall into two simple chemical classes, the **deoxyribonucleic** acids (DNA) and **ribonucleic** acids (RNA). Like proteins, they are linear polymers of subunits. Their individuality also lies in the subunit sequence in the chain.

Nucleotides   The repeating unit or monomer of the nucleic acid is the **nucleotide**, made up of a nitrogenous base (Fig. 2.16), a sugar, and from one to three phosphates. This basic pattern is modified in several ways. The sugar is a pentose—**deoxyribose** in DNA and **ribose** in RNA. The bases can either have one carbon–nitrogen ring (the pyrimidines) or two such rings (the purines). Five different bases are commonly encountered in nucleic acids, only four of them in either type. DNA is assembled in the normal cell from deoxyribonucleotides containing adenine, guanine, cytosine, and thymine (Fig. 2.17). RNA is assembled from ribonucleotides containing adenine, guanine, cytosine, and uracil.

In a condensation reaction, the sugar of the first nucleotide becomes linked to the phosphate of the next. A backbone of covalently linked, alternating sugars and phosphates is formed. The bases project out at an angle from the chain and are free to participate in interactions with other compounds.

Nucleic acid synthesis; Chapter 9, pp. 253–258

Base pairing   The nitrogenous bases possess numerous polar bonds and are capable of forming hydrogen bonds with other polar groups (Fig. 2.18). Most importantly, they bond with each other in a specific pairing mechanism. A purine always bonds to a pyrimidine. The base pairs that function in common nucleic acid interactions are adenine bonded to thymine, adenine bonded to uracil, and guanine bonded to cytosine. Each base has only a negligible tendency to bond to any base other than its normal partner.

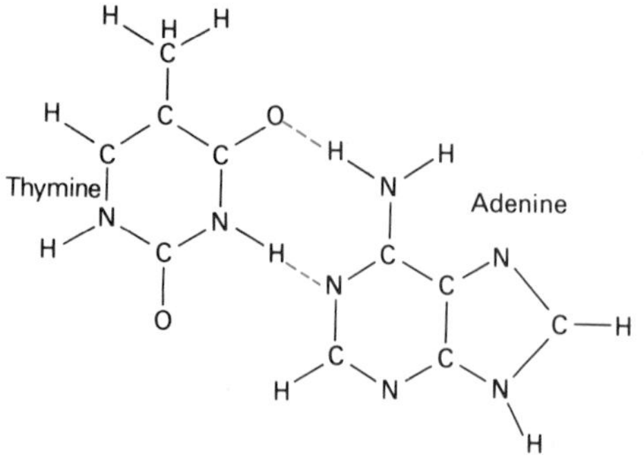

Thymine

Adenine

Uracil

Adenine

Cytosine

Guanine

2.18
Base Pairs When adenine approaches either thymine or uracil at the correct intermolecular distance, two hydrogen bonds may form between the closest, oppositely charged points. Guanine and cytosine can form three bonds; a GC base pair is significantly stronger than an AT-pair. Each pairing arrangement brings together a purine and a pyrimidine, so the pairs all occupy comparable spaces.

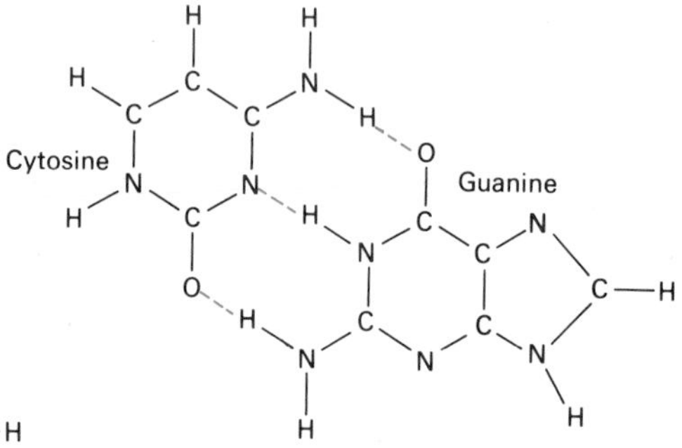

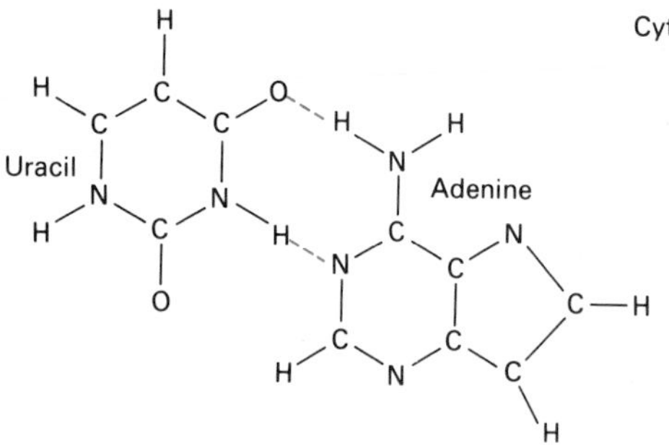

Base pairing is responsible for the secondary structure of the DNA molecule. Two strands of covalently linked nucleotides are twisted around each other in a **double helix** (Fig. 2.19). The two strands are exactly complementary—that is, each base has opposite it on the other strand exactly the base with which it can form stabilizing hydrogen bonds. The core of the helix is a "ladder" of base pairs whose bonds maintain the helical structure.

RNA molecules are usually single-stranded but may have limited regions of base-pairing. However, each type of nucleic acid when single-stranded may base-pair to other nucleic acids if their sequences are complementary. This is a recognition mechanism of utmost importance in molecular genetics.

Cyclic nucleotides   Certain nucleotides function in regulatory capacities (Fig. 2.20). The cyclic nucleotides, among which cyclic AMP is the most studied, are ubiquitous, intracellular regulators. Their functions relate to adjusting the rates of cellular processes to conform to external stimuli. As such their concentrations rise when a cell undergoes a change of function or participates in development or differentiation. Cyclic nucleotides are produced enzymatically from the common ribonucleotides.

Cholera; Chapter 20, pp. 592–593

The ribonucleotide, ATP, in addition to being incorporated into RNA, has a unique role in cellular metabolism. It serves as an energy-storage molecule. In meta-

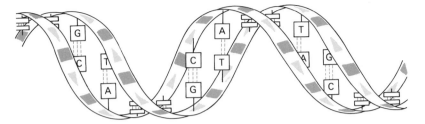

2.19
The Double Helix of DNA   The DNA molecule is composed of two
strands of nucleotides, each of which is covalently linked to its neighbors
above and below. The strand's backbone is alternating deoxyribose and
phosphate groups. The nitrogenous bases project from the backbone. The
two strands are complementary. Each has opposite it on the other strand
precisely the correct base for hydrogen bond formation. Thus base pairs
are formed at regular distances, holding the two strands zipped together.
The two strands coil around each other to form a double helix.

bolism, whenever carbohydrates or fats are broken down, the released energy is
captured in more readily usable form by synthesis of ATP. The stored energy can
then be supplied to drive a large number of different energy-requiring processes such
as biosynthesis, movement, or transport. Other ribonucleotides (GTP, UTP) also
share in this function in limited places in the metabolic picture.

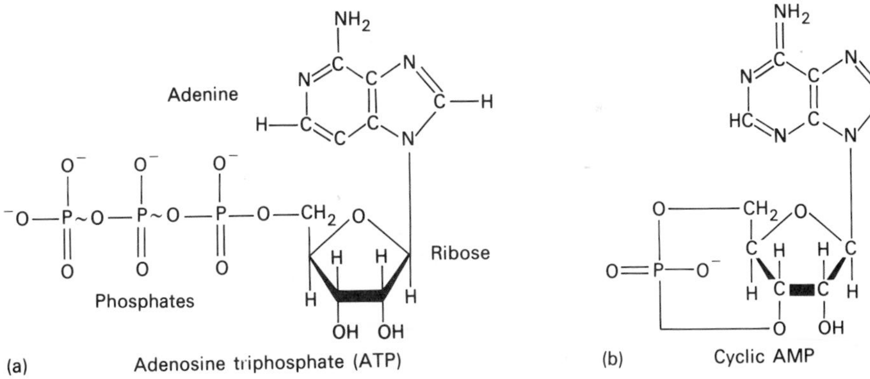

Adenine

O^- O^- O^-

$^-O—P\sim O—P\sim O—P—O—CH_2$

Phosphates

OH OH

(a)   Adenosine triphosphate (ATP)

Ribose

(b)   Cyclic AMP

2.20
Nucleotides in Metabolism   (a) ATP is the universal short-term energy storage
molecule in the cell. Two of its phosphates are attached by unusually energy-rich
bonds ($\sim$). When these are enzymatically broken, the released energy may be used
to drive energy-requiring processes. The molecule is identical to the one used in
RNA synthesis. (b) Cyclic AMP (cAMP) is an intracellular substance that mediates
the cell's response to external chemical signals. It is known to be needed for the
development of fruiting bodies in the slime molds, as well as response to hormones
in human cells.

FROM MOLECULE TO STRUCTURE

Having surveyed the various types of biological molecules to be found in the cell, we can now analyze their specific interactions. These relationships are what make a cell something other than an extraordinarily complex jumble.

Mixtures

We have surveyed biochemistry molecule by molecule, in a sense. This is all well and good from an academic point of view but has only limited relevance to real life. The contents of cells, the makeup of biological fluids such as blood plasma or milk, and the aquatic environment (even clean fresh water) are mixtures in a chemical sense. They contain a variety of compounds in a wide range of concentrations.

Cytoplasm   For an example, let us consider **cytoplasm**, the fluid interior substance of cells. It contains ions such as Na^+, Mg^{++}, and Cl^-. Also present are a variety of small, soluble polar molecules such as the sugars and amino acids. These substances are more or less evenly distributed. The globular protein components form a special type of mixture with water—a thick, viscous, cloudy **colloid**. In a colloid, the particles do not dissolve. Rather they remain evenly distributed without settling out. In many types of cell the colloid is thick, or gel-like, at the periphery and thinner and more fluid in the inner areas. Cytoplasm is, to summarize, a complex collodial mixture of those cell components that are soluble in water or that have hydrophilic surfaces.

Structural units   Cell structures, on the other hand, are assembled from molecules that prefer to associate with molecules other than water. Lipids, for example, are not found in the cytoplasm. Because of their nonpolar nature, they are found primarily deposited in fat inclusions or bound into membranes. When lipids are synthesized in the cell, they migrate to areas in which they can form such hydrophobic interactions.

Cell membrane; Chapter 3, pp. 82–84

Self-assembly   Many simple cell structures on study prove to form by a **self-assembly** process. Look again at how phospholipids in the presence of water form lipid bilayers. Similar patterns of aggregation are seen in the assembly of virus protein coats. TMV (the tobacco mosaic virus of plants) provides a well-studied example. Each TMV particle is composed of a strand of RNA enclosed in a tubular protein coat. When the virus is produced inside the plant cell, the coat is assembled from protein subunits. Each is composed of a sequence of 158 amino acids in an identical globular tertiary form. The subunits have surface configurations such that they can form stable weak bonding patterns if they line up side by side. Some 2,220 identical proteins assemble as they are synthesized and aggregate to form a coiled strand. This in turn forms a hollow tube around the nucleic acid core.

Bacterial flagella; Chapter 3, p. 87

Another self-assembly process is seen in the lengthening of bacterial flagella and eucaryotic microtubules. In each case, the structure is formed from a single type of protein. It elongates as molecules of the structural substance add on to the basal end

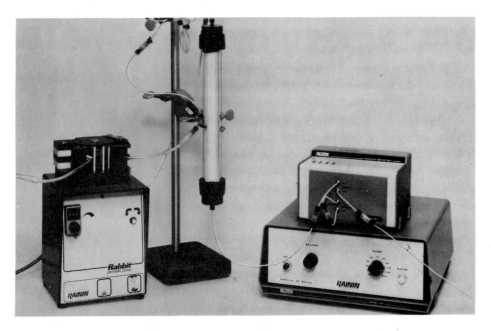

2.21
Liquid Chromatography A liquid that contains a mixture of substances is placed on the top of a column of resin in a glass tube. Buffer is pumped in at a steady rate to carry the substances down through the column. Some adhere more strongly than others in the mixture, so they progress through the resin at differing rates. As the fluid exits from the bottom of the tube, it is passed through a detecting unit that monitors the appearance of various components, and collects the fraction that is desired.

of the structure. Simple self-assembly processes of this sort can be readily duplicated in the test tube (*in vitro)* with the result that a visible structure can be seen to appear before the scientist's very eyes.

Most cell structures are not, however, aggregates of molecules of a single type. They are solid mixtures, just as cytoplasm is a fluid mixture. Cell membranes, for example, contain several different phospholipids, at least four structural proteins plus an undetermined number of enzymatic proteins. The smaller subunit of bacterial ribosomes contains one RNA molecule and 19 different proteins. Scientists have not yet progressed to the point where they can create suitable conditions to allow these structures to assemble *in vitro*. It seems likely, however, that all cell structures do self-assemble *in vivo*. Recognition of the self-assembly principle removes the necessity for thinking about cellular "machinery" to build cell structures. This principle implies that once the cell has manufactured biochemicals of the sorts we have studied, the chemical nature, relative polarity, and surface conformation of the molecules will make it probable that they will end up in the region of the cell in which they will be functional, associated with other substances with which they can effectively cooperate. The apparently random mixture of chemicals that comprises a cell is not then really random at all.

The Separation of Mixtures

Biochemical research deals constantly with the need to **separate** one component from a mixture, or to **assay** it to determine exactly how much is present in the presence of other substances. It would be worthwhile to glance at some of the separation techniques in use.

Filtration   Large aggregates or subcellular structures may often be removed from a mixture by **filtration** on the basis of their size. The mixture is passed through a filter that retains all material above a certain size. Ordinary laboratory filter paper is familiar to most. However, it retains only particles equal or greater in size than the largest cells. Membrane filters have much smaller pores; they will retain bacteria and certain types of macromolecules. Molecular sieve materials have been developed that can remove molecules of any desired size from solution; the process is called **ultrafiltration**.

Filter sterilization; Chapter 25, pp. 732–733.

Electrophoresis   Molecular separation can be made on the basis of differences in charge. A mixture is placed in an electrical field and allowed to diffuse in a buffer solution that conducts current (an **electrolyte**). Molecules will migrate towards the positive pole if they are negatively charged at a rate that indicates how strongly charged they are. Positively charged molecules will move in the opposite direction. This technique called **electrophoresis**, can be used in a variety of sophisticated ways to separate mixtures of enzymes, antibodies, and other large molecules (Fig. 2.22).

Immunoelectrophoresis; Chapter 13, p. 388

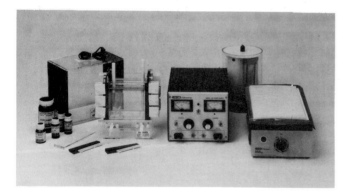

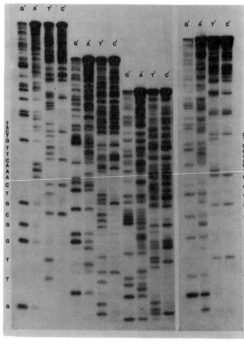

2.22
Electrophoresis A mixture to be separated is spotted onto the surface of a slab or agar gel. The slab is then placed vertically in the center of the glass jar. Buffer is added to fill and an electrical current is passed through. Substances migrate along the gel at different rates of speed depending on their electrical charge. When the gel is removed and stained, the components of the mixture, in this case small fragments of nucleic acid, are spread out. Comparison of the different lines of spots made by different samples will allow the researcher to identify key components by a sort of fingerprint pattern.

Chromatography   Because molecules differ in their polarity, they may be separated by their affinity or lack of affinity for other polar substances. Many forms of **chromatography** use this principle. Most familiar is the paper strip technique in which a mixture is spotted on a supporting piece of filter paper and immersed in a nonpolar solvent. Paper is composed of cellulose, a polar carbohydrate. As the non-polar solvent moves across the surface of the polar paper, it takes along nonpolar molecules at the fastest rate. Semipolar ones move slowly, and strongly polar ones, not at all. By adjusting the relative polarities of the support (which may be paper or cellulose beads or fibers in a column) and the solvent, most mixtures can be resolved.

Centrifugation   **Centrifugation** separates particles and/or macromolecules on the basis of both size and shape. The principle is similar to the spin cycle on the domestic washing machine. As the mixture is spun at high speeds, tremendous forces are generated that propel components toward the outside. With mixtures in test tubes, the larger and heavier materials move to the bottom most rapidly. They sediment first, followed by the sedimentation of progressively lighter materials. The density of the fluid can be increased by addition of sucrose, cesium salts, or other compounds. This slows down migration and increases the apparent difference in size. Separation of even very similar molecules can be made.

These are but a few of the separation techniques available to the researcher interested in teasing out particular pieces of the cellular puzzle. Many separations make use of a sequence of techniques to progressively purify an interesting substance. Very frequently, new discoveries await the development of a new technique. When it appears, researchers all over the world rush to apply it to diverse problems and a flood of new knowledge is released. This has been true in microbiology since its beginnings, when the discovery of the useful properties of agar started the whole era of bacterial study.

SUMMARY

1. All matter is composed of atoms of the various elements. Each element has characteristic numbers of protons, neutrons, and electrons arranged in shells. If the numbers of neutrons vary, then the different atomic forms are isotopes and may be radioactive.

2. The number of electrons in an atom's outer shell determines its valence, or combining potential. Because formation of an outer shell with eight electrons is stabilizing, atoms form bonds in order to share electrons.

3. Oxidation–reduction reactions form ions (charged particles) that in turn can then form ionic bonds. Sharing of electrons between atoms results in covalent bonds. Equally shared electrons yield nonpolar bonds, but unequal sharing gives rise to polarity. These strong bonds hold molecules together. Weak bonds of the hydrogen bond and van der Waals types are responsible for interaction among molecules and for the higher levels of structure of the macromolecules.

4. Most reactions are two-directional and subject to considerable variation both in direction and rate. In order for a reaction to occur, its energy needs must be

met. Its rate will be enhanced by increases in concentration of the reactants, by increase in temperature, and by the catalytic effect of the specific enzyme.

5. Water is an inorganic molecule, but its peculiar properties are essential to the preservation of life. Because the polar water molecules are strongly hydrogen-bonded, water has high surface tension and cohesiveness. Water's heat-absorbing capacity renders it a vital climatic regulator that protects cells and organisms from the effects of uncontrolled temperature variation. Water is a polar solvent and as such transports all the critical small chemical substances within and without cells. It causes ionic and polar covalent molecules to dissociate forming free ions. Of these, the hydrogen ion is of special importance since its concentration determines the pH of the aqueous environment.

6. Carbohydrates are organic substances composed of simple and complex sugars, singly, in pairs, or in long polymeric strands. Assembled by endergonic condensation reactions, polysaccharides serve energy storage and structural functions. Disaccharides are the sugars used for transport; the monosaccharides are used for immediate energy utilization. The carbohydrates are polar, and the smaller ones are readily water soluble. Larger carbohydrates can be immobilized for storage.

7. Lipids are a diverse group of compounds. All share high proportions of hydrogen and carbon and a nonpolar, water-insoluble nature. Triglycerides function as energy-storage substances, waxes as surface protection, phospholipids as membrane structural elements, steroids as hormones, and terpenes as electron transmitters.

8. Proteins are informational polymers in which the sequence of the subunit amino acids is critical. Amino acids may be acidic or basic and polar or nonpolar in nature. They are linked by peptide bonds into a primary structure. Hydrogen-bonding then directs the primary structure into a helical or flat secondary structure. The helical structure will fold to form a globular protein with unique surface characteristics. Several globular proteins may, in turn, associate to form a functional unit. Some proteins have structural roles but most are enzymatic in nature. They carry out catalysis by means of specific surface interactions with their substrates.

9. Nucleic acids are polymers of nucleotides. These monomers contain a nitrogenous base, a pentose sugar, and phosphate. DNA and RNA are both very large macromolecules, and their bases can interact with other nucleic acids by the formation of base pairs. This mechanism is responsible for their ability to store and transmit genetic information. Single nucleotides can serve as energy contributors in driving endergonic reactions. Derived cyclic nucleotides have key roles in modulating cell growth and regulation.

10. In the aqueous environment of life, the interactions of molecules to form structures are conditioned primarily by their ability to mix with water, forming solutions and colloids. If molecules are hydrophobic, they tend to form nonpolar water-excluding structural groups such as membranes. Once molecules have been synthesized, their surface characteristics lead them to associate with other molecules in ways that cause spontaneous self-assembly of cellular structures.

11. Cellular material is a complex mixture of thousands of the substances discussed. Investigators have at their disposal several separation techniques that aid in the study of parts of the mixture. Laboratory methods make use of differences of size, shape, electrical charge, and polarity to resolve mixtures and prepare biochemical materials for study.

Study Topics

1. Construct atomic models of C, H, O, N, P, and S. Which of these tend to gain, lose, or share electrons, and how many? Draw electron dot diagrams of a compound of C and S; H_3PO_4; and nitrogen gas (N_2).

2. Summarize the distinction between ionic, polar covalent, and nonpolar covalent bonds.

3. Describe why the warming of a reaction mixture favors endergonic reactions.

4. What is bond energy?

5. If water were replaced by liquid ammonia in a newly discovered life form, what might be some of the ways in which the new life form would differ from water-based life forms?

6. Distinguish between strong acids, weak acids, salts, weak bases, strong bases, and buffers.

7. Describe how starch, collagen, and DNA all fit the definition of a polymer.

8. Provide four examples of condensation reactions.

9. Compare four separation techniques and the molecular properties they distinguish by.

10. When radioactive isotopes are introduced into a cell, the cell incorporates them along with the normal isotope into its biochemical products. A substance that contains a radioactive atom can be detected by various types of electronic equipment and is said to be labeled. If C^{14} is introduced, what types of compounds might become labeled? If radioactive P^{32} is introduced? If radioactive sulfur is introduced?

Bibliography

General texts

Baker, Jeffrey, and Garland Allen, 1975, *Matter, Energy, and Life* (3rd ed.), Reading, Mass: Addison-Wesley.

Curtis, Helena, 1979, *Biology* (3rd ed.), New York: Worth.

Biochemistry texts

Bohinski, Robert C., 1976, *Modern Concepts in Biochemistry* (2nd ed.), Boston: Allyn and Bacon.

Lehninger, Albert, 1975, *Biochemistry*, New York: Worth.

Stryer, Lubert, 1975, *Biochemistry*, San Francisco: W.H. Freeman.

3

Structure
of Procaryotic
and Eucaryotic Cells

Cell theory; Chapter 1,
pp. 8–9

The discovery of the cell as the basic unit of life occurred, as we have seen, early in the last century. Understanding of the internal construction of cells waited another hundred years for the development of a usable electron microscope. In the last chapter we laid the chemical groundwork to explain the aggregation of macromolecules into cellular structures. In this chapter we will systematically dissect both procaryotic and eucaryotic cells, comparing the ways basic cell functions are accomplished by these molecular aggregates.

MICROSCOPY TECHNIQUES

Microscopy is the essential tool of the **cytologist**—one who studies cell structure. Microscopy in all cases involves the formation of magnified images. Although various means may be employed to produce the image, it must always finally be compatible with the human eye.

The Nature of Electromagnetic Radiation

Images may be formed by various types of **electromagnetic radiation** (EM), all of which have some physical features in common. Radiation is produced when a very hot body emits part of its heat energy in radiant form. Depending on the hotness of the source, radiation with varying energy levels will be emitted. Radiant energy travels in a wave form. Each type of radiation is defined by the distance between peaks (the **wavelength**) and the energy content. As wavelength shortens, the energy content of the radiation increases.

There is a continuous **spectrum** of electromagnetic radiation (Fig. 3.1). A comparatively cool heat source such as an electric stove unit set on the lowest level, around 300°C, will emit **infrared** (IR) radiation that we perceive as heat. If the unit is turned up to the highest setting, it will emit not only an increased amount of IR but also shorter wavelengths from 600–700 nm range. Since this radiation is within the **visible** range, we see the unit glowing bright red. An electric light bulb whose filament is heated to 3,000°C gives off the entire range of visible light containing EM radiations through the wavelengths 400 to 700 nm that we perceive as a white blend. The sun has a surface temperature of one million degrees and is the natural source of the complete EM spectrum. Human beings have designed many devices that emit other portions of the spectrum. It should be noted that all radiation more energetic than visible light is potentially extremely hazardous to life.

All forms of radiation obey certain important physical laws. We will discuss these laws in terms of visible light so that they can be readily related to everyday experience. The speed of light in a vacuum or in outer space is 186,000 miles per second. However, when light travels through a material substance, even a transparent medium such as air, water, or glass, it moves more slowly, with a rate that depends on the substance it travels through.

When light strikes an object, four things may happen to the light. If the object is transparent and the light strikes at a right angle, it will be **transmitted**, passing

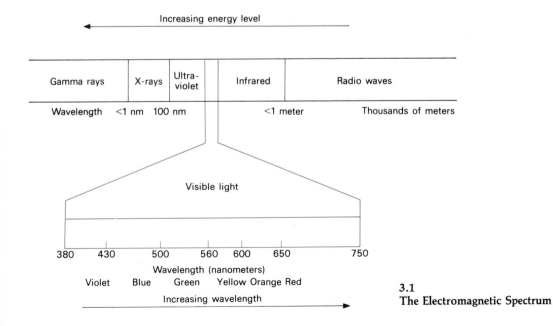

3.1
The Electromagnetic Spectrum

TABLE 3.1
Metric units of length

UNIT	SYMBOL	RELATIONSHIP	EQUIVALENT (METERS)
Meter	m		1m
Centimeter	cm	$\dfrac{1}{100 \text{ m}}$	10^{-2}m
Millimeter	mm	$\dfrac{1}{10 \text{ cm}}$	10^{-3}m
Micrometer	μm	$\dfrac{1}{1000 \text{ mm}}$	10^{-6}m
Nanometer	nm	$\dfrac{1}{1000 \text{ nm}}$	10^{-9}m
Angstrom	$\overset{\circ}{\text{A}}$	$\dfrac{1}{10 \text{ nm}}$	10^{-10}m

The micrometer (μm) is the unit most frequently used in the measurement of microorganisms.

through without affecting the object. If the object is opaque or colored, it will cause complete **absorption** of the light or some portions of the spectrum. Since light is a form of energy, the absorbed energy will have the effect of increasing the kinetic energy of the object. The captured energy may be re-emitted as **fluorescence**, light of a different wavelength. It might cause the object to heat up like a rock in the sun, or the energy might drive chemical reactions such as photosynthesis. Light absorbed by the retinal cells in our eye causes chemical changes. These result in the nerve impulses that our brain synthesizes into an image.

Photosynthesis;
Chapter 8, pp.
239–243

Some light striking an object is **reflected**. Thus even the transparent water of a lake reflects light when its surface is disturbed by the wind.

Last, when light moves from one medium, such as air, to another, such as the glass of a microscope lens, it changes speed. If the light enters a new medium at an angle, its path will be bent. This bending is called **refraction**. Refraction causes a softening or diffusion of images. Each transparent material that light can enter has a characteristic **refractive index** or ability to refract incoming light. The diamond's radiance is caused by its very high refractive index.

The Light Microscope

General considerations   All light microscopes in common use are **compound**—that is to say, they make use of the combined effect of two or more lenses. An image is affected by three parameters that depend largely on the lens system. These parameters are magnification, definition, and resolution (Fig. 3.2).

Magnification is the **orderly** increase of the size of an image in three dimensions. It is a direct result of the curvature of the lenses used.

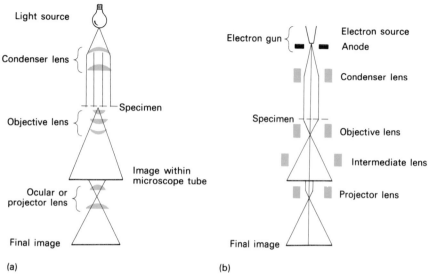

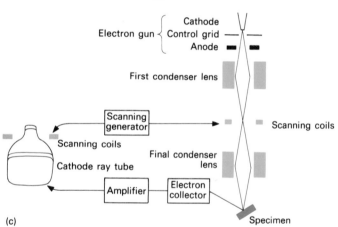

3.2

Schematic Drawings of Microscopes
(a) The light microscope, here inverted to align it with the others, uses visible light that is focused by glass lenses and viewed directly by the eye. (b) The transmission electron microscope (TEM) passes a beam of electrons, focused by electromagnetic lenses, through the specimen. The transmitted image is viewed on an electron-sensitive viewing screen. (c) The scanning electron microscope (SEM) bounces a scanning electron probe off the surface of the specimen. The electron trace is synthesized into a composite picture on the viewing screen.

Although size of the image is important, quality of the image is even more important. An image may be highly magnified without being clear. Clarity or definition partly depends on technical factors such as the lens material and the precision with which the lens is ground.

The major limitation on magnification is contained in the term **orderly** as used in its definition. The wave front of light passing through the object being viewed is disordered somewhat in passing through the lenses. The optical limit of a microscope is technically referred to as its **resolving power**—its ability to take the patterns of light transmitted by two tiny objects and produce two tight patterns of light that can be perceived as separate objects (Fig. 3.3). The smaller such objects may be and still be resolved, the higher the resolving power of the microscopic system. Resolution is physically limited by the wavelength of the type of radiation being used. A

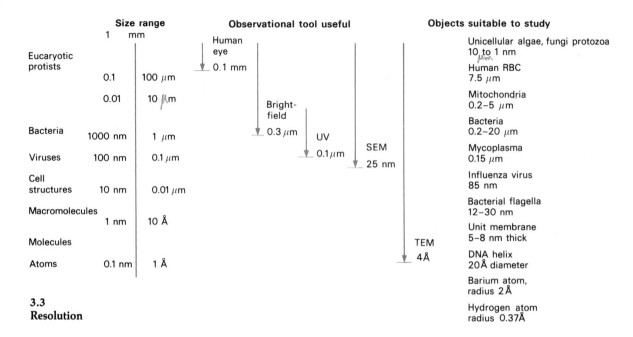

3.3
Resolution

microscope using visible light has a maximum resolution of 0.2μm. Objects smaller than that cannot be clearly seen, no matter what combinations of equipment are used.

Loss and distortion of light due to refraction is minimized by the use of immersion oil. This substance has a refractive index almost identical to that of optical glass. A drop placed between slide and microscope objective significantly improves performance of the **oil immersion** lens.

Introduction of a **monochromatic** glass filter below the slide allows only a narrow portion of the visible spectrum produced by the illuminator to pass through. The image formed is sharper and has more apparent contrast. Because the apparent colors are dependent on the filter, this technique can be used only on unstained materials.

Light microscopes may be equipped with several different types of **condenser**, a device located between the light source and the slide. Condensers redirect the light beam in one of several ways (Fig. 3.4), using systems of shutters and lenses. Each is suitable for certain applications, detailed below.

The bright-field microscope   The illumination for the usual student microscope is supplied through an Abbe condenser that concentrates the light beam for maximum brightness. It is used for viewing nonliving preparations that have been fixed and stained. **Fixation** is the process of denaturing the protein substances of the preparation to render the cell rigidly locked in a more or less natural form. In microbiology, heat is the usual fixative. It has the added effect of attaching the cells firmly to the slide.

Denaturization;
Chapter 2, pp. 52–53

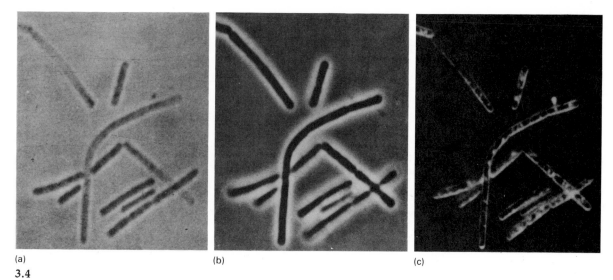

(a)   (b)   (c)

3.4

Light Condensers The Gram-positive bacterium, *Bacillus cereus,* viewed unstained with three different condenser systems. The light source has a large effect on the definition of the image. (a) Bright-field, (b) phase contrast, and (c) dark-field. (× 1,500.)

Microbiological stains Staining is a process designed to render the normally transparent cells opaque to some portion of the visible spectrum. It may be a simple or a complex process depending on the results desired. The stains themselves are synthetic organic compounds of several colors. Due to their ionic nature, they may be acidic, basic, or neutral. Basic dyes, such as crystal violet, safranin, and methylene blue, tend to combine with acidic components of the cell such as nucleic acids and acidic polysaccharides. Acidic dyes such as eosin combine with cellular cations such as basic proteins. In many cases, the exact mechanism of a staining reaction is subject to debate.

    Simple staining procedures require only that a fixed slide be bathed with a single stain, the excess washed off, and the dried slide observed. All organisms will be one color. This type of procedure permits the observation mainly of shape, size, and arrangements or grouping of cells.

    Differential stains involve a multistep process. They yield general morphological data as well as specific data about biochemical features of the cells. This information may be useful in identifying the cells. Certain stains are designed to reveal unusual structural features not readily seen, such as spores, granules, or flagella.

    The **Gram stain**, most familiar and useful of staining protocols, will demonstrate the strategy of differential staining. First, a primary stain is applied—the purple stain crystal violet. It is taken up by all cells indiscriminantly. Second, the dye is locked into place by addition of iodine that enters the cells and forms large complexes with the crystal violet. This locking step is not usual in other staining procedures. Third, a decoloring agent is applied that removes the dye complex from some types of organisms but not from others. Last, a counterstain of a contrasting red color,

TABLE 3.2
Differential staining procedures

STEPS	GRAM STAIN	ZIEHL–NEELSEN ACID-FAST STAIN	WIRTZ–CONKLIN SPORE STAIN
Apply primary stain	Crystal violet 1–3 min	Carbol fuchsin, (reddish-purple) steam 3–5 min	Malachite green, steam 3–5 min
Apply mordant	Gram's iodine 1–3 min	None	None
Decolorize	Alcohol–acetone until no more clear color runs off	Acid alcohol until no more color runs off	None—water rinse to remove excess stain
Counterstain	Safranin (red) 1–3 min	Methylene blue 1 min	Safranin 1–3 min
Observe with microscope	Gram-positive cells are purple Gram-negative cells are pink-red	Acid fast cells are red-purple Nonacid fast are blue	Endospores are green Vegetative cells are red
Report	Presence or absence of single-layer, cell-wall structure	Presence or absence of special lipids in cell wall	Presence or absence of endospores

Each procedure gives important information about particular structural characteristics of bacterial cells by staining one type of structure one color and others the contrasting color of the counterstain.

Cell wall structure; Chapter 3, pp. 76–81

safranin, is applied to stain those cells that were just decolorized. The procedure differentiates bacteria into those that retain the first stain—the purple-colored **Gram positives**—and those that, having been decolorized, now show the second stain—the red **Gram-negatives**. The cytological basis for this reaction will be explained later in this chapter. Compare also the protocols for the **acid-fast** stain and the **spore** stain (Table 3.2).

Negative stains color the background of the slide so that the organisms stand out light in contrast. The most frequent application of negative stains is to demonstrate the bacterial capsule, an external layer that cannot easily be stained directly. It is best viewed as a clear area against a dark background.

Note that preparing single-celled organisms for viewing is much simpler than preparing tissues from normal or pathological human organs. The latter requires fixation, dehydration, slicing or sectioning, mounting, then differential staining by special techniques. We will limit our study of tissue preparation to noting how certain pathogenic microbes are demonstrated to exist in human tissue samples.

The dark-field microscope   The illumination for a dark-field microscope passes through a condenser that cuts off direct light to the slide. The condenser lenses focus the light across the slide at a sharp angle. It does not enter the objective lens unless it is reflected from an object on the stage. Such an object appears brilliantly outlined, a source of reflected light, just as the moon at night appears bright because it is a source of light reflected from an unseen sun. A dark-field microscope is especially useful for viewing of live, motile organisms such as the syphilis spirochete in the fluid sampled from a lesion.

Syphilis; Chapter 21, pp. 630–632

The phase-contrast microscope   This more recent addition to the group of light microscopes makes use of a highly sophisticated system of condensers and special objectives. Remember that as light passes through an object, differences in refractive index cause slight changes in the light path. With the bright-field microscope, these changes are too subtle to be detected. The phase-contrast system allows the comparison of directly transmitted light with refracted light. Differences in refraction are amplified into significant differences in light intensity and a useful image is formed. Phase contrast is employed for viewing wet mounts of live unstained preparations, for observation of bacterial motility, and for viewing delicate intracellular structures in eucaryotes.

Ultraviolet Microscopy

Fluorescence   The use of ultraviolet radiation (UV), shorter in its wavelength than visible light, doubles theoretical resolving power to 0.1μm. However, the human eye has no receptors for UV so we cannot perceive the image directly. We must depend on indirect means of image production. Remember that when radiation is absorbed by certain substances, the energy is immediately re-emitted in the form of a different wavelength radiation, called fluorescence. UV microscopy makes use of fluorescent dyes, similar to those in black-light posters and certain kinds of paints. They absorb 230–350 nm wavelength UV and emit orange, yellow, or greenish light. The coupling of such dyes to cell surfaces results in their outline becoming visible in the UV microscope. Ultraviolet images can also be recorded directly on photographic plates or on UV-sensitive display screens.

Fluorescent antibodies   The most widespread use of the UV microscope combines microscopy with **immunology**—the study of the interaction between antigens and antibodies. Substances can be chemically coupled *in vitro* to fluorescent dyes. They then will attach by a specific lock-and-key mechanism to their complementary structures on the microscopic slide. These structures are accordingly pinpointed and glow with the visible light emitted by the dye. Thus precise chemical identification is combined with microscopy.

Fluorescent antibody techniques; Chapter 13, pp. 391–392

The Electron Microscope

Electron source   Everyone is familiar with the use of X-rays to make unmagnified images of internal body structures. Such pictures require limited focusing and what they reveal is restricted more by the nature of tissues than by the physics of the radiation source. From the early development of X-ray technique, however, came some of the ideas and engineering that went into the development of the electron microscope. Its radiation source is a stream of electrons, with the electromagnetic character of so-called hard or very high-energy X-rays, emitted by a very hot wire in a vacuum. Their wavelength, about 0.05 Å, is so short that the theoretical limit of resolution drops to around 10 Å, more than one hundred times greater than the light microscope.

Operation   Streams of electrons are focused, not by glass lenses, but by electromagnets. The tubular structure through which the electrons pass must be sealed and evacuated to a complete vacuum. The specimen must be prepared in such a way that it is very thin to allow electron passage, and dehydrated and fixed in order to endure the vacuum without distortion. Images consisting of electrons that have passed through a specimen can be made visible by directing them onto an electron-sensitive screen or a plate of X-ray film. In addition to the magnification provided by the instrument, such negatives can in turn be photographically enlarged. This increases the effective magnification still further.

Transmission techniques   Both the original instruments and the majority of modern ones are **transmission** electron microscopes (TEM). They work as described above. The image they produce indicates differences in the electron density of areas of the preparation—denser materials stop more electrons. Special techniques are employed for visualizing various structural features. **Fixation** involves exposing the specimen to chemicals that render all protein features permanent and rigid, hopefully without inducing the formation of spurious features or artifacts.

Internal structures are visible only if the specimen is less than 0.05μm in thickness. In most cases, this will necessitate that the specimen be sliced or cut into **ultra-**

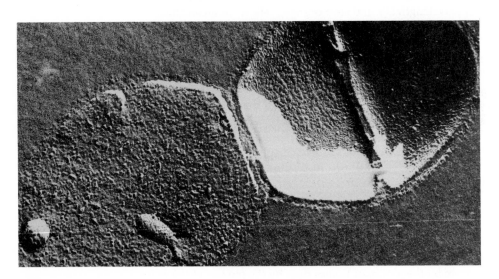

3.5
Freeze-etching Technique   A whole dividing cell of ***Streptococcus faecalis*** was fast-frozen and fractured. Then the fractured surfaced was shadowed. The Gram-positive cell wall shows up as a single layer. The photograph shows the inside face on the upper cell; an exterior face of a neighboring cell is seen below it. The upper cell has been preparing to divide.

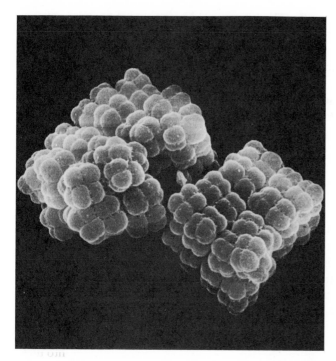

3.6
Scanning Electron Microscope Images This electron micrograph demonstrates the SEM's effective rendering of three-dimensional impressions of cellular interactions. The cells of *Micrococcus rubens,* a Gram-positive coccus, remain attached after division to form regular packets.

thin sections. Imagine a cell of the intestinal bacterium *Escherichia coli,* 1.5μm in length, sliced across into 30 sections 0.05μm in thickness! A quartz or diamond knife is used and the sections are then mounted, as are all EM specimens, on metal support grids.

Surface structure is revealed by **shadowing** the preparation with a stream of electron-opaque heavy metal vapor. The only type of staining used is the application of electron-dense salts or heavy metals that fill in hollows in the specimen. This greatly increases contrast and is called **negative staining**.

In some EM applications, the viewer looks not at the cell itself but at surface replicas. The cell is thinly coated with an electron-transparent plastic or carbon, the cell substance dissolved away, and the remaining replica, in reality a casting of the surface, is shadowed with heavy metal to give contrast for viewing. **Freeze-etching** is an extension of the replica technique in which a deep-frozen specimen is subjected to processes that fracture off certain superficial layers (Fig. 3.5). The exposed surfaces are then replicated, giving views of inner layers of cell walls and membranes. These derived images, like other EM pictures, often are difficult to interpret.

Scanning techniques   Quite recently, we have benefited from the development and widespread use of the **scanning** electron microscope (SEM). These microscopes produce an image of the surface of specimens by electrons reflecting off the surface onto a viewing screen (Fig. 3.6). Materials to be viewed must first be given a protective, electron-dense coat of heavy metal. Thus the procedure at this time remains limited to the viewing of nonliving cells. The images produced are striking, three-dimensional views of surface structures. Because all cells relate to their environments via surface features, the potential of SEM to reveal biologically significant detail is great. SEM permits researchers to view specimens with more clarity than a light microscope affords.

At present, complexities of operation and cost limit the use of electron microscopes to research laboratories. They continue to increase our knowledge of cell structure at a rapid rate.

STRUCTURE OF THE PROCARYOTIC CELL

The bacterial cell uses very simple structures for the performance of basic cell functions (Fig. 3.7). There is no compartmentalization; that is, all portions of the cell are in direct contact with each other. There is but limited specialization. Some structures, such as a cytoplasmic membrane, contribute to the performance of all functions. Thus procaryotic cells have been not only an object of study in their own right but also a source of a rewardingly simple means of answering basic questions about the biology of all cells. Let us use the functional framework of life with which we became familiar in Chapter 1 and extend it, describing the particular structures of the cell involved.

*Characteristics of life;
Chapter 1, pp. 4–5*

Separation from the Environment—the Bacterial Surface

The surface layers of bacteria are as varied as they are complex. We will concentrate on function and avoid some of the complexities of the pure chemistry of these layers.

Capsules and slime layers   Many, perhaps most, bacteria possess amorphous external coats. If the coats are dense, they are referred to as capsules. If they are loosely

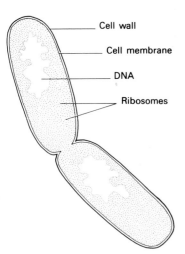

Cell wall

Cell membrane

DNA

Ribosomes

3.7
A Typical Bacterial Cell   The internal structure of a bacterium is simply constructed. A central mass of DNA, the chromosome, is attached to a cell membrane. The cytoplasm is filled with ribosomes. The whole is bounded by a rigid cell wall. Additional structures are found in some, but not all, bacteria.

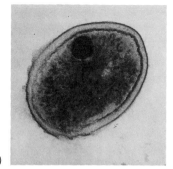

(a)

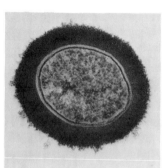

(b)

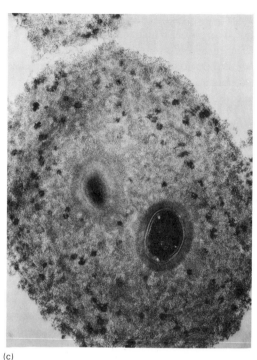

(c)

3.8
Capsule Formation   The Gram-positive coccobacillus *Leuconostoc mesenteroides* (a) possesses the enzyme dextran sucrose that converts sucrose into dextran. This substance accumulates after two hours' growth in a sucrose medium, to form a surface coat (b). The coat is a dextran–dextran–sucrase complex. After 18 hours, a completed capsule of insoluble dextran many times the diameter of the cell has been formed (c). This bacterium is a major slime-producer in the dairy and sugar manufacturing industries.

associated, they are referred to as slime layers. Capsules and slime may be of polysaccharide or protein nature. They originate within the cell and are extruded through the cell wall (Fig. 3.8). Capsules are not readily stained by most dyes. Therefore they are most conveniently revealed by a negative stain procedure that renders the cell colored and the background dark or opaque. The negative stain exposes a transparent zone corresponding to the volume occupied by the capsule around each cell. The amount of capsular material produced by an organism depends on its growth environment. For example, organisms with polysaccharide capsules frequently produce more capsular material in media containing sugar. Organisms that live in the human body can be shown to have more capsular material when freshly isolated than they do after long periods of laboratory growth. In addition, the ability to produce capsular material is a genetic trait that can be lost by mutation or gained by genetic exchange. Capsule-producing colonies of bacteria are called **smooth** strains because of their surface texture. Unencapsulated strains frequently have a dry, irregular surface and are described as **rough**. Genetic variation of capsule production in some bacterial species give rise to many different strains of the organism. These are separated on the basis of small but stable chemical variations in the capsular substances.

The capsule seems to have at least two functions for the producing organism. In disease-causing bacteria such as *Streptococcus pneumoniae*, which causes lung infection, only capsule-bearing strains cause disease. Apparently the capsule protects the bacteria against the destructive phagocytic cells in the body's normal defense

Streptococcal pneumonia; Chapter 19, pp. 554–555

network. Capsules, then, contribute to **pathogenicity**—disease-causing ability. The capsule may also protect the bacterial cell from the harmful effects of drying when it is outside its host's body.

Production of polysaccharide capsular material (dextran) from dietary sucrose occurs in the host's mouth. This production is a result of the activities of *Streptococcus mutans*. The dextran plaque adheres to tooth surfaces, a perfect environment for the acid-forming bacteria responsible for tooth decay. Bacterial slime production is a major nuisance in dairy operations. In the paper industry, up to ten cents of every dollar of production cost is spent for slime control. Capsular material extruded by soil bacteria makes a positive contribution to the normal crumblike texture of topsoil.

Dental caries; Chapter 16, pp. 479–480.

Capsules are frequently antigenic in the host animal. Infection by one strain of a capsular organism may be followed by the appearance of specific antibodies against its unique capsular substance. Such antibodies may protect against future infection by the same agent. The agent may be identified by strain by detecting specific antigen–antibody combination. **Capsular antigens** are also used as immunizing agents to protect against bacterial pneumonias and meningitis.

Pneumococcus vaccines; Chapter 27, p. 799

The cell wall or envelope   Most bacteria are enclosed by rigid external layers that retain their shape when removed (Fig. 3.9). Such a structure is called either a wall or an envelope. Most bacteria produce one of two types (Fig. 3.10), referred to as Gram-positive and Gram-negative because they are revealed by the Gram staining procedure. As knowledge about cell envelopes accumulates, it becomes progressively harder to make sweeping chemical and structural generalizations about them. With few exceptions, all procaryotes contain a unique substance, **peptidoglycan**

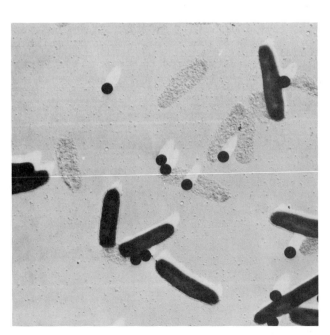

3.9

Inherent Shape   The dark cells are intact cells of the Gram-negative bacillus *Escherichia coli*. The lighter gray rod forms are empty sacculi—cell walls without contents. This picture was made with a shadowing technique. The small, dark spheres are latex beads of $0.3\mu m$ diameter, included as size references. It can be deduced that the empty sacculi are not only less dense but also have been collapsed or have been flattened because they no longer have a shadow. However, the basic shapes of the cells have been retained, even without the contents to round them up.

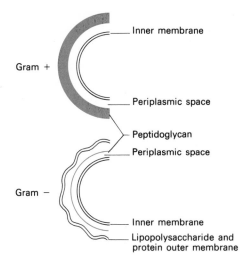

Gram +
— Inner membrane

— Periplasmic space

— Peptidoglycan

— Periplasmic space

Gram −

— Inner membrane
— Lipopolysaccharide and protein outer membrane

3.10
Two Types of Cell Wall   The single, thick, peptidoglycan layer of Gram-positive bacteria is much stronger but less flexible than its Gram-negative counterpart.

(PG), as a wall component. This complex polymer is unknown in eucaryotic cells. The important physical properties of PG are its rigidity and inelastic character. It is responsible for the mechanical strength of the cell wall.

PG is a three-dimensional macromolecule (Fig. 3.11) assembled from a number of highly unusual small units (Fig. 3.12). The carbohydrate portion consists of alternating molecules of the amino sugars N-acetyl muramic acid (NAM) and N-acetyl

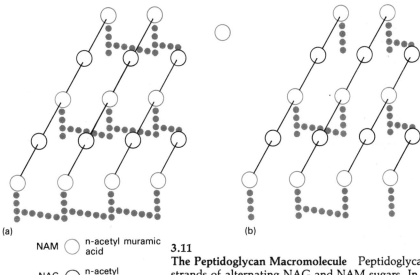

(a)

(b)

NAM ◯ n-acetyl muramic acid

NAG ◯ n-acetyl glucosamine

● Amino acid

3.11
The Peptidoglycan Macromolecule   Peptidoglycan is assembled from linear strands of alternating NAG and NAM sugars. In a Gram-positive cell wall (a), these are completely cross-linked by peptide bridges. In the Gram-negative cell (b), the peptide bridges may not be complete.

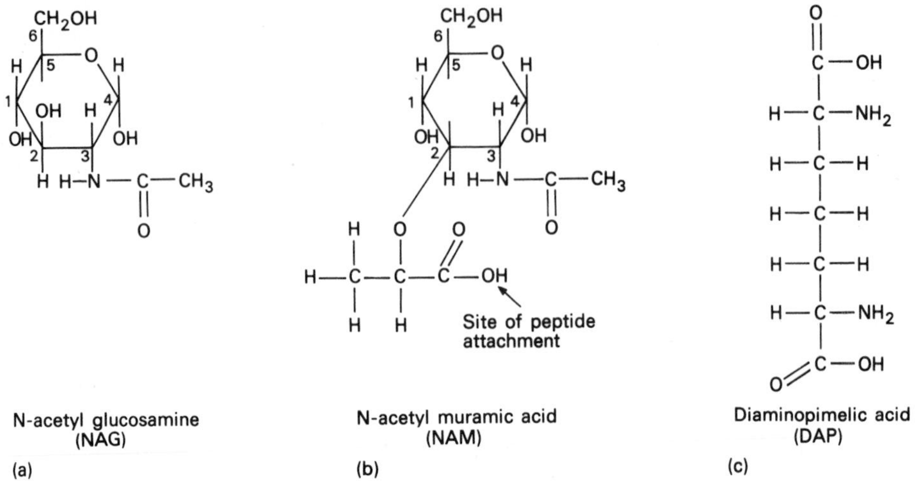

N-acetyl glucosamine
(NAG)

(a)

N-acetyl muramic acid
(NAM)

(b)

Diaminopimelic acid
(DAP)

(c)

3.12

Peptidoglycan Subunits   Peptidoglycan is a complex polymer. Its carbohydrate portion consists of alternating sequences of NAG and NAM, covalently bonded between carbons 1 and 4. These are cross-linked by peptides, with DAP and several D-amino acids present.

Box 3.1

Why Do Antibiotics Work?

Penicillin and tetracycline are widely prescribed to cure or modify bacterial infections. They are not effective against viruses or eucaryotic pathogens such as protozoa and fungi. These limitations are not generally understood by the layperson, who sees them as miracle drugs, good for whatever ails you. This they certainly are not.

The best antibiotics are chemicals that poison uniquely procaryotic functions. They fulfill the dream of a "Magic Bullet," a drug that can be introduced safely into the body to kill only the pathogens at which it is aimed.

Penicillin blocks peptidoglycan synthesis. Because PG has no place in an eucaryotic cell, the drug's effects do not carry over to our tissues. Similarly, tetracycline blocks protein synthesis on procaryotic 30S and 50S ribosomal subunits. It has only a minor effect on the function of our eucaryotic 40S and 60S ribosomes.

However, some other bacterial processes, such as membrane function and DNA synthesis, are not very different from ours. Although there are inhibitory drugs that block these activities, their toxicity to us is almost as high as it is the infectious agent. Thus these drugs have little clinical usefulness.

glucosamine (NAG), joined by glycosidic linkages. These chains are then cross-linked through peptide chains covalently attached to adjacent NAM units. The resulting two-dimensional layer may be linked or laminated to other layers so that all of the PG layer is covalently bonded into one gigantic, continuous molecule. Genetic variation occurs in the exact sequence of the peptide cross-links. In many cases, the peptides contain D-isomers of amino acids, not found in proteins and unique to bacterial cell walls. Many antibiotics such as the penicillins and cephalosporins specifically block PG synthesis thus arresting growth of susceptible bacteria.

Amino acids; Chapter 2, pp. 50–51

Mode of action of antibiotics; Chapter 26, pp. 752–754

The cell wall of Gram-positive organisms   In the Gram-positive cell, PG is the major component of the wall. There are also variable amounts of non-PG polymers such as the **teichoic acids**. Recent advances in electron microscopy have revealed that the wall structure appears to consist of two layers of PG with an area of unidentified material between.

Teichoic acids are phosphate polymers of glycerol or ribitol widely found in Gram-positive cells. Their precise location in the cell wall remains open to debate although they have been demonstrated to be on both surfaces of the wall. These teichoic acid components are of particular interest because, like capsular materials, they are antigens. They show a high degree of strain specificity facilitating bacterial identification by immunological techniques.

In addition to PG and teichoic acids, some Gram-positives such as the streptococci have cell–wall–associated protein M-antigens important in pathogenicity. Both Gram-positives and Gram-negatives may have regularly structured (RS) layers (Fig. 3.13) made up of hexagonally or tetragonally arrayed proteins. Their functions have not been established.

Streptococcal antigens; Chapter 7, p. 504

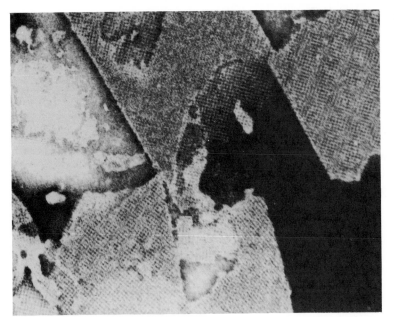

3.13
Regularly Structured (RS) Layers   Protein layers are integral parts of many bacterial cell walls. Often the protein has a regular arrangement, such as the tetragonal arrays seen here. The most striking feature of the tubular fragments in this photograph is that they were re-formed by self-assembly from a purified protein solution derived from cell walls of *Bacillus sphaericus*.

The cell envelope of Gram-negative organisms   Exterior to the cytoplasmic membrane, Gram-negatives typically possess a two-layered envelope structure. The inner of the two layers is composed largely of PG, but the layer is thinner than in the Gram-positives and perhaps less completely cross-linked. The outer layer, frequently called the outer membrane (OM), contains a number of different lipid substances. Among these are phospholipids, lipoprotein throughout the layer, and **lipopolysaccharide** (LPS) substances found mostly on the exterior and interior surfaces.

Endotoxin; Chapter 12, pp. 355–356

The polysaccharide portions of LPS are a class of antigens referred to as somatic or O antigens. The lipid portion (Lipid A) has toxic effects on human circulation when liberated from the cell. This substance, often referred to as **endotoxin**, in large amounts may cause fatal shock. In smaller doses it produces discomfort such as follows immunization for typhoid fever.

Among the proteins of the outer membrane a type called **Protein I** has aroused great interest. It is found in significant amounts and appears to preferentially form clusters. From a number of pieces of evidence it seems likely that passive pores exist for the passage of large molecules through the envelope. Clusters of Protein I molecules (**porins**) or other related substances might form these pores. Protein I also appears to be a receptor site for the attachment and entrance of viruses parasitic on bacteria.

The exterior layers of Gram-negative bacteria therefore differ significantly from the exterior layers of Gram-positive bacteria. Chemical and structural differences

TABLE 3.3
Biological characteristics of Gram-positive and Gram-negative bacteria

GRAM-POSITIVE BACTERIA	GRAM-NEGATIVE BACTERIA
Wall is single layer of peptidoglycan	Wall is typically three layers, inner one is peptidoglycan. Wall deficient bacteria also stain Gram-negative
Shape always rigid	Shape may be rigid or flexible
Rods, cocci	Rods, cocci, spirals, pleomorphic
Spore formation occurs	Spore formation rare
Vegetative cells tolerate dry conditions	Vegetative cells have low tolerance for drying
Growth inhibited by penicillin	Penicillin inhibition not reliable
Wall removed by lysozyme; bacteria form protoplasts	Wall weakened by lysozyme only if peptidoglycan is present; bacteria form spheroplasts
Growth often inhibited by aniline dyes	Dyes not generally inhibitory
Exoenzymes frequently produced; excreted directly	Exoenzymes retained in periplasmic space, released when wall is disrupted
May produce exotoxins	Produce endotoxins; true exotoxins uncommon

are reflected not only in the staining reaction but also in response to antibiotics and environmental challenge (Table 3.3).

The mechanism of the Gram reaction has never been completely clarified. One hypothesis proposes that dye molecules readily enter both types of cell. The iodine mordant forms complexes with the dye within the cell. These complexes are many times larger than individual dye molecules. When a Gram-positive cell containing crystal violet-iodine complexes is treated with the decolorizing agent, alcohol, its cell wall is dehydrated. This causes a decrease in the size of any passages, so the dye complex cannot be easily extracted. A stained Gram-negative cell, however, may have much of its envelope lipid removed by the decolorizer, rendering it porous enough so that the dye-iodine complex will wash out.

The cell wall's function is to provide a mechanical restraint to cell expansion caused by **osmosis**. Without this mechanical restraint, water inflow could rapidly cause the cell to burst. Osmosis is the movement of water between two solutions separated by a membrane such as the plasma membrane of a cell. Water tends to move from the side containing a lesser amount of solute to the side containing a greater amount. Since cells contain quite a bit of dissolved material, if they are placed in fresh water or dilute solution, the direction of water flow will be inward. Unchecked, this would result in bursting of the cell. This is readily demonstrated by placing in water human red blood cells that, like all animal cells, have no wall. On the other hand, cells placed in a strong salt solution may experience water loss—dehydration or drying out—resulting in death or inhibited growth. Because salt inhibits bacterial growth, it is often used in the preservation of foods.

Cell envelope variants   A few types of bacteria differ significantly from the basic Gram-positive and Gram-negative structural models. A brief examination of some of these will shed light on the normal function of a cell wall as well as on the relationship of a cell's structure to the type of environment in which it survives.

L-forms are naturally occurring variants of cell-wall-bearing or normal bacteria. They are isolated primarily from tissues, occasionally in association with disease symptoms. They grow in erratic flexible shapes to form small colonies that require highly specialized growing conditions. The reason for these unusual characteristics is that L-forms lack part or all of their PG or, in some cases, form incomplete complexes. They are thus vulnerable to osmotic destruction and survive best within the osmotically controlled environment of our bodies or on media that are **isoosmotic**, that is, that have the same concentration of solute as found within the cell.

Wall-deficient cells can be produced in the laboratory for study. The techniques of preparation take advantage of the fact that **penicillin**, a clinically useful antibiotic, will kill or inactivate bacteria if it inhibits the synthesis of the PG layer. The enzyme **lysozyme**, isolated from tears, saliva, other body fluids, or egg white, specifically hydrolyzes certain bonds in the PG. Lysozyme can be used to remove the rigid layer; penicillin will prevent its being resynthesized. Removing the PG from a Gram-positive organism results in the formation of a spherical **protoplast** lacking all wall components. Lysozyme in conjunction with EDTA, a chelating agent that increases the porosity of the outer membrane, removes the PG layer from Gram-negatives. This leaves a **spheroplast** that still has the outer membrane portion of the wall.

Such modified cells retain their metabolic capabilities for some period, as long as their structural integrity is protected by an isoosmotic environment. Neither type, however, can multiply. They are extensively used for studying the membrane layers and their transport activities.

Certain groups of bacteria inhabit natural environments with high concentrations of dissolved salts. Marine organisms, frequently Gram-negative, often have internal osmotic pressures close to that of the sea water and envelopes similar to other Gram-negatives. The halobacteria, however, are unusual. These organisms inhabit such environments as the Dead Sea (29–30 percent salts) and pickling brines. They entirely lack PG. Instead, they have a highly specialized protein outer layer; its function is not osmotic support because the cells maintain an isoosmotic interior salt concentration. The protein layer's function is in energy generation. Halobacteria do not tolerate exposure to lower salt concentrations because they have no protection against osmotic lysis.

Mycoplasma; Chapter 19, pp. 555–556

A group of predominantly pathogenic bacteria called **mycoplasmas** illustrate another pattern of adaptation. These organisms, which like L-forms live mostly within animal tissue, entirely lack an envelope. Their outer boundary is the cytoplasmic membrane; its stability depends on incorporation of sterols. Although their membranes have more resistance to osmotic shock than those of other bacteria, they are still too vulnerable to be free-living.

In summarizing our observations on the procaryotic cell wall, then, we can say that the incorporation of PG is a unifying characteristic. The mechanical strength of each organism's cell wall is directly related to the amount of PG that in turn determines the degree of osmotic stress the organism can stand. Cells that lack such osmotic protection occupy ecological niches that are unlikely to expose them to challenge.

Each species of bacteria, as well as the various strains within the species, may possess unique individual patterns of surface antigens, such as the teichoic acids and proteins of the Gram-positives and the lipopolysaccharides of the Gram-negatives. The result is that each strain has a pattern of molecular recognition sites on its surface. In later chapters we will pay particular attention to these markers and their importance in immunology and diagnostic microbiology.

The cytoplasmic membrane   The cytoplasmic membrane is approximately 60 percent protein and 40 percent lipid, the latter being mostly phospholipid. These percentages create an incorrect impression that the membrane is a homogeneous structure with all components evenly distributed. In reality, although the membrane appears to have a relatively uniform thickness of from 5 to 8 nm, it appears that its protein components are far from randomly arranged.

Phospholipids; Chapter 2, pp. 47–48.

The **fluid mosaic** model of the membrane is widely accepted at present (Fig. 3.14). Use your imagination for a moment to visualize a soap bubble revolving in the sunlight. Note the moving patterns of iridescent color on its surface. Soap molecules make up this simple membrane that is held together only by the soap molecules' mutual intermolecular attractions. The molecules are continuously moving around and realigning causing the color patterns change. The soap bubble shows us in a readily visible form what is meant by the term **fluid** when applied to a mem-

3.14

The Fluid Mosaic Membrane Proteins, varying in their polarity, are embedded in a flexible matrix of lipid molecules. Because all are bound together by weak bonds, they are free to slide by one another in a fluid fashion. Some proteins contribute to the structural integrity of the membrane, while others are enzymes that catalyze membrane-localized reactions.

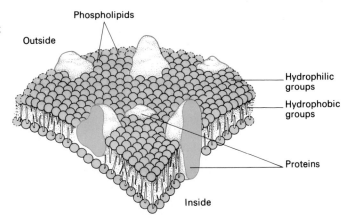

brane. The term **mosaic** refers to the fact that the membrane contains a variety of proteins embedded in its lipid bilayer structure. Some of these proteins are structural. Their function is to contribute to the physical stability of the membrane itself. Others are enzymatic proteins, involved in the catalysis of the numerous reactions for which the membrane is the site. Proteins can be demonstrated on both surfaces and within the interior of the membrane. Cell membranes grow or expand by the insertion of additional molecules of all components into the mosaic.

In many bacterial cells there are structures called **mesosomes.** These are blind infoldings of the membrane, frequently very much coiled. They have received a great deal of attention and many functions have been proposed for them, none of which is universally accepted. The mesosome has been recently described as a structure in search of a function.

The roles of the bacterial cell membrane are many. Three of them are directly related to a cellular contact with the environment. Cell membranes are **semipermeable;** they regulate entrance and exit of all small molecules, with the exception of water that passes freely in both directions. To penetrate the membrane, any compound must be able to passively penetrate the hydrophobic layer or be actively assisted in its migration. Cell membranes then serve to exclude certain substances entirely from the cell and to retain certain others. **Facilitated transport** is the name given to the carrier-assisted movement of some small molecules across the barrier in either direction. The direction of movement is always from an area of high concentration of the substance to one of lower concentration. Such movement is said to go with the **concentration gradient.** It does not require energy expenditure. This process is in some cells responsible for the entrance of certain small polar compounds and exit of waste products.

Active transport is a highly specific directional process in which the cell's energy is expended to move nutrients (sugars, amino acids) and ions into the cell **against** a concentration gradient. The mechanism depends on the presence of membrane-bound transport enzymes called **permeases.** Each permease transports a specific substrate. The active transport mechanism is discussed in Chapter 8.

The cytoplasmic membrane in procaryotes participates in most cell activities. In fact, for the survival of a cell, membrane integrity is the most important need. Cells survive, at least for a while, without cell walls, without genetic material, certainly without organs of locomotion. Disruption of the membrane permeability barrier, however, spells death unless the membrane can be rapidly repaired. Certain antibiotics, such as polymyxin, kill bacteria by disrupting the infectious agents' membrane.

The periplasmic space   Between wall and membrane, the periplasmic space appears to serve as a reservoir. Many bacteria produce hydrolytic enzymes that are specifically extruded from the cell through the membrane. These **extracellular** enzymes are employed to break down neighboring macromolecules into the small subunits that can pass the membrane and enter the cell. In Gram-positive cells, exoenzymes usually pass directly into the medium, whereas in Gram-negative cells, they are frequently retained in the periplasmic space. Conversion of the Gram-negative cell to a spheroplast liberates the exoenzymes.

Exoenzymes; Chapter 12, pp. 352–353

Structures That Capture, Utilize, and Store Energy

High-energy compounds; Table 8.3

Life is sustained by the metabolism of organic compounds with a high-energy content. Specific cellular sites are used for the synthesis and utilization of high-energy molecules.

Energy capture for a procaryote frequently means the capture of solar or radiant energy. The ability to carry out the reactions linking energy capture to synthesis of carbohydrates is found in two major groups of bacteria. The cyanobacteria, also called blue-green algae, contain their light-capturing apparatus in **lamellae,** layers of parallel folded membrane material incorporated into **thylakoids.** The anaerobic photosynthetic bacteria possess various types of chromatophore. These include both lamellae and vesicles (sacs of membrane material filled with liquid). The membranes, however arranged, contain the light-capturing pigments. None contain genetic material or reproduce themselves.

Cyanobacteria; Chapter 4, pp. 103–107

A novel mechanism for the capture of light energy is the cytoplasmic membrane of halobacteria. This "purple membrane" contains **bacteriorhodopsin,** a light-reactive substance chemically related to our retinal pigment. These organisms capture light energy and harness it to a unique membrane transport mechanism for ATP production.

Energy utilization refers to use of the energy potential of organic food molecules for the production of ATP. This energy transfer requires (in different species) various multistep pathways. The enzymes necessary for those steps directly associated with ATP synthesis or **phosphorylation** are incorporated in the cytoplasmic membrane. The membranes of procaryotes probably all contain highly ordered localized enzyme–coenzyme sequences.

ATP structure; Chapter 2, Fig. 2.20

Energy storage occurs when energy-rich substances can be manufactured or assimilated faster than they are needed for growth or maintenance. These substances must be in an insoluble form, often polymeric. Procaryotes exhibit considerable variation in the choice of storage material. It is deposited in **inclusion bodies** or **granules**—irregularly shaped elements usually quite visible microscopically. Inclusions may contain glycogen, other polysaccharides, or lipids such as poly-β-hydroxybuty-

Corynebacteria diphtheriae, Chapter 18, pp. 532–535

rate (PHB). **Volutin,** a polymer of inorganic phosphate, is found in metachromatic (variable-staining) granules. The chemical nature of an organism's storage material is genetically determined. Its identification may be a useful determinative tool.

Granules serve as a metabolic bank account; their appearance and disappearance indicates the adequacy or inadequacy of the organism's energy supply.

The Assembly of Proteins

Protein is quantitatively the most significant organic cell component. Its synthesis in a living cell is continuous, providing for growth, replacement, and adaptive shifts of metabolism. The assembly of each polypeptide requires covalent linkage of a precise sequence of amino acids. The **ribosome** is the major structural participant in this process. It provides a point for simultaneous attachment of a gene copy (in RNA) and the amino-acid carrier molecules (tRNA).

Ribosome function; Chapter 9, p. 255

The structure of the procaryotic ribosome has been exhaustively studied. The complete functional unit is assembled from two subunits. These subunits are frequently described on the basis of their **sedimentation rate.** When placed in an ultracentrifuge and spun, particles move toward the bottom of the centrifuge container at a rate proportional to their density. The Svedberg unit (S) is used to measure behavior in a centrifugal field. The small (less dense) ribosomal subunit is called the **30S particle** and the larger one, the **50S particle.** Separation of the macromolecular components of both particles reveals that each contains one large RNA molecule and a mixture of proteins. Among the latter are the enzymes responsible for establishing peptide bonds between the amino acids.

The cytoplasm of a procaryote contains numerous free ribosomal subunits, but in a healthy cell most subunits will be bound in functional complexes called **polysomes.** A polysome is a group of ribosomes actively translating one genetic message. Some polysomes appear to be directly associated with the cytoplasmic membrane. It is speculated that these polysomes manufacture proteins whose destination is extracellular.

Sources of Genetic Information in the Cell

Procaryotes' intracellular genetic instructions are constructed in a very simple structural format. In the bacteria, all essential genes are contained in a single continuous double helix of DNA in the form of a closed circle. The DNA is not coated with basic proteins and is not enclosed in a membrane. This chromosome structure is usually called a **nucleoid** to distinguish it from the structurally more complex nucleus of eucaryotes. The nucleoid usually appears to be attached to the cytoplasmic membrane or a mesosomal extension. As the cell and its membrane grow, new copies of the nucleoid are produced and spatially separated. A rapidly growing cell at times will have nuclear copying running far ahead of cell division so that as many as four copies can be seen in a single cell.

Most procaryotes also contain accessory DNA-containing structures called **plasmids.** These small, closed circles in the cytoplasm do not contain essential information, as cells can be "cured" of them without losing viability. Their practical

Antimicrobial
resistance; Chapter 26,
pp. 755–756

importance to human beings is considerable, however. Plasmids frequently contain information that allows the bacterial possessor to destroy or otherwise resist the inhibitory effects of antibiotics. Even worse, these resistance plasmids or **R-factors** are readily passed from one strain of organism to another. Plasmids contribute to the ever-worsening problem of bacterial antibiotic resistance.

Recombinant DNA
techniques; Chapter 9,
pp. 276–278

In recent years, laboratory techniques have been developed that allow the enzymatic recombination or splicing of genes from many sources. This recombinant DNA can be "built" into engineered plasmids. When reintroduced to bacteria, these plasmids cause the bacteria to acquire new traits. Because such research is not without hazards, it is done under very strict controls.

Adaptability and Movement

Procaryotes in normal nonlaboratory conditions experience constant small and large environmental shifts. They adapt to these shifts primarily by nonstructural changes such as variations in enzyme patterns.

Movement   Many species are **motile.** They can either move toward or away from stimuli by rotational movement of their flagella.

Flagella have been studied from chemical, physiological, and mechanical viewpoints. Flagella consist of repeating units of the protein **flagellin,** with a weight of 26,000 daltons. Different species make genetically variant forms of flagellin. The variant forms are identified as the **H-antigens.** Flagellin molecules aggregate by a self-assembly process to form fibers; flagella are formed when three fibers twist to form a helix. The helix has a wave form determined by the amino acid sequence of the protein units.

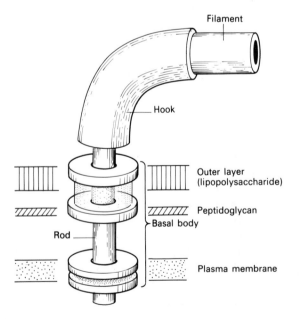

Filament
Hook
Outer layer
(lipopolysaccharide)
Peptidoglycan
Basal body
Rod
Plasma membrane

3.15
Bacterial Flagella   The bacterial flagellum is anchored by means of a complex basal body, which has structural rings that attach to each cell wall layer in a Gram-negative organism. The rings interacting with the cell membrane perhaps assist in transmitting the energy source for flagellar motion.

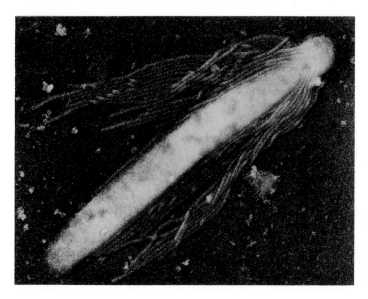

3.16
The Helical Nature of Flagella This rod-shaped bacterium has unusually thick flagella that appear to be spun like strands of yarn. Their arrangement around a "head" is also noteworthy.

Electron micrographs indicate that the flagellum has a hollow core. It is anchored into the cell wall and membrane via a two-part structure consisting of a **hook** and a **basal body** (Fig. 3.15). The hook is composed of subunits of a single protein quite different from flagellin. It is connected via a rod inserted through the cell envelope to a basal structure composed in Gram-negatives of four rings bound to a central rod. From 10 to 13 distinct polypeptides make up the basal structure, the rings of which are bound to each of the envelope layers.

Growth of flagella occurs as flagellin units are added at the end of the filament. Protein molecules move down the hollow core and are added on by the self-assembly process at the tip. Flagellar growth is very rapid; active bacteria from which flagella have been removed may have full-length replacements within 10 to 20 minutes.

Self-assembly; Chapter 2, pp. 58–59

The arrangement of the flagella is characteristic of the taxonomic group to which the organism belongs. The pseudomonads have **polar** flagella located either singly or in clusters (Fig. 3.16). The enterobacteria, on the other hand, have flagella scattered in a **peritrichous** arrangement all over the surface. Most flagellate bacteria are rod-shaped or helical.

Flagellar motion is subject to a fascinating process of direction and coordination called **chemotaxis**. The bacterial cell receives sensory input in the form of chemical signals. These signals cause alterations in the direction of flagellar movement which in turn cause the organism to remain in the vicinity of a desirable chemical source (**positive** chemotaxis) or move away from an undesirable one (**negative** chemotaxis). The cell membrane contains protein receptors that combine specifically with the signal substance. Such signaling, in the presence of the compound adenosyl methionine, produces a change in the direction of rotation of the flagella fibrils. In peritrichous bacteria, such as *E. coli*, counterclockwise rotation of the flagella results in

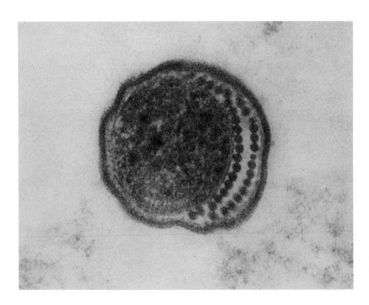

3.17
Flagella in Spirochetes   In spirochetes, the flagella are internal. They are grouped in bundles attached toward the ends of the organism. These bundles wrap around the organism. Thus their contraction produces an undulating motion.

their forming an orderly bundle that propels the cell in a forward direction. Clockwise rotation, on the other hand, separates the flagella into individual uncoordinated fibrils. They produce a tumbling motion that does not advance the cell through the medium. Peritrichous cells then do not start and stop—they either swim or tumble. Polarly flagellar cells appear to swim both forward and in reverse. Flagellar action appears to be continuous as long as an energy source is supplied.

Spirochetes; Chapter 4,
pp. 116–117

Flagella in the spirochetes are chemically and structurally similar to those described above. However, their cellular location is unique (Fig. 3.17). They are attached via a hook and basal body structure near the poles of the cell, just short of the ends. One or more filaments are attached at each end. Each runs toward the opposite end between the cell envelope layers, with their ends overlapping those from the other pole near the middle of the cell. Thus they are not external to the cell at all. In number the filaments range from a single pair to over 100. They are usually grouped in a bundle, the **axial filament**. The mechanical action of the flagella appears to result in the wriggling motions of the spirochetes.

Cyanobacteria move by **gliding** over a surface. In the absence of any apparent external motility structures, the mechanism of this motion has remained unresolved. Freeze-etched electron micrographs reveal a continuous layer of **fibrils** in parallel arrangement, found between the outer membrane and the peptidoglygen layers of the envelope. Contact of fibrils with a solid surface may result in the characteristic forward rotating movement of the gliders.

Vertical movement in a water column is a desirable adaptive mechanism for aquatic organisms, particularly photosynthetic ones. **Gas vesicles,** membranous gas-filled vacuoles, are widespread in cyanobacteria and in two groups of the anaerobic photosynthetic bacteria. They are believed to provide flotation and to assist in vertical movement by providing changes in buoyancy.

Formation of dormant structures   Motility provides means of moving toward or away from environmental stimuli. However, sometimes adverse changes are so major that survival is possible only for those microorganisms that form specialized resistant cells called **spores, cysts,** or **conidia.** The typical endospore is formed by two genera of Gram-positive bacteria, *Bacillus* and *Clostridium.*

Clostridium species; Chapter 23

Sporulation commences in a heathy, adequately nourished vegative cell after the chromosome has been copied. One of the two nucleoids is incorporated into the spore, the other remains outside (Fig. 3.18). Spore formation begins with the appearance of a **spore septum** isolating the spore contents from the rest of the cytoplasm. The spore DNA then directs the rest of the sporulation steps from within the **forespore.** Additional coats are deposited, first the **cortex,** chemically rich in calcium ions and **dipicolinic acid,** then one or two **spore coats** exterior to the cortex. The spore contents include one chromosome, most of the cell's RNA, and many of its proteins condensed and dehydrated. Removal of water arrests metabolic activity.

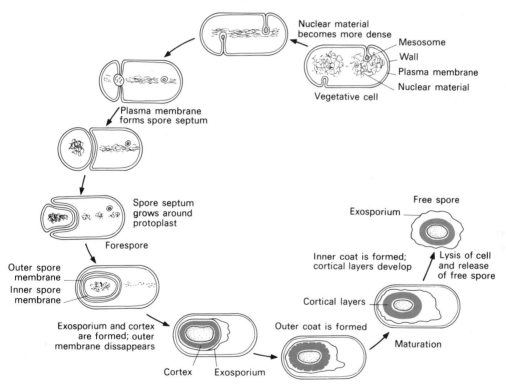

3.18
Sporulation   There are numerous clearly defined stages in the development, maturation, and release of a free spore. The spore's outer layers originate from the cell membrane, which is progressively thickened and reinforced by the deposit of newly synthesized materials. Of the original nuclear copies on the vegetative cell, only the one housed in the spore survives. The other breaks down, after directing certain of the initial stages of sporulation.

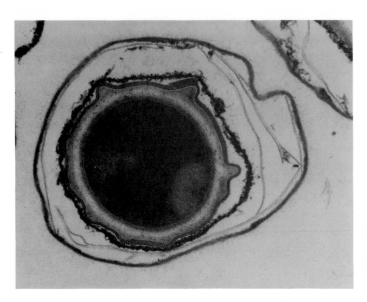

3.19
Bacterial Spores   The protective outer coats surround a core. This core has its own cell membrane and cell wall. The protective layers help a spore survive suboptimal conditions.

Before the sporulation process is complete, the remainder of the vegetative cell autolyses and sloughs off, in some cases leaving a remnant called the **exosporium.** Because of the spore coats, a free spore is highly refractile to light, resistant to heat, and impermeable to chemicals including stains. From the time when normal cell growth ceases until the free spore is released, from six to eight hours will elapse. Sporulation is not a very rapid response, therefore.

Germination follows the spore's contact with certain amino acids, ions, or other chemical signals. The spore loses its resistance and refractility; enzymatic activity begins. **Outgrowth** occurs as swelling due to water uptake and is followed by the splitting of the spore coat and emergence of a vegetative cell. Following a period of readjustment, cell division will begin again. Spore formation is not a reproductive process because each vegetative cell gives rise to only one spore that in turn germinates into just one new cell.

Spore formation allows a bacterium to survive unfavorable conditions but, oddly enough, only if the spore is already formed when the environment starts to deteriorate. Spores are formed only by healthy cells. The signals that start sporulation do not appear to be a "warning" to the cell. Still, the ability to form resting structures contributes greatly to the survival potential of organisms in environments subject to drying and temperature fluctuations because some of the population will at any time be in the resistant state.

Cysts are formed by some gliding bacteria and *Azotobacter.* These structures (Fig. 3.19) lack dipicolinic acid and have lesser degrees of heat resistance than true spores. They do resemble spores in their metabolic inactivity and ability to survive drying. **Conidia** are formed by filamentous bacteria such as *Streptomyces* and *Micromonospora.* They are dormant structures but do not have special stress resistance. Their function is clearly reproductive, since they are formed in large numbers.

Conidia; Chapter 5, pp. 138–141

Attachment structures—the pili   Many strains of Gram-negative bacteria have many small, tubular, surface projections called **pili.** These are formed by aggregation of molecules of the protein **pilin,** (MW 17,000) into a hollow helix. Two types of pilus are recognized. One, the sex pilus, serves for the attachment of two bacteria to each other prior to **conjugation.** This is a process in which portions of donor DNA are transferred to the recipient, perhaps through the tubular pilus. Other types of pilus have been observed to attach bacteria to eucaryotic cells. Attachment of bacteria to red blood cells causes **hemagglutination** or clumping. Pili may be responsible for the formation of **pellicles** or surface films seen when aerobic organisms grow in liquid media, and for the attachment of bacterial viruses to their host cells.

THE EUCARYOTIC CELL

Although the eucaryotic cell is significantly more complex in structure than the procaryotic, it will be dealt with fairly briefly here. This material is covered effectively in most up-to-date general biology texts. These should be referred to whenever necessary. We are most interested in deriving comparisons between the two types. Human infectious disease involves an interaction between a microbial agent—procaryotic, eucaryotic, or viral—and the eucaryotic cells of the host. Thus it is a good idea to have the basic plan of the eucaryotic cell type well in mind.

Separation from the Environment

Cell walls   Algae, fungi, and higher plants all share, with the bacteria, the requirement for a rigid cell wall for osmotic protection. However, eucaryotic cells contain no peptidoglycan. They rely instead on layers of various other polysaccharides, including cellulose and **chitin.** Alginic acid, carrageenan, and agar add rigidity and tensile strength in the algae. Cell walls provide a barrier to the entrance of plant pathogens. They also contribute degradable energy-rich residues in soil that nourish the many groups of soil bacteria. Cell walls are not found in animal cells and tissues or in protozoans (Fig. 3.20).

3.20
A Typical Protist   The cell structure of the familiar *Paramecium* illustrates a protozoan organization. The nucleus and micronuclei are present as separate genetic information structures. Organelles of ingestion (gullet), digestion (food vacuole), and excretion of solid waste (anal pore) are present. Osmotic balance is sustained by excretory contractile vacuoles, while motility is provided by cilia. Mitochondria, endoplasmic reticulum, Golgi bodies, and smaller vesicles are not as readily visible.

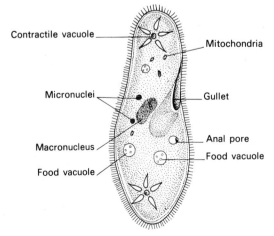

Contractile vacuole

Mitochondria

Micronuclei

Gullet

Macronucleus

Anal pore

Food vacuole

Food vacuole

Cytoplasmic membranes   The cytoplasmic membrane of eucaryotes appears to be chemically and structurally similar to the procaryotic membrane. Animal membranes derive improved physical strength from the incorporation of sterols. However, the function of the cell membrane in higher organisms is comparatively restricted. Its activities are related to passive and active transport, chemical exchange, and chemical signaling. Receptors for a hormone are found on the membranes of its target tissue; immunoglobulin molecules on the membranes of immune lymphocytes provide for specific antigen recognition. All the other membrane functions described for the procaryotes are compartmentalized in cytoplasmic **organelles** in eucaryotes.

Phagocytosis; Chapter 12, pp. 365–368.

     One activity in which the membranes of certain animal cells participate is **phagocytosis,** the surrounding and engulfing of particles. This requires major changes in the shape of the cell and, therefore, in its membrane.

Structures that Capture, Utilize, and Store Energy

Energy capture   In eucaryotes, light energy is captured within a specialized organelle, the **chloroplast** (Fig. 3.21b). These structures are ovals, enclosed by two complete unit membranes. The photosynthetic pigments are located in membrane leaf-

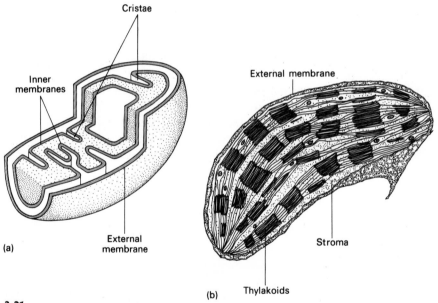

(a)

Cristae

Inner membranes

External membrane

External membrane

Stroma

Thylakoids

(b)

3.21
Mitochondria and Chloroplasts   Both mitochondria and chloroplasts are bounded by outer membranes, while their inner membranes are organized into complex structures. In mitochondria (a), the cristae bear the components essential for ATP synthesis driven by respiration. In chloroplasts (b), the thylakoids, in stacks called grana, bear the photosynthetic pigments needed for the capture of light energy. In both of these organelles, the interior is filled with a thick fluid containing enzymes, ribosomes, and nucleic acids.

Box 3.2

Mitochondria and Chloroplasts—Are They Really Procaryotes?

Eucaryotic chloroplasts and mitochondria both can be described in terms that bring out their similarity to bacteria. They are enclosed in a bilayered membrane; in its inner layers are found enzymes and coenzymes required for electron transport and the manufacture of ATP. They contain DNA in the form of a single, histone-free circle. They transcribe and translate their internal genes with the aid of 30S and 50S ribosomes, using a set of enzymes and transfer molecules completely different from those in the cytoplasm of the cell. Their protein synthesis is inhibited by those antibiotics—chloramphenicol and streptomycin—that stop bacterial protein synthesis. At the same time, these drugs have no effect on the cytoplasmic protein synthesis.

This discrepancy between the biological nature of an eucaryotic cell and its organelles is extremely puzzling. One widely accepted explanation places this problem directly in the middle of an even larger unsolved question—How did the eucaryotes originate in the first place?

The Endosymbiotic Hypothesis proposes that, on one or many occasions, a large procaryotic cell either ingested or was parasitized by other procaryotes (accounts differ on which was the aggressor). The internal procaryotes persisted, and both partners evolved towards a symbiotic relationship—one in which each benefited. The host cell provided shelter, nutrients, and an environment for propagation. The endosymbionts specialized in efficient ATP production and energy capture. This relationship, a stable and mutually advantageous one, has persisted to this day.

Inside each eucaryote, there may be several bacteria—but they are not crying to get out!

lets (lamellae) arranged in flattened disks called thylakoids. Thylakoids, in turn, are arranged in stacks called **grana,** interconnected and floating in the fluid stroma of the chloroplast. Chloroplasts also contain an incomplete genetic system including DNA, ribosomes of the procaryotic type, and associated components of the protein-synthesizing system. Chloroplasts are capable of binary fission within the host cell.

Energy utilization   Energy-yielding metabolism coupled to phosphorylation occurs in eucaryotic cells in two locations. The primary, anaerobic steps take place in the cytoplasm, catalyzed by soluble enzymes. The later phases take place within the **mitochondrion,** another specialized organelle (Fig 3.21a). Mitochondria, like chloroplasts, have two membrane layers, the inner of which is extensively infolded to give projections called **cristae.** Some of the respiratory enzymes are found in the fluid **matrix** that fills the mitochrondrion; others are immobilized in the membrane layers. Like the chloroplast, the mitochondrion contains DNA and protein-synthesizing components and replicates itself.

Energy storage   Plant and animal cells frequently contain loosely organized storage materials—starch or oils in plants, glycogen or fats in animals. These are not membrane-bounded. In human beings, for example, certain tissues are specialized for storage function. Adipose tissue contains cells that can enlarge tremendously as they become filled with stored fats. The liver contains deposits of glycogen. However, accumulation of lipids in other than the expected tissue, due to degenerative disease or genetic error, leads to severe malfunction.

Synthesis of Macromolecules

A complex picture has emerged of the synthesis and transportation of molecules of protein and other macromolecules (Fig. 3.22). The basic protein synthesizing unit is the ribosome, composed of one large and one small subunit. However, eucaryotic ribosomes are larger than procaryotic ribosomes. The subunits are 40S and 60S; when fractioned they contain longer RNA molecules and a larger number of proteins per unit. The polysome is still the basic assembly combination, with genetic messages being translated into protein sequences on groups of ribosomes. However, some polysomes are freely floating in the cytoplasm and seem to make internal pro-

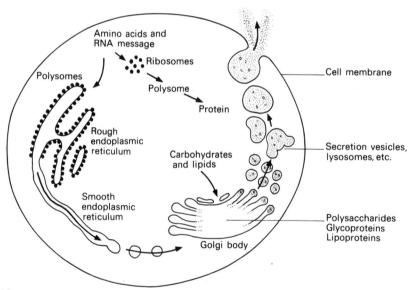

3.22
Protein Processing   Proteins are synthesized when an RNA message, ribosomes, and other factors assemble to form a polysome. When the polysomes are free in the cytoplasm, the protein is released as a cytoplasmic constituent. When polysomes are bound to the endoplasmic reticulum, the protein product enters its lumen, passes through the smooth endoplasmic reticulum, and is packaged into vesicles. These vesicles may transfer the protein to the Golgi body. There the protein may be enzymatically modified. The Golgi body pinches off enzyme-filled lysosomes, which remain inside the cell, or secretory vesicles, which fuse with the cytoplasmic membrane to dump their contents exterior to the cell.

teins, whereas other polysomes are attached to the surface of a deeply folding network of membrane lamellae, the **endoplasmic reticulum** (ER). **Rough** ER, which is ER with associated ribosomes, produces proteins with an extracellular destination. These move toward the outside of the cell by a staging process. They are first packaged in ribosome-free extensions of the ER (**smooth** ER); these then migrate to and fuse with the **Golgi complex,** a stack of flattened oval membrane disks. Within the Golgi complex, large molecules are held in reserve for later use. Proteins to be secreted are often modified by addition of lipid or carbohydrates to form lipoproteins and glycoproteins. Eventually, engorged vesicles of the Golgi complex pinch off, move to the cell membrane, fuse with it, and discharge their contents to the outside. Vesicles pinched off from the Golgi complex, **lysosomes,** also have semipermanent existence as internal enzyme reservoirs. The complex also collects and transports polysaccharides to the developing cell wall in plants.

The structural components of the protein synthesis–secretion function are particularly well developed in those animal cells that specialize in synthesis of export materials. Examples are glandular tissues such as liver and pancreas, and plasma cells that produce large quantities of antibody protein.

Sources of genetic information in the cell   The major source of genetic instruction in an eucaryotic cell is a set of chromosomes located within the cell **nucleus.** The nucleus itself is a sphere surrounded by two layers of **nuclear membrane.** The nuclear interior, however, communicates with the cytoplasm of the cell via numerous nuclear pores that pierce the boundary. Eucaryotic chromosomes are linear structures, rather than circles, and the DNA that contains the genetic code is covered by a coating of basic proteins, the **histones.** These appear to be arranged in a regular pattern as yet not fully understood. Most eucaryotic cells have paired information; they receive in the course of sexual reproduction two chromosomes of each type, one from each parent, usually differing considerably in the exact details of the instructions. Such a cell is genetically a **diploid.** Bacteria that have only one parent and one chromosome (or multiple copies of it, all identical) are normally **haploids.** Certain rare recombination events may produce **partial diploid** bacteria.

In most eucaryotes, a spherical membrane-bounded organelle, called the **nucleolus,** lies within the nucleus. The nucleolus also contains DNA. The genes involved are those concerned with the manufacture of ribosomes.

When viruses infect an animal cell, most of them locate and are replicated within the nucleus in which enzymes for copying DNA and RNA are found. Viral genes may become permanently incorporated into the host cell DNA and affect certain cell functions. As mentioned earlier, both chloroplasts and mitochondria also contain DNA and incomplete genetic systems. There are many unanswered questions concerning the interactions of these heterogeneous segments of DNA with the nuclear chromosomes.

Viral replication; Chapter 11, pp. 319–323

Adaptability and Movement

The possible adaptations of the eucaryotic cell to changing environments are far too varied to be covered adequately here. Let us look briefly at one relevant organelle,

TABLE 3.4
Structure and function in procaryotic and eucaryotic cells

FUNCTION	PROCARYOTIC STRUCTURE	EUCARYOTIC STRUCTURE
Boundary layers	Capsule or slime layer	No common equivalent
	Cell wall containing peptidoglycan	Cell wall in algae, fungi contains cellulose
	Cell membrane-fluid mosaic type	Basically same
Energy capture, utilization, and storage	Thylakoids, chromatophores	
	Cell membrane	
	Inclusions of glycogen, starch, volutin, PHB	Inclusions of glycogen, starch, fats, and oils
Macromolecular synthesis	30S and 50S ribosomal subunits	40S and 60S ribosomal subunits, endoplasmic reticulum, Golgi complex
Genetic information	Circular nucleoid	Membrane-bounded nucleus
		Nucleolus
	Plasmids	Organelle DNA
Movement	Flagella of protein fibrils	Flagella and cilia of "9 + 2" microtubular type
	Gas vesicles	Pseudopodia
		Gas vesicles
Adaption	Endospores, cysts, conidia	Cysts
	Pili	No equivalent
	No structural equivalent	Lysosomes

the **lysosome.** These membrane-bounded sacs contain colloidal mixtures of hydrolytic enzymes. Hydrolases are potent catalysts for the breakdown of macromolecules into subunits. Proteases, lipases, and nucleases are examples; their names indicate the substrate they attack. These enzymes in a free cytoplasmic form would be incompatible with cellular integrity. Thus they are retained in an encapsulated form to be released only as needed.

3.23
Eucaryotic Flagella   In the flagellar shaft (a), there are nine pairs of microtubules, each of which has two rows of projections. These are *dyneium* proteins, actively engaged in the ATP-dependent motion of the microtubules. There is also a central pair of microtubules. Thus the arrangement is called the "9+2" pattern. In the basal body (b), the architecture is somewhat changed.

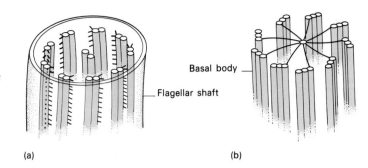

(a)   Basal body
Flagellar shaft
(b)

Some tissues need remodeling during growth and also during healing and the formation of scar tissue. Remodeling in tissues, as well as in our homes, requires that destruction occur first, followed by reconstruction. Lysosomal enzymes are released to carry out the molecular wrecking.

Lysosomes and other cellular granules are largely responsible for the effects of inflammation and some degenerative tissue changes. Most direct interception of microbial invaders within our bodies is carried out by phagocytic white blood cells, whose lysosomal enzymes destroy the ingested invaders.

Inflammation; Chapter 12, pp. 368–369

Movement   Eucaryotic cells have mechanisms for relocating cell contents, resulting in internal movement and the change of cell shape. Three locomotor structures coexist. The **microfilaments** are slender protein fibrils that appear to have contractile properties. They are active in cytoplasmic streaming, contraction of muscle, and the separation of developing daughter cells in mitosis. **Microtubules** are hollow cylinders assembled by the helical stacking of molecules of the protein **tubulin.** Intracellularly, they form the mitotic spindle. Their major role is external; they compose the organs of true motility, eucaryotic flagella and cilia. These structures, arising from a membrane-bound basal body, are composed of a circle of nine double microtubules surrounding a pair of single microtubules. This **"9 + 2" arrangement** occurs in all flagella (Fig. 3.23) and in cilia that are just short flagella. Many single-celled algae and certain protozoa have one or two flagella. A uniform covering of short cilia with coordinated movement is seen in protozoa such as *Paramecium* as well as in the ciliated epithelium that lines the human respiratory tract.

A third type of structure called the **intermediate** fiber has recently been demonstrated.

SUMMARY

1. An appreciation of the structure of the cell developed concurrently with the increase in sophistication of the microscope. Bright-field microscopes allowed study of shapes and the larger organelles of eucaryotes. Their use led to the development of many clever techniques for the differential staining of cell structures. Dark-field and phase contrast optics were developed later and are now widely used for the study of unfixed living cells. With the development of ultra-violet microscopy came an increase in resolution. In addition, techniques were created that coupled microscopic study with the biochemical and immunological probing of the cell surface.

2. The electron microscope was developed to take advantage of the much increased resolving power of shorter wavelength radiation. Transmission electron microscopes create images of very thin sections of cells or of surface replicas. A variety of techniques have been developed to provide images with the desired degree of detail. Scanning electron microscopes use deflected electrons to assemble an image of surface details that appears to be three-dimensional.

3. Electron microscopes provided convincing evidence that there were two different types of cell. A combination of microscopic and biochemical techniques has provided us with detailed structural information about procaryotes and eucaryotes.

4. The procaryotic cell has a number of surface layers. The capsule is a loosely organized external layer, antigenically unique, that serves protection functions. The cell wall or envelope contains a layer of the complex polymer peptidoglycan that exists as one massive covalently linked molecular enclosure. It provides protection against osmotic shock and confers shape on the cell. While in Gram-positive bacteria the peptidoglycan sac comprises the entire cell wall structure, in the Gram-negative bacteria and cyanobacteria it is sandwiched between an outer membrane and an inner one. The outer membrane contains a mixture of substances, notably the lipopolysaccharides or endotoxins. The inner membrane is equivalent to and is usually called the cytoplasmic membrane. Cytoplasmic membranes are fluid mosaics of structural and enzymatic proteins interacting with phospholipid and other lipid components. Membranes carry out the regulatory functions associated with differential permeability; they contain enzymatic and carrier proteins necessary for passive and active transport. Between the membrane and the peptidoglycan layer is found a storage reservoir, the periplasmic space. When the cell wall is weakened or absent, bacteria lose their characteristic shape and become osmotically fragile. All of the surface layers contain unique chemical substances, antigens, by which they can be chemically recognized both *in vivo* and *in vitro*.

5. Energy capture in procaryotes takes place in extensions of the cytoplasmic membrane, thylakoids in the cyanobacteria and chromatophores in the true bacteria. Energy utilization and ATP production take place in specialized regions of the cytoplasmic membrane. Energy storage involves irregular aggregates of high-energy compounds within the cytoplasm.

6. Protein synthesis occurs both in the cytoplasm on free-floating polysomes and on membrane-bound polysomes. Genetic information is contained in the circular DNA double helix called a nucleoid. Nonessential information may also be present in the form of extrachromosomal elements called plasmids.

7. Motility is conferred by contractile protein fibrils. Flagella are hollow tubes formed as three fibrils assemble to form a helix. They rotate in response to chemical stimuli in the environment, allowing partially coordinated movement toward or away from stimuli.

8. The true spore is a highly refractile, resistant structure formed by some bacteria during conditions of healthy growth. If exposed to adverse conditions, spores survive when vegetative cells in the population die. When the surroundings become hospitable again, the spore will germinate and commence active growth.

9. In eucaryotic cells each function is to a large degree localized in individual specialized organelles. These are enclosed in membranes that partially isolate them from the cytoplasm. The cytoplasmic membrane's key function is to regulate permeability and transport. The mitochondrion serves as the site of energy generation; the chloroplast carries out photosynthesis. Protein synthesis takes place on polysomes, and a complex packaging network of membranes captures and moves those proteins and other macromolecules designed for exterior use.

10. The nucleus of eucaryotes is surrounded by a double membrane penetrated by numerous pores. Within the nucleus, genetic information is organized in linear chromosomes. The DNA is enclosed in a regular coating of histones. In most eucaryotic cells, all genetic information is present in duplicate because chromosomes are paired.

11. Eucaryotic movement structures are either contractile microfilaments or rigid, hollow microtubules. Their flagella or cilia are composed of microtubular bundles.

12. The lysosome is of great importance in defense against disease; it also may simultaneously be a major contributor to inflammation and other symptoms of disease.

Study Topics

1. Identify the portions of the EM spectrum used in each type of microscope. What type of viewing apparatus is needed to see each type of image?

2. Express the size of the human red blood cell in mm, nm, and angstrom ($\overset{\circ}{A}$) units.

3. Mitochondria and bacteria are about the same size and have the same type of ribosomal subunits. What might this suggest about a common origin?

4. Compare the various types of cell-wall deficient bacterial cells to normal procaryotes.

5. What are the functions of the procaryotic cell membranes?

6. Describe how bacteria move in response to chemical stimulation.

7. Compare the structure and characteristics of a bacterial spore with those of the vegetative cell.

8. Describe the process whereby eucaryotic cells produce macromolecules and move them to the exterior of the cell.

9. Microtubules play a crucial role in several eucaryotic functions. Describe them.

10. What role does the lysosome play in the life of eucaryotic organisms?

Bibliography

Books

Lowry, A. G., and P. Siekevitz, 1979, *Cell Structure and Function* (3rd ed.), New York: Holt, Rinehart and Winston.

Review articles

Adler, J., 1975, Chemotaxis in bacteria, *Ann. Rev. Biochem.* **44:** 341–356.

Salton, M. R. J., and P. Owen, 1976, Bacterial membrane structure, *Ann. Rev. Microbiol.* **30:** 451–482.

Silverman, M., and M. I. Simon, 1977, Bacterial flagella, *Ann. Rev. Microbiol.* **31:** 397–420.

Singer, S. J., and A. L. Nicholson, 1972, The fluid membrane model of the structure of cell membranes, *Science* **175:** 720.

4

A Survey
of the Microbial World

THE microbial world contains phenomenally diverse organisms. It is endlessly fascinating for those patient enough to devote the time to the study of these organisms. This chapter will describe each of the major subdivisions of the procaryotes; special mention will be made of one or more important species in each group. We will build on the conceptual framework already acquired to enlarge your understanding of the biology of microorganisms and their place in the ecosystem.

BASIC RULES OF CLASSIFICATION

The science of biological classification is called **taxonomy.** Its history is old and honorable, dating back at least to the Greek philosopher Aristotle. The desire to group things together is a basic human drive—it recognizes the advantages of generalizing information. The first workable plant and animal taxonomies were prepared by the Swedish scientist Carolus Linnaeus in 1753 and 1758. All succeeding classification schemes have been progressive modifications of these originals. Taxonomists today have available all the tools of biochemistry, and sophisticated computerized analysis. Even so, classification schemes for the microscopic life forms are still far from perfect. They are constantly being changed as new information comes to light and old criteria are reexamined.

TABLE 4.1
Microbial classification

A BACTERIUM	A HUMAN BEING	
Kingdom Procaryotae	Kingdom	Animalia
Division II	Phylum	Chordata
Part 8	Subphylum	Vertebrata
Family Enterobacteriaceae	Class	Mammalia
Genus *Escherichia*	Subclass	Eutheria
Species *coli*	Order	Primates
	Family	Hominidae
	Genus	*Homo*
	Species	*sapiens*

Classification of the bacteria is, at present, less formalized than that of the animals because there is not enough information on bacteria to justify development of a complete taxonomic scheme. Fewer subdivisions are used because the subdivisions or taxa between kingdom and genus are in a state of flux.

The Binomial System

A newly described organism always receives a two-part name—a **binomial** designation. The first portion, capitalized, is the **genus** name (equivalent to our family surnames) and the second, lower cased part is the **species** name (equivalent to our first or given name). The names are usually derived from Latin or Greek descriptive terms. Newly proposed bacterial names are subjected to the scrutiny of the International Code of Nomenclature for Bacteria. Before a bacterial strain is recognized as being truly new, it must be established that it is not merely a variant of an existing, already named organism.

Groups of similar genera are arranged in families, and groups of families constitute orders. This type of system is a **hierarchical** one (Table 4.1).

Purpose of Classifying Organisms

The goal of classification efforts is to group organisms that are closely related and similar in most of the more important characteristics. This is a **natural taxonomy;** it should also indicate evolutionary relationships. Unfortunately, this goal is particularly difficult to achieve among the microorganisms. Their great biological age perhaps accounts for the fact that they have evolved extremely complex interrelationships. Organisms that are structurally very similar may be so different functionally that their actual relatedness is doubtful. Conversely, two organisms of very different morphology may share some unusual function so that it is difficult not to believe that they are closely related.

The euglenid group of algae, which possesses features of both plant and animal cells, are claimed by both botanists and zoologists. The world of microorganisms is full of similar enigmas. In any case, some things can be learned about *Euglena* by considering it as a plant, and other things can be learned by considering it as an animal. Classification, for all its difficulties, allows us to organize what knowledge we possess, and to formulate the most sensible questions to ask in order to get more knowledge.

Some Criteria by Which Classification Is Done

Among macroscopic organisms, descriptions of species are almost entirely based on structural or gross anatomical criteria. The species, basic unit of taxonomy, is composed of all individuals capable of fertile interbreeding. These criteria are clearly visible and usually unambiguous.

The classification techniques developed for bacteria that do not reproduce sexually are necessarily more subtle. Before a newly discovered protist can be classified, we may need electron microscope studies of its cellular ultrastructure; extensive investigation of its metabolic patterns; and determination of the chemical characteristics of its DNA, ribosomal RNA, and other cellular constituents. In **numerical taxonomy,** these data are subject to a computerized analysis.

With each group of organisms, we will introduce the criteria most frequently used for their identification.

Classification of the Procaryotae

All life forms can be separated into the five kingdoms—Plants, Animals, Fungi, Protists, and Procaryotes. The *Procaryotae* consist of two divisions, the cyanobacteria (Division I) and the bacteria (Division II). In discussing these we will observe the almost endless variations that evolution has played on the simple cell theme. Each variation in turn, has a particular place in the ecosystem for which it is fitted. It will be valuable for you to think about the role of each type of organism. Why is each so well adapted to survive from the early stages of biological history, in the face of competition from more advanced life forms?

THE CYANOBACTERIA

For about 150 years the cyanobacteria were classified by the taxonomic rules of botany as plants. Since the early 1960s they have been shown conclusively to be a highly specialized group of bacteria. The information about them is far from complete, partly because isolating them in pure or **axenic** culture is quite difficult. However, certain key structural and functional characteristics are now clearly outlined.

Biology of the Cyanobacteria

All cyanobacteria are photosynthetic and produce molecular oxygen as a by-product. Some have the ability to live and grow, although poorly, with organic food sources in the dark. They are widely distributed in fresh and salt water as well as in soil and may inhabit hot springs and saline lakes. They contribute to the productivity of these environments by fixing CO_2 into carbohydrates via photosynthesis. In addition, many members of this group enzymatically fix molecular nitrogen (N_2) into an organic form, the amino acid glutamine, that is useful to plants. Cyanobacteria require no vitamins, and use nitrate or ammonia as a nitrogen source. They use organic nutrients only as a source of carbon and not for energy. Some cyanobacteria produce odorous substances that alter the potability of drinking water; others release neurotoxins poisonous to animals. Cyanobacteria also possess another striking characteristic—the capacity to move over a solid surface by gliding. It is believed by many, perhaps most, biologists that a primitive cyanobacterial cell was the ancestor of the chloroplast of photosynthetic eucaryotes.

Gliding mobility; Chapter 3, p. 88

Structures of the Cyanobacterial Cell

Gram-negative cell wall; Chapter 3, pp. 80–81

The cell wall   The cyanobacteria possess a typical Gram-negative cell wall. Two groups also produce a hollow exterior structure that surrounds the cell and binds groups of cells into long strands. This **trichome,** which is usually open-ended, breaks apart to allow for reproduction.

Photosynthetic apparatus   All photosynthetic organisms contain combinations of light-absorbing pigments built into membranous structures. One or more types of chlorophyll are used as the principal energy-capturing core; accessory pigments fur-

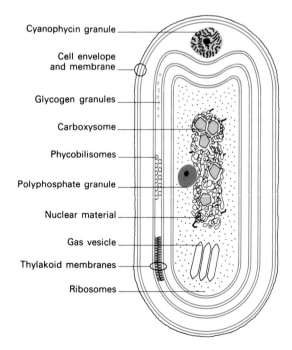

Cyanophycin granule

Cell envelope and membrane

Glycogen granules

Carboxysome

Phycobilisomes

Polyphosphate granule

Nuclear material

Gas vesicle

Thylakoid membranes

Ribosomes

4.1
The Structure of the Cyanobacterial Cell   The interior of the cyanobacterial cell is somewhat more complex than that of the bacteria because it contains much membranous material and numerous inclusions. (Courtesy of Drs. R. Y. Stanier and G. Cohen-Bazire, Phototrophic prokaryotes: The Cyanobacteria, *Ann. Rev. Microbiol.* **31:** 236. Reproduced, with permission, from the Annual Review of Microbiology © 1977 by Annuals Reviews Inc.)

ther expand the light-capturing potential. Cyanobacteria contain only chlorophyll a. Their accessory pigments include the yellow to red carotenoids and the **phyco-bilins**, a group of biliproteins that absorbs red or blue light. The light-capturing apparatus (Fig. 4.1), is constructed of **thylakoids**, layers of unit membrane that lie inside of and are connected with the cell membrane. The thylakoid membrane contains the chlorophyll, while the phycobilins are arranged in **granules** between the thylakoid layers. Together these carry out the "light" reactions of photosynthesis producing ATP. (See Chapter 8.)

Granules   Many types of cytoplasmic granules may be seen. The storage granules may contain either glycogen, volutin, polyphosphate, or **cyanophycin**, a polypep-tide containing aspartic acid and arginine. **Carboxysomes** contain enzymes for the incorporation of CO_2 into carbohydrate, the "dark" reactions of photosynthesis. **Gas vesicles**, presumably for buoyancy, are also frequently found.

Four Groups of Cyanobacteria

Confusion plagues the classification of cyanobacteria. Stanier and Cohen-Bazire have, on the basis of developmental criteria, placed these organisms in four groups (Table 4.2). This appears to be a simple, if not universally satisfactory, arrange-ment.

The Chroococcaceans   These are unicellular rods or cocci, looking much like bac-teria. They are mostly nonmotile and reproduce by simple binary fission or bud-ding. One genus, *Synechococcus*, is the single most studied of the cyanobacteria.

TABLE 4.2
Major subdivisions of the cyanobacteria

GROUP	MORPHOLOGY	MOTILITY	DNA BASE RANGE* MOL % G AND C	REPRODUCTIVE METHOD	NITROGEN FIXING
Chroococcacean	Unicellular rods or cocci	Mostly immotile	35–71	Binary fission or budding	No
Pleurocapsalean	Single cells in fibrous coat	Baeocytes may show gliding motility	41–47	Multiple fis-sion yielding baeocytes	No
Oscillatorian	Single vegetative cells in trichomes	Gliding motility	41–67	Fragmentation of trichome	No
Heterocystous	Trichomes with both vegetative cells and heterocysts	Gliding motility	38–47	Fragmentation of trichome	Yes

*Data derived from Stanier and Cohen-Bazire, 1977, explained later in this chapter.

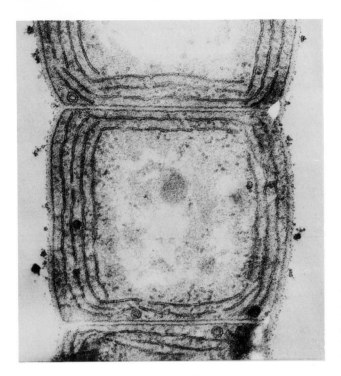

4.2
Anabaena, a Filamentous Cyanobacterium
Strands of vegetative cells such as these are interspersed with nitrogen-fixing heterocysts (not
shown). This organism is a major contributor to
summer algal blooms in fresh waters.

The pleurocapsaleans   In these organisms, daughter cells produced by binary fission remain glued together in small aggregates by a fibrous wall outside the regular
wall. They also reproduce, alternately, by a process in which a mother cell will
undergo multiple fission. The tiny daughter cells, called **baeocytes,** may number
from four to 1,000. They are released by rupture; on release they exhibit gliding motility. *Dermocarpa, Xenococcus,* and *Pleurocapsa* are among the form genera in this
group.

The oscillatorians   These organisms form trichomes and therefore are usually observed in water samples as long strands. These trichomes reproduce by fragmentation; within the trichome elongation is a result of binary fission. *Oscillatoria,* ubiquitous in fresh water, can be used to demonstrate the ability of the trichome to
glide, probably by means of fibrils. *Spirulina* and *Pseudoanabaena* are other genera.

Heterocyst formers   These form trichomes in which the ordinary vegetative cells
are interspersed with specialized cells called heterocysts (Fig. 4.2). Their function is
nitrogen fixation. The nitrogen-fixing enzymes are rapidly inactivated by oxygen
and therefore require an anaerobic intracellular environment. These cells have extra-
thick cell walls and reduced amounts of photosynthetic pigment. They carry out
only a portion of the photosynthetic process and do not produce oxygen. Sufficient

energy in the form of ATP is captured to drive the endergonic process of nitrogen fixation. This results in the manufacture of the amino acid glutamine. A tubular neck connects the heterocyst to adjacent vegetative cells. Through pore openings, glutamine passes into the nonnitrogen-fixing vegetative cells to nourish them. There is probably no DNA in the heterocyst, which does not divide. *Anabaena* and *Nostoc* are common heterocyst formers in fresh water.

Role of Cyanobacteria in the Environment

Although cyanobacteria cause no known human diseases, their contribution to our existence through their ecological functions is enormous. Cyanobacteria contribute a very significant part of the fixed nitrogen essential for plant and animal life. They are a part of the **phytoplankton**—floating microscopic plant life—and constitute an important base for aquatic food chains. They are consumed in large quantities by aquatic animals. Their oxygen production adds substantially to the total atmospheric oxygen. Approximately three billion years ago their appearance converted the earth from an anaerobic to an aerobic habitat.

Nitrogen fixation; Chapter 6, pp. 166–167

Cyanobacteria may be found in soils, especially moist, warm soils such as those found in greenhouses. They produce a substance called **geosmin** that is responsible for the characteristic "soil smell." They form crusts on hot, dry soils and over rocks; in shallow waters they form multicolored mats, often in association with the eucaryotic algae. A number form a symbiosis with fungi, producing the unusual partnership of the **lichen**. (See Chapter 6.)

THE BACTERIA

The bacteria, descendants of the earliest forms of life, have survived and prospered. Representatives of this group occupy all areas of the globe, including such unlikely spots as Antarctic ice masses, the Dead Sea, and steaming hot geysers. Their diversity of form and function is great.

Criteria Used in the Classification of Bacteria

Cellular morphology   Once a newly isolated organism is obtained in pure culture, the microbiologist describes its **shape, size,** and the characteristic **groupings** in which it appears. A **bacillus** (plural, bacilli) is a rod-shaped or sausage-shaped cell; a **coccus** (plural, cocci) is a spherical cell; and a **spirillum** (plural, spirilla) is a helical, rigid rod (Fig. 4.3). These shapes must be accurately described by the acute observer. In addition, cell groupings (clusters, chains, pairs, tetrads) are significant. Cell length and/or diameter can be measured by the use of a calibrated micrometer scale mounted within the microscope ocular. Not all the cells in a population will be exactly alike, but a composite picture of the average cell can be prepared.

In conjunction with the microscopic examination, a determination of the Gram reaction is made. This may be followed up by biochemical analysis of cell-wall com-

Staining methods; Chapter 3, pp. 69–70

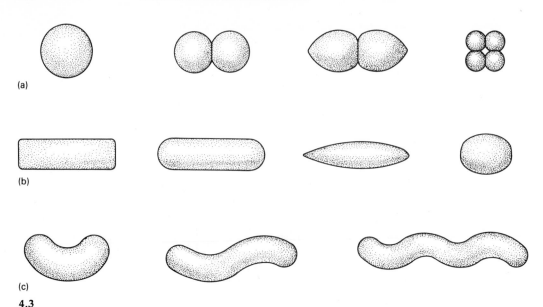

(a)

(b)

(c)

4.3
Bacterial Cell Shapes   (a) Cocci (spheres) may occur singly, in pairs as diplococci or paired lancets, and in tetrads. (b) Bacilli (rods) may be square or round-ended. If the ends are elongated and pointed, the bacteria are fusiform. If they are short and fat, the bacteria are coccobacilli. (c) The spirilla are helically curved. A half-turn of the helix produces the common vibrio. One or several full turns may be seen in the spirals.

ponents. Special structures such as endospores, capsules, and granules are also important to demonstrate by special staining methods.

Motility   The presence, number, and arrangement of flagella or the presence of nonflagellar motility are important determinative traits (Fig. 4.4). In special cases, measurements of the length and wave form of the fibril and biochemical studies of the flagellar proteins have been used.

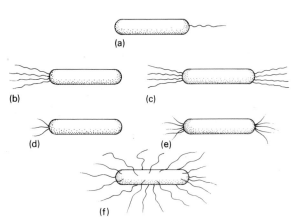

(a)

(b)   (c)

(d)   (e)

(f)

4.4
Distribution of Bacterial Flagella   Flagella may be located in polar groupings (a–e) singly, or in tufts, at one or both ends. In other bacterial forms, the flagella are found over the entire surface(f). The exact numbers and groupings may be determined with a flagella stain, using the light microscope, or with electron microscopy.

Mode of division   Most bacteria reproduce by binary fission, but other methods, such as budding, formation of reproductive spores, and fragmentation are seen. The actual plane of division plays a role in establishing cell groupings.

Binary fission; Chapter 7, pp. 203–205

Nutrition   The nutritional classification of an organism depends on the energy sources it uses. Let us establish a common terminology. You are already familiar with the biological distinctions between autotrophs and heterotrophs. These terms have been further refined in microbiology to distinguish among the varied modes of inorganic and organic energy gathering. The microbiologist recognizes four nutritional categories:

Photoautotrophs—able to grow with light as the energy source in a strictly inorganic medium using CO_2.

Photoheterotrophs—obtain energy for growth from light, but use organic carbon compounds as a source of carbon.

Chemoautotrophs—able to obtain energy by the oxidation of inorganic compounds; they use CO_2 as their main carbon source.

Chemoheterotrophs—obtain energy by the oxidation of organic compounds. These are also the main sources of carbon.

The last category can be further divided into those called **saprophytes,** that use organic compounds obtained as wastes or residues from dead organisms, and those called **parasites** that either prefer or are required to live at the expense of a living organism. All microbial pathogens are chemoheterotrophic.

Metabolism   Metabolic patterns in bacteria are varied and complex, but all generate electrons. One question of particular importance is precisely what substance is used as the final electron acceptor (for redox reactions) in the metabolic pathways. Bacteria that use oxygen or another inorganic substance as electron acceptor are **respiratory;** those that use only organic electron acceptors are **fermentative.** Organisms such as *Bacillus subtilis,* that require oxygen are **strict anaerobes;** they are always respiratory. Organisms, such as *Escherichia coli,* that survive adequately both with and without oxygen are **facultative anaerobes;** they are often both respiratory and fermentative. Organisms, such as *Clostridium perfringens,* that do not use oxygen and tolerate exposure to it poorly are **strict anaerobes;** they are fermentative and, in many cases, respiratory by means that do not use oxygen. We will discuss the metabolic bases for these adaptations in Chapter 8.

Respiration fermentation; Chapter 8, pp. 236–239

Inclusions   The subject of cellular energy storage was covered in the last chapter. In many bacterial groups, knowledge of the chemical nature of the storage substance is a useful piece of identification. For example, the formation of poly-β-hydroxybutyrate or its lack is an important key for identification of species of *Pseudomonas.*

DNA   The genetic material of each species is unique. However, the more closely related two species are, the more similar their DNA material is. One useful analysis is

TABLE 4.3
Criteria useful in classification and identification of unknown bacteria

MICROSCOPIC
MORPHOLOGY

Cell shape
Cell size
Groupings
Internal structures
Appendages
Staining reactions

BIOCHEMICAL
DETERMINATIONS

Cell-wall materials
Cellular lipids
Capsular materials
Pigments
Cellular storage inclusions
Characteristic antigens

ENVIRONMENTAL
RELATIONSHIPS

Atmospheric gases required
pH tolerance
Temperature range
Osmotic tolerance
Antibiotic sensitivity
Pathogenicity
Symbiosis
Habitat from which isolated

APPEARANCE OF GROWTH

Location in liquid culture
Colonial morphology
Pigmentation
Rate of growth

METABOLIC PATHWAYS

Carbon sources used
Nitrogen sources used
Sulfur sources used
Fermentation potential
Respiratory potential
Characteristic end products

MOLECULAR GENETICS

DNA base composition
Nucleic acid hybridization
Base sequences (tRNA, rRNA)
Genetic recombination
Amino acid sequences of polypeptides
   such as cytochromes

DNA; Chapter 2, pp. 55–56

the determination of the **base ratio** of the purified DNA from a culture. Remember that in DNA, A always pairs with T, and G with C. Thus molar amounts of A and T will in all cases be identical as will the amounts of G and C. However, different organisms show wide fluctuation in the ratio of GC pairs to AT pairs. An accurate measurement of the mols percent of the DNA bases represented by G and C combined is an almost universal part of the identification process today (Table 4.3). Similarity of base ratios implies relatedness only if other characteristics are also very close. Hybridization techniques now available tell to what degree DNAs from two organisms are identical in information content—the most critical measurement of relatedness. These techniques will be discussed in Chapter 11.

This summary of determinative criteria is of necessity incomplete. The reader can most quickly get an idea of the methods of bacterial classification by leafing through the primary reference on bacterial classification, *Bergey's Manual of Determinative Bacteriology*. This compendium, first published in 1923, is now in its eighth edition (1974). In each edition, major rearrangements have occurred. A summary is presented in Table 4.4 on pages 111–112.

It is still not possible to construct a cohesive natural classification for the procaryotes. Thus the editors of *Bergey's Manual* have taken a provisional approach, dividing procaryotes into 20 parts. The first of these is Division I, the cyanobacteria that we have already reviewed. The 19 groups in Division II constitute the bacteria. Although the *Bergey's Manual* system is cumbersome, it is the most authoritative. The manual is readily available in most microbiology libraries, clinical libraries, and laboratories.

The Major Subdivisions of the Bacteria

The phototrophic bacteria: Part I This group contains all the photosynthetic bacteria. These organisms, found in anaerobic aquatic habitats, have a more primitive photosynthetic apparatus than do the cyanobacteria, algae, and higher plants (Fig. 4.5). They use light energy solely for synthesis of ATP; wavelengths in the near infrared range (700–900 nm) are absorbed. The hydrogen for synthesis of their carbohydrate is obtained from H_2 gas, H_2S, or organic compounds, not water (H_2O).

Bacterial photosynthesis; Chapter 8, pp. 239–243.

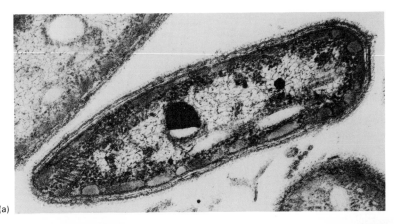

(a)

4.5
Some Autotrophic Bacteria
(a) *Pelodictyon clathratiforme;*
(b) *Rhodopseudomonas spheroides;*
(c) *Ectothiorhodospira mobilis;* (d)
Nitrobacter agilis; and (e) *Thiobacillus neopolitanus.*

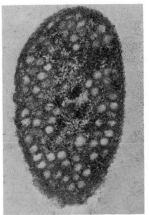

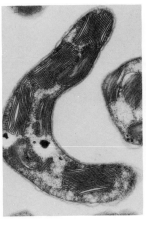

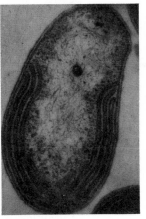

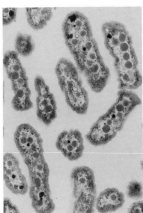

(b) (c) (d) (e)

TABLE 4.4
The characteristics of the major subdivisions of bacteria

NUMBER	TITLE OF BACTERIAL GROUP	CELL SHAPE AND GROUPING	MOTILITY
1	Phototrophic	Rods, cocci, spirals occur singly	Flagella
2	Gliding	Single rods	Gliding
3	Sheathed	Rods inside sheath	Flagella
4	Budding	Single rods, cocci, also rosettes	Flagella
5	Spirochetes	Single flexible spirals	Axial fibrils
6	Spiral and curved	Rigid curved rods, occur singly	Flagella
7	G^- aerobic	Single rods, coccobacilli	Flagella
8	G^- facultative anaerobes	Single rods	Flagella
9	G^- anaerobic	Single rods, straight, curved, pointed	Usually none
10	G^- cocci and coccobacilli	As described, single and pairs	Usually none
11	G^- anaerobic cocci	Single cocci or pairs, masses	Usually none
12	G^- chemoautotrophic	Rods, spirals, cocci, single and aggregates	Flagella
13	Methane-producing	Variable shapes, seen singly	Flagella
14	G^+ cocci	Cocci in pairs and in chains	Usually none
15	Endospore-forming	Rods and cocci, single and in chains, form spore	Flagella
16	G^- asporogenous rods	Rods, single and in chains, do not form spores	Flagella
17	Actinomycetes	Irregular rods, club-shaped, some form mycelium	Usually none
18	Rickettsias	Tiny, irregular	None
19	Mycoplasmas	Tiny, irregular, form filaments, pseudomycelium	None

GRAM STAIN	NUTRITION	OXYGEN NEED	CELL DIVISION
Negative	Photoautotrophic; photoheterotrophic	Anaerobic	Binary fission, budding
Negative	Chemoheterotrophic	Aerobic	Binary transverse fission
Negative	Chemoheterotrophic	Aerobic	Binary fission
Negative	Chemoautotrophic; Chemoheterotrophic	Anaerobic and Aerobic forms	Budding
Negative	Chemoheterotrophic	Facultative anaerobic	Transverse fission
Negative	Chemoheterotrophic	Aerobic	Binary fission
Negative	Chemoheterotrophic	Aerobic	Binary fission
Negative	Chemoheterotrophic	Facultative anaerobic	Binary fission
Negative	Chemoheterotrophic	Strictly anaerobic	Binary fission
Negative	Chemoheterotrophic	Aerobic	Binary fission
Negative	Chemoheterotrophic	Strictly anaerobic	Binary fission
Negative	Chemoautotrophic	Aerobic	Binary fission
Positive Positive Negative	Chemoautotrophic; chemoheterotrophic	Strictly anaerobic	Binary fission
Positive	Chemoheterotrophic	Aerobic facultative anaerobic	Binary fission
Positive	Chemoheterotrophic	Strict and facultative aerobes and anaerobes	Binary fission
Positive	Chemoheterotrophic	Facultative anaerobes	Binary fission
Positive Negative	Chemoheterotrophic	Aerobic, facultative anaerobic	Budding, fragmentation
Negative	Intracellular parasitic	Aerobic	Binary fission, elementary bodies
Negative	Chemoheterotrophic; parasitic	Facultative anaerobic	Elementary bodies

Thus they produce elemental sulfur or other end products rather than oxygen; their environment remains anaerobic. Most photosynthetic bacteria also assimilate organic compounds, both during photosynthetic (light) periods and while respiring in the dark. These organic compounds may be used as carbon, hydrogen, or energy sources under varying conditions. Phototrophic bacteria are of various shapes, including rods, cocci, spirals, and budding forms. Most reproduce by binary fission and are motile.

There are two families of purple bacteria, the rhodospirillaceae and the chromatiaceae. These contain bacteriochlorophylls a and b, contained in lamellae or tubules continuous with the plasma membrane.

The green bacteria, or chlorobiaceae, contain bacteriochlorophylls c, d, and e, with some a. These pigments are found in discontinuous membrane vesicles. Many green bacteria are planktonic in lakes; there is incomplete evidence that they might have been ancestral to the cyanobacteria.

The gliding bacteria: Part 2   These unusual organisms are common inhabitants of soil and areas exposed to animal wastes. They are unicellular, Gram-negative rods or filaments; most are motile by means of gliding. *Cytophaga* is a soil and water organism that lives on polysaccharides such as cellulose, chitin, and agar. Certain related organisms cause diseases in fish. *Beggiatoa* (Fig. 4.6a) is a filamentous glider found in environments rich in H_2S; it accumulates intracellular sulfur granules.

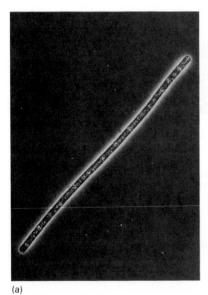
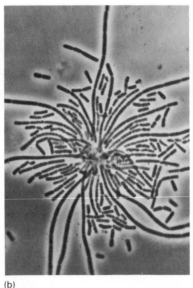

(a)   (b)   (c)

4.6
Some Aquatic Bacteria   (a) *Beggiatoa* sp., a filamentous sulfur bacterium. Bright areas in the cell are sulfur granules. (b) *Leptothrix* sp. (c) *Caulobacter crescentus*, two-stalked cells and a swarmer.

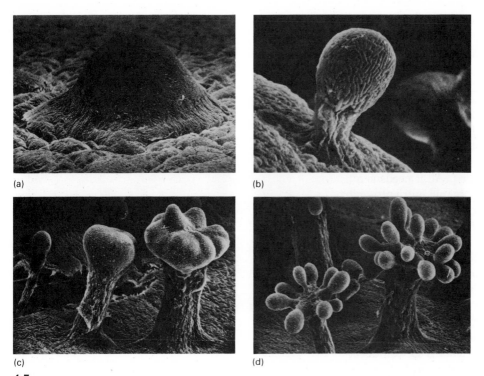

(a)   (b)   (c)   (d)

4.7
Chondomyces crocatus (a) In fruiting, cells first aggregate and mound up. (b) Then they start to migrate upward forming a stalk and a head. (c) A slime coat obscures the cellular detail. (d) The mature fruiting body bears golden yellow sporangia.

Leucothrix, also filamentous, is found in association with marine seaweeds; only certain cells glide.

The **myxobacteria** exhibit a complex developmental cycle. Under conditions of nutrient restriction, the vegetative colony will begin swarming. Individual cells glide and converge in the colony center. The aggregate (Fig. 4.7) becomes elevated forming an involved structure, often brightly pigmented. The cells within this **fruiting body** become resistant resting cells called **myxospores.** The gliding bacteria such as *Myxococcus* and *Chondromyces* have been extensively studied in an effort to learn more about cell-to-cell signaling and differentiation.

The sheathed bacteria: Part 3 This is a diverse assortment, probably not closely related. The rod-shaped, Gram-negative cells form filaments surrounded by sheaths. The sheaths may become impregnated with oxides of iron or manganese.

Within the sheath, which resembles a tubular capsule, binary fission gives rise to motile swarmer cells, which exit from the sheath's ends. Each swarmer may originate a new filament. *Sphaerotilus* and *Leptothrix* (Fig. 4.6b) are widely distributed in fresh and salt waters, usually in association with sewage pollution.

Box 4.1

Living on Air

Hyphomicrobium vulgare, whose species name means "common" in Latin, is a very undemanding microorganism. It will develop as a film on the surface of liquids, such as tap water, to which nothing at all in the way of nutrients has been added. All it needs is a source of the usual inorganic nutrients—nitrogen, sulfur, phosphorus, and air. It is not photosynthetic, so without organic carbon, how is it living?

The conclusion that this organism is an autotroph of some sort has been proved wrong. *Hyphomicrobium* must have organic carbon, preferably in the form of single-carbon compounds such as methyl alcohol, formaldehyde, or even cyanide. However, its ability to concentrate dilute nutrients is so well developed that it gets sufficient nourishment from volatile substances present in the air in a few parts per billion concentration. These organics form a very dilute solution in the growth fluid.

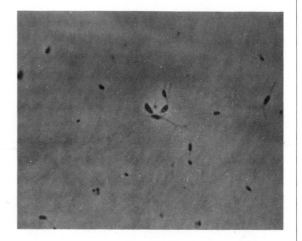

Since such volatiles are in ample supply in the somewhat odorous atmosphere of many laboratories, these bacteria are quite content to live on air.

The budding and appendaged bacteria: Part 4   The organisms in this group are placed together because they undergo unequal cell divisions, and the daughter cells originate from one end or an extension of the cell. Most organisms excrete polymeric strands of material that twist to form stalks. These serve as holdfasts to attach the cell to a solid surface.

Hyphomicrobium forms long, slender extensions called hyphae; the daughter cell, with a flagellum, develops at the tip, then separates. *Caulobacter* (Fig 4.6c) bears a stalked holdfast, attached to a solid surface. The holdfasts may attach to many other cells, forming a rosette. *Caulobacter* is common in tapwater. All budding and appendaged bacteria are aquatic.

Treponema; Chapter 21, pp. 630–631

Leptospira, Chapter 21, p. 624

The spirochetes: Part 5   Spirochetes possess a flexible cell wall. They are Gram-negative but, because of their slenderness, they often are poorly visible under the light microscope. Using flagella-like axial filaments, they have an undulating motility (Fig. 3.17). The genus *Spirochaeta* are free-living, aquatic organisms, but the four other genera are inhabitants of animals. Three genera, *Treponema* (syphilis, yaws), *Borrelia* (relapsing fever), and *Leptospira* (leptospirosis) contain species highly pathogenic to human beings. Spirochetes can be fermentative or respiratory. Several, including the agent of syphilis, are obligate parasites, requiring living organisms as a habitat; they have defied all efforts to culture them in nonliving media. Certain *Borrelias* have been grown *in vitro*.

The spiral and curved bacteria: Part 6   One difference between this group and the preceding one is that these organisms are true spirilla; they are rigid helices. Their flagella are located on the cell surface. *Spirillum*, (Fig. 4.8) *Aquaspirillum*, and *Oceanospirillum* are predominantly aquatic. The genus *Campylobacter* is a fairly recently described group, microaerophilic or anaerobic, implicated in reproductive and intestinal infections in human beings and in domestic animals. *Bdellovibrio* parasitizes bacteria of other species. Unicellular motile cells enter the host bacterium and multiply in the periplasmic space. They may attack a variety of other Gram-negative species. A very long, multinuclear spiral cell is formed, which subsequently fragments into numerous progeny.

Campylobacter; Chapter 20, p. 594

Gram-negative, aerobic rods and cocci: Part 7   The organisms of this group are all strictly respiratory chemoheterotrophs, fermentation does not occur. Most members possess the respiratory enzyme **cytochrome oxidase** and are said to be **oxidase positive.** For further discussion it will be well to subdivide the group.

Cytochromes; Chapter 8, p. 226

The pseudomonads (Family I) have polar flagella, are chemoheterotrophs, and are extraordinarily versatile in their ability to utilize compounds for energy. The genus *Pseudomonas* includes many water and soil organisms, as well as several species that are occasional pathogens of human beings (*P. aeruginosa, P. pseudomallei, P. cepacia*) and plants (*P. solanacearum*).

4.8
Spirillum   The cell body is helically curved as shown by the shadowed effect in this electron micrograph.

The *Azotobacter* group (Family II) contains organisms that are nitrogen-fixers. They are free-living and inhabit soil, water, and leaf surfaces. Some species form resistant cysts as resting stages.

The rhizobium group makes major contributions to soil fertility. *Rhizobium* forms **symbiotic** (mutually helpful) associations with leguminous plants such as beans, peas, and alfalfa. The bacteria–plant partnership forms **nodules** on the root network and in these nodules nitrogen is fixed by the bacteria and absorbed by the plant. Legume crops significantly increase the available nitrogen in soils. The discovery of the "enriching" character of leguminous crops led to the introduction of crop rotation, a major agricultural advance in human history. *Agrobacterium* forms root tumors called crown galls. It has been established that the tumorigenic factor in the bacterium is present in a plasmid.

Plasmids; Chapter 9, pp. 274–276
Symbiosis; Chapter 6, pp. 159–162

The most interesting features of the **halobacteria** are specializations that allow them to function in highly saline environments. Their enzymes require high salt concentrations for catalytic activity. A high concentration of carotenoid pigments in the membrane affords protection against sunlight. When grown under conditions in which oxygen is limited, they produce bacteriorhodopsin, a light-sensitive pigment, that establishes an ionic gradient by releasing hydrogen ions (H^+) from the cell on illumination. Ionic gradients can be used by this organism to generate ATP, capturing solar energy in chemical form.

Three pathogenic genera are included in Part 7 because they are Gram-negative and aerobic. They differ from the preceding groups in many ways especially in that they are usually associated with warm-blooded animals. Members of the genus *Brucella* cause brucellosis and spontaneous abortion in domestic animals and occasionally in people. In the genus *Bordetella* is found the organism that causes whooping cough. The genus *Francisella* causes tularemia in human beings and animals.

Bordetella; Chapter 19, p. 552

Gram-negative, facultatively anaerobic rods: Part 8

In this group are placed those organisms capable of both fermentative and respiratory metabolism. Almost all of these lack the enzyme oxidase found in strict aerobes. If flagella are present, they are usually peritrichous (Fig. 4.4f).

The enterobacteria are probably the best understood of all microorganisms. It is said that we know more about *Escherichia coli* than we do about any other living thing including ourselves. Because of the presence of many pathogens in this group, techniques for identification and classification have been brought to a very high level of sophistication. Some members of the enterobacteria (*Enterobacter, Citrobacter, Proteus*) are components of the **normal flora** or resident microbial population in the mammalian intestine. They may cause opportunistic disease in a weakened host. Others, such as the *Salmonella* and *Shigella* genera, cause serious epidemic diseases including typhoid fever and bacterial dysentery. *Klebsiella* may cause pneumonia. One group, *Erwinia*, is found in association with plants, sometimes as harmless and sometimes as pathogenic flora. The enterobacteria are discussed in detail in Chapter 20.

Normal flora; Chapter 6, pp. 176–180; Chapter 12, p. 346

The vibrios (presently classed with the Part 8 organisms) have much in common with the pseudomonads (such as polar flagella). Most of the species are curved rods.

Some, like *Vibrio cholerae*, the agent of cholera, are pathogens. Others are fresh water and salt-water organisms that occasionally cause disease (*V. parahaemolyticus*). *Photobacterium* is a luminescent organism that gives off yellowish light when respiring with sufficient oxygen. Most luminescents are marine; they are usually associated with fish and may be the "battery" in the fish's luminescent organs and markings. The light-generating mechanism uses an enzyme called **luciferase,** also found in fireflies.

Loosely associated with other Part 8 organisms are some human pathogens including *Hemophilus* (pneumonia, ear infections), *Pasteurella* (bubonic plague), and *Streptobacillus* (rat-bite fever).

Hemophilus, Chapter 19, p. 555

Gram-negative, anaerobic bacteria: Part 9   Anaerobic organisms by definition tolerate oxygen poorly. Thus they can be found in specialized environments. The organisms in this group are isolated from intestinal contents, deep tissues in animals, or microenvironments such as the groove between the teeth and the gums in the human mouth.

The *Fusobacterium* and *Bacteroides* genera are a major component of human oral flora and feces. When trauma allows them to escape from the bowel into the body cavity, *B. fragilis* and *B. melanogenicus* cause complications such as abscesses, surgical infections, and peritonitis. Organisms of this group frequently infect bites and other wounds. For successful culture they require special techniques of isolation and maintenance (See Chapters 7 and 23.)

Wound infections; Chapter 23

The anaerobic sulfate-reducing bacteria *Desulfovibrio* and others are found in aquatic, intestinal, or food spoilage environments. One genus, *Desulfomaculatum*, forms spores.

Gram-negative cocci and coccobacilli: Part 10   The Gram-negative cocci (Fig. 4.9) are predominantly aerobic and respiratory, and only one genus, *Neisseria*, is fer-

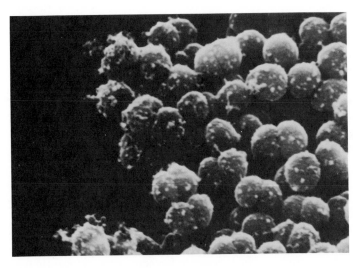

4.9
Neisseria   The doubled effect of the diplococci is evident in this scanning electron micrograph.

N. gonorrhoeae;
Chapter 21, pp.
628–629
N. meningitidis;
Chapter 22, pp.
659–660

mentative. The enzyme oxidase is generally present. The genus *Neisseria* contains two pathogenic species (*N. gonorrhoeae* and *N. meningitidis*) that we will study in detail at a later time. *Branhamella* and *Moraxella* are normal human flora, occasionally opportunists. *Acinetobacter*, which is rod-shaped, is oxidase negative and moves with a twitching or jumping motility. It is pathogenic only in debilitated individuals.

Gram-negative, anaerobic cocci: Part 11

Dental plaque; Chapter
16, pp. 475–479

Gram-negative, anaerobic cocci: Part 11   This group contains only four species. *Veillonella* is a parasitic component of the normal intestinal flora of warm-blooded animals and is also found in dental plaque. It is parasitic but not individually associated with disease. It may be detected as part of a mixed infection.

The chemoautotrophic bacteria: Part 12   Metabolism of the chemoautotrophic type of bacteria, in which all the energy necessary for cellular activities can be extracted from redox reactions between inorganic chemicals, is a feat unique to bacteria. Strictly aerobic chemoautotrophs depend on sources of reduced nitrogen or sulfur compounds, which they oxidize to make ATP. Many have membranous lamellae containing the necessary enzyme–coenzyme sequences (Fig. 4.5d).

Nitrifying bacteria may utilize ammonia (*Nitrosomonas, Nitrosococcus*) or nitrate (*Nitrobacter, Nitrococcus*) in aerobic, energy-yielding processes that lead to the production of ATP. In small numbers, they are common in soil and water. Their sequential activities yield nitrate, a plant nutrient. Sulfur-oxidizers such as *Thiobacillus* (Fig. 4.5e) can utilize the reduced inorganic sulfur compounds, with sulfate as an end product. Some tolerate extremely acidic or hot conditions. *Sulfolobus*

Thermophiles; Chapter
7, p. 195

lives in sulfur-rich, hot springs. The iron bacteria produce sheaths that are encrusted with iron and manganese compounds.

Chemoautotrophy can be viewed as an evolutionary experiment, one that was successful enough so that its possessors have survived, but not profitable enough so that it became incorporated into the eucaryotic forms.

The methane-producing bacteria: Part 13   The Methanobacteriaciae are strictly anaerobic and are more exquisitely sensitive to traces of oxygen than are any other anaerobic bacteria. Most are chemoautotrophic, reducing CO_2 with H_2 gas to form methane. This reaction yields energy and provides the cell with organic carbon for growth. Some are chemoorganotrophic, combining H_2 with simple organic compounds such as acetate. Their natural distribution is very restricted. They are commonly found, when special anaerobic isolation techniques are used, in the rumen

Rumenbacteria;
Chapter 6, pp. 176–177

and intestinal tract of animals, in composting manure, in anaerobic sludge digesters, and in oxygen-free sediments. Their commercial use for the production of methane gas is being exploited.

The Gram-positive cocci: Part 14   This group contains both aerobic and facultatively anaerobic members. The genera *Micrococcus* and *Staphylococcus* primarily occur in tetrads (fours) and regular grapelike clusters. They are both respiratory and fermentative, and contain catalase. Most are harmless members of the normal flora or soil and water forms. *Micrococcus*, which does not ferment, is often brightly pig-

Box 4.2

The Methanogenic Bacteria—Another Kingdom?

The methanogens are a morphologically diverse group—rods, cocci, spirals, and packets. All share the physiological traits of strict anaerobiosis and methane production from carbon dioxide and hydrogen gas. Recent studies have turned up a number of important points in which these organisms differ from other bacteria. They have cell walls, and most stain Gram-positive; however, they completely lack peptidoglycan. Their systems for electron transport by coenzymes are atypical—they lack cytochromes, but have two unique carriers, coenzymes F_{420} and M. All previously studied autotrophic organisms use a metabolic pathway called the Calvin cycle for incorporating CO_2

into carbohydrates; these organisms lack the necessary enzymes, so they are fixing carbon dioxide by an unknown method different from that used by all other autotrophs. Analysis of their ribosomal RNAs reveals a base sequence notably different from that of other bacteria. These distinctions have led to the suggestion that the methanobacteria, so well suited to the primordial atmosphere thought to have been present during the early stages of biological evolution, may be little-changed descendants of some of the first life forms to arise. Some authors have suggested that their uniqueness may be sufficient to justify their placement in a separate, sixth kingdom.

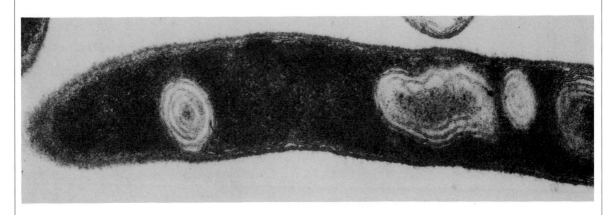

Methanobacterium thermoautotrophicum   This electron micrograph shows the cell wall, cytoplasmic membrane, and introcytoplasmic membranes of this unusal autotroph.

mented with yellow to red carotenoids. Some species form cubical packets (Fig. 3.6). *Staphylococcus aureus* causes severe infections in human beings with some fre-

Staphylococcus aureus; Chapter 17, pp. 501–504

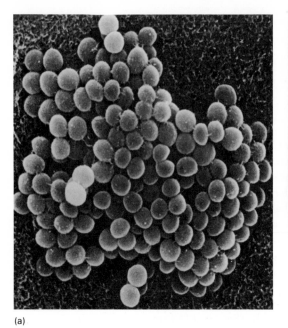

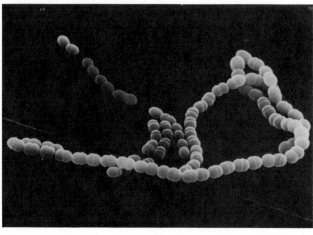

(b)

4.10
Gram-positive Cocci   (a) Clusters of *Staphylococcus*, a common inhabitant of human skin. (b) Chains of *Streptococcus faecium*, a common inhabitant of the human bowel.

(a)

quency; *S. epidermidis* is a normal skin organism (Fig. 4.10). All of these organisms are quite salt-tolerant.

The streptococci characteristically grow in chains, especially if in liquid media (Fig. 4.10). They are fermentative and make lactic acid as the main end product of carbohydrate metabolism. They do not respire and have no catalase. Streptococci are found in normal oral flora and bowel contents of warm-blooded animals. They are a major part of the bacterial population of milk and cause souring. *Streptococcus pyogenes* (Group A), is a serious pathogen, causing so-called strep throat; *S. agalactiae* (Group B) causes infant mortality. An important determinative trait is their **hemolytic** effect (destruction of red blood cells). Surface antigens are used to place the streptococci in groups.

Streptococcus pyogenes; Chapter 17, pp. 504–507

Peptococcus, which is not completely anaerobic, does not usually grow in chains. It is weakly fermentative and produces products other than lactic acid. Along with *Peptostreptococcus*, it has been isolated from normal human urogenital, respiratory, and intestinal tracts as well as from sites of infection. *Ruminococcus* is found in the rumen of cattle and sheep and the cecum of other herbivores; since the bacteria can digest cellulose they are crucial in enabling their hosts to live on hay and grasses.

The endospore-forming rods and cocci: Part 15   Formation of true heat-resistant refractile endospores is a trait restricted to a relatively small number of bacterial species (Fig. 4.11). The mainly aerobic genus *Bacillus*, widely distributed in soil and vegetation, is frequently recovered in bacteriological sampling of dry or dusty environments in which the spores survive indefinitely. Most members of this group are

Spore stain; Table 3.2

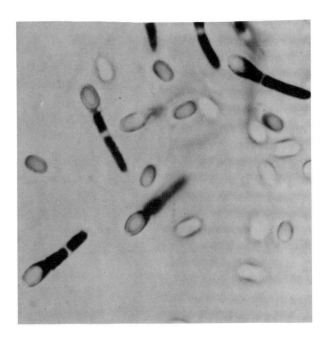

4.11
Spore Formation *Clostridium pectovorum,* an anaerobic spore former, stained to reveal the polar spores that distend the ends of the cells. Location of the spore within the cell as well as its size can be important clues in identification of these species.

strongly Gram-positive and actively motile with peritrichous flagella. Only one species, *Bacillus anthracis,* is normally pathogenic for animals and human beings; it is facultatively anaerobic. Many *Bacillus* strains produce extracellular hydrolytic enzymes such as amylase and penicillinase. Others yield antibiotics, including bacitracin and polymixin.

Anthrax; Chapter 17, p. 506

The genus *Clostridium* contains strict anaerobes, usual inhabitants of intestinal tracts of animals, animal wastes, and sewage. They are also found in waterlogged (airless) soils, black muds, and decomposing organic material. Clostridia are strongly fermentative, do not respire, and lack catalase. All remain harmless to human beings unless unusual events—puncture wounds (tetanus) or consumption of improperly canned foods (botulism)—expose human beings to their protein toxins. The butyric acid group of clostridia produce commercially valuable organic compounds such as acetone and isopropanol. *Cl. pasteurianum* is a nitrogen fixer.

Tetanus, Chapter 23, pp. 676–678
Botulism; Chapter 20, pp. 609–610

Gram-positive, asporogenous rods: Part 16   The genus *Lactobacillus* comprises most of this group. These organisms are characterized by their very exacting nutritional requirements for preformed amino acids and vitamins. This restricts their natural distribution to nutritionally complete environments such as foods and intestinal contents. Many of them are associated in some way with milk, dairy products, or the fermentation industries. Their anaerobic, fermentative metabolic activities are used by human beings for time-honored methods of food preparation such as the manufacture of cheeses (*L. casei*), yogurt, sauerkraut, and kefir (*L. bulgaricus*), the pickling of vegetables, and the rising of sourdough types of bread. In many of these applications a stable mixed culture, often with streptococci, is present.

The lactobacilli are also universally found in digestive tracts, and in the oral and vaginal flora of human beings. As part of the normal flora in these areas, they have poorly understood, but apparently positive, roles in maintaining health.

The genus *Listeria*—aerobic to microaerophilic, catalase-positive rods—are provisionally grouped in Part 16. *L. monocytogenes*, widespread in the flora of animals and their wastes, is an occasional but serious pathogen, especially in newborn infants.

Actinomycetes and related organisms: Part 17   This group is the largest and most diverse of the 19 bacterial subdivisions. All members possess a Gram-positive type of cell wall, although the actual staining reactions may be unreliable. They are all more or less irregular in cell morphology. Although basically rod-shaped, they may become clubbed, coccoid, branched, or filamentous.

The coryneform group are aerobic, nonmotile, club-shaped, small rods often showing pronounced granules. The members are divided into animal pathogens, plant pathogens, and nonpathogenic forms. Human beings carry nonpathogenic diphtheroids such as *Corynebacterium xerosis* and *C. pseudodiphtheriticum* among their normal flora. *C. diphtheriae* is responsible for a serious human disease, diphtheria. The genus *Arthrobacter* characteristically undergoes a shape change from rod to coccus as cultures age. Among this genus there are soil and water organisms.

Diphtheria; Chapter 18, pp. 531–535

The actinomycetes all tend to form branching filaments that in some families develop into a dense mat called a **mycelium.** Many form spores or conidia at certain growth phases. They are relatively slow-growing.

Some form only transitory or short branching filaments. Cell-wall composition is an important determinative trait. The genus *Mycobacterium* deserves mention because it contains human normal flora and the agents of tuberculosis and leprosy. Mycobacteria are readily identified by a positive acid-fast differential stain

Acid-fast stain; Table 3.2

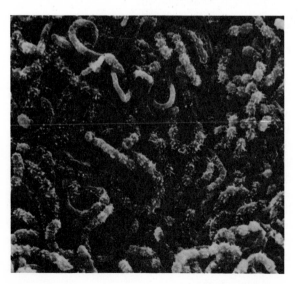

4.12
Streptomyces   Growth in this group of bacteria is characteristically mycelial. Spores, which are reproductive, are formed in large numbers at the ends of specialized, coiled structures.

because their cell walls contain lipids that allow them to resist decolorization with acid alcohol.

The anaerobic genus *Actinomyces* is found in the oral cavity; *A. israelii* may cause oral or respiratory lesions. *Bifidobacterium*, which often has branched cells, is an important first colonist of the infant intestine. *Nocardia*, which forms mycelial filaments, is found in soil and in animal flora. *N. brasiliensis* causes a disease, mycetoma.

In the mycelial actinomycetes *Streptomyces* and *Micromonospora*, colonial growth often looks much like that of true fungi (Fig. 4.12). However, these are procaryotes and thus true bacteria. Reproductive spores (**conidia**) are produced on aerial filaments of many types. They are major components of the aerobic bacterial flora of soil. Most are saprophytic and some are actively helpful. The streptomycetes are important as producers of such antibiotics as streptomycin, chloramphenicol, and the tetracyclines.

Within Part 17, the diversity of cell form and reproductive methods is considerable. Efforts to clarify the natural relationships of this mixture of bacterial types to each other and to their eucaryotic look-alikes in the fungi are continuing.

Source of antibiotics; Chapter 26

The rickettsias: Part 18 This section contains two types of organisms that are sometimes called deficient bacteria. The rickettsias and the chlamydias both are clearly cellular, procaryotic organisms. However, each has lost in the process of evolution some of the essential capabilities for independent free-living existence. These organisms are **obligate intracellular parasites.** That means that they can live and multiply only within and at the expense of another intact cell, more particularly the eucaryotic cell of a vertebrate or invertebrate animal. Most can be cultivated and studied only in live animals and tissue cultures. However, *Bartonella* can be propagated in media to which blood or body fluids have been added.

The Rickettsiales, including the genera *Rickettsia*, *Bartonella*, and a number of other poorly described genera, have a typical Gram-negative cell wall, cell membrane, and cytoplasmic components. They multiply by binary fission. The exact nature of their cellular deficiency is not entirely established. In some cases, it has been suggested that their cell membranes are imperfect permeability layers: that is, the cells are leaky. All members are animal parasites and many cause serious disease. *R. prowazekii* causes typhus and *R. rickettsii* causes Rocky Mountain spotted fever. The rickettsial diseases are characteristically transmitted by the bite or the feces of an **arthropod vector** such as a tick or louse.

Rickettsial diseases; Chapter 24, pp. 697–702

Arthropod vectors; Chapter 15, Table 15.2

The genus *Chlamydia* at present contains only two known species, *C. trachomatis* and *C. psittaci* (Fig. 4.13). Until studied by electron microscopy, they were thought to be large viruses. However, they are cellular in nature. Their cell walls are of a Gram-negative multilayered type. The deficiency that limits them to a parasitic reproductive pattern appears to lie in their energy metabolism. When separated from host cells, they can metabolize (but not reproduce) if a supply of high-energy compounds, including ATP, is provided. They seem to depend on the host for a supply of ATP and other high-energy compounds.

Their reproductive cycle consists of an alternation of two cell types. The **elementary body** is the small, rigid, infective form. It is taken into the host cell by

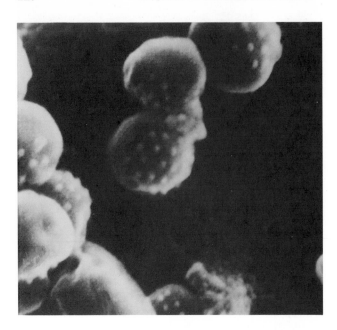

4.13
**Scanning Electron Micrograph of *Chlamydia
psittaci* on the Surface of Cultured Mammalian
Cells**   The arrays of regular projections are characteristic of the chlamydial surface.

phagocytosis and then develops into a larger, thin-walled, noninfectious **initial
body.** This continues to enlarge, then divides intracellularly to organize new
daughter elementary bodies. Eventually the cell ruptures and releases these daughter
bodies. Chlamydiae cause trachoma (an eye disease), psitticosis or parrot fever, and
other diseases.

The mycoplasmas: Part 19   We have discussed previously these bacteria deficient
in cell walls from a structural point of view. Mycoplasmas are presently included as
Part 19 of the bacteria, but some evidence indicates that they may be different
enough to deserve classification as a third division of the procaryotes, the class
Mollicutes.

Mycoplasmas are very small, ranging from 125–250 nm in diameter. Thus they
are at about the limits of resolution for the light microscope. Cellular form is highly
pleomorphic and filaments are often seen (Fig. 4.14). Although they reproduce by
binary fission, they also produce elementary bodies and spherical reproductive
bodies arising by fragmentation of filaments.

The outer cell boundary is a three-layered membrane. Cells stain Gram-
negative although peptidoglycan is lacking. Cholesterol is a component of the membrane and is required for growth. Colony formation is often difficult to observe, as
colonies are very tiny. Frequently, growth on solid media is shown as a film with
spots of precipitated matter. Species of the genus *Mycoplasma* are recovered from
body cavities and fluids of animals. In human beings, *M. pneumoniae* is associated
with respiratory disease. The *Ureaplasma* genus (T-mycoplasma) is being increasingly implicated in venereal disease, spontaneous abortion, and human neonatal
infections.

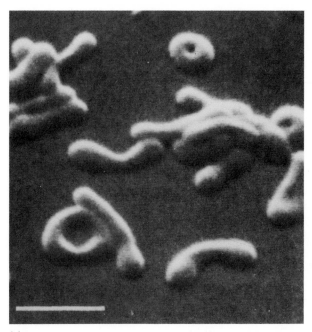
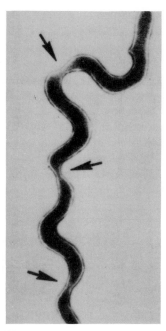

(a)   (b)

4.14
Mycoplasmas   (a) *Mycoplasma meleagris* cells. The mycoplasmas are the tiniest of the bacteria. Because they lack cell walls, their shapes are very irregular. (b) *Spiroplasma citri*. This filamentous form exhibits constrictions, the sites at which it would have soon divided.

The genus *Acholeplasma* lacks cholesterol in its membrane and has more susceptibility to osmotic shock. Some of these organisms are free-living while others are parasitic. They have been extensively used in studies of the structure and function of biological membranes. Other recently discovered genera include the plant pathogens, *Spiroplasma* and *Thermoplasma*.

Role of the Bacteria in the Environment

When reading through a survey of the bacteria such as the above, several impressions strike us. The bacteria are certainly diverse in habitat and in the metabolic modes that suit them to their environment. Clearly, bacteria are found in many places that are successfully inhabited by no other life form. It is also clear that most bacteria are useful members of the biosphere. Although some are producers, most are **decomposers**, responsible for removal and breakdown of waste materials. Only a small minority of genera produces disease. However, we tend to zero in on the occasional species that is pathogenic.

Decomposers; Chapter 6, pp. 164–165

The taxonomic treatment of the bacteria is far from settled. Probably the reader will feel that Linnaeus was closer to the truth than modern scientists are when he lumped all bacteria together in a group he named **Chaos**!

SUMMARY

1. Taxonomy is the process of classifying organisms. It involves discerning fundamental similarities and differences, and separating these from superficial resemblances. In the case of the procaryotes, these judgments are often not easy. The classifications arrived at are not often completely satisfactory.

2. The discovery of a new organism is followed by its careful examination and description. The results will show either that it is identical with a previously described and named species, or that it differs from all established classifications. If the differences are considered sufficient, an international jurisdiction will approve the assignment of a new name. These procedures have largely eliminated the wild proliferation of names that occurred earlier in the century. Each organism bears a binomial designation.

3. The procaryotes are now divided into two groups, the cyanobacteria and the bacteria. The four groupings of cyanobacteria include unicellular and filamentous forms, soil and aquatic inhabitants, and heterocyst-forming nitrogen fixers. All carry out aerobic photosynthesis and produce oxygen as a by-product.

4. The bacteria are divided into 19 groups. Among them are found microorganisms adapted to all extremes of environment. Every possible means of energy generation, every available carbon source and nitrogen compound is utilized by one or another bacterial group. Organisms pathogenic to human beings are found in many groups, but on the whole pathogenic bacteria are in the minority.

Study Topics

1. What is a natural classification? Find some examples, in the chapter, of natural groupings and of unnatural ones.

2. What are the differences between the cyanobacteria and the photosynthetic bacteria?

3. List the photoautotrophic, chemoautotrophic, and some chemoheterotrophic bacterial genera.

4. Outline the primary differences between the pseudomonads and the enterobacteria. Refer to *Bergey's Manual*, if it is available, to augment your answer.

5. Compile a list of the strictly anaerobic bacterial genera discussed.

Bibliography

Cyanobacteria

Carr, N. G., and B. A. Whitton (eds.), 1973, *The Biology of Blue-Green Algae*, Oxford: Blackwell Scientific Publications.

Fogg, G. E.; W. D. P. Stewart; P. Fay; and A. E. Wolsby, 1973, *The Blue-Green Algae*, London: Academic Press.

Stanier, R. Y., and G. Cohen-Bazire, 1977, Phototrophic procaryotes: the cyanobacteria, *Ann. Rev. Microbiol.* **31**: 225–274.

Bacteria

Brock, Thomas D., 1979, *Biology of Microorganisms* (3rd ed.), Englewood Cliffs, N.J.: Prentice-Hall.

Buchanan, R. E., and N. E. Gibbons (eds.), 1974, *Bergey's Manual of Determinative Bacteriology* (8th ed.), Baltimore: Williams and Wilkins.

Stanier, R. Y.; E. A. Adelberg; and J. L. Ingraham, 1976, *The Microbial World* (4th ed.), Englewood Cliffs, N.J.: Prentice-Hall.

5

The Protista and Fungi

There are several groups of eucaryotic microorganisms, including the algae, protozoa, and fungi. Although these are all microscopic forms, not very much larger than the procaryotic microorganisms, their eucaryotic cell structure has important effects on their biology. Some of these effects include the use of sexual as well as asexual modes of reproduction, complex life cycles, slower growth rates, and more elaborate structural adaptations.

Structure of eucaryotic cell; Chapter 3, pp. 91–97

THE EUCARYOTIC ALGAE

Introduction

Algae are photosynthetic protista, a large group of organisms that usually receive little attention from biology students. This is unfortunate because these organisms manufacture a significant portion of the annual global production of organic material. Along with the cyanobacteria, eucaryotic algae provide the nutritional basis for all aquatic life and in so doing replenish global oxygen supplies. There are six groups of algae. The groupings have been established primarily on the basis of cell coloration and morphology.

The Subgroups of Algae

The green algae or chlorophyta   The first land plants are usually considered to be descended from ancestors of the green algae. Green algae occur in unicellular, paired, or filamentous forms similar to those of other organisms we have already considered (Fig. 5.1). Certain groups, experimenting with multicellular life, have developed structural arrangements new to our discussion. Spherical colonies containing vegetative and reproductive cells are found in *Volvox*. In *Chlorococcus*, vegetative cells are frequently multinucleate. The multinucleate cell, called a **coenocyte**, undergoes mitosis (nuclear division) several times in succession without undergoing cytokinesis (cell division). Prior to asexual reproduction, the coenocytic cell completes cytokinesis that allows it to release a number of small independent cells. **Ulva,** the sea lettuce, develops a macroscopic structure, the **thallus,** composed of a sheet of cells two cells thick. The cells divide in two planes, causing the thallus, which superficially resembles a leaf, to grow to a visible size. Other cells develop into an attachment structure, the **holdfast,** by which the organism adheres to submerged rocks, pilings, and so forth. This algae we can clearly identify as plant-like.

(a)

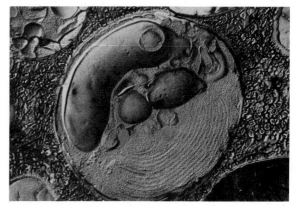

(b)

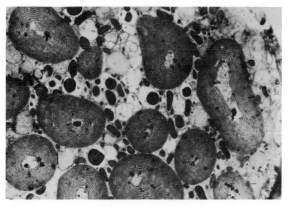

(c)

5.1
Green Algae   (a) A scanning electron micrograph of *Pediastrum*, showing a microcolony of neatly arranged cells. (b) A freeze-fracture electron micrograph shows the interior structure of *Cyanidium*. The chloroplast is the layered structure in the lower half of the cell. (c) The interior of *Glaucocystis* is shown by a stained thin section. In this species, there are numerous chloroplasts, the large, concentrically layered bodies.

Green algae have been intensively studied in an attempt to understand the evolution from unicellular to multicellular modes of life and from water to land existence. In conjunction with their evolutionary developments, these algae show a progressive restriction of flagellar motility. Motility is possessed by all cells in the more primitive algae, but only by reproductive cells in the more advanced forms.

In the green algae, we encounter for the first time one of the most important traits distinguishing eucaryotes from procaryotes. This is the occurrence of true sexual reproduction. We define sexual reproduction as a process in which there are two parental cells. Each parental cell contains paired chromosomes (each is diploid). The parental cell undergoes a type of cell division called **meiosis,** in which the pair of chromosomes are separated and one of each pair of chromosomes is installed in a special **haploid** reproductive cell called a **gamete.** Fusion of haploid gametes produces a diploid **zygote** and results in a new individual with a genetic combination that differs from those of both of the parents. All sexually reproducing organisms exhibit **alternation of generations.** That is, in the life cycle diploid forms alternate with haploid forms. In the lower life forms, including the protists and fungi, the haploid phase predominates. All protists reproduce asexually; almost all reproduce

TABLE 5.1
The major subdivisions of eucaryotic algae

DIVISION	COMMON NAME	MORPHOLOGY
Euglenophyta	Euglenids	Unicellular
Chlorophyta	Green algae	Unicellular One group is multicellular
Chrysophyta	Diatoms	Mostly unicellular
Pyrrophyta	Dinoflagellates	Unicellular
Phaeophyta	Brown algae	Multicellular, large, plantlike
Rhodophyta	Red algae	Mostly large, multicellular, some unicellular

The botanical taxonomy of the eucaryotic algae rests largely on coloration of the organism. Other clear distinctions can also be made. Wherever two or more characteristics are noted, they appear in approximate order of importance.

sexually as well. The development of sex paved the way for rapid evolutionary change. The algae remained aquatic until certain of their descendants developed capacities for reproduction out of water and thus invaded the land.

Green algae are extremely widely distributed. They grow on snow and ice and on the shady sides of tree trunks. They grow on the fur and feathers of certain animals and birds giving them a greenish color. They also grow in the more expected soil and water habitats.

The euglenids or euglenophyta   *Euglena* is claimed by both botanists and zoologists. We will consider it only once, as an alga. All euglenids are unicellular, photosynthetic, and motile (usually by means of two flagella). In these respects, they resemble certain of the green algae. However, they have a gullet; they take in and digest particles of organic material. Furthermore, they cannot survive in a completely inorganic medium and thus are heterotrophic. That is why the zoologists claim them. To confuse the issue further, euglenids store a glucose polymer called paramylum that is dissimilar from both plant starch and animal glycogen. Most euglenids are fresh-water forms and reproduce asexually.

CELL WALL	STORAGE	REPRODUCTION	HABITAT
None	Paramylum Fat	Asexual—fission	Fresh water
Cellulose Pectin	Starch	Asexual—fission Zoospores Sexual fusion	Fresh water Soil
Siliceous test	Oils	Asexual—fission	Fresh water Salt water Soil
Silica plates Cellulose	Oils	Asexual—fission Sexual—fusion	Salt water
Cellulose Alginic acid Pectin	Laminaran Mannitol Fat	Sexual—zoospores Asexual	Salt water Intertidal and surface
Cellulose Pectin	Starch	Sexual-specialized sex organs Asexual	Salt water Below mean low tide

Box 5.1

Colonial Life

Volvox is a colonial green algae. A single colony may consist of from 500 to 50,000 individual photosynthetic cells arranged in a large hollow sphere. Each individual cell is coated by a gel-like substance but is connected to other cells by bridges of cytoplasm permitting chemical communications. Each cell bears two flagella on the outer-facing surface, and the colony is motile, rolling slowly but smoothly through the water with a directed spin.

In the most advanced species, individual colonies are either male and female. The male colonies develop special sperm-bearing structures and release many haploid sperm. The single egg is produced in the female colony; fertilization is internal and results in the eventual release of haploid swimmer cells that may originate new colonies.

The *Volvox* cells have an unusual dependence on each other, since reproduction, either sexual or asexual, can be carried out only by the intact colony. As a colony all cells coordinate their movements towards the available light and orient the cell mass for optimum illumination.

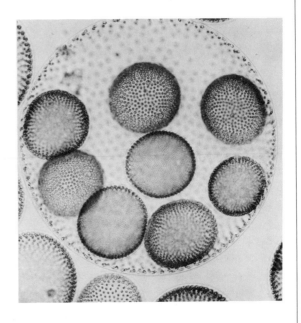

Cooperative function has reached a very high level in these algae. The next evolutionary step is to the formation of true tissues, at which point organisms will enter the higher kingdoms of plants and animals.

The diatoms or chrysophyta   Diatoms are unicellular forms in which the cell is encased in a hard, silica-impregnated **frustule** that looks like tube or a petri dish or a box with a lid (Fig. 5.2). Diatoms are very beautiful; their skeletons, being almost indestructible, are preserved in vast fossil deposits of diatomaceous earth. This material is useful as an abrasive powder.

The diatoms are a significant division of the marine-water and fresh-water producers.

Golyaulax; Chapter 20, p. 610

The dinoflagellates or pyrrophyta   The dinoflagellates are, like the euglenids, unicellular and biflagellate. Their distribution and ecological importance are quite similar to those of the diatoms. These organisms are unique in that each cell is armored with interlocking plates (Fig. 5.3). Most of these organisms, of which more than 1,000 species are known, are photosynthetic; however, some are also hetero-

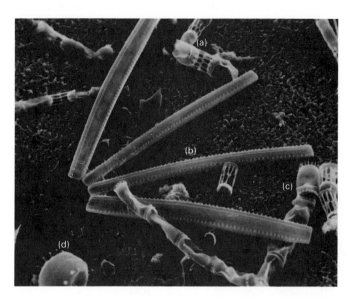

5.2
Diatoms Marine plankton may be collected by towing a fine-meshed plankton net through the water. This sample, obtained in Narragansett Bay, Rhode Island, contains four diatoms: (a) *Skeletonema costatum;* (b) *Thalassionema nitzschioides;* (c) *Detonula confervacea;* and (d) *Thalassiosira nordenskjoldii.*

trophic. They are an important component of the marine phytoplankton (floating plants).

The brown algae or phaeophyta   Brown algae are true seaweeds. All are marine and widely distributed around the coasts of the world's oceans. They occupy the intertidal zone and shallow ocean waters to a depth limited by effective light pene-

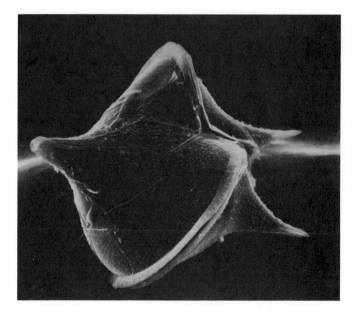

5.3
A Dinoflagellate *Peridinium,* from Florida waters (× 1,225). This organism has two flagella, one of which provides forward motility, while the other, housed in the beltlike groove surrounding the cell, creates rotary motion. The cell is enclosed in armor-like plates.

5.4
Seaweeds Exposed at High Tide Two species of brown algae are attached to the upper and middle levels of rock. At the water line, mossy red algae are attached. These need a habitat in which they are submerged almost all of the time. The brown algae here can tolerate some exposure to sunlight and air during a portion of the tidal cycle.

tration (Fig. 5.4). *Sargassum* is a floating seaweed. The Sargasso Sea is an area of the North Atlantic in which many thousands of square miles of ocean surface are covered by this algae. The plant body or thallus of the giant kelps such as *Laminaria* or *Macrocystis* may be many meters long and many hundreds of kilograms in weight. Their holdfasts are similar to, but not homologous to, plant parts such as roots.

Brown algae in nature are "grazed" by many of the marine animals. The polysaccharides in brown algae are a nutrient source for heterotrophic marine bacteria. Although we are more familiar with the brown algae than with the green algae, their actual importance in the marine ecosystem is much less than that of the microscopic green algae, diatoms, and other unicellular forms. Brown algae are used for fertilizer, stock feed, and as a source of food thickeners and stabilizers like the alginates.

Alginates; Chapter 2, Table 2.8

The red algae or rhodophyta There are some indications that the red algae might be an evolutionary link with the ancestral procaryotic cyanobacteria. They resemble cyanobacteria in that they possess phycobilin pigments that absorb short wavelength blue light with very high efficiency. Since the short wavelengths of light penetrate deepest into water, red algae can capture sufficient light energy for growth at greater depths than other algae. By growing at lower levels in the euphotic zone they occupy an ecological niche in which they do not compete with the brown seaweeds. Red algae may be unicellular, filamentous, or thallus-producing.

Ocean zones; Figure 6.6

The coralline forms deposit calcium carbonate; they are the major builders of coral reefs in warm seas. *Chondrus crispus*, the Irish moss, is used in gelled puddings and carrageenan is extracted from it for use in chocolate milk. *Porphyra* species are used by the Japanese in dishes such as sushi. *Gelidium* is the source of the agar bacteriologists find indispensable.

Bacteriological media; Chapter 7, pp. 192–193

THE FUNGI

Introduction

Fungi have characteristics intermediate between algae and protozoa and are given an individual kingdom designation. Fungi have cell walls, but they are heterotrophic in nutrition, usually dependent on the absorption of small organic molecules by diffusion. In most cases, they are saprophytic. They produce and secrete highly active digestive enzymes and are responsible for the rapid deterioration of organic materials such as wood, fabrics, and leather in a damp environment or in contact with soil. The fungi, with the bacteria, make up the ecological class of **decomposers,** organisms that return scarce elements to the soil by releasing them from animal and plant remains. Some fungi are parasitic; they live in and upon living plants and animals in which they sometimes cause disease. Most fungi are respiratory in nutrition, but some, especially the yeasts, are facultatively anaerobic. These have the option of metabolizing fermentatively. Many fungi tolerate high osmotic pressures and survive in concentrated solutions of salt and sugar. In addition, they readily tolerate extremes of pH. Their abililty to adapt to extreme heat or cold is less than that of certain bacteria, however.

Structural Characteristics

The fungal cell is typically eucaryotic. It is surrounded with a rigid cell wall composed of cellulose and chitin. Two growth types are distinguished, molds and yeasts. In the dimorphic fungi, these two growth types are to some degree interconvertible. **Molds** are fungi that form **hyphae,** long threads or filaments. These may or may not have crosswalls called **septa.** If septa are present, however, they are not complete but are perforated by pores. Thus all the cytoplasm of the fungus is continuous. Fungal cells are multinucleate or coenocytic. Hyphal growth usually branches and becomes extensive enough to form a visible mat, or **mycelium.** The mycelium may be subterranean, within tissue, or may spread upon the surface of the material the fungus is utilizing for growth. Some portions of the mycelium are vegetative and serve to gather nutrients. Other portions give rise to asexual and sexual reproductive units. Since all of the cytoplasm of the mycelium is continuous, it may be regarded as a single huge coenocyte. Cytoplasmic streaming and diffusion carry nutrients from one area to another. In mushrooms, most familiar of fungi, portions of the mycelium, instead of being simply a tangled mat, are organized into a solid fruiting body. The rest of the mycelium, in the soil or in wood, nourishes the visible portion. Mycelial threads readily penetrate devitalized plant and animal tissue and, in some cases, enter living cells.

    Yeasts (Fig. 5.5) are unicellular fungi, usually oval in outline. They reproduce by budding, a highly organized asymmetric form of cell division preceded by mitosis. Binary fission of procaryotes, by contrast, is symmetrical and does not require mitosis to subdivide cellular components. Whereas many species of yeast are always unicellular, others are **dimorphic** or **biphasic** and can convert to mycelial growth in the appropriate environment.

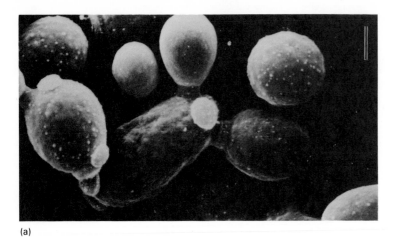

(a)

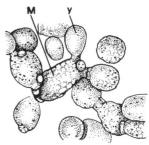

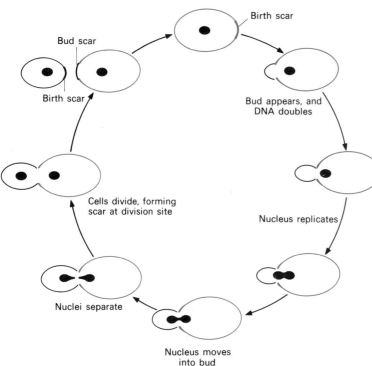

(b)

5.5
Budding Yeast Cells   (a) *Candida lipolytica* is shown actively growing. Young, smooth buds (y) are forming from a mature, rougher-surfaced cell (M). Each will contain a nucleus formed by mitosis. After the bud separates, a collar-like bud scar will be left on the parent cell, matched at the arrow by a birth scar on the bud cell. (b) A schematic drawing of the budding process in yeast.

Reproduction   Fungi have evolved many different reproductive structures for the production of spores. Increase in numbers of organisms depends primarily on asexual reproduction, almost continuous in most forms. Sexual reproduction occurs less consistently. The reproductive structures produced are the criteria used in establishing subdivisions of fungi.

Asexual. Fungal spores (Fig. 5.6) are small, modified cells. Their chemical composition is not significantly different from the parent mycelium. They have less refractility and resistance than does the bacterial endospore; they are metabolically inactive. Since one individual (one mycelium) can produce many spores, the role of the spore is clearly reproductive. Remember that in bacteria the spore, rather than being an instrument of reproduction, serves solely as a resting structure.

Asexual spores arise by pinching off portions of the coenocytic cell containing one nucleus. All are genetically identical. Many fungi are extremely prolific, and visible clouds of thousands of spores become airborne when the mycelium is disturbed (Table 5.2).

The **zoospore** produced by the lower fungi is a motile, flagellate cell. A zoospore has a striking resemblance both to certain unicellular algae and to the reproductive forms of some protozoa, a resemblance that emphasizes again the close relatedness of all the protists.

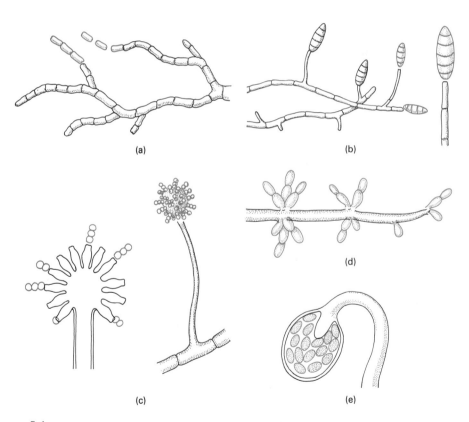

5.6
Asexual Spores of Fungi   (a) Arthospores; (b) aleurospores; (c) conidiospores and conidia; (d) blastospores; and (e) sporangia with sporangiospores.

TABLE 5.2
The major groups of fungi

CLASS	MICROSCOPIC MORPHOLOGY	METHOD OF ASEXUAL REPRODUCTION
Lower fungi	Nonseptate mycelial fungi Some unicellular species	Sporangiospores Zoospores Conidiospores
Ascomycetes	Septate mycelial fungi Yeasts	Conidiospores Budding
Basidiomycetes	Septate mycelium May form complex fruiting bodies	Conidiospores (rare)
Deuteromycetes	Yeasts Dimorphic fungi Septate mycelial fungi	Conidiospores Arthrospores Blastospores Chlamydospores Budding
Myxomycetes	Plasmodial or unicellular amoebae Form aggregations	Spores formed in fruiting body

Sexual. All sexual reproductive modes require preliminary fusion of haploid nuclei from two different sources. In the protista and fungi, for lack of identifying marks, these two sources are often referred to simply as plus (+) and minus (−) strains rather than male and female. After fusion, the diploid nucleus undergoes meiosis, producing haploid spores (Fig. 5.7). The fungal mycelium is predominantly haploid; only certain cells or structures have a transitory diploid chromosome number.

Performance of sexual reproduction requires contact and conjugation between two mating types. Many fungi rarely "meet their match" and so are not often observed in the sexual act. These fungi have proved difficult or impossible to classify and they have been placed in the catchall category of **fungi imperfecti** or deuteromycetes. It appears likely that they are asexual members of other established groups. Periodically, a species will be observed carrying out sexual reproduction. Then it is officially assigned to its correct group. Systematic application of ultrastructural and biochemical analysis to the fungi may allow conversion to a more sensible classification system in the future.

METHOD OF SEXUAL REPRODUCTION	COMMON HABITAT	ROLES IN ENVIRONMENT
Zygospores Oospores Flagellate gametes	Soil Water Plants	Saprophytes—decay, spoilage Parasites—diseases of plants, fish
Ascospores	Foods Soil Plants	Saprophytes—cause fermentation Parasites—diseases of plants
Basidiospores	Soil Plants Trees	Saprophytes—decompose forest litter Parasites—invade and kill trees
None yet observed	Soil Plant tissue Animal tissue	Saprophytes—decomposition in soil Parasites—plant disease Important fungus diseases of human beings
Zoospores	Soil Manure	Saprophytes—decompose wastes Parasites—live by absorbing other microorganisms

5.7
Sexual Spores of Fungi (a) The zygomycetes among the lower fungi form a zygospore following fusion of haploid hyphae of two mating types. The hard-coated prickly zygospore contains a diploid sexual cell. (b) The definitive structure for the ascomycetes is the ascus, a sac that contains four or eight haploid ascospores. (c) In basidiomycetes, hyphal fusion produces a swollen basidium. Its diploid nucleus undergoes meiosis, producing four haploid basidiospores that project from the surface. When the basidiospores mature, they are discharged.

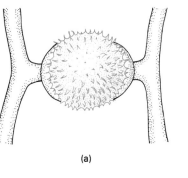
(a)
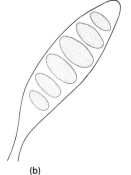
(b)

(c)

The Classes of Fungi

The lower fungi   The primarily aquatic fungi are primitive and have nonseptate mycelium, some unicellular motile forms, and asexual spores produced in a sac from which they burst when mature. The **chytrids** such as *Allomyces* and *Blastocladiella* are unicellular through much of the life cycle, although a coenocytic thallus may form to give rise to a new generation of motile, uniflagellate zoospores (Fig. 5.8). The oomycetes are mildew and blight organisms, usually aquatic, also multiplying by means of zoospores. Many are highly destructive pathogens; *Pythium*, for example, is a pathogen of plants and *Saprolegnia* is a pathogen of fish.

The zygomycetes reproduce sexually by the fusion of two gametangia, one from a parent of each mating type. A zygospore forms at the point of fusion.

A familiar example is *Rhizopus nigricans*, the common black, bread mold (Fig. 5.9). It also has an asexual reproductive cycle, during which spores are formed in **sporangia.** These spherical black balls are just visible, readily seen when you look at a piece of moldy bread. Other zygomycetes include *Phycomyces*, *Pilobolus*, and *Cunninghamella*.

The ascomycetes   Ascomycetes form a separate mycelium; some are dimorphic and are also found in the yeast form. The morels, cup fungi, and truffles form very solid mycelial masses that closely resemble true mushrooms. Ascomycetes have no motile forms.

Asexual reproduction.   The mycelium forms brushlike **conidia** that pinch off strings of **conidiospores.** In yeasts, budding serves the same function.

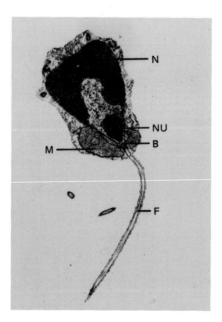

5.8
Zoospores   *Blastocladiella emersonii*, a chytrid fungus, forms large numbers of motile zoospores within a thick-walled sporangium. The free zoospore shown here is motile by means of a single flagellum (F) arising from a basal body (B) that passes through a large mitochondrion (M). The zoospore's nucleus (N) and nucleolus (NU) are also clearly seen in this thin section. (× 10,260)

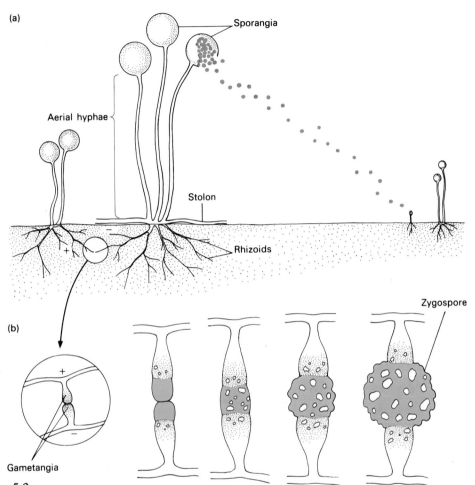

(a)

Sporangia

Aerial hyphae

Stolon

Rhizoids

+

Zygospore

(b)

Gametangia

5.9
Rhizopus **Life Cycles**   (a) Asexual reproduction. A mycelium arises from the outgrowth of
a fungal spore. Once the mycelium reaches a certain size, aerial hyphae arise, bearing
sporangia filled with haploid spores. On dissemination, these spores give rise to new
mycelial nets. (b) When the subsurface rhizoids of a plus and minus type met, they fuse
with resulting zygospore formation in sexual reproduction.

Sexual reproduction. Cell fusion produces a diploid nucleus. This undergoes
meiosis, yielding four or eight haploid ascospores held like peas in a pod within a sac
or **ascus.** The ascus eventually ruptures to release the **ascospores.** *Neurospora
crassa,* the pink bread mold, is widely used in genetic research. Each spore in its
ascus represents a discrete meiotic product. The spores can be removed by microsur-
gery and separately cultured. In this way genetic recombination, gene interaction
and linkage arrangements can be studied. Human beings derive sustenance and

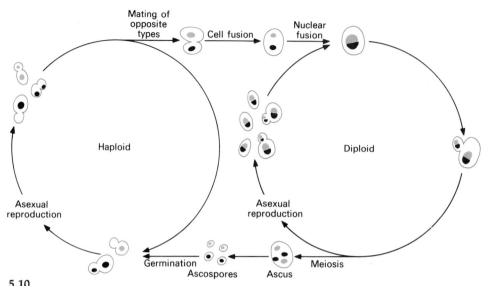

5.10
Asexual and Sexual Reproduction in an Ascomycete Yeast   *Saccharomyces cerevisiae* is
the yeast used for baking and brewing purposes. When rapidly multiplying in bread
dough, it replicates by budding, asexually. Only when two opposite types meet and fuse
does nuclear fusion produce a diploid. Diploids may reproduce asexually by budding, or
they may undergo meiosis to yield haploid ascospores.

Alcoholic fermentation;   cheer from the products of *Saccharomyces cerevisiae* (Fig. 5.10). Special strains of
Chapter 8, p. 236   this ascosporogenous yeast are employed in bread making and in the production of
alcoholic beverages.

Ascomycetes are extremely widely distributed saprophytes and parasites. They
are found in soil (*Chaetomium*), dung (*Sordaria*), in mycorrhizal association with
the roots of plants (*Endothia*), on grains (*Claviceps*), and in the oceans (*Lepto-
sphaeria*), where they are the most common type of fungus.

The basidiomycetes   True mushrooms and bracket fungi are the reproductive fruiting bodies of the basidiomycetes. Less conspicuous members, the rusts and smuts,
are economically important plant pathogens.

People are fascinated by mushrooms, by their sudden appearance and rapid
growth, by their often fantastic shapes and colors, by the folklore surrounding the
few poisonous species, and by the delicate flavors of the edible forms. Because they
are so fascinating, mushrooms are the best known fungi. Each mushroom is a highly
organized extension of its underground vegetative mycelium. The mushroom, a
sexual reproductive body, develops rapidly because the mycelial cytoplasm is continuous and nutrients move to the site by streaming. Each basidiomy-

cete species develops best in certain types of soils or in association with the litter from specific types of trees or grasses.

Asexual reproduction. The vegetative portion of the fungus is an extensive mycelium. In mushrooms this may spread out a number of yards in the soil or permeate the roots and vascular system of a large tree. Conidiospores are asexually produced. The mycelium continues to expand and ramify, with some cells dying and being replaced by others. Thus the mycelium, but not individual cells, may be extremely long-lived. Digestive enzymes liberated by the fungal mycelium may kill the tree and decompose any organic material the mycelium invades. A crumbling fallen tree trunk is usually totally penetrated by fungal hyphae responsible for its disintegration.

TABLE 5.3
Some fungi of economic importance

CLASS	GENERAL NAME	ECONOMIC EFFECT
Lower fungi (six classes)	*Mucor* spp.*	Cheese manufacture; food spoilage
	Rhizopus spp.*	Food spoilage; black mold on bread
	Saprolegnia spp.*	Parasitic on fish
	Pythium spp.*	Damping off of vegetable and flower seedlings
	Plasmoparium spp.*	Downy mildew of grapes
	Phytophthora spp.*	Potato late blight; caused famine in Ireland (1845)
Ascomycetes	*Neurospora crassa*	Chestnut blight; eliminated tree from the United States; Dutch elm disease; bread mold, genetics research
	Saccharomyces cerevisiae	Beer brewing, bread making, and wine making
	S. cerevisiae var. *ellipsoideus*	
	Saccharomyces spp.*	Fermented milks such as kefir, koumiss
	Morchella esculenta	Edible morel; prized, wild mushroom
Basidiomycetes	*Agaricus campestris*	Commercially raised, edible mushroom
	Amanita spp.*	Poisonous or hallucinogenic mushrooms
	Polyporus spp.*	Bracket mushroom—rots live trees, destroys lumber
	Puccinia graminus	Black stem rust of wheat
	Cronartium ribicola	White pine blister rust
Deuteromycetes	*Penicillium notatum*	Penicillin manufacture
	Botrytis cinerea	Disease of grape vines
	Trichothecium roseum	Rots apples and other fruit
	Trichoderma viride	Ammonification in soil
	Fusarium spp.*	Wilt disease of tomatoes

*spp. = multiple species, names not significant or unknown.

Sexual reproduction. A basidiomycete becomes ready to fruit when portions of two haploid mycelia make contact and fuse (Fig. 5.11). Mycelial growth and fruiting are seasonal phenomena. Certain mushrooms appear at certain seasons during suitable weather, usually after a rainy spell. A **dicaryon** (cell with two separate haploid nuclei) results from mycelial fusion. Dicaryons develop to form the fruiting body. Under the cap of the mushroom are gills or pores bearing **basidia**. Within these, dicaryons undergo meiosis. As the mushroom ripens, clouds of variously colored haploid basidiospores are released. Each basidiospore may germinate and initiate the growth of a new mycelium.

The rusts and smuts lack the classic fruiting body; they multiply in soil and on woody and soft plants, causing great economic and nutritional losses. Black stem rust of wheat is a disastrous basidiomycete disease, causing extensive crop losses. The life cycle of the causative agent *Puccinia graminis* is complex. Asexual spores are responsible for wheat damage. However, continuation of the infection from season to season depends on a sexual cycle. Basidiospores are produced in soil and then develop into the infective spore form on the leaves of barberries. Control of the

Box 5.2

The Once-stately Elms

Over the past 30 years, the eastern and central United States have suffered the loss of many millions of large and graceful shade trees. Many an Elm Street or Elm Plaza no longer deserves its name. The loss of the elms has greatly increased the harshness of much urban life.

This sorry victim succumbed several years ago to the lethal effects of Dutch elm disease, caused by the ascomycete *Ceratocystis ulmi*. Perhaps it was already weakened by the ascomycetes *Gnomonia ulmea* and *Systremma ulmi*.

Fungi also will carry out the removal of the dead tree from the landscape. Its trunk and bark have been penetrated by the mycelium of a basidiomycete, the bracket fungus *Pleurotus ostreatus*, that is fruiting prolifically, using nutrients extracted from digested wood. Chunks that have fallen to the ground will be finally destroyed by the dryrot fungus *Merulius lachrymans* and a variety of other soil fungi.

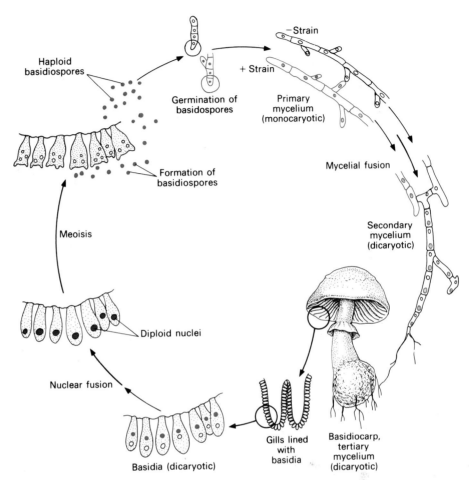

Haploid
basidiospores

Germination of
basidospores

−Strain

+ Strain

Primary
mycelium
(monocaryotic)

Mycelial fusion

Formation of
basidiospores

Secondary
mycelium
(dicaryotic)

Meoisis

Diploid nuclei

Nuclear fusion

Basidia (dicaryotic)

Gills lined
with
basidia

Basidiocarp,
tertiary
mycelium
(dicaryotic)

5.11
Sexual Reproduction in the Basidiomycetes The reproductive sequence involves fusion
of the + and − strain mycelium to yield a dicaryotic mycelium. This develops into a
fruiting body. Shown here is an *Agaricus*, a common, edible, meadow mushroom. The
gills beneath the cap produce the spores following a nuclear fusion and meiosis sequence that recombines the parental genetic material. If you place a ripe mushroom cap
on a piece of paper overnight, you will see a spore print corresponding to the gill
arrangement when you lift the cap in the morning. Each of these basidiospores can give
rise to a new mycelium.

problem depends on destruction of barberries. Wheat varieties genetically resistant
to rusts and smuts are under development.

The deuteromycetes Deuteromycetes, or Fungi Imperfecti as they are also called,
are placed in an artificial classification based on the absence of a trait (sexual repro-

duction) rather than its presence. These are species for which sexual stages have not yet been described. Almost all of the human pathogens are in this classification.

Asexual reproduction is sufficiently efficient as to make the use of sexual methods unnecessary. *Aspergillus* and *Penicillium* form conidia of a sort that suggest strongly that they are really ascomycetes. The yeasts such as *Candida* and *Torulopsis* bud. Dimorphic fungi such as *Coccidioides* form chlamydospores, arthrospores, and conidia. Penicillin, the first antibiotic to have wide clinical usefulness, is a natural product of the fungus from which its name is derived.

Fungal diseases of human beings are quite widespread, particularly in subtropical and tropical areas. Many of the diseases tend to be chronic and refractory to treatment. Parasitic fungi growing at the expense of the living host cause a variety of destructive effects. The major **mycoses** (fungal diseases) are summarized in Table 5.4.

Yeasts are fungi in which the unicellular form is dominant. In the dimorphic fungi, the unicellular form may alternate with a mycelial phase; in others the yeast phase is exclusively present. Some yeasts such as *Candida* may under favored conditions produce pseudohyphae (tubular extensions). These do not develop into mycelial growth.

Candida; Chapter 17, pp. 510–512

TABLE 5.4
Some fungus diseases of human beings

LOCATION OF DISEASE	GENUS NAMES	MORPHOLOGY	DISEASE	SYMPTOMS
Superficial skin layers	*Epidermophyton* *Microsporum* *Trichophyton*	Mycelial fungi, form conidia	Ringworm of scalp, beard, skin, groin	Limited, usually to moist areas Some contracted from animals Recurrent in nature Cause inflammation, itching, peeling of skin layers
Deep tissues	*Coccidioides*	Dimorphic fungi	Coccidioidomycosis	Lung infection, usually mild, occasionally severe, caused by fungi in dust
	Histoplasma	Dimorphic fungi	Histoplasmosis	Lung infection, mild to severe, transmitted by bat or bird dung
	Blastomyces	Dimorphic fungi	North American blastomycosis	Lung infection, mild to severe, life-threatening if spread
	Sporotrichum	Dimorphic fungi	Sporotrichosis	Subcutaneous nodules, ulcers, lung involvement. Acquired by puncture wound
	Cryptococcus	Yeast	Cryptococcosis	Lesions in lung, skin, and central nervous system. Transmitted by pigeon droppings
	Candida	Dimorphic fungi	Candidiasis Moniliasis Thrush	Usually nonpathogenic. Causes disease in throat, genital areas in very young or old, or those on long-term antibiotic therapy

The slime molds   Four other groups of protists are not considered true fungi because they exhibit gliding motility and clearly organized behavioral patterns more reminiscent of protozoa. The plasmodial slime molds are coenocytic, forming a multinucleate film of growth called a **plasmodium.** The plasmodium expands and swarms over the growth surface as a thin film. By means of a chemical signal the cytoplasm aggregates in one area and becomes formed into a complex, elevated fruiting body. Inside, the plasmodial material organizes into numerous flagellate, uninucleate spores. On release they are distributed, each then forming another plasmodium.

The cellular slime molds such as *Dictyostelium* remain as individual cells that lack a cell wall and strongly resemble amoebas. These also aggregate, and the individual cells construct an elaborate fruiting body with a central cellulose shaft. Inside the fruiting body, individual cells acquire a cellulose layer and are eventually released as spores. Each spore develops into a cell. Slime mold developmental cycles are reminiscent of the fruiting cycle of the procaryote myxobacteria. They have been studied to elucidate the possible role of cyclic AMP and other cellular messengers in modulating the aggregation and fruiting behaviors.

cAMP; Chapter 2, p. 56

THE PROTOZOA

Introduction

Protozoa are best described as one-celled animals (Fig. 5.12). Their cell structure is the most sophisticated of the protists; they have the unicellular design evolved to its maximum capabilities. All groups of protozoa move, and they are classified primarily by their means of locomotion. All are heterotrophic, but some also have chloroplasts and carry out photosynthesis. Some ingest other living organisms and are true **consumers.** Others take in dissolved nutrients and are thus saprophytic or parasitic. Certain protozoan species are the largely uncontrolled agents of the world's worst infectious diseases.

Consumers; Chapter 6, pp. 164–165

Divisions of the Protozoa

There are approximately 25,000 species included in the protozoa. They are separated into four subgroups—Sarcodina, Mastigophora, Ciliata, and Sporozoa.

The Sarcodina   The Sarcodina (Fig. 5.12) move by formation of **pseudopodia,** flexible extensions of the cytoplasm. Active cytoplasmic streaming in conjunction with the stretching of the cell membrane allows pseudopodia to advance and retract. Pseudopodia may be used for a type of creeping motion, but primarily serve as food-capturing devices. The extension of several coordinated pseudopodia allows amoebas to surround and engulf other smaller organisms that are then digested and assimilated.

All members of the Sarcodina, like all protozoa, lack a cell wall and have flexible exterior membranes. True amoebas have no distinct shape but rearrange their cytoplasm continuously. Other Sarcodinas such as *Arcella* or *Difflugia* produce

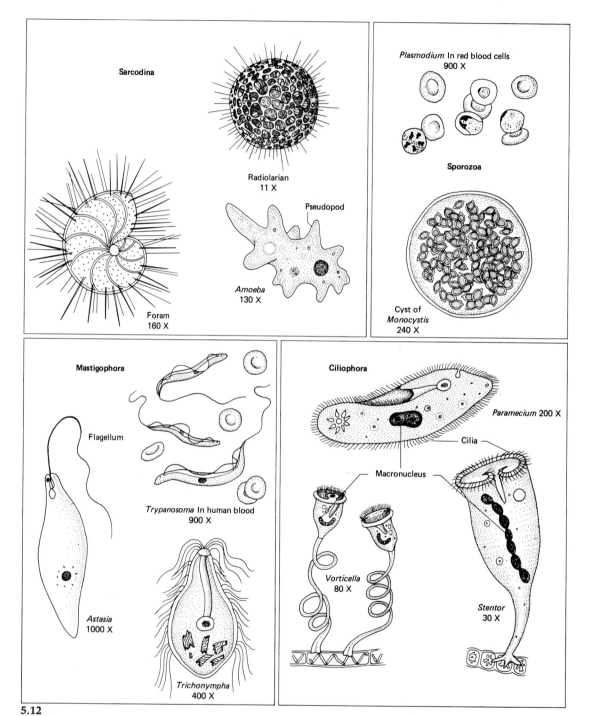

5.12

Representatives of the Four Groups of Protozoa   These eucaryotes are classified on the
basis of their means of locomotion, shown in these diagrams.

TABLE 5.5
The classification and characteristics of the protozoa

CLASS	LOCOMOTION	REPRODUCTION	NUTRITION
Mastigophora	Flagella, usually paired	Asexual—longitudinal fission	Some are photosynthetic All are heterotrophic Absorb or capture food
Sarcodina	Produce pseudopodia Ameboid motion	Asexual—binary fission	Absorption of dissolved materials Phagocytic capture of prey
Ciliophora	Many cilia, in bands or over all of surface	Asexual—transverse fission Sexual—exchange of nuclear material	Capture and digest other unicellular organisms as prey
Sporozoa	Usually none Some stages may have flagella	Asexual—multiple fission Sexual—in host organism Micro- and macro-gametes form spores at some point	Usually parasitic, absorb dissolved materials from host

external shells or skeletons, constructed primarily of inorganic materials excreted or glued together. The organism extrudes its pseudopods through openings in the casing.

The fresh-water protozoans must cope with the problem of life in a medium more dilute than their cytoplasm. Water continually enters the cell by osmosis. All have mechanisms for collection and evacuation of the unwanted water. These simplest excretory systems are usually called **contractile vacuoles**. They are well developed in the fresh-water Sarcodinas.

Reproduction in this group is asexual, with nuclear division being followed by equal cytoplasmic division. Some forms, under adverse conditions, form a resting stage called a **cyst**, more resistant to drying and other stresses than is the vegetative cell or **trophozoite**.

The amoebas. Most amebic species are free-living aquatic and soil organisms. Some are found as apparently harmless inhabitants of the animal gut. However, several species are major human disease agents. *Entamoeba histolytica* causes amebic dysentery, an often chronic, debilitating disease. Like all protozoan diseases, it is most commonly found in subtropical and tropical countries. Other amebic species also cause intestinal disease and one, *Naegleria fowleri*, causes a rare but usually fatal form of encephalomeningitis.

Amebic dysentery; Chapter 20, pp. 602–605

The foramenifera. Foramenifera produce a true shell, made of calcium salts. It is highly structured and resembles a miniature snail shell. *Globigerina* is present in

5.13
Scanning electron micrograph of the exoskeletons of radiolarians.

enormous numbers in nutrient-rich areas of sea water. When it dies, its shells contribute to the formation of ocean sediments, trapping substantial amounts of calcium and phosphorus. Forameniferan shells compose the chalk of the White Cliffs of Dover.

The radiolarians. These beautiful creatures form delicate exoskeletons of silica compounds with radially arranged spikes (Fig. 5.13). They are also marine and, like the forameniferans, serve as a mainstay food source for larger marine animals and at death contribute their skeletons to bottom sediments.

The heliozoa. These are fresh-water forms, some of which resemble the radiolarians. They construct compact casings out of gummy substances and sand grains.

The Mastigophora   The Mastigophora possess from one to six flagella, inserted in a pore, and powered by the chemical reactions taking place in a **kinetoplast.** In *Trypanosoma,* the flagellum is not free, but anchored along the length of the cell by an undulating membrane.

Photosynthetic forms. Zoologists include in the Mastigophora those photosynthetic protists with heterotrophic nutrition or lacking cell walls. We have already discussed the euglenids and the colonial forms such as *Volvox* among the green algae, and the dinoflagellates as a separate algal group.

Trichomoniasis; Chapter 21, p. 635

Giardiasis, Chapter 20, pp. 605–606

Nonphotosynthetic forms. Of the obligate heterotrophs—those that lack photosynthetic capability—four genera cause serious human infection. *Leishmania* species, transmitted by the bite of infected flies, cause superficial and systemic ulcerations. *Trypanosoma* species, transmitted by the tsetse fly, cause sleeping sickness. *Trichomonas vaginalis* causes a genitourinary infection usually spread by sexual intercourse. *Giardia intestinalis* causes various forms of intestinal discomfort.

The flagellates characteristically reproduce asexually. Many form cysts; several undergo cyclic morphological changes.

The Ciliophora Ciliophora have cell surfaces with bands of cilia or are entirely covered by short cilia. The cilia arise from basal structures interconnected by fibrils. Thus these organelles beat with a coordinated direction and rhythm. A ciliate can respond to chemotactic stimuli and is able to start, stop, or move in any direction. Most are successful free-living forms; *Paramecium* is a well-known example. They capture other unicellular organisms as prey and ingest them via the oral groove or gullet. Undigested material leaves by way of the anal pore. Capturing techniques include the use of **trichocysts,** sticky threadlike harpoons that are discharged to entangle the prey. The suctorians have tentacles.

Reproduction. Both asexual and sexual reproduction occur. Ciliophora possess a **macronucleus** and a **micronucleus,** both diploid. The former is functional in asexual reproduction, which is a form of transverse fission. The micronucleus functions in sexual reproduction, giving rise to haploid daughter nuclei that are exchanged between the two partners. This process has been extensively studied in *Tetrahymena.*

Endosymbionts. Some ciliates, including *Paramecium,* contain other organisms as intracellular dwellers. Symbiotic relationships between ciliates and photosynthetic algae, nonphotosynthetic bacteria, and other protozoa are known to exist. These relationships are often studied as models for the type of interaction between cells that might originally have given rise to the eucaryotic cell type.

Endosymbiosis; Chapter 3, p. 93; Chapter 6, pp. 177–178

The Sporozoa In general, adult members of this group lack means of locomotion, although immature forms and gametes are occasionally motile by means of flagella, flexion, or gliding. The group name reflects the fact that spores are formed following sexual reproduction. The sporozoans are not generally capable of independent life and are parasitic. They receive dissolved nutrients and environmental protection from a host. Frequently they cause the host to be seriously diseased. Complex life cycles are usual, and sexual and asexual reproduction may alternate, each taking place in a different host.

Sporozoan diseases. Malaria, in human beings and other warm-blooded animals, is caused by the genus *Plasmodium.* The life cycle of this parasite in human beings, detailed in Chapter 24, involves an asexual cycle in which reproduction of the parasite results in rupture of red blood cells giving rise to fever and chills. The parasite, after being ingested along with the infected human blood by the mosquito, can then complete a sexual cycle in the insect's gut. Mature infective parasites are then transferred to the mosquito's salivary gland. When the mosquito bites a new human host, the mature parasite is transported via the saliva. Malaria is partially controlled by the use of drugs both for those who are exposed and those who develop the disease. Drainage and spray programs help control the insect vector. Still, as many as 300 million people a year, principally children, contract the disease and around 1.5 million die as a result.

Malaria; Chapter 24, pp. 709–713

 Two other human parasites are considered to be provisionally members of the sporozoa. *Toxoplasma gondii* causes an infection that is mild or asymptomatic in

Toxoplasmosis; Chapter 21, pp. 639–640

TABLE 5.6
Some of the medically important protozoan parasites of human beings

CLASS	GENUS AND SPECIES	NAME OF DISEASE
Mastigophora	*Giardia lamblia*	Giardiasis
	Leishmania donovani	Leishmaniasis
	L. tropica	
	L. brasiliensis	
	Trichomonas vaginalis	Trichomoniasis
	Trypanosoma gambiense	Trypanosomiasis
	T. rhodesiense	
	T. cruzi	
Sarcodina	*Entamoeba histolytica*	Amoebiasis
	Naegleria fowleri	PAME
Ciliophora	*Balantidium coli*	Balantidiasis
Sporozoa	*Plasmodium vivax*	Malaria
	P. malariae	
	P. falciparum	
	P. ovale	
	Toxoplasma gondii	Toxoplasmosis
	Pneumocystis carinii	pneumonitis

adults but can cause severe congenital defects in infants. *Pneumocystis carinii* causes a severe pulmonary infection in weakened persons.

SUMMARY

1. We have surveyed the eucaryotic cellular microorganisms found in the protista and fungi. They exhibit a tremendous variety of structures and metabolic abilities, and make many contributions to the environment. As these microorganisms evolved, certain characteristics appear to have been passed from one group to another. Other characteristics appear to have evolved independently in several groups at once. Thus it is not possible at this time to make ironclad conclusions about their mutual relationships. This fact stands in the way of a complete natural classification.

2. Classification techniques themselves evolve. Early taxonomists relied heavily on physical description. Later, basic analysis of nutritional traits was attempted and, as in the dispute over the placement of *Euglena*, gave answers that were in contradiction to the first approach. More recently, biochemical characterization of cell components and ultrastructure determination with the electron microscope have been added. As might be expected, the new information has raised almost as many questions as it has settled.

SYMPTOMS AND COMMON NAME		ADDITIONAL HOSTS
Diarrhea in some cases		None
Kala-azar, an infection of the reticuloendothelium		Mammals
Oriental sore, cutaneous		Sand fly
Espundia, mucous membranes		Monkeys and dogs
Commonly causes itching and discharge		None
Sleeping sickness—blood and lymphatic tissues		Monkeys, cats, and dogs
Chaga's disease—blood		Tsetse fly
Amebic dysentery, chronic and acute		None
Primary amebic meningoencephalitis		None
Mild to severe diarrhea		Swine
Vivax tertian fever	Affects blood cells and liver; causes fever, chills; often fatal if not treated	Anopheles mosquito
Malarial quartan fever		
Falciparum subtertian fever		
Ovale fever		
Mild in adults, may cause severe CNS damage in infants		Cat, cow, other mammals
Usually seen in weakened individuals only		

3. The algae are photosynthetic eucaryotes. In addition to the true nucleus, they possess subcellular organelles. One of these is the chloroplast, in which photosynthesis takes place. These interesting organisms may have motility by flagella. Some are multicellular, forming an undifferentiated cell mass called a thallus. Others are coenocytic, with many cells merged within one outer boundary. Sexual reproduction occurs, and some cells are diploid.

4. The groups of algae are separated on the basis of their photosynthetic pigments, anatomy, and habitat. The unicellular forms along with the cyanobacteria make up the phytoplankton. Others live in soil and on damp surfaces. Macroscopic forms, the seaweeds, are harvested for extraction of useful polysaccharides and as food or fertilizer.

5. Fungi are eucaryotic and heterotrophic. They are related to algae because they possess a cellulose-containing cell wall and are frequently coenocytic. They resemble protozoa in that they are nutritionally dependent. Some fungi also produce ameboid or flagellate forms that resemble protozoa. In their environmental roles they resemble the bacteria in that they function as decomposers and occasionally as pathogens.

6. The classes of fungi are divided on the basis of the means of sexual reproduction. Cell structure may be mycelial, yeastlike, or dimorphic. Both asexual and sexual reproduction occurs, and the resulting spore types are both diverse

and numerous. In one group, the deuteromycetes, sexual reproduction has not been observed. This group includes all the species that are pathogenic for human beings. Slime molds stand apart from the four groups of true fungi. They have elaborate developmental cycles leading to the production of fruiting bodies.

7. Protozoa are animal-like in nature. Their eucaryotic cells, very complex in design, do not have cell walls. Their shape may be completely changeable, as are the shape of amoebas, maintained by a stiff gel-like exterior layer of cytoplasm, or enclosed in a mainly inorganic shell or skeleton. Sexual reproduction is seen in two of the four groups. Many form resting bodies called cysts.

8. Most protozoa are free-living components of the aquatic zooplankton, feeding on each other and the phytoplankton and in turn nourishing the larger animals. Others are parasitic and occupy the intestinal tracts and orifices of human beings and animals. Although most of these do no appreciable harm, a few cause diseases characterized equally by their seriousness, wide distribution in certain areas of the globe, and the lack of effective measures to control them.

Study Topics

1. How do the eucaryotic algae differ from the cyanobacteria?

2. Why is the classification of *Euglena* debatable?

3. Fungi, like plants, have cell walls. Why are they given the status of a separate kingdom?

4. What are the types and functions of fungal spores?

5. What effect did the development of sexual reproduction have on the evolution of early eucaryotes?

Bibliography

Algae

Pickett-Heaps, J. D., 1975, *Green Algae*, Sunderland, Mass.: Sinauer Associates.

Trainor, F. R., 1978, *Introductory Phycology*, New York: Wiley.

Villee, C. A., 1977, *Biology* (7th ed.), Philadelphia: W.B. Saunders.

Fungi

Ainsworth, G. C., and G. R. Bisby, 1971, *A Dictionary of the Fungi* (6th ed.), Kew, England: Commonwealth Mycological Institute.

Alexopoulos, C. J., 1979, *Introductory Mycology* (3rd ed.), New York: Wiley.

Burnett, J. H., 1968, *Fundamentals of Mycology*, New York: St. Martin's Press.

Protozoa

Grell, K. G., 1973, *Protozoology*, Heidelberg: Springer-Verlag.

Sleigh, M., 1973, *The Biology of Protozoa*, Amsterdam: Elsevier.

6

Microorganisms
in Natural Environments

Microorganisms play essential roles in the earth's ecology. In the last chapter, as each group was introduced, major environmental contributions were noted. We will now develop an understanding of the natural laws that shape and maintain the stable, life-supporting biosphere we take for granted. We will discover how microorganisms participate in all of the natural cycles; we will become aware of the complex partnerships between microbial species and higher organisms. These interactions are essential for normal function of plants and animals; disturbed, they give rise to disease.

THE ECOSYSTEM CONCEPT

What Is an Ecosystem?

A powerful concept in modern biology is that of the **ecosystem.** An ecosystem is the sum total of organisms of all types living in a defined area in addition to those portions of the nonliving or inorganic environment with which they interact.

An ecosystem is more than the sum of its parts. Researchers can isolate each organism, count its numbers, and study its biology in isolation. They can determine amounts of minerals in the soil, and measure temperatures. Yet even when all this is done, the ecosystem is not fully comprehended. The functioning ecosystem is not adequately defined by a series of tests done on single species at one moment in time. It is a dynamic equilibrium in which each component constantly and subtly affects all others.

The organism in isolation is not the same as the organism in its natural environment. Clearly, this presents ecologists with a serious dilemma.

Homeostasis   Many ecosystems, especially those that have been undisturbed for a period of time, are highly stable. We perceive that stability when we walk in a deep pine woods and realize that it must have looked much the same a hundred years ago as it does today. More scientifically, if numbers and distributions of organisms and nutrients are determined at points over time, there will be but small changes. This stability is deceptive, however. Ecosystems are not changeless. Their equilibrium is the very active balance of an acrobat on a tightrope, whose position is maintained by a continuous sequence of hundreds of tiny, interactive adjustments.

In an ecosystem, stability is achieved because a series of checks and balances guarantees that no one species will increase its numbers to an extent that would threaten the survival of the others (Fig. 6.1). The proliferation of each species is normally checked by the availability of suitable nutrients and interactions with other species. Population expansion is usually contained, sooner or later, by **negative feedback.** This means that, as the numbers increase, so also does some negative factor in the ecosystem. Using rabbits (notorious proliferators) as an example, increase will usually cause two negative results—a shortage of food suitable to rabbits and an increase in the number of rabbit predators such as foxes and owls. Both of these results serve to reduce the rabbit population. Thus the numbers of rabbits tend to oscillate slightly, with a slight increase followed by a slight decrease. As you walk in the woods, you would probably not be aware of the variation from year to year.

Microbial populations behave in the same way, but we are not aware of them at all, until we start doing microscopic analysis. Negative feedback operates equally effectively to control microbial populations.

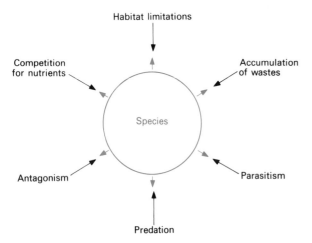

6.1
Population Control Factors   Each reproducing species tends to increase its population size. However, many pressures limit this tendency. Habitat limitations and the biological activities of other species in the community counteract growth of one group at the expense of the others.

Diversity Ecosystem homeostasis depends heavily on regulatory interactions among species. The greater the variety of organisms present in an environment, the more checks and balances will operate and the more sensitive will be the "fine tuning." Stable ecosystems are usually so diverse that it is next to impossible to map out every interaction. Each organism's growth is subject to a large number of different influences.

When an ecosystem is severely disturbed, it will return slowly to the stable state. Using the forest example, let us imagine the severe disturbance a fire can cause. It leaves the soil bare; perhaps only soil microbes are still present. The area will be recolonized by first one, then several types of small plants. These will modify conditions enough to allow the development of shrubs and brush, and these will soon give way to an increasing diversity of trees. As plant life develops, animals will return. Microbial species will thrive on the growing supply of plant and animal wastes. After dozens, maybe hundreds, of years, the ecosystem will have returned to its prefire state.

A newborn human being is normally devoid of microbial organisms. Yet its skin surface rapidly undergoes a colonization process similar to reforestation, as many species of nonpathogenic normal flora take up residence. Within a very short time, the baby's skin is a stable ecosystem in which many microbial species coexist under conditions such that their multiplication is restricted and the baby's health is maintained.

In areas of the human being in which microorganisms are present, there is always a diversity of species. In other areas microbes are absent. There is no such thing as a normal ecosystem inhabited by just one microbial species. Should one species establish an abnormal dominance, the stable person–microbe interaction will break down; the organism will usually cause pathological changes in the affected area.

Replacement flora, Chapter 26, p. 769

Symbiosis It is time to explore more closely the types of interactions among species, commonly referred to as **symbiosis**—living together. **Neutralism,** perhaps the rarest form of symbiosis, is a condition in which two species, such as owls and mushrooms, have no demonstrable effect on each other. **Competition** is a general phenomenon among species that have similar food or habitat requirements, as various types of algae compete for the supply of available light.

In **mutualism** or **synergism,** two or more species have evolved a relationship that is mutually advantageous. An intriguing example is found in the luminous organs of deep-water fish. Living well below the limits to which sunlight penetrates water, fish still need to recognize potential food, sexual partners, and predators. Each fish species has specially patterned receptacles on its head and feelers that harbor populations of photoluminescent (light-emitting) bacteria. The bacteria receive nutrients from the fish, using some for growth and others to produce light. The fish recognize each other by exhibiting light signals. The bacterial metabolism generates the light. This interaction, like many mutualisms, is obligatory in that the fish without the bacteria or the bacteria without the fish do not normally survive.

Luminescent bacteria; Chapter 4, p. 119

Lichens are a much-studied mutualistic form. The encrustations of lichen, often seen on rocks, consist of two microbial species, one a photosynthetic alga and the

Box 6.1

Lichens

Lichens are small, plantlike, spreading growths seen on such inhospitable surfaces as exposed stone, bark, and bare soils. They were once considered to be a type of plant, but are now understood as a symbiotic partnership between a fungus, usually an ascomycete, and a cyanobacterial, or green, algal species. In nature, the two partners are completey interdependent. The photosynthetic partner provides fixed carbon in the form of glucose/and or sugar alcohols. If it is a cyanobacterium, it may also provide amino acids. In compensation, the fungus provides inorganic nutrients, such as phosphate and metal ions, extracted from the underlying substrate. The algae, which are very sensitive to drying, are protected by being enclosed in a protective network of fungal mycelia.

The fungal and phototrophic partners can be cultivated independently in the laboratory. They cannot be induced to reassociate unless the growth conditions are so bad that neither can make it alone. The symbiosis clearly arises by necessity, not choice.

The growth rates of lichens are exceptionally slow, but the uniquely efficient mutual support system allows lichens to succeed in areas, such as north of the Arctic Circle, in which there is no other permanent form of life. The reindeer lichen is the basis of the food chain for both reindeer and caribou, and thus traditionally for Eskimos and Laplanders.

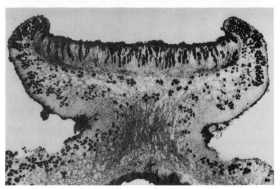

Urban dwellers may never have seen lichens. Because of the lichens' exposed growth locations, they are exquisitely sensitive to the toxic effects of pollutants both in air and rainwater. As the zone of pollution expands from an urban center, lichen life disappears.

other a thick-walled fungus. The algae live within the fungus and nourish it. The fungal cell wall in turn provides the protection necessary for algal survival, enabling lichens to survive on the hostile surfaces of bare stone or tree bark. Some lichen partnerships have been successfully resolved, and the alga and fungus cultivated separately. However, this is possible only in the laboratory—in nature, the partners are inseparable.

Commensalism is an interaction in which one partner benefits and the other is not appreciably affected. The growth of algae on the bark of trees is an example. The surface of the bark is damp and nutrients leach out of it to feed the algae, which are in a most favorable position to obtain light. The presence or absence of algae seems to make no difference whatever to the tree.

Antagonism in symbiosis occurs when one organism exerts an inhibitory or repellant effect on others, usually by the release of a special chemical. The antagonistic effect gives its possessor an edge in competing for space or nutrients.

Medical practice was revolutionized when microbiologists discovered and developed antibiotics for therapeutic use. The original drugs were natural antagonistic substances produced by soil fungi and streptomycetes to inhibit bacterial competitors (Fig. 6.2). In the ecosystem of fertile soil the chemicals reduced the growth of other decomposers, increasing the share of the food supply remaining for the drug-excretors. Drugs such as penicillin and streptomycin continue to act as bacterial antagonists when given to patients with pneumonia or tuberculosis.

Antibiotics; Chapter 26

All heterotrophic organisms need organic foods manufactured by other organisms. If they wait until the source organism is dead, they are called saprophytic or saprozoic. If they do not wait for the source organism to die, they may be classed as parasites, or if they cause its death, as predators.

Predators characteristically pursue and consume organisms (prey) smaller than themselves. Most protozoans derive a majority of their nutrition by consuming bacteria, just as the multicellular animals graze or hunt. Predatory behavior is also seen in some fungi. Some soil fungi form hyphal snares to capture nematodes (roundworms) that they then digest. Population size of a predator species is clearly limited by the number of prey organisms.

Protozoan nutrition; Chapter 5, Table 5.5

Parasites are usually smaller than their **host** organisms. The interaction between the two is frequently very subtle. Typically, the parasite does not kill the host outright, as a predator does. For example, the wheat rust fungus weakens the wheat plant and reduces its capacity to produce grain. *Entamoeba histolytica*, the protozoan parasite of human beings, causes diarrhea and ulceration but rarely death.

Host–parasite interactions seem to evolve to permit continued feeding and reproduction of the parasite, while keeping the host alive. After all, in most cases, if the host dies, so also do the parasites. Most serious students of symbiosis propose that relationships evolve over time from an initial state that is essentially predatory

Host-parasite relationship; Chapter 12, pp. 346–347

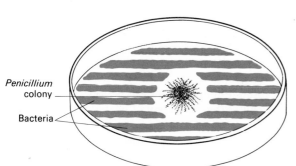

6.2

Antagonism Parallel streaks of a penicillin-susceptible bacterium were inoculated onto a nutrient agar medium. Then a single inoculation of *Penicillium* mold was made in the center. Following incubation, bacterial growth is inhibited over the agar zone containing antibiotic produced by the central mold colony.

Penicillium colony

Bacteria

(one member is slowly killed) to commensalism (both partners or their progeny survive indefinitely).

All infectious diseases, by definition, can be classified as parasitic relationships. However, they vary tremendously in their severity. The average survival time of the infected human host without medical intervention differs greatly. Recovery from infectious disease usually, but **not always**, means that the parasite has been eliminated and the relationship has terminated. The parasite species' survival is dependent on its "jumping ship" to a new host before the host eliminates it.

Microenvironments

We may make generalizations about macroscopic environments such as garden soil or the human mouth. Analyses of these areas give us very broad data that are certainly useful. However, careful study will reveal that both environments are really subdivided into many smaller differing **microenvironments.** In a microenvironment, conditions may be totally different from the "general" situation. For example, a tiny crumb of soil may have different microbial populations on each of its faces because of differences in availability of foods and oxygen. The flora on the aerobic surface of a tooth will be almost entirely different from that in the anaerobic crevice between that same tooth and the surrounding gum. This diversity of environments contributes in turn to a stabilizing diversity of microorganisms.

THE FLOW OF ENERGY AND MATERIALS

Within an ecosystem, organisms live together in mutual dependence. The structured nature of their relationships derives from the fact that organisms pass energy sources and crucial elements from one species to another. **No organism** can permanently fulfill all of its biological needs for itself.

The One-directional Flow of Energy

Life-sustaining forms of energy   Although energy is available in many forms, organisms on this planet have evolved, so far as we know, the ability to use only certain types of energy and energy-yielding chemical reactions. The primary energy source for all ecosystems is light, the visible portion of the solar electromagnetic spectrum. We do not know of organisms that use nonvisible wavelengths. Sunlight, we recall, is used by phototrophs, adapted and distributed to capture visible light efficiently under a variety of conditions. Although sunlight is an almost unlimited energy source, the amount actually captured is restricted by climate and availability of key nutrients. Captured energy is used to fix carbon—to synthesize organic compounds. Solar energy not directly captured in photosynthesis energizes the cycle of water evaporation and precipitation, causes the movement of heated air masses, and warms the atmosphere.

Reduced inorganic chemical compounds may be oxidized with a resulting release of energy. Chemoautotrophs also use the energy so gained to fix carbon.

Electromagnetic spectrum; Figure 3.1

Chemoheterotrophs satisfy their energy requirements by assimilating organic materials produced by phototrophs and chemoautotrophs. When these organic materials are metabolized, part of the energy released is used by the chemoheterotrophic organism to manufacture its own chemicals.

Note that whether the energy source be light, reduced inorganic compounds, or organic compounds, part of the energy is always used for the same purpose—making needed organic compounds. These new compounds are themselves stored-energy sources. Energy is also used for movement and active transport.

All organisms are in the business of converting energy from one form to another. This conversion proceeds under the limitation of the First Law of Thermodynamics that states that in any process energy can be neither created nor destroyed. An organism's growth and reproductive process is therefore strictly limited by the amount of energy it is able to capture.

Inefficiency of energy transformation   The Second Law of Thermodynamics states that energy exchanges can never be perfectly efficient. This means that whenever an organism uses energy, it can retain only a fraction of the starting amount of energy locked in the new organic compounds it has made with the energy. What happens to the rest of the energy? It is released in disordered useless forms, primarily as heat. Dissipated energy cannot be used, either by the producing organism or any other, to run a biological process.

The one-way flow of energy   In a typical **food chain**, a phototroph producer captures a certain amount of energy from the sun. It successfully transfers some fraction, usually quite small, of this into the chemical bonds of new plant material. The phototroph is grazed on by an animal consumer whose metabolism releases the stored chemical energy in the plant tissue for purposes of making animal tissue. Because the animal's metabolism is inefficient, the amount of added animal tissue will be much smaller than the amount of plant tissue consumed to build the animal tissue. At the end, the dead tissues and wastes of plant and animal are left, with some residual energy content, to the microorganisms (decomposers). These progressively break down the residues to obtain energy and nutrients for microbial growth. At the point at which all organic compounds have been broken down, then all the sun's energy has been converted to entropy and returned to the environment in an unusable form. The uses of energy lead inexorably to energy's dissipation. The energy is converted from one form to another, and the final form is useless.

The cyclical flow of matter   The conversions of matter do not produce loss. The biologically important elements—carbon, nitrogen, sulfur, and phosphorus—are chemically transferred from one compound to another. But all C,N,S, and P compounds are potentially useful to some organism. In a complete ecosystem no biologically vital element is ever left unused for long. **Nutrient cycles** are sequences of biologically mediated chemical conversions that transform elements from one molecular form to another. Each organism within a cycle takes a key element from the environment in one molecular form and returns it to the environment in another molecular form, as a waste product. The waste product is in turn taken up by an-

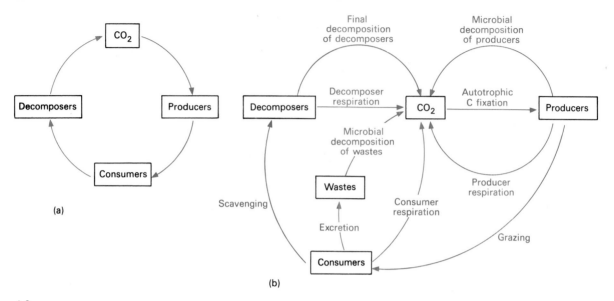

6.3
The Carbon Cycle   (a) The basic flow of carbon atoms is from the atmosphere through producers to consumers to decomposers and back again. However, this view is vastly oversimplified. (b) Individual carbon atoms removed from the atmosphere may return to it more or less indirectly. All fixed organic matter, including the decomposer cells themselves, is eventually broken down.

other organism for which it is usable, then returned to the environment in yet a third form. A sequence of such transfers results in the regeneration of the element in its original form. Cycling is of crucial importance because the essential elements are relatively scarce in the earth's crust and atmosphere. If they were progressively tied up in unavailable forms, life would soon cease to exist.

The carbon cycle   Carbon makes up the backbone of organic molecules. No organism can manufacture a single new molecule, organic or inorganic, without a carbon source. Organic carbon sources are, of course, also energy sources.

The main source of carbon is atmospheric CO_2 (Fig. 6.3). **Carbon fixation** is the energy-driven binding of carbon atoms from atmospheric CO_2 into organic carbon compounds. **Producer** organisms are photo- and chemoautotrophs than can fix carbon. **Primary production** results in an increase in the biomass of producer species.

The rest of the carbon cycle is a progressive "unfixing" of carbon atoms during which organic compounds are used sequentially by various members of the food chain resulting in the release of CO_2. Most of the unfixing process results from a complex series of metabolic reactions called **respiration.** These reactions break down carbohydrates to form CO_2. There is a simultaneous release of energy, recaptured by making the energy-storage molecule ATP.

Respiration; Chapter 8,
p. 238

Producers also respire, releasing some of the carbon they fix. Plants, for example, respire in the dark. Consumers—grazers, predators, and parasites—respire, breaking down most of the more readily metabolized organic compounds. Residues of fixed carbon are left in the tissues of organisms that die, and in the wastes of animals. Such unutilized matter is collectively called **detritus.** We have already made it clear that a major role of bacterial and fungal decomposers is to clean up these leftovers. Decomposers are most important not because of the sanitary or aesthetic value of their work but because decomposition completes the unfixing of carbon. The work of decomposition, carried out by bacteria and fungi, has two steps: (1) **Dissimilation,** the breaking down of macromolecular structures, that would lead, for example, to the disappearance of a tree stump as a recognizable object, and (2) **mineralization,** the conversion of organic elements to their inorganic state. The two steps are clearly continuous.

Some carbon atoms will be retained in a fixed form for a very long time if they are in refractory compounds. Resistant plant polymers, such as lignin, may persist for a long time, particularly under anaerobic conditions. Peat is an accumulation of undecomposed material preserved by overlying bog water. Dried peat is used for fuel in some areas. Coal, oil, and natural gas production occurred in the past when accumulated partially decomposed organic matter was overlaid with water and rock before the carbon was mineralized. Combustion of these fossil fuels rapidly completes the return of this carbon to the atmosphere.

Some synthetic organic compounds such as plastics and pesticides are *nonbiodegradable*—a term that tells us that decomposers possessing the right enzymes to decompose the compound are rare or nonexistent. Such synthetic substances break down extremely slowly, mostly as a result of nonbiological processes. They comprise some of our more severe environmental pollutants.

Sometimes genetic changes in organisms do occur, allowing them to metabolize a previously unavailable organic compound. Bacterial species have been genetically "engineered" to break down spilled crude oil at an accelerated rate. Such genetic changes may also be undesirable from the human point of view; certain strains of the gonorrhea organism recently acquired the enzymatic capacity to degrade penicillin.

Penicillinase—positive strains; Chapter 21, p. 628

Energy flow is linked to the carbon cycle. As carbon is fixed, energy is captured in the organic compounds needed by heterotrophs. As heterotrophs metabolize, the energy is released through the process of entropy. The energy rule of "diminishing returns" tells us that the biomass of consumers must be less than that of producers. The universal need for energy sources ensures that almost every organic compound will eventually be used by some organism needing to get at not only the compound's carbon atoms but also its precious supply of stored energy.

The nitrogen cycle All organisms require nitrogen to form proteins and nucleic acids. Both inorganic and organic forms of nitrogen can be used. The interconversions among inorganic forms and the movement of the element from free to fixed status are more complex than those in the carbon cycle.

Let us survey the nitrogen needs of different groups of organisms. Plants characteristically need nitrate ion (NO_3^-) that they incorporate into new plant tissue

TABLE 6.1
Microbial participants in the nitrogen cycle

PROCESS	SOME MICROBIAL PARTICIPANTS
*Nitrogen fixation	Bacteria—*Rhizobium, Azotobacter, Beijerinckia,* and others Cyanobacteria—All heterocyst-formers such as *Anabaena, Nostoc*
Assimilation	All organisms; not dependent on microorganisms
Ammonification	Aerobic—*Bacillus,* most aerobic species Anaerobic—*Clostridium,* most anaerobic species Urea breakdown—*Proteus*
*Nitrification	Nitrite formation—*Nitrosomonas, Nitrosospira* Nitrate formation—*Nitrobacter, Nitrococcus*
Denitrification	*Pseudomonas, Bacillus* (some species)

*At each stage, certain microbial species participate. They are essential for the asterisked processes.

concurrently with their carbon-fixing operations. Animals need organic nitrogen, principally bound as the $-NH_2$ group of amino acids. Different groups of microorganisms can use all the forms of nitrogen, and are primarily responsible for interconverting the forms. Yet in the biosphere, all of these nitrogen compounds are scarce. The only bountiful nitrogen compound is the diatomic gas N_2. This substance is useless to plants and animals alike!

Nitrogen actually moves in two interlocked cycles (Table 6.1). One is an atmospheric cycle involving nitrogen gas (Fig. 6.4). The gas is removed from the atmosphere and fixed into nitrogenous organic compounds by **nitrogen-fixing** microor-

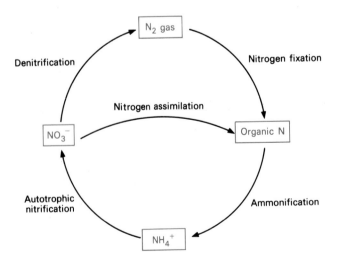

6.4
The Nitrogen Cycle The global supply of fixed forms of nitrogen is very limited. Because many different types of organisms seek to use every atom of fixed nitrogen, cycling tends to be rapid. Nitrogen gas is removed from the atmosphere by microbial action. This beneficial activity is supplemented somewhat by the creation of nitrogen oxides by lightning discharges.

ganisms. There are free-living forms such as the cyanobacteria and *Azotobacter*. Other nitrogen-fixers live in highly specialized mutualistic relationships with the root tissues of host plants. The bacterium *Rhizobium* forms nodules on the roots of leguminous plants (peas, beans, clover, alfalfa) in which it fixes gaseous nitrogen into a nongaseous organic product, glutamic acid. The glutamic acid is passed on to the plant and also excreted into the environment. Traces of nitrogen are also converted to nitrate by lightning discharges, and tons of inorganic nitrate fertilizer are made by industrial processes. Even so, it is estimated that over 90 percent of the annual global nitrogen supply for plant growth is fixed by microorganisms.

Nitrogen gas is also returned to the atmosphere by microbial reduction of nitrates. **Denitrification** seems wasteful because it strips nitrogen from the biosphere. Yet the magnitude of the effect is small and rarely compromises soil or water fertility. It most often occurs in environments, such as water-logged soil, in which oxygen for bacterial respiration is limited.

The second cycle is that of combined nitrogen. Nitrate ion is readily assimilated by plants and microorganisms to build tissue. The immobilized organic nitrogen is passed around among producers, consumers, and decomposers in the food chain.

As wastes are produced and decomposers work, amino acids are broken down to release ammonium ion (NH_4^+). This process of ammonification yields a product useful to microorganisms but not to plants or animals. Some nitrogen will be lost to the atmosphere as ammonia gas.

Closing the cycle requires another microbial process, **nitrification.** Bacteria such as *Nitrosomonas* and *Nitrobacter* oxidize ammonium to nitrate. This is a chemoautotrophic process; the organisms derive energy while producing a critical plant nutrient.

In summary, five major nitrogen conversions occur in the nitrogen cycle. Microorganisms participate in all of them, and are essential to three—nitrogen fixation, nitrification, and denitrification. Since suitable nitrogen compounds are a requirement for life, microbial activities provide the foundation for the productivity of the biosphere.

The sulfur cycle   Sulfur is an essential component of living cells. It is found predominantly in the amino acids cysteine and methionine. The quantitative need for sulfur is smaller than that for nitrogen, and the reserves of the element in nature are more often adequate to sustain full productivity. The sulfur cycle (Table 6.2) is similar to the nitrogen cycle except for the fact that the free form (S) does not physically leave the biosphere as does nitrogen gas (Fig. 6.5).

Sulfate ion ($SO_4^=$) is the form of sulfur assimilated by plants, animals, and most microorganisms to be converted into organic sulfur compounds. These are passed through the food chain. The sulfur is finally released by microbial decomposition in the form of the bad-smelling gas, hydrogen sulfide. H_2S is sequentially reoxidized to elemental sulfur and then to sulfate by energy-yielding reactions carried out by chemoautotrophic bacteria such as *Beggiatoa* and *Thiobacillus* (Fig. 4.5e). Rhodospirillum, in **sulfate oxidation,** oxidizes H_2S and elemental sulfur during anaerobic photosynthesis. The opposite process, **sulfate reduction,** occurs in

TABLE 6.2
Microbial participants in the sulfur cycle

PROCESS	SOME MICROBIAL PARTICIPANTS
Sulfate assimilation	All organisms; not dependent on microorganisms
*H_2S release	Facultative anaerobes: *Enterobacteria*
	Strict anaerobes: *Clostridium, Bacteroides*
*Sulfur oxidation	Accumulate sulfur—*Beggiatoa, Thiothrix*
	Produce sulfate—*Thiobacillus, Thiomicrospira, Sulfolobus*
	Photosynthetic—*Rhodospirillum, Chlorobium*
*Sulfate reduction	Spore-formers—*Desulfatomaculum*
	Nonspore-formers—*Desulfovibrio*

*Microorganisms (chemoautotrophs and anaerobes) are essential for these processes.

anaerobic environments; bacteria such as *Desulfovibrio* use sulfate in place of oxygen as an electron acceptor yielding H_2S. These sulfate losses are similar to the nitrate losses in the denitrification step of the nitrogen cycle and occur for much the same reason, inadequate aeration of the environment.

Other elements   Oxygen, phosphorus, iron, calcium, and other elements are of primary importance in all ecosystems. Their transformations do not, however, form closed cycles like the carbon or nitrogen cycles. Except for oxygen, these elements are generally derived from mineral deposits. Their availability is often limited to the amounts dissolved in water. Microbes in some cases play in important role in their transformations.

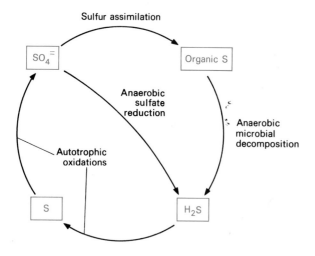

6.5
The Sulfur Cycle   Many steps in the sulfur cycle resemble similar biological events in the nitrogen cycle. A difference is that whereas anaerobic denitrification releases N_2 from the pool of fixed nitrogen, anaerobic sulfate reduction converts it to a less useful form that is still directly available for some biological use.

AQUEOUS ENVIRONMENTS

Life forms evolved in water. Water masses still cover approximately 70 percent of the planet. Thus aqueous ecosystems, both fresh water and salt water, are the dominant ones in global importance.

Biological Activities

Classes of organisms   In a water ecosystem, **plankton** are small floating organisms, primarily algae, protozoans, and the larvae of multicellular animals. These are consumed by the swimming animals (**nekton**). Plant and animal species attached to the bottom or found in the sediments are classed as **benthos**. Bacteria and fungi are found both in the water column and in great numbers in the sediments in which the detritus from all other organisms is deposited.

Limits on productivity   Productivity (Table 6.3) depends on both planktonic and benthic algae and cyanobacteria. It is limited primarily by available light. Clear water absorbs light, and suspended particles reduce the transparency of water, so

TABLE 6.3
Microbial contributors to the major ocean zones

ZONE	MICROBIAL CONTRIBUTORS	PREVALENCE
Intertidal	Producers—cyanobacteria, unicellular and multicellular algae	Many
	Consumers—protozoa, bacteria	Many
	Decomposers—bacteria and fungi—aerobic	Many
Neritic euphotic zone	Producers—photosynthetic diatoms, dinoflagellates	Many
	Consumers—forameniferans, radiolarians	Many
	Decomposers—bacteria and fungi—aerobic	Moderate
Oceanic euphotic zone	Producers—photosynthetic diatoms, dinoflagellates	Few
	Consumers—forameniferans, radiolarians	Few
	Decomposers—bacteria—aerobic	Few
Aphotic zone	Producers—	None
	Consumers—protozoa	Very Few
	Decomposers—bacteria—aerobic	Very Few
Sediments	Producers—bacterial chemoautotrophs in certain locations	Limited
	Consumers—detritus feeders	Few
	Decomposers—bacteria, aerobic and anaerobic	Many

The oceanic euphotic zones and underlying aphotic zones are with few exceptions very underpopulated. They are often referred to as biological deserts. Near the shore and over sediments the water is enriched by essential nutrients, and adapted organisms thrive.

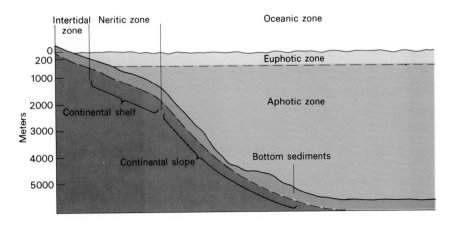

6.6
Ocean Zones The ocean zones are shown with approximate depth readings. Actually, ocean zones vary widely around the globe. In an intensely murky area of surface water, the euphotic layer may be very thin. Life is found in every zone; the intertidal, euphotic, and bottom sedimentary zones support the greatest biological activity.

light penetrates only the top layers of deep waters (Fig. 6.6). Thus productivity is limited to shorelines and surface waters. The zone in which light is sufficient for photosynthesis is the **euphotic** zone.

Nitrogen. In addition to light, algae also need sources of carbon dioxide, fixed nitrogen, and phosphate in order to grow. Water dissolves small but usually adequate amounts of CO_2 from the air; since algal growth is a shallow water phenomenon, it is rarely limited by inadequate supplies of carbon. Nitrogen in a freshwater environment may be supplied by nitrogen-fixing cyanobacteria, or washed in with rainwater runoff that picks up fertilizers, agricultural wastes, and effluents from human sewage disposal systems. In the oceans, all of these contribute, but the role of the cyanobacteria is proportionately much more important.

Phosphorus. Phosphate enters the biosphere as a result of the weathering of rocks. It is tied up in the form of calcium salts in tests, shells, bones, and teeth. These structures are resistant to decomposition and tend to accumulate in sediment. The phosphate is only slightly water soluble; thus it is "trapped" and only slowly released. It frequently is the scarcest, "most wanted" nutrient in water. Phosphate supplements enter water from fertilizer runoff and domestic waste waters containing detergent products and human waste. In fresh water, nitrogen and phosphorus supplements usually derive from these sources. The oceans, however, receive their major phosphate additions through a process known as **upwelling.** In certain areas, ocean currents rise to the surface bringing water from very deep zones. This water has had thousands of years to dissolve phosphate from the ocean floor sediments, mainly composed of phosphate-rich tests of foraminiferans. When the enriched water comes to the surface, it supports dense algal populations. Upwelling areas are the world's richest fishing zones.

Foraminiferans;
Chapter 5, pp. 151–152

Temperature. Temperature affects productivity. Algal growth slows as temperatures drop. In fresh waters, particularly in temperate zones, summer brings high levels of productivity but winter reduces productivity to near zero. In the oceans,

however, the annual variation in water temperature is comparatively slight. Therefore the seasonal changes do not greatly affect algal growth except when daylight
periods are much shortened in winter.

Productivity, then, clearly varies greatly from one area to another, depending
on physical conditions such as temperature and light, and chemical factors such as
availability of nitrogen and phosphorus. Aqueous production, however, is estimated to average 0.5 g dry organic matter/square meter/day worldwide. Marine
ecosystems alone are responsible for about 44 percent of the gross annual global
production.

Algal blooms   When a population of algae increases with explosive speed, we know
that a stimulation to its growth has been suddenly supplied. Warming conditions or
the addition of a needed nutrient are among such stimulations. Algal blooms have
undesirable side effects. In lakes, the consumer population never increases fast
enough to graze down the algal population to a suitable level. Then, when the algae
begin to die off, the water mass must absorb a large mass of dead algae settling to the
bottom. The decomposers usually quickly exhaust the dissolved oxygen supply in
metabolizing the dead algal cells. A sediment of partially decomposed, objectionable
algae accumulates; water animals may become suffocated from lack of dissolved
oxygen.

Geochemical Activities

Biological activities in aqueous habitats have built up extensive geological deposits
of economically important minerals. Sedimentary accumulation of diatom frustules
has produced diatomaceous earth, in the same way that deposits of forameniferan
tests gave rise to chalk and limestone. Sulfur domes contain elemental sulfur supposed to have originated from the activities of sulfur-cycle bacteria. Petroleum reserves represent hydrocarbon substances formed abiologically from the stored oils
in collected dinoflagellates.

Diatoms; Chapter 5, p. 134

Dinoflagellates; Chapter 5, p. 135

Many areas of the ocean floor are strewn with **manganese nodules,** irregularly
shaped, metallic globules made up of precipitated atoms of iron, manganese, and
other rarer metals. In the near future, these will be commercially mined to extract
their extremely valuable mineral content. Microorganisms are essential to the precipitation of metals from sea water in the form of nodules.

TERRESTRIAL ENVIRONMENTS

Although the land is, geologically speaking, a relatively recently exploited habitat
for life, it supports the majority of the higher plants and animals. Terrestrial habitats
derive their nutrients from soil, and organisms live either in or directly on the soil
surface. It is estimated that average terrestrial productivity is from 1 to 10 g dry
weight of organic matter per square meter per day. Let us examine the ecological
characteristics of the two parts of a terrestrial ecosystem, the soil and the air.

Composition of Soil

Soils are composed of inorganic and organic components. The mineral portion is derived from weathered rock, with particles in descending order of sizes called rock, gravel, sand, silt, and clay. These particles pack irregularly, leaving spaces for penetration of water and air within the soil. The mineral content of the soil is clearly dependent on the rock from which it is derived. The inorganic portion of soil is added to from below by the fragmenting of the underlying rock and occasionally from above by the deposition of sediment as in riverbeds. The thickness of soil in an area depends on rates of weathering, sediment deposit, and protection from erosion provided by plant cover.

Most soils are seen to have definite layers (Fig. 6.7a). **Subsoil** is principally inorganic, has few nutrients, and supports little life. **Topsoil** has a significant population, contains and recycles nutrients, and is the fertile zone. Over time, subsoil matures into topsoil as biological activities increase its proportion of **humus,** a type of amorphous organic matter composed primarily of those plant residues most resistant to decomposition. In temperate climates, active plant growth in a grassland will annually add to the soil more organic matter than is completely mineralized. The topsoil thus gains in humus content and the topsoil layer becomes thicker. Humus

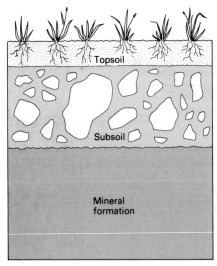

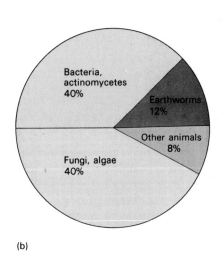

(a)   (b)

6.7
The Terrestrial Ecosystem   (a) A typical soil horizon (vertical analysis) shows a thin layer of topsoil. The topsoil, which contains decomposing organic material and the soil organisms, supports terrestrial life. The roots of plants are largely confined to the topsoil. The underlying subsoil is rocky, and poor in humus and nutrients. (b) Less than 1 percent by weight of topsoil is living matter. Once plant roots have been removed, most of the remaining life forms are microorganisms. The proportions of various life forms in a typical soil sample are shown here.

TABLE 6.4
Microbial populations of soil

TYPE OF ORGANISM	ROLES	NUMBERS/GRAM
Bacteria—Autotrophs/heterotrophs Aerobes/anaerobes Cellulose digesters Protein digesters Sulfur oxidizers Nitrifiers, denitrifiers, Nitrogen fixers	Decomposers, mineral cycles, plant symbiosis	2.5 billion (direct count) 15 million (dilution plate)
Actinomycetes—Filamentous procaryotes *Nocardia, Micromonospora, Streptomyces*	Decomposers	700,000
Fungi—Mostly mycelial forms Primary global habitat for fungi	Decomposers, mycorrhizal associations	400,000
Algae—Limited to lighted surface Green algae, diatoms, cyanobacteria	Producers— minor importance	50,000
Protozoa—Flagellates and amoebas— mainly consume bacteria	Consumers	30,000

One gram of fertile garden soil taken from a well-drained surface location might well contain this extraordinary diversity of microorganisms.

improves soil structure, making it more readily infiltrated by the roots of the plants. Humus also improves soil's water-retentive capacity and provides a reserve supply of organic matter that supports a rich and varied microbial population (Table 6.4).

Soil Organisms

It is seldom possible to determine accurately exact numbers of each type of organism in soil. However, there is a direct relationship between increasing soil populations and improved soil fertility. The reason for this relationship is implicit in what we already know of the carbon, nitrogen, and sulfur cycles. Microorganisms, by recycling wastes, provide plants with the elements they need in the forms in which they need them—thus the greater the number of microorganisms, the stronger the plants.

Bacteria in a rich garden soil may number up to 2.5 billion per gram. These include aerobic spore formers, anaerobic organisms, heterotrophic decomposers adapted to a variety of substrates, and autotrophic species. In the work of decomposition the bacteria are aided by hundreds of thousands of fungi (Fig. 6.7b). Significant numbers of algae are found on the surface of bare soil; their numbers decline in planted areas because of shading and competition for nutrients. Protozoa are present in large numbers. They are predatory on the soil bacteria, and may serve to limit their numbers.

Box 6.2

Mycorrhizae

Mycorrhizae or fungus roots arise from colonization of the fine roots of plants by beneficial soil fungi. This relationship, which is mutualistic, is seen in the vast majority of economically important crop plants. The mycorrhizal association contributes to increased plant vigor and rapid growth. In plant growth, the limiting factor is the availability of ions such as phosphate, zinc, copper, molybdenum, and ammonium, as well as water. Increased root surface area improves absorption; the fungal hyphae appear to function as a kind of accessory root.

The ectomycorrhizal fungi are found in association with particular tree species, such as pine, hemlock, oak, birch, and poplar. Stimulated by chemicals secreted by the tree roots, the hyphae form a fungus mantle over the surface of the feeder roots. When tree seedlings are grown in nurseries in sterilized soil, they benefit from artificial inoculation with the appropriate fungus. Colonized seedlings develop larger root systems and grow several times as fast, both while in the nursery and after being moved to the plantation area.

Endomycorrhizae are formed when the fungus actually penetrates the root. This process does not noticeably change the appearance of the root. It significantly improves the root's capacity to extract water and scarce nutrients such as phosphate from the soil. This relationship is seen in most annual crop plants, but it has been largely ignored as a means of improving crop fertility because, until recently, there have been ample supplies of relatively cheap inorganic fertilizers. As the price of artificial soil supplements climbs, agriculturalists are looking toward increasing crop yield by assuring complete fungal colonization. Field experiments in several deficient soil areas have demonstrated 20 percent or greater yield in-

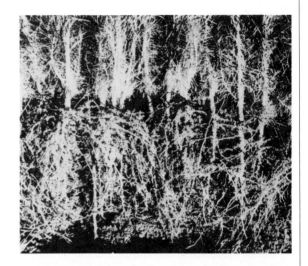

creases in potatoes, corn, wheat, and barley when either the soil or the plants were inoculated with the appropriate fungus. Fungal symbionts, then, may provide a means of obtaining adequate crops on some of the world's marginal soils.

The Rhizosphere

The zone immediately surrounding the root network of a plant is termed its **rhizosphere.** This microenvironment sustains dense microbial growth. The apparent reason is that the plant excretes useful organic substances into the rhizosphere. The relationship may be mutually beneficial in that many soil microorganisms in turn produce plant nutrients.

Many higher plants develop an intimate relation with a fungal species in which fungal mycelium, called a **mycorrhiza,** infiltrates the roots of the plant. Many of the participating fungi are basidiomycetes, and their characteristic fruiting bodies develop year after year in association with a certain type of tree. The association is a mutualistic one as the fungal hyphae absorb nutrients from the tree's vascular system while they assist the tree in the transport of water and dissolved minerals from the soil.

Root nodules formed by bacteria on legumes are a major site of nitrogen fixation in the soil (Fig. 6.8). Legumes excrete organic nitrogen into the soil to nourish surrounding organisms. Also, if left to decompose in the soil, they make a long-range contribution to its nitrogen reserves.

The variety among types of soils and their resulting life-support potential is very wide. Soil type, in conjunction with climate–temperature range and rainfall, determines soil productivity. At the moment, the most productive global areas are all approaching maximum agricultural utilization. Thus there is a growing interest in the development of farming on marginal lands. An appropriate use of microbial enrichment can be of great help.

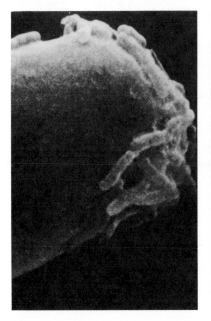

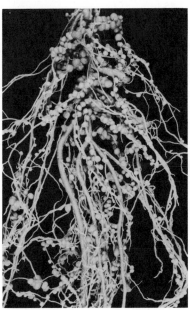

(a)   (b)

6.8
Symbiotic Nitrogen Fixation
(a) *Rhizobium trifolii* adhering to a clover root hair tip (× 7,500). Leguminous plants produce chemicals called lectins. These substances are believed to crossreact with receptors on both root hairs and bacterial surfaces. The first step in establishing a symbiosis is this special chemical binding of the partners. (b) When symbiosis is established, areas of swollen tissue (*root nodules*) form on the plant. Within these nodules, bacteria reduce the nitrogen gas (present in well-aerated soil) to amino groups that are incorporated in amino acids.

Atmospheric Microbiology

Airborne infection;
Chapter 15, p. 443

The air, although essential for life, is not itself a habitat. Organisms enter the air either from aqueous or terrestrial habitats. Even the upper layers of the earth's atmosphere contain some microorganisms, but these are all simply travelers, cells that have been picked up by air currents. Air dispersal spreads microbial species. It aids in colonization of natural habitats and in the dissemination of infectious disease.

Most organisms enter the air through the production of **aerosols,** clouds of microscopic droplets of microbe-bearing fluid. These droplets are sufficiently tiny so that they do not readily settle out, but remain airborne for an indefinite period. Significant aerosol production occurs not only in the obvious events of coughing or sneezing but also from wind-whipped water surfaces such as the crests of waves.

ANIMALS AS MICROBIAL HABITATS

Normal flora   The body surfaces of animals constitute a major habitat of microorganisms. The term **normal flora** is used to categorize a microbial species that is most freqently found in association with human beings or other animals, and more frequently than not in the absence of disease. This heavily qualified definition says basically that the organism in question is adapted to life in or on the animal. However, it is not primarily parasitic because the host usually sustains no harm from its presence. In most animals, probably human beings as well, the normal flora is essential to life.

Ectosymbiosis. In ectosymbiosis the microbial partner lives on the surface of tissues rather than within cells of the host. The surfaces may be exposed, such as skin or mucous membranes, or protected, such as the lining of the gut. An amazing example of a protected surface is the digestive system of ruminant animals such as cows, sheep, goats, and other grazing animals that derive their nutrition from grass, hay, and foliage—materials that are rich in cellulose and other polysaccharides. The main productivity of plants yields cellulose. In fact, it is estimated that at least 50 percent of the total global biomass is in the form of cellulose. But cellulose can be hydrolyzed only by cellulase enzymes. And cellulolytic enzymes, needed to unlock that 50 percent of the total energy supply, are found among heterotrophs **only** in bacteria and fungi.

Rumen microflora;
Chapter 4, p. 120

Ruminants have evolved a four-chambered stomach; the first two portions comprise the **rumen,** a giant fermentation vat with the capacity of about 100 liters (Fig. 6.9). In the rumen, microorganisms multiply to a density of around ten billion cells per milliliter. These organisms constitute a complex mixed flora, strictly anaerobic. One group digests the cellulose and other polysaccharides by producing large amounts of bacterial enzymes. The resulting sugars are then fermented by other microorganisms into simple fatty acids. The fatty acids are absorbed and distributed as an energy source to the tissues of the ruminant. The digestive process also permits large-scale multiplication of protein-rich bacterial cells. Since the rumen bacteria can assimilate ammonia to make amino acids, protein can be made from "waste" forms of nitrogen. As the bacterial cells reach the lower stomach regions, they are digested by the ruminant's own proteolytic enzymes to provide its amino

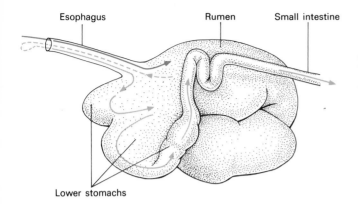

Esophagus   Rumen   Small intestine

Lower stomachs

6.9

The Digestive System of a Cow A cow while grazing rapidly shreds and swallows grasses and other green plants. This ingested matter passes into the rumen in which fermentation occurs. The carbohydrates are microbially converted to fatty acids, which are absorbed. The fermented mixture is regurgitated in small masses (cuds), chewed again, and reswallowed. The food goes into the lower stomachs in which protein digestion and amino acid absorbtion take place.

acid requirements for growth. The next time you eat a hamburger, pause to thank the rumen microflora!

Endosymbiotic Relations

Endosymbionts are organisms that live within the cell or cells of the host. Their presence may introduce a major genetic change in the host. The protozoan *Paramecium's* eucaryotic cells often contain procaryotic endosymbionts such as the **kappa** particle. *Paramecia* that contain kappa particles are killers—they manufacture killer particles and liberate them into the medium. These particles have a toxic effect on sensitive strains of *Paramecium.* The killer is a whole kappa particle. The kappa particle is not essential to the host—it can be eliminated by chemical treatments without harming the *Paramecium.* It is clearly not harmful to the host, and may be helpful since it serves to eliminate the competition. The relationship between the protozoan and the bacteria-like endosymbiont is a mutualistic one.

Insects often have microbial endosymbionts. As animal species, insects have a wide variety of nutritional requirements. Yet many insects occupy an ecological niche in which their diet is highly restricted and nutritionally incomplete. For example, termites eat only wood—not a very well-balanced diet. The cultivation of a gut population of bacteria is the solution to their problem.

In one relatively simply adaptation, plant-eating insects harbor large numbers of bacteria or fungi, usually of one species, in specialized cells. In some cases, these cells form a special organ, a **mycetome.** The endosymbionts produce needed amino acids from the insect's food materials. Very few of the microbial partners have been successfully cultured *in vitro.* In insects "cured" of their endosymbionts, vitamin and amino acid deficiencies develop unless the insects are fed a complete diet.

The true complexity of mutualistic relationships can really be appreciated by considering the lowly termite. Few of us pause to ask how a termite survives, since it clearly survives extremely well. Careful examination of the gut of the termite reveals that it contains flagellate protozoa, sometimes constituting as much as a third of the weight of the animal. But do the protozoans digest the wood cellulose?

No. Instead they are host to a population of ectosymbiotic spirochetes attached to their surfaces, and are host as well to endosymbiotic bacteria. The latter are believed to be the real source of the cellulase enzyme. Thus wood digestion depends on bacteria within protozoans within termites!

Human Normal Flora

The human animal has a microbial flora of great complexity. The flora has been extensively studied, with the revelation that our relationship to our flora is mutualistic. This implies that the flora does something essential for us. Recognition of the beneficial roles of normal flora in maintaining health, however, has paralleled discoveries of its ability to cause disease in some unusual circumstances. That implies a parasitic relationship. The true picture, then, is one to which no single label can be universally attached.

Colonization of the human body is external. It is restricted to the skin, intestinal tract, and orifices that open to the outside, such as the mouth, upper respiratory tract, vagina, and lower urinary tract. Microorganisms are not normally found in the lower respiratory tract, blood, tissues, and upper regions of the urinary and genital tracts. The specific normal flora of each area is presented in subsequent chapters.

Although a newborn baby is microbiologically sterile, its entrance into a world in which all the human beings that care for it have normal flora results in its being rapid colonized by the same organisms. This process is essential for the baby's continued normal development.

Roles of our normal flora   We can get a good idea of the effects of normal flora by studying experimental animals reared in such a way that they are **germ-free.** To be germ-free the newborn animal must be delivered by surgery and transferred immediately to a sterile environment in which it will spend its entire life. It breathes sterile air, eats sterile food, and is handled only by sterile-gloved hands (Fig. 6.10). Colonies of germ-free animals have been established for the study of biological functions unobscured by the contributions of the normal flora. The immune systems of germ-free or **gnotobiotic** animals do not develop at the normal rate. Since they have nothing to protect themselves against, they do not express their genetic potential to produce immune responses. Introduction of normal flora organisms rapidly kills germ-free animals. Development of appropriate body defenses in the normal animal thus appears to be dependent on the stimulus provided by continuous intimate contact with normal flora microorganisms.

Immune response; Chapter 13

The use of antibiotics gave us another insight into the beneficial role of the normal flora. Particularly in the early years of the "antibiotic era," patients received very large doses of antibiotics for long periods. Clinicians observed that these patients frequently developed troublesome secondary infections. The causative organism was often the "normal" yeast *Candida albicans.* This organism is a minor component of normal flora in most areas of the body. Antibiotic use was repressing other portions of the flora and allowing the yeast to gain the upper hand.

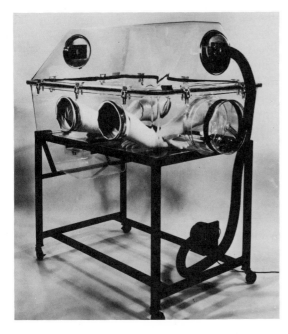

6.10
An Isolation Unit for the Handling of Germ-free Animals
Germ-free animals, delivered aseptically by surgical means
and reared in a sterile environment, have no natural flora.

When we remove the normal flora by disinfection or antibiotic use, we remove one
ecological control that reduces the potential for disease. **Opportunistic** disease organisms frequently proliferate as a **replacement flora.** The effect can be compared with
plowing a field in early spring, then leaving it uncultivated. The result will be a field
full of weeds! Bare soil, like uncolonized skin, is an invitation to uncontrolled
growth of whatever organism gets there first. The normal flora excludes potential
pathogens from promising sites of colonization; competition for food and energy
sources limits populations. The mucous membranes and the skin provide microbes
with only a limited amount of nutrients. Multiplication of the normal flora exhausts
these nutrients, leaving little for a potential pathogen to live off. Antagonism plays a
role in exclusion, too, as some normal floras produce by-products strongly inhibitory to invading organisms.

Normal flora of the
skin; Chapter 17, pp.
499–501

Our gut flora   In our gut, microorganisms are present in huge numbers—up to 100
billion organisms per gram of feces in the lower bowel. It has proven difficult to establish exactly what these organisms do for us. It is clear enough what we do for
them—we provide them with warmth, moisture, and a nutritionally complete diet.

Normal flora of the GI
tract; Chapter 20, pp.
580–586

Bowel organisms produce vitamins, but which vitamins and how much of
them, in normal circumstances, is not clear. Because starving people develop vitamin deficiencies, the microbial supply is clearly not adequate in the absence of a
poor human diet. Furthermore, in certain circumstances it appears that the bowel
flora uses up more vitamins than it produces.

Gut bacteria in human beings do not digest cellulose as they do in cows and termites. Their role in supplying digestive enzymes to assist with other types of food is

uncertain. Germ-free animals encounter no difficulty in digesting a normal diet. Yet human babies who do not acquire the *Lactobacillus* organism as the dominant member of their infant gut flora have grave difficulties digesting milk; the problem responds to the administration of a culture of living *Lactobacillus* organisms. Only one point is unequivocal—the gut flora is responsible for the production of gas in the process of digestion!

Human beings cannot live a healthy life in a normal world without their normal floras. We must consider them to be mutualistic partners of ours and learn to respect their importance.

THE ECOLOGY OF ARTIFICIAL ENVIRONMENTS

All ecosystems change in response to the same set of requirements for energy, nutrient elements, and space, yet some are much more stable than others. Certain manufactured ecosystems reveal very clearly that the artificially simplified environment is inherently unstable. Consider a well-publicized example of the effects of human ecological intervention on the macroscopic level.

The Monoculture Model

In agriculture, farmers attempt to simplify the population structure of a field to the point at which it supports only one producer species, such as corn or potatoes. Consumers are eliminated and decomposers ignored. The resulting **monoculture** (a term very similar in meaning to the microbiologist's **pure culture**) is not nutritionally self-sustaining because nutrients are not recycled and fertilizers must be added in large amounts. Insect populations are not checked by the usual interspecies interactions and therefore multiply to the point at which they are destructive pests that must be controlled with chemical poisons. How different a picture this is from the stable grassland ecosystem of the uncultivated field, that needs neither fertilizer nor pesticides. Yet the human population must feed itself, and it cannot eat grass!

Human beings, of necessity, have created many highly specialized environments in which the relationships among the people, plants, animals, and microorganisms have been radically changed. The resulting instability constantly challenges our efforts to control the system we have built. Let us examine two radically different ecosystems with the eye of the microbial ecologist.

The Hospital Environment

A hospital, or any other resident health facility, constitutes a highly selective environment for both human beings and microorganisms. The hospital is partially isolated from other environments. Staff, new patients, and visitors come and go with great frequency. Yet the physical structure itself, with its contents—people and equipment—is largely separated from other exterior factors. It has little flora, unless you count the lettuce in the kitchen and the florist's bouquets; one hopes there will be no insects or other animal life.

Specialized microbial population   Each patient, as well as each "healthy" staff person or visitor, brings to the hospital his or her own normal flora. This personal flora may be added to or modified in hospitals by organisms derived from other sources. Organisms are shed from human beings and their clothes and may become airborne.

Microbes can multiply only when they have a source of nutrients and moisture. A properly managed hospital will provide little opportunity for microbes to multiply in such favored habitats as food, water reservoirs, and human wastes. Most actual multiplication will take place in the one habitat that cannot be eliminated—people.

The hospital microbial population is not disproportionally large in numbers. However, it is exposed to strong environmental pressures that selectively allow only the fittest to remain. A new bacterium, being introduced to the hospital, will encounter a barrage of disinfectants. If it survives them and finds a location in which to multiply, it will frequently be exposed to one or more antibiotics (Fig. 6.11). Contact with the antibiotic has a selective effect by which the resistant members of the microbial population quickly become dominant. Thus our newly introduced bacterium will join the hospital's resident microbial population only if it survives exposure to disinfectants and antibiotics.

Specialized host population   Whereas hospital microorganisms are "tough," their hospitalized hosts are correspondingly weak. The person admitted to a health-care facility is almost by definition in some sort of biological trouble. Restorative procedures such as surgery, intravenous feeding, and strong medications that suppress

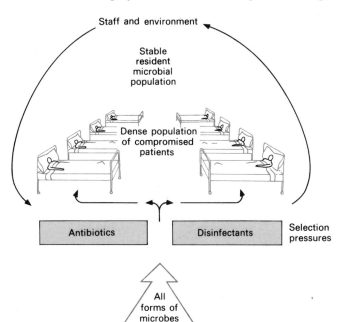

6.11
Microbial Ecology of a Hospital Environment
All microorganisms may enter the hospital, but rigorous selection keeps the total numbers low and promotes the survival of antibiotic-resistant forms. These forms colonize the hospital staff and are rapidly transmitted to each entering patient, sometimes with unfortunate results.
Within hours of admission, each entering patient is often colonized with the resident flora.

body defenses may at least temporarily make the situation worse. The normal human being and his or her flora exist in a precise balance, severely threatened as the patient becomes **compromised.** Hospitalized individuals therefore have a high incidence, perhaps as much as 5 percent, of hospital-acquired infection (6.11). Some of these infections are **exogenous**—the organism came from the environment or someone else's flora—and some are **endogenous**—the patient's own normal flora, no longer controlled by natural competition or the patient's defenses, invaded the patient's tissues. Antibiotics contribute to endogenous infections because they inhibit the antibiotic-sensitive portion of the patient's flora thus giving free rein to the antibiotic-resistant members such as the surface fungi.

An ecological conclusion　　The point at issue is that this distressing situation is inherent in the artificial nature of the hospital environment. Strong and intelligent efforts at hospital infection control can and must minimize its effects, but the problem will never entirely go away as long as an inherently unstable balance between microbe and human beings is maintained within the walls of a high population density institution. A full discussion of the problem and its solutions will be presented in Chapter 28.

Hospital infection control; Chapter 28

The Activated Sludge Process

Sewage　　In developed areas of the world, human waste is disposed of as sewage, a solution/suspension of body wastes, wash water, and waste water from various sources. It is noxious because it contains infectious microorganisms. It is offensive because it smells.

Approaches to sewage treatment　　There are several goals of sewage treatment. One is to render it noninfective. Another is to mineralize all of the organic nutrients to their inorganic state. A third is to minimize its negative aesthetic value. All of these goals can best be served by setting up a specialized ecosystem in which decomposer organisms are provided with the optimum conditions to do their work. Decomposition can occur both aerobically and anaerobically, but it is more rapid and more complete when oxygen is available. In addition, aerobic decomposition does not yield smelly end products. These considerations are built into the design of the activated sludge system Fig. 6.12.

Sewage treatment systems; Chapter 15, pp. 456–458

Function of the activated sludge process　　When sewage is first piped to the treatment plant, it is allowed to settle (**primary treatment**), so that the sludge, containing the majority of the problem material, will separate out. Then the sludge is activated by **secondary treatment.** It is converted to **floccules**—porous, gelatinous aggregates of bacteria, fungi, protozoa, and sludge material. As the microorganisms grow at the expense of the sludge, the floccules grow, then fragment. Active aeration is provided by pumps. An extremely rapid breakdown of the organic wastes occurs with a minimum of smell. After only a few hours, a much reduced amount of residue is removed. This material is odorless, noninfective, and nontoxic; it is frequently dried

TABLE 6.5
Microbial colonization of a baby

	TIME	MICROBIAL POPULATION
PRENATAL	Embryonic and fetal period	All tissues sterile in normal fetus
	After rupture of amnion	Fetus contacts mother's vaginal flora; fluids bathe skin, are inhaled and swallowed
NEONATAL	Eight hours after birth	Skin: Vaginal flora has been or is being eliminated; skin already colonized by normal skin flora from mother and attendants
		Oral cavity: Some streptococci and lactobacilli present
		GI (gastrointestinal) tract: *Lactobacillus* and *Bifidobacterium* present
	1–2 weeks after birth	Skin and oral flora well developed
		GI tract acquiring facultative anaerobes such as *Escherichia coli* and *Streptococcus faecalis* from baby's adult attendants
POSTNATAL	Solid food introduced	GI tract: Strict anaerobes (*Bacteroides* spp.) displace most of previous flora
	Teeth erupt	Oral cavity: Strict anaerobes colonize gum region around teeth

Essentially all of a baby's normal flora is acquired from parents or other human contacts. It is sequentially developed; some species succeed others as the diet changes, while certain groups await the appearance of the appropriate micro-environment.

and sold as fertilizer. Some portion of the sludge is reserved and reintroduced as a starter to get the next batch going.

The ecosystem analyzed As an ecosystem, we can view the activated sludge process as one in which a constant more-than-adequate supply of energy and nutrients is available. Selection of organisms occurs in favor of those that multiply using the supplied nutrients most rapidly in the existing circumstances. This selection competes out the pathogenic organisms that are adapted to other conditions. The useful population maintains itself in the sludge and is continuously reintroduced from batch to batch (Fig. 6.13). Sewage plant workers make no effort to select the organisms present, but simply adjust the conditions so that those organisms that do the job best are favored and undesirables are eliminated.

Growth rate; Chapter 7, p. 205

Instability in this system comes from variations in the water supply. Large amounts of toxic industrial wastes or sudden reductions in the volume of waste will unbalance the ecosystem. This will cause its waste-consuming efficiency to fall off temporarily.

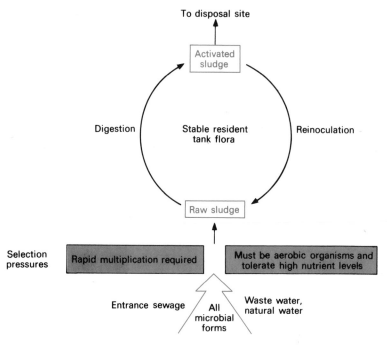

6.12
Microbial Ecology in an Activated Sludge System All forms of microorganisms enter the system, but stringent selection pressures allow only certain forms to persist in the activated sludge while others die out. The desired resident flora is retained by continuous reinoculation of new batches of raw sludge.

People cannot treat sewage; only microorganisms can. The activated sludge process shows us a model of how ecological principles can be made to operate in our favor, and our unavoidable partnership with microorganisms can be made actively beneficial.

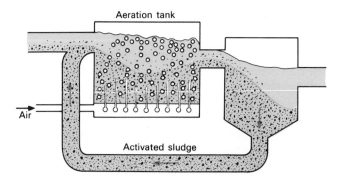

6.13
The Activated Sludge Process The sludge derived from primary treatment undergoes rapid, aerobic decomposition that reduces its bulk. After treatment, some of the sludge may be reintroduced as a starter culture.

SUMMARY

1. The evolution of microorganisms has led to their diversification; they occupy all possible ecological niches. Furthermore, they perform unique environmental roles, carrying out specifically microbial functions. The planetary life-support system for higher plants and animals is completely dependent on microbial contributions.

2. An ecosystem is the sum total of all living and nonliving contributions to an area's life-support network. An ecosystem is stable if it contains diverse organisms interacting to limit population growth and nutrient consumption to sustainable levels. Ecosystem homeostasis is a dynamic equilibrium. It becomes more difficult to maintain as species diversity is reduced. When ecosystems are radically simplified, they become inherently unstable. Undisturbed, ecosystems return to maximum diversity and maximum stability.

3. Symbiotic relationships among species are the major means of establishing homeostasis. These relationships may be positive, neutral, or negative. Frequently they are extremely complex; most are impossible to study accurately *in vitro*.

4. Energy supplies are necessary for the growth and survival of all organisms. Primary productivity depends largely on the photosynthetic capture of sunlight. The captured energy is passed on as organisms consume and are consumed in the food chain. However, because energy exchanges are always inefficient, each succeeding link in the food chain has available to it less energy than the previous one. Microbial decomposition eventually lowers the useful energy content of organic material to zero.

5. Elements move in cycles in which they are converted from inorganic to organic form (fixed) and then returned to the inorganic form (mineralized). Producer organisms fix carbon derived from atmospheric CO_2 into new biomass. This material in turn feeds consumers. The leftovers are used by decomposers so that all carbon returns eventually to the atmosphere as CO_2. Nitrogen, as N_2 gas, is removed from the atmosphere by nitrogen fixers and converted into amino acids, useful to all forms of life. Ammonification, the breakdown of nitrogenous wastes, yields a nitrogen form useful only to microorganisms. Certain autotrophic bacteria use ammonia as an energy source for growth and, in the process (nitrification), oxidize it to nitrate, the prime nitrogen source for plants. Nitrogen returns to the atmosphere as N_2 gas only when certain bacteria metabolize under anaerobic conditions. Sulfur is most available to organisms in the form of sulfate. However, in metabolism sulfur is converted to a reduced form. When decomposition releases sulfur, it appears in the form of free H_2S gas. This is reoxidized by autotrophic bacteria.

6. Fresh and salt waters constitute the largest ecosystems. Main producer organisms are the floating microscopic plankton. Some consumers are microscopic; they become food for swimming animals such as fish and whales. Much life in deeper water is found in the bottom sediments. The benthic organisms break down all undigested detritus completing the cycling of nutrients. Aquatic productivity varies greatly from area to area. Photosynthetic organisms need light, nitrogen, and phosphorus, and are limited by whichever one runs out first. Human activities have enhanced the productivity of some water masses, with unfortunate results. Aquatic microorganisms, over long periods of time, have produced deposits of important minerals and petroleum.

7. Terrestrial environments are an interaction between soil and air. Thirty percent of the globe is covered with soils of varying degrees of fertility. Soil fertility is dependent on the same elements needed in water, and these elements are converted to useful forms largely by microorganisms. A soil's agricultural potential improves with increasing numbers of microorganisms and amounts of humus to feed them. The greatest biological activity is found in the rhizosphere, the soil immediately around the roots of plants, in which mutually beneficial symbiosis among plants and soil microorganisms is important.

8. Animals, including human beings, serve as a specialized microbial habitat. Normal flora microorganisms have a close association with their hosts. The symbiosis is usually mutually beneficial and only rarely results in damage. The cow's rumen is a site for an ectosymbiosis. Microorganisms multiplying in the digestive tract hydrolyze cellullose, converting it first to sugars and then to organic acids that the cow's circulatory system distributes to its tissues. The bacteria multiply, producing amino acids that will also in time be used by the cow. Insects frequently maintain endosymbiotic colonies of bacteria in the gut to nourish them. Our gut flora is ectosymbiotic. Its actual contribution to our well-being is unclear.

9. All human surfaces and orifices have specialized normal floras. The organisms are generally ones that do not flourish elsewhere; they are adapted to the human surface. They provide a nonspecific stimulus necessary for the normal development of our immune systems; they also exert a competitive and exclusionary pressure on other microorganisms that protects us against invasive pathogens.

10. Artificial environments frequently contain a very specialized, highly selected flora. In hospitals, disinfection and antibiotic use selects for a resident flora resistant to these normal control measures. The resident human population of hospitals is unusually susceptible to disease, so hospital-acquired infection is a very severe problem. In this case, the ecosystem structure has some unavoidable negative effects. In the activated sludge system for sewage treatment, the ecosystem continually selects for microorganisms that will decompose sewage at the maximum speed and with the minimum production of unpleasant end products. This system not only gets the job done, but also competes out pathogenic organisms. The residual material is no longer a health hazard.

Study Topics

1. Explain why species diversity is a population-regulating device.

2. Provide additional examples of each of the various categories of symbiosis.

3. Describe the major features of a host–parasite relationship.

4. List the inorganic forms of carbon, nitrogen, sulfur; list as many organic forms as you can (Chapter 2).

5. Ecologists do not pay much attention to cycles of hydrogen and silicon. Why not?

6. Explain the ecological limitations on aquatic productivity. What direct effect do these have on human interests?

7. Why are airborne microorganisms important?

8. Compare the contribution of the gut flora of cows, insects, and human beings to the health of each host.

9. From your reading so far, what are some of the negative effects of antibiotic use in specific environments?

10. Why are artificial environments less stable ecologically than natural ecosystems?

Bibliography

General environmental concepts

Odum, Eugene P., 1979, *Fundamentals of Ecology* (3rd ed.), Philadelphia: W. B. Saunders.

Turk, Amos, et al., 1978, *Environmental Sciences*, Philadelphia: W. B. Saunders.

Specialized environments

Alexander, Martin, 1977, *Introduction to Soil Microbiology* (2nd ed.), New York: Wiley.

Bartha, R., in preparation, *Microbial Ecology*, Reading, Mass.: Addison-Wesley.

Mitchell, Ralph, 1974, *Introduction to Environmental Microbiology*, Englewood Cliffs, N.J.: Prentice-Hall.

Rueble, John L., and Donald H. Marx, 1979, Fiber, food, fuel, and fungal symbionts, *Science* **206:** 419–422.

Sieburth, John McNeill, 1979, *Sea Microbes*, New York: Oxford University Press.

Slanner, F. A., and I. A. Carr (eds.), 1974, *The Normal Microbial Flora of Man*, New York: Academic Press.

7

Studying Microbial Pure Cultures

During the late nineteenth century, microbiology's Golden Age, the major bacterial diseases were identified one after another. When their causative agents were isolated, most of these diseases were quickly brought under control. The early microbiologists under the influence of Koch demonstrated the primary importance of isolating individual species from the mixtures always found in exposed body regions. As we will see, characterizing an organism depends on getting it into pure culture and maintaining it in a hospitable environment.

Isolation of new species is a continuous process. One would think, looking over the microorganisms presented in Chapter 4, that every possible type of protist has already been discovered. Yet through the efforts of a legion of microbiologists, more than 25,000 in the United States alone, major new groups and species are constantly coming to light. New discoveries are usually the result of a directed search, using specialized isolation techniques.

ISOLATION TECHNIQUES

A fisherman goes out on a lake for a day's sport, taking full tackle. Depending on which type of bait or lure he selects, he may catch anything from a sunfish to a salmon, if he's skillful and lucky. Microbiologists may take their samplers to the

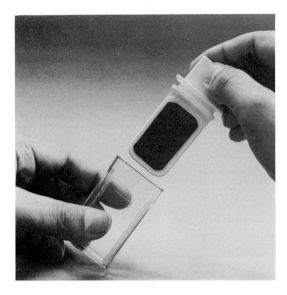

7.1
Sampling Techniques   There are many different ways to obtain microbiological samples. Shown here is one of the simplest, a medium-impregnated paddle. This is supplied sterile and can be dipped into a water sample on location or in the laboratory. The system absorbs a fixed quantity of the sample. As colonies develop on the surface of the medium, they can be counted, giving a quantitative measurement of the number of bacteria per milliliter of sample.

same lake and recover anything from cyanobacteria to anaerobic, spore-forming bacteria, depending on where they sample and what they do with the samples back in the lab (Fig. 7.1). Let us look at some of the procedures used to deal with microbially rich mixed samples from various sources.

Physical Separation

The first step is to physically separate one organism from another so that each will have room to multiply unhindered and can be separately studied. The discovery of agar as a gelling agent made this possible. After 1.5–2.0 percent agar is added to a nutrient medium, the medium is then boiled to dissolve the agar and finally heat-sterilized. Bacteria, fungi, and some other protists can be readily cultivated on this surface. Aseptic conditions are maintained by enclosing the agar in a dish or test tube. When a single cell is placed on an agar surface, its multiplication results in a heap of cells, a **colony,** all derived from a single parental cell. This colony is a **pure culture.** Geneticists call it a **clone.**

Agar; Table 2.8

Considering that soils can contain 2.5 million bacteria per gram, and human intestinal contents, 100 million, it is clear that the microbiologist is usually faced with the problem of the needle in the haystack—how to pick out individuals from among astronomical numbers. Most isolation techniques include a method for severely limiting the number of organisms picked out in a single operation. A standard petri dish full of agar medium cannot readily support the growth of more than about 300 isolated colonies.

The streak-dilution technique   The most commonly used isolation technique, especially in clinical situations, is the **streak plate.** A sterile nichrome loop or probe is dipped into the material to be sampled; it is then applied to the plate so that a very

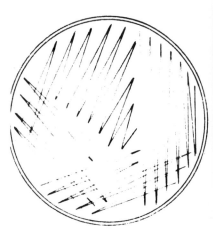
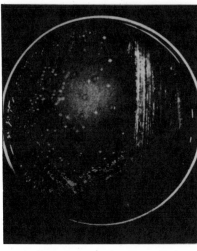

7.2
Streak Plate   This is perhaps the first technique each new microbiologist learns. A skillfully streaked agar plate gives good separation of the different types of microorganisms in a mixture and allows their transfer to pure culture as desired.

small amount is transferred. The **inoculum** is then spread about in a pattern that guarantees progressively thinner application. Following incubation, isolated colonies appear in the thinnest areas (Fig. 7.2). In thicker areas, the colonies run together—are **confluent**—and cannot be readily studied. Since it is rarely possible to guess exactly how many organisms a sample contains, the streak plate allows the worker to prepare, in one plate, a range of dilutions so that isolated colonies will be obtained almost every time. Streak plates are suitable for the isolation of many types of bacteria and fungi; they are not useful for some mycelial fungi and the extremely motile **swarmers** that fan out through the thin film of water on the agar surface. In the clinical laboratory, streak plates are used routinely for the study of throat cultures.

Proteus mirabilis;
Figure 20.4

Pour plates   Although agar melts at almost boiling temperatures, it can be cooled to about 45°C without solidifying. An inoculum can be added to the liquid medium, thoroughly distributed, and the mixture poured into a sterile petri dish to harden. If skillfully done, this process will "trap" single cells in the gelled agar in which they will develop into isolated colonies. Because the whole volume of the agar is used rather than the surface, a pour plate can contain more colonies than a streak plate. Also, because all colonies are roughly equidistant, they may be counted. The pour plate may be used in water analysis and urinalysis to determine the numbers of colony-forming microorganisms in a given volume of sample. Some disadvantages are that strict aerobes will grow only on the surface, and it is difficult to sample submerged colonies for further study.

Liquid dilution   A liquid inoculum, such as a drinking water sample, can be sequentially diluted 1:10, 1:100, or more in sterile water, then measured quantities can be transferred to sterile media. This technique can be used to count the numbers of bacteria/mL in the sample. Carried to greater lengths, it may be used to establish pure cultures in broth.

Slide culture   Microscope slides can be coated with medium and inoculated with fungi. The inoculated slide is incubated in an aseptic environment. When growth appears, it is particularly convenient to study. This technique is useful for the recovery and identification of human dermatophytes (Fig. 7.3).

Clean or media-coated slides have been used in marine microbiology to isolate organisms, when the slides are dipped directly into the water. Since open sea water contains relatively few organisms—sometimes less than one per mL—there is no need for dilution. Microorganisms simply attach directly to the slides and grow *in situ.*

Enrichment Culture

In employing the enrichment culture technique the microbiologist places a sample in a closed container and provides conditions that favor the emergence of one species or group of species. The medium may be continuously circulated. All the microorganisms introduced grow to the limits of their ability to adapt to the adjusted ecosystem. The most favored organism, often a minor component of the original microbial population, gradually becomes dominant. Competition progressively eliminates those organisms that multiply less rapidly. Over a period of time the population becomes simplified. The surviving organisms can then be isolated for individual study.

Microorganisms that decompose oil in nature are few in number. Thus spilled petroleum persists in water and creates severe and prolonged ecological disasters. Random plating of sea water would not be likely to reveal or recover these very uncommon species. However, an enrichment culture with oil residues as the major

7.3
Slide Culture   A slice of sterile agar medium is aseptically sandwiched between a sterile glass slide and cover slip. A fungal inoculum is introduced at the center of each side of the agar block. Then the sandwich is sealed with sterile wax to prevent its drying out. As the fungi grow, their mycelial and reproductive structures can be readily observed through the cover glass "window."

energy-yielding nutrient will probably develop a petroleum-decomposing flora. The oil-digesting organisms can then be isolated and their potential in oil-spill control evaluated.

Types of Media

Selective media   Selective agents are chemicals that inhibit most microorganisms, allowing only certain types to grow. When sodium tetrathionite is added to media for the examination of fecal samples, it inhibits the normal gut flora but permits the growth of the resistant pathogens *Salmonella* and *Shigella*. Other selective agents include salts, dyes, antibiotics, and specific enzyme inhibitors. Selection vastly improves the microbiologist's chances to find important minor members of a complex microbial ecosystem.

Differential media   Media may include pH indicators or color reagents. A colony carrying out a particular reaction will change the pH of the adjacent medium, and the pH indicator will change color (see Table 7.3) After incubation, the positive colony may itself be a different color or have a visible zone of different colored

Box 7.1

First, Catch Your Hare . . . .

There is an old recipe for jugged hare that begins, "First, catch your hare," putting first things first. Many of us can see why it is useful—even important—to study individual microorganisms, but it is a little hard sometimes to see why there should be so much fuss made about which medium and what formulation is used. However, primary isolation is the equivalent of catching the hare, and it depends entirely on having the right medium.

When the initial Legionnaire's disease outbreak occurred, months elapsed before the causative bacterium was identified, for lack of a suitable medium. It was finally found by inoculating suspect samples into guinea pigs. Later, a highly enriched medium was developed that would support the organism once it was isolated. *Legionella* could then be isolated from normally sterile specimens, such as pleural fluid in which it was the only organism. However, isolation from the sputum or lung tissue—clini-

cally more important specimens—was still impossible. Three years went by before a workable semiselective medium was developed to allow its primary isolation from fluids and tissues containing other organisms.

The problem is that *Legionella pneumophila* is a highly fastidious, slow-growing organism, usually found in a mixed flora with numerous much less fussy species. When all are cultured, without selection, *Legionella* will be overgrown.

A charcoal-yeast extract agar supplemented with two antibiotics vancomycin and polymixin, has recently been proposed. The growth of enterobacteria and yeasts is sufficiently inhibited so that *Legionella* could be recovered after a two-week incubation period. This formula is still far from perfect, but it will be worked on and improved. The ultimate goal is to be able to isolate this bacterium from soil and water, its apparent natural source.

media around it. Negative colonies will remain undistinguished. Blood agar is a rich nutrient medium to which whole blood has been added. When a colony produces hemolysins, enzymes that attack eucaryotic cell membranes, it will have a visible zone of destroyed red blood cells around it. The red cells are thus the indicators.

Hemolysis in identifiation; Chapter 17, p. 502

Culture Maintenance

By one or a combination of these isolation techniques, the microbiologist obtains pure primary cultures of desired organisms. These can be maintained and subcultured indefinitely for identification and study and are called **stock cultures. Subcultures** are derived from primary cultures by aseptic transfer of a portion of the growth. Watchfulness is needed to prevent introducing extraneous organisms.

Active microorganisms cannot survive indefinitely in a closed area. They must be transferred periodically to a fresh medium or they will die out. Another storage problem is that cultures may develop altered characteristics on prolonged subculture.

Preservation in an inert state is preferable. Cultures may be placed in a state of suspended animation by freeze-drying or **lyophilization** (Fig. 7.4). In this procedure, small volumes of an active culture are placed in glass vials. These are flash frozen to a very low temperature. Then the cells remain firmly frozen until all the water is gone. Then the glass vials are sealed shut under vacuum. Lyophilized cultures may be stored indefinitely at room temperature because cellular activity is completely arrested in the absence of water.

Compare with cold effects in microbial killing; Chapter 25, p. 731

Cultures may also be stored for long periods under sterile mineral oil. Simple refrigeration is not reliable for long periods for most organisms.

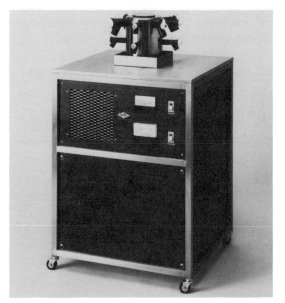

7.4

Lyophilizing Apparatus A sample to be lyophilized is first flash-frozen in a dry-ice acetone bath or by other means. It is swirled while freezing to obtain a thin shell of frozen material on the surface of the container. Then the vessel is attached to one of the small ports on the upper portion of the machine. A powerful vacuum pump causes the water in the sample to sublime, that is, to convert from the solid frozen state directly to the gaseous state. The water vapor is then pulled away. The specimen dries without thawing, and the dry powder remaining is sealed in the vessel while still under vacuum.

Stock culture collections   Reference collections of known microorganisms are maintained by many research facilities and universities. The American Type Culture Collection (ATCC) is the most comprehensive. It maintains stock cultures of all generally accepted species and subspecies of bacteria and many other types of microorganisms. The original isolate or a later, completely typical isolate, serves as the **reference culture.** All later isolates must be compared to it to verify their identification. Reliable stock cultures can be purchased from the ATCC or other stock culture collections.

PROVIDING THE OPTIMAL PHYSICAL ENVIRONMENT

The physical and inorganic growth environment must be adjusted to approximate the ecological realities of a microorganism's native habitat. After all, polar bears require air conditioning to live in zoos in hot climates. One cannot remove a bacterium from a hot sulfur spring and expect it to do well in the 37°C incubator designed for our body flora.

Temperature Ranges

In nature, microorganisms survive and multiply at temperatures from − 10°C to as high as 90°C (Table 7.1). Each individual species multiplies over a range rarely exceeding 30°C. Three major groups have been identified. The **psychrophiles,** or cold-loving organisms, multiply within the range − 10°C to 20°C. Their natural habitats are the colder waters and soils; most marine organisms are psychrophilic. Since they do not multiply at human body temperatures, none is pathogenic for human beings. However, some are important pathogens of fish and other aquatic animals. Furthermore, psychrophiles multiply at refrigerator temperatures (4–10°C) and cause substantial economic loss due to spoilage of fresh foods.

TABLE 7.1
Temperature ranges for growth of selected bacterial species

GROUP	ORGANISM	MINIMUM, °C	MAXIMUM, °C
Psychrophiles	*Bacillus globisporus*	− 10	22
	Vibrio marinus	< − 5	22
	Xanthomonas pharmicola	0	40
Mesophiles	*Escherichia coli*	10	42
	Neisseria gonorrhoeae	30	40
	Hemophilus influenzae	25	40
Thermophiles	*Lactobacillus delbruckii*	20	53
	Bacillus coagulans	25	68
	Bacillus stearothermophilus	35	75

Source: Adapted from R.Y. Stanier; E.A. Adelberg; and J. Ingraham, 4th ed., 1976.

Mesophiles have growth optima in the middle of the temperature range. An organism is officially classed as a mesophile if its optimum growth temperature falls within the range 20°C to 50°C. No organism, of course, grows equally well over the whole range. Mesophiles are isolated from warmer soils and waters, animal wastes, and from the bodies of animals. Our normal flora and pathogens are mesophiles, growing best over a narrow range from 35–40°C. Their growth usually noticeably declines as the temperature rises to 40°C, or 104°F.

Thermophiles grow between 40°C and 90°C, with optima usually between 50°C and 60°C. They are isolated from hot environments—areas with volcanic activity, heaps of decomposing material, and heat-producing industrial processes. They are capable of causing rapid spoilage in cooked foods that are held at warm temperatures.

Physiological mechanisms, especially enzyme function, determine an organism's temperature restrictions. Enzymatic activity increases as the temperature, and thus the kinetic energy of molecules, is raised. Beyond a certain temperature, increasing random molecular movement disrupts the weak bonds that maintain the enzyme's secondary and tertiary structures. The enzyme is then denatured and becomes nonfunctional. Weak bonds also dissociate at very cold temperatures. However, the physical stability of enzymes from different organisms varies. Psychrophiles have enzymes with maximum activity rates at low temperatures and rapid inactivation at 30°C. Mesophilic enzymes are maximally active at around 35°C and are denatured as the temperature approaches 50°C. Thermophilic enzymes are maximally active at 50–60°C; they resist denaturation at the temperature of boiling water. Adaptation to extreme environments requires the genetic capacity to produce highly specialized enzymatic proteins.

Enzyme denaturation; Chapter 2, p. 53

Osmotic Pressure

All cells gain or lose water by osmosis depending on the relative concentrations of dissolved substances within and without the cell. When the exterior environment is hypertonic, or contains more dissolved materials than the cell's cytoplasm, it will lose water to the environment. The cell will shrink; it will become dehydrated. The loss of water reduces, then stops, cellular activities.

When the exterior environment is hypotonic, more dilute than the cytoplasm, water enters the cell. The cell swells; if the influx of water exerts greater pressure than the cell wall's mechanical strength can sustain, the cell lyses. Less disastrous stresses reduce the cell's normal rate of function.

The osmotic pressure of a fluid depends on the concentration of dissolved particles, both inorganic ions—**crystalloids**—and macromolecules—**colloids.** A cell can do little about the external environment. However, the cytoplasmic osmotic pressure may be balanced to that of its environment by varying the intracellular potassium concentration.

Salinity is the term used to indicate the concentration of dissolved ions in a fluid. Sodium and chloride ions predominate in natural waters. Organisms that tolerate a wide range of salinity are **euryhaline.** Organisms, usually marine, that

Marine microbiology; Chapter 6, pp. 169–171

7.5
The Limitations of Pickling   Although preserving with salt or vinegar, or a combination of these, is useful, many molds can grow in a saline or acetic environment. Pickles, especially if they are exposed to air, will spoil just as these have.

tolerate only slight fluctuations, are **stenohaline.** The true **halophiles** live in environments containing 15–30 percent dissolved salts. The structural adaptations that permit this have already been noted (Chapter 3).

Microorganisms of the human gut do not tolerate elevated salt levels well, a characteristic that contributes to their rapid disappearance from sewage effluents released into the ocean. However, the skin flora, exposed to the salts provided by sweat, are very salt-tolerant. The staphylococci tolerate 10 percent NaCl, a level almost high enough to classify them as halophiles.

Bulk quantities of NaCl, alum, and other salts have traditionally been used to protect meats and vegetables from microbial decomposition. Similarly, high concentrations of sugar preserve fruits. Osmotically preserved foods, however, are eventually degraded by growth of remaining tolerant organisms (Fig. 7.5). Salting and preserving by sugar have largely been replaced or supplemented by more absolute means of eliminating microorganisms, such as canning.

pH of the Environment

The pH range of natural waters is surprisingly wide. Drainage waters from mineral deposits and abandoned mines may be between pH 2 and 3. Clean, fresh-water sources average about pH 6, the oceans, about pH 8. The alkali lakes, containing high concentrations of basic salts, may have pH values as high as 10. Selected types of microorganisms can initiate and continue growth across this entire pH range.

Chemiosmosis; Chapter 8, pp. 226–228

The pH within the microbial cell remains at about 7.5 independent of the external pH. Thus a cell in an environment other than pH 7.5 experiences a pH gradient across its membrane. In many organisms, growing at acid pH, this gradient is a source of energy used to generate flagellar movement and/or ATP. Ability to tolerate extreme gradients may lie in the structure of the membrane.

Microbial metabolic activities can cause accumulation of acidic end products. Alternatively, the release of ammonia in the breakdown of proteins may raise the pH. These pH changes may arrest the microorganism's growth eventually. The sulfur-oxidizing autotroph *Thiobacillus* yields sulfuric acid as a metabolic end product; it can readily lower the pH of its environment to 1 or less and still maintain active growth.

Sulfur cycle; Chapter 6, pp. 167–168

In preparing media for a particular organism, two aspects of the pH issue must be considered. The medium pH should be similar to the natural habitat. If the organism tends to modify the starting pH to a degree that inhibits its growth, then buffer substances must be included. Amino acids and proteins are excellent natural buffers because of the amphoteric nature of amino acids. If these are not present, then inorganic salt mixtures or synthetic organic substances must be incorporated.

Gaseous Atmosphere

Although most microorganisms in their natural setting grow in the presence of either gaseous or dissolved air, some do not. They are adapted to areas with altered concentrations of oxygen, carbon dioxide, and other gases. These atmospheres must be recreated in the lab by special means.

Strict aerobes   These organisms require a continuous supply of molecular oxygen for growth. They are not capable of extracting sufficient energy from the environment for growth without the use of oxygen-dependent metabolism. However, they frequently survive for periods of time without oxygen and they usually grow at reduced rates when oxygen is limited. On solid media, oxygen is available. However, when grown in liquid, these organisms tend to rapidly exhaust dissolved oxygen supplies. Liquid cultures should be aerated or shaken to obtain optimum growth.

Facultative anaerobes   These organisms have a versatile metabolism that allows them to multiply both with and without oxygen. However, they use energy sources much more efficiently when oxygen is present, so aerobic growth is more rapid. Facultative organisms can make the shift from aerobic to anaerobic function very quickly.

Microaerophiles   Microaerophiles prefer an atmosphere with a lowered oxygen concentration and increased amounts of carbon dioxide. This type of atmosphere is found in the tissues of higher animals. As tissues carry out respiration, they use up the dissolved oxygen supply in body fluids, replacing it with waste CO_2. Most microaerophiles are members of animal flora. There is a simple way to provide the right gas mix in the lab. The cultures are placed in a gallon jar, a candle is inserted and lighted, and the lid put on. As the candle burns out, most of the oxygen will be consumed and replaced by CO_2. More accurately controlled gas mixtures are obtained in special incubators supplied with a bottled gas mixture. This type of system is frequently used for incubation of primary throat cultures, since it promotes the development of the streptococci. Microaerophilic conditions are also needed to isolate the gonorrhea organism.

Neisseria gonorrhoea; Chapter 21, pp. 628–630

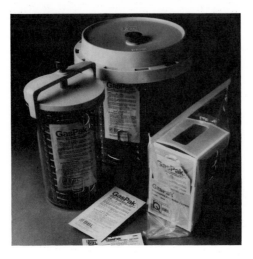

7.6

Anaerobic Incubation   Plates or tubes to be incubated anaerobically are placed in this vessel. Foil packets containing a hydrogen-generating system are activated, and the jar is then sealed. A catalyst in the jar's lid promotes the formation of water vapor from the oxygen in the vessel and the generated hydrogen. This reaction removes all oxygen from the air sealed in the jar.

Strict anaerobes   These organisms do not grow in the presence of molecular oxygen. Some can tolerate exposure to small amounts of O_2, others are almost instantaneously killed by minute traces. In all cases, O_2 is a toxic substance that inhibits growth because the enzyme systems of the bacterium cannot assimilate it in harmless forms.

It is not an easy technical feat to exclude oxygen totally from a microbial culture. At present, most anaerobic cultures are incubated in airtight jars or incubators (Fig. 7.6). Either a catalytic reaction removes oxygen from the interior or an oxygen-free atmosphere is introduced. For successful recovery of strict anaerobes from nature, anaerobiosis must begin with the sampling technique. As the sample is taken, transported to the lab, and placed in culture, it must **at no time** come in contact with the air. Correct procedures are of critical importance, for example, in the attempt to isolate anaerobic pathogens from human wound infections. If they are not followed, the sample may arrive at the laboratory with all the anaerobes in a nonviable state. Only the clinically less significant aerobes will survive. Such mishandled cultures will yield no useful information; worse, they may be actually misleading. Anaerobic techniques are examined in detail in Chapter 23.

Anaerobic culture;
Chapter 23, pp.
673–674

Light Sources

Normal cultivation of photosynthetic bacteria, cyanobacteria, and algae requires a source of light. Although photosynthesis operates primarily on the energy provided by the red and blue portions of the visible spectrum, white light sources are usually employed to simulate natural lighting.

Light exposure has little effect on nonphotosynthetic microorganisms unless the source is direct sunlight. The ultraviolet component of direct sun exposure, without intervening glass, is lethal to many bacteria and fungi. Light exposure may induce pigment formation, a characteristic used to identify certain mycobacterial groups.

Atypical mycobacteria;
Table 19.2

PROVIDING THE OPTIMAL NUTRITIONAL ENVIRONMENT

Chapter 6 introduced the ecological concepts of energy flow and nutrient cycles in the intact environment. Each microorganism is adapted to use specific groups of nutrients available in its particular niche. The nutrient environment must be recreated or improved on when the organism is lifted from its habitat into the artificial environment of the laboratory pure culture.

Carbon Sources

Carbon compounds are needed by cells for two purposes—to assemble into new organic compounds and, in the case of heterotrophs, to metabolize as energy sources. Some microorganisms, such as *E. coli*, are capable of fully satisfying both requirements with a sugar such as glucose (Table 7.2). Members of the bacterial genus *Pseudomonas* are noted for their ability to use any one of more than a hundred simple organic compounds as their sole carbon and energy source. This implies an amazing array of metabolic abilities; organisms such as these are enzymatically constructing all cell components from glucose and inorganic salts.

Proteins can also be metabolized for basic growth needs, but carbohydrates and lipids are preferentially used by many organisms. Microbiological media frequently provide sugars and starches.

Nitrogen Sources

Some microorganisms can derive nitrogen atoms for cellular proteins and nucleic acids by reducing nitrate ions to amino groups, or by the direct incorporation of ammonium ion. The acquired nitrogen can then be used to construct all needed nitrogenous compounds. Other organisms are enzymatically less well endowed and require organic nitrogen. They may need one, several, or all of the 20 amino acids. Highly **fastidious** organisms may need practically every cellular building block molecule handed to them in order to survive.

TABLE 7.2
Comparison of a minimal defined medium and a complex medium

GLUCOSE MINIMAL MEDIUM		NUTRIENT BROTH (COMPLEX)	
Glucose	5 g	Beef extract	3 g
NH_4Cl	1 g	Peptone	5 g
K_2HPO_4	1 g	Water	1 liter
$MgSO_4 \cdot 7\,H_2O$	10 mg		
$CaCl_2$	10 mg		
Trace elements (Mn,Mo,Cu,Co,Zn, etc.)	Less than 1 mg each		
Distilled water	1 liter		

Hydrolysis; Chapter 2,
p. 45
In bacteriological media, hydrolyzed proteins are regularly used. The native protein is predigested into a mixture of small peptides. This renders them easily soluble and also allows their ready assimilation by the cell. Such substances (peptones, tryptones, or trypticases) are often supplied both as an amino nitrogen source and as the carbon-energy source.

Phosphorus Sources

Practically all microorganisms utilize inorganic phosphate ion. Nucleic acids may be hydrolyzed to nucleotides that provide an additional source.

Sulfur Sources

Amino acid structures;
Chapter 2, p. 51
Many microorganisms assimilate sulfur for the synthesis of cysteine and methionine by the direct reduction of sulfate ion. Others, unable to carry out these steps, need to be supplied one or both of these amino acids.

Inorganic Ions

Sodium, potassium, and chloride ions must be supplied according to the organisms' salinity needs. Smaller amounts of metallic ions, the ones dieticians refer to as minerals, are also of critical importance. Magnesium, manganese, copper, calcium, zinc, and others are required for all cellular life. In defined media, these metals are added in specific amounts. In complex media, sufficient amounts creep in as impurities in previously mentioned constituents.

Iron deserves particular mention. Although the quantity needed is tiny, any condition that reduces iron's availability may severely limit microbial growth. It is essential as a cofactor in microbial energy generation systems.

Vitamins

Many microorganisms synthesize vitamins themselves and thus do not require an exogenous supply. Others need one or more of these growth factors. When vitamins are required, the amount of growth a culture achieves is limited by the quantity of vitamin provided. Thus accurate measurement of microbial growth in a medium makes possible accurate determination of its vitamin content. This technique, the **microbiological assay,** is very widely used to quantitate vitamin preparations. Its advantage over straight chemical assay is that it specifically assays only biologically active vitamin molecules.

In culturing microorganisms, a crude mixed vitamin source is often included. Yeast extract, prepared by extracting the water-soluble cytoplasmic components of yeast cells, is most commonly used.

Selective and Differential Substances

In addition to the basic nutrient profile, media used for primary isolation, identification, or environmental screening frequently contain the selective and differential substances mentioned earlier (Table 7.3).

TABLE 7.3
The composition of mannitol salt agar

COMPONENT	AMOUNT	PURPOSE
Beef extract	1 g	Carbon, nitrogen, and energy source
Protease peptone	10 g	Carbon, nitrogen, and energy source
NaCl	75 g	Selective agent; allows only salt-tolerant growth
Mannitol	10 g	Differential agent; sugar fermented by *Staphylococcus aureus* but not by harmless organisms
Agar	15 g	Solidifies medium
Phenol red indicator	0.025 g	Turns yellow around colonies that have produced acid by fermenting mannitol
Water	1 liter	Solvent

A selective and differential formula used for the isolation and preliminary identification of *Staphylococcus aureus*.

Enrichment

Particularly for the cultivation of the normal and pathogenic symbionts of human beings, best results are obtained if tissue or tissue fractions are offered. Whole or hemolyzed defibrinated blood, blood serum, or specific blood factors are routinely added to media used for isolation and maintenance of human bacteria. For example, media for the isolation of *Hemophilus influenzae* contain hemolyzed blood, plus added X-factor (hemin) and V-factor (the coenzyme NAD). Chopped meat, beef brain/heart extract, and others are occasionally used.

H. influenzae; Chapter 19, p. 555

GROWTH AND MULTIPLICATION IN PURE CULTURES

The isolated pure culture may be used for a wide range of studies. Most fundamental are those that determine its growth rate under differing sets of environmental conditions.

Quantitation of Cell Numbers

In the quest for suitable counting methods, there are two major requirements—accuracy and convenience. Numerous techniques, none perfect, are in use. Some determine the number of cells only, without being able to distinguish between the living and the dead. Others assess viability—active cellular metabolism and reproduction. These fail to enumerate organisms which may be present, very much alive, but unable to reproduce under the preselected conditions.

Counting cells   The **direct count** involves placing a suspension of microorganisms or a bacteria-containing fluid such as milk into the sample well of a special microscope slide. A calibrated cover slide engraved with guidelines is used. The observer can count all cells seen and, by a simple calculation, figure out organisms present per mL of the test fluid sample. The observer cannot tell reliably if the organisms are alive.

A suspension of organisms can be chemically analyzed to determine its **total protein** content. These data can be correlated with other observations to give an estimate of the cell number. The method is quite tedious.

A suspension of organisms can be dried in an oven, then weighed with an analytical balance to give a **dry weight**. This method, also time-consuming, has many sources of inaccuracy.

Most laboratories use a **turbidity** method for routine quantitation. A suspension of cells is placed in an optically clear test tube called a **cuvette**. The cuvette is then placed in the path of a beam of monochromatic light. Light rays hitting cells are deflected or absorbed and do not reach a photoelectric detector. The more cells there are, the less light is transmitted by the suspension. Electronic **colorimeters, spectrophotometers,** or **nephelometers** are employed to take these readings. Turbidimetry is extremely convenient; competently done it is quite reproducible. Both living and dead bacteria deflect light. Thus a single reading gives no information about viability. However, growing cultures show steady increases in optical density, and the turbidimetric method is effectively used for following growth of a culture.

Assessing cell activities   The **plate count** is used to determine the number of cells capable of reproducing. Measured amounts of three or more dilutions of sample are mixed with cooled, melted agar medium and incubated. The resulting colonies are counted and reported as **colony-forming units** (CFUs). A CFU may be either a single cell or a clump.

Alternately, the sample may be pulled by a vacuum through a microbial filter (Fig. 7.7). Microorganisms are retained on the filter surface. The filter, placed onto the surface of the appropriate agar medium, supports the growth of all colonies that result; these can be readily counted.

Cellular ATP assay provides an extremely precise measurement of the amount of metabolizing cell mass present. One very sensitive method makes use of the fact that extracts from the luminescent organ of the firefly emit light of an intensity exactly proportional to the quantity of ATP provided. These extracts contain **luciferin** (the light-emitting substance) and **luciferase** (the firefly enzyme that catalyses the coupled breakdown of ATP and light emission). Sensitive photoelectric equipment records the intensity of light emission.

Photoluminescence in nature; Chapter 6, p. 159

Gas exchange techniques quantitate cellular activity by following the amount of oxygen taken up or the amount of radioactive CO_2 given off. In the experiments used to search for life on Mars, sensitive instruments were positioned to detect $C^{14}O_2$ releases from radioactive substances mixed with the Martian soil. The results were negative.

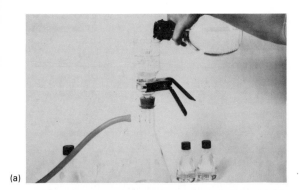

(a)

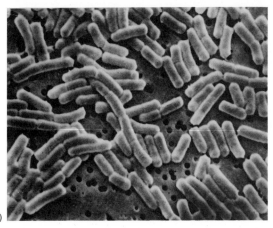

(b)

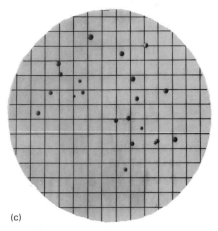

(c)

7.7

Membrane Filtration   This technique may be used either for sterilizing fluids or for collecting microorganisms from samples. (a) Solutions of certain biochemicals such as urea will be degraded if exposed to the heat of the autoclave for sterilization. Here the prepared solution is poured into a holder above the membrane filter, which is held in place by a metal clamp. The hose on the bottom flask goes to a vacuum line. The vacuum causes the fluid to flow through the filter into the lower vessel, where it is aseptically collected. (b) Retained bacteria on the filter surface will develop into colonies (c) if the filter is placed on top of a filter paper pad moistened with an appropriate medium.

Automated quantitation   A new wave of research and development has produced automated, computer-assisted systems designed to instrumentally detect and quantitate bacterial growth in normally sterile clinical samples such as blood, cerebrospinal fluid, and urine (Fig. 7.8). They are also being adapted to detect growth or inhibition of a bacterial isolate in the presence of antibiotics. These systems are currently receiving large-scale clinical trials. Although they are very costly, after further testing and refinement they may eventually replace most manual microbiology techniques in large-volume routine applications.

Cell Division

In unicellular procaryotes, the usual reproductive process is **binary fission,** a division into two (Fig. 7.9). It appears quite simple and orderly, but we are far from understanding all of the details. Let us look briefly at what happens to three cell components, the cytoplasm, the nucleoid, and the cell wall or envelope.

Metabolic activity in the cytoplasm is continuous whenever the cell is growing. There is constant accumulation of inorganic ions, small organic molecules, and

7.8

The Automicrobic System for Automated Urine Culture   A small volume of a urine specimen is automatically distributed into a number of tiny receptacles (wells) in a plastic holder. Each receptacle contains a different highly selective medium that will allow the growth of one group of the organisms found in urinary tract infections. Any isolate will grow only in one or a few wells. After incubation, the receptacles are automatically scanned for turbidity, the results analyzed by a computer program, and the identification printed out on a card.

macromolecular species such as enzymes, RNA molecules, and ribosomal subunits. Rising intracellular concentrations of certain substances may signal the onset of cell division. Roughly speaking, the cytoplasmic contents will double prior to division.

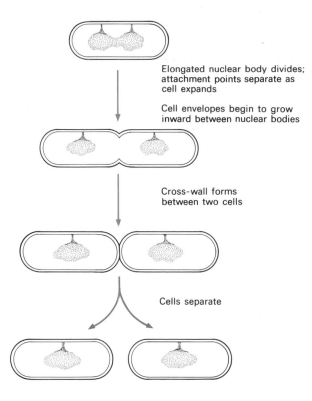

Elongated nuclear body divides; attachment points separate as cell expands

Cell envelopes begin to grow inward between nuclear bodies

Cross-wall forms between two cells

Cells separate

7.9

Binary Fission   Cell division occurs at the site between two nuclear bodies. Ingrowth of the new wall material progresses to separate the two daughter cells. If well nourished, the cell continuously enlarges during the cycle.

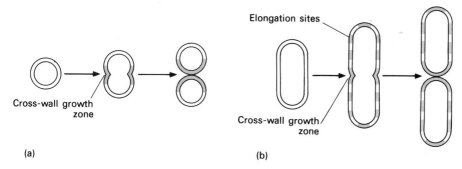

7.10
The Areas of Cell Wall Growth   (a) In a spherical coccus, new cell wall is added at
the central zone where the cross-wall forms during cell division. Each daughter cell
keeps the coccus shape. (b) In a rod-shaped bacillus, new cell wall is added by two
mechanisms. A large area of new wall originates from the added cross-wall.

Bacterial genetic material is contained in a ring-shaped DNA double helix, the
nucleoid. The ring is enzymatically duplicated once in each division cycle. The nucleoid is attached to the cell membrane at one point; the new copy becomes attached
at an adjacent point. Cell expansion spatially separates them. It appears that the cell
can divide itself only in the region between two nucleoid attachment sites. Cell walls
must expand if the cell is to grow (Fig. 7.10). Expansion requires that the peptidoglycan layer be enzymatically opened and new polymer added. In rod-shaped bacteria this occurs at multiple points on the cell surface, leading to the elongation of a
cylinder. In cocci, it occurs only in a central zone. Both rods and cocci separate into
two daughter cells by the progressive formation of a double-thickness layer of cell
wall material (Fig. 7.11). When complete, the two layers pull apart. The daughter
cells may completely separate or remain aggregated in pairs, tetrads, clusters, or
chains.

> Bacterial nucleoid;
> Chapter 3, p. 85

> Peptidoglycan; Chapter
> 3, Fig. 3.11

   The three aspects of cell division are not sequential; rather, they are synchronized. Cells in culture usually divide on their own individual timetable. When one
observes a slide of an actively growing culture, the cells will be different sizes, since
some have just divided, while others are about to divide. Special techniques can induce all cells to divide at about the same time, causing **synchronous growth.**

Growth Kinetics

Binary fission has an inevitable result. Every division doubles the number of cells. In
an actively growing population, few cells die. Growth of this sort is expressed
mathematically as a **geometrical progression.** This is a number series in which each
number is derived from the one previous by multiplying it by a certain factor, an
exponent. In a **binary** series the multipier is 2. Thus increase follows the (infinite) series: 1,2,4,8,16,32,64,128,256,. . . . Note that each increase equals the effect of all
previous increases. This is exponential growth. When a single bacterial cell is placed
in a suitable sterile medium, cell numbers will increase exceedingly rapidly.

> See also surface to
> volume ratio, Chapter
> 1, pp. 11–12

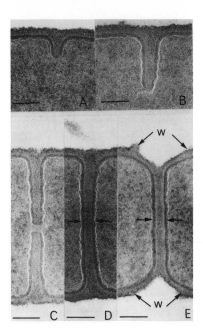

7.11
Cross-wall Formation   This sequence of TEM photographs simulates the formation of the cross-wall between two daughter cells. The wall (w) grows inward from both sides (A–C), and finally closes (D). As the cell begins to tear apart, small tabs of wall material remain to mark the point where ingrowth originated (E).

Generation time   Under controlled conditions, one can determine the length of time that elapses between the time a new cell arises until it in turn divides to give rise to two cells. This is the **generation time** or doubling time (Table 7.4). An average generation time of a heterotrophic mesophile is 20 minutes. However, intervals as short as 12 minutes have been seen. This means that the total number of individuals in such a population will double in 12 minutes. Minimum generation time is determined by the genetically fixed rates at which different species can assimilate nutrients. However, the time needed for doubling can be extended by any environmental change that makes circumstances less favorable—altered temperatures, less favored nutrients, or shortage of a key element. Soil and water organisms persist under adverse circumstances by their ability to adapt to generation times of days or weeks. In nature, organisms rarely approach their maximum growth rate.

In exponential growth, cell numbers rapidly become very large. Thus they are more conveniently expressed as powers of ten. It is more convenient to say that a certain culture contains 2.5×10^8 cells/mL than to write out 250,000,000. Bacterial growth is plotted graphically on semilog plots.

Closed System Growth Curves

When an inoculum is introduced into a test tube or flask filled with fresh, liquid, nutrient medium, the resulting population dynamics are predictable. They are also totally divorced from the realities of population growth in a balanced natural habitat. Consider that the pure culture in a test tube has no competition, that nutrients are not recycled, that wastes are not removed.

TABLE 7.4
Maximum growth rates

ORGANISM	DOUBLING TIME
Bacillus stearothermophilus	8.4 min
Escherichia coli	21 min
Vibrio marinus	81 min
Mycobacterium tuberculosis	360 min (6 hr)
Nitrobacter agilis	1,200 min (20 hr)

Source: Data adapted from R.Y. Stanier; E.A. Adelberg; and J. Ingraham (1976)

At their optimum growth temperature, and during the log phase of growth, these organisms attain maximum growth rates with the average cell completing a round of division in the indicated time

Box 7.2

Growth Rate and Pathogenesis

Cystitis—infection of the bladder—is very common in adult females. By far the commonest infecting agent is the Gram-negative bacillus *Escherichia coli*. This bacterium, like the other microorganisms causing cystitis, is presumed to originate in the bowel and become transferred to the urinary tract from there. It has not been clear why *E. coli* is so clearly better adapted to the role of bladder pathogen than its other enterobacterial cousins.

In an effort to answer the question, a group of investigators built an artificial model of the human lower urinary tract. It was supplied with sterile urine, flowing through at the physiologically correct rate. Pure cultures of various bacteria previously isolated from urinary tract infections were obtained and their generation time in urine measured.

The superior adaptation of *E. coli* became clear. In urine it showed generation times of 21–22 minutes. Other less common urinary tract pathogens (*Serratia marcescans*, 34–35 minutes, and *Pseudomonas aeruginosa*, 50 minutes) were much slower growing.

When mixed cultures containing *E. coli* and another organism were introduced into the artificial bladder model, *E. coli* outgrew 16 different partners 50 to 100-fold in 24 hours.

Thus if *E. coli* alone were introduced into the bladder, it would be able to grow and multiply at a rapid rate that could keep it from being "washed out" or overwhelmed by host defenses. If it were introduced as one of a group of contaminants, which is more likely, its more rapid growth rate would explain its greater likelihood of persisting.

Source. J. D. Anderson; F. Eftekhar; M.Y. Aird; and J. Hammond, 1979, *J. Clin. Microbiol.* **10:** 766–771.

In a laboratory pure culture we have a simplified state that cannot be compared to an ecosystem. It is a closed system because neither materials nor energy flow through it.

The growth curve has four distinct phases (Fig. 7.12). When first introduced to the fresh medium, the inoculated cells adjust to the new environment, primarily by producing new enzyme molecules. Only minimal growth and essentially no cell division occurs during this **lag phase.** The lag phase is further extended if the inoculum is added to refrigerated medium, if the cells are in poor physiological condition, or if they are being transferred from a very different medium. It is very short if log phase cells are used to inoculate prewarmed medium of a similar formula.

Once adaptation is completed, growth and cell division begins. If cell numbers are determined at frequent intervals and plotted on a semilog plot, they will generate a sharply rising straight line indicating exponential growth. This phase is called the **log** or **exponential phase.** Calculations made by analyzing the slope of the straight portion of the line yield the generation time achieved.

As cell numbers increase geometrically, so also does nutrient consumption and waste material production. In a closed system, this inevitably brings exponential growth to an end. The rising straight line tails off into a horizontal line on the graph.

Stationary phase is the term for this horizontal portion of the graphed sequence of events. In stationary phase, cell numbers do not change. Some cells continue to divide using the dwindling nutrient supply. The equal number of cells that die may release some nutrients. Most cells simply survive; microscopic observations show that they become progressively more abnormal structurally.

Although compounds can be used and reused as carbon or nitrogen sources, every time they are used as an energy source inefficient conversion results in dissipation of some energy. The basic media constituents are reused until their useful energy content is very low.

Microbial killing; Chapter 25, pp. 723–724

Death phase succeeds stationary phase because energy sources are inadequate to maintain cell integrity. Aerobes respond to declining energy supplies and low intracellular ATP levels with **autolysis,** enzymatic destruction of their own envelope

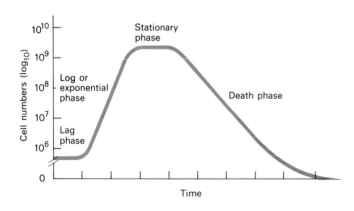

7.12

Closed System Growth Curve When an inoculum is introduced to a container of liquid medium at zero time, the phases of growth succeed each other as shown here. The duration of the phases and the slopes of the exponential growth and death phase lines will vary with the organism and the conditions. No "average" times can be suggested. However, log phase is usually measured in hours, while stationary and death phases may last several days. Cell numbers are plotted as logarithms to the base 10. That is, a sample containing 250 million cells would be plotted as $2.5 \cdot 10^8$.

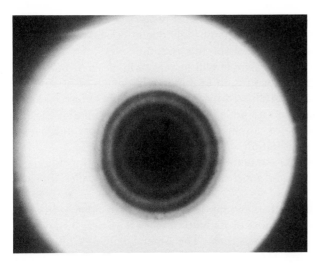

7.13
Colony Zonation Many organisms in colonial growth exhibit periodic oscillations in metabolic function. As the colony grows outward, these oscillations may become visible as rings. *Pseudomonas nigrifaciens*, a marine bacterium, produces a blue-black pigment. As colonies develop, zones of pigment production alternate with colorless growth. The outermost ring—the newest growth—is colorless.

layers. In a surprisingly short time intact cells may become very scarce. Anaerobes may produce toxic end products that eventually not only arrest growth but kill.

Death is also exponential. A very few living cells may persist into the tail of the curve, long after the majority have failed.

Kinetics of Colony Growth

Zones representing each phase of growth are present within a developing colony. As the colony develops, the initial cell divides to form multitudes. The newer cells heap up and push outward; the colony expands radially. Cells left behind in the center quickly suffer nutrient deprivation and waste buildup. The center of a 24-hour-old colony will contain cells in a stationary or death phase while the periphery contains young, log-phase cells. Colony centers often collapse due to autolysis. Sometimes colonies show concentric rings of growth or swarming (Fig. 7.13). The signaling that modulates these phenomena is not understood.

Bacterial colonies do not expand indefinitely. They grow to a predictable size, then stop. Mycelial fungus colonies expand indefinitely. The surface patterns, outline of colony margin, consistency of growth and translucency, opaqueness, or pigmentation are characteristic of a given species and serve as useful identification marks.

SUMMARY

1. To acquire accurate, detailed knowledge about microorganisms, it is necessary to isolate them from all other microbial members of their ecosystem. Once isolated, they are called pure cultures. These are maintained, subcultured, and used for systematic investigation.

2. Successful isolation often requires first a physical process of dilution because most microbial habitats contain vast numbers of individuals. Pour plates and streak plates solidified with agar are used for this purpose.

3. In some cases, the original mixed sample will be placed in an environment with added nutrients that promote the growth of a particular group of organisms. The preliminary enrichment is then followed by isolation, with an enhanced probability of success. Alternatively, an added selective agent may inhibit the growth of a dominant group of flora; a minor component will prosper and become sufficiently numerous for isolation.

Many primary isolation media contain dyes or pH indicators to quickly show up colonies with certain desired biochemical traits. These are called differential media.

4. Once pure cultures are obtained, they may be subcultured to other media. To maintain a stock culture without loss of its natural characteristics, it may be stored under mineral oil, or in lyophilized form. The American Type Culture Collection (ATCC) is one of many organizations that maintain and supply stock cultures and reference strains of microorganisms.

5. The physical growth conditions supplied for a pure culture should approximate the natural habitat. Temperature is an important variable. Each species has a defined growth range, below which metabolism is too slow to produce growth, and above which enzyme losses due to heat denaturation cause death.

6. Cellular organisms maintain a constant internal osmotic pressure. When placed in hypertonic or hypotonic media, they vary in their ability to sustain life in the presence of an osmotic gradient. Loss of water to the exterior reduces rates of cellular metabolism; gain of water results in swelling or bursting.

7. The hydrogen ion concentration of the environment—its pH—may be as low as 2 and as high as 10. Yet most bacteria maintain an interior pH of about 7.5. Many organisms terminate their own growth by adversely altering the pH of the medium with acid or alkaline end products. Buffering agents may be added to compensate for this tendency.

8. Microorganisms vary widely in the concentration of oxygen and carbon dioxide needed and tolerated. Incubator atmospheres can be adjusted to provide exactly the optimum gas mixture to favor a given group.

9. The nutritional environment must supply all key elements in the form needed. Most microorganisms use many small organic compounds interchangeably as carbon sources. However, requirements for one or many amino acids are common. These are usually supplied by hydrolyzed protein and also serve as carbon/energy source. Phosphorus is usually supplied in the form of phosphate ion and sulfur as sulfate ion; however, sulfur-containing organic compounds may be required for some strains. Inorganic ions are added to fulfill osmotic and transport requirements; these often contain sufficient trace metal contamination to make specific additions of iron, copper, or zinc unnecessary. Vitamins or growth factors are often added in the form of yeast or organ extracts. Nonspecific enrichments such as blood often promote faster or more ample growth.

10. Cell division results from coordinated increases in cytoplasmic contents, duplication and movement of the nucleoid, and formation of membrane and wall partitions. Its maximum rate is genetically fixed, but it is retarded by suboptimal growth conditions. The observed time needed for a complete round of cell division is the generation time.

11. In the closed system growth curve seen in the ordinary liquid culture, four sequential phases are observed. The lag phase, just after inoculation, corresponds to a period of internal readjustment. Once prepared, cells begin rapid exponential growth that continues through the log phase. As the environment becomes poor in nutrients and rich in waste products, net growth stops and the population moves into stationary phase. Further deterioration of the medium triggers the onset of an exponential death phase. Similar phases are observed, in less clear form, in growth on solid media.

Study Topics

1. Design an isolation method for the recovery of thermophilic, anaerobic, nitrogen-fixing bacteria.

2. A microorganism is causing serious mortality in schools of North Atlantic salmon. You wish to isolate and study it. How would you proceed?

3. Explain the basis of the antimicrobial preservative effects of salt (brining), acid (pickling), sugar (preserving), and drying as food storage methods.

4. Strict anaerobes require complete absence of molecular oxygen. Look up methods for creating laboratory environments that are oxygen-free.

5. Using a standard reference manual, look up the formulations of blood agar, eosin methylene blue agar, Sabouraud's dextrose agar, and Simmons citrate agar. What organisms are supposed to grow on each? Determine the contribution each component makes to the isolation, differentiation, and maintenance of organisms.

6. A mesophilic organism is inoculated into a nutrient broth tube. Describe what is probably happening to an individual cell during each growth phase.

7. The kinetics of the closed-system growth curve make periodic subculture of stock cultures necessary. Why? What storage methods avoid this requirement?

Bibliography

Lennette, E. H.; A. Balows; W. J. Hausler, Jr.; and J. P. Truant, 1980, *Manual of Clinical Microbiology* (3rd ed.), Washington, D.C.: American Society of Microbiology.

Pramer, David, and E. L. Schmidt, 1964, *Experimental Soil Microbiology*, Minneapolis: Burgess.

————, 1953, *Difco Manual* (9th ed.), Detroit: Difco Laboratories.

Rodino, A. G., 1972, *Methods in Aquatic Microbiology*, Baltimore: University Park Press.

8

Bioenergetics

A microbial cell, or any cell for that matter, can be compared to the type of factory or mill that was common throughout the nineteenth century. Situated near the falls on rivers, mills produced power for spinning and weaving fabrics or assembling shoes, extracting it from the energy source supplied by the river. The power so generated was then directly conveyed to the manufacturing process by an amazingly intricate series of belts and pulleys.

Within the cell, one part of the metabolic network is devoted to the process of extracting energy from nutrients and converting it into immediately useful forms. The other part of the metabolic network is a manufacturing section in which the energy is directly applied to synthesis of useful products. This chapter will deal with the energy conversion section, saving the manufacturing section for the next chapter. Please bear in mind, however, that this separation is purely artificial. You should be aware of the connections of each concept to those previously mastered.

NATURE AND FUNCTION OF ENZYMES

Protein structures;
Chapter 2, pp. 52–54

As we begin the study of metabolism, our first major investigation must be into the nature of enzymes. By reviewing the discussions on proteins in Chapter 2, you may refresh your memory about the details of their biochemical structure as globular proteins.

Classification of Enzymes

Each enzyme may be classified most usefully in direct relation to what it does. Classification of enzymes on the basis of their size, shape, or amino-acid composition is not generally useful.

Exoenzymes contrasted with endoenzymes   One clear division reflects whether the enzyme customarily catalyzes reactions inside or outside of the cell (Fig. 8.1). Most microorganisms have a rigid cell wall and membrane barrier allowing passage of only small molecules. External digestion of macromolecules is commonplace, particularly in free-living species. **Exoenzymes** are those produced within the cell but functioning exterior to the membrane. They are excreted through the membrane and, in Gram-positive bacteria, they are also then passed through the wall. In Gram-negative species they are usually retained in the periplasmic space. The characteristic function of exoenzymes is hydrolysis of high molecular weight polymers into monomers or dimers. Certain exoenzymes produced by pathogenic bacteria and fungi hydrolyze essential tissue components, contributing to the disease.

Bacterial cell walls, periplasmic space; Chapter 3, pp. 82–84

   Endoenzymes are retained entirely within the cytoplasm. They are the catalytic agents of the metabolic processes. Some are freely suspended in the fluid contents of the cell. Others are embedded in cell structures such as the membrane and ribosomes.

Essential enzymes contrasted with optional enzymes   Highly variable environments impose constantly changing demands on the cellular machinery. Some enzymes are

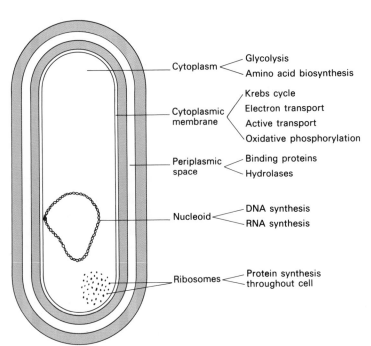

Cytoplasm — Glycolysis / Amino acid biosynthesis

Cytoplasmic membrane — Krebs cycle / Electron transport / Active transport / Oxidative phosphorylation

Periplasmic space — Binding proteins / Hydrolases

Nucleoid — DNA synthesis / RNA synthesis

Ribosomes — Protein synthesis throughout cell

8.1
Bacterial Enzymes   Bacterial enzymes are produced to function in specific locations. In a Gram-negative organism (as pictured), the digestion of macromolecular food stuffs takes place in the periplasmic space. Hydrolases release small molecules that are retained by binding proteins that are actively transported across the membrane to the inside. In Gram-positive bacteria (not shown), the hydrolases are released into the environment as true exoenzymes. Once in the cytoplasm, substances may be broken down by cytoplasmic enzymes, or routed to nucleic acid or protein synthesis at the designated spots.

so essential that the cell produces and uses them under all conditions. An enzyme that is always present is a **constitutive** enzyme. Its concentration may vary, but it is always readily detectable. The cell is permanently committed to maintaining a supply of a constitutive enzyme. Enzymes that make ATP are constitutive. Cell functions can usually be sustained by a few molecules of each of the essential enzymes, replaced on a regular basis.

Enzyme synthesis demands cellular energy and materials. It is costly, in other words. Most microorganisms do not maintain continuous supplies of enzymes for which they have not an immediate need.

Inducible, or optional, enzymes are produced only when the cell is exposed to the enzyme's specific substrate. On exposure to the substrate, genetic expression results in the cell's immediately producing the enzyme(s) for its metabolism. The process is very rapid; measurable enzyme appears within a couple of minutes. Once the substrate is exhausted, no more enzymes will be produced.

Another form of optional enzymes, **repressible** enzymes, are ones whose production is turned off when they are in adequate supply. Remember that the function of a catalyst is to convert A to B. When a cell has sufficient amounts of substance B being provided by the existing enzyme molecules, then repression turns off the manufacture of the enzyme that makes B. When a cell stops making new molecules of an enzyme, the existing ones will remain and continue their function until they are degraded. The genetics of enzyme regulation are discussed in Chapter 9.

Enzyme regulation;
Chapter 9, pp. 262–265

The main classes of enzymes   Enzymes are formally classified according to the type of reaction they catalyze. An international classification scheme divides all enzymes into six classes (Fig. 8.2). Within the classes individual enzymes are given numbers and systematic names. Most of the enzymes with which we will be concerned also have common names of long standing. The common names will be used in the metabolic descriptions.

Oxidoreductases are those enzymes that catalyze redox reactions. This type of reaction was introduced in Chapter 2. In metabolism, energy useful to the cell is invariably derived from the oxidation of an energy-rich nutrient.

Redox reaction;
Chapter 2, p. 31

Transferases move specific functional groups from one molecule to another. Among the groups frequently moved are amino, methyl, and phosphate groups.

Hydrolases catalyze hydrolytic cleavage of the covalent bonds that link monomers into polymers. Most exoenzymes are clearly in this group. These enzymes are also referred to as digestive or degradative enzymes. Amylase is a simple hydrolase; it breaks down starch into maltose. Other hydrolases break down proteins, lipids, nucleic acids, and other high molecular weight substances.

Lyases also separate groups from the parent molecular compound. However, these enzymes do not add water across the bond to be broken; that is, they are not

8.2
Enzymes Can Catalyze Six Basic Operations   Reactions presented in this chapter are offered as examples. In any of the pathways of metabolism, several different types of operations are performed in sequence. This makes a pathway similar to an assembly line.

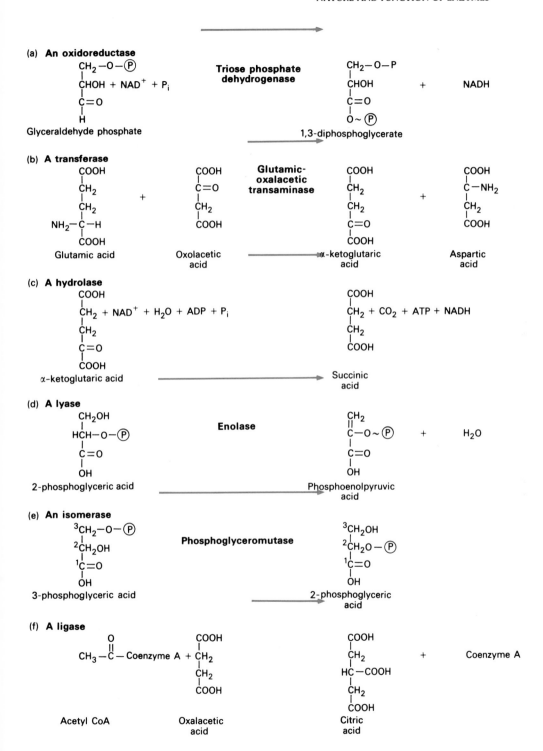

hydrolytic. Thus when the bonds are broken to remove a grouping, double bonds form to satisfy the valences of the remaining atoms. Groups commonly removed are amino groups, released as ammonia; carboxyl groups that give rise to CO_2; and water.

Isomerases carry out atomic rearrangements within molecules. These result in the reshuffling of atoms without any being added or subtracted. The empirical formula for a compound operated upon by an isomerase remains the same. Only the structural formula is changed. Polarities, strengths, and energy contents of various

TABLE 8.1
Important coenzymes of metabolism

TYPE OF FUNCTION	NAME OF COENZYME	GROUP TRANSFERRED	VITAMIN COMPONENT
Electron carriers	Nicotine adenine dinucleotide (NAD)	Two electrons, one proton	Niacin (B_5)
	Nicotine adenine dinucleotide phosphate (NADP)	Two electrons, one proton	Niacin
	Flavin mononucleotide (FMN)	Two electrons, two protons	Riboflavin (B_2)
	Flavin adenine dinucleotide (FAD)	Two electrons, two protons	Riboflavin
	Ubiquinone	Two electrons, two protons	Related to K
	Cytochrome	One electron	—
	Ferredoxin	One electron	—
Group transfer carriers	Thiamine pyrophosphate (TPP)	Aldehyde	Thiamin (B_1)
	Lipoamide	Acetyl	—
	Coenzyme A	Acetyl	Pantothenic acid
	Biotin	CO_2	Biotin
	Tetrahydrofolate	1-carbon, such as methyl	Folic acid
	S-adenosyl methionine	Methyl	—
	Methylcobalamin	Methyl	Cobalamin (B_{12})
	Pyridoxal phosphate	Amino	Pyridoxine (B_6)
	Uridine diphosphate (UDP)	Glucose	—
Energy donors	Adenosine triphosphate	Phosphate	—
	Guanosine triphosphate	Phosphate	—
	Uridine triphosphate	Phosphate	—

bonds may be significantly altered, however. Isomerases convert one hexose to another, for example, by rearranging the positions of alcohol, aldehyde, and keto groups.

Ligases link things together. They catalyze the condensation processes that assemble macromolecular polymers. Similar terms are synthetase and polymerase. Their actions thus are exactly opposed to those of the hydrolases.

Nonprotein components of enzymes   In many enzymes, the complete functional unit is a chemical hybrid. It is assembled from one or more polypeptides (apoenzymes) plus a nonprotein component. In many instances, the nonprotein substance is an organic molecule related to the nucleotides. Such enzyme helpers are correctly referred to as coenzymes (Table 8.1). They may be loosely or tightly attached to the apoenzyme to make the functional catalytic unit, the holoenzyme. The specific function of the coenzyme is to reversibly accept, hold, or contribute single atoms, electrons, or small groups of atoms. Because these diffuse away rapidly in the fluid medium of the cell, a mechanism is needed to convey them efficiently to the enzyme that requires them as substrates.

Cofactors are inorganic metal ions, the minerals referred to in our nutritional discussions in Chapter 7. Cations of iron, magnesium, calcium, and other elements may be bonded to the reactive center of an enzyme and be necessary for its correct combination with substrates.

Minerals; Chapter 7, p. 200

How Enzymes Catalyze Reactions

Before reading this section, it would be helpful to review the section on chemical kinetics in Chapter 2. This gives the conceptual framework from which we will develop an understanding of enzyme function.

Chemical reactions; Chapter 2, pp. 34–36

Enzyme specificity   An enzyme, as a three-dimensional globular protein with an intricately sculptured surface configuration, can form close contact only with molecules whose surface shape and charge distribution is complementary. The enzyme holds the substrate at a specific location on its surface called the active site. Some enzymes are very highly specific and catalyze only a single reaction in vivo. Frequently, by altering laboratory conditions, they can be "tricked" into catalyzing closely related reactions in vitro. Other enzymes, less specific, will perform a given reaction on all generally similar molecules.

Our use of the term specificity is intended to convey the fact that, in the healthy functioning cell, there is no appreciable chance that an enzyme will cause random changes in cellular materials other than those that are its own natural substrates.

Enzyme inhibitors   Let us, however, consider some unnatural situations in which enzyme specificity is confounded. Imagine that there might be natural or artificial substances closely similar to enzyme substrates that a cell was highly unlikely to encounter. Its enzymes in consequence would not be designed to differentiate between natural substrate and the other substance. An unnatural contact with the artificial substance would result in the enzyme's accepting it as a substrate as readily or more avidly than it would accept the real thing (Fig. 8.3) This acceptance would block the enzyme from its normal activities. Such is the role of the **competitive inhibitors.** We are all familiar with the dangers of carbon monoxide (CO) poisoning. The gas CO is a competitive, unnatural substrate for hemoglobin. It is not normally encountered, but when it is CO attaches more avidly than oxygen to the oxygen-carrying portion of the hemoglobin molecule, blocking hemoglobin's normal oxygen-transport function. In sufficient concentration, it is lethal. In bacteria, sulfa drugs act as competitive inhibitors for the enzymatic processes that synthesize folic acid, an essential vitamin. A competitive inhibitor may dissociate from the enzyme later, allowing it to resume functioning.

Folic acid pathway;
Figure 26.2

Noncompetitive inhibitors combine with the enzyme irreversibly to effectively prohibit catalysis. Cyanide ions are noncompetitive inhibitors of key respiratory enzymes.

Enzyme efficiency   Under optimum conditions, each enzyme molecule can catalyze a certain number of reactive operations per second. The measured velocities of certain reactions approximate one million individual catalyses per second. Even relatively sluggish enzymes crank out ten thousand or so! It is clear why a cell needs only a few molecules of each enzyme at any time.

Energy and reactions;
Chapter 2, p. 55

Enzymes remove energy barriers   The features of endergonic and exergonic reactions have already been discussed. Remember that reactions proceed spontaneously when they are exergonic and result in a net release of energy.

Reactions occur between compounds; within the compounds each atom is in a stable-bonded relationship with the others. In order for reaction to occur, bonds must be broken, and breaking bonds requires energy. Thus compounds such as paper (cellulose) and oxygen (gas), which potentially react with energy release, may be put together without reacting.

Activation energy must be supplied to raise the average energy level of the molecules to the point at which bond breakage and subsequent reaction is probable (Fig. 8.4). The combination of enzyme with substrate appears to remove or lower the activation energy barrier. It probably alters the nature of the substrate bonds to render them highly unstable. Once the bond of the substrate(s) is broken, the enzyme holds the portions and orients them to promote the formation of the desired new bonding combinations.

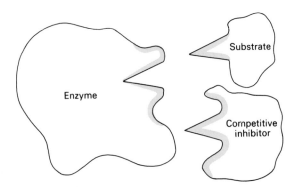

8.3
Enzyme Inhibitors   An inhibitor binds to an enzyme as shown by

$$E + I \rightleftharpoons EI,$$

forming an enzyme–inhibitor complex. A competitive inhibitor binds to an enzyme's active site as well as, or better than, the natural substrate. It not only displaces substrate but it also tends to remain bound longer. It can, however, dissociate if the equilibrium is favorable, that is, if all unbound inhibitor is removed.

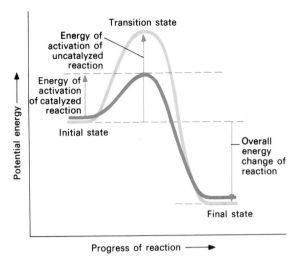

Potential energy →

Transition state

Energy of activation of uncatalyzed reaction

Energy of activation of catalyzed reaction

Initial state

Overall energy change of reaction

Final state

Progress of reaction ⟶

8.4

Activation Energy The molecules that enter into a reaction are in a stable state. They require an input of activation energy to render them less stable. The presence of an enzyme significantly reduces the amount of energy needed to start a reaction. After the reaction, the producers are in a second stable state. Their energy has been lowered and released to the environment in a useless form.

This process is summarized in the **Michaelis–Menton** equation.

$$E \ + \ S \ \rightleftharpoons \ ES \ \rightleftharpoons \ E \ + \ P.$$

enzyme / substrate   enzyme–substrate / enzyme / product
complex

The enzyme's activity is based on the formation of the specific, unique enzyme–substrate complex, a union of extremely short duration. The equation also points out that enzymatic catalyses, like most other chemical reactions, are reversible. Within the cell, they operate primarily in the forward direction with rapid net production of product (P). Since the product is useful, it is consumed, and does not accumulate to a level that would encourage a reverse reaction.

Control of Enzymatic Reaction Rates and Activity

Reaction Rate   Enzymatic reactions are accelerated by increasing the substrate concentration to saturation so that each enzyme molecule works at maximum speed. Increases in temperature also accelerate reaction rates up to a point; beyond that point high temperatures inactivate enzymes.

Each enzyme has an optimum pH for maximum activity; this is based on the charge characteristics of the amino acids that comprise its active site. A value of pH under or over the optimum causes less effective substrate binding.

Feedback inhibition   Often, accumulation of product causes a shutdown in enzyme activity. This effect, called **feedback inhibition,** is comparable to the interaction between a thermostat and a household heating system. As heat (the product) accumulates, the thermostat shuts off the heat-producing unit (the furnace). After the heat dissipates, the thermostat reactivates the furnace. In a cell, the accumulation of ATP, a highly desired product, signals a slowdown of the steps leading to ATP pro-

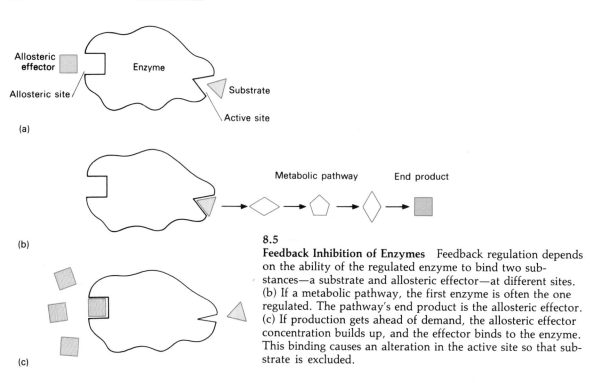

(a)

(b)

(c)

Metabolic pathway   End product

8.5
Feedback Inhibition of Enzymes   Feedback regulation depends
on the ability of the regulated enzyme to bind two substances—a substrate and allosteric effector—at different sites.
(b) If a metabolic pathway, the first enzyme is often the one
regulated. The pathway's end product is the allosteric effector.
(c) If production gets ahead of demand, the allosteric effector
concentration builds up, and the effector binds to the enzyme.
This binding causes an alteration in the active site so that substrate is excluded.

duction by inhibiting certain enzymes. Once the ATP supply has been reduced, ATP
synthesis is reactivated.

The mechanism for feedback inhibition of enzymes is an **allosteric** interaction.
A product combines with the enzyme at other than its active site, changing its three-
dimensional shape so that combination with substrate is no longer possible (Fig.
8.5). The blocking molecule stays there until it is removed for cellular use. Its removal turns the enzyme back on.

Control at the genetic level   Cells have highly sophisticated genetic mechanisms
that determine whether or not the cells will manufacture a specified enzyme. Induction and repression control enzyme activity by controlling the presence of enzyme.
To return to the heating system analogy, it would be like deciding that heat is
needed, then installing a furnace, which you subsequently remove when the house
gets warm enough. Clearly, the response time is longer than with a feedback system.

Metabolic pathways   Enzymes and enzymatic reactions are isolated only in the test
tube or in textbooks. In the cell, all are linked in pathways. In a metabolic pathway,
the products of one enzymatic reaction become the substrates for the next. In
straight pathways such as glycolysis, a sequence of enzymatically catalyzed steps results in the conversion of one substance into another. In branched pathways, an

intermediate may be taken up as a substrate by either of two enzymes and thus be converted into several different things. Cyclical pathways tend to regenerate at least one of their starting materials. Simultaneously, they may provide starting materials for other pathways or yield waste of end products. Enzymatic reactions within pathways are rate-regulated by the steps that precede and follow them.

OXIDATION, ENERGY RELEASE, AND ENERGY CAPTURE

Energy is the ability to do work to drive a process in a desired direction. A cell's energy source allows it to create internal order and maintain its integrity. In order to be so used, the work force must be released from a storage form, then quickly repackaged in small molecular packages with a minimum of waste. The physical laws regulating energy behavior in the universe, which we saw regulating ecosystem function in Chapter 6, also establish the ground rules for cellular metabolism.

Laws of thermodynamics in ecology; Chapter 6, pp. 162–163

The Laws of Thermodynamics in Metabolism

The first law of thermodynamics   The Law of Conservation of Energy states that whenever energy is converted from one form to another, all the energy can be accounted for afterward. No energy is created or destroyed. Thus equations for chemical reactions, involving changes in the bond energy of the components, can and should be balanced, not only in terms of the atoms of each element but also in terms of the units of energy contained in each molecular entity.

In the exergonic reaction,

$$C + O_2 \rightarrow CO_2 + energy,$$

energy values on the left side of the arrow, corresponding to the energy contents of elemental carbon and molecular oxygen, will be exactly equal to the sum of the two energy sources on the right—the bond energy of CO_2 plus the released energy. The exact magnitude of all four values can be experimentally determined; such information is available in reference tables. Thus it is relatively easy to find out the theoretical energy yield from the conversion of any biologically useful nutrient into another. For example, when glucose is oxidized to CO_2, it will release 686,000 calories per mol in the process. Synthesis of ATP from ADP and inorganic phosphate ion requires about 8,000 calories per mol. Thus oxidation of one mol of glucose will supply enough energy, theoretically, to put together 85 mols of ATP.

The second law of thermodynamics   It doesn't work that way, however. Accurate measurements of ATP yields per mol of glucose rarely indicate a yield as high as 38 mols, usually somewhat less. The reason is the second law of thermodynamics, which states that as energy is changed from one form to another, some portion of the energy will be inevitably converted from an ordered, useful form called **free energy** (G) into a form unavailable to do work, **entropy.**

Coupling of Exergonic and Endergonic Reactions

Exergonic reactions give off energy; endergonic reactions proceed only if they are supplied with an energy source. The emergence of living organisms was dependent on the development of a way of coupling the two. This is the **coupled reaction;** an exergonic and an endergonic process are linked by sharing an enzyme and/or its products and reactants. A coupled reaction can be visualized as similar to the relationships among an automobile engine, its fan belt, and its battery. Oxidation (the combustion of gasoline) causes release of energy and the rotation of the crankshaft. By means of a fan belt, some of this energy is transmitted via an alternator to the storage battery. When the fan belt breaks, the linked processes of energy release and energy storage are uncoupled and the battery does not charge.

An endergonic reaction may be symbolically represented by

$$A + B + energy \rightarrow C + D.$$

This reaction represents an increase in the free energy (G) of the products over that of the reactants. The change in G (ΔG) is positive. Let us use an arbitrary value of 100 units of energy for ΔG in this reaction.

A simultaneous exergonic reaction,

$$X \rightarrow Y + Z + energy,$$

has a decline in the free energy of the system, a negative value of ΔG ($-\Delta G$). Let us set an arbitrary value of 200 units of energy for $-\Delta G$ in this reaction. Both reaction sequences occur at the same site; they are catalyzed by the same enzyme or enzyme-coenzyme complex. Summing the process,

$$A + B + X \rightarrow C + D + Y + Z,$$

we get an equation that accurately represents the chemistry of the process but no longer shows the energetics.

If a coupled reaction is to proceed, the $-\Delta G$ (energy output) of the exergonic reaction must be greater than the ΔG (energy demand) of the endergonic reaction. Thus for the coupled reaction, we see that the net free energy change is

-200 units (exergonic reaction)
$+100$ units (endergonic reaction)
———————
-100 units (coupled reaction).

Thus the coupled process itself has a net $-\Delta G$ value. Note that this shows that more energy must be supplied than is actually needed; waste is inevitable under the second law.

In biological energy-manipulating processes, the exergonic component is most often an oxidation process. The endergonic component is usually a **phosphorylation** process, the addition of a phosphate group.

Biological Oxidations

Oxidation is the loss or removal of electrons from a substrate. The oxidized substance is frequently referred to as the **electron donor.** The electrons are usually passed to another substrate, the **electron acceptor.** The latter acquires electrons and is said to be reduced. Clearly, the electron acceptor has a higher affinity for the electrons than does the donor.

Redox couples   Let us look at the simple reaction

$$\frac{1}{2} H_2 \xrightarrow{\text{oxidation}} H^+ + e^-.$$

reduced   oxidized.

The paired chemical forms $\frac{1}{2} H_2/H^+$ are called a **redox couple.** The reduced substance appears to the left, the oxidized substance to the right. In any redox reaction, redox couples can be similarly identified (Table 8.2).

Electrode potentials   The process of oxidation–reduction involves the movement of a current of electrons. In physics, currents of electrons are measured as voltage. It is

TABLE 8.2
Redox potentials (E_o' or Eh) of biologically important redox couples

ELECTRON DONOR (REDUCED COMPOUND)	ELECTRON ACCEPTOR (OXIDIZED COMPOUND)	E_o' IN VOLTS
H_2O	$\frac{1}{2} O_2 + 2H^+$	0.82
Ferrous ion (Fe^{++})	Ferric ion (Fe^{+++})	0.77
Cytochrome $A_{reduced}$	Cytochrome $A_{oxidized}$	0.50
Nitrite ion (NO_2^-)	Nitrate ion (NO_3^-)	0.42
Cytochrome $C_{reduced}$	Cytochrome $C_{oxidized}$	0.22
Quinone$_{reduced}$	Quinone$_{oxidized}$	0.10
Cytochrome $b_{reduced}$	Cytochrome $b_{oxidized}$	0.07
Succinate	Fumarate	0.03
$FADH_2$	FAD	− 0.03
Lactate	Pyruvate	− 0.19
NADH or NADPH	NAD^+ or $NADP^+$	− 0.32
Ferredoxin$_{reduced}$	Ferredoxin$_{oxidized}$	− 0.43
Chlorophyll$_{reduced}$	Chlorophyll$_{oxidized}$	− 0.61

Negative values indicate strong reducing potential. Electrons flow from negative to positive with a release of free energy proportional to the change in E_o' (or Eh) value.

readily possible to measure the voltage or "electron pressure" of the electrons flowing from one half of the redox couple to the other. This value, E_o' (or Eh), is a characteristic of each redox reaction.

Electron transfer   The oxidized member of one redox couple may be reduced not only by its "normal" coupled partner but also by the reduced member of another couple. Electrons may be passed from the donor of system A to the acceptor of system B if system B has a higher (more positive) E_o' value (Table 8.2). This transfer of electrons is associated with a decline in free energy; that is to say, the transfer is exergonic. By comparing the electron potential of the donor system with that of the acceptor system, one can derive the change in E_o' value, and calculate the exact amount of free energy made available. Calculations show that the transfer of a pair of electrons from the $NADH^+/NAD$ couple to the $H_2O/\frac{1}{2}O_2 + 2H^+$ couple yields sufficient energy to drive the endergonic synthesis of three molecules of ATP.

Phosphorylation

ATP is the energy currency of the cell—like our paychecks, earned only to be too soon spent. It is formed from the nucleotide diphosphate ADP by the addition of an inorganic phosphate group, indicated by the symbol P_i (Fig. 8.6)

$$\text{ADP} + P_i + \text{energy} \quad \overset{\text{transferase}}{\rightleftharpoons} \quad \text{ATP} + H_2O$$

The left-to-right direction of this reversible reaction is endergonic, energy-storing; it has a $+\Delta G$. For the right-to-left reaction all the opposite statements are true. In particular, ATP breakdown is often the driver in coupled reactions in which energy must be supplied.

We will be using the term **phosphorylation** for a variety of reactions in which phosphate groups are added to molecules, including ADP (Table 8.3). Phosphorylation always increases the compound's free energy.

8.6
Phosphorylation   The bond that attaches the third phosphate to the ATP molecule is unusually rich in energy and unstable. It is readily hydrolyzed to release its energy. In a normal cell, this would never happen except when coupled to an endergonic process.

TABLE 8.3
High-energy compounds

COMPOUND	WHERE FOUND	ΔG IN KCAL/MOL
Phosphoenolypyruvate	Glycolysis—donates $\sim$ P* to make ATP at Step 9	− 14.8
Diphosphoglycerate	Glycolysis—donates $\sim$ P* to make ATP at Step 6	− 12.0
Creatine phosphate	In muscle tissue—provides energy for contraction	− 10.3
Pyrophosphate	Cleavage provides energy to polymerize carbohydrate	− 8.0
ATP,GTP,UTP,CTP	Cleavage drives cellular transport, movement, synthesis	− 7.3
Glucose-1-phosphate	Provides energy for glycogen synthesis	− 5.0
Glucose-6-phosphate	Glycolysis—intermediate	− 3.3

* $\sim$ P = High-energy phosphate group.

The amount of free energy released when these compounds are hydrolyzed increases as the negative value of ΔG increases. Compounds with a − ΔG of 5.0 or greater have a high phosphate group-transfer potential and actively drive cellular reactions.

Substrate-level phosphorylation   The phosphorylation process can take place in two different ways, at distinct cellular locations. **Substrate-level phosphorylation** takes place in the cytoplasm, not in association with any cell structures. The energy released by substrate oxidation is used directly, at the reaction site, to phosphorylate ADP to ATP. This process is comparatively simple. It is universal in cellular organisms. **Oxidative phosphorylation** is carried out in the presence of a membrane-bound electron transport system. Electrons derived from the oxidized substrate are passed via electron transport coenzymes to an inorganic electron acceptor, typically oxygen. In transit, the electrons lose energy because they are passed from one redox couple to another with steadily increasing E_o' values. This released energy is coupled to oxidative phosphorylation. The sequential electron transfer necessitates an orderly, physical arrangement of the transfer coenzymes.

Electron transport pathways   Coenzymes specialized for the transport of electrons and protons make up the unambiguously named **electron transport system** (ETS) (Fig. 8.7). The first coenzyme, the one whose redox couple has the lowest E_o' value, is NAD^+. In the oxidized form this substance, nicotine adenine dinucleotide, has a single positive charge in the reactive site. On reduction it picks up two electrons and one proton (H^+) to become the electrically neutral NADH.

The two electrons are passed from the reduced NADH to the oxidized form of the next coenzyme, FAD, named flavin adenine dinucleotide. In becoming reduced,

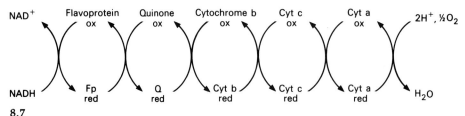

8.7

Electron Transport System   A linear model of electron transport represents the passage of a pair of electrons, transferred from one carrier to another at the point where the paired curved arrows are tangent. No attempt is made to indicate what happens to the protons. The electrom finally unite with a pair of protons (2H$^+$) to reduce an oxygen atom to H$_2$O.

FAD accepts two electrons and a proton from NADH, and takes on a second proton from the environment to become FADH$_2$. The transfer is catalyzed by an enzyme.

At the next step, the paired electrons and protons reduce a **quinone,** a vitamin K derivative. From this point on, electrons are transported separately from protons.

After the quinone stage, **cytochrome** is the next electron acceptor. The cytochromes are chemically similar to chlorophyll and have either an iron or a copper atom as the electron acceptor portion. They do not carry protons. There are many different cytochromes; in microorganisms there may be none, one, or several in a sequence. In a more highly developed ETS, the electrons may pass from ubiquinone to cytochrome b, cytochrome c, and a complex of cytochromes a and a$_3$.

At the end of the electron transport sequence, the pair of electrons have a much reduced energy level. Cytochrome oxidase plays a role in forming a low-energy waste product H$_2$O. Hydrogen peroxide (H$_2$O$_2$) may be formed as a highly toxic intermediate. In aerobic organisms the enzyme catalase breaks down H$_2$O$_2$. Strict anaerobes usually lack catalase, so molecular oxygen promptly poisons them.

Bacteria exhibit great variation in the electron transport mechanisms they possess. This outline presents the most extended developed sequence; bear in mind that many common species have less complex schemes. Eucaryotes, however, show a high degree of uniformity in their electron transport systems.

How is the free energy given off by the electrons used to make ATP? Oddly enough, this extremely important question has resisted at least 40 years of dedicated and ingenious investigation. There are at least four hypotheses currently in use. Rather than reviewing them all, let us look only at the most recent, the one that seems to answer the most questions.

The chemiosmotic hypothesis   This hypothesis proposes that as electrons move down the electron transport chain their free energy is used to extrude protons (H$^+$) through a membrane to the outside (Fig. 8.8). In bacteria, they are extruded through the cell membrane; in eucaryotic cells electrons are pushed out through the mitochrondrial membrane. Thus the medium outside contains a higher concentration of

H^+ than the interior. In other words, a pH gradient is created. Like all gradients, this represents an energy potential. At other points in the membrane, specific sites contain membrane-bound ATPases. At these sites, H^+ can reenter the cell, perhaps in an energy-rich bonding to a carrier. It moves by diffusion down the concentration gradient. The free energy of the pH gradient is harvested at these sites to phosphorylate ADP to ATP. Support for the chemiosmotic hypothesis comes from the fact that phosphorylation occurs only in structurally intact membrane. If the membrane is modified to prevent a proton gradient from developing, phosphorylation will stop even though electron transport continues.

Ion gradients are used similarly in the energy-gathering purple membrane of the halophilic bacteria. These organisms use solar energy to pump out the protons.

Halobacteria; Chapter 4, p. 118

In summary, electron pairs are passed across a sequence of coenzyme carriers and finally to oxygen, with a progressive release of free energy. This aspect of the process is called electron transport. The released free energy is used to expel protons to the outside across the cell membrane. At certain locations on the membrane, these protons reenter, providing energy to make ATP. This final step is the part that is called oxidative phosphorylation.

The two processes can be uncoupled so electron transport proceeds independently of oxidative phosphorylation but the resulting energy production is then not captured for subsequent use. Oxidative phosphorylation, however, cannot proceed by itself—it is endergonic and totally dependent on its energy source.

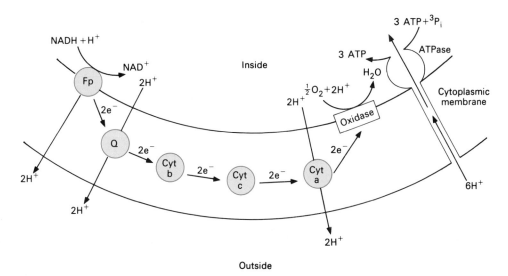

8.8

A Model of Chemiosmotic ATP Production   The electron transport coenzymes are oriented in the membrane mosaic so as to be able to extrude protons. During the passage of $2C^-$ from NADH to oxygen, three pairs of protons may be extruded. At the same time, protons reenter through an ATPase gate to drive ATP synthesis.

Box 8.1

Ionic Gradients and Membrane Functions

When an ionic gradient is established on either side of a semipermeable membrane, it has the potential to do work. The chemiosmotic hypothesis explains the relationship between electron transport and oxidative phosphorylation by proposing that, during electron transport, the free energy of the electrons is used to pump protons (H^+) to the exterior of the membrane. The proton's energy is called the **proton-motive force** (PMF). These protons can then fall back down a concentration gradient to provide the energy to make ATP.

However, ion gradients have another equally important set of uses. When the proton is pushed out with its positive charge, it also establishes an electrical charge gradient—outside positive, inside negative. Positive charges (H^+) accumulate outside, and an equal number of negative charges, mainly hydroxyl ions (OH^-), accumulate inside. Ionic and electrical gradients are believed to provide the driving force for active transport.

Positive ions can enter the cell directly because of the charge gradient, in the presence of

a suitable carrier (a). Negative ions can enter with a proton, as a neutral salt (b). Uncharged molecules can also be carried along with a proton (c), in a process called **symport.** By a type of revolving door arrangement (d), the entrance of a proton can be linked to the eviction of an unwanted material (**antiport**).

In aerobes, electron transport, then, provides the energy for transport. However, fermenting or anaerobic microorganisms must use up cellular ATP to create their ion gradients in the absence of electron transport. In respiring cells, protons that perform transport functions have used up their PMF and cannot drive phosphorylation. Thus transport does use up ATP, either directly or by subtracting from the oxidative phosphorylation yield.

Another energy-requiring, membrane-associated function is motility. It is not clear what role, if any, ion gradients play in motility, but it is suggestive that all of the chemotaxis receptor proteins studied to date are also directly involved with active transport.

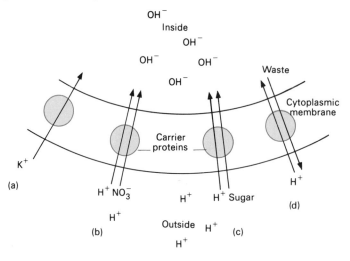

METABOLIC SOURCES OF ELECTRONS

Electrons are the primary source of metabolic chemical energy. Heterotrophs break down organic compounds to obtain the electrons to power phosphorylation. This process is called **catabolism.** Note that we are shifting our viewpoint—now we are looking at nutrients as substrates to be oxidized, to give up those all-important electrons. Carbohydrates and fats are the most commonly used fuels and the most commonly stockpiled reserve energy materials. Proteins are most important as structural and functional cell components; however, they too may be used as energy sources when they are plentiful or when carbohydrate or fat starvation threatens the cell.

Utilization of Carbohydrates

Cellular metabolism of carbohydrates starts with glucose. Polysaccharides are hydrolyzed to yield hexoses; those hexoses other than glucose are enzymatically transformed into glucose prior to the start of glucose oxidation.

From glucose to pyruvate   Let us attempt to think of intermediary metabolism—the interconversion of all the cell's small organic molecules—as a road map. Viewed as a whole, the thing is just a maze of lines. Closer examination shows that all points are connected and that there is a reason for the existence of each road.

When a cell attacks a glucose molecule, three results are desirable. The glucose should be oxidized, since that gives the cell some electrons to use. The glucose should provide some energy directly for substrate-level phosphorylation, if possible. The glucose should be subdivided into smaller molecules that might be useful for building blocks. As we explore the metabolic roads, attempt to keep your eye on three things—the electrons, the newly formed molecules of ATP, and the carbon atoms. It will help you figure out what the cell is accomplishing along the way.

The first stage in the cell's attack on glucose converts some or all of it to pyruvic acid, a three-carbon compound. Three pathways have evolved for the performance of the process. No one organism has all three, but many have two options. We will discuss glycolysis, the most nearly universal pathway.

Glycolysis   The Embden–Meyerhof pathway, more frequently referred to simply as **glycolysis,** is a nine-step process (Fig. 8.9). In Step 1, a glucose molecule is phosphorylated by transfer of a phosphate group from a donor ATP molecule. This raises the energy level of the glucose considerably. It becomes less stable; it is said to be **activated.**

In Step 2, the internal arrangement of atoms is shifted and the carbonyl group ($C=O$) is moved from carbon 1 to carbon 2. This creates an alcohol grouping at the upper end of the glucose molecule.

Step 3 is similar to Step 1 in that a phosphate group is added to activate the molecule. With these three sequential changes, the glucose molecule has been altered so

8.9
Glycolysis

The steps of glycolysis

Glucose

Step 1 Hexokinase — ATP → ADP

Glucose 6-phosphate

Step 2 Phosphogluco-isomerase

Fructose 6-phosphate

Step 3 Phospho-fructokinase — ATP → ADP

Fructose 1,6-diphosphate

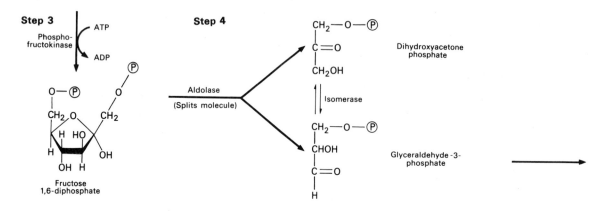

that the middle carbon-to-carbon bond (between 3 and 4) has been weakened. It can now be enzymatically broken without a prohibitively expensive input of activation energy.

In Step 4 the glucose is cleaved into two almost identical halves. Both three-carbon molecules, glyceraldehyde-3-phosphate and dihydroxyacetone phosphate, are readily interconvertible. Glyceraldehyde-3-phosphate is the substance that continues in the glycolytic pathway; dihydroxyacetone phosphate is shunted to a different pathway for the manufacture of fats. The **relative** amounts of the two three-carbon products at this stage will vary with the cell's immediate needs. The phosphate bonds in both compounds are only moderately energetic.

Step 5 is a complex reaction in which we see very clearly the direct coupling of exergonic and endergonic events. Glyceraldehyde-3-phosphate is oxidized, and one electron-proton pair (a hydrogen atom) is removed. Simultaneously, another hydrogen atom is removed from a molecule of inorganic phosphate. The pair of electrons reduce a molecule of NAD^+ to NADH. The resulting free energy is conserved by a substrate-level phosphorylation process that adds a phosphate group to the carbon backbone. The product—1, 3-diphosphoglyceric acid—has now two phosphates. One is a high-energy group ($\sim P$), the other is not.

Step 6 is a transfer reaction. The accumulated bond energy of the high-energy phosphate group in 1, 3-diphosphoglyceric acid is greater than that of the high-energy bond of ATP. Thus a phosphate may be moved from one to the other as an exergonic reaction. This is the first payoff in the form of accumulated ATP.

Steps 7 and 8 are rearrangements; they convert the remaining low-energy phosphate group to a high-energy group. This is done by first moving the phosphate, then removing the molecule of water. These steps put a double bond right next to the phosphate, a very unstable, highly activating configuration. Phosphoenolpyruvate (PEP) contains the highest-energy phosphate group of any naturally occurring organic molecule.

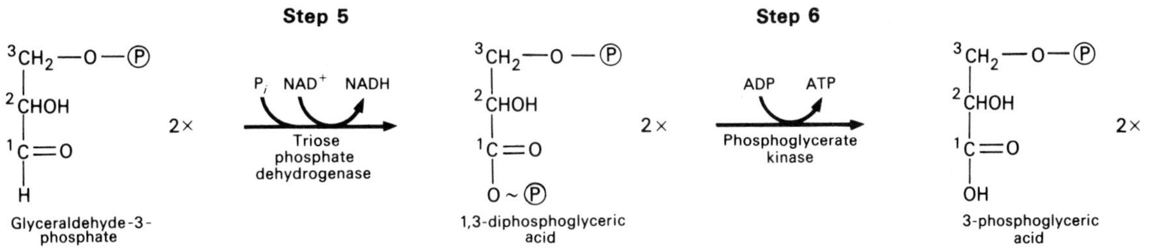

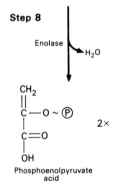

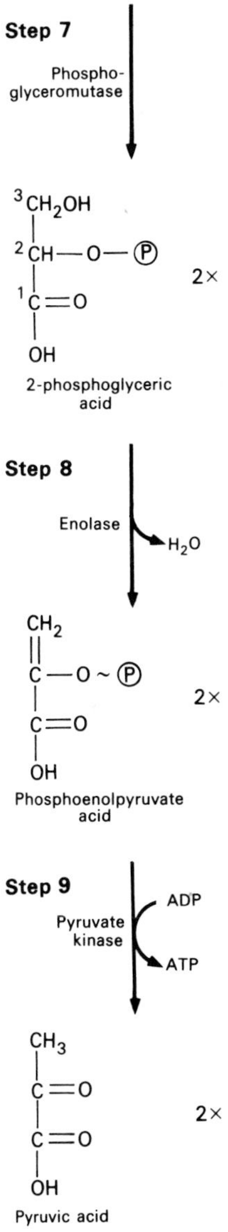

Step 5

$^3CH_2$—O—Ⓟ
^{2}CHOH
^{1}C=O
H

Glyceraldehyde-3-phosphate

2×

P$_i$ NAD$^+$ NADH

→ Triose phosphate dehydrogenase

$^3CH_2$—O—Ⓟ
^{2}CHOH
^{1}C=O
O~Ⓟ

1,3-diphosphoglyceric acid

2×

Step 6

ADP ATP

→ Phosphoglycerate kinase

$^3CH_2$—O—Ⓟ
^{2}CHOH
^{1}C=O
OH

3-phosphoglyceric acid

2×

Step 7

Phosphoglyceromutase

^{3}CH$_2$OH
^{2}CH—O—Ⓟ
^{1}C=O
OH

2-phosphoglyceric acid

2×

Step 8

Enolase

H$_2$O

CH$_2$
‖
C—O~Ⓟ
C=O
OH

Phosphoenolpyruvate acid

2×

Step 9

ADP

Pyruvate kinase

ATP

CH$_3$
C=O
C=O
OH

Pyruvic acid

2×

Step 9 is like Step 6—it is a transfer reaction, with a net $-\Delta G$, in which ATP is made. The reaction yields pyruvic acid (pyruvate). This three-carbon compound still contains over 90 percent of the original energy of the glucose molecule. It can be further metabolized via several pathways.

Summarizing glycolysis:

$$\text{glucose} + 2P_i + 2ADP + NAD^+ \xrightarrow{\text{9 enzymes}} \text{2 pyruvic acid}$$
$$+ 2NADH + 2H^+ + 2ATP.$$

We can see that a **net** yield of 2ATP is achieved. Furthermore, glycolysis is a source of electrons that may be directed to electron transport.

Complete Oxidation of Pyruvic Acid

Large yields of usable phosphorylating energy become available to the cell that completes the conversion of pyruvic acid to CO_2 in **respiration**. A respiring cell must have an electron-transport chain and a supply of electron acceptors to receive the low-energy electrons.

Oxidative decarboxylation   This is a single irreversible reaction of great importance (Fig. 8.10). It uses a set of three coenzymes, thiamine pyrophosphate (TPP), lipoamide, and Coenzyme A (CoA), in its complex atomic transfers. The name of the reaction clearly describes what happens. Pyruvic acid is oxidized, and one hydrogen atom is removed. Energy made available by the oxidation is used to **decarboxylate**, that is, to remove a CO_2 group. The remaining two carbon fragments make up an **acetyl** group. The acetyl group is accepted by Coenzyme A, becoming acetyl CoA. Two hydrogen atoms are made available—one derived from the pyruvic acid, and one from the coenzyme. Both electrons and one proton are accepted by

$$\text{Coenzyme A-SH} + \underset{\substack{\text{Pyruvic}\\\text{acid}}}{\overset{\displaystyle CH_3}{\underset{\displaystyle COOH}{\overset{|}{\underset{|}{C=O}}}}} \xrightarrow[]{NAD^+ \quad NADH} \underset{\substack{\text{Acetyl}\\\text{CoA}}}{\overset{\displaystyle CH_3}{\underset{\displaystyle CoA}{\overset{|}{\underset{|}{\underset{S}{\overset{|}{C=O}}}}}}} + CO_2$$

8.10
Oxidative Decarboxylation   When CO_2 is removed from pyruvic acid, Coenzyme A becomes bound via a divalent sulfur atom. This is a high-energy bond that will provide part of the energy for the introduction of the acetyl group into the Krebs cycle.

NAD^+ and it is reduced to NADH. The acetyl group is then introduced into the Krebs cycle, the last stage in glucose oxidation.

The Krebs cycle   Cyclic pathways regenerate at least some of their starting materials. The **Krebs** or **citric acid cycle** has two starting materials, the two-carbon

8.11
The Krebs Cycle: Fumaric Acid

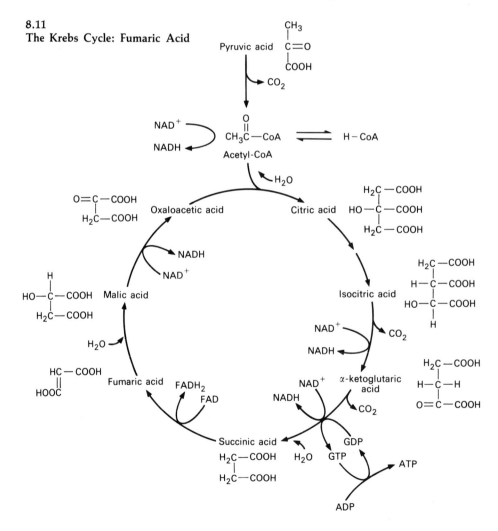

acetyl group and the four-carbon compound **oxalacetic acid.** One turn of the cycle dissimilates two carbons (equivalent to an acetyl group) as CO_2. Simultaneously, it regenerates one unit of oxalacetic acid (Fig. 8.11). The energy-yielding strategies of the reactions are similar to those seen in glycolysis.

Step 1 bonds the entering acetyl group with the oxalacetate molecule, an energy contribution, and Coenzyme A is released. Steps 2 and 3 are rearrangements that prepare the carbon chain for cleavage.

Step 4 combines oxidation of the substrate, isocitrate, with cleavage of the carbon chain; CO_2 is released and NAD is reduced. The other product is a five-carbon compound, **α-ketoglutaric acid,** a key metabolic intermediate. In addition to its role in the Krebs cycle, it can also be converted into an amino acid. It provides a bridge between carbohydrate and protein metabolism.

In Step 5, oxidation is again associated with decarboxylation. However, in this case the oxidation energy is used to attach a Coenzyme A unit to succinyl CoA via a high-energy linkage. Substrate-level phosphorylation occurs when the coenzyme is exergonically removed and the released energy is conserved by the manufacture of ATP. In Step 6, another pair of electrons are removed. This pair is at a slightly lower energy level than electron pairs previously removed. Thus it enters the electron-transport system at an intermediate point, carried by FAD.

Steps 7 and 8 further oxidize and rearrange, so the remaining four-carbon fragment reappears as oxalacetate.

The Krebs cycle dissimilates two carbons and provides four pairs of electrons for phosphorylation per turn. One high-energy molecule, GTP, is formed directly. In summarizing glucose oxidation as an **electron-providing** mechanism, we see that glycolysis provides two pairs of electrons and two pyruvates. Oxidative decarboxylation of the two pyruvates yields two pairs of electrons and two acetyl groups. The Krebs cycle consumes the acetyl groups and yields eight pairs of electrons and fully oxidized carbon (CO_2).

Fat Hydrolysis

Fats are also metabolized primarily for the production of electrons. Fats are **triglycerides**—three fatty acids bonded to glycerol. Before oxidation, the fatty acids are separated from the glycerol by **lipases.**

Triglycerides; Chapter 2, pp. 46–47

Fatty acid oxidation   The β-oxidation process removes two-carbon segments from the long carbon chains of fatty acids (Fig. 8.12). Before each two-carbon unit can be separated off, two pairs of electrons are removed, one pair by NAD and one by FAD. Then the carbon-to-carbon link is broken and a free acetyl group is available to be attached to Coenzyme A and diverted into the Krebs cycle.

Glycerol oxidation   The glycerol backbone is converted into glyceraldehyde-3-phosphate, enters glycolysis, and is broken down in the standard fashion.

Fats are more highly reduced organic molecules than are carbohydrates. Thus on oxidation they yield more pairs of electrons per gram, and thus more energy.

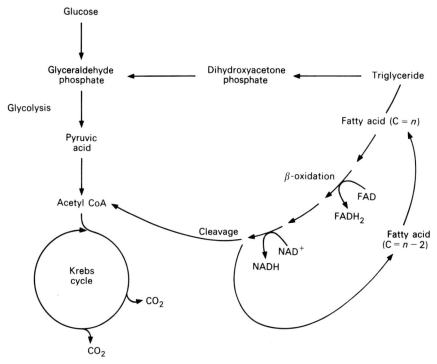

8.12

Fat Breakdown   The fatty acids are enzymatically separated from the glycerol backbone. One round of β-oxidation of a C–18 fatty acid would retain 16 carbons; the second round, 14 carbons, and so on.

This is the explanation for the high calorie value of fats and their efficiency as energy-storage materials.

Amino Acid Breakdown

Amino acids; Chapter 2, Figure 2.13

When a hydrolyzed protein is provided to organisms as the sole source of carbon, nitrogen, and energy, they use amino acids effectively as a source of energy for phosphorylation. They deal with the amino acid as if it is composed of two parts, an amino group and a carbon chain. When the amino group has been disposed of, the carbon chain can be utilized via reactions that terminate in the Krebs cycle.

Deamination   Enzymatic removal of the amino group leaves a **keto acid** (Fig. 8.13). Thus deamination of glutamic acid (an amino acid) gives α-ketoglutaric acid, the five-carbon intermediate of the Krebs cycle. Other amino acids are deaminated to other useful organic acids. The free amino groups form ammonium ion in solution. This nitrogenous waste is toxic because it raises the pH of the medium. Terrestrial animals convert amino nitrogen to urea, a less toxic waste.

Deamination

$$NH_2 - \overset{\overset{\displaystyle H}{|}}{\underset{\underset{\displaystyle CH_2}{|}}{C}} - COOH \xrightarrow{\text{Phenylalanine deaminase}} O = \overset{}{\underset{\underset{\displaystyle CH_2}{|}}{C}} - COOH \quad + \quad NH_3$$

(a) Phenylalanine

Phenylpyruvic acid

Transamination

$$\underset{\substack{\text{Glutamic} \\ \text{acid}}}{\overset{\displaystyle COOH}{\underset{\underset{\displaystyle COOH}{|}}{\underset{\underset{\displaystyle NH_2 - CH}{|}}{\underset{\underset{\displaystyle CH_2}{|}}{CH_2}}}}} \quad + \quad \underset{\substack{\text{Oxalacetic} \\ \text{acid}}}{\overset{\displaystyle COOH}{\underset{\underset{\displaystyle COOH}{|}}{\underset{\underset{\displaystyle CH_2}{|}}{C=O}}}} \xrightarrow{\substack{\text{Glutamic-} \\ \text{oxalactic} \\ \text{transaminase}}} \underset{\substack{\alpha\text{-ketoglutaric} \\ \text{acid}}}{\overset{\displaystyle COOH}{\underset{\underset{\displaystyle COOH}{|}}{\underset{\underset{\displaystyle C=O}{|}}{\underset{\underset{\displaystyle CH_2}{|}}{CH_2}}}}} \quad + \quad \underset{\substack{\text{Aspartic} \\ \text{acid}}}{\overset{\displaystyle COOH}{\underset{\underset{\displaystyle COOH}{|}}{\underset{\underset{\displaystyle CH_2}{|}}{HC-NH_2}}}}$$

(b)

8.13

(a) The amino group may be removed from an organic acid, leaving a carbon backbone to be broken down further. Phenylalanine deaminase is characteristic of *Providencia,* an enteric organism. (b) Transamination can create new amino acids at the expense of old ones. In this case, glutamic acid is used as the amino group donor. The amino group is transferred to oxalacetic acid, yielding respectively alpha-ketoglutaric acid and aspartic acid.

Transamination   Amino groups can be enzymatically transferred from one amino acid to another. Pyridoxal phosphate serves as the amino carrier coenzyme for this type of reaction. The donor compound, on losing the amino group, becomes an organic acid suitable for metabolic breakdown. Transamination functions both as a way to prepare amino acids for breakdown and, in reverse, as a means of synthesizing amino acids as needed.

Utilization of carbon chains   The organic acid residues of amino acids are broken down to enter metabolism as pyruvic acid, acetyl CoA, or some other by now familiar intermediary. Dissimilation and energy harvest proceed exactly as if the material had originated from glucose or fat metabolism (Fig. 8.14).

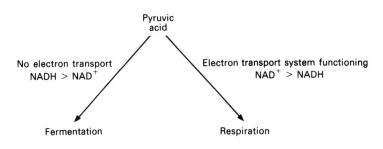

8.14

A bacterium that can both respire and ferment switches from one to the other on the basis of the ratio of oxidized to reduced NAD. Reduced NADH is required for fermentation, and oxidized NAD$^+$ is required to carry out oxidative decarboxylation and Krebs cycle reactions. If electron transport is operative, NAD$^+$ levels will be high; if it stops, NADH will accumulate, pushing the equilibrium toward fermentation.

PATTERNS OF HETEROTROPHIC ELECTRON USE

So far, we have analyzed the processes whereby energy-carrying electrons can be removed by oxidation from reduced organic nutrients such as carbohydrates and fats. In addition, we have examined substrate and oxidative phosphorylation, by which free-energy release is coupled to the manufacture of high-energy phosphate compounds. Let us now put these two parts of the metabolic map side by side and link up the main routes of heterotrophic energy utilization.

Fermentation   Fermentation is a method for producing ATP without oxygen. Since the atmosphere in which procaryotic life evolved was anaerobic, this comparatively simple mechanism is presumed to have appeared first. It persists to this day (in a much restricted form) in higher organisms, including ourselves. Certain groups of bacteria can subsist entirely on the very small ATP yield from fermentation.

Oxygen relationships; Chapter 7, pp. 197–198

A fermentative pathway begins with the glycolytic breakdown of glucose to pyruvic acid. Remember that glycolysis does not require oxygen; neither is it inhibited by it. Thus glycolysis by itself is neither aerobic nor anaerobic. In addition to yielding two pyruvic acid molecules and two ATP molecules, glycolysis also yields two molecules of reduced NADH.

NADH can be valuable to a cell as an electron source. However, it can also be a liability. In the reduced form, it cannot accept electrons; thus NADH is useless to keep glycolysis going. Consider for a moment that there is only a limited number of molecules of this key coenzyme per cell. The cell must have a means of off-loading the electrons from NADH in order to stay in business.

Fermentation reoxidizes NADH to NAD^+. The basic pattern is that NADH starts out reacting with pyruvic acid. The electrons from the NADH are transferred to the pyruvic acid that becomes an **organic electron acceptor**. The reduced compound or mixture of compounds that results is the **terminal** electron acceptor in the fermentation process. The electron acceptors (or the compound or mixture) still contain much of the original free energy of the glucose molecule. Fermentation processes yield **only** the two units of ATP derived by substrate-level phosphorylation in glycosis. When fermentation is occurring, no use is made of electron-transport pathways and oxidative phosphorylation; thus the electrons have no energy value.

Specific pathways of fermentation are characterized by a great diversity (Fig. 8.15). The lactic acid fermentation is shared by bacteria such as the streptococci and higher organisms. Pyruvic acid is converted quantitatively to lactic acid, a toxic substance. As it accumulates in the medium, the pH drops and growth is eventually arrested.

Yeasts; Chapter 5, p. 148

The alcoholic fermentation is characteristic of the yeasts. Pyruvic acid is converted to a mixture of ethanol and CO_2. Yeast fermentations have been exploited by human beings for thousands of years. Flour mixtures are leavened, or lightened, by the formation of expanding bubbles of CO_2 within the dough; the small amount of alcohol is driven off in baking. Still (noneffervescent) wines are prepared in ways that let the carbon dioxide escape, while beers and champagnes make use of both alcohol and carbonation.

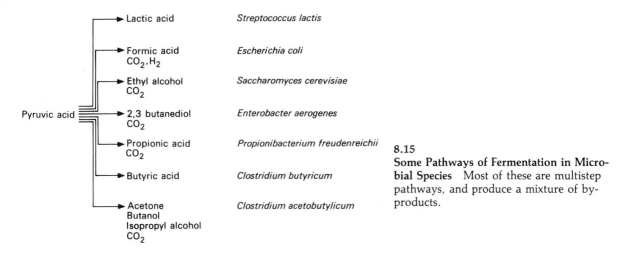

	Lactic acid	*Streptococcus lactis*
	Formic acid CO_2, H_2	*Escherichia coli*
	Ethyl alcohol CO_2	*Saccharomyces cerevisiae*
Pyruvic acid	2,3 butanediol CO_2	*Enterobacter aerogenes*
	Propionic acid CO_2	*Propionibacterium freudenreichii*
	Butyric acid	*Clostridium butyricum*
	Acetone Butanol Isopropyl alcohol CO_2	*Clostridium acetobutylicum*

8.15
Some Pathways of Fermentation in Microbial Species Most of these are multistep pathways, and produce a mixture of byproducts.

The pathways of fermentation are genetically determined. Pinpointing fermentative end products can be useful in identification. Alcohols and acids released by fermenting bacteria are of great industrial usefulness. A technology has been built up around the large-scale culture and purification of metabolic end products (Fig. 8.16).

8.16
Bulk Fermentation Equipment For commercial production, sophisticated engineering is used to provide the exact conditions for maximum yields. Pure stock starter cultures, themselves the product of genetic selection or engineering, are supplied when the fermenter is loaded.

Respiration   Respiratory metabolism derives electrons from reduced organic compounds, primarily sugars and fatty acids, and funnels them via an electron-transport system or cytochromes to an **inorganic electron acceptor.** Pathways that make use of electron transport are efficient energy-gatherers. Electrons are acquired through any of the oxidation pathways previously described and any oxidizable substrate may donate electrons.

Aerobic respiration. True **aerobic respiration** occurs when the terminal electron acceptor is oxygen. This process is found in strict aerobes and facultative anaerobes. Electrons are derived from glucose or other organic electron donors and passed to oxygen. CO_2 (the most oxidized form of carbon) and water are the only end products.

As each pair of electrons passes through the electron-transport system, the energy release is sufficient for three phosphorylations. One glucose molecule may (at maximum metabolic efficiency) be broken down with the subsequent recapture of energy in 38 high-energy bonds of ATP. This gives a calculated efficiency of 40 percent. The rest of the energy is dissipated as increased random molecular movement or heat. An actively decomposing pile of organic matter, such as compost or imperfectly dried hay, will release sufficient heat to ignite.

Anaerobic respiration. Certain bacteria use inorganic ions other than oxygen, such as nitrate (NO_3^-), to accept electrons (Table 8.4). This is called **anaerobic respiration.** The mechanisms are very similar to those of aerobic respiration. A partial or full electron-transport chain is required, with specific enzymes. If it is able to carry out anaerobic respiration, a cell deprived of oxygen is still able to obtain a high-energy yield from sugars.

Nitrogen and sulfur cycles; Chapter 6, pp. 165–168

Nitrate and sulfate reduction are essential links in the elemental cycles of nitrogen and sulfur in nature.

Many groups of bacteria can dispose of electrons by more than one route. For example, the aerobic genus *Pseudomonas* uses both aerobic and anaerobic respiration. Similarly, the strictly anaerobic *Clostridium* carries out fermentation and also respires anaerobically with nitrate. If two options are available, then the direction of metabolism will be determined by environmental conditions. Respiratory pathways

TABLE 8.4
The terminal electron acceptor in respiration

CONDITIONS	EQUATION
Aerobic	$2H^+ + 2e^- + \frac{1}{2}O_2 \rightarrow H_2O$
Anaerobic	$2H^+ + 2e^- + NO_3^- \rightarrow NO_2^-* + H_2O$
	$2H^+ + 2e^- + SO_4^= \rightarrow SO_3^= \dagger + H_2O$
	$8H^+ + 8e^- + CO_2 \rightarrow CH_4 + 2H_2O$

*Further reduction of nitrite (NO_2^-) to N_2 occurs in some species: this is denitrification.

†The sulfite ion ($SO_3^=$) is usually sequentially reduced through a number of steps to elemental sulfur (S) or H_2S with the addition of more pairs of electrons.

always take precedence, when feasible, because the energy yields are as much as 20 times higher. Bacteria retain a degree of flexibility in their metabolism that has been lost by higher organisms such as ourselves. In that flexibility lies their adaptability to a wide range of environments.

AUTOTROPHY

Autotrophs are self-feeders. In metabolic terms, they are independent because they synthesize their cellular ATP using electrons derived from a nonorganic source. The energetic electrons, on passage through a special form of electron-transport system, can phosphorylate. They also can be used to fix carbon, by reducing it from the oxidized form CO_2 to the more reduced hexoses. In discussing autotrophy, then, we will deal first with the method of ATP generation and second with the methods of CO_2 fixation, an aspect of biosynthesis.

The Photoautotrophs

Different patterns of photosynthesis have evolved during biological history and are still present. The true bacteria carry out **anoxygenic** photosynthesis, a process that does not evolve oxygen gas. Many photosynthetic bacteria are also anaerobic. Some species switch over to heterotrophic nutrition under suitable conditions.

Photosynthetic bacteria; Chapter 4, pp. 111–114

   The cyanobacteria carry out the more complex **oxygenic** photosynthesis; they split water molecules and give off molecular oxygen. Most cyanobacteria are obligate autotrophs.

The capture of energy   A photosynthetic energy-capture system or **photosystem** has three major components. The **antenna of light-harvesting pigments** absorb specific wavelengths of electromagnetic energy. Because different species possess pigments covering various absorption ranges, they utilize different portions of the visible spectrum.

   Photochemical reaction centers contain the pigment chlorophyll. This substance, which appears green, reflects green light and absorbs red and blue wavelengths.

   There are a number of different **chlorophylls**; all contain the **tetrapyrrol nucleus**, a structure found also in heme compounds and cytochromes. In chlorophyll, the metallic ion in the core is magnesium. When a chlorophyll molecule is struck by a **photon** (discrete bundle of light energy) the absorbed energy ejects an electron. The released electron reduces electron-transport coenzymes.

   In close association with the chlorophyll molecules are specialized **electron-transport chains**. The first coenzyme is an iron-containing substance called **ferredoxin**. Cytochromes are also present.

The light reactions—cyclic and noncyclic phosphorylation   The process of photosynthesis can be generally summarized by the equation

$$CO_2 + H_2X \longrightarrow C_6H_{12}O_6 + X.$$

Box 8.2

The Pasteur Effect and the Art of Making Home Brew

Many people try to make their own beer or wine at some point or other in their lives. After all, making their own home brew would be cheap. The result should be much better than the commercial product, shouldn't it?

For every amateur home-brewer venturing out, it is a good bet that there will soon be a jug, barrel, or plastic garbage can filled with nonalcoholic, vilely sour, undrinkable fluid. The Pasteur effect has struck again.

The yeast *Saccharomyces*, like so many other facultative anaerobes, possesses both fermentative and respiratory mechanisms. When making home brew the goal is to encourage fermentation, which produces alcohol and CO_2, and to shut off respiration, which uses up the sugars without giving the desired results. In order to do this, the conditions must be completely anaerobic.

The Pasteur effect, named for its discoverer, is the yeast's prompt shift from anaerobic to aerobic growth when air is introduced. Few novice home-brewers exclude all air, so their product contains little alcohol. They also tend to use baker's yeast (not a pure culture) and less than sterile conditions. As a result they introduce unwanted organisms, such as *Acetobacter*, that will oxidize any alcohol to acetic

In traditional wine making, the inoculum is the bloom on the grape, a mixture of yeast cells and bacteria. The juice extraction and fermentation processes are not carried out under aseptic conditions. Somehow or other, though, the product is more highly valued than wine made in the huge, American, stainless steel wineries.

acid if air is present. Other contaminants produce a variety of other unpleasant by-products.

Most fledgling brewers give up in disgust at this point. If they go on, they succeed by mastering three basic microbiological techniques—providing the right growth conditions, using a good stock culture, and maintaining clean conditions.

H_2X is the compound that serves as a source of reducing power. The equation, however, covers two quite separate processes, the light reactions and the dark reactions. The so-called light reactions are those that are directly dependent on a supply of photons and occur only while light is provided. They generate ATP and reducing power (Fig. 8.17).

Light reactions take place in photosystems. Photosystems are always embedded in membrane structures, to assure the correct spatial orientation of the components.

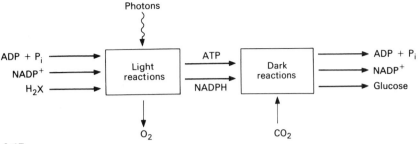

8.17
The photosynthetic process consists of two separate reaction complexes. The light and the dark reactions are shown here as "black boxes," with only the inputs and outputs specified. Light reactions are membrane-bound, whereas dark reactions, which need no light but depend on the products of light reactions, are carried out by soluble enzymes.

The light reactions coordinate these processes:

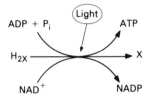

Thus light energy drives **photophosphorylation** and **photoreduction**. $NADP^+$ (nicotine adenine dinucleotide phosphate) is the coenzyme that functions as an electron carrier in most reactions of autotrophic metabolism.

Photosystem I. There are two different types of photosystem. Photosystem I is the only type found in the true bacteria. The high-energy electrons ejected from chlorophyll pass through a series of carriers and eventually return at a low-energy state to the chlorophyll molecule. This gives rise to the name **cyclic phosphorylation** (Fig. 8.18a). In the process the harvested energy is used to make ATP.

However, since the electrons return to chlorophyll, and not enough energy is captured to split water, reducing power is derived from another source. In bacteria, hydrogen atoms are commonly obtained from H_2 or from the oxidation of hydrogen sulfide (H_2S) to sulfur.

Photosystem II. This system contains chlorophylls that are able to transfer a larger portion of the photon's energy. This provides enough potential to break down water into protons, electrons, and oxygen gas. In the higher plants, eucaryotic algae and cyanobacteria, Photosystems I and II are both found working together. The electrons evicted from the chlorophyll reduce the NADP and are replaced by electrons

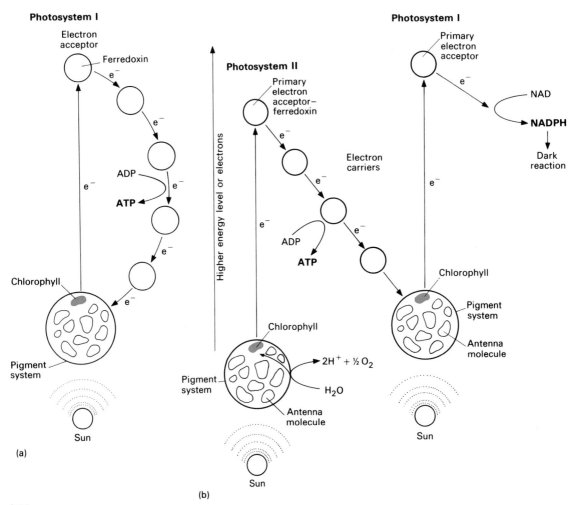

8.18

Phosphorylation   (a) Cyclic phosphorylation. When only Photosystem I is present, as in the photosynthetic bacteria, electrons return to the reactive chlorophyll molecule. (b) Noncyclic phosphorylation. When Photosystem II is added, water is split, and its electrons are pulled away by the chlorophyll to replace the electrons ejected by light. The free electrons pass to Photosystem I when another input of light energy activates their passage to NADP$^+$.

from the split water. The process is called **noncyclic phosphorylation** (Fig. 8.18b). It has three results: ATP is synthesized, NADP is reduced, and oxygen is released.

The reduced materials from which photosynthetic bacteria derive electrons are stable only in anaerobic environments because they are spontaneously oxidized in the presence of free oxygen. Therefore, these bacteria are restricted to environments in which anaerobic conditions coincide with light penetration. The cyanobacteria

Algae; Chapter 5, pp. 130–136

and the eucaryotic photosynthesizers evolve oxygen, live in an aerobic environment, and contribute directly to its aerobiosis.

The dark reactions Carbon fixation is endergonic. A highly oxidized material, CO_2, must be reduced; bonds must also be forged between carbons to form carbon chains. The reducing power supplied by light-generated NADPH carries out the first task, and light-generated ATP carries out the second.

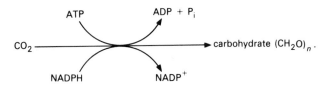

$$CO_2 \xrightarrow{\text{ATP} \quad \text{ADP + P}_i \quad \text{NADPH} \quad \text{NADP}^+} \text{carbohydrate (CH}_2\text{O)}_n.$$

However, the chemistry is far from simple. An elaborate reaction network, the **Calvin cycle**, incorporates each new carbon atom into preexisting carbohydrate molecules. **Carbon dioxide fixation** occurs when CO_2 reacts with the five-carbon sugar, **ribulose diphosphate** (RuDP). An unstable six-carbon product immediately separates into two molecules of the three-carbon compound 3-phospho-glyceric acid (3-PGA). A portion of the 3-PGA is combined to form fructose-6-phosphate. This early intermediate in glycolysis can be converted into glucose and the other useful carbohydrates. Fructose-6-phosphate is the exportable product of the Calvin cycle. The five-carbon RuDP acceptor is regenerated in a complex series of interchanges, utilizing the other portion of the 3-PGA.

The fixation of one molecule of CO_2 requires that three molecules of ATP and two molecules of NADPH be supplied by the light reactions. The enzymes of the Calvin cycle are found in the **carboxysomes** in procaryotes, and in the **stroma** of the eucaryotic chloroplast.

Carboxyomes; Chapter 4, pp. 104–105

Chemoautotrophs

Chemoautotrophy resembles photoautotrophy in many ways. Like photosynthesis, it can be divided into two processes. First, ATP and reducing power is generated in conjunction with specialized electron-transport systems embedded in membrane structures. By oxidation of an inorganic substrate, large amounts of NADPH are obtained. Some of the electrons can be passed through the respiratory electron-transport chain to yield ATP; others can be used as the source of reducing power in CO_2 fixation. Table 8.5 presents the energy-yielding reactions of some typical bacterial species. These reactions are of major significance in nutrient cycling, and have already been discussed in that connection in Chapter 6. Reduced forms of hydrogen (H_2), sulfur, iron, and nitrogen are the four inorganic energy sources exploited.

The Calvin cycle provides the means for carbon fixation in chemoautotrophs as well. Many such organisms can be shown to possess carboxysomes. Many questions relating to the ability of these organisms to use organic compounds as carbon sources remain unanswered.

TABLE 8.5
Energy-yielding reactions of some chemoautotrophs

GENUS	REDUCED SUBSTRATE	REACTION	OXIDIZED PRODUCT
Hydrogenomonas	H_2 gas	$H_2 + \frac{1}{2}O_2 \rightarrow H_2O$	H_2O
Thiobacillus	H_2S	$H_2S + \frac{1}{2}O_2 \rightarrow H_2O + S$	$S, SO_4^=$
		$S + 1\frac{1}{2}O_2 + H_2O \rightarrow H_2SO_4$	
Gallionella	Fe^{2+}	$2Fe^{2+} + \frac{1}{2}O_2 + H_2O \rightarrow 2Fe^{3+} + 2OH^-$	Fe^{3+}
Nitrobacter	HNO_2	$HNO_2 + \frac{1}{2}O_2 \rightarrow HNO_3$	HNO_3

SUMMARY

1. Cells are energy-converting systems that capture or extract energy from many sources, then redirect it toward their own growth and maintenance needs.

2. Each chemical reaction is catalyzed by a particular, unique enzyme. Enzymes are classified by the type of reaction they perform, their location, and the cellular regulation of their activity rate and synthesis.

3. A catalyst facilitates a reaction, causing it to occur more quickly. An enzyme's catalytic ability depends on the highly specific fit between enzyme and substrate. This brings substrate molecules in close proximity, markedly increasing the probability of a reaction. Also, for new compounds to be formed, the original compounds must be altered to expose combining sites; this means that covalent bonds must be broken, a process that requires energy. However, enzymes can alter the chemical environment around the substrate to reduce the activation energy so that the reaction may proceed.

4. Enzyme activity is controlled in two ways. Existing enzyme molecules can be controlled or stopped by feedback inhibition. The cell's genetic information may be regulated to start or stop the manufacture of new enzyme molecules.

5. The energy-utilizing mechanisms of cellular life channel energy from exergonic reactions to endergonic reactions. This activity utilizes coupled reactions in which the exergonic portion drives the endergonic portion. These processes are always less than completely efficient.

When heterotrophic organisms oxidize organic compounds, electrons are removed enzymatically and passed via coenzymes such as NAD^+ to a terminal electron acceptor. In passage, the electrons give off much free energy.

The released free energy then drives phosphorylation at the substrate level or via membrane-bound, electron-transport systems. The formation of the high-energy bond of ATP conserves the free energy.

6. The cell derives electrons from carbohydrates and fats. Proteins and other organic substances may also be oxidized. Glycolysis partially oxidizes and cleaves the glucose molecule to yield pyruvic acid. Then, if a supply of NAD^+ is pres-

ent, pyruvic acid will be oxidatively decarboxylated to an acetyl group, then oxidized via the Krebs cycle to CO_2.

Fats are cleaved to yield glycerol and fatty acids. The glycerol enters the glycolytic pathway; fatty acids are oxidized to remove acetyl groups that enter the Krebs cycle.

Proteins are hydrolyzed into amino acids. The amino group is removed by deamination or transamination, leaving an organic acid that directly or indirectly enters the Krebs cycle.

7. In anaerobic conditions, facultative and strictly anaerobic bacteria utilize pyruvic acid as an organic electron acceptor in fermentation. Gases, organic acids, and alcohols are produced as end products. The pyruvic acid cannot then be used for further energy extraction.

8. Respiration requires a functioning electron-transport mechanism. The electrons are transferred to oxygen (aerobic respiration) or to some other inorganic acceptor (anaerobic respiration). As long as the electron acceptor supply is adequate, cells will respire since oxidative phosphorylation can occur. If the electron acceptor supply is exhausted, NADH builds up, causing the metabolic equilibrium to favor fermentation. Many bacteria and yeasts switch readily from respiration to fermentation as the environment alters.

9. Photosynthetic organisms capture light energy. Chlorophyll molecules absorb light energy and emit electrons. As the emitted electrons move through electron-transport systems, phosphorylation occurs. Simultaneously, light energy also drives the endergonic reduction of coenzymes such as NADPH. The ATP and the reducing power generated are then utilized via the Calvin cycle to fix CO_2 into glucose.

10. Chemosynthetic organisms derive energy from reduced inorganic compounds, oxidizing them so that the released energy drives phosphorylation and coenzyme reduction. The Calvin cycle carries out carbon fixation in these organisms too.

11. All these metabolic pathways have a common pattern. High-energy electrons are removed from a chemical donor, transferred to an acceptor at a lower energy level, and the energy differential captured by the synthesis of ATP. Although the sources of electrons, the nature of the final acceptor, and the means of phosphorylation may vary, the basic pattern remains the same.

Study Topics

1. To which of the six functional classes of enzymes does each of the glycolytic enzymes belong?

2. When electrons are passed from a NADH/NAD$^+$ couple to a lactate/pryuvate couple, as in fermentation, is energy released or required?

3. There is an unexplained discrepancy between the data given in Table 8.2 and the order of the electron transport carriers in Fig. 8.11. Find it. If you have access to biochemistry references, try to seek out the explanation.

4. Using the data in Table 8.2, predict whether a bacterium respiring anaerobically with nitrate captures more or less energy than when it respires aerobically with oxygen.

5. Why does phosphorylation always occur simultaneously with or shortly after a redox reaction?

6. Describe why a pH gradient across a membrane can serve as a source of free energy.

7. Among present forms of life, the glycolytic pathway is used almost exclusively. The very similar Entner–Doudoroff pathway is used in only one procaryotic group. Both probably originated very early in evolution. Provide an explanation for the present situation. (References required.)

8. The Krebs cycle is a pathway for the oxidative breakdown of carbohydrates. Krebs cycle enzymes may be present and functional in organisms growing anaerobically without carbohydrates. Why?

9. If an aerobically growing organism is deprived of carbohydrates, its Krebs cycle can be kept operative. The correct organic intermediates are obtained from what sources?

10. Facultative anaerobes function both fermentatively and respiratively. Under most conditions both necessary enzyme sets will be present. The ratio of oxidized to reduced NAD is the "signal" that directs metabolism into one of the two paths. Predict the direction of metabolism when NAD^+ is much larger than NADH.

11. Describe how the energy of the photon is captured and converted to chemical bond energy in the light reactions.

12. In what ways does photosynthesis in the true bacteria differ from photosynthesis in cyanobacteria?

13. What aspect of metabolism is common to all autotrophs? Explain its operation.

14. In heterotronic respiration, fermentation, and in photoautotrophy and chemoautotrophy energy is captured as electrons move from an electron donor to an electron acceptor. Compare these four processes by completing the following equation for each, identifying A,B,C,D, and E.

$$\underset{A}{\text{electron source}} + \underset{B}{\text{electron acceptor}} \rightarrow \underset{C}{\text{oxidized products}} + \underset{D}{\text{reduced products}} + \underset{\text{Compound E}}{\text{energy storage.}}$$

Bibliography

General texts

Gottschalk, Gerhard, 1979, *Bacterial Metabolism*, New York: Springer-Verlag.

Stryer, Lubert, 1975, *Biochemistry*, San Francisco: W.H. Freeman.

Review articles

Fraenkel, D.G., and R.T. Vinopal, 1973, Carbohydrate metabolism in bacteria, *Ann. Rev. Microbiol.* **27**: 69–100.

Haddock, B.A., and C.W. Jones, 1977, Bacterial respiration, *Bacterial Review* **41**: 47–99.

Harold, F.M., 1977, Ion currents and physiological functions in microorganisms, *Ann. Rev. Microbiol.* **31**: 181–203.

Parson, W.W., 1974, Bacterial photosynthesis, *Ann. Rev. Microbiol.* **28**: 41–59.

9

Biosynthesis
and Genetic Control

Growth, repair, and reproduction all are processes requiring the creation of order among atoms, molecules, and structures. As ordering activities, they have two requirements. First, they occur only with continuous input of metabolic energy and progress at a rate directly determined by the rate of biological ATP generation. Second, they require the expression of a genetic blueprint.

In this chapter, we shall look at how the major classes of biochemicals are synthesized, emphasizing the synthesis of protein, the primary gene product. We shall see how microorganisms adjust their patterns of gene expression in response to environmental changes. The mechanisms for genetic change in procaryotes will be examined. These include mutation occurring within a single cell, as well as means of exchanging genetic materials among cells. Newly arising genetic variations in bacteria are rapidly expressed. This implies that we will have a changing relationship with microorganisms as our species and the bacteria co-evolve.

BIOSYNTHESIS OF STRUCTURAL AND RESERVE MATERIALS

It is estimated that about 150 different small organic molecules are needed for cellular growth and reproduction. These small molecules are the building blocks of the polysaccharides, lipids, proteins, and nucleic acids.

Biosynthesis of Carbohydrates

By reviewing Table 2.7, you will recall that microorganisms manufacture polysaccharides from a number of different simple sugar units and use them for a variety of purposes. Let us look at polysaccharide assembly methods.

Energy storage;
Chapter 3, pp. 84–85

Synthesis of glycogen, an intracellular reserve   Glycogen and poly-β-hydroxybutyrate are the most frequently encountered bacterial energy-storage materials. Glycogen is composed of only one type of monomer—glucose—and is a **homopolymer**. The strategy used for its synthesis is typical of cellular polymerizations.

A condensation reaction that covalently links two units is strongly endergonic. Thus one or both of the units must first be activated. They can either be phosphorylated by reaction with a high-energy compound such as ATP or become attached via a high-energy bond to a carrier coenzyme.

Glycogen is assembled by addition of glucose–phosphate units to the growing polymer. Prior to condensation the unit to be added reacts with ATP.

$$\text{glucose-1-phosphate} + \text{ATP} \rightarrow \text{ADP-glucose} + \text{PP}_i \text{ (pyrophosphate)}.$$

This is followed by the polymerization step

$$\text{ADP-glucose} + \text{glycogen}_n \rightarrow \text{ADP} + \text{glycogen}_{n+1},$$

where n stands for the number of glucose units in the strand. The initial phosphorylation of glucose-1-phosphate also required the expenditure of one ATP (not shown). Thus the addition of one monomer unit to a polymer requires the energy equivalent of two ATPs. This is generally the case for all polymerizations.

Other polysaccharides   Other types of cellular polysaccharide find use in capsules, the carbohydrate portion of lipopolysaccharides, and glycoproteins.

Biosynthesis of Lipids

Triglycerides   Neutral fats are formed by linking glycerol molecules, derived from glycolysis intermediates, with three fatty acids. The commonly occurring fatty acids have even-numbered carbon chains. They are synthesized by a reductive pathway that adds two carbon atoms at a time. The stepwise addition reactions involve adding an activated acetyl group, followed by two reduction steps. NADPH serves as the electron donor in this process; it is the carrier used in all biosynthetic reductions. NAD serves as electron acceptor/donor in energy-yielding metabolism.

Phospholipids   The phospholipid backbone is the three-carbon compound 3-glycerophosphate. Two fatty acids are added to the nonphosphate bearing carbons. The phosphate group may then have polar groups such as choline or ethanol added to form the various phospholipids.

9.1
Of the 20 amino acids, 19 are assembled using carbon backbones derived from carbohydrate metabolism. Only histidine arises from other pathways. If the cell is not receiving sugars in the nutrient media, the glycolytic and Krebs cycle intermediates may be in short supply. However, these substances can often be synthesized from other cellular constituents. Many organisms are genetically lacking the ability to synthesize all of the enzymes needed to make 20 amino acids.

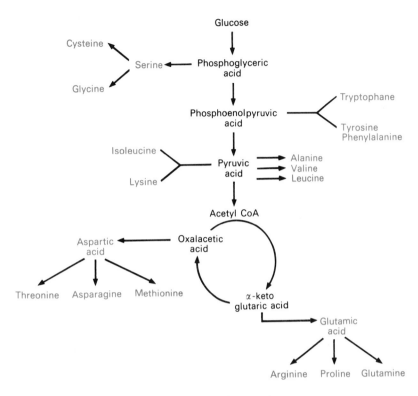

Biosynthesis of Peptidoglycan

The amino sugars   N-acetylglucosamine (NAG) and N-acetylmuramic acid (NAM) are both derived from fructose-6-phosphate, an early glycolysis intermediate. The amino groups are donated by glutamine and the acetyl groups are donated by acetyl CoA. NAG and NAM are transferred across the cell membrane in an activated form, then covalently linked to form the repeating network of the polymer.

Structure of PG; Figure 3.10

The peptide crosslinks   Before crossing the membrane, five amino acids are added one at a time to the sugar units. The peptide bonds formed are an exception to the normal cellular mechanism for creation of linkages between amino acids since they are not under direct genetic control. However, certain types of transfer RNA are involved in the assembly process. Once the subunits are in place exterior to the membrane, the peptides may or may not be enzymatically cross-linked to the adjacent peptide to complete the three-dimensional meshwork of peptidoglycan.

Biosynthesis of Amino Acids

A cell needs 20 amino acids in order to assemble proteins. The nutritionally self-sufficient bacteria, of which *E. coli* is an example, can synthesize all of these from intermediates produced in the course of glucose metabolism and from an inorganic nitrogen source that can be transformed into amino groups. There are six pathways that result in the synthesis of the 20 compounds. These are summarized in Fig. 9.1. The pathways themselves are quite intricate. Research leading to the clarification of these pathways and analysis of their regulation has been one of the more rewarding projects to occupy microbiologists in the molecular era.

Krebs cycle; Chapter 8, pp. 232–233

Note that Krebs cycle intermediates are essential for the manufacture of eight amino acids. This provides further evidence of the indispensable nature of the Krebs cycle. If an organism is actively using amino acids, withdrawal of incompletely oxidized intermediates from the Krebs cycle will considerably reduce the energy yield from carbohydrate oxidization.

Only one amino acid, histidine, is produced by means that are largely independent of carbohydrate metabolism. Its biosynthesis is closely linked to that of the purines; it receives only its amino group from glutamine.

Biosynthesis of Nucleoside Triphosphates

Purines and pyrimidines   The bases themselves are assembled from carbon and nitrogen atoms derived from various sources. The core of single-ringed pyrimidines contains four carbon atoms and two nitrogen atoms. These derive from aspartic acid (3C, 1N) to which a carbamyl group (1C, 1N) is added. The resulting unit, uracil, is modified to yield cytosine and thymine.

Purines contain two heterocyclic rings with a core of nine atoms, of which four are nitrogen and five are carbon. Their assembly is extremely complex and requires the methyl-transporting coenzyme, **folic acid**. Both adenosine and guanine are derived from an intermediate—inosine—that also appears as a "rare base" in certain nucleic acids.

The pentoses   Ribose is produced in metabolism from other sugars. It is incorporated into nucleotides as ribose-5-phosphate, an activated form.

Assembly of nucleotides   We may view a cell as needing eight different nucleotides for synthesis of DNA and RNA (Fig. 9.2). All eight are needed in the triphosphate form. Four of them (ATP, GTP, CTP, and UTP) must contain ribose for incorporation into RNA. The other four (dATP, dGTP, dCTP, and dTTP) must contain deoxyribose for incorporation into DNA.

In synthesis of both purines and pyrimidines, a ribose bearing one phosphate group is attached early in the assembly process before the base itself has been completed. Then a second phosphate is added, and then a third. As a final step, some of the units bearing T will have the ribose sugar enzymatically reduced to deoxyribose; thus the units will become deoxyribonucleotides.

BIOSYNTHESIS OF INFORMATIONAL MACROMOLECULES

We have, so far, accounted for the synthesis of many different cellular components. Their manufacture is directed by the catalytic properties of specific enzymes, under indirect genetic control. The polymers we are about to consider—proteins and nu-

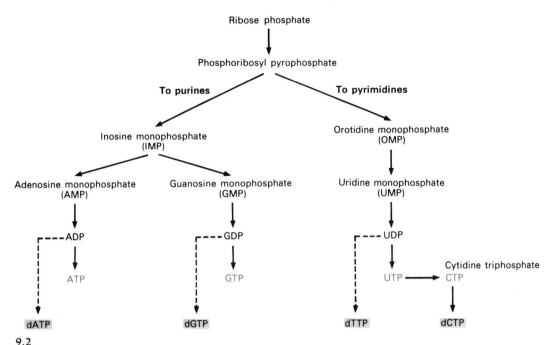

9.2
Nucleotides   Ribose is the starting material for the assembly of the nucleotides. In *E. coli*, the nucleotides appear first in the ribonucleotide monophosphate form (AMP, GMP, UMP). These are then phosphorylated again to the diphosphate form (ADP, GDP, UDP). These three compounds are then modified further to convert the ribose sugar to deoxyribose. Further modification yields the deoxyribonucleotide triphosphates dATP, dGTP, dCTP, and dTTP.

cleic acids—are all unique, ordered sequences. Because they are sequences of variable units, they are chemically heterogenous and are called **heteropolymers**. Proteins and nucleic acids must be assembled with constant reference to a pattern or **template** that is physically present at the site of assembly. They are called **informational** because a source of information is needed for their manufacture.

The Bacterial Genome

The **genome** of an organism is the sum total of all of its genes and thus the source of all information for macromolecular assembly. Those familiar with classical or Mendelian genetics will recognize the term **genotype**, describing the genetically determined characteristics of an organism. **Phenotype**, on the other hand, is the observed appearance and behavior of the organism; this is determined not only by genes but also by the interaction of organism and environment.

In the period from 1940 on, geneticists have been coming to understand the structure and function of the bacterial genome. From these studies have come great advances in molecular genetics. Bacteria have proved to be apt tools, primarily because of the comparative simplicity of their nuclear material and its patterns of expression.

Organization   Essential genetic information in procaryotes is contained in a single, circular, double helix of DNA. This DNA molecule has associated with it molecules of basic proteins and large numbers of inorganic cations; these have the function of neutralizing DNA's strong anionic character. Bacterial DNA can be extracted from the cell in a form that is chemically pure and physically intact, not without difficulty

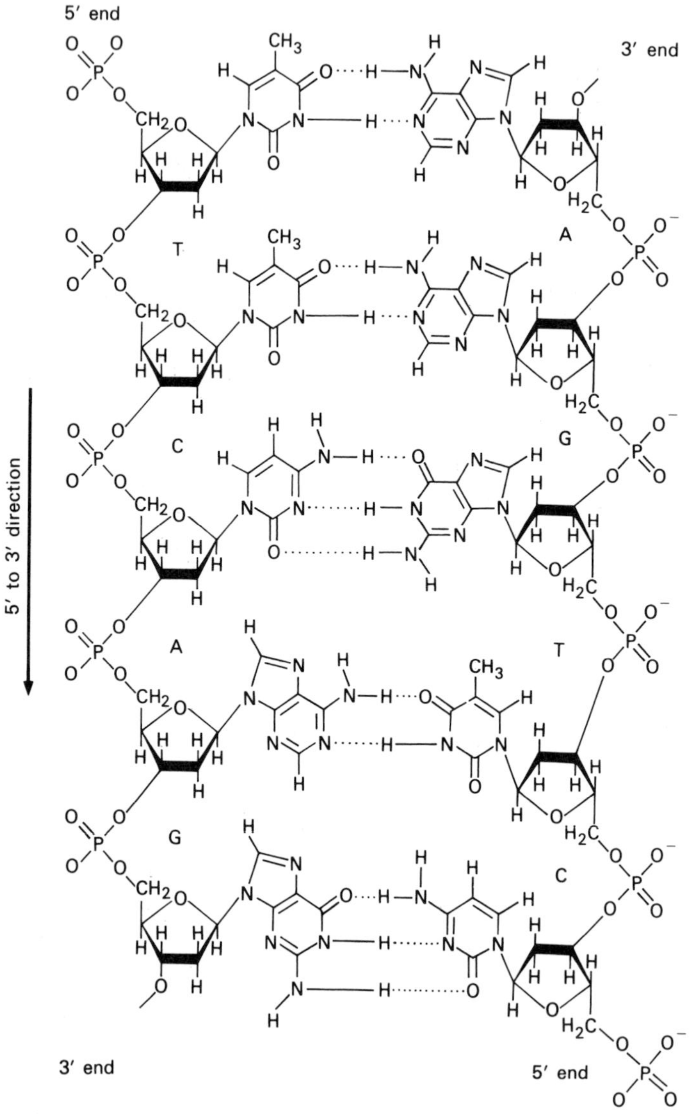

9.3
DNA   A DNA double helix has two strands with opposite polarities. Since DNA can be synthesized only from 5′ → 3′, the left-hand strand here would be copied from bottom to top, and the right-hand one from top to bottom.

but certainly far more readily than from higher organisms. Before proceeding further, it would be helpful to reexamine the information about the nucleic acids given in Chapter 2.

Nucleic acids, Chapter 2, pp. 55–56

DNA replication   In order to discuss the replication process, it will be necessary to get the directional nature of nucleic acids firmly in mind (Fig. 9.3). In a single strand of DNA, there is one end bearing a 5′ phosphate group; the other end has an exposed 3′ OH group on the pentose. The positive **polarity** of this strand—the 5′-3′ direction—runs from top to bottom. Its paired complement, on the other side, runs in the opposite direction, and is **antiparallel**. The significance of strand polarity is that DNA synthetases assemble DNA **only** from 5′ to 3′, in other words, by adding nucleotides to exposed 3′ OH ends. Thus the enzymes cannot move down both strands copying both simultaneously in the same direction. The situation is far more complex.

In order to replicate DNA, the helix must first be unwound. It has been suggested that this process is catalyzed by the enzyme **DNA gyrase** (Fig. 9.4). Then the hydrogen bonds holding the base pairs together must be broken. It appears that multiple molecules of a **helix-destabilizing protein** (HDP) are required. The related activities of unwinding and helix separation create a **replication fork**, and it is at this fork that the actual copying of the DNA may take place. In *E. coli*, at least 13 enzymatic proteins are associated with the fork.

The actual replication is asymmetrical. Copying of the "old" 3′-5′ strand is straightforward. As each "old" base is exposed, a "new" nucleotide base-pairs to it and is enzymatically linked by a condensation reaction to the preceding "new" base. The strand grows then in the 5′-3′ direction.

The "old" 5′-3′ strand presents a problem, however, since it must also be copied with the correct polarity. In fact, this **lagging strand** is copied in a discontinuous manner. At certain sites copying begins, with the first segment of the copy being assembled from RNA bases. The RNA **primer** section is followed by a longer true DNA copy.

9.4
At the growing point or replication fork, enzymatic reactions supply energy to unwind the disestablished helix. One strand, the old 3′ → 5′, is copied directly in DNA. The other, the old 5′ → 3′, is copied in fragments as unwinding progresses. Each fragment starts with a RNA primer section. Further out from the fork the primer is enzymatically removed, and the gap filled in with complementary deoxyribonucleotide.

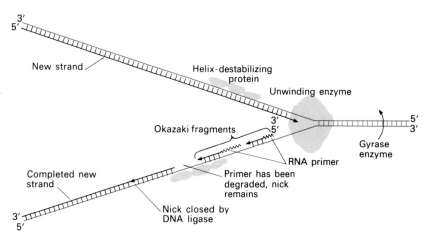

Then the RNA primer sections are degraded or "erased" and their place filled in with the correct DNA bases by the repair enzyme **DNA ligase**. This apparently wasteful process is explained by the need for complete copy fidelity. The beginning of the copy is where errors are most likely to occur. This erasing mechanism may have arisen as a fail-safe procedure to replace the error-prone portions "just in case."

The replication fork moves completely around the circular chromosome of a procaryote. DNA synthesis is **semiconservative**; each of the two daughter strands contains one "old" and one "new" strand. The replication process places heavy energy demands on the cell.

DNA modification   We have discussed the assembly of DNA from the four familiar deoxyribonucleotides as it actually occurs. However, following synthesis of the new DNA strands, certain sequences of bases may be enzymatically modified by addition of methyl groups, for example. These chemical changes do not affect base-pairing capacity. The changes executed are unique to the species in whose cells they occur; modification marks the DNA as uniquely *E. coli* DNA or *S. pneumoniae* DNA. Such modifications serve a role in controlling DNA transfer among species.

The Eucaryotic Genome

Structural features   Despite recent progress, much remains unclear about the structure of eucaryotic nuclear material. The chromosomes appear linear during their condensed phase in mitosis. When uncoiled, they appear to be organized in regular repeating units called **nucleosomes**. A nucloesome contains a length of DNA of about 140 base pairs, and this is coated by four histones, probably two molecules each, to form an octamer. A nucleosome does not, generally speaking, equal a gene, since the information content would suffice only for the smallest proteins.

Replication   DNA replication in eucaryotes is restricted to those times in the cell cycle during which the DNA is not condensed, and thus to times during which the cell is neither prepared for dividing nor actually dividing. There are portions of the DNA that are **early** replicating and others that are **late** replicating. The replication process is semiconservative, and similar enzyme activities to those of procaryotes are present. However, much of the replication process, which is obscured by the presence of the highly organized histone packing, remains to be elucidated.

Transcription

To transcribe, in common terms, means to make a copy of something. Normally, one makes copies only of that portion of a document for which there is an immediate need. Thus in the cellular process of transcription, RNA copies are made of those parts of the DNA genome necessary for current cell activities.

The classes of RNA   There are three major classes of RNA transcript copied from the DNA. They differ in their size, in the number of copies of each in existence at any one time, and in their role in protein synthesis. The three major classes are messenger RNA (mRNA), ribosomal RNA (rRNA), and transfer RNA (tRNA).

Messenger RNA (mRNA). This contains the largest number of different (unique sequence) RNAs, yet it makes up only about 5 percent of the total cell RNA. Most of the total length of transcribed DNA yields RNA copies of the mRNA type. Their molecular size is large and heterogeneous, averaging 1,500 bases in length. They are predominantly single-stranded. Each DNA gene, when copied, yields a unique mRNA. Thus the sequence of one message differs from all other mRNAs except those copied from the same gene. Messenger RNA provides the direct sequence instructions to select, line up, and link together individual amino acids.

Ribosomal RNA (rRNA). There are three different rRNA molecules, many hundreds of copies of each per cell. rRNA is always the most abundant type of RNA in the cell. Each cell contains multiple DNA sequences to produce rRNA transcripts; the cell mass-produces this substance.

It appears that a single, long RNA molecule is transcribed. It is then cleaved to yield two parts—a 16S and a 23S portion— by a process called **posttranscriptional modification**. The 16S portion becomes associated with molecules of 19 different proteins to form the small ribosomal subunit (30S). A 23S rRNA plus about 35 proteins form the 50S or large ribosomal subunit, in association with a small 5S RNA (Fig. 9.5). The sizes indicated are those for the procaryotic cell. Eucaryotic rRNAs and ribosomes are larger. Ribosomal aggregates provide attachment sites for the molecular participants in the protein-synthesizing apparatus.

Transfer RNA (tRNA). Molecules of tRNA are the smallest among RNA molecules, averaging 75 nucleotides in length. It is not clear how many different types are present in a given cell but on theoretical grounds there must be at least 20 and no more

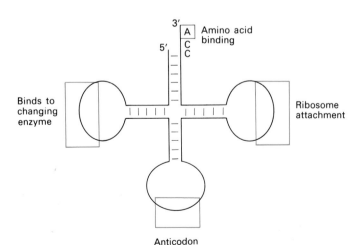

Anticodon

9.5

A Transfer RNA A transfer RNA has a base-paired secondary structure, producing a clover-leaf profile. Four functional sites are present. The amino acid binds via a carboxyl group to the 3′ terminal of adenine. The right-hand arm is important in binding to the 50S ribosomal subunit. The anticodon, a 3-base sequence unique to each different tRNA, binds to the messenger codon. The left-hand arm identifies the individual tRNAs as substrates to the charging enzymes that attach the amino acid.

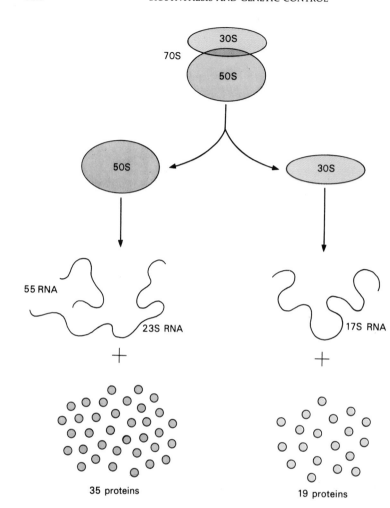

9.6
sRNA Molecule Each ribosomal
subunit is a complex of a long, linear
sRNA molecule in association with
numerous proteins. It is not a smooth
oval body, however, as shown here.
Electron micrographs reveal pockets of
various sizes and shapes, whose significance is unclear. Possibly some such
surface features might form specific
combining sites.

than 61. Experimental results point toward the higher number. Transfer RNA is the
second most abundant RNA type. Each tRNA molecule has a complex, three-dimensional, secondary structure (Fig. 9.6). All share key base sequences, but each also
has a unique three-base sequence called the **anticodon.** Transfer RNAs are acceptors
for activated amino acids, which they deliver to the appropriate coded locations in
the growing polypeptide chain. They also function in bacterial cell-wall synthesis
and in a variety of regulatory roles.

RNA synthesis   RNA synthesis is a localized process on the DNA genome. At a
given point in time, single genes or groups of genes will be transcribed. The copies
(mRNA, rRNA, or tRNA) are manufactured by the same chemical steps with the
same complex enzyme.

The DNA molecule serves as the template for aligning RNA nucleotides. The direction of synthesis is 5'-3', just as with DNA synthesis. However, the directional problem of nucleic acid synthesis is resolved because only one DNA strand is copied into RNA. The complementary DNA strand does not simultaneously serve as a template.

The enzyme responsible for transcription is **DNA-dependent RNA polymerase**. Its name tells us that it uses DNA as a template and makes RNA polymer copies. It is a quaternary structure composed of four permanently associated polypeptides and two temporarily associated regulatory proteins. The RNA polymerase binds to the beginning of genes at an initiation or **promoter** region by means of the **sigma** subunit of the enzyme (Fig. 9.7). Once binding occurs, sigma dissociates. The DNA is locally unwound, then RNA nucleotides are base-paired and condensed into place. A second regulatory protein factor, **rho**, seems to recognize the stop signals or **terminators** at the ends of genes and causes the incorporation of new bases to cease.

After synthesis is complete, **modification** of the products occurs, with cleavage into smaller units, addition of methyl groups, or modification of the bases.

The control point   If a cell is to express a gene, that gene must first be transcribed. Thus if the gene is not to be expressed, the most efficient way to prevent expression is to prevent transcription. As we shall see, genetic regulation centers on the control

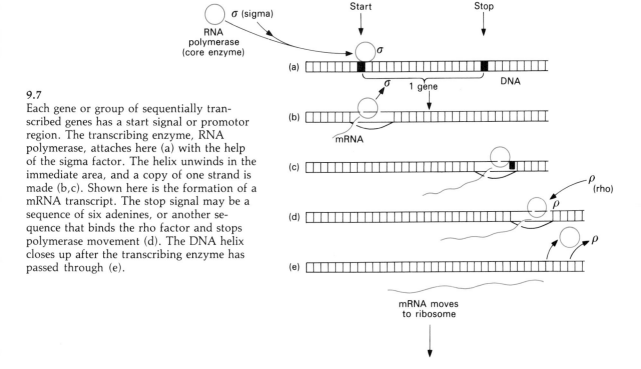

9.7
Each gene or group of sequentially transcribed genes has a start signal or promotor region. The transcribing enzyme, RNA polymerase, attaches here (a) with the help of the sigma factor. The helix unwinds in the immediate area, and a copy of one strand is made (b,c). Shown here is the formation of a mRNA transcript. The stop signal may be a sequence of six adenines, or another sequence that binds the rho factor and stops polymerase movement (d). The DNA helix closes up after the transcribing enzyme has passed through (e).

of transcription and, more specifically, on the accessibility of the promoter region to which RNA polymerase must attach to initiate transcription.

Translation

Translation poses two problems. First, the information encoded in unique sequences of nucleotides must be converted into a quite different chemical form—unique sequences of amino acids. This translation process, like those associated with verbal languages, requires a "dictionary," that is, a fixed statement of equivalents. We call this the **genetic code.** Second, enzymatic and structural mechanisms must bring the message, the amino acids, and the synthetase enzymes together to catalyze the formation of peptide bonds.

The genetic code   The essential features of the genetic code were first deduced on purely logical grounds. There are four bases available. These must be arranged to provide unequivocal instructions for about 20 amino acids. If the bases are arranged in all possible groups of three, there will be 4^3, or 64, possible three-letter "words," more than enough to convey the 20 needed pieces of information. In a sequence of experiments, each of the three-letter message units or **codons** was assembled from ribonucleotides. Each was then tested for its ability to bind one of the 20 amino acids, in a test-tube simulation of protein synthesis. From these results, the genetic code (Table 9.1) was constructed. The code is a universal biological language in all living things.

TABLE 9.1
The genetic code

FIRST LETTER	MIDDLE LETTER							
	U		C		A		G	
U	UUU	Phe	UCU	Ser	UAU	Tyr	UGU	Cys
	UUC	Phe	UCC	Ser	UAC	Tyr	UGC	Cys
	UUA	Leu	UCA	Ser	UAA	Stop	UGA	Stop
	UUG	Leu	UCG	Ser	UAG	Stop	UGG	Trp
C	CUU	Leu	CCU	Pro	CAU	His	CGU	Arg
	CUC	Leu	CCC	Pro	CAC	His	CGC	Arg
	CUA	Leu	CCA	Pro	CAA	Gln	CGA	Arg
	CUG	Leu	CCG	Pro	CAG	Gln	CGG	Arg
A	AUU	Ile	ACU	Thr	AAU	Asn	AGU	Ser
	AUC	Ile	ACC	Thr	AAC	Asn	AGC	Ser
	AUA	Ile	ACA	Thr	AAA	Lys	AGA	Arg
	AUG	Met	ACG	Thr	AAG	Lys	AGG	Arg
G	GUU	Val	GCU	Ala	GAU	Asp	GGU	Gly
	GUC	Val	GCC	Ala	GAC	Asp	GGC	Gly
	GUA	Val	GCA	Ala	GAA	Glu	GGA	Gly
	GUG	Val	GCG	Ala	GAG	Glu	GGG	Gly

An RNA message is an **unpunctuated sequence** of three-base codons. As each codon in the linear message is translated, the equivalent amino acid will be incorporated into a linear protein. Thus the information in the message is **colinear** with the information in the protein product. If a message contains 1,500 bases, it will code a protein the length of 500 amino acids. Any change in the sequence or number of mRNA bases will produce a corresponding change in the sequence of amino acids in the protein product.

Amino acid activation   We have seen previously that biosynthesis of any polymer requires that the subunits be activated first. Amino acids are no exception. Activation is a two-step process.

The first step is nonspecific; that is, all amino acids are handled in the same way. They react with ATP with the liberation of PP_i. By the action of a transferase they become covalently linked to the AMP residue via a high-energy bond. This bond will provide the stored energy to form this amino acid's peptide bond to its neighbor in the protein.

The second step is highly specific. The cell contains an array of **charging enzymes**, probably as many as there are different tRNAs. These highly specific enzymes recognize two substrates—a particular amino acid–AMP molecule and the correct tRNA. The charging enzyme then moves the amino acid from the AMP molecule to the terminal adenine unit of the tRNA. Because of the recognition system, it is not likely that a tRNA, under normal conditions, could become charged with any amino acid other than the correct one.

Step 1: amino acid + ATP → aminoacyl-AMP + PP_i.
Step 2: aminoacyl-AMP + tRNA → aminoacyl-tRNA + AMP.

Thus the message moves over the ribosome, being decoded one codon at a time. Each entering tRNA attaches to the A-site (Fig. 9.8), has the existing peptide trans-

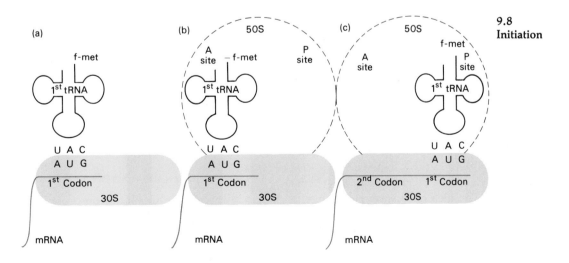

**9.8
Initiation**

9.9
Elongation

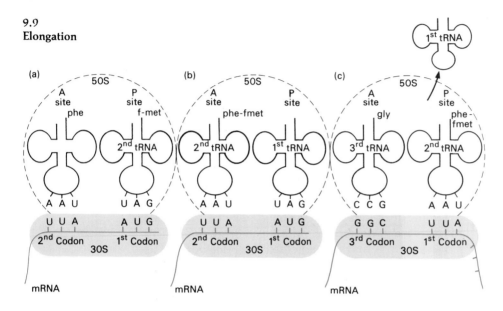

ferred to it from the tRNA in the P-site, then moves to the P-site itself as the discharged tRNA vacates it (Fig. 9.9). The ribosome may be viewed as the fixed point across which a series of codons and charged tRNAs moves (Fig. 9.10). Several initiation complexes can form on one message. Once the complex has moved sufficiently far down the message, a new start will be made. One mRNA can be completely decoded a number of times and direct the synthesis of several molecules of its product (Fig. 9.11).

Protein modification   After synthesis, the polypeptide may be cleaved, disulfide bonds may be formed, and/or carbohydrates or lipids added.

A comparison of procaryotic and eucaryotic protein synthesis   Eucaryotic protein synthesis is generally similar to procaryotic protein synthesis. However, procaryotic ribosome structures, and some of their associated protein factors, differ from those of the eucaryotes. These differences are great enough to allow a number of substances (chloramphenicol, tetracycline, etc.) to inhibit procaryotic protein synthesis. These substances have only minimal effect on eucaryotic protein synthesis. Chloramphenicol and tetracycline are antibiotics extremely effective in combatting bacterial infections.

Antimicrobial drugs;
Chapter 26

PHENOTYPIC CHANGES IN PROCARYOTES

The idea of phenotypic change, a change in the **pattern of expression** of genes, is clear when we consider everyday examples. Nutrition and exercise affect growth

**9.10
Termination**

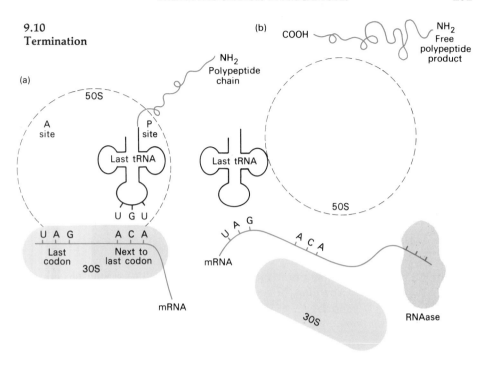

(a)

(b)

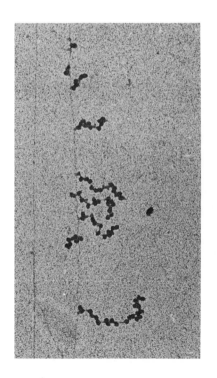

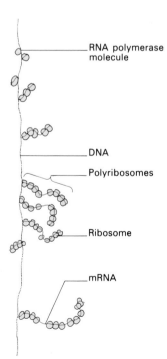

RNA polymerase molecule

DNA

Polyribosomes

Ribosome

mRNA

**9.11
The Expression of a Bacterial
Gene** Nine molecules of RNA
polymerase have attached themselves sequentially to this gene.
Moving along it, RNA transcripts
have been produced and attached
to the strands at approximately
right angles. The longest one, at
bottom, was the first to begin synthesis. (See *Science* **169**: 392, 1970,
reference) As each mRNA is synthesized, protein synthesis initiates
and ribosomes attach. They move
from the free end of the RNA in
toward the gene, their movement
just about keeping up with the
rate of the RNA growth. As the
ribosomes move inward, new
ribosomal complexes initiate at the
free end. The "oldest" RNAs
have the most ribosomal complexes.

rates and development of animals. Stunting effects in plants follow their exposure to environmental pollutants. In discussing phenotypic changes, we assume that the genetic material and its information content are unchanged; what is altered is how the genes are expressed.

Responses to the Environment

Morphology and staining   During different growth phases, bacteria may exhibit variable forms. In response to nutrient deprivation, aging, and other events, motility may be lost, cell shape may become irregular, and the integrity of the peptidoglycan layers of the wall may weaken, causing changes in the Gram reaction. Like all phenotypic changes, they are temporary; they may be reversed by transfer of the bacteria to new media.

Stationary phase; Chapter 7, p. 308

Colonial morphology   Alterations in colony characteristics reflect the moisture content of the media and the availability of suitable substrates for capsule synthesis. Crowding, of course, markedly changes appearance of colonies, a result of nutrient shortage.

Enzyme activity   As discussed in Chapter 8, the activity of existing enzyme molecules may be markedly reduced by feedback inhibition. The inhibitory effect is quickly relieved if the enzyme's product is again needed.

Feedback inhibition; Chapter 8, pp. 219–221

Regulation at the Transcriptional Level

By controlling the initiation of transcription, the cell can control the **amount** of gene product present from none to moderate to maximum. You are already familiar with the terms **inducible**, **repressible**, and **constitutive** enzymes. Let us examine some hypotheses about the mechanism of transcriptional control in bacteria.

Enzyme classes; Chapter 8, pp. 214–217

Functional subdivisions of the DNA   By accurate measurement of the length of a bacterial cell's DNA, it has been calculated that the DNA is theoretically sufficient to code for about 5,000 average-size proteins. As our knowledge of the actual composition of this bulk DNA has improved, it has become clear that there are at least four functional types of DNA, only two of which code for proteins.

The first type composes the genes for rRNA and tRNA. These genes are transcribed, but their RNA products are not intended to be translated: instead, they assist in the translation of the mRNAs.

A second type of DNA is not transcribed into RNA at all. This untranscribed DNA has a "punctuation" role as spacer, starter areas such as the promoter, and stop signals.

There is an undetermined number of **regulatory genes**—genes that when transcribed and translated yield a protein product, a **regulator**, whose cellular role is to influence the transcription of other genes.

The fourth type, constituting the majority of the DNA, contains **structural genes** that on expression give rise to enzymatic proteins and structural materials for cell growth.

Arrangement of genetic regions   In bacterial chromosomes, genetic functions seem to be arranged in groups. Reading along the DNA in the direction of transcription, there is first a **promoter** to which RNA polymerase attaches. Then comes an **operator region** to which regulator proteins may attach, blocking the forward movement of the RNA polymerase. Last comes the structural gene or genes. Transcription of structural genes occurs only if the enzyme passes over at least two control points. This linear sequence—control elements plus "controlled" structural gene(s)—is called an **operon**. Operons may be controlled by a repressible or inducible response to the environment.

Repressible systems   These systems regulate—that is, they repress—when the end product of the enzymatic process they control builds up (Fig. 9.12). For example, the amino acid, tryptophane, appears as the last product in a sequence of enzymatically catalyzed reactions. In elevated concentrations, it combines specifically with the regulator protein for the tryptophane operon. The combined unit attaches to the pro-

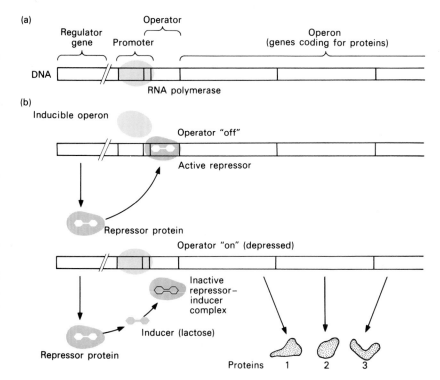

9.12
The Lactose Operon   The lactose operon is inducible. When lactose enters the cell, it binds to the repressor protein, removing the barrier to transcription. Structural genes are transcribed, producing proteins necessary for lactose transport and utilization. Once the last molecules of lactose, including those attached to the repressor, are used up, transcription stops.

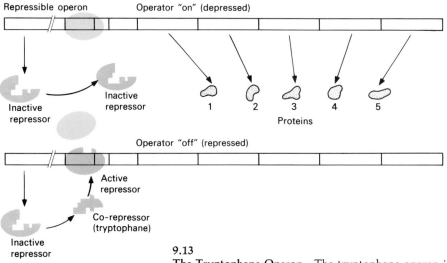

9.13

The Tryptophane Operon   The tryptophane operon is repressible. A regulatory gene produces a specific repressor. This substance is inactive unless it combines with tryptophane, the end product of the biosynthetic pathway and corepressor. Together, these block transcription of the enzymes of this operon. Once cellular tryptophane levels go down, the repressor–corepressor complex dissociates, and transcription proceeds with the appearance of the five enzymes of the tryptophane pathway.

moter and blocks further expression of the **trp** genes. This results in a gradual reduction in the level of tryptophane synthesis as existing enzyme molecules wear out and are diluted by cell division without being replaced. However, as tryptophane levels fall to a low point, the regulator-bound amino acid molecules will be removed. This frees the promoter; transcription begins, biosynthetic enzymes are made, and amino acid production rises again. Repressible systems regulate biosynthetic pathways and respond to intracellular signals. In combination with feedback inhibition, operative at the cytoplasmic level, they provide coordinated control over biosynthesis.

Inducible systems   Inducible systems are turned on for transcription in the presence of a newly presented substrate (Fig. 9.13). In the case of the highly studied lactose operon in *E. coli*, the appearance of the disaccharide signals induction. The inducer (lactose) combines with the specific regulator protein. In this case, the combination inactivates the regulator so that it **cannot** block the RNA polymerase. Transcription is initiated, the structural proteins for lactose dissimulation are promptly made, and they function until the inducer molecules are consumed. At that point, the regulators become free to prohibit transcription.

Lactose, source; Table 2.7

Catabolite repression   In the induction process, after the removal of the repressor molecule, another step is needed before transcription will occur. A protein factor and the cyclic nucleotide **cyclic AMP** must also attach to the operon. Levels of

9.14
When an organism is grown in a medium containing two energy sources, such as glucose and lactose, it will utilize one first, then the other. The growth curve may show two shoulders. In Phase I, logarithmic growth on glucose is taking place. Cellular ATP levels are high, and cAMP levels are low. In Phase II, the glucose in the medium has been exhausted. Cellular ATP levels have fallen but cAMP is high. Without an ATP supply, growth is temporarily arrested. High cAMP levels, however, permit synthesis of 1-galactosidase, the enzyme needed to metabolize lactose. In Phase III, the cells have shifted over to lactose utilization and once again are synthesizing adequate ATP for logarithmic growth.

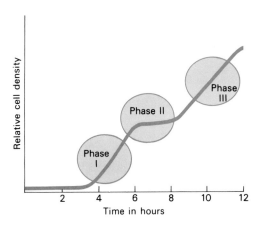

cAMP in the cell are always inversely related to levels of ATP. That is, in a well-fed cell, ATP supplies are high, and available cAMP will be low. If cAMP is low, induction will not occur even if the inducer is present (Fig. 9.14). This interesting mechanism prevents a cell from "making a pig of itself" by oxidizing nutrients for which it has no immediate need. Any nutrient that is a preferred ATP source will thus exert **catabolite repression**, preventing the mobilization of other nutrients until it is exhausted.

A number of metabolic pathways have been exhaustively studied in *E. coli* and other bacterial species. In one study with *E. coli*, the amount of 140 different proteins was determined under different growth conditions. Of these, 38 were inducible or repressible—their level depended on the chemical nature of the medium. The rest (102) of the proteins varied only with the growth rate and were presumed to be constitutive. There must be many different control modalities. The basic pattern of events can be summarized as follows. A simple chemical substance can combine with a regulator. The act of combination either turns off the system (repression) or removes the obstacle to the system's function (induction) at the level of the promoter region of an operon.

Eucaryotic regulation   Nothing comparable to an operon system has yet been demonstrated in eucaryotic cells. It appears that the structural features of the nucleosome particles might have an important effect on whether genes were or were not available for transcription. In higher organisms, hormones clearly have a great influence on gene activity. In all probability, there is a large number of intricate, interlocking control mechanisms.

GENOTYPIC CHANGE IN BACTERIA THROUGH MUTATION

Mutation results when the base sequence of DNA undergoes a change. It is basically a chemical event. Mutation acquires genetic significance if the alteration creates a protein with altered function. Mutation may yield a varient phenotype. This will be a permanent condition since the underlying **genotypic** change can be reversed only by a statistically unlikely reverse mutation.

Mutation can be **induced** or caused to occur at an enhanced rate by contact with a chemical or with radiation. **Spontaneous** mutations by contrast are those in which there is no apparent cause. This may be a false distinction because **mutagenic** chemicals and low-level radiation are now and have always been part of the natural environment. Responding to social concern about the long-term effects of technology on genetic inheritance, researchers turn to the bacteria to study how mutation occurs and what conditions increase or decrease mutation rates.

The Biochemical Basis of Mutation

The DNA replication process is normally extremely accurate; we have discussed the means by which proofreading in the error-prone initiating sequences is carried out. The spontaneous rate of copying error is about one error per 100 million bases copied. **Point mutations**—changes in a single base—occur principally during the copy process. They occur when an incorrect base is inserted (**substitution**) or when an uncoded base is inserted by the enzyme complex (**addition**) or a base is not inserted as coded (**deletion**).

Substitution mutations occur as a result of temporarily incorrect base-pairing. (Remember that the addition of each base, like any enzyme-catalyzed reaction, requires only microseconds.) Such an error has the potential for changing one codon. Genetically, the change will be significant if this results in a different amino acid being coded for. The rest of the protein will be unchanged unless the mutation creates a stop signal. In that case, an abbreviated protein will be made.

Addition and deletion mutations seem to occur as a result of spatial disruptions of the DNA helix. If the helix is stretched, an additional base may be inserted opposite the gap. It is probably the function of accessory protein factors in replication to prevent this.

However, when this type of mutation occurs, its results are always extensive. Remember that the message is coded without punctuation. Because the message is always read in three-base units, these mutations cause a **frame shift** of reading. From the error point to the end of the message all codons are read off-center, giving a different message and a useless protein. The translation machinery will faithfully translate any message, whether it "makes sense" or not.

Mutagenic Agents

If exposure to chemical or physical conditions increases the observed rate of mutation, the conditions are called mutagenic. Mutation is entirely random; there is little evidence that one portion of the genome is more susceptible than another. The techniques of mutagenesis can be used to produce mutant organisms with desired traits. Mutagenesis was pioneered by the fruit fly geneticists, then applied to yeasts and bacteria. **Selection** techniques were developed to permit the recovery of the sought-after mutant strains.

Chemical mutagens   A number of mutagenic substances are **base analogs**. These are chemical derivatives of the normal purine and pyrimidine bases. They pair incorrectly for a sufficient length of time to cause the cell to synthesize DNA with errors. The mutations are randomly spread through the genome. If the dose of the drug is high enough, the mutation load will be lethal. Base analogs are widely used in the treatment of rapidly growing tumors, since their effect is to kill or inactivate reproducing cells. They also injure normal mitotic tissues as well.

Some other substances, including dyes such as acridine, are mutagenic because they become inserted or **intercalated** into the DNA helix causing the spatial distortions that lead to frame shifts.

Radiation mutagenesis   All electromagnetic radiations with wavelengths shorter than visible light will cause mutation. So also will particulate radiation—streams of high-energy subatomic particles released in atomic reactions.

Ultraviolet radiation forms covalent linkages between adjacent thymines, and these cause incorrect copying. As UV is a component of sunlight, this explains part of the lethal effect of sunlight on some bacteria as well as some of the deleterious effects of sunlight on overexposed human skin.

Electromagnetic spectrum; Chapter 3, p. 65

High-energy radiations are **ionizing**; they break bonds and create environments in which incorrect bonding will occur. They may actually snap DNA helixes. Their effect is a general inactivation of all cellular macromolecules.

Production and Selection of Mutants

Exposure   Because different mutagenic agents produce fairly specific types of mutants, the choice of agent is dictated by the results sought. Exposure to the agent is usually at rates that kill most of the population outright and thus induce mutation in almost all of the survivors. Following exposure, the cells may be treated in ways that minimize repair of genetic damage. For example, cells treated with UV will show a high rate of recovery if exposed to visible light, which activates a DNA-repair enzyme. The repair does not occur if the cells are kept in the dark for a period. The exposed population is then plated out under conditions that select the desired mutants.

Selection techniques   The exposed population contains, in a bacterial sample, a huge number of cells. Most will have been badly enough damaged so that they are incapable of reproducing under any conditions. Those mutants that are still viable have been altered in all conceivable ways. Certain types of selection are easy. Selection for a chloramphenicol-resistant mutant of a chloramphenicol-sensitive microorganism involves plating the exposed population on media containing the drug. All colonies that develop are derived from resistant mutants.

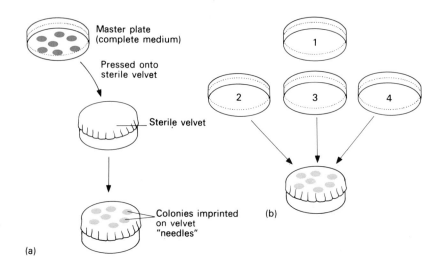

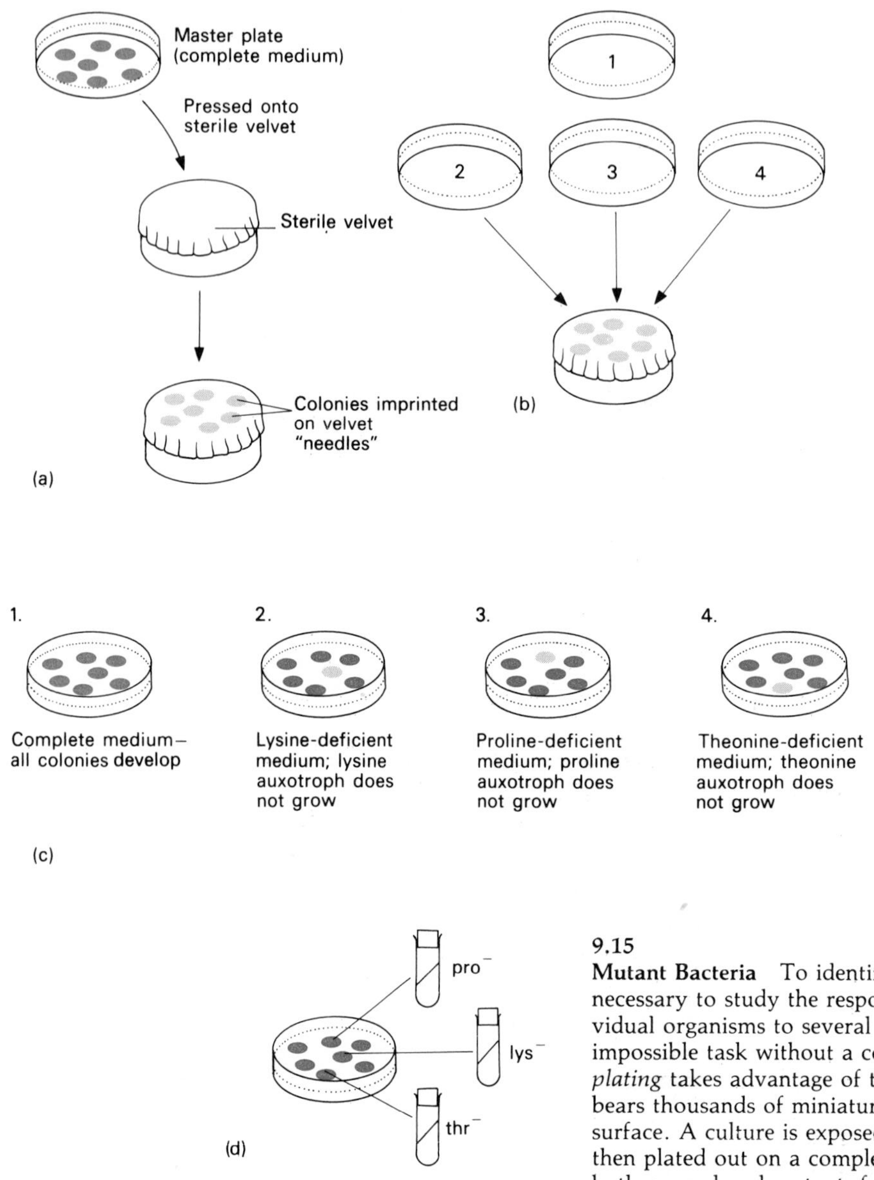

(a)

(b)

1.

Complete medium—
all colonies develop

2.

Lysine-deficient
medium; lysine
auxotroph does
not grow

3.

Proline-deficient
medium; proline
auxotroph does
not grow

4.

Theonine-deficient
medium; theonine
auxotroph does
not grow

(c)

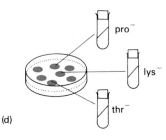

(d)

9.15

Mutant Bacteria   To identify mutant bacteria, it is often necessary to study the response of thousands of individual organisms to several nutrients. This would be an impossible task without a convenient technique. *Replica plating* takes advantage of the fact that velvet fabric bears thousands of miniature inoculating probes on its surface. A culture is exposed to a mutagenic treatment, then plated out on a complete medium so all organisms, both normal and mutant, form colonies. Prints of the colonies are then transferred to the sterile velvet (a) and it in turn may be used to inoculate a number of selective media (b). Mutants are those that do *not* appear on the selective media (c). Individual colonies of bacteria auxotrophic for three different amino acids are seen here. They can be isolated to pure culture on complete media (d).

One widely used class of mutants is that called **auxotrophs**. In the case of a non-fastidious organism such as *E. coli*, the normal or **wild-type** organism has no specific nutritional requirements for amino acids or vitamins. Auxotrophic mutants are those that have lost the function of one or more biosynthetic pathways through genetic changes that render an enzyme dysfunctional. Auxotrophs must be supplied the corresponding nutrients to grow. Isolation of such mutants can be done using the **replica-plating** technique (Fig. 9.15).

Study of mutants has several main uses. First, measurements of mutation rates in the presence of chemicals allows analysis of the chemical's mutagenic potential. Second, mutagenesis reveals which aspects of cell function are essential; these are functions for which no viable mutants can be recovered. Third, cataloging of mutants for a given function allows analysis of the number of genes involved and the order in which their products participate in metabolic pathways. Last, mutations introduced into a strain of bacteria (or other organisms) serve as **markers**; that is, they identify that cell and its progeny, thus permitting the analysis of **genetic crosses** or **recombinations**, as the microbial geneticists call them. The work we are about to discuss on gene exchange in bacteria is done with mutant strains containing readily identifiable markers.

GENETIC RECOMBINATION IN PROCARYOTES

Sexual contrasted with asexual reproduction In eucaryotic organisms, sexual reproduction is common. In the higher forms of life, it is used almost exclusively. However, it was long believed that the simplest organisms, the bacteria, reproduced exclusively by the asexual means of binary fission. This has been shown to be untrue.

Sexual reproduction in eucaryotic microorganisms; Chapter 5, p. 132

Let us compare the genetic consequences of the two approaches to reproduction. Sexual reproduction results in the formation of a zygote in which the genetic material is equally derived from each of two parents. The zygote is **diploid**; that is, each piece of information is present in at least two copies. Each new singleton individual is a **recombinant**—its genes are derived from two sources—and unique.

Asexual reproduction produces new individuals that are exact copies of their single parent, with change from generation to generation occurring solely on the basis of mutation. There is no genetic exchange or recombination between individuals. The cell produced by asexual reproduction is normally **haploid** and has only one copy of each information unit. In a population of bacteria there appear to be vast numbers of organisms that are genetically identical. The appearance of a single cell with new traits can be spotted with the right techniques.

Beginning in the 1940s, a series of rarely occurring genetic exchange processes was identified in bacteria. All result in the formation of recombinants and are thus sexual processes. These processes have provided much information about the structure, function, and regulation of bacterial genetic material.

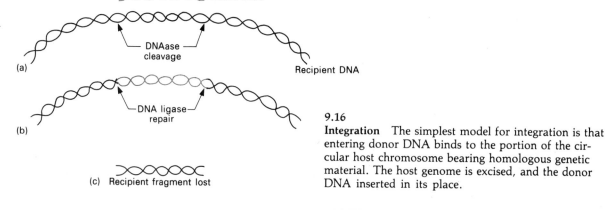

(a) Donor DNA

DNAase cleavage

Recipient DNA

(b) DNA ligase repair

(c) Recipient fragment lost

9.16

Integration   The simplest model for integration is that entering donor DNA binds to the portion of the circular host chromosome bearing homologous genetic material. The host genome is excised, and the donor DNA inserted in its place.

The Integration of New Genetic Material

The key issue in bacterial recombination is that new genetic material becomes permanently integrated into the recipient cell's genome, so the genes can be passed to the cell's progeny.

Compatibility   In legitimate recombination, donor and recipient must be closely related, the same species or in rare cases closely similar ones. Remember that DNA is chemically tagged with species—specific side-chains. Donor DNA, modified differently from recipient DNA, will be immediately destroyed by highly specific DNAases called **restriction enzymes**. These recognize "foreign" base sequences in an unrelated DNA and hydrolyze the DNA at those sites.

Recognition of complementary regions   Usually only a portion of a donor cell's genome is introduced. The donor and recipient single strands recognize each other by their complementary base sequences. These allow them to pair up or hybridize (Fig. 9.16); the pairing may not be perfect. This initial recognition and pairing is essential before integration; it guarantees that exact joining will occur, without loss of critical portions of genes.

Integration   The exact mechanism of integration remains in doubt. Once introduced, donor DNA separates into single strands, as does the recipient DNA in the region that carries the same genetic information. In one model, the complementary single strands of both donor and recipient DNA are broken or "nicked" enzymatically. The free ends of the hybrid double strand (not perfectly complementary) are rejoined covalently to the main molecule by repair enzymes. Subsequently, the recipient strand in the hybrid region is digested away and replaced by a sequence fully complementary to the donor strand. The final integrated portion is genetically of the donor sequence. The recipient cell remains haploid, having but a single copy of the relevant information. It has simply had some of its genes replaced by genes from another source.

In other types of recombination events, a length of double-stranded donor DNA is added directly into the double-stranded recipient chromosome. This is not a replacement—the recipient cell becomes a **partial diploid**, having two genes for certain functions.

Permanence of recombinations   Once introduced, genetic material becomes part of the covalent structure of recipient cell DNA. It will be replicated right along with the rest and passed on to all progeny in cell division. Unless the material has been introduced by the specialized mechanisms of virus infection, its presence in the cell is as permanent as that of the original DNA. However, DNA introduced by viruses in **transduction** can be induced to **deintegrate** by various means and to leave the cell along with the virus.

Transformation

Transformation occurs when short (10^5 to 10^6 base pairs in approximate length) pieces of DNA from a dead donor cell are taken up by a living recipient cell and integrated into its chromosome (Fig. 9.17). The process was first observed in the "transformation" of unencapsulated nonlethal strains of the pneumococcus into encapsulated or smooth, lethal strains. This was the first demonstration of recombination in bacteria. Identification of the "transforming factor" as DNA was a milestone in molecular biology.

The transforming process can transfer roughly 20 genes at a time; these may be from any portion of the donor chromosome. All donor genes are transmitted with equal frequency. In order for the donor material to pass the recipient cell wall, the latter must be in a certain phase of the division cycle (**competent**). Transformation has been observed in five major bacterial groups *in vitro*. It is presumed that it also occurs *in vivo* during simultaneous growth of two strains in natural environments. It could also occur in mixed infections of a human host.

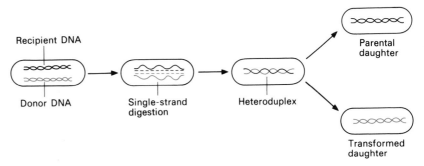

Recipient DNA

Donor DNA

Single-strand digestion

Heteroduplex

Parental daughter

Transformed daughter

9.17

Transformation   In transformation, a portion of double-stranded donor DNA is released from a dead cell and taken up by a competent living recipient. One strand of each DNA is hydrolyzed, and a heteroduplex formed. When each strand is copied, and cell division occurs, one daughter will have a parental genome, the other a transformed genome.

Conjugation

The conjugation process is a true mating between two bacterial cells, one of which serves as a donor or "male" and the other as a recipient or "female." Both cells survive the process; only the recipient is genetically changed.

Mating types   Maleness is a trait conferred by the presence of a special segment of DNA containing at least 11 genes. In its simplest form, this DNA segment exists as an independent closed circle in the cytoplasm. In this form it is called an **F-particle**, or F-factor, and cells that have it are designated F^+. F^+ cells are male; they can transmit the F-factor to female (F^-) cells that thus become F^+ in turn.

Conjugation with transfer of bacterial DNA occurs when the F-factor is integrated into the chromosome of the male bacteria. In this form the male cells are called **Hfr**. They transmit their chromosomal material to F^- cells with a high frequency of recombination.

The F-factor is a type of transmissible plasmid. The term **plasmid** is used for circular gene sequences that are extrachromosomal and replicate autonomously in the cytoplasm. F is one of a group of plasmids that also functions as **episomes**, becoming

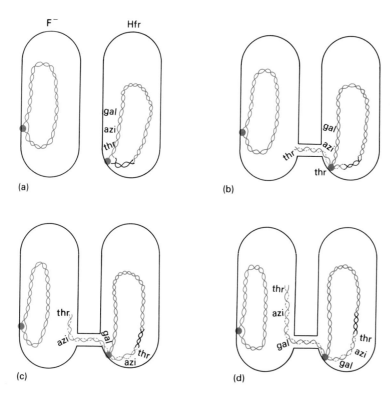

(a)

(b)

(c)

(d)

9.18

Conjugation   (a) The Hfr donor cell in conjugation bears genes for threonine synthesis (thr), azide resistance (azi) and galactose utilization (gal). They are organized in this order following the Hfr region (dark helix). (b) During conjugation, the donor helix separates, and one strand enters the recipient, where it is copied (dotted line). The other strand, remaining in the donor, also is copied. (c) With the passage of time, first thr, then azi, then the gal gene are transferred. Only if the entire circle was transferred would the recipient F become Hfr itself (d).

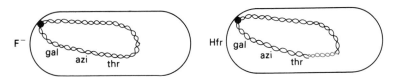

9.19

Interrupted Conjugation   If conjugation was interrupted after Fig. 9.18, the recipient cell would integrate the donor genes, and become thr + , azi + , and gal + , just like the donor.

integrated into the chromosome proper and being replicated in synchrony with it. In the Hfr state, F is an episome.

Hfr on occasion deintegrates and may resume autonomous cytoplasmic existence as the F-factor. If an error is made in the excision process, some normal nuclear material may also be removed from the bacterial chromosome and become part of the F-factor. Such nuclear genes will be transferred with the F-particle. Such "hybrid" F-factors are called F'.

Time-dependent transfer   Mating between Hfr and F⁻ cells requires the formation of a conjugation tube through which DNA may pass (Fig. 9.18). The tube is formed from pili produced by the male partner. The majority of the genes in the F sequence code for pilus formation.

The donor chromosome breaks at the point next to the site of Hfr integration. One single strand of the DNA passes into the recipient while the other, remaining behind, is used as a template for a new complementary strand. As the donor strand enters the recipient, it too is used as a template to produce its complement.

Transfer continues as long as the cells remain attached and genes are transmitted in a linear order. If uninterrupted, the entire donor chromosome will be transferred, concluding with the Hfr genes. Only then will the recipient become Hfr itself (Fig. 9.19). By correlating genes transmitted with elapsed mating time, it has been possible to map the genes of *E. coli*.

The introduced donor DNA is integrated into the host DNA by a replacement mechanism. Both donor cell and recipient emerge from the process with a single complete circular DNA molecule. Any fragments of recipient DNA that have been replaced in the chromosome are lost through dilution or are degraded.

Sexduction   Mating between F' and F⁻ cells results in the transfer of the F-genes plus a small portion of bacterial DNA extracted from the main chromosome. This is called sexduction. However, the genes that are introduced are not integrated by replacement. Thus the recipient will possess two copies of certain genes, its own plus those introduced with the F'-factor. These cells and their progeny are **partial diploids**.

Transduction

In transduction, bacterial DNA is carried from donor bacterium to recipient bacterium as an almost accidental passenger molecule within a virus particle. Transmission of bacterial DNA occurs as a side effect of virus infection of the recipient. Transduction cannot be adequately discussed without dealing with the mechanisms of virus infection. Therefore we will defer it until Chapter 10.

Plasmids

Characteristics of plasmids   Plasmids are closed circles of extrachromosomal DNA found in the cytoplasm of many strains of bacteria (Fig. 9.20). The genetic material they carry is not essential to the host organism's existence. However, plasmid genes may modify the host's phenotype in important features.

Some plasmids are nontransmissible and are propagated only by replication within the host cell. Plasmid replication is autonomous. It may be well synchronized with that of the host DNA, in which case each daughter bacterial cell will receive in division one plasmid copy. If plasmid replication is comparatively more or less rapid than host cell chromosomal copying, some daughter cells may receive more than one copy, or not receive a copy and become "cured."

Transmissible plasmids contain, as part of the genetic sequence, information similar to that contained in the F-particle. This allows ready transfer of plasmids

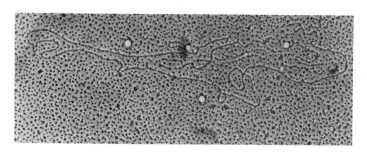

9.20
An electron micrograph of an R-factor from *Hemophilus influenzae* with an interpretive drawing. Some portions of the circle are double-stranded while others, imperfectly paired, form single-stranded loops. Ap is the ampicillin resistance region, and Tc is the tetracyline resistance region. This organism's resistance to ampicillin is posing an increasing clinical problem.

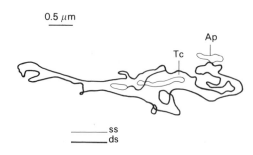

Box 9.1

Transposons and Illegitimate Recombination

Bacterial conjugation and transformation involve legitimate or homologous recombination. One genetic region from the recipient is exchanged for another from the donor, both having similar if not identical genetic information. The preliminary recognition step requires extensive base-pairing between the two DNA sequences.

Illegitimate recombination events are those occurring between DNA segments from unrelated sources. These would have little or no base-sequence homology. Such events can be engineered in the laboratory with plasmids. They occur freely in nature, directed by transposable genetic elements or **transposons**.

A transposon is a DNA strand that consists of one or more transferable genes, sandwiched between control elements. The control elements are unusual DNA base sequences that form inverted repeats of each other. Insertion sequences are recognized in turn by a unique enzyme, the **transposase**, that cleaves DNA at these points. The gene for the production of transposase is also carried by the transposon. In addition the transposon carries a **repressor** gene that regulates transposase activity; thus it controls the transfer of the transposon from one DNA-containing genetic unit to another.

Transposons may be found in either chromosomal or plasmid DNA in bacteria. They may move antibiotic-resistance genes from the plasmid to the chromosome or to the DNA of a bacterial virus. Transposons may move such resistance genes from one plasmid to another within the bacterial cell, thus assembling groupings of resistance factors that may make the host bacteria **multiply drug resistant**. They may transfer penicillin resistance from one bacterium to another of an unrelated species. Transposons may promote rearrangement of or deletion of genes on the chromosome. These sequences must be matched by complementary sequences in the insertion site of the recipient. One transposable element, Mu, even has an alternate existence as a free virus.

In higher organisms, transposons may serve as switches, turning existing genes in the DNA on or off. Certain genetic events were once viewed as inherently abnormal, such as gene fusion, inversion, or deletion. It is increasingly common to find that these events are actually an essential part of gene function. An example is antibody synthesis, where gene rearrangement is a necessary prerequisite to the development of immunity. Since transposons also are known in higher organisms, their role in this process is being investigated.

Transposons are clearly of great importance for the natural creation of genetic diversity. Their role in the evolution of new genetic capabilities is just beginning to be assessed.

---

from one bacterium to another via pili. Plasmid transfer is very similar to conjugation. For example, the R or **resistance factors** of the coliform organisms contain genes for transfer (**RTF** or resistance transfer factor), plus **R-genes** that confer resistance to being killed by antibiotics and heavy metals. A given R-factor may have, in addition to the RTF segment, R-genes for up to eight different antibiotics, sulfona-

Antibiotic resistance; Chapter 26, pp. 755–756

mide drugs, and four heavy metals. Possession of such a plasmid obviously renders a bacterial strain very difficult to control by the conventional means.

The changing nature of plasmids   Because plasmids are not essential to their cells, they are not under the same selection pressures as the chromosome. Random changes in the information content of plasmid genes will not kill the host cell in most imaginable circumstances. Thus plasmid genes exhibit great variability. When two different plasmids of compatible types enter one cell, they can recombine. Some plasmids also integrate into the host chromosome as episomes. It has also been shown that plasmid genes are occasionally excised from one plasmid and incorporated into another under the control of **transposons**.

In fact, it is becoming clear that former concepts of the genetic stability and integrity of chromosomes are somewhat superficial. Segments of genetic material are constantly being shuttled around among chromosomes, plasmids, and viruses by a variety of transfer mechanisms.

Recombinant DNA research   Plasmids have two characteristics that make them exceptionally useful in research. First, groups of genes can conveniently be added to them by biochemical means. Second, as long as they possess a transfer factor, they can be readily introduced into host bacteria. Then, as the bacterial population grows, the plasmids will be replicated too (Fig. 9.21). The types of research that can be done by using genetically engineered plasmids are almost unlimited.

The plasmid can be constructed to contain DNA sequences for which biochemical and/or genetic information is needed. It is then introduced into a bacterial host. As the bacteria grow and multiply, they will produce large quantities of plasmid DNA. This can be harvested, and the interesting DNA sequences obtained in bulk. In effect, one sets up a bacterial factory to make specific DNA to order.

It is also possible to establish plasmid-containing bacterial lines in which the plasmid contains not bacterial but plant or animal DNA sequences. Such host bacteria will express these genes, making plant or animal proteins. Mammalian genes for insulin production have been placed in a plasmid, and the plasmid introduced into bacteria. These bacteria then produce human insulin. The present techniques of obtaining insulin require extraction from animal pancreas. The cost of this process is high, and the supply of insulin thus produced is barely adequate to meet a growing demand. In addition, many diabetics become allergic to the "foreign" animal protein. When human insulin produced by bacteria becomes available, the supply and allergy problems will be solved.

Because of scientific and public concern, recombinant DNA work is performed under strict containment guidelines. The fear is that a genetically "new" bacterium–mammal hybrid with unpredicted pathogenicity might escape into the environment.

A useful biologic containment procedure is the use of "crippled" host cells—mutants that have become totally dependent on specialized laboratory growth conditions. Should these "escape," they would not be able to replicate in the environment or in warm-blooded animals. Engineering containments involve several levels of aseptic techniques, enclosure hoods for transfers, and isolation units.

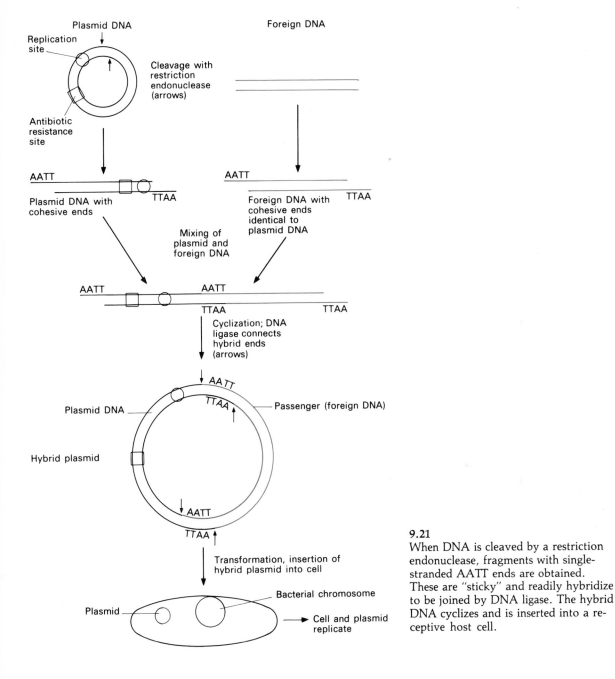

9.21
When DNA is cleaved by a restriction endonuclease, fragments with single-stranded AATT ends are obtained. These are "sticky" and readily hybridize to be joined by DNA ligase. The hybrid DNA cyclizes and is inserted into a receptive host cell.

Recombinant DNA tools promise to significantly enhance the already major contributions bacteria can make to human welfare. It remains to be seen what the social and economic costs of such advances will be.

Box 9.2

Patenting Life

The essence of scientific endeavor over the years has been a spirit of community, based on the free exchange of ideas, techniques, and conclusions. The occasional researcher who was secretive or refused to share materials was regarded with disapproval.

There is a time-honored story about a West Coast virologist who isolated a new genetically interesting strain of bacteriophage. Wishing to exploit its potential for his own ends, he refused an East Coast scientist's request for a stock strain with an abrupt letter. Smiling slightly, the recipient went into his laboratory, found a flask containing a culture of the host bacterium, and gently tapped the offending letter over the broth. You can easily anticipate the outcome of this story—abundant replication of the desired virus in its new home!

A new restrictive atmosphere is entering molecular biology. Many large corporations have formed research groups to exploit the new recombinant DNA techniques. Some potentially profitable substances are already being made—insulin, growth hormone, and interferon—and many others are being sought. In view of the large profits at stake, these laboratories operate with tight security (autoclaved mail?). Patent protection is being sought for key new bacteria strains.

The issue of whether or not a life form can be patented is complex in both legal and philosophical arenas. The original United States patent law of 1790 limited the patent right to anything that was "a manufacture or composition of matter" clearly invented by the person seeking the patent. Can we say that a researcher has "invented" a new life form?

The 1930 Plant Patent Act extends the patent right along these lines by allowing the patenting of certain new plant strains. In 1978, a Court of Customs and Patent Appeals ruled that a newly developed, oil-degrading bacterium could be patented by its developer, A. Chakrabarty, and General Electric. The court said that "the natural and commercial uses of biologically pure cultures of microorganisms. . . are much more akin to inanimate chemical compounds. . . than to horses and honeybees." Do you agree?

The patent decision was appealed to the United States Supreme Court in 1980. Their final ruling was that patent rights could be awarded for microorganisms. This ruling opens up vast new avenues of investigation that will be made more inviting by the potentials for large corporate profits. In the next few years, there will probably be several hundred such patent applications.

SIGNIFICANCE OF GENETIC CHANGE IN BACTERIA

Constant genetic exchange among bacterial chromosomes, plasmids, and bacterial viruses leads to the appearance of new bacterial strains. Their altered characteristics may have major effects on human health. New strains can be developed as tools for industry or environmental cleanup.

Genetic Contributions to Pathogenicity

Antibiotic resistance   When a new antibiotic is introduced into clinical practice in a locality, the target bacteria are effectively controlled at first. One or two years later there may be an outbreak of infections by resistant bacterial strains. This shift results from **selection** of resistant cells from a predominantly sensitive population; only the resistant members will survive and multiply in the presence of the antibiotic. The genetic basis for antibiotic resistance differs from species to species. The commonest way for a bacteria to become penicillin-resistant is to acquire genes permitting the manufacture of a **penicillinase** enzyme. Penicillinase genes may be chromosomal, transferred from one strain to another via transduction, or plasmid-associated and transmitted via conjugation.

The R-factors or resistance plasmids were first discovered in *Shigella*, the causative agent of dysentery, in Japan. When first identified in 1953, only about 0.2 percent of *Shigella* strains harbored R-factors. Due to the transmissibility of plasmids, and a widespread use of antibiotics that favored the spread of resistant bacterial strains, the incidence had risen to 58 percent by 1965. Furthermore, the R-factors had also been transferred to the closely related genera *Proteus* and *Salmonella*.

Capsule formation   Capsular layers are a requirement for pathogenicity in species such as *Streptococcus pneumoniae*. Capsular production may be altered by transformation, as discussed previously, as well as by mutation.

Capsules; Chapter 19, p. 554

Cell–surface antigens   Cell–surface materials such as the M-protein of *Streptococcus pyogenes* may be virulence factors that directly increase pathogenicity. Others stimulate development of functional host immunity. Genetic changes that result in antigen changes permit the altered bacterial strain to evade host defenses established against the parent strain. Changes in antigens can occur by all types of genetic change. An example of the great diversity that may develop is seen in the genus *Salmonella*. When classified by antigenic analysis, in excess of 400 distinct species can be cataloged.

M-protein; Chapter 17, p. 504

Growth rate changes   Successful pathogenicity requires a growth rate fast enough to allow multiplication to outstrip the body's defense efforts. Mutation produces nonpathogenic, slow-growth strains of *Salmonella*. On the other hand, enhanced virulence in *Brucella* and *Francisella* may be caused by mutation-induced resistance to the growth repression exerted by host metabolic products.

Toxin production   The elaboration and release of protein **exotoxins** is the major aspect of pathogenicity in several bacterial diseases, notably diphtheria, tetanus, and botulism. For most of toxins studied, the necessary genetic information comes from integrated bacterial virus genes. For others, the genetic information is plasmid-borne. In both cases, the genes are readily lost, so that **nontoxigenic** strains are common.

Exotoxins; Chapter 12, pp. 352–353

Economic Importance of Bacterial Genetics

Penicillin production;
Chapter 26, p. 769

Industrial applications   The use of bacteria in industrial production of valuable biochemicals constitutes a multimillion dollar business. The efficiency of production can be greatly enhanced by induction and selection of **high-yield mutants**. Genetic and cultural selection techniques have been used with *Penicillium notatum*, the fungus that yields penicillin. When first used in 1941, only four units of penicillin/mL of culture fluid could be obtained. Present strains produce up to 50,000 units/mL. In the future, we will see plasmids used to induce the ever-cooperative bacteria to produce in bulk substances normally elaborated only by plants or animals in trace amounts.

Environmental applications   Many microorganisms, notably *Pseudomonas* and its relatives, decompose hydrocarbons including crude oil. The natural rates of decomposition are quite slow. Some genes for breakdown of specific hydrocarbons such as camphor or octane seem to be plasmid-associated. Recently, researchers using the tools of recombinant DNA production have designed plasmid-carrying marine bacteria with greatly increased catabolic activity against crude oil. They metabolize actively in cold and saline ocean environments in which most oil spills of concern occur. These bacteria are currently under evaluation for their ability to deal with the major ecological threat of oil in the global ocean.

SUMMARY

1. Cells use the chemical raw materials from catabolic metabolism plus the stored energy in ATP and other high-energy compounds to carry out biosynthesis leading to growth, repair, and reproduction.

2. The biosynthetic pathways are the primary means by which the genetic information of the cell is expressed.

3. Random-sequence polymers are assembled without specific genetic blueprints. An activation step comes first, in which the monomer becomes attached via a high-energy bond to an intermediate. Breakage of this bond energizes the second step that links the new unit to the growing chain. The production of peptidoglycan for cell wall presents a complex example of the principle.

4. Informational macromolecules are heteropolymers with variable units whose sequence is not random but genetically specified. The primary sequence blueprint is DNA; the information is amplified by template-directed copying of DNA into RNA, and RNA into protein.

5. The bacterial genome is a single closed circle of double-stranded DNA. It is replicated semiconservatively, and the circle is copied completely from beginning to end. In eucaryotes, with far more complex chromosomal structure, replication is discontinuous.

6. Transcription of the DNA results in the appearance of three classes of RNA molecules, mRNA, rRNA, and tRNA. The synthesis of RNA is a highly regulated copying of one of the two DNA strands in specific regions only.

7. Translation of the genetic code, contained in mRNA codons, results in the assembly of a polypeptide. Activated amino acids are carried by specific tRNAs that base-pair with message codons on ribosomal surfaces. Transferase enzymes simultaneously link the amino acids together and release the tRNAs.

8. Bacterial cultures growing under altered or adverse conditions normally show considerable phenotypic variation. This is temporary and reversible variation caused by regulation of gene function at the level of transcription. By controlled access of RNA polymerase to genes, the cell regulates the formation of mRNA and thus the synthesis of the protein in question. The genetic controls work in conjunction with enzyme-level feedback controls in most cases. Methods of regulation in eucaryotes remain largely unelucidated.

9. The genomes of bacteria undergo permanent genetic change through mutation and genetic recombination of the bacterial chromosome with DNA from donor sources. Mutation may be induced by exposure to chemicals or radiation; it usually reflects a single-base change in the DNA sequence for a particular gene. Sexual recombination in bacteria is a rare event in nature as well as in the laboratory. When donor DNA is introduced into a recipient cell, the DNA will remain undigested only if it is of the correct modification type. Portions of the donor DNA will become integrated into the host genome and will probably affect the host cell's phenotype.

10. There are several means of recombination. Transformation occurs when short pieces of DNA, liberated from a dead donor cell, are taken in and integrated. Transduction occurs when a bacterial virus, as a side effect of its infection of the bacterial host, introduces bacterial genes abstracted from its last host. Conjugation is a true mating between two bacteria, one of which, the donor or male, possesses the sex or transfer F-factor integrated (as Hfr) into its chromosome. A portion of all of the genome will be transmitted as a single strand.

11. The F-factor is a plasmid or extrachromosomal genetic element. A typical plasmid contains transfer genes plus other genes related to such functions as antibiotic resistance or bacteriocin production. Plasmids, which are often transmissible to other bacterial strains, confer potential survival advantages on the host cell. Plasmids may be enzymatically reconstructed in the laboratory to join any genetic material, no matter what its origin or information content, with transfer genes that permit introduction of the plasmid into a bacterial cell. The host cell will then reproduce the plasmid DNA and express its genes, producing the designated proteins.

12. There are practical consequences of the low-frequency gene transfer among bacteria, viruses, and plasmids. Such genetic exchanges determine many aspects of bacterial pathogenicity as well as bacterial susceptibility to standard drug therapy. Furthermore, by careful selection of recombinants or mutants, bacterial strains can be found for industrial or antipollution applications.

Study Topics

1. Identify the building blocks for biosynthesis of each of the major types of cellular macromolecule. Locate the source of these building blocks in metabolism.

2. What is the ATP cost of synthesizing a glycogen molecule from 5,000 glucose units? What is the ATP cost of synthesizing a polypeptide from 5,000 unactivated amino acids? (The conversion of ATP $\rightarrow$ AMP is equivalent to 2ATP $\rightarrow$ 2ADP, of GTP $\rightarrow$ GDP is equal to ATP $\rightarrow$ ADP.)

3. What is the purpose of DNA modification and restriction enzymes?

4. Describe sequentially how a peptide is elongated.

5. Differentiate between the mechanisms, speed, and duration of regulation due to feedback inhibition and to genetic repression.

6. How does the presence of ample amounts of an energy source such as glucose prevent a cell from inducing synthesis of enzymes to metabolize a second substrate such as lactose?

7. Differentiate among the biochemical types of mutation on the basis of the severity of their results.

8. What is the difference in the type and degree of gene transfer from cells with F^+ and with Hfr to recipients?

Bibliography

Books

Goodenough, U., and R.P. Levine, 1974, *Genetics*, New York: Holt, Rinehart and Winston.

Jackson, David A., and Stephen P. Stick (eds.), 1979, *The Recombinant DNA Debate*. Englewood Cliffs, N.J.: Prentice-Hall.

Mandelstam, J., and K. McQuillen, 1973, *Biochemistry of Bacterial Growth* (2nd ed.), New York: Wiley.

Stryer, Lubert, 1975, *Biochemistry*, San Francisco: W.H. Freeman.

Watson, J.D., 1975, *Molecular Biology of the Gene* (3rd ed.), Menlo Park, Calif.: Benjamin/Cummings Publishing Company.

Article

Devoret, Raymond, 1979, Bacterial tests for potential carcinogens, *Scientific American* **241**: 40–49.

10

The Study
of Viruses

Early in this century, accumulated evidence hinted at the existence of a form of life smaller than any known bacteria. These small agents appeared to cause disease in animals and cellular breakdown in bacteria as the agents multiplied. Laboratory techniques, such as electron microscopy, nucleic acid biochemistry, and cell culture, have allowed us to see and study the viruses. It is now clear that viruses infect every form of cellular life, from the simplest bacteria to the most highly evolved plants and animals (Table 10.1). Intracellular infection is the only means by which viruses propagate; the virus particle apart from its host cell is incapable of essential life functions. Virus particles are not cells.

In this chapter we will discuss the biological nature of the virus particle, structure and classification of viruses, the strategy of virus replication within the host cell, and virus life cycles in bacteria, plants, and insects.

VIRUSES AS BIOLOGICAL ENTITIES

Discussions of the viruses often conclude with the question: "Are viruses living or not?" Let us dispose of this question at the beginning. Viruses are certainly living in

TABLE 10.1
A general scheme for classification of viruses

NUCLEIC ACID	CAPSID SYMMETRY	SIZE OF CAPSID IN nm	ENVELOPE PRESENT	BACTERIAL VIRUS EXAMPLES	ANIMAL VIRUS EXAMPLES	PLANT VIRUS EXAMPLES
RNA	Helical	17.5 × 300	−			Tobacco mosaic virus
		9	+		Myxovirus	
		18	+		Paramyxovirus	
	Polyhedral	20–25	−	f_2		
		28	−		Enterovirus	Bushy stunt virus
		70	−		Reovirus	
DNA	Helical	5 × 800	−			
		9–10	+	Filamentous phage fd	Poxvirus	
	Polyhedral	22	−	$\phi \times 174$		
		45–55	−		Polyomavirus	
		60–90	−		Adenovirus	
		140–200	+		Herpesvirus	
	Complex	Head 95 × 65	−	T-even series		
		Tail 17 × 115				

Essential criteria for life; Chapter 1, pp. 4–6

at least one sense of the word, if not in others. In practical terms, we all perceive that if viruses were **not** living, if they did not replicate efficiently, we would be less actively concerned with them. However, virus replication **does** occur, causing severe economic loss in crops, and common, sometimes fatal, diseases in animals including human beings. Viruses play an important role in the ecological balance of nature as population-regulating parasites.

The Nature of Viruses

If viruses are not cells, what are they? It is difficult for us to think of any thing as living that is not cellular. However, remember that any cell's entire structural and functional capacity, as well as its reproduction, is encoded in the nucleic acid genome. The cell is constructed and maintained solely on the basis of that information.

Viruses are simply **genomes** surrounded by one or more exterior coats. The genome of a virus is a group of genes that have evolved to provide only a minimum of necessary information. Viral genomes need code for only three functions: (1) infection of a host cell, (2) redirection of the host cell's metabolism to the production of virus-specified macromolecules, and (3) escape from the host cell. The viral particle has bypassed the requirement for maintaining its own cellular systems for energy production and biosynthesis. As a parasite, it does none of the work, but simply takes advantage of the generosity of its host.

Viruses are **obligate intracellular parasites.** No true virus can be cultivated on cell-free laboratory media. Our success in studying them depends entirely on providing the appropriate physical conditions and host-cell system. For the hepatitis viruses, these requirements were so exacting that laboratory isolation was impossible until very recently.

When the virus takes over the host cell's metabolism, it uses the cell's own reserves of ATP and building block molecules for viral synthesis. Cellular macromolecules, including the DNA, may be degraded. An infected cell is usually significantly harmed by viral infection. The severity of the effect varies considerably. Some infections are **inapparent** or **asymptomatic** either because of small numbers of viruses or conditions that hold virus multiplication in check. Some infections are **cytolytic**—the host cell dies. Viral infection may be rapidly lethal to an organism if it causes overwhelming destruction of essential tissues.

Distribution of Viruses

All forms of life have viral parasites. Although the viruses of bacteria and mammals were the first to be studied, investigation of various subdivisions of the plant and animal kingdoms demonstrates that viral infection is ubiquitous. Economic and health considerations have directed the most intensive research toward the crops and domestic animals raised by human beings, and toward human beings themselves.

Most of our basic knowledge of viruses was gained from the study of the bacterial viruses. This knowledge has been crucial to successful investigations of the much more intricate relationships between the animal virus and its host.

Origin of Viruses

Although viruses are "simple" particles composed of genome and coverings, their total dependence on cellular life for reproduction almost makes it certain that they evolved after cellular forms. A useful hypothesis for their origin postulates that an extrachromosomal group of genes, like a plasmid, became able to leave the cell that produced it and to enter other cells. Plasmids, as we have already seen, do this under specialized conditions. Viruses, however, go one step further. They engineer large-scale manufacture of new virus in the invaded cell. When a transmissible plasmid becomes self-replicating, it becomes virus-like.

Chemical analysis of viral nucleic acids indicates the probability that viruses arose from host cellular material on a number of separate occasions. Since the time of origin, hosts and viruses have co-evolved. There are complex interactions between host and viral genomes, with each having inherited capabilities to regulate the function of the other. Careful investigation now detects viral nucleic acid in a latent state in virtually all tissues of certain mammalian species such as mice, and possibly of human beings. This is striking evidence of a "special relationship." Certain questions: "Which genes are ours and which are viral?" and "Which genes are normal and which abnormal?" are becoming very difficult to answer.

Importance of Viruses

We can identify several ways that viruses are biologically important. When their genes are incorporated into the host-cell genome in a semipermanent state, they cause phenotypic changes in the host. Bacterial strains acquire the capacity to produce new proteins, including deadly toxins. Animal cells undergo growth changes that cause them to become cancerous.

Genetic recombination; Chapter 9, pp. 269-274   As viruses complete a round of replication in one host cell and move to another, they may carry with them and introduce genes from the first host. They thus can serve as vectors of genetic recombination, and cause the second host to become genotypically altered.

They can also destroy the host. It is sobering to note that whereas we have effective therapy to cure patients with the other microbial diseases, we can do little to assist the patient who is fighting a severe virus infection. On the bright side, vaccination has been extremely effective in preventing such viral infections as smallpox, polio, and measles. Such notable exceptions as the influenza virus have largely frustrated immunological control efforts to date.

STRUCTURE OF VIRUSES

Resolution; Chapter 3, pp. 66-68   The virus particle in its extracellular condition is called a **virion**. The largest type of animal virus, the pox virus, is just within the theoretical limit of resolution of the light microscope (0.2 mμ) but other viruses are not visible. Structural observations of virus particles are done by electron microscopy. Virus particles have unique architectural patterns that suit them for the role of infective agents (Fig. 10.1).

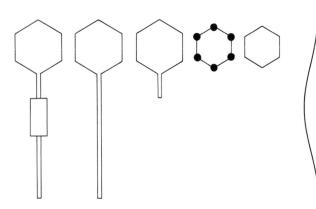

10.1
Structure of Bacterial Viruses   Bacterial viruses range in structure from tiny and simple to large and complex. All consist of a nucleic acid core surrounded by a protein capsid. The capsid has evolved to protect the viral genome in transit between host cells. The capsid also aids in the adsorption of viral particles to the host cell.

Viral Nucleic Acids

All cellular organisms contain both DNA and RNA. Virions, on the other hand, contain only one of the nucleic acids, either DNA or RNA. The nucleic acids may be double-stranded DNA (D2), single-stranded DNA (D1), double-stranded RNA (R2), or single-stranded RNA (R1). To biologists, accustomed to thinking that a gene must be double-stranded DNA, the discovery that viruses could successfully code their genetic information in three alternative fashions required some rethinking. The replication and expression of each of these genetic systems results from specific interactions with the host.

Nucleic acids of bacterial viruses contain modified bases such as HMC— **hydroxymethylcytosine.** Like the modified bases of the bacterial DNA itself, these protect the viral DNA (on infection) from degradation by restriction enzymes in the host. Modified bases do not seem to be a feature of plant or animal viral nucleic acids.

Each virus has a fixed amount of nucleic acid corresponding to a fixed number and sequence of genes. Some tiny bacterial viruses contain only four genes but the poxviruses contain around four hundred genes.

Viral Architecture

With the exception of the **viroids** of plants, all virus particles have an architecture based on the association of the anionic nucleic acids with a surrounding coat of basic or cationic proteins.

The capsid Usually, all of the protein units that make up the capsid are identical—the product of one viral gene—but in some capsids several different proteins are present. The coat proteins self-assemble as the virus is being replicated, enclosing the nucleic acid in a structure with one of three regular symmetries, helical, spherical, or polyhedral. Many virions that appear to be spherical on closer inspection show a regular crystalline structure, a polyhedron. The most frequent type is a 20-sided geometric figure called an **icosahedron.** Alternatively, the nucleic acid may be coiled up inside the protein units forming a hollow cylinder. This helical arrangement was first described for the tobacco mosaic virus (TMV) and is also observed in certain bacterial and animal viruses. In certain animal viruses combinations of polyhedral and helical symmetry are seen. Several types of accessory structure may also be present. These viruses are said to be **complex.**

Many animal viruses have layers external to the protein layer covering the nucleic acid. In these animal viruses, the nucleic acid–protein complex is referred to as the **nucleocapsid** or sometimes the **core.**

Accessory Structures

Tails and fibers Some bacterial viruses have elaborate structures to attach to the cell wall of the host and to inject their nucleic acid forcibly into the cell (Fig. 10.2).

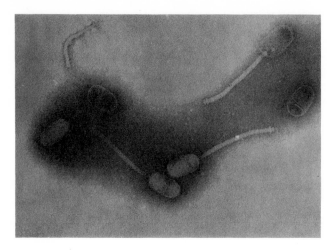

10.2
A Characteristic Bacteriophage   A mycobacteriophage is a phage that parasitizes *Mycobacterium*
species. The staining technique reveals the long
tail, its hollow shaft, and a double baseplate at the
site where it attaches to the host bacterium.

Tail fibers form highly specific weak-bonded attachments to receptor sites on the
host surface. The hollow, contractile **tail** is the tube through which nucleic acid can
be injected.

Envelopes   Many animal viruses have **envelopes,** layers exterior to the nucleocapsid. These envelopes are primarily lipoprotein in character but they may also
contain enzyme molecules. Envelope material is usually derived directly from the
host-cell membrane, as the virus leaves the host cell. In many cases, special proteins,
coded for by viral genes, are manufactured by the host cell and inserted into the envelope as it is formed. These added proteins may form spikes or projections called
peplomers on the exterior surface of the envelope.

Envelopes are not characteristic of bacterial virions and are found only rarely
on plant virions.

VIRAL CLASSIFICATION

It has proved difficult to develop classification schemes for the viruses because the
basic information needed requires much sophisticated and time-consuming research.
Provisional schemes have been constructed for the viruses of bacteria, plants, and
animals. A partial, unified classification appears in Table 10.1.

Useful Criteria

Perhaps the most important single characteristic of the virus is its nucleic acid content—chemical composition, weight in daltons, and conformation. The nucleic acid
may be linear, circular, or a group of small segments.

The symmetry of the virus is also considered. In icosahedral viruses, exact information about the arrangement of protein units into the crystalline structure is useful.

For the plant and animal viruses, another important characteristic is the **means of transmission**—the method by which the virus particles get from one host to another. Many plant viruses and a significant group of animal viruses are spread by an **insect vector**. The vector carries the virus either on its mouthparts or within internal organs and introduces it by biting or chewing the new host.

Discovery of New Viruses

To a much greater extent than other areas of microbiology, virology is still in the early phases of development. In consequence, new viruses are frequently being discovered and existing information about familiar viruses is being continuously reexamined. Thus we can realistically expect that current classification schemes will undergo future revision.

BASIC FEATURES OF VIRAL REPLICATION

Although there are many different patterns of virus–host interaction, it is possible to make some general statements about the process. You can then keep these general concepts in mind as you proceed, in this chapter and the next, to study specific examples.

The Strategy of the Virus

When a virus introduces a group of its genes into a functional host cell, it initiates a series of events under **viral** control. Certain host-cell activities will be bypassed or inactivated. Viral genes will direct the host cell to manufacture viral nucleic acids and proteins. These macromolecules will be assembled into new virions and released.

Note that the primary difference between normal host cell and infected host cell is the presence of a new group of genes. Viral genes are duplicated, transcribed, and translated by the host cell in exactly the same way the host cell processes its own genes. In studying viral replication we shall draw on the understanding of genetic expression developed in the previous chapter. We shall concentrate on thinking in terms of molecular information transfer. In the cases of RNA viruses, there are some new and unique variations on the theme.

The Functions of a Genome

Let us clarify our views on what viral genomes do. Any organism's genome must do two things, either simultaneously or sequentially. First, it must serve as the template

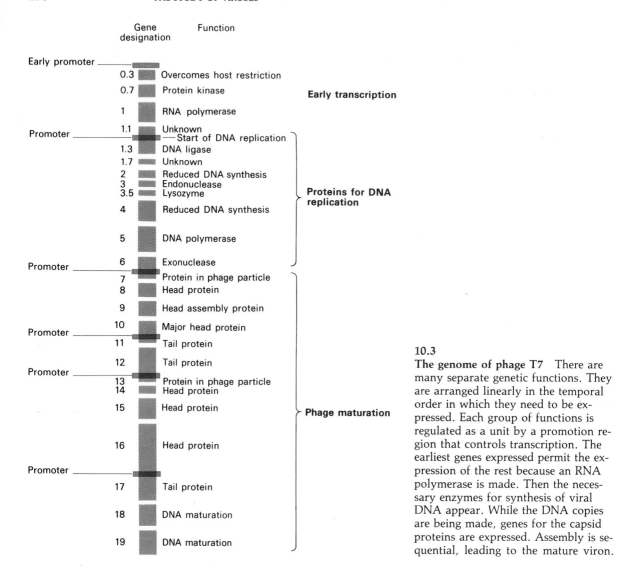

Gene designation   Function

Early promoter

0.3 Overcomes host restriction
0.7 Protein kinase

Early transcription

1 RNA polymerase

Promoter

1.1 Unknown
     Start of DNA replication
1.3 DNA ligase
1.7 Unknown
2 Reduced DNA synthesis
3 Endonuclease
3.5 Lysozyme

4 Reduced DNA synthesis

5 DNA polymerase

Promoter

6 Exonuclease

Proteins for DNA replication

Promoter

7 Protein in phage particle
8 Head protein

9 Head assembly protein

Promoter

10 Major head protein

11 Tail protein

12 Tail protein

Promoter

13 Protein in phage particle
14 Head protein

15 Head protein

16 Head protein

Promoter

17 Tail protein

18 DNA maturation

19 DNA maturation

Phage maturation

10.3

The genome of phage T7   There are many separate genetic functions. They are arranged linearly in the temporal order in which they need to be expressed. Each group of functions is regulated as a unit by a promotion region that controls transcription. The earliest genes expressed permit the expression of the rest because an RNA polymerase is made. Then the necessary enzymes for synthesis of viral DNA appear. While the DNA copies are being made, genes for the capsid proteins are expressed. Assembly is sequential, leading to the mature viron.

to direct its own replication; that is, it must provide new copies of the blueprint. DNA viruses make DNA copies, while RNA viruses make RNA copies. In virus infections, the single entering genome is usually copied many times. Hundreds of new viral genomes are produced to incorporate into hundreds of new virions; this contrasts with the single copy made in each cell cycle of the dividing bacterium or mitotic human cell.

Second, the same genome, during the same period, is serving as the template for protein synthesis (Fig. 10.3). DNA viruses are copied into messenger RNAs that are then translated. RNA viruses serve **directly** as messengers for protein synthesis.

Products of Viral Genes

What kinds of proteins could be useful to a virus?

Replication proteins   The products of some viral genes are proteins that assist directly in the process of viral replication. Some viral proteins inactivate host functions. These proteins are repressors that block host genes and viral DNAases that preferentially hydrolyze host DNA. Other virus proteins regulate the order in which the virus genes are expressed so that the replication activities occur in the correct time sequence. There may be special synthetic enzymes, such as the one produced by bacterial viruses to convert cytosine to hydroxymethylcytosine.

The proteins above are "early" proteins. They can be detected in the host cell quite early in the infection process. They do not become incorporated into the completed virion. Their functions relate entirely to the replication process itself; when replication is complete they have no further function.

Structural proteins   The other major group of viral gene products are those proteins that will make up the external structures of the new crop of virions. These include the proteins of the capsid, and of the tail and fibers if they are present.

Animal viruses with envelopes commonly carry genes to encode for certain proteins to be incorporated in the envelopes. Two of these, seen in the influenza virus, are the **hemagglutinins** (red blood cell-aggregating proteins) and **neuraminidases** (enzymes that break down the mucus that protects the respiratory epithelium from invasion). These proteins are inserted into the envelope as peplomers in the last stages of the replication cycle.

Structural proteins are largely "late" proteins. They appear when the replication process is well advanced. They are then incorporated into the virions, assembling around the completed genomes.

Repressor proteins   In many virus life cycles, introduction of the genome into the host cell is followed by a prolonged latent period. To maintain latency, one viral gene must be expressed to produce a repressor protein. The repressor in turn **prevents** expression of the virus's other potentialities as long as it is present.

Defective viruses   A complete infective virus must contain genes to code for all proteins needed in its replication cycle, although not all viruses need every protein mentioned above. Through mutation or inaccurate replication, viruses have arisen lacking one or more of the required genetic functions. They are called **defective** viruses. Defective viruses can infect but cannot complete replication. For them to replicate successfully, they must infect a susceptible host cell simultaneously with a **helper virus** that carries in its own genome the genes the defective virus is missing. In **combined infection** the host cell will produce a mixture of both viruses—helper and defective. The progeny of the defective virus is not "cured"; it will still be defective.

THE BACTERIAL VIRUSES

It often surprises people to learn that bacteria, which are often wrongly regarded primarily as parasites, have viruses parasitic on them. However, virus infection is widespread among bacterial and cyanobacterial genera. These viruses, called **bacteriophages,** exhibit host specificity that usually restricts them to infecting a single species (Fig. 10.4). Because viral replication cycles in bacteria are short and the "test

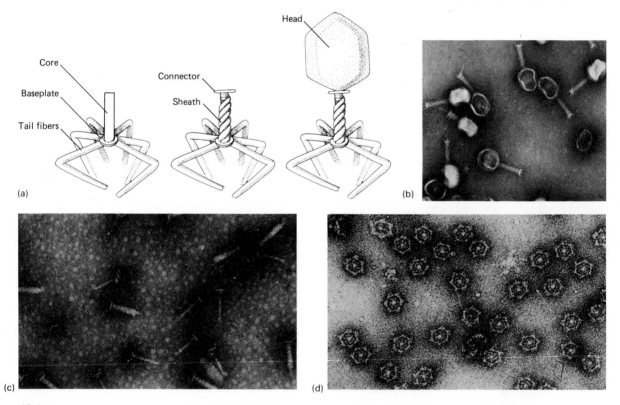

10.4

(a) T4 phage. This virus, parasitic on *Escherichia coli,* has an extremely complex structure for its tiny size. The polyhedral head is attached to a contractile tail, which is composed of a connector, a hollow core wrapped around with a sheath of contractile proteins, and a baseplate bearing tail fibers. (b) A group of T4 phage. Those with light-appearing (electron-dense) heads are packed with DNA, while those with darker heads are empty. (c) T4 tails with and without sheaths. (d) Baseplates, produced as intermediates in the assembly of the virus tail, about 500 Å in diameter. Each baseplate is assembled in the host cell from multiple units of two different proteins, one forming an inner ring (darker) and the other an outer ring (brighter).

animal" is so readily available, the bacterial viruses have been exhaustively re-searched. There are six structural types and three types of life cycle.

Cytolytic Viral Infections

The Gram-negative bacillus *E. coli* is infected by a number of different types of phage. Let us first consider the replication of Type I phages of the T-even group (T2, T4, and T6) (Fig. 10.4). The T-even phages are large, complex viruses with highly developed hollow tails and fibers structures, attached to an icosahedral "head." The head contains double-stranded DNA; in T4, the DNA encodes approximately 75 genes.

The replication cycle of any productive virus consists of three phases, **initiation, replication,** and **release** (Fig. 10.5).

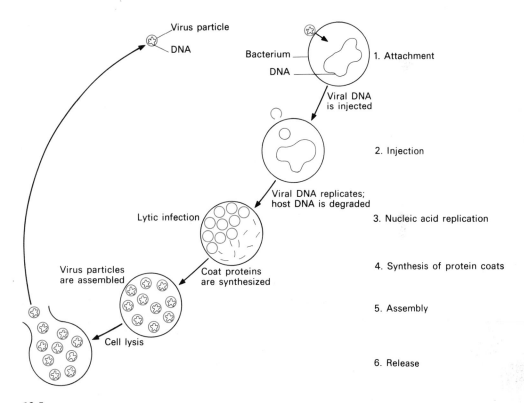

10.5
The Lytic Cycle   The bacteriophage attaches to specific chemical sites on the surface of the coccus. The actual means of penetration may depend on lysozyme from virus or host. Once inside, bacterial metabolism is effectively halted by destruction of the host genome. First DNA, then coat proteins appear; they then self-assemble and the host cell lyses. The free virions may promptly reinfect if a supply of susceptible host bacteria is present.

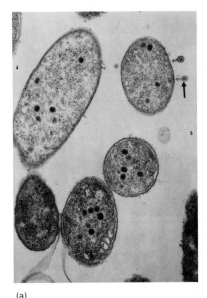

(a)

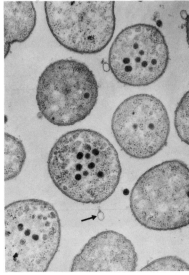

(b)

10.6
Replication of a Lytic Bacterio-phage (a) The absorption process. Two T4 phage attached to the outer membrane of a cell of *Escherichia coli*, cut in cross section. If you look closely, you can see the fine tail fibers and will detect that the tail sheath has contracted prior to the injection of the DNA. (b) A similarly sectioned cell bearing an empty phage head and showing numerous dense bodies within the cytoplasm. These are developing phage heads.

Initiation The first step in viral infection, the one that governs the specificity of virus–host relationships, is the firm **attachment** of the virus to the exterior surface of the bacterial cell (Fig. 10.6). Host-cell surfaces have genetically determined receptor sites. Since bacterial viruses have only one attachment site, a genetic alteration of the cell surface can render the bacterium resistant to infection.

After the virus attaches itself, it penetrates the bacterial cell wall. The contractile proteins of the virus tail coil and inject the viral DNA into the bacterial cytoplasm. Small amounts of protein are also injected. However, most of the viral protein (capsid, tail, and fibers) remains outside and serves no further function.

Once its nucleic acid is inside the cell, a virion is no longer infective for other cells. Furthermore, after the empty capsids fall off the host cell, there is for a time no demonstrable evidence of viral infection. The virus has entered the so-called **eclipse** phase that marks the end of initiation.

Replication The replication period begins when the viral DNA enters the host. Early proteins are made by transcription and translation of host genes. The host-cell DNA is degraded, thus manufacture of host-cell gene products stops. Viral DNA is manufactured—about 200 copies of the viral genome per infected cell. Nucleotides scavenged from the degraded host DNA are used after enzymatic modification. The cytoplasm of the host cell now looks decidedly abnormal.

Expression of late genes then begins, and viral structural proteins begin to accumulate. Capsids assemble around viral genomes. Immature viral particles can now be seen with the electron microscope. **Maturation** of the viruses is complete about 25 minutes after initiation.

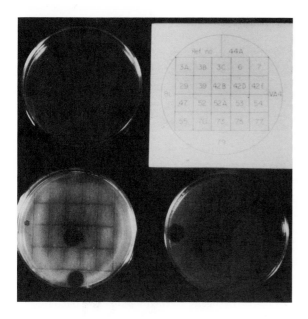

10.7
Bacteriophage Typing   Drops of bacteriophage suspensions are placed on a lawn of *Staphylococcus aureus* confluent growth. The host bacteria are susceptible only to one or a few virus strains. These infect, lyse the host cells, and leave a plaque where there is no bacterial growth. The *S. aureus* is then reported by its **phage type.** Type 80/81 is a strain notorious for lethal skin infections of newborns.

Release   The host-cell contents have been largely replaced by mature viruses. Lysozyme weakens the host-cell wall causing autolysis. On rupture of the wall, the virions are released. Little is left of the host cell except an empty wall-membrane ghost.

Since viral replication (from 1 to 200 in 25 min) rapidly outdistances bacterial replication (from 1 to 2 in 20 min) a cytolytic virus will devastate a population of susceptible bacteria. This effect is the basis for the **plaque count.** A suspension of virus to be counted is plated in known concentration on a **lawn** of sensitive bacteria, confluent growth on a petri dish of nutrient agar. Each virus and its progeny clears a visible hole or plaque in the lawn (Fig. 10.7). By counting these, the number of infective virus or **plaque-forming units** (PFU) can be determined. In addition, since virus–host attachment is highly specific, it is possible to use phage as an identification tool. Each *Staphylococcus aureus* subgroup is lysed by different strains of staphylophage. The unknown bacterial strain is plated onto nutrient agar and incubated overnight. Then drops of different phage suspensions are placed on this lawn in a pattern. By observing which phages form plaques, the bacterial strain can be assigned precisely to a subgroup. **Phage typing** is of particular use in hospital epidemics of staphylococcal infection. By pinpointing the exact strain, one can determine which of several carriers is the source of the outbreak.

Staphylococcus typing; Chapter 17, p. 502

Temperate Viruses

Not all virus infections lead to immediate cell destruction. In a noncytolytic infection, initiation is followed by a latent phase of indeterminate length called **lysogeny.**

Bacteriophage **lambda** (λ) is a widely studied lysogenic virus of *E. coli*. The initiation steps of infection by the lysogenic virus are as already described. Once in the cell the viral DNA may cause one of two possible outcomes. Less commonly, it might cause a cytolytic sequence, as outlined previously. Usually the virus DNA is converted into a **prophage.** When this occurs, few viral genes are expressed and new mature phages are not produced. Since a prophage does not damage its host, it is said to be **temperate.**

Prophage replication   A prophage, which is similar to an episome, is viral DNA integrated covalently into the bacterial chromosome (Fig. 10.8). The prophage state requires the expression of **repressor genes** in the viral genome. Protein repressors block the expression of the rest of the viral genes. The host cell remains healthy and growing. As it replicates its chromosome prior to division, the prophage is replicated too. Each bacterial daughter cell receives a chromosome that includes prophage. The progeny make up a clone of latently infected cells.

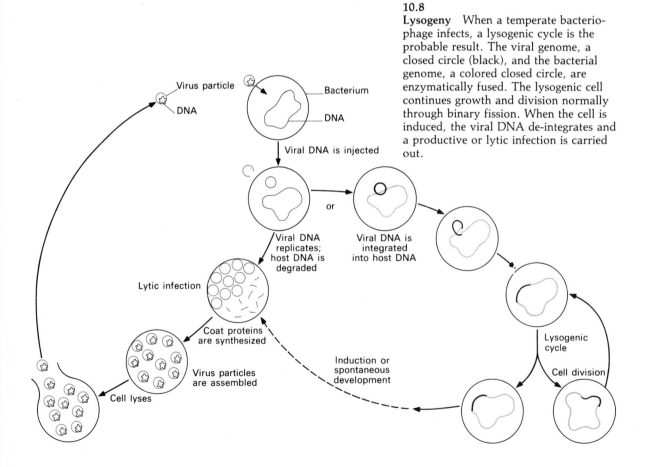

10.8
Lysogeny   When a temperate bacteriophage infects, a lysogenic cycle is the probable result. The viral genome, a closed circle (black), and the bacterial genome, a colored closed circle, are enzymatically fused. The lysogenic cell continues growth and division normally through binary fission. When the cell is induced, the viral DNA de-integrates and a productive or lytic infection is carried out.

Lysogenic conversion   Genes controlled by prophage may be expressed, and the gene products will cause phenotypic change in the host bacterium. This is called **lysogenic conversion.** For example, when the Gram-positive rod *Corynebacterium diphtheriae* contains a prophage, the bacterium releases diphtheria toxin. The toxin produces the clinical manifestations of diphtheria. Without the prophage, *C. diphtheriae* is harmless. Lysogenic conversion yields an altered phenotype as long as the prophage is present. Many species of bacteria are normally lysogenized. Thus certain bacterial characteristics, long regarded as "normal," can be shown on elimination of the prophage to be due to conversion.

Diphtheria toxin; Chapter 18, p. 533

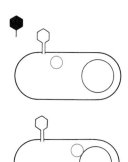

Productive replication   A small fraction of lysogenized cells spontaneously undergo lysis with the appearance of infective virions. The percentage of productive infections can be increased to almost 100 percent by exposing the lysogenized culture to radiation and certain chemicals. Lysis is **induced** when the host cell fails to continue to make repressor protein. Without repressor, the viral genes needed for productive replication become activated in their proper sequence. The events that follow resemble the cytolytic infection of *E. coli* by T-even phages described earlier. The viral DNA deintegrates from the host chromosome and is then copied and assembled with viral proteins into mature infective viruses and released.

To summarize, the lysogenic phage can replicate in two ways. It may integrate into the host, with its genes being copied and incorporated into the numerous bacterial progeny, or it may carry out cytolytic productive infection.

Transduction   **Transduction** is genetic recombination by transfer of genetic material from one bacterium to another by a virus vector (Fig. 10.9). From time to time, as a prophage detaches from the host chromosome at the beginning of the productive cycle, a small error by the excising enzymes removes with it some adjacent bacterial DNA. These bacterial genes will be copied as the viral genome is copied, and placed in all the mature virions produced in that cell. As these virions infect subsequent hosts, they again integrate as prophages. The second host's chromosomes become mosaics of their own DNA, DNA from the virus's previous bacterial host and from the virus's DNA itself. The transduced bacterium will be a **partial diploid;** it contains two genes for the transduced functions. If the transduced gene is different from the host-cell gene, the bacterium may exhibit a new phenotype. A bacterium unable to metabolize galactose (gal⁻) may be transduced by a virus carrying a bacterial gene

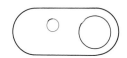

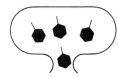

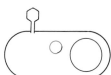

10.9
Transduction   Transduction requires three stages. First, a temperate phage establishes integration in a bacterial host. (1–3) Second, the prophage de-integrates, taking with it some bacterial DNA, (4) and carries out a productive cycle. The first host is destroyed. (5) All the virus progeny carry bacterial genes. Third, the "hybrid" virus infects a second host and integrates. (6–7) This host is now transduced. Its genome contains not only the viral prophage but also bacterial genes from the first host.

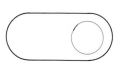

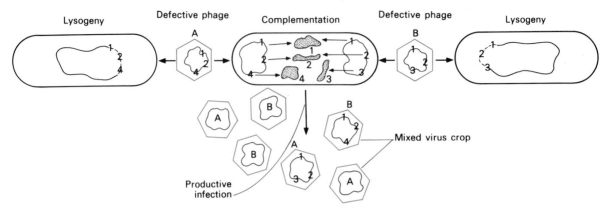

10.10

Complementation   Defective phage A possesses Genes 1, 2, and 4, but lacks 3. When it infects, it can integrate, but cannot carry out a productive infection. This is because Gene product 3 would be missing and the viral capsid could not be assembled without it. Defective phage B lacks Gene 4, thus a different product is missing. During a combined infection, all four gene products are produced. Complementation provides all necessary proteins to assemble the two types of complete virus progeny. But, since the defective viral genomes are copied exactly as they are, the viral progeny will remain defective.

conferring the ability to use galactose (gal$^+$). The host becomes a partial diploid heterozygous for this function (gal$^+$/gal$^-$) and it will be able to use galactose.

Because the amount of DNA that fits into a virus capsid is limited, any bacterial DNA included is balanced by phage DNA left out. Thus transducing phages are usually defective because they lack one or more key viral genes. They may lysogenize the next host adequately, resulting in transduction. But they can never complete a productive cycle unless they are "rescued." **Rescue** occurs when the host is infected by a **helper** virus that supplies the missing genes (Fig. 10.10). This process is called **complementation**.

The Filamentous Phages

The filamentous phages have single-stranded DNA that associates with capsid proteins in a helical symmetry, forming a semirigid helical structure.

When phage **fd** infects *E. coli*, replication of new virions begins. However, replication is very slow and regulated so that only one virion is produced at a time. Single copies of viral DNA are made; phage coat proteins are synthesized and accumulate at the cell membrane. The phage DNA is extruded through the bacterial membrane, acquiring its protein covering in the process. Production of new phage particles requires only a small percentage of the host cell's energy reserves and does

Box 10.1

Temperate Phage and Toxin Production

In diphtheria, botulism, and scarlet fever, many or all of the disease symptoms are caused by the pharmacological effect of exotoxins released by the infecting bacteria. In each of these diseases, toxin is produced only by strains of bacteria bearing a temperate phage.

Diphtheria has been the most intensively studied. Toxigenic *Corynebacterium diphtheriae* synthesize and excrete a protein (62,000 daltons) that enters the human cell and blocks protein synthesis by inactivating one of the elongation factors. Clinically, the toxin hits the heart, kidneys, and nervous tissue hardest; fatalities are usually the result of heart failure.

Corynephage β resembles the lambda phage of *E. coli*. It contains a double-stranded DNA genome, linear in the free form but capable of becoming circular and integrating with the bacterial genome. The toxin is coded for by the viral **tox** gene, which is regulated by promoter and operator regions. Toxin may be produced by viral genes both in the integrated and free form.

Toxin production is repressed by addition of as little as 350 μm of iron to the medium. It has been shown that the host bacterium makes a protein repressor that combines with iron, then blocks transcription of the tox gene.

This control would, unfortunately, not operate in the infected host. Normal human plasma contains on the average only 18 μm iron because serum also contains iron-binding proteins. Thus toxigenic *C. diphtheriae* are able to make their virally coded toxins unimpeded in the human body.

---

not block host-cell processes. Virus production can continue indefinitely because the host cell does not die.

Host–Virus Interaction

Note that one host, *E. coli*, can be infected by viruses that cause all three types of infections. One virus, **lambda,** can undergo two different processes of replication in the single host. This flexibility is due to interaction of the genetic regulation systems of the bacteria and virus. There is information feedback between two genomes sharing a single cell. Geneticists largely understand these complex interactions for the three infections presented here. However, there are many weighty unanswered questions about animal virus infections. For example, why does the identical virus that causes measles in one child cause pneumonia in a second and a fatal neurological degenerative disease in a third? The answer lies not just in what the virus does to the host cell, but also in what response the host makes.

Measles virus; Chapter 17, pp. 517–519

TABLE 10.2
Representative plant viruses

NAME	SYMMETRY	SIZE IN Å	ENVELOPE
Tobacco etch	Helical	200 × 7,000	−
Soybean mosaic	Helical	200 × 7,000	−
Alfalfa mosaic	Ellipsoid	180 × 300	−
Potato yellow dwarf	Helical	750 × 3,800	+
Southern bean mosaic	Icosahedral	280	−
Turnip yellow mosaic	Icosahedral	280	−

THE PLANT VIRUSES

Virus infection is widespread in the plant kingdom (Table 10-2). Its impact on human beings is significant. Crop failure caused by virus damage can lead to hunger and economic disaster.

Structure of Plant Viruses

All plant viruses so far discovered contain RNA as their genetic material. Their capsid structure is frequently of the helically assembled rodlike type, as typified by tobacco mosaic virus (TMV) (Figs. 10.11 and 10.12). However, other agents, such as

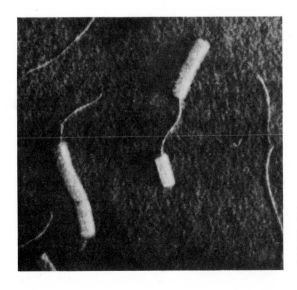

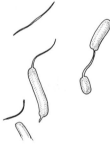

10.11
Tobacco Mosaic Virus The tobacco mosaic virus causes leaf damage to tobacco and tomato plants. Its helical structure is assembled as the long strand of RNA binds and coils with more than 2,000 protein molecules.

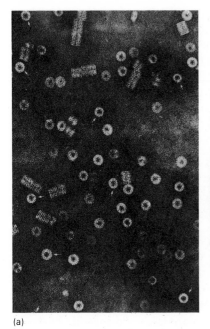

(a)   (b)

10.12
TMV Replication (a) Short segments of TMV virus are seen both from the side and in cross section. The pattern of protein coils around a central dark core is clearly visible. (b) In an infected plant cell, huge regularly arranged aggregates of the virus develop. The linear viral structures here occupy a large portion of the cell's area.

tobacco necrosis virus, are icosahedral. Envelopes are seen in a small number of plant viruses.

Mechanism of Infection

Plant viruses appear to lack specialized mechanisms for penetration of the plant cell wall. Thus infection probably always follows mechanical damage to the plant cell to permit entry of the virus. Such damage occurs as a result of bad weather, chewing insect attack, and cultivation by human hands and machinery. Studies show that plants in the field are often well "dusted" with virus particles: the primary barrier to infestation is the physical integrity of the plant itself. Thus explosive multiplication of insect pests in a crop area may precipitate an outbreak of viral infection to add to the damage caused by insects.

Types of Viral Damage

Once inside plant cells, viruses migrate through tiny pores called **plasmodesmata**, perforations in the cell wall that connect the cytoplasm of adjacent plant cells. Virus particles may also enter the plant's vascular system and be distributed with the sap. Plants, in contrast to animals, have no specific defense mechanisms to control the internal spread of infections. Thus plant infections usually progress throughout the life of the plant.

Viruses appear to interfere with plant photosynthesis and respiration. They may block water and nutrient transport. The regulation of flowering and fruit or seed development may be altered. Leaves become spotted, streaked, thickened, or curled. Under the microscope, tissues appear distorted and characteristic inclusion bodies may be seen.

The virion yield in plant infections is much higher than in bacteriophage infections, up to 60 million virions per cell. Tobacco mosaic virus may constitute as much as 10 percent of the dry weight of tobacco leaf. When a smoker drops tobacco scraps in the garden, the virus has been known to transfer its attention to such crop plants as the tomato.

Box 10.2

Controlling Insect Pests with Microorganisms

For every human being on the globe there are about one billion insects. At least 3,000 insect species are in direct competition with people for food crops, annually reducing human food supplies about 30 percent. Because of the tons of food they devour, insects are a major contributor to world starvation.

Chemical pesticides have many disadvantages. They kill beneficial insects and are toxic to animals and human beings. Because they require advanced technology and oil for production, their price is skyrocketing. Detailed training is needed in order to administer the pesticides safely and legally.

Biological control employs bacterial and viral pathogens of insect species. The bacterium *Bacillus thuringensis* (BT) kills the larval or caterpillar stage and can control the tent caterpillar, gypsy moth, and cabbage worm. BT cells, which are mass-produced commercially now, contain toxic protein crystals. The crystals, ingested by the voraciously feeding larvae, destroy their intestinal tracts and kill them. The protein is not toxic to other insects or vertebrates.

Baculoviruses are large rod-shaped DNA viruses pathogenic to insects. To use these for pest control, an ingenious method has been developed. Pheromones (female insect sex attractants) are used to attract and trap male insects. These males are dosed with virus, then released. Before becoming incapacitated, the males mate with wild female insects, infecting them too. The females in turn infect the eggs they lay. Thus the virus is spread throughout all members of the insect population. Baculoviruses successfully control corn earworm, and are being studied for their effect on African armyworm, pine sawfly, and Mediterranean cotton worm.

Microbial insect parasites are highly host-specific, and do not attack beneficial pollinating insects, animals, or humans. They are relatively cheap to produce, requiring simple technology and little energy. They are readily stored as dry crop dusting powders. Users do not need extensive protective gear or sophisticated instruction. With these advantages, biological insect controls probably will come to have an important part in pest control programs.

Viroids

Much interest has attended the discovery of a group of plant viruses, the **viroids,** the simplest self-reproducing agent yet found. They differ from "standard" virus in that they consist **only** of the nucleic acid genome and lack external structures such as capsid, envelope, or tail. Although they are widespread in plant disease, they have not yet been found in animal disease although it has been suggested that some of the elusive hepatitis agents may be viroids.

INSECT VIRUSES

Insects play an important role in spreading virus disease among plants and animals. In this role they are called **vectors.** Insect vectors are discussed in Chapter 15. Insects are themselves also parasitized by viruses important in the natural regulation of insect populations.

Insect vectors; Chapter 15, pp. 460–461

Recently it has been shown that viral insect pathogens have promise as biological controls for insect pests in agricultural settings. A number of insect viral preparations have been developed to spray or dust on pest-infested crops. Because these viruses have great host specificity, they do not damage helpful insects such as pollinators. However, to date it has proved difficult to obtain federal licensing for commercial use of viral insecticides.

SUMMARY

1. Viruses are an ubiquitous, highly specialized life form that completes a life cycle by means of obligate intracellular parasitism.

2. The free virus particle, or virion, consists of a nucleic acid genome, either DNA or RNA, enclosed in a protein capsid. Exterior to the capsid may be other structures essential to the infective process, such as the tail of bacterial viruses and the envelope of animal viruses.

3. The viral genome contains a minimum of genetic material. Only those genes needed to specify its own replication and direct the host cell to make new viruses are present. The host cell provides the raw materials, the enzymatic machinery, and the metabolic energy. Viral replication may result in the complete destruction of the host cell, reversible damage, or no apparent ill effect.

4. The cytolytic bacteriophages, such as the T-even group, carry out a lytic cycle of infection. Infection is initiated. Then there is an eclipse phase during which early proteins are synthesized and viral DNA is manufactured. Later, there is a maturation phase during which capsid proteins are made and the virus particles are assembled. The cycle terminates with digestion of the bacterial cell wall and the release of viruses.

5. Lysogenic viruses, as exemplified by phage lambda, normally do not destroy the host. Instead the viral genome integrates into the circular bacterial chromosome to form a prophage. Most, or all, of the viral genes are repressed. The prophage is replicated along with the bacterial genome and is passed to all the host cell's progeny. The prophage may cause new phenotypic traits in the infected strain.

6. On occasion the viral genome directs a productive lytic cycle yielding free virus. Portions of the bacterial DNA may be included in the new virions and transmitted to new bacterial hosts. This is called transduction.

7. Filamentous phage carries out an unobtrusive form of replication. New viral particles are made and released at a slow rate maintaining normal function in the host cell.

8. Plant viruses always contain RNA. They infect plant cells following mechanical injury to the plant. The infections are slowly progressive.

9. Insects serve as vectors for viral and other infections.

10. Many virus diseases pathogenic to insect species are known.

Study Topics

1. What aspects of **cellular** function must have evolved to a functional level prior to the appearance of viruses?

2. In what ways are viruses more complex than plasmids? Refer to Chapter 9.

3. Summarize the genetically coded functions a virus might be expected to have.

4. How do viruses cause their host cell to become phenotypically altered?

5. How do viruses cause their host cell to become genotypically altered?

6. Under what conditions is infection of a host cell by a virus very difficult to detect?

7. Describe how simultaneous infection of a host cell by a defective virus and a helper virus results in production of both types of virions.

8. What are the general characteristics of viral infection of plants?

9. In what two ways are the interaction of insects and viruses important?

Bibliography

Brock, Thomas D., 1979, *Biology of Microorganisms* (3rd ed.) Englewood Cliffs, N.J.: Prentice-Hall, Inc., Chapter 10, pp. 309–349.

Hughes, S. S., 1977, *The Virus: A History of the Concept*, London: Heinemann.

Watson, J. D., 1976, *The Molecular Biology of the Gene* (3rd ed.) Menlo Park, Calif.: Benjamin/Cummings Publishing Company.

11

Animal
Viruses

Animal viruses cause many of the major unconquered infectious human diseases—perhaps even that most feared of all diseases, cancer. Animal viruses exert their effects on eucaryotic cells, about which we lack much basic knowledge. Thus our comprehension of the ways in which the animal host may respond to viruses is still incomplete. Studies of virus-infected tissues, at first frustrating, are now beginning to clarify the basic function of both normal and abnormal cells. There are reasonable hopes for control of viral killers in the near future.

CHARACTERISTICS OF ANIMAL VIRUSES

Animal viruses share in the general characteristics of viruses, as described in Chapter 10. Their specialized features will be discussed here. Like other viruses, the animal virus particle consists of a genome (D2, D1, R2, or R1) and a protein coat or capsid, and sometimes an envelope.

Genome   In viruses with envelopes in which the protein layer surrounding the nucleic acid is **not** the external layer, the complex of genome and capsid material is

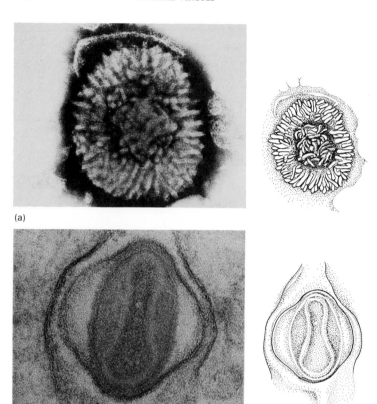

(a)

(b)

11.1
Pox Viruses The pox viruses are large, almost visible with the light microscope. The nucleic acid and protein are coiled up as a nucleocapsid within the lipid-soluble envelope.

called the **nucleocapsid** or simply the **core** (Fig. 11.1). Within the nucleocapsid, the nucleic acids appear to be "packaged" with the protein in a variety of ways. The genome of animal viruses containing DNA does not contain unusual bases; in fact, its physical resemblance to the chromosomes of the host cell is considerable. The DNA of the polyoma virus has been used as a model to study the DNA-histone interaction in the nucleosome, the subunit of eucaryotic chromatin.

Nucleosomes; Chapter 9, p. 254

Some of the RNA viruses, such as the reoviruses, possess a genome made up of several unconnected segments of RNA. This situation poses interesting questions as to the correct sorting and assembly of virus progeny in the host cell.

The possession of an RNA genome, single or double-stranded, requires RNA viruses to utilize unique enzymatic methods for copying and expressing their genetic material. Several quite separate patterns have evolved, all without parallel in cellular organisms.

Envelopes The larger animal viruses all possess envelopes. As previously mentioned, the envelope is lipoprotein in nature. The source may be nuclear membrane (herpesviruses) or cytoplasmic membrane (poxviruses). In either case, the viral envelope shares many physical and chemical characteristics with the host's own cells. This may confuse the body's immune mechanisms in their search for "foreign" invaders, in certain diseases.

However, the envelope may also contain up to 30 percent unique viral proteins. These are produced by the host cell in response to viral infection and incorporated into the envelope at the time when it is wrapped around the emerging virion. The viral proteins may appear as more or less pronounced spikes or **peplomers.** The external layers of the virus are essential for the physiochemical attachment with host cells preceding successful infection. Envelope antigens also stimulate the body's immune defenses leading to protective immunity.

Classification

The natural taxonomy of animal viruses is still very far from complete. We will use the formulation of the International Committee for the Nomenclature of Viruses. Please realize that these subdivisions are likely to undergo future change.

At present, taxonomists recognize seven genera of DNA viruses and 15 genera of RNA viruses. These are summarized in Tables 11.1 and 11.2. A very brief description of each group follows.

The DNA Viruses

Papillomaviruses are very small, unencapsulated virions. They cause benign tumors in various animals and warts in human beings.

Polyomaviruses also cause tumors but, so far, only in artificial laboratory situations. Recently, the JC strain has been implicated in degenerative brain disease in human patients receiving immunosuppression therapy.

Adenoviruses occur in a large number of immunologically distinct groups, or **serotypes,** and have pronounced spikes on the exterior of the capsid. They are found in many warm-blooded species. In human beings, they are agents of respiratory diseases of the "cold" variety. However, certain strains, when inoculated into newborns of species other than the natural host, cause malignant tumors.

Herpesviruses are currently receiving intense research efforts. They can cause a wide range of infections, from a malignant tumor (Marek's disease of chickens) to a generalized contagious disease (chickenpox), to diseases with intermittent symptoms (cold sores or fever blisters). Herpesviruses have been implicated in four different types of human cancer. They have a mysterious ability to remain latent but intact within the body for indefinite periods, and then reactivate.

The **iridovirus** group contains viruses of insects and nonhuman vertebrates. The polyhedrosis viruses are being tested and promoted as control agents for insect pests.

Poxviruses are the largest animal viruses and structurally the most complex. They cause disease in a wide range of species. They may be associated with generalized skin rash and respiratory involvement (smallpox) or may cause single or mul-

TABLE 11.1
The DNA viruses

GENUS	GENOME	SHAPE	SIZE (nm) NUCLEOCAPSID	EXAMPLES
Papillomavirus	D2, circular	Icosahedronal	55	Rabbit papilloma Human papilloma ("warts")
Polyomavirus	D2, circular	Icosahedronal	45	SV40, passenger virus of monkeys Mouse polyoma virus JC virus, human agent of PME
Adenovirus	D2, linear	Icosahedronal	70–80	Human adenovirus ("colds") Three types are strongly oncogenic
Herpesvirus	D2, linear	Icosahedronal, enveloped	100	Herpes simplex Types I and II Varicella-zoster Epstein–Barr virus
Iridovirus	D2, linear	Icosahedronal, some enveloped	190	Tipula virus
Poxvirus	D2, linear	Brick-shaped	$300 \times 240 \times 100$	Smallpox Cowpox Molluscum contagiosum
Parvovirus	D1, linear	Icosahedronal	20	Viruses of rabbits, dogs Adenovirus-associated viruses Hepatitis B

tiple benign tumors of skin (rabbit fibroma and the human venereal disease molluscum contagiosum).

Parvoviruses are very tiny and contain single-stranded DNA. Hepatitis B virus has recently been placed in this group. A severe and highly communicable disease of dogs is caused by parvovirus.

The RNA Viruses

Enteroviruses are very tiny and are found in the intestines of human beings and animals. They include the poliovirus group; the echovirus group, less commonly associated with overt disease; and the coxsackieviruses, which may cause influenza-like

TABLE 11.2
The RNA viruses

GENUS	GENOME	SHAPE	SIZE (nm) NUCLEOCAPSID	EXAMPLES
Enterovirus	R1	Icosahedronal	20–30	Polio virus Coxsackie virus
Rhinovirus	R1	Icosahedronal	20–30	Over 100 types, human "colds" Foot-and-mouth disease
Calcivirus	R1	Icosahedronal	20–30	Swine disease
Alphavirus	R1	Spherical, with envelope	50–60	Equine encephalitis virus Sindbis virus
Flavivirus	R1	Spherical, with envelope	40–50	Dengue Yellow fever
Myxovirus	R1, 7 segments	Spherical, with envelope	80–120	Influenza types A, B, and C
Paramyxovirus	R1	Spherical, with envelope	100–300	Mumps Parainfluenza Measles Distemper
Coronavirus	R1	Spherical, with envelope	80–120	Respiratory virus ("colds")
Arenavirus	R1	Spherical, with envelope	85–120	Lymphochoriomeningitis; Lassa virus
Bunyaweravirus	R1	Spherical, with envelope	90–100	Insect-borne group
Retrovirus	R1, 4 segments	Spherical, with envelope	100–120	C-type leukosis-leukemia-sarcoma virus B-type mammary tumor virus
Rhabdovirus	R1	Bullet-shaped, with envelope	175 × 70	Rabies
Reovirus	R2, 10 segments	Icosahedronal	70–80	Reovirus Type 1
Orbivirus	R2, 10 segments	Icosahedronal	50–60	Bluetongue virus
Rotavirus	R2	Spherical	< 36	Gastroenteritis agents

symptoms, myocarditis, or central nervous system disease. Enteroviruses are stable at a pH value 3. Thus they can be transmitted via food or liquids and pass through the highly acid stomach.

Rhinoviruses, of which more than 100 serotypes are known, are the usual agents of the common cold, although upper respiratory disease with coldlike symptoms may also be due to other genera of viruses. Foot-and-mouth disease is a severe rhinovirus disease in cattle.

Calciviruses cause severe generalized skin eruptions in domestic animals.

Alphaviruses and the following group are also known as **arboviruses.** They are arthropod-borne, transmitted by insect vectors. The arboviruses may cause inapparent infections in birds or other species; on transmission to the horse or human being they cause encephalitis.

Flaviviruses are tick- or mosquito-transmitted viruses. They cause generalized disease with rash (dengue), liver destruction (yellow fever), or central nervous system effects (encephalitis).

The rubella virus, which causes german measles, is loosely classified with the two groups above on the basis of structure, although it is not insect-borne.

Myxoviruses or influenza viruses contain a seven-part R1 genome in a helical nucleocapsid. Their envelopes contain two proteins, hemagglutinin and neuraminidase, synthesized under viral genetic control. As new strains of influenza virus evolve, the antigenic nature but not the function of these proteins change. Because each new strain has a new exterior antigenic surface, our old immune defenses are periodically rendered obsolete.

Paramyxoviruses contain a unitary R1 genome. The group includes the mumps virus, several groups of parainfluenza virus, and perhaps the measles virus and the respiratory syncytial virus.

Seen through the electron microscope, **coronaviruses** have a crownlike appearance. The human strains of the group cause "common colds." The animal strains may infect many systems.

Arenaviruses include the lymphochoriomeningitis (LCM) virus of mice, a persistent viral infection. In human beings, one type causes the highly fatal Lassa fever.

Bunyaweraviruses, a "supergroup", is an assemblage of more than 100 insect-borne viruses, most of them found in Africa.

Retroviruses are a heterogeneous group, many of which are **oncogenic,** or tumor-causing. One subgroup, the C-type particles, causes malignant leukemias and sarcomas in rodents and birds. The B-particles cause mammary gland tumors in mice. A third group causes progressive neurological degenerative diseases of sheep. All possess a unique enzyme called the **reverse transcriptase.** This enzyme allows them to make DNA copies of the RNA genome. The process is a backwards information flow; thus the **retro** virus name.

Rhabdovirus particles are bullet-shaped and have envelopes. The most important, from the human point of view, is the rabies virus.

Reoviruses contain a ten-piece genome within two concentric capsids. Reoviruses are widespread in warm-blooded animals but are not generally associated with disease.

Orbiviruses are associated with severe generalized disease in nonhuman animals.

Rotaviruses, recently classified agents, are a major cause of diarrheal illness in infants up to two years of age.

Unclassified viruses   Several important human disease agents have not yet been classified, among them Hepatitis A and the one or more so-called non-A, non-B, agents of hepatitis. The subacute spongiform encephalopathies (slow, neurological, degenerative diseases) are all very similar: the agents seem closely related to the retroviruses.

Species and Tissue Specificities

In nature, most viruses cause disease in only one species of animal. They are also most often restricted to one type of tissue. However, there are plenty of exceptions to both rules.

Species specificity   The polioviruses and rhinoviruses are maintained in nature exclusively in the human population. Epidemics develop as a result of person-to-person transmission. However, one can experimentally infect nonhuman primates with polio by an artificial route (injection) using large doses of the virus. Influenza has long been considered an exclusively human virus. It now seems likely that natural infections of rodents and of swine occur, and that these animals harbor virus of the A subgroups between human epidemics.

A few years back, an all-out campaign was started to eradicate yellow fever by means of mosquito control and vaccination. However, it was discovered that the virus also infects wild monkeys in jungle areas in which insects cannot conveniently be wiped out. The monkeys constitute a permanent reservoir of the virus, and new cases will continue to occur whenever unvaccinated persons enter the jungle. Eradication could occur if and only if the impossible goal of 100 percent vaccination could be achieved and maintained indefinitely. By contrast, smallpox has no nonhuman reservoir. Through quarantine and vaccination, human smallpox has been eradicated worldwide. Smallpox will not recur unless laboratory-maintained virus stocks were to be released.

Vaccination; Chapter 27, pp. 792–806

The rabies virus (Fig. 11.2) is unusually unselective. It can infect both wild and domestic mammals, including dogs, cats, foxes, cows, skunks, raccoons, and bats. Human beings are also infected when bitten by a sick animal. Rodents are the only common animals that seem to have species immunity to this virus.

Tissue specificity   Rhinoviruses attack only the superficial epithelium of the upper respiratory tract. They are not found in other tissues, apparently because their replication cycle is very sensitive to temperature. They will infect and replicate at 34°C, the temperature of the nasopharynx, but not at 37°C, the temperature of the rest of the body. By contrast, the virus of Creutzfeld–Jacob disease has been recovered from lymph nodes, liver, kidney, spleen, lung, cornea, and cerebrospinal fluids of victims.

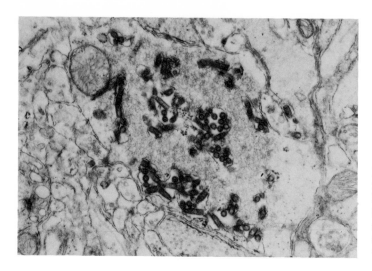

11.2
Rabies Virus in Mouse Brain   Clusters of rabies virus are shown within a cell from the brain of an experimentally infected mouse. Some viral particles are seen in cross section and appear round. Others, seen sideways, show the characteristic bullet shape of rabies virus.

In most cases, although a virus may be found in several tissues, its replication usually damages one type. The severity of the disease depends on the ability of the affected tissue to replace itself. Compare the effects of epithelial destruction by the influenza virus (in which the damage is repaired within one or two days) with that of nerve cell destruction by the poliovirus (in which replacement does not occur and paralysis is permanent).

Virus binds to tissue when viral surface sites are compatible with the glycoproteins of the host-cell membrane. Each type of tissue in the human body has a characteristic set of surface glycoproteins. The virus forms a semi-permanent, weak-bonded interaction with the target cell.

Modification   Viruses may be adapted to grow in tissues different from those in which they grow in nature. Pasteur was the first to demonstrate this fact in developing the rabies vaccine. He inoculated infective material into the brains of laboratory rabbits (not natural hosts). As each rabbit died, part of its brain was removed and transferred to the brains of healthy rabbits. The virus lost virulence (became **attenuated**) and, after many transfers, this rabbit-brain material no longer caused rabies in either susceptible animals or in human beings. Instead, it stimulated immunity. Modern rabies vaccines contain modified virus adapted to duck embryo or human fibroblast cell cultures.

Modification is a selective adaptation. Mutant viruses, if any are present, may have an advantage in the unnatural host. In most cases, a mutation that augments the virus's replicative ability in the new host renders it less adapted to damage the original host. The adaptation takes place slowly, requiring many passages in the unnatural host. These slowly acquired characteristics cannot be reversed rapidly. Thus

Stock strains; Chapter 27, pp. 795–796

such adapted strains can be used safely as the source of live virus for vaccines. When introduced to the natural host, they can no longer cause typical disease; their side effects will be limited to mild, nonspecific symptoms.

TECHNIQUES FOR ISOLATING AND PROPAGATING VIRUSES

Any virus needs living cells in which to grow. Virus laboratories have developed four different ways of supplying living cells to growing viruses: **cell culture, chick embryo, organ culture,** and laboratory **animal inoculation.**

Cell culture By providing suitable conditions, cells derived from mammalian tissues can be encouraged to survive and carry out mitosis after removal from the animal to laboratory glassware (Table 11.3).

Sources of tissue. Animal tissues are of four types: epithelial, connective, muscle, and nerve. The tissues share the same genetic material, but have undergone **differentiation** during embryonic life. As a result, the various cell types exhibit specialized functions. The highly differentiated muscle and nerve cells have permanently lost their ability to divide. Such cells cannot normally be cultured.

TABLE 11.3
Some representative cell culture lines and their uses

	LINE	SOURCE	CELL TYPE	USED TO
Primary cultures	HA	Normal human amnion	Mixed	Isolate and maintain adenovirus, coxsackie, herpes, reovirus, rubella, vaccinia, varicella, etc.
	CE	Normal chick embryo	Mixed	Isolate and maintain coxsackie, herpes, influenza, poxviruses, etc.
	AGMK	African green monkey kidney	Mixed	Isolate and maintain adenovirus, coxsackie, herpes, influenza, measles, polio, etc.
Serial cultures	3T3	Normal mouse embryo	Fibroblast	Assay transformation by oncogenic viruses
	HeLa	Human cervical cancer	Epithelium	Isolate and maintain most viruses of human disease
	L	Mouse connective tissue	Fibroblast	Isolate and maintain herpes, influenza, and measles
	WI38	Normal human skin	Fibroblast	Isolate and maintain most human viruses, produce virus for vaccines

TABLE 11.4
Growth characteristics of cell lines

LINE	SHOWS GROWTH LIMITATION	PERSISTENCE	KARYOTYPE	SPECIAL TRAITS
AGMK	Yes	Most if not all cells fail to reproduce after a few passages	Normal diploid	Like most normal tissues, does not persist in cell culture; new tissue sources always needed
HeLa	No	Cells of this line have been in serial culture since 1952	Heteroploid—numbers of chromosomes vary widely	Tumor tissue—has traits of transformed cell
WI 38	Yes	Stable in culture since 1962	Normal diploid	Normal in all respects except it has immortality

Epithelium or connective tissue cells retain the capacity to divide but at a fixed rate under regulatory controls. Thus the epithelium of our skin, cornea, and mucous membranes is replaced at a steady rate. Bone marrow, a connective tissue, matures the various types of blood cells and releases them into the circulation. Mitotic tissues of this type can be cultured, and in culture they will demonstrate **growth limitation**. That is, the cells in the vessels will divide only until a certain population density is reached. Then they will stop. Fibroblasts, cells derived from connective tissue, grow in a **monolayer**—a sheet only one cell thick. If there are only a few cells in a vessel, they divide regularly, their membranes undulate, and the cells stretch across the glass. Whenever these movements bring them into contact with a neighbor cell, motion ceases and cell division stops. Growth limitation is a property of normal tissue, in culture and (in all probability) in the body too.

Tumors may arise in any type of tissue. A tumor is a tissue that exhibits unusually rapid growth. When tumor tissues are placed in cell culture, they do not show growth inhibition. Cells multiply all over each other to form heaps many cells thick. One of the major markers of the tumor cell in culture, as opposed to the normal cell, is its inability to regulate growth.

Primary cell cultures. These result when a sample of tissue is taken aseptically from an animal, minced, treated with an enzyme to separate the cells, placed in sterile glassware, provided with medium, and incubated (Fig. 11.3). The tissue taken may be normal or abnormal. The resulting culture will consist of a mixture of cell types.

A useful primary cell culture can be created from human amnion tissue, collected following normal deliveries. It is used to cultivate a large variety of human pathogenic viruses. Another is African green monkey kidney, which also supports the growth of most human viruses.

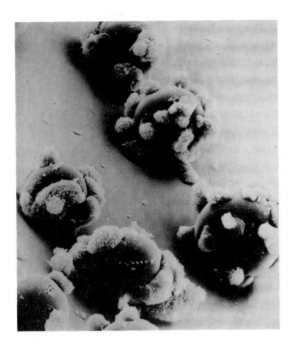

11.3
Cell Culture   Human cells can be cultured in large numbers when they have a large surface area for attachment. One way of providing a large area is by suspending inert microscopic beads in the culture fluid. Here cells of the Vero line are seen attached to a bead.

Serial cell cultures.   If the primary culture is harvested, and the cells transferred to a new container, a secondary culture arises from those cells that are still dividing. If the subculture process is continued for an extended period of time, one of two things happens. Most frequently, the culture dies out. After a period of time, all the cells fail to divide. In other cases, stable cell lines capable of indefinite serial propagation emerge. These "immortal" cell clones, such as HeLa or L-cells, are usually abnormal or tumor-like. With special care, cell lines have been selected which both carry out serial growth and exhibit normal growth limitation. An example is WI-38, a cell line derived from normal human skin. WI-38 has normal growth limits and normal chromosomes. It is used both for isolation and maintenance of human viruses in clinical applications, and also for bulk production of vaccine viruses (Chapter 27).

Production of antiviral vaccines; Chapter 27, p. 794

Culture conditions.   Cell culture growth media provide a complete set of amino acids, vitamins and other growth factors. Most cells in culture also require serum. To prevent microbial contamination, antibiotics are included. The culture fluid pH must be regulated to within 0.2 pH units, in many applications. In the intact animal, pH is controlled by the natural buffering action of the bicarbonate ion system. *In vitro* this is simulated by providing an incubation atmosphere containing CO_2. Organic buffers may also be added; a pH indicator, usually phenol red, is included to give direct visual monitoring of the medium pH. The medium is changed at regular intervals to replace nutrients and remove cell wastes. All manipulations are carried

11.4
Large-scale Production of Cultured Cells   This rack system
supports many large cell culture vessels and rolls them for media
circulation and aeration.

out under very strict, aseptic conditions (Fig. 11.4). The usual incubation temperature for tissues and viruses of human origin is 37°C.

Virus may be inoculated into the culture media of a healthy cell culture. The
virions will infect the cells and replicate, sometimes causing characteristic **cytopathic
effects** (CPE) in the cells (Fig. 11.5). Mature viruses may be harvested from the culture media or the cells.

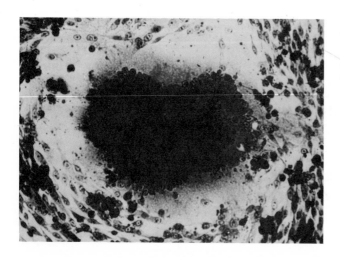

11.5
Virus infection may cause visible changes in cultured host cells. Varicella-zoster virus infecting
human melanoma cells causes them to heap up and
form a compact syncytium.

11.6

Chick Embryo   The fertilized egg contains a developing chick embryo, enclosed in an amniotic sac. The yolk sac, which fills much of the egg, contains its food supply. The albumin or "white" of the egg supplies amino acids for growth. The chorioallantoic membrane is the tough inner membrane. When the inoculum is injected, the needle passes though the blunt end of the egg, where the air space is located. Then virus may be dropped on the chorioallantois, injected into the yolk sac, or introduced into the amnion. Each virus has a preferred site for replication.

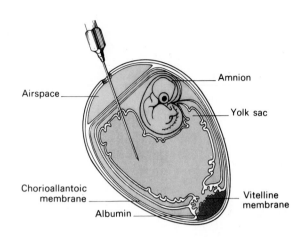

Airspace — Amnion — Yolk sac — Chorioallantoic membrane — Albumin — Vitelline membrane

The chick embryo   A fertilized hen's egg (if uncracked) is a very convenient, sterile package of cells. As the embryo develops, the embryonic membranes that form resemble cell monolayers (Fig. 11.6). The embryo itself contains rapidly dividing cells of each tissue type, in the early stages of differentiation.

Early in the development of virology, fertilized eggs were widely used, both for isolation and identification of virus from clinical samples and for vaccine production. At present, eggs are still used to isolate influenza virus from the respiratory secretions of patients during epidemics. The influenza vaccines, yellow fever vaccine, and rabies vaccine are produced in embryonated eggs (Fig. 11.7). Persons allergic to eggs may experience side effects when they receive these vaccines.

Virus replication begins when the inoculum containing virus is injected into a chosen location in the egg. Viral multiplication will cause visible changes after a

11.7

Viral Vaccine Production   Vaccine production requires the mass inoculation of large numbers of fertilized eggs. This is done with strict aseptic precautions, both to maintain the eggs free from contamination and also to protect the workers.

11.8
Research Animal Facility   Many research and diagnostic procedures require a source of standardized, pathogen-free animals. Here mice are raised and maintained in large numbers. Note that the cages are covered with air filters and that the technician is transfering a mouse with forceps to avoid transfer of microorganisms.

time. Most characteristic are small zones of cell destruction called **plaques** on the embryonic membranes. Death of the embryo will eventually occur.

Virus is harvested by opening the egg, removing the affected tissue, mincing it, and separating the virus particles from the debris.

Organ culture   When the maintenance of an intact segment of an organ is desired, the explant may be embedded in a serum clot on a chemically inert grid. The tissue is moved regularly to new clots of serum.

Animal inoculation   Cultivation of virus in intact animals is very expensive. For both humanitarian and practical reasons, it is used only when other techniques are not available or suitable. Current virology applications often involve the use of newborn animals. Newborns are highly susceptible because their natural defenses are undeveloped.

In order to render experimental results as clear as possible, the genetic variability between individuals may be minimized by using **inbred** animals (Fig. 11.8). An inbred line is the result of many generations of brother–sister mating. All individuals in an inbred line are genetically identical. Because of this genetic uniformity, tumors or normal tissues can be transplanted from one individual to the next without the interference of rejection mechanisms.

REPLICATION OF ANIMAL VIRUSES

The characteristics of cytolytic and temperate infections were discussed in Chapter 10. In this section, we will examine how these cycles occur within the specific framework of an animal cell. Most of the detailed observations discussed were obtained by the study of infected cells in culture. We cannot view these events at all clearly in the intact animal.

The Initiation of Infection

Initiation requires two sequential events, attachment and penetration, that exhibit certain differences from the events occurring in the bacterial model.

Attachment A virus's species and tissue affinities are determined by its attachment to liproprotein or glycoprotein receptors on the host cell.

Certain viruses, for example, the influenza viruses and adenoviruses, will cause the agglutination or clumping of red blood cells—**hemagglutination.** This is a nonspecific attachment because the RBC does not serve as a host cell for viruses. It demonstrates, however, the presence of numerous receptor sites on the virus particle; one virion links to two or more RBCs, forms cross-connections, and thus makes clumps.

The receptors on the host-cell surface are probably not immobile. Because the cell membrane is a fluid mosaic, receptors may at one time be exposed on the surface and at another time be concealed. This could affect the susceptability of host cell to virus.

Penetration Virions are actively taken into the host cell, by an invagination of the cell membrane, with their outer layers intact. This contrasts with the bacteriophage model in which nucleic acid is injected into the cell and the capsid is excluded.

The virus nucleic acid is uncoated by enzymes either of host or viral origin. A poliovirus is uncoated as it passes the cell membrane; larger viruses are uncoated within the cell, perhaps by lysosomal enzymes. It is unclear how viral nucleic acid escapes enzymatic attack.

Fluorescent antibodies; Chapter 13, pp. 389, 391–393

The viral nucleic acid may remain in the cytoplasm to direct viral replication (poxviruses) or move to the nucleus (herpesviruses). At the spot at which viral replication and assembly occurs, a dense aggregate of viral material referred to as an **inclusion body** may develop. Inclusion bodies can be specifically identified by fluorescent antibody techniques, a valuable diagnostic approach.

DNA virus replication Recall that bacterial viruses are capable of cytolytic, lysogenic, or steady-state replication. These patterns are also seen in animal viruses. The cycle carried out depends on virus–host-cell interaction. Thus a virus may be lytic in a host of one species (a **permissive** host) and form a provirus in another (a **nonpermissive** host). In a **semipermissive** host it may be lytic in some percentage of cells or under certain physiological conditions while remaining latent in adjacent cells or under other conditions.

Cytolytic replication of DNA viruses proceeds in permissive cells according to the general plan laid down in Chapter 10. The phases are synthesis of early proteins, replication of the DNA, synthesis of late proteins, assembly, and maturation. Many copies of all viral materials appear to be made, and the assembly process is somewhat inexact, so that such infections yield a high proportion of incomplete particles—empty capsids, structurally deficient particles, or noninfective virions. Viral DNA synthesis can occur either in the nucleus or in the cytoplasm; host-cell DNA is not usually destroyed.

Polyoma viruses, in nonpermissive cells, carry out temperate infection similar to bacteriophage lysogeny. After penetration and uncoating, the DNA is copied, and one or more viral genomes are covalently integrated into the host chromosomes. No further steps are taken toward viral replication. In contrast to the bacterial system, however, there is no evidence of specific repressors. The integrated

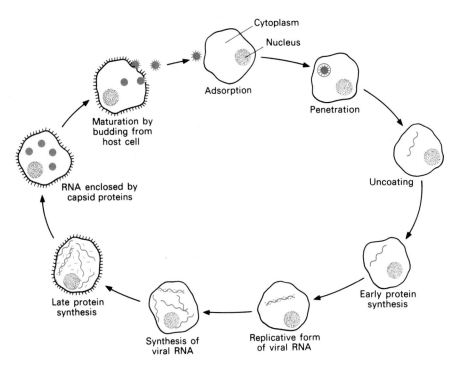

11.9
Replication of Influenza Virus   The replication of influenza virus illustrates many of the features of animal virus interaction with host cells. The virion is taken into the cell intact, then uncoated, releasing the single-stranded RNA genome, which is in seven segments (only one is shown here). A replicative form (Rf) (color) is made for each segment. The Rf in turn templates the synthesis of new genomes. As the late proteins are synthesized, they come to lie just beneath the cell membrane. Virions exit by budding through the membrane. A segment of the host cell's membrane becomes their envelope. It is chemically modified by the inclusion of the viral proteins hemagglutin and neuraminidase. *In vivo*, the virus production is eventually brought to a halt by interferon. Virus replication does not lyse the cell.

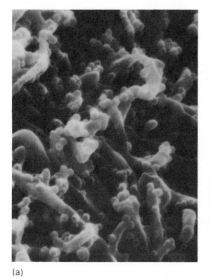

(a)

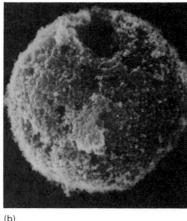

(b)

11.10
Virus Budding (a) Eleven hours after infection with influenza these monkey kidney cells seen with the SEM show many finger-like protrusions from which viruses are exiting. Numerous tiny round to oval bodies are free viruses. (b) An infected cell 24 hours after infection, detached from the growth surface. Its membrane is completely covered with viral buds.

viral genome, or *provirus,* is transmitted to subsequent generations of cells by mitosis. It may cause changes in the host cell phenotype (transformation) or be completely inapparent.

Steady-state replication occurs in persistent infections.

Replication of RNA viruses The RNA viruses are unique forms of life because their nucleic acid is RNA, not DNA. They have at least five variant replication cycles. We will deal with just two of these.

Poliovirus has a single-stranded RNA genome that can serve directly as a messenger RNA in that it has the correct polarity for translation. However, in order to produce copies of this molecule for new genomes, the viral RNA must be used as a template to make complementary copies, **replicative forms** (RF). These RFs can serve as templates to make copies of the original viral RNA. Viruses that make replicative intermediates have genes that code for a special enzyme, **RNA-dependent RNA polymerase.** In their replication cycle, one of the first early proteins to be made is the polymerase. Then come replicative forms, then new genomes. Expression of late genes leads to appearance of viral coat proteins, assembly, and maturation (Figs. 11.9 and 11.10).

The retroviruses can integrate their genomes with that of the host. You may wonder how a RNA genome could integrate with a host chromosome containing DNA. The answer, discovered quite recently, was thoroughly startling to many scientists. In retroviruses, the flow of genetic information is reversed from "normal," and RNA is used as a template for the synthesis of DNA.

When a retrovirus carries out productive infection in a permissive cell, the single-stranded RNA genome is copied by the unique virus-coded enzyme, RNA-

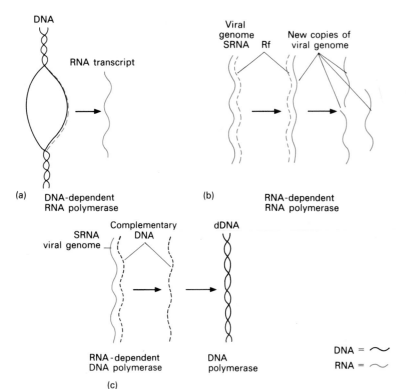

11.11
Information Flow (a) The normal cellular enzyme DNA-dependent RNA polymerase expresses genes by transcription of RNA. It may transcribe either host genes or viral genes if the genome is DNA. (b) To copy a viral RNA genome, RNA-dependent RNA polymerase (an enzyme coded for only by viral genes) is needed. It operates in two stages, first making the complementary sRNA copy, a replicative factor or Rf, then via Rf making many new copies of the viral genome. (c) In the retroviruses, the reverse transcriptase, RNA-dependent DNA polymerase, makes a single-stranded DNA copy of the viral genome. DNA polymerase completes the job by adding the second strand to the DNA copy. The finished dDNA can then be integrated into host chromosomes.

dependent **DNA** polymerase, called the reverse transcriptase (Fig. 11.11). The first single-stranded DNA copy, which is complementary to viral RNA, is copied again by another enzyme to yield the usual double-stranded type of DNA. This DNA then serves two functions. It operates both as a template for the production of new virus RNA genomes and also as a template for the production of viral messenger RNA, thus directing viral protein synthesis. However, it is not a part of the final assembled virus.

Productive retrovirus infections are uncommon. Most cells infected with these agents are nonpermissive; that is, they yield no infectious virus. The DNA duplex derived from the viral genome is covalently integrated into host DNA as a provirus. It may change the cell's behavior causing it to become cancerous.

Release of viruses   Some of the viruses without envelopes are directly cytolytic. Release occurs as the membrane disintegrates.

Viruses with envelopes are released by **budding,** and this may or may not destroy the cell. It appears that viral glycoproteins enter the plasma membrane, displacing normal cellular membrane protein, and a special viral protein coats the inside of this prepared membrane area. Then the viral nucleocapsid attaches to the inside of the region, the viral glycoproteins push through to the outside of the membrane where they appear as spikes, and the membrane pouch pinches off.

Some viruses pass directly from one cell to a neighboring cell through adjacent membranes and have no extracellular phase.

ACUTE DISEASE PATTERNS

Much of the foregoing discussion has focused on the effects of viruses on single cells or cell cultures *in vitro*. Now we shall look at some basic patterns of actual disease in animals. Specific diseases will be analyzed in later chapters, in conjunction with our survey of medical microbiology.

Localized Infections

Certain viral infections are restricted to one type of tissue. Viruses can be demonstrated in only one area, and damage is also restricted to that area. Some of the viral infections of the respiratory and gastrointestinal tracts are of this type.

Let us use influenza as an example. The virus, usually acquired by inhalation, becomes attached to specific receptors on the epithelial surface of the upper respiratory tract. There is a very short incubation period of up to three days, as cellular infection occurs. Many thousands of viral particles are released per host cell. These infect adjacent tissue, creating foci of infected and dead cells. Tissue damage leads to the secretion of fluids, inflammation, and fever. Eventually, the respiratory epithelium is **desquamated;** that is, the surface layer of cells is lost. The exposed surface may be colonized by bacteria that occasionally cause secondary pneumonia, the cause of death in the fatal cases. The acute infection comes to an end when the supply of target tissue is exhausted or the body successfully fends it off.

Influenza; Chapter 19, pp. 562–565

Systemic Infections

Measles is a systemic or generalized infection. The virus, acquired by inhalation, infects the respiratory mucosa without symptoms, and then passes into the lymphoid system. An **incubation period** of 10 to 12 days intervenes between exposure and the appearance of symptoms. During this time the virus is actively multiplying in lymphoid tissue. At the end of the incubation period, the virus is disseminated to body surfaces via the blood stream (**viremia**), and a rash suddenly appears on the skin and mucous membranes. Simultaneously toxic symptoms, such as fever, also appear. At this point, virus replication is almost at an end. Many of the symptoms, such as the skin lesions, are caused principally by the body's defense systems attacking those cells that harbor virus. The symptoms subside over a period of days and, in most but not all cases, the virus is completely eliminated from the body.

Measles; Chapter 17, pp. 517–519

Mechanisms of Cell Death

Cells are damaged or die in the infected animal for one or many reasons. Virus infection usually shuts down the synthesis of host-cell protein, RNA and DNA. Mitosis is usually prevented and, in some cases, chromosomes become markedly abnormal. The lysosomes and cell membrane become leaky, and the cell may autolyze. Certain viral proteins are directly toxic. Some viruses cause cells to fuse, forming **syncytia** or **giant cells,** that are nonfunctional.

Responses of the Animal to Infection

Mammalian species have several protective responses to viral infection. Nonspecific defense mechanisms operate against all types of pathogenic microorganisms. Phagocytic cells (neutrophils and macrophages) can phagocytose virus particles and digest them. Circulating and secreted antibodies specific for the virus can neutralize viral infectivity. Specially sensitized lymphocytes and macrophages of the cell-mediated immune (CMI) system can carry out attack on the viruses or on the cells that harbor them. These mechanisms are discussed in Chapters 12 and 13.

Interferon   The **interferons** are a special class of proteins produced by virus-infected cells; their function is to interfere with the viral replication cycle in neighboring cells (Fig. 11.12). They play a major role in the arrest of acute short-term (under one week) infections.

There are three types of interferon, produced by fibroblasts, leukocytes, and lymphocytes, possibly with slightly different effects. All are proteins with weights between 15,000 and 30,000 daltons. Interferon can also be produced by bacterial cells containing plasmids bearing the human interferon gene.

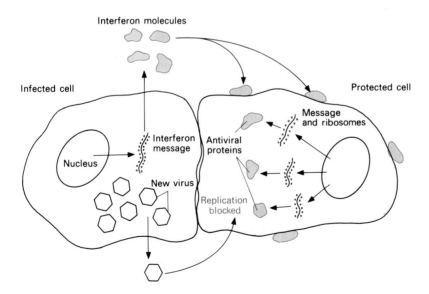

11.12
When a cell is infected, viral nucleic acids derepress the gene(s) for interferon production. The interferon message is translated and its product released, along with new virus particles. When interferon interacts with the cell membrane of neighboring cells, they are induced to produce antiviral proteins. When the second cell is infected, the viral replication cycle is blocked.

Box 11.1

What Price Interferon?

Interferon is currently one of the scarcest, most sought-after, and most expensive biochemicals. This is because it can engage in not only antiviral but also antitumor activity. Because interferons inhibit cell division, and enhance macrophage and lymphocyte killer activity, they have great promise in cancer chemotherapy. Very small-scale clinical trials have been done, limited because there is very little interferon available. A few patients with osteogenic sarcoma, lymphoma, Hodgkin's disease, melanoma, breast cancer, and leukemia have obtained variable levels of benefit from interferon therapy. The results are promising, but far from conclusive. Sometimes patients that were responding well to interferon therapy have had to be taken off the drug and put back on regular therapy because the supply of interferon ran out.

To expand the scope of the research, in 1978 the American Cancer Society put up $2 million for purchase of interferon to give to investigators for expanded clinical trials. This amount will buy only enough interferon to treat about 150 patients. Why is interferon so expensive? Although all mammalian species produce it, only human interferon is functional in human beings. It must be produced by living human cells. Until recently, the only source

was leukocytes obtained from the blood of donors. Since these cells do not divide, their yield is tiny. Now a method is available for culturing human fibroblasts that make interferon. Interferon produced by any method to date is very impure. This makes it difficult to say that any effects, good or bad, are due solely to the interferon in the preparation.

Several investigators are attempting to engineer a plasmid that will produce interferon. As we go to press, two research groups have reported success in construction of plasmids bearing interferon genes. Right now, there is no proof that host bacteria will make interferon or, if the interferon is made, that it will be biologically active.

This and other lines of interferon research are big gambles. However, the stakes are high. If interferon does work as a treatment for cancer, we will, for the first time, be attacking cancer with the body's own weapons. It is likely that side effects of interferon will be less severe and perhaps more controllable than the side effects associated with the use of drugs or radiation. Many knowledgeable researchers hope that interferon will also work better than conventional therapy, at least for some patients.

Interferon appears rapidly, within 24 hours of infection. When a cell is infected by a virus, the viral nucleic acid depresses a gene for the synthesis of interferon. The protein is produced and released from the infected cells at about the same time the virus particles mature. The interferon then binds to the membrane surface of the neighbor cells, but apparently does not enter the cells. Its binding triggers depression of other genes within the neighbor cell. New intracellular gene products appear; their combined effect is to block viral synthesis of protein and thus replication.

It is generally accepted that interferon synthesis is the means by which the body brings acute viral infections, such as the common cold and influenza, under control. That interferon synthesis is the means is indicated because the disease is cured before significant levels of immunity are present. Interferon's role in longer viral infections is not so clear, but preliminary research data show that interferon has therapeutic promise against rabies, hepatitis B, arthropod-borne encephalitis, and generalized herpes infections.

Elimination of virus In longer-term infections, during which there is an incubation period and/or generalized distribution of the virus, there is sufficient time for the specific immune response to develop. Both antibodies and immune cells appearing after ten days to two weeks are available to assist the nonspecific interferon and phagocytes. In persistent viral diseases, however, virus is not eliminated effectively by any combination of these defenses.

PERSISTENT DISEASES

There are several virus diseases (some common, some rare) in which virus persists in the tissues for extended periods of time, and the body's defenses are not successful in eliminating it (Table 11.5). Symptoms may occur sporadically or not until many years after an infection.

Latent Infections

Latent infections are characterized by intermittent episodes of acute cellular involvement and active virus replication. These episodes are separated by long periods during which all cells appear normal and infective virus is not demonstrable. The classic example is cold sores caused by herpes simplex virus Type I. The virus is usually contracted in childhood by direct contact and causes an ulcerative sore on the lip. The sore heals and a latent stage results. Unusual stress such as fever, sunburn, or other events can reactivate the virus, and another sore will result. The state of alternating activity and latency will persist for life. During latency, it is believed that the virus, perhaps in a provirus state rather than as an intact virion, harbors in sensory nerve ganglia serving facial skin. Antibodies to the virus are usually present; they probably cannot eliminate the virus because of its protected location in the nerve.

Chronic Infections

In chronic infections the virus is actively replicating and can usually be demonstrated in tissue fluids. Disease symptoms, on the other hand, are usually absent and will appear only in certain circumstances. Viral production is of the steady-state type. An immune response is present, but it fails to reject the virus.

TABLE 11.5
Persistent infections

TYPE	VIRUS	HOST	LOCATION OF VIRUS	DISEASE AND SYMPTOMS
Latent	Herpes simplex	Human beings	During attack, in epithelium Between attacks, in ganglion cells	Recurrent fever blister on lip Activated by numerous stresses
	Varicella-zoster	Human beings	During attack, in epithelial cells Between attacks, in ganglion cells	Shingles—widespread, very painful eruption over face and/or trunk
Chronic	SV 40	Monkey	Kidney	None
	Hepatitis B	Human beings	Liver	Usually none, found in recovered cases that act as carriers
	Epstein–Barr virus	Human beings	Lymphoid tissue	Usually none, occasionally causes infectious mono-nucleosis; associated with Burkitt's lymphoma and nasopharyngeal carcinoma
	Lymphocytic choriomeningitis virus	Mouse	Widespread	Glomerulonephritis, immune complex attack on kidney
Slow	Creutzfeld–Jacob virus	Human beings	Central nervous system and other sites	Creutzfeld–Jacob disease, a type of subacute spongiform encephalopathy (SSE) with progressive central nervous system degeneration
	Measles virus	Human beings	Central nervous system	Subacute sclerosing panen-cephalitis (SSPE) years after acute measles
	JC group human polyoma virus	Human beings	Central nervous system	Progressive multifocal leukoencephalopathy (PME), usually follows immunosuppressive therapy

Hepatitis Type B (serum hepatitis) becomes a chronic infection in about 5 per-cent of those who recover from the first, acute attack. The serum of such persons is infectious because it contains both viable and defective virus particles—as many as a million infective units per milliliter. These persons are carriers. Most of them even-tually stop producing virus.

Hepatitis Type B; Chapter 20, pp. 598–601

Other chronic infections are asymptomatic in early life, but develop into fatal diseases later. These may be **immunopathological.** The animal's immune system destroys the infected tissue and thus kills the animal.

Malignant diseases of virus origin will be discussed in the next section.

Slow Infections

In **slow-virus disease,** viral replication occurs as a continuous but slow process. An extended time—measured in years in human beings—elapses before the cumulative viral damage is sufficient to be clinically noticeable. These diseases are progressive and invariably fatal. Most of them are manifested as central nervous system degeneration. Fortunately, they are also quite rare.

One rare human disease is called subacute sclerosing panencephalitis (SSPE). This disease occurs some years after recovery from attacks of measles. The measles virus has been demonstrated microscopically in brain biopsies from fatal cases and isolated from these samples. Abnormally high levels of antimeasles antibodies are found in these patients, but the antibody is clearly not protective. There are no good answers as to why certain individuals develop SSPE after measles. With the advent of mass antimeasles vaccination, it is expected that the incidence both of acute measles and of SSPE will drop.

Persistent infection is currently receiving great attention as a possible causative mechanism for diabetes, multiple sclerosis, and certain degenerative diseases of connective tissue.

ONCOGENIC VIRUSES

Lately, cancer investigation has dominated health-related research. A tremendous amount of information has been gathered, showing that cancer is not one disease, but many. There is clearly neither a single cause nor a single cure. The section that follows analyzes the role of certain viruses in laboratory animals. Information that supports the hypothesis that specific viruses may contribute to some types of human cancer will be presented.

Cancer is a **multifactorial** disease. That is, many factors in combination increase its incidence and, by inference, cause the disease. Among these factors are diet, radiation, chemicals, age, ethnic group, geographical location, immune status, and viruses. At the present time, it is impossible to separate these sufficiently to make unambiguous statements about what causes cancer.

The Nature of Cancer

Cancers in intact animals   A cancer or **neoplasm** is a group of cells that proliferates at a rate greater than that of its surrounding tissues. A **tumor** is a solid or localized mass. **Benign** tumors remain intact, growing in one locale only. They are dangerous if they distort organs, undergo necrosis and release toxic substances, exert pressure on major arteries, or cause intracranial pressure as a brain tumor might. Those cells that break away from the mass are not capable of establishing themselves in other areas of tissue and are unable to grow there.

Malignant tumors shed cells (**metastasize**) as they grow. These cells are not inhibited from growth in other tissue sites. Instead they "seed" new tumor foci throughout the body, primarily in lymphoid tissues. These **metastases** become numerous. As they grow, they have deleterious effects on normal organs and, if unchecked, can cause death.

Not all cancers are solid masses. **Leukemias** consist of (1) proliferating, abnormal, lymphoid cells, with aggregates in bone marrow, and (2) many circulating forms. The latter frequently attack normal tissue.

Distribution   Several hundred different types of cancer have been described in human beings. Most types of tissue can undergo the mysterious changes that cause them to become cancerous, and thus to become initiators of the disease.

The Transformation of Cultured Cells

The laboratory model for the study of the cancerous change is **transformation** that occurs when normal cultured cells undergo an observable metamorphosis into cells with altered characteristics. Table 11.6 lists a number of the abnormal traits acquired in cell transformation.

TABLE 11.6
Altered properties of cells transformed by oncogenic virus

Growth changes	Grow to high cell-population densities
	Grow in less oriented fashion
	Grow in media containing agar
	Require less or no serum in media
	Cell motility not inhibited by contact with neighbor cells
	Chemical communication with other cells reduced or absent
	Form tumors when injected into susceptible animals
Surface changes	Agglutinated by plant lectins
	Up to 30% change in membrane glycoprotein composition
	Acquire tumor-specific transplantation antigen (TSTA)
	Acquire renewed expression of fetal antigens
	Microtubular activity reduced
Evidence of virus present	Virus-specific antigens in nucleus (T-antigen)
	Virus DNA sequences detected by hybridization
	Virus mRNA sequences detected by hybridization
	Virions produced under certain conditions
	Reverse transcriptase present (retroviruses)

These traits are shared by cells containing provirus of DNA or RNA virus origin. A single transformed cell will not necessarily show all of these traits simultaneously.

Major features of the transformed cell are unresponsiveness to normal growth limitation, lack of cell-to-cell communication, ability to multiply in an altered environment, (that is, without serum or in the presence of agar), and the appearance of novel antigens on its surface or in its nucleus. Let us stress that although scientists can observe and measure the change, they still do not know exactly what **causes** these differences.

Assays for transformation   Normal cultured animal cells are exposed to **carcinogenic** (cancer-causing) energy or matter, such as radiation, hazardous chemicals, or viruses. The cells are then placed in culture media that either lack serum or contain semisolid agar. Only transformed cells can successfully initiate growth and form clones under these conditions. Those clones of cells that appear are examined to see if they demonstrate transformed characteristics.

Assay for carcinogenesis   The **Ames test** is a recently developed, convenient assay using bacterial cells to measure the carcinogenic potential of chemicals. The test determines how much any chemical increases the rate of mutation of a single gene in *Salmonella typhimurium.* Mutagenic potential of a chemical in bacteria has been demonstrated to correlate well with carcinogenic potential in animals, and it is vastly quicker and less expensive to evaluate.

Oncogenic DNA Viruses

About one quarter of the known animal viruses cause cell transformation in one system or another. Many fewer cause naturally occurring tumors. Five of the seven genera of DNA viruses contain members that cause abnormal cell proliferation (Table 11.7).

Papillomaviruses   These cause benign tumors in human beings and rabbits. In the rabbit, such benign growths frequently develop into malignant carcinomas later in life. Papilloma viruses cause a low frequency of transformation of cells in culture.

Polyomaviruses   The SV40 virus was first discovered as a passenger virus in monkey kidney cell cultures used in the production of Salk polio vaccine. The procedure used to inactivate the poliovirus in the vaccine did not kill the more resistant SV40 virus. Millions of people received vaccine containing live SV40 virus. It was simultaneously shown that, although SV40 had no visible effects on normal monkey cells (which is why its presence went unsuspected so long), it transformed cultured mouse, hamster, or human cells. However, these *in vitro* results do not necessarily correlate with *in vivo* situations. There is no evidence at all linking SV40 to cancer in humans. Most of our basic information about the transformed cell came from study of cultured cells transformed with SV40 and mouse polyoma virus. The genome of cells transformed by either of these viruses contains integrated DNA with base sequences homologous to that of the transforming virus. This indicates that the trans-

TABLE 11.7
Some oncogenic DNA viruses

GENUS	VIRUS	CELL CULTURES TRANSFORMED	CAUSES TUMORS IN ANIMALS	IMPLICATED IN HUMAN TUMORS
Papillomavirus	Rabbit papilloma	None	Wild and domestic rabbits	No
	Human papilloma	Human being (low efficiency)	No	Causes benign warts
Polyoma	Mouse polyoma	Hamster Rat	Newborn mice Newborn rats Newborn hamsters	No
	JC virus (human)	Human being Hamster Monkey	Hamsters Owl monkeys	No
	BK virus (human)	Human being Hamster Monkey	Hamsters	Isolated from brain sarcoma
	SV 40	Human being Mouse Hamster	Hamsters	No
Adenovirus	Types 12,18,31 human	Hamster, rabbit, rat, human	Newborn hamsters Newborn mice Newborn rats	No
Herpesvirus	Marek's disease virus	None	Chickens	No
	Epstein–Barr virus	Human being	No	African Burkitt's lymphoma? Nasopharyngeal carcinoma? Hodgkin's disease?
	Herpes simplex Type I	None	None	Benign tumor and (rare) malignancy of lip?
	Herpes simplex Type II	Hamster	Mice, hamsters	Cervical carcinoma?
Poxviruses	Yaba poxvirus	None	Monkeys	No
	Rabbit fibroma	Rabbit?	Rabbits	No

forming agent is present as a provirus. Messenger RNA made in the transformed cell has base sequences complementary to viral DNA. These mRNA molecules indicate that viral genes are being expressed in the transformed cell. Polyoma-transformed cells also contain a new nuclear protein, the **T-antigen.** Cell surfaces may acquire new tumor-specific transplantation antigens (TSTA). It is believed that these are products of viral genes.

However, cells containing provirus need not always be transformed. Transformed cells have been observed to revert to the normal phenotype while still retaining the proviral DNA.

Human polyoma JC virus is implicated in a noncancerous, slow-virus disease, **progressive multifocal leukencephalopathy** (PML). JC virus is extremely widespread. Sixty-five percent of the children develop antibodies against it by their early teens. The disease, however, is extremely rare and seems to occur only in persons who have undergone immunosuppressive therapy.

Viruses of the BK group are also extremely common and usually cause no disease. However, they have been isolated from human brain tumor tissue. DNA hybridization experiments also reveal base sequences homologous to BK virus in cellular DNA samples from other human tumors. The significance of these observations is unclear at present. The simple **presence** of a virus in tumor tissue cannot be interpreted to mean that the virus caused the tumor.

There are literally hundreds of research reports documenting viruses, or viral products, in tumor cells of one sort and another. There are equal numbers of reports documenting the same factors in normal tissues. No one is sure how to interpret much of this data.

Adenoviruses   These viruses are primarily respiratory pathogens. There are 31 serological subgroups of human adenoviruses. Subgroups 12, 18, and 31 are highly oncogenic. They cause tumors in newborn hamsters and mice and transform hamster, rat, rabbit, and human cells in culture. Covalently integrated adenoviral DNA is found in transformed cells in which viral gene transcripts and virus-coded T-antigens are present. Adenoviruses probably do not cause neoplasms in human beings.

Herpesviruses   These are the only DNA viruses that cause naturally occurring malignancy. Marek's disease of chickens is caused by a herpesvirus. Here the role of the virus is not in doubt. It is highly infectious and can cause neoplasms that can kill up to 70 percent of a flock within a few weeks. Vaccination now effectively controls Marek's disease.

Proof that the herpesvirus is the **etiological** or **causative** agent was completed by classic methods. The virus was isolated and introduced into healthy chickens in which it produced the disease. The virus was then reisolated. Note that this method of proof is ethically impossible for human cancer viruses until we have a sure-fire cure for cancer.

The Epstein–Barr virus (EB) is ubiquitous; about 90 percent of adults have anti-EB antibodies in their serum which indicates exposure to the virus. A significant number of these adults actively shed virus in saliva, and is it suspected that most, if not all, antibody-producing or **seropositive** people contain occasional, circulating lymphocytes that contain the viral genome. All strains of the virus so far isolated appear to be basically the same.

This is the "normal" background—a single virus is found in the integrated form in many normal individuals. However, EB virus has also been implicated in a variety of clinically pathological states.

The EB virus is frequently isolated from malignant tissues in African Burkitt's lymphoma (ABL) and nasopharyngeal carcinoma (NPC) among the Southern Chinese. It has been implicated in Hodgkin's Disease. EB also causes infectious mononucleosis (IM), a nonmalignant hyperplasia of lymph nodes, in Caucasians. One way of viewing IM is as a type of leukemia in which the immune system effectively controls the malignancy and eradicates it. IM is fatal only in immunodeficient individuals. The evidence that EB virus causes Burkitt's lymphoma is circumstantial, but very strong. In each of these diseases, the picture is clouded by the fact that EB virus is not always found, and other viruses are occasionally found. The conditions may all stem from differing relationships of this virus to the lymphoid cells of genetically distinct individuals, affected also by environmental variables.

If an infective agent causes a disease, then it must be possible to show how it is transmitted. Person-to-person transmission is called **horizontal** transmission. Viruses that can be transmitted horizontally are **exogenous** viruses because they are at some point exterior to the body. It has not been established how ABL and NPC are transmitted. Folklore has long labeled infectious mononucleosis the "kissing disease" and it may well be true that it is spread by kissing but this has not been proven. An EB virus vaccine is under development. The reduction of incidence of these diseases by the vaccine will be a strong indication that EB virus is the prime cause of the diseases.

Means of disease transmission; Chapter 15, pp. 440–450

Herpes simplex Type II, genital herpes, is transmitted sexually. Increasing incidence of genital herpes lesions parallels increasing incidence of cervical cancers in the United States, but not in Japan. This constitutes grounds for suspicion, but not proof that herpes causes cervical cancer. Intensive research is in progress.

Genital herpes; Chapter 21, p. 625, Table 21.6

The poxviruses   All poxviruses cause cellular proliferation in the skin lesions. Poxviruses of monkeys and rabbits cause localized benign tumors. This group is not considered oncogenic for human beings.

RNA TUMOR VIRUSES

The retroviruses are subdivided into four groups, two of which are oncogenic (Table 11.8). Unlike the DNA viruses, oncogenic RNA viruses cause malignancy in natural circumstances. Although retroviruses are strongly suspected of being implicated in human cancer it has not been proven that they are.

Leukemia, leukosis, and sarcoma viruses   These are a large, serologically related group of retroviruses. The electron microscope shows that, as they bud from the cell, a **C-type particle** is formed and extruded (Fig. 11.13). This group is often classed

TABLE 11.8
Some oncogenic RNA viruses

SUBGENUS	VIRUS	CELL CULTURES TRANSFORMED	CAUSES TUMORS IN	IMPLICATED IN IN HUMAN TUMORS
A (C-type particles)	Avian leukosis group: Rous sarcoma virus	Chicken Rat Mouse Hamster Human being	Chickens, other birds, hamsters, all ages Newborn mice Newborn guinea pigs Newborn monkeys	No
	Murine leukoviruses: Moloney leukemia virus Gross leukemia virus Friend leukemia virus Rauscher leukemia virus	Mouse, when coinfected with other virus only	Mice (certain strains of mice only)	No
	Feline leukemia virus	None	Cats	No. Extensive studies show no connection between feline and human disease C-type particles under study in human leukemia
B (B-type particles)	Mouse mammary tumor virus: Bittner milk factor	None	Female mice (certain strains of of female mice?)	Mammary cancer?; perhaps related virus C-type particles also implicated

as Type C viruses, and unidentified viruses seen in tissue are also so classified if they exhibit the characteristic particle morphology.

The Rous sarcoma virus (RSV), which causes malignant disease in chickens, is the most intensely studied. Like all retroviruses, its core contains the enzyme reverse transcriptase. Most strains of RSV are defective, incapable of productive infection. On infection, the viral genome is integrated in the form of a DNA copy with a very high efficiency of transformation. Transformed cells are always malignant when injected into susceptible animals. Infective virus is produced rarely, only when "complete" virus infects or the defective virus coinfects with helper strains. The virus is transferred from the hen to her egg, or from generation to generation. This is called **vertical** transmission, and viruses transmitted this way are called **endogenous** viruses. They are apparently never external to the body. These viruses have a long latent period, thus relatively few birds develop sarcomas until old age. Active early tumor development occurs only when virus is introduced into young uninfected chicks or newborn mammals; in both kinds of animals an active immune defense is absent.

The murine leukemia viruses (MLV) can be studied in intact mice because the viruses are readily transplantable among members of the inbred strains of mice. There are major differences in susceptibility among strains. Mice of the AKR strain develop disease many more times frequently than mice of the C57Bl strain. In this case, as with the EB virus in humans, it is clear that the oncogenic potential of a virus depends heavily on the genetics and physiology of the host.

Among the viruses found in human cells, not one has yet been implicated as a cause of leukemia. However, continuing investigation may reveal that one or more viruses are involved in the disease.

Mammary tumor viruses   Mammary tumor viruses (MTV) compose the second retrovirus subgroup. As they bud from the host cell, they form a so-called B-particle, which differs morphologically from the C-particle in that the core is eccentrically placed in the envelope. MTV is transmitted vertically via egg and/or sperm in mice of certain strains. Large numbers of infectious virions are shed in the milk, and these particles can infect mice of other strains (Fig. 11.14).

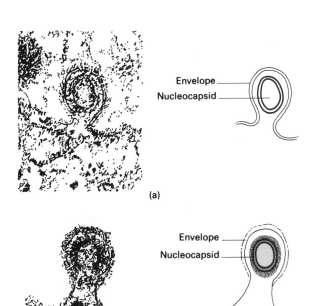

(a)

(b)

Envelope

Nucleocapsid

Envelope

Nucleocapsid

11.13
Retroviruses   (a) A B-type particle budding from a cell. Both membrane layers have an equal electron density. (b) A C-type particle budding from a cell. C-particles have an electron-dense inner layer.

Box 11.2

Nucleic Acid Hybridization

The feature that renders a given sequence of DNA or RNA unique is its base sequence. A single strand of nucleic acid will be able to hydrogen-bond to any other strand that has a complementary base sequence by forming A–T or A–U and G–C bonds. Either DNA–DNA hybrids or DNA–RNA hybrids can be made. This ability of a nucleic acid to seek out and "recognize" strands with matching information is the basis for some ingenious and very powerful research approaches.

In order to use DNA in a hybridization experiment, it must first be changed from the normal double-stranded duplex form to single strands. This can be done by heating the solution, then rapidly cooling it. Heat separates the strands by breaking hydrogen bonds. Rapid cooling prevents their coming back together.

The DNA to be used may be radioactively labeled by providing the cell that synthesized it with thymine containing tritium (H^3). RNA can be labeled by use of H^3 uracil.

Single-stranded unlabeled DNA is immobilized on an inert membrane filter. Single-stranded labeled DNA or RNA from a source to be examined for complementary sequences is placed on the filter. After a time, unbonded nucleic acid is washed away. The radioactivity retained on the filter is quantitatively counted. Since residual radioactivity results from nucleic acid sequences that have effectively base-paired with the membrane-bound DNA, the amount of radioactivity indicates the amount of genetic relatedness between the two nucleic acid samples.

APPLICATION 1—DETECTION OF PROVIRUS DNA IN TRANSFORMED CELLS.

DNA extracted from transformed cells is immobilized on a membrane and exposed to highly radioactive viral DNA. Annealing of labeled DNA to cellular DNA occurs if the cellular DNA contains integrated provirus DNA. DNA from uninfected cells does not bind viral DNA.

APPLICATION 2—DETECTION OF VIRUS-CODED RNA IN CELLS FROM MALIGNANT HUMAN BREAST TUMORS.

The RNA genome is extracted from mouse mammary tumor virus particles. In the test tube, it is used as a template to make a DNA copy, using the reverse transcriptase enzyme. Then, mRNA is extracted from cells of malignant tumors. Control mRNA extracts of normal breast tissue are also prepared.

The viral DNA is allowed to hybridize with both types of mRNA. Detectable specific hybrids are formed between viral DNA and tumor cell mRNA, but not with mRNA from normal cells. This result shows that the human malignant tissue contains nucleic acid with sequences that produce messages very similar to MTV and provides some of the circumstantial evidence for the role of B-type particles in human breast cancer.

Investigation of human breast cancer   Malignancies of the breast occur in about one out of 25 American women. Statistically, there is a strong familial correlation among affected females. In view of the clear evidence of viral etiology in mouse mammary cancer, several investigators have been comparing the biopsies of and the milk from normal and malignant human breasts. The evidence is exciting but inconclusive. The milk from cancerous mammary glands may contain viral particles morphologically similar to MTV; these particles contain reverse transcriptase. Tissue from malignant, but not from normal, glands contains RNA sequences that hybridize with MTV DNA. However, there is no evidence that milk is infective.

Endogenous C-type particles   The major complicating factor in the investigation of oncogenesis by the RNA viruses is the discovery that uninfected "normal" tissues in culture often produce C-type particles. Accidental viral contamination has been ruled out. Over the last few years, these observations have been repeated with various types of tissue from a number of species. Thus it is highly likely that much normal tissue contains integrated viral DNA derived from the reverse transcription of viral RNA. These proviruses can be **induced** to undergo productive infection. The induction rate is greatly increased by treating cells with known carcinogenic chemicals or radiation. Yet, these are not oncogenic viruses because they are not associated with any detectable level of transformation of cells in culture or tumor development in intact animals. In reality, researchers have no clear indication as to what these viruses do to or for the cells that harbor them. Endogenous retroviruses have been apparently maintained in vertebrate cells through millions of years of evolution. This may indicate some positive adaptive value of the viral presence. The viral genome may contain some useful, if not essential, information.

Compare induction of prophage; Chapter 10, pp. 296–297

One possible explanation for viral oncogenesis is that the endogenous C-type particles occasionally may give rise to oncogenic C-type tumor viruses by a process similar to transduction in bacteria. The induction process results in the excised provirus carrying host genes. These excess host genes, on introduction into the next host cell, cause regulatory imbalances sufficient to transform the host cell.

The ubiquitous distribution of endogenous C-particles is another reason why the mere discovery of virus in malignant tissue is insufficient to claim viral etiology for the cancer. We need evidence that is more convincing. Probably the question we should be asking is not the simple one: "Do viruses cause cancer?" but the more involved one: "Are viruses essential for the origin and development of human cancer when other causative factors are also present?"

The Oncogene Hypothesis

At this time there are a number of strands of evidence that very much need to be tied together into some sort of coherent theory. We know that naturally occurring cancers are caused by environmental hazards, such as irradiation and chemical carcinogens. We know that integration of viral genetic material is the essential first step in cell transformation, but that the integration of viral nucleic acid does not inevitably lead to transformation. We know that retrovirus genes appear to be ubiquitous in mammalian cells. We know that administration of radiation, chemicals, or physiological changes enhance the rate of viral expression and/or development of transformation (in cells with provirus). The **oncogene hypothesis**, first formulated in 1969 and still considered valid, unifies the accumulated knowledge as it relates to RNA oncoviruses.

It is proposed that all cells contain in their chromosomes informational DNA sequences that correspond to the complete genome of a retrovirus. This information is transmitted vertically via meiosis to the next generation and via mitosis to all cells of the individual. Some **virogenes** may play an essential role in normal development. Other **oncogenes,** normally repressed, can cause the cell to acquire the transformed phenotype if derepressed. Environmental carcinogens may destroy the essential cellular control over the oncogene and cause induction and transformation.

Gene repression; Chapter 9, pp. 263–264

If this hypothesis is sound, then preventing cancer due to RNA viruses would hinge on preventing oncogene expression. Apart from minimizing environmental hazards, it is unclear at present how this could be done. Since transmission is vertical, vaccination would not prevent acquisition of the virus, although it might amplify the body's ability to obliterate any neoplasms that did develop.

With DNA viruses such as the herpesviruses, the causative agent appears to be horizontally transmitted by contact among individuals. The role of environmental factors in activating a provirus state seem to be relatively less important than are social factors that permit transmission. Vaccination against transmission is apparently feasible and might be truly preventive.

The conclusion that emerges is that rational clinical approaches will have to be tailored to each different type of cancer. An estimated one out of every four people in the United States will contract cancer during his or her lifespan. The battle against cancer is a battle well worth fighting.

LABORATORY IDENTIFICATION AND DIAGNOSIS

Highly specialized techniques are needed for the identification of viruses and the diagnosis of the causative virus from a clinical sample. These tests are not routine in many clinical laboratories. Viral samples are usually sent to specially equipped reference laboratories.

Viral Identification

Unknown viruses may be sought in samples from normal and ill individuals, food and drink, and environmental materials. The sample is first treated to remove extraneous material and to reduce or eliminate microbial contamination.

Identification by cultural characteristics   The sample may be inoculated into a vigorous cell culture. If the cells are of a type that support the replication of the virus, they may, over the course of hours or days, show visible changes as a result of viral parasitism. These **cytopathic effects** (CPE) are sometimes highly specific and provide a strong presumptive identification of the virus present. However, some viruses do not cause unambiguous or discernible CPE; although they are replicating, the host cells are not morphologically changed, and other techniques will be needed to identify the viruses.

Immunological pinpointing   Viruses within cells may not be visible, yet at certain points in the replication cycle large amounts of virus-specific protein will be present, usually in aggregates. Viral proteins are strongly antigenic and combine avidly with antibody. Fluorescent antibody techniques can be used to confirm the presence of specific viral protein in cells.

The complete isolation and characterization of a virus from a sample is a lengthy process, often requiring two weeks. Most viral diseases have run their course, and the patient has recovered before the identification is completed. Thus these techniques are most useful in studies of the spread of a particular virus in populations and to confirm a diagnosis for the physician's information.

Serological Studies of Individuals

Almost all virus infections of an immunologically normal person will stimulate production of virus-specific antibodies. An increase in serum concentration of such antibodies is evidence of recent virus infection.

Paired samples   The antibody level will be low or nonexistent prior to infection, but it will start to rise during infection and continue to rise for several weeks. If possible, two samples two weeks apart should be taken and tested. A **rising titer,** or concentration of antibodies, is present if the second sample is more strongly positive than the first. This is evidence of concurrent or very recent infection. A shift from low titer to high titer is called **seroconversion.**

Types of Serological Tests Used

Virology labs use several testing procedures. These include complement fixation (CF), hemagglutination inhibition (HI), radioimmunoassay (RIA) and immunoelectrophoresis. The theoretical basis of these tests is presented in Chapter 13.

Serological tests; Chapter 13, pp. 385–391

Diagnosis via serology is comparatively rapid. If suitable serum samples are available, the results may be reported after a few hours. However, because antibody appears many days after the onset of infection, such results are also primarily confirmatory.

CONTROL OF VIRAL INFECTION

Although viruses differ greatly from the cellular pathogens, the methods used for their control are in many ways similar.

Prevention

Of all the pathogenic organisms, the viral group has been most successfully controlled by vaccination. Many of the serious acute viral diseases—smallpox, measles, yellow fever, dengue, and rabies, to name a few—can be effectively prevented by immunization. Postexposure protection of nonimmunized individuals by administration of preformed antibodies or drugs is usually much less successful.

Therapy

Many virologists have investigated interferon as a therapeutic agent. It has not been used widely in the United States. Only when large amounts of interferon in an active, purified form become available for clinical trial will it become clear whether or not interferon is generally useful.

Several chemotherapeutic drugs (Chapter 26) are in occasional use. All are competitive inhibitors of nucleic acid synthesis and have significant toxicity for patients. So far, they are licensed only for restricted applications or life-threatening conditions such as herpes encephalitis in infants. There is no "miracle drug" in sight for the common cold.

*Antiviral drugs;
Chapter 26, pp.
781–783*

SUMMARY

1. The animal viruses have an acellular structure and an obligate intracellular parasitic mode of replication. They range from the very tiny, simple polyoma viruses to the large, complex poxviruses. Many animal viruses are enclosed in a characteristic envelope derived from host-cell membranes.

2. Viruses are classified by several features. There are seven groups of DNA viruses and 14 groups of RNA viruses; most contain serious human pathogens. Each virus has a characteristic species and tissue specificity. These can be modified by artificially manipulating the conditions of viral growth.

3. Viruses are most commonly grown and maintained on cell cultures. Serial cell cultures have been derived. These are stable cell lines in which the replication of a given virus can be studied in a controlled fashion. The fertilized egg is also widely used for cultivation of viruses and vaccine production. Live animals may be inoculated with virus if no *in vitro* system is available or if the pathogenesis of the viral disease is to be studied.

4. Animal viruses have several patterns of replication. They attach to the host cell by very specific weak bond interactions. Subsequently, the host cell ingests the virus and enzymatically uncoats it. Cytolytic replication, integration as a provirus, or steady-state replication may ensue, depending on the genetic and physiological regulation systems of virus and host.

5. If new viruses are produced, they are released either by lysis of host cell or by budding through the membrane. In the integrated state, both DNA and RNA viruses can *transform* cells to an uncontrolled growth pattern.

6. The replication of RNA viruses requires unconventional information transfer, either from RNA to RNA or from RNA to DNA. Unique enzymes are coded for by the viruses to execute these syntheses.

7. The more familiar viruses, such as influenza, chickenpox, and polio, usually cause acute disease. Virus multiplies to large numbers in certain types of tissue, and tissue destruction causes symptoms. After a predictable length of time, body defense systems put a stop to viral increase, damaged tissue may or may not be replaced, and the victim recovers. Interferon, a cellular protein produced in response to viral infection, appears to play a major role in recovery.

8. Persistent infections are those in which the virus is present in the body for months to years. The normal defense mechanisms fail to eliminate it. Virus may be shed asymptomatically, or cause sporadic episodes of cell damage, or multiply very slowly with progressive and cumulative degeneration.

9. Virus-caused cancers in animals are a special case of persistent infection. Integration of an oncogenic virus as a provirus can cause the cell to become transformed, to lose its normal growth limitation, and to express abnormal characteristics. Transformed cells, or transforming viruses, injected into susceptible animals lead to tumor growth. The integrated viral genomes that cause malignant growth, called oncogenes, may be ubiquitous in animal cells. Circumstantial evidence points to viruses as contributing to several types of human cancer.

10. Viruses are identified in the laboratory by observation of characteristic patterns of cell damage in culture, and by their interaction with specific antibodies. Clinical applications involve isolation and identification of virus from patient samples, or characterization of antiviral antibodies produced by the patient.

11. Many viral infections may be prevented by the standard methods of sanitation and vaccination. In certain diseases, such as influenza, the administration of drugs during the brief incubation period may have a preventive or modifying effect. Therapy of viral infection is in its infancy. A few drugs are available, but their high toxicity has limited their useful application.

Study Topics

1. What are the practical implications of host modification of a viral strain?

2. Evaluate the significance of growth limitation as a cellular characteristic both in cell culture (*in vitro*) and in the animal (*in vivo*).

3. What are interferons? How do they differ from antibodies? What is their role in defense against viral infection?

4. Why is the enzyme RNA-dependent DNA polymerase called the reverse transcriptase?

5. Compare the means by which a DNA virus and an RNA virus are converted to the provirus state.

6. Identify some disease states in which the virus is not eliminated from the body. What are the consequences?

7. Compare the acute and the persistent disease caused by measles virus. (See also Chapter 17.)

8. Vaccination may be a practical preventive approach for tumors caused by "contagious" cancer viruses like Marek's disease, but not for those caused by viral oncogenes. Why?

Bibliography

Books

Fenner, Frank; B. R. McCauslan; C. A. Mims; J. Sambrook; and David O. White, 1974, *The Biology of Animal Viruses* (2nd ed.), New York: Academic Press.

Reviews and articles

Baron, S., and F. Dianzani (eds.), 1977, The interferon system: a current review to 1978. *Tex. Rep. Biol. Med.* **35:** 1–573.

Epstein, M. A., and B. G. Achong, 1977, Recent progress in Epstein–Barr virus research, *Ann. Rev. Microbiol.* **31:** 421–446.

Pimentel, Enrique, 1979, Human oncovirology, *Biochem. Biophys. Acta* **560:** 169–216.

Schlessinger, David (ed.), 1978, Human papovaviruses. A series of original papers in *Microbiology—1978*, Washington, D.C.: American Society for Microbiology.

II

HOST–PARASITE INTERACTIONS

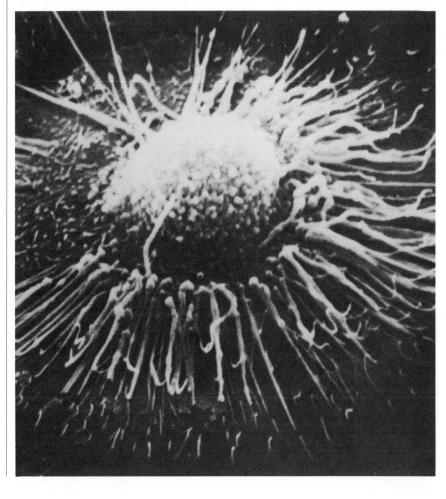

A Scanning Electron Micrograph of a Macrophage

12

The Host–Parasite Relationship

Infectious disease arises when the delicate natural balance between the microbial world and the human being is disrupted. Consider for a moment how you are overwhelmingly outnumbered by the microorganisms in your environment. Your skin carries thousands of organisms per square centimeter. Each time you inhale, you may draw a million organisms into your upper respiratory tract. Saliva often contains tens of thousands of bacteria per milliliter. In your intestinal tract the microbial population may actually outweigh the food being digested. Some fraction of the organisms we contact have the genetic potential to cause disease. That we are not overwhelmed by infection shortly after birth is because of a very complex, interacting, system of defenses. These defenses have arisen as the vertebrate animals evolved, and have reached their highest development in the mammals (Fig. 12.1).

THE STABLE STATE: A HEALTHY BALANCE

Population Dynamics

In Chapter 6, we systematically evaluated the features regulating population size of each species within a given environment. Among similar species, such as heterotro-

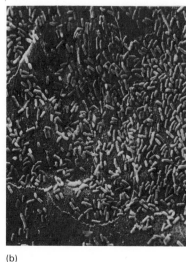

12.1
Body Surface and Flora These scanning electron micrographs show the wrinkled epithelial surface, strands of mucus, and adherent organisms, predominantly yeasts (a) and bacteria (b). Every surface of the human mucous membranes is subject to microbial colonization.

(a)   (b)

phic microorganisms, competition for nutrients is the dominant regulatory factor. Between unlike organisms, such as microorganisms and animals, the parasitism of the small on the large and the predation of the large on the small are most apparent. A critical aspect of the mammal's defense against the microbe is phagocytosis, which may be viewed as a specialized form of predation.

The Normal Flora

Population factors;
Chapter 6, pp. 158–159

An animal's normal flora contains certain species maintaining themselves within a range of population density acceptable for the host's continued health. There is no evidence that possession of a normal flora carries any biological penalty as long as the population is controlled.

At the same time, the microbial partners must maintain a sufficient density to reliably perpetuate themselves. The survival of both partners (host and flora) is crucial; if either partner succeeds in eliminating the other, it will also die. If the host eliminates the normal flora, it will be overwhelmed by opportunistic species. If the flora kills the host, its source of nutrients is gone. For the flora, a dead host is not as adequate an environment as a live one. A mutual restraint on aggressiveness benefits both.

The Evolution of Host–Parasite Interactions

Reaching a mutually satisfactory condition appears to require long periods of microbe–host interaction, while both simultaneously undergo gradual genetic change. Medical history provides examples, such as the mutual adaptation of the syphilis organism and the European population over the last 500 years. Syphilis, when it first came to Europe, was so virulent that it weakened and killed its victims within a very short period. Since the disease is spread by sexual contact, the most aggressive strains incapacitated their hosts before the host could transmit them many times and thus perished with the host. Less virulent strains allowed their host a longer period of grace in which to disseminate them, and so were rapidly selected. At the same time, to a less significant extent, the selective pressure of the newly introduced disease resulted in a more resistant human population. In our time, untreated syphilis (very rare) will take many years to kill its host.

Syphilis; Chapter 21, pp. 630–632

It is possible that all highly lethal pathogens may gradually evolve to perfectly domesticated symbionts. The disease agents we will study are those such as syphilis that have progressed only part way along that path.

MICROBIAL PATHOGENICITY

A microbial pathogen has unique features that permit it to exist in a parasitic relationship. These may be destructive enzymes, toxins, surfaces resistant to host defenses, or the ability to multiply in a location in which host's defenses do not operate. When such virulence factors are present, some percentage of encounters between the organism and the host will result in disease.

Infection

Infection occurs when a microorganism gains access to an area of the body, and multiplies. If no clinically significant damage results, it is an **inapparent, subclinical,** or **asymptomatic** infection. But if detectable alteration in normal tissue function results, we call it a **disease** (Table 12.1).

Note the three successive events required—gaining access, multiplication, and interference. The host defenses can intervene by preventing any stage.

Pathogenicity   Pathogenicity means simply the ability to cause disease. There is no organism so pathogenic that it causes disease 100 percent of the time. There are, on the other hand, many nonpathogens that do cause disease in very extraordinary circumstances (Table 12.2). The microbiologist regards as a "true" pathogen a microbial species that causes disease in an initially normal, healthy individual. The term **opportunistic pathogen** is used for agents that can initiate disease only in a severely compromised host. This is a person with a wound, underlying disease, or other predisposing factor.

Virulence   Virulence describes the degree of pathogenicity, the efficiency with which the agent attacks. Virulence can be quantitated by determining the **morbidity rate**—the percentage of those exposed who develop disease—or the **mortality rate**—the percentage of those with disease who die.

TABLE 12.1
Frequently used terms applied to microbial diseases

Acute disease	A disease with a short, well-defined course
Chronic disease	A disease of slow progress and long or indeterminate length
Local infection	An infection confined to a limited area, such as the bladder
Focal infection	Infection in a single area of infected tissue, such as an abscess that sheds organisms that infect distant areas
Systemic infection	An infection in which the infectious agent is widely disseminated throughout the body
Primary infection	An initial infection in a healthy individual
Secondary infection	Infection with another agent that follows a primary infection, usually occurring only because the host has been weakened
Mixed infection	An infection, e.g., dental caries, in which the symptoms are produced by the combined efforts of several species of bacteria
Latent infection	An infection in which the agent is present but not multiplying and in which symptoms are absent, e.g., quiescent intermittent periods between herpes cold sores
Inapparent infection	An infection in which the organism does not attain sufficient numbers to cause clinical symptoms
Bacteremia	A condition in which bacteria are being transported by the blood but are not multiplying in blood
Viremia	A condition in which viruses are transported by blood
Septicemia	A condition in which microorganisms are multiplying in blood
Communicable disease	A disease that can be transmitted from one individual to another
Contagious disease	An older term for communicable disease, implies ready transmission
Exogenous disease	A disease in which the infective agent comes from outside the body
Endogenous disease	A disease in which the agents are those usually present in the body; occurs in compromised hosts

In the laboratory, virulence may be measured by determining critical inoculum size. If the organism is highly virulent, a very small number of cells or virus particles suffices to cause disease. A single particle of certain *Rickettsia* can initiate fatal disease, whereas a million cells of *Salmonella typhi* may be required to cause a case of typhoid fever. By administration of graduated doses of agent to groups of test ani-

TABLE 12.2
Varying degrees of pathogenicity of bacteria

PATHOGENICITY	EXAMPLES OF SPECIES AND CONDITION
Pathogenic—initiates disease in normal, nonimmune individuals	*Bordetella pertussis*—whooping cough
	Mycobacterium tuberculosis—tuberculosis
	Neisseria gonorrhoeae—gonorrhea
	Streptococcus pyogenes—acute pharyngitis
	Yersinia pestis—bubonic plague
Frequently opportunistic—initiates disease in comprised hosts such as those with wounds or who are alcoholic. Can cause secondary infections	*Bacteroides fragilis* ⎤ anaerobes that *Clostridium tetani* ⎦ infect wounds
	Klebsiella pneumoniae—may cause pneumonia principally in alcoholics
	Pseudomonas aeruginosa—colonizes burns
	Streptococcus pneumoniae—infects aged persons or persons with lung disease
Rarely opportunistic—initiates infection only in severely compromised hosts such as premature babies, or persons with immunodeficiency or unusual structural defects	*Propionibacterium acnes*—causes infections in patients with implanted prosthetic heart valves
	Staphylococcus epidermidis—causes infections arising at site of intravenous catheters
	Streptococcus agalactiae—subacute bacterial endocarditis in patients with abnormal heart valves
	Flavobacterium meningisepticum—meningitis in premature infants

mals the ID_{50} (infectious dose or number of particles required to infect 50 percent of host group) or LD_{50} (lethal dose or number required to kill 50 percent) can be assayed. This information can be of practical importance, because certain pathogens, such as the diphtheria organism, have strains of variable virulence.

Disease Transmission

Each microbial disease is transmitted by a cycle in which the agent leaves one host, survives temporarily in the exterior environment, then enters and establishes itself in the next host. Careful analysis of this circuit is an activity in **epidemiology**—the study of occurrence of disease in populations. The key factors, summarized here, will be presented in depth in Chapter 15.

Epidemiology; Chapter 15, pp. 437–440

    A **communicable** disease, such as measles, is transmitted from one individual to another. Most microbial diseases are more or less infectious. There are a small number of **noncommunicable** microbial diseases, of which tetanus is an example. The

Tetanus; Chapter 23, pp. 677–679

TABLE 12.3
Site of replication of microbial parasites

GROUP	SPECIES OR GROUP	EXTRACELLULAR	FACULTATIVELY INTRACELLULAR	OBLIGATELY INTRACELLULAR
Bacteria	*Bacillus anthracis*	Yes		
	Clostridium, all species	Yes		
	Corynebacterium diphtheriae	Yes		
	Neisseria gonorrhoeae	Yes		
	Pseudomonas aeruginosa	Yes		
	Staphylococcus aureus	Yes		
	Streptococcus, all species	Yes		
	Vibrio cholerae	Yes		
	Yersinia pestis	Yes		
	All spirochetes	Yes		
	Brucella, all species		Yes	
	Francisella tularensis		Yes	
	Mycobacterium, all species except *leprae*		Yes	
	Mycobacterium leprae			Yes
Fungi	Dermatophytes	Yes		
	Candida albicans		Yes	
	Coccidioides immitis		Yes	
	Histoplasma capsulatum		Yes	
	Other systemic fungi		Yes	
Rickettsia	All species			Yes
Chlamydia	All species			Yes
Viruses	All groups			Yes

host contracts tetanus from introduction of bacterial spores from soil or manure; it is not possible, in any ordinary way, to contract tetanus from an ill individual.

Entrance to the body   Microorganisms enter the body by various routes. Food and drink enter the gastrointestinal tract; air containing droplets of respiratory secretions enters via the nose and mouth. The skin, urogenital tract, or wounds provide other routes of entry. For each disease there is one or more likely routes.

Once the organism has entered, it must form an attachment to an epithelial surface in the respiratory tract, intestine, or wherever infection occurs. Pili are known to play a role in attachment of *Neisseria gonorrhoeae* to the genital mucosa. Normal flora may be protective, in part because it covers up attachment sites.

Site of replication In the body, some organisms multiply outside the cells, either in the spaces between cells or in body fluids. Others multiply primarily within cells (facultative intracellular parasites) or only within cells (obligate intracellular parasites) (Table 12.3). Organisms exterior to cells are comparatively readily destroyed by serum factors or phagocytes. Intracellular parasites can usually be eradicated only by destroying the host tissue that is harboring them (Fig. 12.2).

Exit from the body During infection, certain body areas contain living organisms. Body fluids, secretions, or excretions in contact with those tissues will become contaminated. Saliva may be contaminated with hepatitis virus, and urine with the leptospirosis agent.

Survival outside the host Transmission necessarily requires that the agent be outside the host for a period of time. Agents such as the mycobacteria, some viruses, and spore-forming bacteria can survive in an inert state for prolonged periods. Their transmission can be much delayed or very indirect. The syphilis organism, on the other hand, rapidly loses viability when removed from the human mucous membranes. For that reason, it is transmitted only by direct contact between individuals.

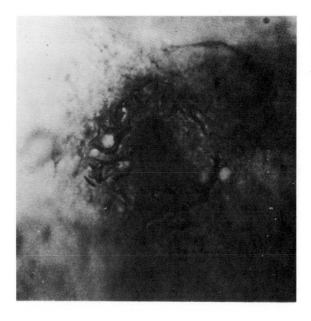

12.2
Intracellular Parasitism The acid-fast *Mycobacterium tuberculosis* is a facultative intracellular parasite. *In vivo* it is characteristically seen within cells. Here huge numbers of the slender bacilli pack peritoneal macrophages.

Microbial Pathogenicity Factors

One microbial disease differs from another in its symptoms and outcome because each pathogenic agent has different physiological weapons. In interacting with our tissues, they carry out characteristic activities (Table 12.4). Observation of the results—the symptoms—contributes to accurate diagnosis that may be confirmed by laboratory procedures.

Invasiveness   It will be remembered that the agent must first gain access without being intercepted and destroyed by host defenses. Invasive pathogens can successfully enter normal tissues or, once introduced via a wound, penetrate underlying tissues. In some cases, the mechanism by which invasion occurs is known, but in others it is not.

The chemistry of the pathogen's exterior layers may be important. Many bacteria produce capsules or slime layers when multiplying in the host. Capsular polysaccharides and proteins, such as the M-protein of *Streptococcus pyogenes*, may actively prevent ingestion by phagocytes.

Cell wall components may also confer resistance. The Gram-positive bacteria are not readily lysed by phagocytes without specific antibodies. The unusual lipids of the mycobacterial cell wall that render it acid-fast also render it resistant to phagocytic action. Bacteria of this type can be seen not only surviving but also multiplying inside phagocytes.

Invasion of the intestinal tract is possible if the agent can pass through the acid environment of the stomach without loss of viability. The protein of enterovirus capsids is not sensitive to acid hydrolysis. Many protozoan parasites enter as cysts—resting forms that resist acid attack in the stomach but are later uncoated by the digestive enzymes in the intestine.

Protective effect of stomach acid; Chapter 20, p. 582

Enzymes produced by the agent are used for tissue penetration. The influenza virus carries **neuraminidase** on its envelope. This enzyme attacks neuraminic acid, the key polymer in mucus, and by hydrolyzing it reduces the viscosity of the mucus coat. This is believed to allow the virus easy access to the surface epithelium of the mucous membrane to which it must adhere before infection can be initiated.

Invasiveness factors are characteristic of bacteria that produce exoenzymes. Many examples have been studied *in vitro*, but it is not always clear what the exoenzymes contribute to *in vivo* invasiveness. All exoenzymes hydrolyze cellular and extracellular macromolecules. Some enzymes lyse specific groups of cells (hemolysins), and others open a path between cells (hyaluronidase) or remove fibrin clots (fibrinolysins). Two of the most invasive bacterial species, *Streptococcus pyogenes* and *Clostridium perfringens*, have multiple enzymatic potentials.

Many invasiveness factors are specifically directed against or opposed by key host defenses. After studying these later in this chapter, you should return to this section and look for connecting links between attack and defense.

Microbial toxins   Some microbial invaders interfere with host function by making a **toxin.** A toxin is a poison—a chemical substance that reversibly or irreversibly

TABLE 12.4
Pathogenic species and their invasiveness factors

TYPE OF PRODUCT	SPECIFIC FACTOR	SPECIES OR GROUP	PATHOGENIC EFFECT
Capsule	Type-specific polysaccharides	*Streptococcus pneumoniae*	Inhibits phagocytic adherence and capture
		Klebsiella pneumoniae	
		Bacillus anthracis	
		Yersinia pestis	
Cell wall component	M-protein	*Streptococcus pyogenes*	Inhibits phagocytic digestion
	Waxes, mycolic acids	*Mycobacterium tuberculosis*	
Extracellular enzyme	Collagenase	*Clostridium perfringens*	Permits invasion of connective tissue
	Coagulase	*Staphylococcus aureus*	*In vivo* effect uncertain; may inhibit phagocytosis
	DNAase; streptodornase	*Staphylococcus aureus*	Reduces viscosity of purulent discharges improving bacterial mobility
		Streptococcus pyogenes	
	Hemolysin	*Staphylococcus aureus*	Bacterial mobility. Lyses red blood cells.
		Streptococcus species	*In vivo* importance uncertain
		Others	
	Hyaluronidase	*Streptococcus pyogenes*	Permits invasion of epithelial tissue
	Kinase; streptokinase	*S. pyogenes*	Activates plasma fibrinolysin and dissolves clots
		Staphylococcus aureus	
	Lecithinase	*Clostridium perfringens*	Destroys cell membranes, especially red blood cells
	Leukocidin	*Staphylococcus aureus*	Lyses leukocytes; *in vivo* importance uncertain
	Lipase	*Staphylococcus* species	Breaks down lipids
	Protease	*Clostridium* species	Breaks down protein; permits attack on muscle tissue

alters a key physiological reaction. Toxin may be taken up and distributed to host tissues by circulation. Colonization of a small area of the body by a completely non-invasive organism may result in a lethal outcome. *Corynebacterium diphtheriae*

Box 12.1

Why Do I Feel So Bad When I Get an Infection?

Here are some capsule answers.

Inflammation
Release of mediators from your damaged tissue have mobilized defenses at the invaded site.

Headache
Probably the inflammatory mediators are dilating blood vessels in your brain.

Aches and pains
The mediators are affecting your joints and muscles.

Fever
Pyrogenic substances released by white blood cells are acting on the thermoregulatory region of your hypothalamus to increase heat production.

Swollen glands
Your lymph nodes are enlarged. Drainage of lymph carrying mediator substances causes enlargement and increased cell proliferation in lymph nodes "downstream" of the infection. Also, some agents multiply in lymphoid tissue causing enlargement, e.g., as in infectious mononucleosis.

Sore throat
You have swollen lymph nodes in the pharyngeal area or, less commonly, microbial denuding of pharyngeal lining.

Rashes
Capillary hemorrhage, attack by immune system on cells harboring an intracellular parasite, or small areas of cytolytic activity (pox) may cause spots or vesicles on your skin.

Vomiting
You have ingested food containing toxins, which activate the brain's vomiting reflex center. Vomiting ejects contaminated food.

Diarrhea
Toxins are causing body fluids to pour into the lumen of your intestine, accelerating peristalsis and yielding watery stools. In viral infections, destruction of the intestinal epithelium has a similar effect.

Coughing
You are trying to rid yourself of excessive mucus, caused by bacterial products altering permeability of respiratory epithelium.

Runny nose
You are experiencing excessive fluid loss by virus-damaged epithelium. This brings protective plasma factors to the surface.

---

grows only on the superficial epithelium of the nasopharynx. Yet the toxin it produces at that site is absorbed and can cause fatal degeneration of heart, liver, kidneys, and nerve tissue.

Exotoxins are proteins; they are synthesized within the cell and released into the environment in which they diffuse freely. They have highly specific modes of action and have pronounced effects in microgram amounts (Table 12.5). Between 200 and 400 molecules of diphtheria toxin will kill a human cell. Most exotoxins are inacti-

TABLE 12.5
Pathogenic bacteria that produce exotoxins

ORGANISM	DISEASE	COMMON NAME OF TOXIN	SPECIFIC ACTIVITY
Bordetella pertussis	Whooping cough	—	Necrotoxin: destroys upper respiratory tract epithelium
Corynebacterium diphtheriae	Diphtheria	Diphtheria toxin	Inhibits protein synthesis; damages heart, liver, nerve tissue
Clostridium botulinum	Botulism	Botulin	Blocks transmission of nerve impulses. Causes flaccid paralysis
Clostridium perfringens	Food poisoning	Enterotoxin	Acts on vomit reflex center of brain
Clostridium tetani	Tetanus	Tetanus toxin	Blocks neuromuscular junction. Causes rigid paralysis
Escherichia coli, some strains	Diarrhea	Enterotoxin	Causes water and electrolyte loss via intestinal tract
Staphylococcus aureus, certain strains	Food poisoning	Enterotoxin	Activates vomit reflex center
Streptococcus pyogenes, certain strains	Scarlet fever	Erythrogenic toxin	Necrosis of capillaries, gives rash
Shigella dysenteriae	Dysentery	—	A neurotoxin; causes water and electrolyte loss across intestinal wall
Vibrio cholerae	Cholera	Choleragin	Alters cAMP activity in intestinal lining causing rapid water and electrolyte loss by diarrhea

vated by heat. As proteins, they are antigenic, and the body reacts to exposure by making protective neutralizing antibodies. The first immunizations to be made safe and effective were those that used active diphtheria and tetanus toxins chemically detoxified and converted to **toxoids.**

Toxoids; Chapter 27, pp. 787, 793–794

Antibodies produced by an experimental host, against a toxin or toxoid, can be collected and used therapeutically. A serum product containing antitoxic antibodies is called an **antitoxin.** If administered to a person poisoned by the toxin, before irreversible effects have occurred, the antitoxin may arrest the disease. To arrest the disease, it is not necessary to get rid of the bacterial producer, just to neutralize the toxin.

Antitoxins; Chapter 27

Endotoxins are a group of lipopolysaccharide substances produced by Gram-negative bacteria. The lipopolysaccharide material is part of the cell wall, liberated after bacterial death and disintegration. Although there are small chemical differences among endotoxins, useful in identification of Gram-negative species, all endo-

TABLE 12.6
Some biological effects of endotoxin

EFFECTS	MECHANISM
Temperature elevation	Causes release of endogenous pyrogens from PMNs
Pain	Inflammatory response at site of infection or injection of vaccine
Increased phagocytosis	Activates alternate pathway of complement
Rash (meningitis)	Causes capillary fragility resulting in tiny hemorrhages
Endotoxin shock	Decreased contractility of arteries produces drop in blood pressure with serious or fatal damage to vital organs

toxins have the same nonspecific mode of action (Table 12.6). Large (milligram) doses are necessary to cause death. Small doses cause rather general symptoms of fever, aches, and malaise. Large doses, usually a result of Gram-negative septicemia, result in septic or **endotoxin shock.**

Endotoxin is weakly antigenic. Apparently exposure to endotoxin does not give much protective immunity. Immunity to endotoxins may give rise to damaging hypersensitivity reactions (Chapter 14).

NONSPECIFIC RESISTANCE

Let us first establish the distinction between **nonspecific resistance** and **immunity.** Defenses that protect the host in general against all microorganisms are **nonspecific.** Effective nonspecific resistance does not depend on prior exposure to the agent in question. It does not require development of lymphocyte populations. The role of the nonspecific defenses is to protect against first attacks by the widest possible variety of agents.

Specific or **immune** defenses, by contrast, are directed against individual bacteria, viruses, fungi, or other agents. They require development of specifically programmed lymphocytes. They arise only after the body's lymphoid tissues have experienced contact with the agent, and then only after a five- to ten-day induction period. Thus immune defenses control infection in the later stages of the first attack, and defend against second or subsequent attacks by the same agent.

The immune system cooperates with and potentiates the nonspecific defenses. The primary role of immunity can be interpreted as "sharpening the weapons" of the nonspecific systems or targeting them against the challenge of the moment. The two systems are, in fact, inseparable.

Immune system;
Chapter 13

Innate or Genetic Resistance

In discussing viruses, we noted the species specificities of certain viral agents. Although the physiology of all mammals is much alike, it is also sufficiently different that human beings cannot be infected by canine or feline distemper, hog cholera, mousepox, or even most of the diseases of nonhuman primates. The exact nature of the barriers to cross-infection is not known.

There are a few diseases humans **can** acquire from animals that are extremely serious exceptions to species resistance. These diseases include rabies, bubonic plague, and viral encephalitis.

General Health Factors

Although most people can readily define "disease," they have a much harder time coming up with an accurate definition of "health." Good health is a protective state. Departure from optimal physical condition clearly predisposes to infection.

Nutrition   Individuals with malnutrition, in particular with inadequate dietary protein and vitamins, are severely compromised and susceptible to infection. Infections, once acquired, are much more likely to be severe. Measles is rarely fatal in individuals living in developed countries. However, in South America, mortality caused by measles in children from one to four years of age—the age group most poorly nourished—is 100 to 400 times higher than it is in the United States. Many of the children who do recover suffer permanent disability. A cyclical pattern of poverty—malnutrition–infectious disease–economic unproductiveness—is characteristic of large areas of today's world.

Measles control; Box 27.2

Stress   Severe physical or emotional stresses, including fatigue, anxiety, or depression, can produce the "stress syndrome." Stress can result in enhanced production of adrenalin and altered hormone levels. The condition is clearly associated with higher attack rates for several infectious diseases. Stress is the major contributing factor in the development of "trench mouth" or acute necrotizing ulcerative gingivitis (ANUG) in college students during exam periods. *In vitro* studies have demonstrated reduced activity of both the nonspecific and immune defenses during periods of stress.

Active necrotizing ulcerative gingivitis (ANUG); Chapter 16, pp. 482–483

Age   The incidence of infectious disease is higher in the very young and the very old. In the young child the immune system is not yet fully potent, and in the older individual its effectiveness wanes along with the effectiveness of all other systems.

Age-related incidence of disease; Figure 15.6

Underlying conditions   A preexisting disease state may create predisposing conditions (Table 12.7). The individual with structural abnormalities of the heart has a vastly increased risk of bacterial endocarditis following dental procedures or minor

TABLE 12.7
Relationship of existing disease state to risk of infection

DISEASE STATE	DEFECT IN DEFENSES	ASSOCIATED INFECTIONS
Rheumatoid arthritis	Reduced chemotaxis of phagocytes	Skin infections, pneumonitis
Myelogenous leukemia	Few functional granulocytes	Infections of pharynx, rectum, urinary tract; septicemia
Lymphosarcoma	Affects antibody synthesis	Pneumonia, skin and urinary tract, septicemia
Hodgkin's disease	Affects cell-mediated immunity (CMI)	Infections by intracellular parasites
Renal failure	Chemotaxis, affects CMI	Many different infections and infective agents
Alcoholism	Swallowing reflex, leukocyte function. If cirrhosis of liver occurs, phagocytic clearance also declines	Pneumonia, usually due to aspiration
Diabetes	Granulocyte function	Urinary, soft tissue, skin and lung infections; septicemia
Staphylococcus aureus is the primary agent
Malnutrition	Phagocytosis, complement (C') activity, CMI	Skin, respiratory and gastro-intestinal infections; septicemia
Burns	Destroy normal skin structure, phagocytic action
Reduced pulmonary function	Skin colonization, pneumonia and septicemia
Cystic fibrosis, chronic lung disease	Poor clearance of mucus, reduced phagocytic function	Pneumonia, septicemia

Source: Compiled from A. von Graevenitz, 1977, The role of opportunistic bacteria in human disease, *Ann. Rev. Microbiol.* **31**: 447–471.

Compromising factors; Chapter 28, pp. 814–816, Box 28.1

traumas that introduce small numbers of bacteria into the blood stream. Neoplasms of blood-forming or lymphoid tissues cause general immunodeficiencies that leave the patient poorly protected. A bedridden patient has a significantly increased risk of pneumonia. Then, too, many therapeutic measures—intravenous catheters, chemotherapy, surgery—taken to help such patients carry risk factors themselves.

Climate   Sometimes the effect of climate is obvious. For example, certain areas of Africa were referred to, in the colonial period, as "white men's graveyards" because of the high yellow fever mortality. These areas have an ideal climate to hatch huge swarms of the mosquito vectors.

Opinions differ about the effect of climate and seasonal changes on the various epidemic respiratory diseases. That there is such an effect is unarguable. Public health data collected over the years show an annual increase in influenza in January through March. However, the increase is noted both in warm and cold parts of the country. Certainly **cold** doesn't cause "colds." But intensive chilling prior to exposure to a "cold" virus may increase the likelihood that the exposure will result in disease.

Influenza; Chapter 15, Figure 15.8

Occupation   Public health surveillance and appropriate medical treatment have eliminated most infectious occupational diseases. However, brucellosis is still frequently diagnosed in persons in meat-packing occupations. The risk of rabies is high for veterinarians, game wardens, and dog officers. Although the fact is little mentioned, persons working in health care, offering personal services to ill individuals, have the highest occupational risks of all. The risk is fortunately offset by the fact that health professionals are taught how to protect themselves, as well as how to prevent spread of disease to others.

Urbanization   The spread of epidemic diseases requires a high-density, susceptible population. Most epidemics are spread by direct, interpersonal contact. Thus urbanization per se increases risk. In addition, life in crowded or decayed areas adds dirt, pest infestation, and stress to directly challenge the dweller. Diseases such as diphtheria, almost totally eradicated elsewhere, have their last stronghold in such environments.

An adequately fed, stressfree, uncrowded life that carefully avoids risky occupations, bad habits, and unfavorable climates will reduce, but not eliminate, infection.

Anatomical and Physiological Barriers

If microorganisms are unable to enter through a wound, they must enter the body via one of its orifices or must penetrate its skin. Each point is guarded by anatomical barriers, the body's first line of defenses. Local secretions create an unfavorable physiological environment for microbial growth or simply wash away invaders. Later in this book, each body system is considered as a unit, starting with its normal anatomy, physiology, and protective mechanisms.

Anatomical protection   The central nervous system has the most developed anatomical protection of any body area (Fig. 22.2). The skeletal features (skull and spinal column) that enclose these vital structures not only protect against mechanical damage but also prevent microbial invasion. Pathogens that penetrate overlying tissues do not often penetrate bone. The three layers of membrane that enclose the nerve tissue present a further barrier.

Figure 22.2

In respiration, inhaled air enters through the nose. The nasal passage (see Fig. 18.1a) is a tortuous pathway among the highly convoluted **nasal turbinates,** bony

Figure 18.1a

protrusions covered with mucous membrane. Entering air must follow a turbulent path that directs it against a sequence of moist, sticky surfaces with the functional characteristics of old-fashioned flypaper. Dust particles and microscopic matter are trapped there. Unless the air is extremely heavily laden with particles, or the humidity is so low as to actually reduce the moistness of the mucosal surfaces, few particles reach the lungs.

Physiological protection　　The eye is exposed to extensive environmental contamination, but few agents actually initiate infection there. The steady flushing action of tears removes invaders every few seconds. Tears, like many other body fluids, contain **lysozyme,** the enzyme that hydrolyzes the peptidoglycan of the bacterial cell wall.

The vaginal secretions have an acid pH of from 4.4 to 4.6 that renders the vaginal area inhospitable to microorganisms that might intrude from adjoining skin and intestinal tract. Although the organisms are not killed, they cannot grow sufficiently well to compete with the area's acid-tolerant, normal flora.

Anatomical and physiological barriers can be adversely affected by poor health. For example, vitamin deficiencies can cause cracking and ulceration of skin and mucous membranes that then present no barrier at all to microorganisms. Some birth-control pills render vaginal secretions less acid (pH 5–6) increasing the likelihood of gonorrhea.

Role of Blood Components in Body Defense

Figure 24.1

An adult human has about five liters of blood circulating through arteries, capillaries, and veins (Fig. 24.1). Simultaneously, a variable amount of tissue fluid and lymph follows a one-way return path via lymphatic ducts and through lymph nodes, to rejoin the blood. Blood and lymph transport oxygen and nutrients, remove wastes, and regulate temperature. In addition, they carry antimicrobial biochemicals and phagocytic cells, the body's internal second line of defense.

Formed elements of the blood　　Blood consists of cellular elements suspended in a fluid matrix, the plasma (Fig. 12.3). Blood cells are of three types. Red blood cells (RBCs or **erythrocytes**) are the most numerous. They have no nuclei. They are formed in the bone marrow, released into the blood, and then circulate for many days without division or change. Their function is oxygen transport; they have no known role in body defense.

White blood cells (WBCs or **leukocytes**) are colorless unless stained. When blood smears are stained with Wright's stain, a mixture of acid and basic dyes, the leukocytes can be differentiated into five types. Three are **granulocytes,** the neutrophils, basophils, and eosinophils. These cells have irregular, often lobed, nuclei and numerous, and clearly visible, cytoplasmic granules. Under the conditions of the Wright stain, the granules in each of the three types are stained a different color.

Neutrophils or **polymorphonuclear** leukocytes (PMNs) are the most numerous of white blood cells. Their granules pick up both acidic and basic dyes and thus are

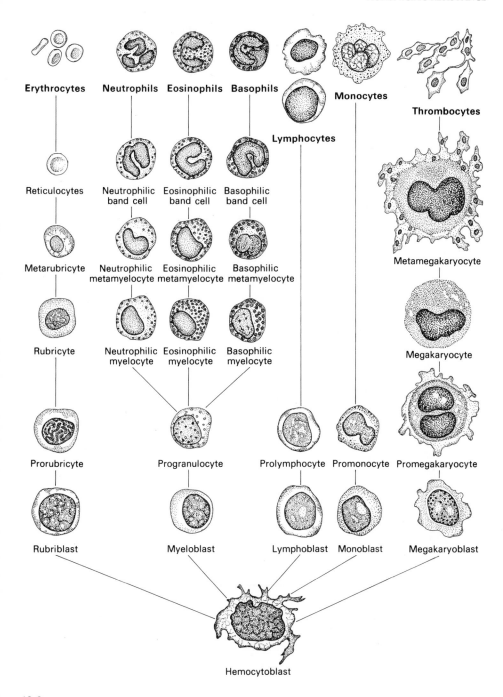

Erythrocytes   **Neutrophils**   **Eosinophils**   **Basophils**   **Monocytes**   **Thrombocytes**

Lymphocytes

Reticulocytes   Neutrophilic   Eosinophilic   Basophilic
              band cell   band cell   band cell

Metarubricyte   Neutrophilic   Eosinophilic   Basophilic   Metamegakaryocyte
        metamyelocyte metamyelocyte metamyelocyte

Rubricyte   Neutrophilic   Eosinophilic   Basophilic   Megakaryocyte
          myelocyte   myelocyte   myelocyte

Prorubricyte   Progranulocyte   Prolymphocyte Promonocyte Promegakaryocyte

Rubriblast   Myeloblast   Lymphoblast   Monoblast   Megakaryoblast

Hemocytoblast

12.3
Development of the Blood Cells in Human Beings   A population of stem cells, located in the bone marrow, gives rise to several differentiated types of mature blood cells.

stained a neutral grayish-pink color. Their nuclei frequently have three or four lobes separated by constrictions. About 100 billion per day are produced in bone marrow. Thirty times that number is reserved in the marrow to be rapidly released in the case of infection. This release accounts for the great increase in circulating white blood cells (**leukocytosis**) seen in certain infections such as acute appendicitis. In the event of a severe infection, these reserves may become exhausted and **leukopenia,** low WBC count, will result. Leukopenia is a diagnostic sign in typhoid fever and influenza. PMNs are mobile phagocytic cells. They respond chemotactically to substances released by damaged cells and migrate to the scene of injury or infection at which they ingest foreign particles.

Allergy: Chapter 14, pp. 411–416

Basophils contain granules that are stained deep blue-purple by the basic dyes. Although basophils are not numerous, they and the **mast cells** in connective tissue are responsible for the unpleasant or dangerous effects of the Type I hypersensitivities (anaphylaxis and allergy).

Eosinophils contain granules that collect the acid stain eosin, and become a vivid orange-red. Eosinophils are few in number, except during allergic flare-ups and certain parasitic infestations. It is believed that the eosinophil's role is to counteract the physiological effects of the basophils and control the inflammatory response.

There are two groups of **agranulocytes,** the lymphocytes and the monocytes. Granules, although present, are not deeply stained, and the nucleus is a single body.

Lymphocytes originate from lymphoid tissues. The characteristic form in the blood is the **small lymphocyte.** This cell has a large, darkly staining nucleus and a small halo of light cytoplasm. All lymphocytes participate in specific immunity. Although all look much alike, they actually belong to many distinct functional classes, each with a special role in the immune response.

Monocytes are quite large, and their nuclei are oval or horseshoe-shaped. They are actively phagocytic, like PMNs. Circulating monocytes later differentiate into the **macrophages** in the tissue fluid, body spaces, and aggregates along the blood and lymphatic circulations.

Platelets or **thrombocytes** are cell fragments that arise from **megakaryocytes** produced in bone marrow. They contain a key triggering substance, **thromboplastin,** that initiates blood clotting. Because platelets are extremely fragile, they readily break down on contact with damaged tissue. Clotting occurs when the platelet releases factors that react with plasma-clotting factors.

The clotting mechanism is an important nonspecific defense. Within tissue, clotting walls off a traumatized area and restricts the spread of all but the most invasive infectious agents. Surface clots form the structural framework for scab formation, sealing off raw areas while tissue replacement is carried out. Platelets also release inflammatory substances.

Blood plasma and serum   **Plasma** is the whole fluid matrix of blood with cells removed. **Serum** is the fluid left after whole blood has been allowed to clot and the clot has been removed. Therefore serum is plasma with the clotting proteins removed. Both fluids contain many proteins that have protective roles. The most important group of serum proteins is the immunoglobulins (Igs), effectors of specific immunity.

Plasma contains two proteins, **prothrombin** and **fibrinogen,** that participate in the stepwise formation of clots, in cooperation with the platelet factors. **Lysozyme,** mentioned as a component of tears, is also found in other body fluids, including plasma. Its ability to significantly damage cell walls of bacteria in the blood is not established.

Interferon produced by virus-infected cells finds its way not only into adjacent tissue but also into the blood stream. It may serve to restrict viral multiplication in distant tissues; it also is a regulator of immune function.

Interferon; Chapter 11, pp. 324–327

Opsonins are a loosely defined group of plasma proteins that increase the effectiveness with which the phagocytic cells adhere to particles to be ingested. The adherence may be nonspecific (not dependant on previous immunization) or specific. Specific opsonins are actually antibody molecules.

Complement (C′) is a group of 11 proteins. On signal, these accumulate sequentially to form an aggregate. Certain of the proteins act enzymatically on the others, releasing biologically active fragments. The complement network kills invaders by assisting phagocytosis and catalyzing cell lysis (Fig. 12.4).

12.4
When the Gram-negative bacterium *Escherichia coli* is exposed to specific antibody and complement, lesions are formed in the cell wall. Each hole is caused by a single complement complex. There are many thousands of lesions on this cell, yet probably only one would be sufficient to inactivate and kill the bacterium.

Box 12.2

The Complement Pathways

The complement system is composed of at least 11 serum proteins. Once triggered, these react sequentially. Complement complexes function enzymatically; each complement factor is the activator for the next one in the sequence.

The key step in the complement network is the cleavage of C_3. C_{3a} promotes inflammatory changes useful in fighting infection, and C_{3b} promotes adherence of phagocytes and continuation of the complement pathway. C_3 may be activated by the **alternate pathway** (shown in color) which does not require antibody. Nonspecific microbial substances such as polysaccharides or endotoxin can trigger the production of a C_3 activator, $C_{3b}B$. The **classic pathway** (shown in color) requires specific antigen–antibody combination leading to the formation of C_{142}, the classic C_3 activator.

Once C_3 is activated, then other complement factors can be added. C_5 is also cleaved, and the C_{5a} fragment functions with C_{3a} while the C_{5b} fragment, plus $C_{6,7,8,\text{ and }9}$ combine to form a complex that can lyse cell membranes and some bacterial cell walls.

At each step, the activated complex can activate more than one molecule of the next component. A chain reaction is set in motion. Thus a single molecule of antibody bound to the surface of a cellular antigen may lead to that cell's eventual death.

The complement system functions both in nonspecific and immune defense. It contributes essential parts to two of the main defensive actions of the body—phagocytosis and cytolysis of invading microorganisms.

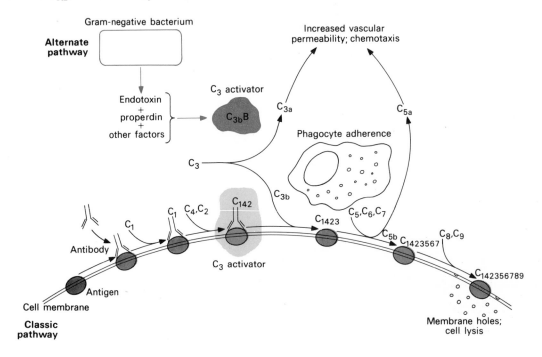

1. *Trypanosoma rhodensiense.* (Carolina Biological Supply Company.)

2. *Penicillium,* hyphae in orange peel. (Carolina Biological Supply Company.)

3. *Aspergillus fumigatus.* (Carolina Biological Supply Company.)

4. *Saccharomyces,* budding. (American Society for Microbiology Slide Collection.)

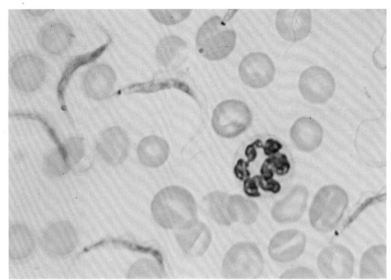

1.

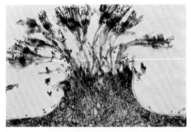

2.

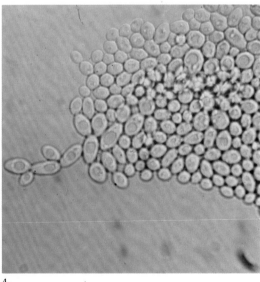

3.

4.

5. Nitrogen-fixing bacteria.
   (Carolina Biological Supply
   Company.)

6. *Penicillium* rotting a lemon.

7. A common microbial disease,
   black knot of plum.
   (Carolina Biological Supply
   Company.)

8. Fungal decomposition of
   compost; note cloud of
   Aspergillus spores, (American
   Society for Microbiology
   Slide Collection.)

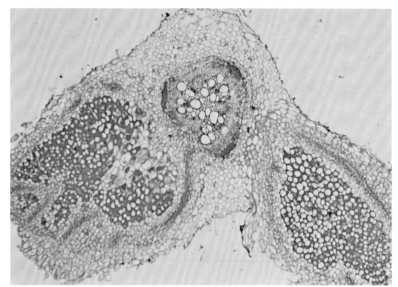

5.

6.

7.

8.

9. *Vaccinia* virus multiplying on chick allantoic membrane. (American Society for Microbiology Slide Collection.)

10. Urease test. (American Society for Microbiology Slide Collection.)

11. *Clostridium perfringens,* stormy fermentation of milk. (American Society for Microbiology Slide Collection.)

12. Simmon's citrate test. (American Society for Microbiology Slide Collection.)

13. Methyl red test. (American Society for Microbiology Slide Collection.)

14. Voges-Proskauer test. (American Society for Microbiology Slide Collection.)

15. Nitrate reduction. (American Society for Microbiology Slide Collection.)

16. Sugar fermentation. (American Society for Microbiology Slide Collection.)

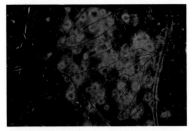

9.

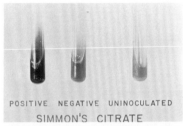

POSITIVE   NEGATIVE   UNINOCULATED
UREASE TEST

10.

11.

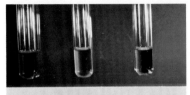

POSITIVE   NEGATIVE   UNINOCULATED
SIMMON'S CITRATE

12.

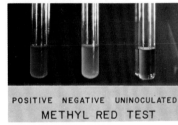

POSITIVE   NEGATIVE   UNINOCULATED
METHYL RED TEST

13.

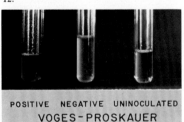

POSITIVE   NEGATIVE   UNINOCULATED
VOGES-PROSKAUER

14.

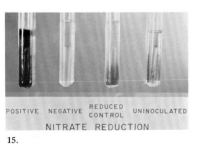

POSITIVE   NEGATIVE   REDUCED CONTROL   UNINOCULATED
NITRATE REDUCTION

15.

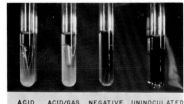

ACID   ACID/GAS   NEGATIVE   UNINOCULATED
SUGAR FERMENTATION

16.

Photographs 17–20 (Courtesy of Don Valencia, Antibodies Incorporated) show autoantibodies, a major feature of autoimmune disease. Here the FA technique is used to detect intracellular structures diagnostic of specific autoimmune conditions.

17. Antinuclear antibody.

18. Antinucleolar antibody.

19. Antimitochondrial antibody.

20. Antimitotic spindle antibody in dividing cells.

21. The FTA-ABS test showing fluorescent *Treponema pallidum*, a positive diagnosis for syphilis. (Courtesy of Curtis E. Miller, M.D., Beckman Diagnostics Operations, Fullerton, California 92634)

22. Positive tuberculin test 48 hours after injection of old tuberculum. (American Society for Microbiology Slide Collection.)

23. Langhans giant cell. (Courtesy of Dr. Alwin H. Warfel from Sloan-Kettering Institute for Cancer Research and of the *American Journal of Pathology* 93: 753–780, 1978.)

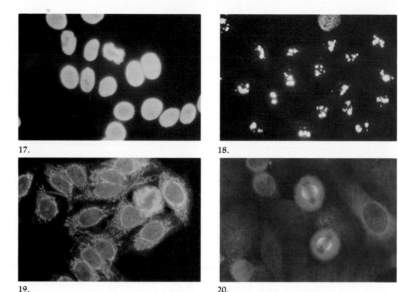

17.

18.

19.

20.

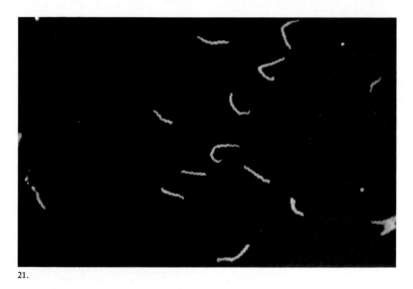

21.

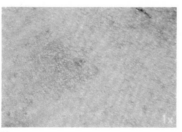

22.

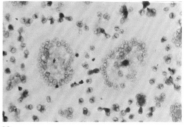

23.

The cascade-like reaction sequence can be triggered by nonspecific activation (the so-called **alternate pathway**) or by antigen–antibody combination, the **classic pathway.**

Transferrins are proteins that bind free iron atoms by electric attraction. Iron is an element absolutely essential to both microbial and human growth. Transferrin in serum, lactoferrin in milk, and conalbumin in egg whites act to tie up the body's iron reserves in a form unavailable to microorganisms. The amount of free iron in the plasma of normally nourished humans is 10^8 times less than what is required by bacteria. In infection, the host wages a war with the bacteria for the possession of the crucial iron atoms. The human body increases the amount of iron-binding proteins. If the host is successful in withholding iron, the microorganism's growth potential is greatly reduced. Interestingly enough, the addition of excessive iron supplements to the diet of pregnant women, neonates, or malnourished children may actually **increase** infections. In one study, when human neonates received daily doses of iron from birth, the incidence of Gram-negative infection increased more than six-fold.

The Reticuloendothelial (RE) System

Let us now look at how phagocytic cells are integrated into the body as a whole and how they form the body's second line of defense.

The RE network   The RE system consists of tissues with an underlying network of fibers, a **reticulum.** Fixed phagocytic cells inhabit the mesh. Cells of other types may group around it. Cells of the monocytic line, both the circulating monocytes and the macrophages, comprise the RE system (Table 12.8). Although this is called a "system," it is not, like the gastrointestinal system, a single, anatomically continuous unit. The cells are spread throughout the body and some migrate from one place to another.

RE cells are all more or less actively phagocytic. The **fixed cells** include the **Kupffer cells** in the liver (Fig. 12.5) and **dendritic macrophages** in the lymph nodes. Blood and lymph flow over and through strategically located aggregations of phagocytic cells. These carry out a routine cleansing activity called **blood clearance.** Small numbers of microorganisms introduced into the blood are rapidly removed. They cause no further problem unless they then resist phagocytic killing.

Wandering cells migrate through tissue, passing into and out of the circulation. Wandering cells police areas external to the circulation.

Role of the PMNs   The PMNs, which arise from a different stem cell line, are not usually included in the RE system. However, they carry out the same type of work. Their surveillance is normally restricted to the blood. However, trauma sets in motion localized circulatory changes. Both PMNs and monocytes are then able to leave capillaries rapidly by wriggling through the capillary wall (Fig. 12.5), which is only one cell thick. They then migrate to the site of trauma.

TABLE 12.8
Cells of the reticuloendothelial system

	CATEGORY	LOCATION
Fixed	Reticulum cells—make up a supporting structural network	Spleen
		Thymus
		Lymph nodes—dendritic macrophages
		Bone marrow
	Endothelial cells—line blood and lymph sinuses in glandular tissue	Liver—Kupffer cells
		Spleen
		Lymph nodes
		Adrenal gland
		Pituitary gland
Wandering	Histiocytes—cells in tissue and body cavities	Skin
		Peritoneal cavity
		Alveoli of lung
		Uterus
		Thoracic cavity
		Bladder
	Circulating cells	Monocytes in blood; enter tissue in inflammation

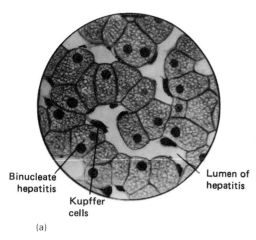

Binucleate
hepatitis

Kupffer
cells

Lumen of
hepatitis

(a)

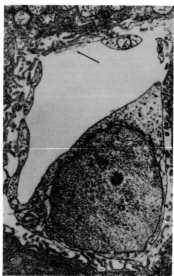

(b)

12.5
(a) The Kupffer cells of the liver are spaced along the sinusoids through which circulating blood percolates. These phagocytic cells remove particulate debris from the blood. (b) The large triangular Kupffer cell seen here occupies much of the area of the sinusoid passage.

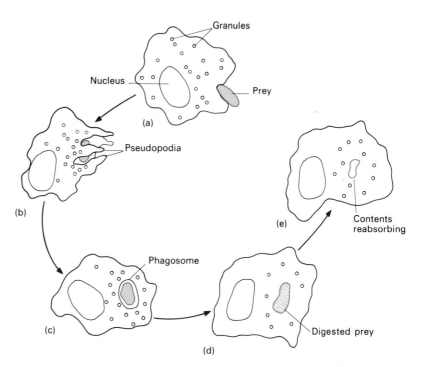

12.6
The prey must first attach positively to the surface of the phagocyte membrane (a) Attachment is assisted by complement and/or antibodies. Numerous finger-like pseudopodia (b) surround the prey. Their membranes fuse to enclose the prey in a phagosome. (c) When the enzymes from numerous granules are dumped into the phagosome, the prey will often be digested (d) and the fluids and solutes reabsorbed (e).

The process of phagocytosis   The phagocytic cell first establishes effective contact with the particle to be ingested. Phagocytes will attach to microorganisms, inanimate charcoal or quartz particles, and whole or fragmented dead cells. Attachment is enhanced by **opsonization,** coating the particle with opsonins, complement, and/or antibody. Capture may depend on the phagocyte's trapping the particle against a surface, such as the lining of a blood vessel.

Ingestion of the particle occurs by ameboid movements (Fig. 12.6). Cytoplasmic extensions surround the adhering particle. Where they meet and fuse, the prey becomes enclosed in a **phagosome,** a vacuole bounded by a membrane layer derived from the cytoplasmic membrane of the phagocyte.

Digestion or killing of the prey is enzymatic. The granules of phagocytic cells are lysosomes and other enzyme-containing vesicles. Granules are found in large numbers in PMNs. Macrophages normally contain some granules but produce many more as a result of **activation** by contact with foreign material.

Lysosomes; Chapter 3, pp. 96–97

Granules move to and fuse with the phagosome, forming a **phagolysosome.** Certain granules contain a **myeloperoxidase** enzyme complex. In the presence of chloride ion, this complex releases hydrogen peroxide, rapidly lethal to many microbes. Other granules contain **acid hydrolases.** In their presence the particle may rapidly disintegrate. Eventually, the phagolysosome moves to the cell boundary and discharges spent enzymes and undigested material. When phagocytes retain undigested material, their further activity is hindered.

The lysosomal enzymes cannot kill such pathogens as *Neisseria gonorrhoeae* and *Mycobacterium tuberculosis.* The bacteria survive, multiply, and are transported with the migrating phagocyte. Specific immunologically activated macrophages, with vastly enhanced digestive activity, are required to finally destroy them.

In the end, elimination of an infectious agent will usually depend on successful phagocytic action or complement-mediated lysis. The critical importance of the RE cells is made clear in the **Chediak-Higashi** syndrome in which phagocytic cells do not develop. Although the immune system appears normal, such individuals are unable to repel microbial invasion, and suffer repeated and severe infections.

The Inflammatory Response

Inflammation is the body's response to trauma or infection. It is characterized by localized or general swelling (tumor), reddening (rubor), elevated temperature (calor), and pain (dolor). Inflammation is a protective mechanism designed to arrest the spread of infection. At the same time, it has the potential to cause tissue damage, great discomfort and, in extreme circumstances, death.

Initiating causes   The inflammatory response is typically initiated by mechanical damage (such as a small cut) to tissue. When cells of any type are traumatized, their released contents have inflammatory effects. A cytolytic virus infection, cell lysis by complement, or contact of cells with toxin will all initiate inflammation.

Development of the inflammatory state   The chemical mediators of inflammation include **histamine,** which has a vasodilator effect, **prostaglandins,** which increase vascular permeability; **kinins,** which cause smooth muscle contraction; and **chemotactic substances,** which attract PMNs. Mediators are released in amounts related to the extent of the damage. They make the capillary bed in the vicinity much more permeable. Fluids bearing serum proteins leave the circulation and enter the area. Dilated capillaries give the reddened appearance; the fluid that pours into the tissue causes swelling and pain.

When **pyrogens** are present, the temperature will be elevated. Pyrogens may be **exogenous** substances, such as endotoxin derived from invading bacteria, or **endogenous** proteins released from phagocytic cells. The heating may be localized or, if enough pyrogen is present, cause a generalized increase in the body temperature, **fever.**

Large numbers of PMNs migrate to the area and pass through the capillary wall into the tissue. If inflammation persists, macrophages will appear somewhat later.

Phagocytic cells ingest and digest tissue debris and microorganisms, their cellular respiration increases, and they produce hydrogen peroxide and superoxides that kill most ingested microorganisms. The PMN often dies as a result, releasing more of the chemical mediators that augment the response. Platelets, breaking down in response to the mediators, precipitate clotting. This occurs both over the surface of a wound and surrounding the periphery of the traumatized area. **Fibroblasts** in connective tissue also contribute to a walling-off process.

If the microorganisms are few in number or quickly subdued, the inflammation rapidly subsides. When no more cellular damage is occurring, no more mediators are released. The mediators are quickly removed from the circulation by the liver. Healing follows, as the fibrin strands of the clot become the network for the replacement of tissue.

If, however, the microorganisms resist phagocytosis, increasing numbers of phagocytes are themselves lysed in the battle, releasing higher levels of mediators. The capillaries in the area will break down and surrounding tissue will die. The region will become a hollowed-out **necrotic** area. It contains no living cells, only PMNs in varying stages of degeneration. Fibrin deposition completely isolates the area from all surrounding tissue, contact with the circulation is cut off, and the area becomes anaerobic. The result is an **abscess.** Abscesses are eventually resolved when erosion of overlying tissue leads to drainage. The abscess is the typical lesion of *Staphylococcus aureus* infections; in it, organism and host defenses fight each other to a standstill.

Chronic inflammation occurs in hypersensitivity and autoimmune diseases, as well as in chronic low-grade infections. Progressive tissue destruction, scarring, and loss of function may occur in the kidney or lung.

Protective value   Inflammation brings together the main defensive components—phagocytes and plasma factors—on the scene of battle. Elevated temperature is believed to increase metabolic rates of phagocytes while slowing down replication of invading agents. Deposition of fibrin encloses all pathogens except those with fibrinolytic enzymes. Working together, the defensive factors will have success where separately they would fail.

Inflammation is a useful response in another way too. In an incipient infection, the reddening, swelling, and pain alert us to the threat and direct us to take steps. We can assist the body in disposing of the invaders by disinfecting or draining the area, or by administration of antibiotics.

Anti-inflammatory drugs such as the corticosteroids are available, and they are essential in the treatment of some chronic conditions. They should not be used indiscriminately because they compromise the body's most important defensive responses. When steroids are given, they may make the patient feel better fast, as the discomforts of inflammation subside. In one early study, pneumonia patients receiving steroids sat up and started asking for their dinners, exclaiming how marvelous they felt. However, alert clinicians studying the patients' X-rays noted that the lung involvement was getting rapidly worse. They withdrew the medication. Although the patients then felt subjectively much worse again, the infection was arrested and they recovered.

Immunosuppressive therapy; Chapter 14, p. 411

SUMMARY

1.   We live our lives in ecological balance with microorganisms that surround and colonize us. Most of these organisms are not harmful. Others can cause dis-

ease if our defenses are lowered, while yet another and by far the smallest group cause disease in healthy human beings.

2. Virulence can be expressed in terms of the number of organisms necessary to initiate disease, the attack rate, or the mortality rate.

3. Infectious disease occurs when the organism enters the host's tissues, multiplies at a rate greater than the rate at which the host defenses can remove it, and manufactures chemical substances that upset the host's homeostatic mechanisms or that damage tissues.

4. The infectious process depends on the means by which the organism is transmitted, the organism's ability to attach itself to tissue, the extracellular or intracellular location of replication, and the agent's invasiveness.

5. Toxins can also damage the host. Exotoxins have very specific effects on key functions such as neurological activity and protein synthesis. Endotoxins have the ability to cause vascular changes, inflammatory changes and, in high dosage, fatal shock.

6. The host has both nonspecific and specific immune defenses. The nonspecific defenses are constantly active in surveillance; they take care of most microbial invasions. The general state of the individual's health has a profound, if incompletely understood, effect on the efficiency of defense.

7. The first line of defense is made up of the anatomical features that reduce access by microorganisms and the physiological conditions that produce a chemical environment unsuited for rapid multiplication of invaders.

8. Blood cells and plasma factors are important internal defenses. All of the white blood cells have special functions in defense. Some are phagocytic, others mediate or control inflammation, and a third group mediates specific immunity. Plasma proteins include immunoglobulins, clotting factors, lysozyme, and interferon.

9. The proteins of the complement series combine to promote chemotaxis, inflammation, immune adherence of antigen–antibody complexes to phagocytes, and lysis of foreign cells. Opsonins are a group of proteins that promote phagocytosis.

10. In the living animal, phagocytosis removes the organism that has passed the external barriers and established itself inside. This requires attachment of phagocyte to prey, ingestion, exposure to lysosomal enzymes, killing, and digestion. The normal phagocytic cells—PMNs and macrophages—work effectively only in the presence of plasma factors such as complement (C') and the opsonins. Their efficiency is increased by active immunization.

11. The inflammatory response is a generally encountered effect of infection. It delivers a high concentration of plasma factors and phagocytes to the infected area. Raised temperature may inhibit microbial growth; clot formation restricts microbial migration.

12. Individuals receiving anti-inflammatory drugs have increased susceptibility to infection.

Study Topics

1. In what ways does the normal flora constitute a defense against disease?

2. Starvation leads to a progressive loss of blood serum proteins. How does this loss contribute to increased susceptibility to disease?

3. In what ways do pathogenic microorganisms evade the phagocytic action of PMNs and the RE system?

4. What are the differences between PMNs and macrophages?

5. Determine how each of the symptoms mentioned in Box 12.1 is related not only to pathogenic effects of the microbial agent but also to defensive efforts of the host. What happens if we give medication for "symptomatic relief" of coughing, runny noses, inflammation, fever, or vomiting?

Bibliography

Books and review volumes

Burnet, F. M. (ed.) 1975, *Immunology*. A collection of readings from *Scientific American*. San Francisco: W. H. Freeman.

    Relevant to this chapter:
    DeDuve, Christian, 1973, The lysosome, pp. 134–142;

    Mayer, Manfred M., 1973, The complement system, pp. 143–145;

    Mayerson, H. S., 1963, The lymphatic system, pp. 123–133.

Leeson, T. S., and C. R. Leeson, 1970, *Histology* (2nd ed.), Philadelphia: W. B. Saunders.

Roitt, Ivan, 1977, *Essential Immunology* (3rd ed.), Oxford: Blackwell Scientific Publications.

Smith H., and J. H. Pearce (eds.), 1972, *Microbial Pathogenicity in Man and Animals*. Cambridge: Society for General Microbiology.

    Relevant to this chapter:
    Glynn, A. A., Bacterial factors inhibiting host defense mechanisms, pp. 75–112;

    Hirsch, J. G., The phagocytic defense system, pp. 59–74;

    Savage, D. C., Survival on mucosal epithelia, epithelial penetration and growth in tissues of pathogenic bacteria, pp. 25–58;

    Stoner, H. B., Specific and nonspecific effects of bacterial infection on the host, pp. 113–128.

Articles

Lachman, P. J., 1975, Complement. In P. G. H. Gell; R. R. A. Coombs; and P. J. Lachman (eds.), *Clinical Aspects of Immunology* (3rd ed.), Oxford: Blackwell Scientific Publications.

Von Graevenitz, A., 1977. The role of opportunistic bacteria in human disease, *Ann. Rev. Microbiol.* **31:** 447–471.

Weinberg, Eugene D., 1978, Iron and infection, *Microbiol. Revs.* **42:** 45–66.

13

The Immune Response

Specific immune responses are characteristic of vertebrate animals. Immune responses reach their highest development in the warm-blooded birds and mammals, in which the evolution of a four-chambered heart permits a high-pressure blood circulation. Because fluid is continuously extruded from capillaries into tissues, a lymphatic system has evolved to collect and recycle the lost fluids. Lymphoid tissues became more widely distributed and important as they acquired an increased range of functions. The immune responses are an intricate communication system in which **lymphocytes** differentiate into many functionally distinct, mutually cooperating, defensive weapons.

THE BASIC FRAMEWORK OF IMMUNE FUNCTION

Because the immune response is extremely complex, we must approach it with a clear view of its actual functions. Then we can discuss how populations of cells become committed to carrying out those functions.

The Contribution of Specific Immunity to Defense

Having studied the impressive array of nonspecific defenses, one might ask what is left for immunity to do. To learn the answer, we have only to study the dilemmas experienced by individuals with genetic deficiencies in the immune system. Infants with complete immunodeficiency can be kept from inevitable death only by being isolated in a germ-free environment.

The immune system, often referred to as the **third line of defense,** is an essential companion to the nonspecific defenses. It improves the specificity with which non-specific defenses home in on invaders. The classic antibody-mediated, complement pathway creates **immune adherence** of phagocytes to foreign cells and assists in their lysis. The immune system also increases the aggressiveness of macrophages, enabling them to kill intracellular pathogens.

Nonspecific defenses in and of themselves are relatively ineffective against most Gram-positive bacteria, bacteria with capsules, bacteria with waxy walls such as the tuberculosis organism, and most intracellular parasites. Infections with these organisms may become established and progress until the body mounts an immune response.

Specific immunity is thus essential for successful recovery from many microbial diseases. It also defends against successive invasions by the agent.

The Two Types of Immune Response

There are two fundamentally different modes of immune response. In each, a different subpopulation of lymphocytes responds to the presence of foreign material (Fig. 13.1). **Antibody-mediated immunity** (AMI) results if the lymphocyte population produces immunoglobulin antibodies. These are found largely in the blood serum. Thus they are also called **humoral,** a word derived from the old-fashioned term "humor" for body fluid. Antibodies are also found in body secretions. The antibody

13.1

The Two Types of Lymphocytes   In a large population of identical-looking lymphocytes, some portion are the thymus-derived T-lymphocytes. Their reactions to foreign antigens result in cell-mediated immunity. Others are B-lymphocytes. When these react to antigen, the outcome is antibody production. Specific antibodies and specifically activated immune cells work together to inactivate or eliminate invading microorganisms or their products. It is essential to have both the AMI and the CMI systems functioning to have normal resistance to infectious disease.

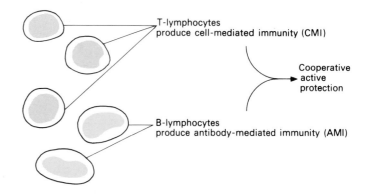

response is what most people think of as "immunity." It can also be temporarily acquired by **passive transfer** of immune serum to a nonimmune individual.

Cell-mediated immunity (CMI) results if lymphocytes acquire the capacity to kill the target themselves, on contact, or to activate macrophages to do so. This type of immunity cannot be transferred by serum, but only by injection of immune lymphocytes.

When a human being is naturally exposed to a new immunizing agent, or **antigen,** both antibody-mediated and cell-mediated immune responses develop to variable degrees. In consequence, two types of defensive weapons or **effectors** will appear—the immunoglobulins, effectors of AMI, and the immune lymphocytes, effectors of CMI. Each has a specific protective value when human health is challenged.

THE LYMPHOID TISSUES

Anatomy of the Lymphoid System

In human beings, the lymphoid system weighs only about two pounds. At any given moment it may consist of about 10^{12} (one trillion) lymphocytes and 10^{20} molecules of various antibodies.

Lymphoid tissues are intimately associated with the reticuloendothelial (RE) system. In most lymphoid organs, often referred to as lymphoreticular tissues, the

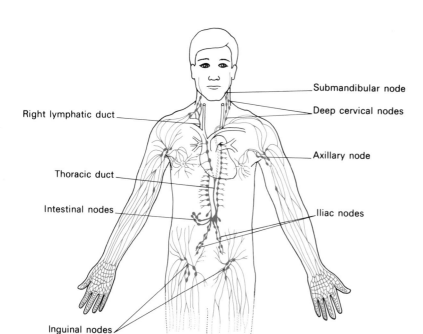

Right lymphatic duct

Thoracic duct

Intestinal nodes

Inguinal nodes

Submandibular node

Deep cervical nodes

Axillary node

Iliac nodes

13.2
The Human Lymphatic System
Lymphoid tissues are composed of lymphocytes. Such organs as the thymus, lymph nodes, spleen, and tonsils are composed partly of lymphocytes and these are considered lymphoid organs. The lymphoid tissues are in communication through both the blood and lymph circulations. Both lymphocytes and lymphocyte products circulate among the lymphoid organs and throughout the body.

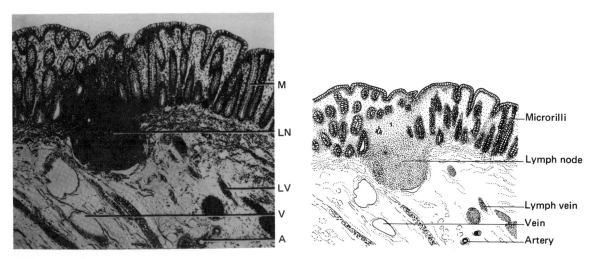

Labels: M, LN, LV, V, A

Labels: Microrilli, Lymph node, Lymph vein, Vein, Artery

13.3
Lymphoid Tissue in the Intestine One of the locations in which much lymphoid tissue is exposed to environmental antigen is the intestine. In this cross section of the small intestine, aggregates of lymphocytes called Peyer's patches are seen. These bodies are stimulated by antigenic material in the intestinal contents, such as microorganisms causing gastrointestinal damage. The lymphoid patches produce and secrete IgA antibodies that protect the mucosal surface from invasion.

lymphocytes and the macrophages of the RE system are in close proximity or in direct cell-to-cell contact. This facilitates cooperation between the two systems.

A lymphoid organ characteristically contains large numbers of lymphocytes in various stages of maturity, very loosely retained in a **reticulum** or mesh.

Primary lymphoid organs The **bone marrow** and the **thymus gland** are sites for active proliferation of immature lymphocytes. These progeny cells all undergo further differentiation before they actually contribute to an immune response.

Secondary lymphoid organs These contain both immature and mature lymphocytes. The **lymph nodes** (Fig. 13.2) and the **spleen** are regular, highly organized structures, encapsulated by a membranous covering. **Unencapsulated** lymphoid tissue is found along the gastrointestinal tract in the **tonsils,** in the **Peyer's patches** (Fig. 13.3) in the wall of the small intestine, and in the **appendix.** Diffuse lymphoid tissue occurs along the respiratory tract. Immune responses can occur in any of these areas.

Early Differentiation of Lymphoid Cells

In the last chapter, a scheme (Fig. 12.3) was presented showing development of all blood cell lines from a single **hematopoietic stem cell.** We will now follow the pathways of lymphocyte development.

TABLE 13.1

A comparison of the characteristics of B- and T-lymphocytes of mammals

CHARACTERISTIC	B-LYMPHOCYTE	T-LYMPHOCYTE
Surface or membrane markers	Fc receptor Complement binding site	Theta antigen Ly antigen
Surface receptor to bind specific antigen	Specific immunoglobulin	Unknown
Site of embryonic processing	Bone marrow or gut-associated lymphoid tissue	Thymus
Percentage of circulating lymphocytes	20%	80%
On antigenic stimulation differentiate into:	Plasma cells Memory cells	Helper cells Suppressor cells Cytotoxic (killer) cells Delayed hypersensitivity cells Memory cells
Soluble products	Immunoglobulins	Lymphokines
Primary function	Antibody-mediated immunity (AMI)	Cell-mediated immunity
Secondary function	Regulate cell-mediated immunity (CMI)	Regulate antibody-mediated immunity

Cell differentiation results when groups of cells, under different patterns of genetic control, express different phenotypes. Differentiation is a gradual process of **commitment**: certain functions are developed to their fullest, while others are turned off. The signals that direct differentiation may be surface stimuli, such as cell-to-cell contact, or diffusible chemical signals. Differentiating cells "learn" by the company they keep. An embryonic lymphocyte that lodges in one organ will be processed one way; a similar cell passing through another organ will be directed toward a divergent path.

B-cells and T-cells   Mature lymphocytes are of two types (Table 13.1). The B-cells give rise to antibody-mediated immunity. The T-cells give rise to cell-mediated immunity. Each group also regulates the other group's functions, which complicates the issue.

Origin and location of B-cells   The embryonic lymphoblast stem cell in birds migrates to an intestinal pouch called the **Bursa of Fabricius** in which it is processed into a mature B-cell. Mammals have no such organ, and the "Bursa analog," the exact site of B-cell processing in mammals, is still not known. Bone marrow, or gut-associated lymphoid tissue (GALT) are the candidates for its location. From birth on, some B-cells reside in special areas within all the lymphoid tissues. About 20 percent of circulating lymphocytes are B-cells.

Origin and location of T-cells　　T-cell progenitors migrate through, in sequence, the fetal liver, spleen, and then the thymus gland (Fig. 13.4). The thymus is an unusual organ. Its main function is to establish adult immune responsiveness; this work is over at adolescence.

Lymphocytes that migrate to the thymus are processed as T-cells. Rapid lymphocyte proliferation occurs there through childhood. The cells released from the thymus (**thymocytes**) are committed T-cells but nondividing and nonfunctional. They migrate to T-cell areas in the other lymphoid organs, where they are in close proximity to B-cells. By adolescence, the secondary lymphoid tissues are fully stocked with T-cells. About 80 percent of circulating lymphocytes are T-cells.

Lymphocyte traffic　　To the immunologist, nothing is more frustrating than the fact that lymphocytes just will not stay put and be studied. They temporarily lodge in

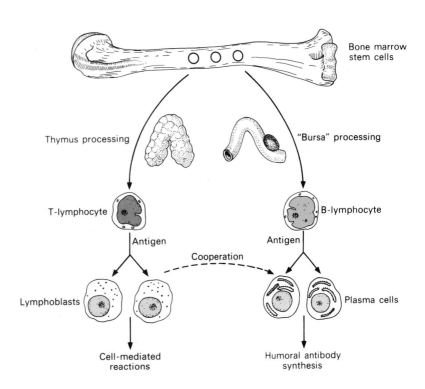

13.4
Lymphocyte Processing This diagram shows a commonly accepted hypothesis about lymphocyte processing that states that primitive lymphoid cells originate in the fetal bone marrow. Randomly, some pass through the thymus gland and undergo a "processing" that converts them to T-lymphocytes. Others pass through gut-associated lymphoid tissue and become immunocompetent B-lymphocytes. Antigen stimulation will cause either type to divide and change into the enlarged blast or plasma cell forms.

Bone marrow stem cells

Thymus processing

"Bursa" processing

T-lymphocyte

B-lymphocyte

Antigen

Cooperation

Antigen

Lymphoblasts

Plasma cells

Cell-mediated reactions

Humoral antibody synthesis

the reticular network of a lymphoid organ, then are freely washed out and carried by blood or lymph to the next lodging point. Many lymphocytes complete a circuit from a lymph node through the circulation and back to the node again in about two days. This mobility brings lymphocytes and antigenic stimulants together and provides the opportunity for the cell-to-cell communication needed to mount and control immune response.

MECHANISM OF THE IMMUNE RESPONSE

When contact with a foreign body is made by a lymphocyte (either a B-cell or a T-cell), a change occurs in the lymphocyte that results in its being tailored to combat that particular foreign body. To put it another way, during contact, certain chemical markers of the foreign body are recognized by the lymphocytes and stimulate defensive reactions in them.

Antigens

Unique chemical markers identify the surfaces of all cells. All cells in your body have some markers in common. An individual has only about one chance in 30,000 of encountering another human being with the same marker pattern, unless that individual has an identical twin. Differentiated cells also have tissue-specific markers. Your own cells are unambiguously labeled **self**. Only in pathological states can your lymphocytes react against your own cells.

Markers not a normal part of your tissue are **nonself**. Lymphocytes will react to nonself markers. We will operationally define an **antigen** as a substance against which an individual mounts an immune response.

Antigens are usually very large molecules or cells. Proteins and polysaccharides are strongly antigenic, and nucleic acids and lipids are weakly antigenic. The surface of an antigen has one or more chemically defined sites, the **antigenic determinants** or **haptens**. The antigenic determinant is also the combining site for the immune effector. The rest of the antigen is the **carrier**. Some antigens have multiple antigenic determinant sites. These antigens tend to be "stronger" or provoke a greater response.

Small molecules may also become antigenic. A penicillin molecule, for example, is a potential antigenic determinant. In order to signal lymphocytes to respond, the small unit must be coupled to a carrier, usually a protein.

Antigens vary in the number of combining sites on their surfaces to which immune effectors can attach. Soluble antigens such as protein molecules may have few sites; cellular antigens such as bacteria have sites all over their surfaces. The **valence** of an antigen is the number of its combining sites.

Immune Tolerance

There are many potential antigens in your body—the self-antigens—that your immune system does not react to. This **tolerance** develops during fetal life, in the period before the immune system is functional. Any antigen to which fetal lymphoid tissue is exposed becomes tolerated; that is, the individual can never mount an immune response against that antigen. After birth, however, antigen contact provokes immunization.

Clonal selection theory   Many hypotheses have been advanced to explain immune responsiveness. Studies of immune tolerance led to Burnet's Clonal Selection Theory, which has survived 20 years of probing and questioning.

Literally billions of lymphocytes are produced during fetal development. The theory postulates that a genetic mechanism renders each lymphocyte capable of participating in an immune response to a single antigenic determinant. Each fetal lymphocyte bears on its surface receptors specific for the antigen it recognizes. The fetal lymphocytes are in an intermediate state of development—they can **recognize** antigens with their cell surface receptors but they cannot **respond** with cell division.

Recognition and contact with self-antigens in the fetal state—when the only source of antigens is the fetus itself—destroys the lymphocyte (Fig. 13.5). This is **negative selection**. Toleration is induced by eradicating all potential responders. The tolerant state is permanent because the remaining lymphocytes are nonresponders, unable to attack the self-antigens.

After birth, contact with an antigen stimulates the lymphocyte. When it attaches to an antigen, the lymphocyte begins rapid division (**blastogenesis**) and gives rise to many daughter cells. This is **positive selection**. The resulting **clone** of lymphocytes all express one genetic potential and thus one antigenic reactivity. They become the effector cells for a single specific immune response. Unselected lymphocytes remain present but inactive.

The main commonsense objection to the Clonal Selection hypothesis is that it requires a very large number of genetically variant lymphocytes, probably about 100,000 but perhaps as many as a million, to account for responsiveness to all possible antigens.

It would be possible, but not probable, that each lymphocyte had a million different genes for immune responses, only one of which would be operative per cell. A more likely mechanism would be something that caused fetal lymphocytes to undergo random somatic mutation in the gene(s) regulating their immune response. Each lymphocyte might emerge with a different allele of the gene or genes.

Such mutational enzymes have recently been discovered. They can be shown to selectively randomize genetic sequences in the immune response genes. These enzymes function only during the fetal period and, so far as is known, only in lymphocytes.

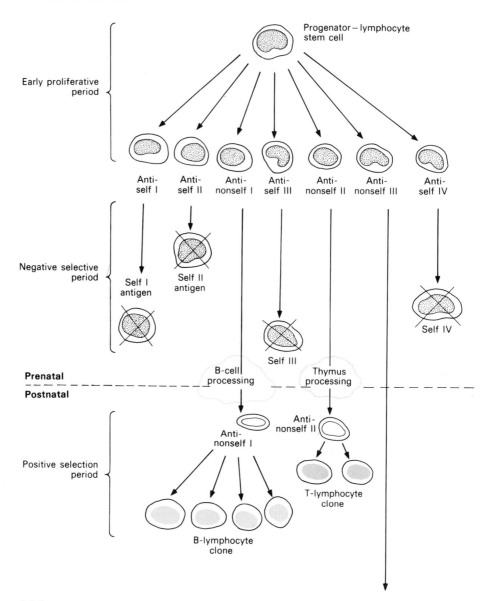

13.5

Clonal Selection   From a lymphocyte stem cell arise many thousands of different anti-body-labeled fetal lymphocytes. During the negative selection period, any of these that binds antigen (of necessity a *self*-antigen) will die. Those reactive with nonself-antigens survive and undergo either B or T processing.

After birth, a human being begins to have contact with nonself-antigens such as mi-croorganisms. Some of these contacts will positively select B or T cells, promoting prolif-eration, clone formation, and AMI or CMI, respectively. Other lymphocytes may never contact their antigen and remain unstimulated throughout life.

ANTIBODY-MEDIATED IMMUNITY

The outcome of an antibody-mediated immune response is the production of immunoglobulin antibody with highly specific antigen-binding potential. The antigen-binding site is a unique, three-dimensional structure.

Structure of the Immunoglobulins

There are five classes of immunoglobulins. All are quaternary proteins, assembled from four or more polypeptide chains held together by disulfide bonds. Each of the five classes of immunoglobulins is a different combination of polypeptides. The structural details are summarized in Table 13.2 and Fig. 13.6.

Protein structure; Chapter 2, pp. 51–54

Light chains and heavy chains Treatment of antibody molecules with reducing agents breaks disulfide bonds. In the case of immunoglobulin G(IgG), this treatment yields four separate chains, two of which are **light chains** and two, the longer **heavy chains.** Light chains are of two types, designated kappa or lambda. The Ig product of one clone of antibody-forming cells will contain two kappas or two lambdas but not one of each. Each class of immunoglobulin has its own type of heavy chain.

Papain fragments Digestion of intact IgG with the proteolytic enzyme papain cleaves the four-chain unit roughly in the middle. This yields two antibody-binding fragments (Fab) and one crystallizable fragment (Fc). Fab fragments recognize antigen but Fc fragments have no affinity for antigen. The Fc regions bind antigen–antibody complexes to complement, to phagocytes, or to cell surfaces.

TABLE 13.2
Physiochemical characteristics of immunoglobulins

	IgG	IgA	IgM	IgD	IgE
Light chains	Kappa or lambda 2	Kappa or lambda 2	Kappa or lambda 2	Kappa or lambda 2	Kappa or lambda 2
Heavy chains	Gamma 2	Alpha 2	Mu 10	Delta 2	Epsilon 2
Basic 4-chain units	1	1 in serum 2 in secretions	5	1	1
Antigen-binding sites	1	1	5 (10)	2	2
Other peptides	—	J-chain secretory unit	J-chain	—	—
Percentage of total immunoglobulin	80	13	6	1	0.002

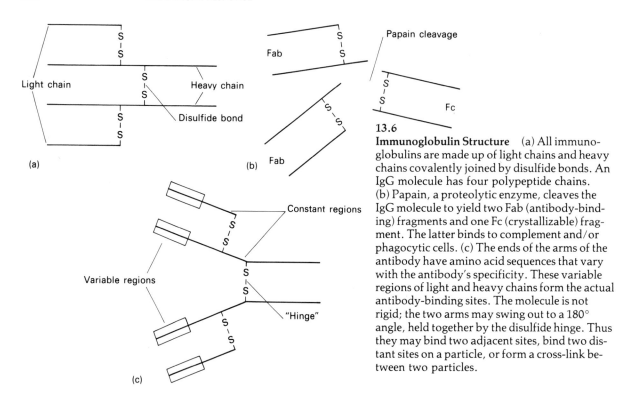

13.6

Immunoglobulin Structure (a) All immunoglobulins are made up of light chains and heavy chains covalently joined by disulfide bonds. An IgG molecule has four polypeptide chains. (b) Papain, a proteolytic enzyme, cleaves the IgG molecule to yield two Fab (antibody-binding) fragments and one Fc (crystallizable) fragment. The latter binds to complement and/or phagocytic cells. (c) The ends of the arms of the antibody have amino acid sequences that vary with the antibody's specificity. These variable regions of light and heavy chains form the actual antibody-binding sites. The molecule is not rigid; the two arms may swing out to a 180° angle, held together by the disulfide hinge. Thus they may bind two adjacent sites, bind two distant sites on a particle, or form a cross-link between two particles.

Variable and constant regions The researcher can obtain large amounts of a single human antibody for biochemical analysis as a result of a neoplastic disease, **myeloma.** This is a bone marrow tumor in which one clone of B-cells undergoes wild proliferation. The resulting pure or **monoclonal** Ig is produced in excess and appears in the urine of patients as the **Bence-Jones protein.** The protein can be collected and subjected to amino-acid sequence analysis. Each individual patient produces a different one. **Hybridoma** cells in culture are now being extensively used to provide monoclonal antibody also.

Comparison of many different monoclonal Igs shows that the antigen-binding portions of the Fab regions are **variable,** and each antibody has a different amino-acid sequence. The rest of the Ig unit, including the rest of the Fab fragment and the Fc region, is the **constant region.** All antibody specimens in a given class will have the same amino-acid sequence in those regions.

Antigen-binding capacity In the variable regions, interactions between the three-dimensional folded patterns of the light and heavy chains create uniquely shaped

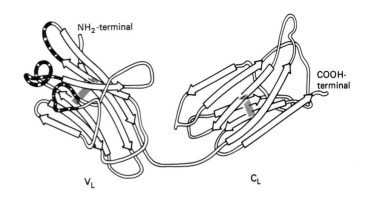

13.7
Detail of a Light-Chain Arrows show original direction of synthesis of the polypeptide from NH_2-terminal to COOH-terminal. Dark bars represent disulfide bonds. The three dotted loops constitute the antibody-bonding portion. This chain would normally function with a heavy chain. The variable region of the heavy chain also contributes to the antibody-binding site.

grooves (Fig. 13.7). The antigen fits into these, very much as a substrate fits into the surface of an enzyme. IgG, IgA, IgD, and IgE each have two binding sites per molecule, and IgM has five sites.

The closer the fit, the stronger the binding power or **avidity** of antibody and antigen. When your body experiences a new antigen, different clones of antibody-forming cells may respond. These may produce Igs with different amino acids in the binding site and thus with varying avidities.

Induction of Antibody Synthesis

Route of antigenic presentation When antigen enters or is introduced into the body, it rapidly encounters lymphoid tissue. If it enters by skin break or injection, it is carried with tissue fluid to the nearest lymph node. If antigen is introduced into the blood, it is filtered out by the spleen. If it penetrates the respiratory mucosa, diffuse lymphoid aggregates capture it; and if it invades the gut, it is picked up first by the Peyer's patches.

Evidence of response When antigen is captured, visible changes begin in the lymphoid tissue. **Germinal centers** develop in the cortex of the tissue (Fig. 13.8). Here B-cells with receptors specific for that antigen divide and differentiate into the larger **plasma cells.** Within one to six days after antigen contact, one or more such centers have developed. Specific antibody is detectable in quantity in the blood within one to two weeks later. The number of antibody-producing plasma cells increases for several weeks, then declines. Although cell division slows, the germinal center persists. It is considered to contain **B-memory cells.**

On second exposure to the same antigen, the response, although similar, is much more rapid. Because of the existence of B-memory cells, new plasma cells and antibodies appear in much shorter time. This is called the **amamnestic** response. The

Vaccination schedules; Chapter 27, pp. 796–809

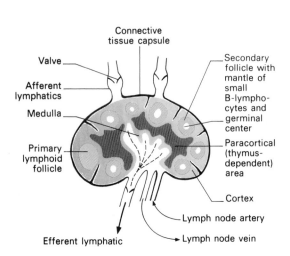

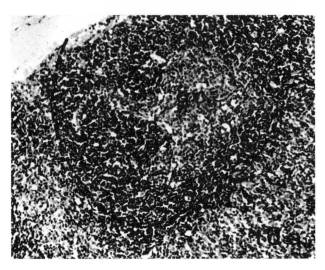

13.8

The Lymph Node   Lymph nodes have both lymphatic and blood circulation. Thus substances can come to them by both arterial blood and afferent lymphatics. Once in the lymph node, antigenic particles will be trapped by the spongelike structure of the organ. The particles may be phagocytosed. The exterior or cortical layers of the node contain lymphoid follicles in which B-lymphocytes are found. On antigenic stimulation, a primary follicle develops into a larger secondary follicle. This has a germinal center in which active cellular proliferation yields plasma cells. The plasma cells release their product—antibody—that passes into the venous circulation. The paracortical layers contain T-lymphocytes. These proliferate on antigenic stimulus and may leave the node and travel in the circulation to the site of an infection. The photograph shows an active germinal center containing enlarged cells (light area) in the center of a secondary follicle.

final antibody level is higher and persists longer (Fig. 13.9). This is the rationale behind booster shots.

Cellular activities   Response seems to require the cooperation of three types of cells, the B-cell, the T-cell, and the macrophage.

Macrophages of the lymphoid reticulum are considered to **present** the antigen to the lymphocytes. There is no agreement at the moment as to the actual manner of antigen presentation.

What is the actual stimulus to antibody production? Both B-cells and T-cells have surface receptors that enable them to recognize the antigen to which they are able to respond. In the B-cell, the receptor is the specific antibody the B-cell is programmed to make. The T-cell receptor is probably also immunoglobulin or a closely related molecule.

When antigen is presented to B-cells, it is bound and held by existing antibody molecules inserted into the fluid mosaic membrane. These antigen–antibody complexes then sink through the membrane into the interior of the cell. Shortly there-

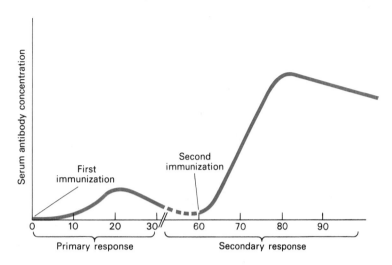

13.9
The Anamnestic Response Immunizations are often given in series of two or more doses. When the first dose is given, a mild antibody response is noted at just about six to eight days. Response peaks at about three weeks, then gradually declines. When the second or other subsequent doses are administered, there is a significant increase in antibody titer within about three days. The peak antibody titer achieved may be as much as 100 times that obtained with the first dose, in some cases.

after, the stimulated cell starts to divide. **Blastogenesis,** or division, is the primary evidence that a lymphocyte has become committed to immune response.

To summarize, then, we can visualize that antigen is captured in a lymphoid tissue and offered to a large number of B-cells as they migrate through. Each B-cell has a different genetic potential for antibody synthesis and is wearing a sample of its wares on its surface. Any that are able to bind the offered antigen will be stimulated to division. They will settle down and populate a germinal center. Via mitosis, they pass on the ability to produce the antibody to two groups of progeny, plasma cells (committed to all-out antibody synthesis) (Fig. 13.10) and memory cells (primed to begin synthesis quickly at a later time).

Thus antigen "selects" a B-cell, which gives rise to a clone. Many cells may be selected, and not all of them will produce exactly the same antibody. The antibodies will differ in amino-acid sequence and in avidity, but they will all bind the antigen in some way.

B–T cooperation Persons with genetic deficiency in T-cell functions cannot readily make antibody. Recall that antigens consist of an antigenic determinant (the hapten) and a carrier. The carrier must be bound to a helper T-cell, and the hapten to the B-cell receptor, before the B-cell is activated. This requirement for contact with two cells is called a "second signal." It is rather like those fail-safe systems by which two keys must be simultaneously turned to launch an atomic missile.

Laboratory Tests of the AMI Response

The presence of antibody in an individual's body fluids can be taken as evidence of exposure to the antigen. Detecting the presence of specific antibodies (qualitative

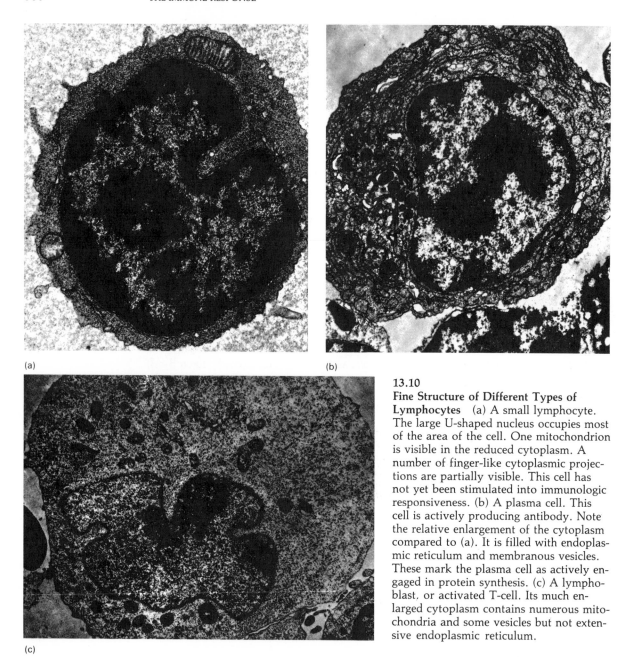

(a)

(b)

(c)

13.10
Fine Structure of Different Types of Lymphocytes (a) A small lymphocyte. The large U-shaped nucleus occupies most of the area of the cell. One mitochondrion is visible in the reduced cytoplasm. A number of finger-like cytoplasmic projections are partially visible. This cell has not yet been stimulated into immunologic responsiveness. (b) A plasma cell. This cell is actively producing antibody. Note the relative enlargement of the cytoplasm compared to (a). It is filled with endoplasmic reticulum and membranous vesicles. These mark the plasma cell as actively engaged in protein synthesis. (c) A lymphoblast, or activated T-cell. Its much enlarged cytoplasm contains numerous mitochondria and some vesicles but not extensive endoplasmic reticulum.

testing) and determining their concentration or **titer** (quantitative testing) is often an aid to clinical diagnosis. **Serology** is the study of antibody in serum.

Combination with a known antibody unequivocally identifies an antigen. This can occur in a mixture, structure, or tissue sample without the requirement for extraction and purification. Antigen–antibody reactions can also be an extremely sensitive means of quantitating substances such as enzymes or hormones. Immunological techniques are extensively used in basic biological research. Any immunological test must render the antigen–antibody complex, if it is formed, visible or otherwise detectable. There are literally dozens of approaches.

Agglutination tests   When the antigen is a large particle or a cell, and it complexes with antibody, clumping or agglutination occurs. The clumps are large enough to see with the naked eye (Fig. 13.11). Any test, such as blood grouping, in which the red blood cell is the antigen, is of this type. Bacterial cells are also agglutinated by their antibodies. Soluble antigens may be attached to inert latex beads, thus rendering them "particulate." Certain rapid screening pregnancy tests make use of latex bead antigens. One may also attach the antibody to the latex bead.

Some viruses such as influenza agglutinate red blood cells nonspecifically. Antiviral antibodies produced by patients will attach to the virus particles preventing hemagglutination. The **hemagglutination inhibition** test is widely used in virology laboratories to assay antiviral antibody.

Hemagglutinins;
Chapter 11, p. 319

The red blood cell may also be used, exactly like the latex beads mentioned above, to carry an antigen and render it agglutinable. A test antigen such as thyroglobulin is attached to the surface of the red blood cell after the latter has been treated with tannic acid. Exposure of antigen-bearing RBCs to antithyroglobulin antibody leads to agglutination of the red cells. This is called passive or **indirect hemagglutination.**

Slide-type tests give a rapid qualitative result indicating the presence or absence of antigen or antibody. Titration-type tests are used for quantitation. In these tests,

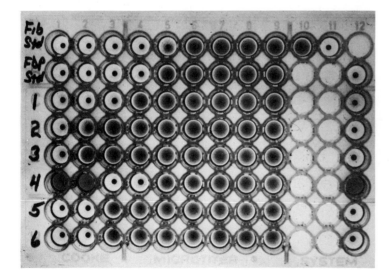

13.11
Hemagglutination The microtiter technique is widely used for quantitative analysis of either antigens or antibodies. The test is performed in a plastic tray containing 96 small wells, which function as tiny "test tubes." Antigen or antibody standards are presented linked to the surface of red blood cells, which serve as inert carriers. When antigen and antibody combine, the red blood cells settle out in a layer covering the entire bottom of the well (top line, tubes 6 through 10). If antigen-antibody combination has not occurred, the red blood cells settle directly to the center bottom of the well. This shows up as a small, round button (tubes 1 through 4, top row). Controls and standards are included in each plate.

the serum is **serially diluted** so that each sample in a series contains only half as much serum as the previous one. This may be done in test tubes or in calibrated wells (microtitration). The higher the antibody level in the original serum, the higher the titer or dilution that will give a positive result.

Precipitation tests   Many antigens, such as proteins, are soluble macromolecules. When they combine with antibody, they form a macromolecular aggregate or precipitate that does not settle out of suspension readily; it is apparent only as a cloudiness. To be able to be seen, the antigen and antibody must meet in a small area where the aggregates are trapped. **Immunodiffusion plates** allow antigen and antibody to diffuse toward each other out of wells cut into flat slabs of highly purified agarose gel. A line of precipitate forms in the gel at the point at which antigen meets homologous antibody. If the antigen is part of a mixture, preliminary **immunoelectrophoresis** may be used to separate the components on the basis of electric charge. Then the resolved mixture can be exposed to antibody and components identified by precipitation (Fig. 13.12). With closely controlled conditions, concentrations can be calculated from the size of the precipitation zones.

Neutralization tests   The biological role of antibody is, after all, the neutralization of the pathogen. Thus the effectiveness of the patient's serum in protecting a live animal, or reducing viability of the agent, is a useful measure of immunity.

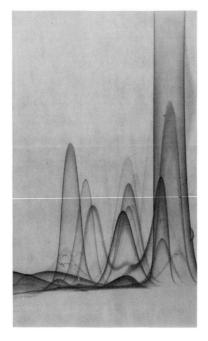

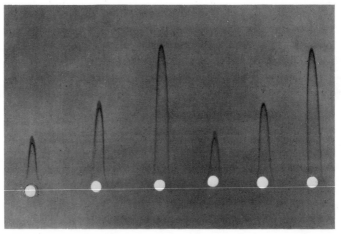

13.12
Immunoelectrophoresis   When a serum sample containing a mixture of immunoglobulins is placed in an electric field, the substances migrate from left to right. After this step, the antibody is exposed to antigen and the resulting pattern is made visible by a staining procedure.

TABLE 13.3
Immunological techniques

TEST		IDENTIFIES OR QUANTITATES	
Technique	Method used	Antigen: infective agent	Antibody against agent of
Agglutination	Slide test	*Salmonella* species	Syphilis
	Tube titration		Cholera
			Whooping Cough
	Hemagglutination inhibition		Measles
			Rubella
	Indirect hemagglutination		Plague
			Pseudomonas infection
			Amebic dysentery
Precipitation	Immunodiffusion		Blastomycosis
			Influenza
	Immunoelectrophoresis		Candidiasis
			Trichinosis
Neutralization	Virus		Respiratory infections (viral)
			Polio
	Toxin		Diphtheria
			Tetanus
Complement fixation			Primary atypical pneumonia
			Histoplasmosis
			Herpetic infections
Radioimmunoassay	Direct	Insulin	
		Growth hormone	
		Hepatitis B-virus	
ELISA	Indirect		Rubella
			Measles
Fluorescent antibody	Direct	*Treponema pallidum*	
		Varicella-zoster virus	
	Indirect	*Neisseria meningitidis*	
		Plasmodium species	
Immune electron microscopy	Direct examination of stool	Hepatitis A-virus	
		Other gastrointestinal virus	
		Other gastrointestinal bacteria	

TPI test; Chapter 21, p.
632

Virus neutralization tests assess the ability of serum, when preincubated with virus particles, to block their infectivity on subsequent injection into a test animal. **Toxin neutralization** determines the ability of antitoxin-type antibodies in a patient's serum to protect a test animal from the effects of infected toxin.

The *Treponema pallidum* immobilization test (TPI) is used to identify the syphilis spirochete in the exudate from suspected lesions. This agent moves with an undulating motion. When exposed to specific antiserum, its motion stops. Unrelated spirochetes that might be part of the normal flora are not affected.

Complement-fixation tests   Antigen–antibody complexes, if IgM or IgG is the antibody, fix complement. *In vivo*, complement fixation leads to more effective phagocytosis or cell lysis. *In vitro*, the amount of complement fixed can indicate the amount of antibody present in a serum sample. The technique is based on competition—the competition of two antigen–antibody pairs for a limited amount of complement.

The first pair, the **test system,** can be any antigen–antibody pair of interest. Let us say that we are looking for antibody to rubella virus. The test antigen will be purified virus and the test antibody will be a patient serum in which there is an unknown amount of antirubella antibody. Complement fixation by this pair is not directly visible. The second pair, or **indicator system,** is sheep red blood cells (antigen) and antisheep hemolysin, an antibody produced by rabbits against sheep red blood cells. When complement is fixed by this pair, the RBCs hemolyze.

The test is done in stages. First, the test system—rubella virus and patient serum—is allowed to react. The reaction fixes an amount of complement proportional to the amount of AG–AB complex formed. Depending on the amount of antibody in the unknown sample, none, some, or all of the complement will be consumed. Then the indicator system is added. Any unfixed complement will react with the sheep cell–hemolysin complex and lyse cells. The amount of lysis indicates the amount of complement left over from the first step. The more lysis, the less antibody there was in the patient sample. This test method, although complex, is extremely sensitive and reliable in trained hands.

Radioimmunoassay   RIA is also a competition test system. Its extreme sensitivity comes from the fact that one can accurately measure very small amounts of radioactive iodine (I^{125}). RIA was first developed for the detection of minute amounts of hormones. It can be used to detect any substance to which antibody can be produced in a test animal.

The first step is exposure of specific antibody to a sample containing an unknown amount of antigen. Then I^{125} labeled antigen is added to combine with any uncombined antibody. The resulting complexes are precipitated and their radioactive count determined. Since antigen from the "unknown" competes with labeled antigen, the more the unknown antigen, the lower will be the final radioactivity.

Enzyme-linked immunosorbant assay   ELISA is a competition system too. The antigen is "labeled" by linkage to an enzyme, such as peroxidase (Fig. 13.13). Antibody absorbed to a surface is allowed to react with unlabeled antigen from the

13.13

Enzyme-linked immunosorbent assay This test works by the combination of three reactants to make a kind of "sandwich." On the left, antigen (first reactant) binds specific antibody (second reactant). Nonspecific antibody does not bind and is washed away before the next step. The third reactant is antihuman IgG that has been conjugated to the alkaline phosphatase enzyme. This IgG binds to the previously bound antibody, forming a three-part complex. Then the enzyme's substrate, p-nitrophenyl phosphate, is added. Bound enzyme is revealed by the formation of brightly colored product, p-nitrophenol. On the right, in the absense of specific antibody, no enzyme IgG conjugate is bound, and thus there is no conversion of substrate to product. Although this test has many steps, it can be performed rapidly and is extremely accurate.

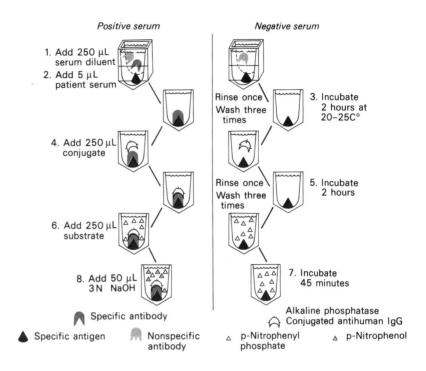

Positive serum   Negative serum

1. Add 250 µL serum diluent
2. Add 5 µL patient serum

Rinse once Wash three times

3. Incubate 2 hours at 20–25C°

4. Add 250 µL conjugate

Rinse once Wash three times

5. Incubate 2 hours

6. Add 250 µL substrate

7. Incubate 45 minutes

8. Add 50 µL 3 N NaOH

Specific antibody

Alkaline phosphatase
Conjugated antihuman IgG

Specific antigen   Nonspecific antibody   p-Nitrophenyl phosphate   p-Nitrophenol

---

unknown sample, then later with the enzyme-labeled antigen. Last, the total enzyme activity of the complex is assayed. Since unlabeled antigen from the unknown competes with enzyme labeled antigen, the more the unknown antigen, the lower the final enzyme activity of the complex. ELISA, very similar in design to RIA, is easier to use because it does not require the iodine isotope that has a very short half-life and is tricky to handle. The techniques are relatively easy to learn, and equipment costs are modest.

Immune Microscopy **Fluorescent antibody** techniques make use of fluorescent dyes coupled to antibody to locate antigen with the ultraviolet microscope.

Fluorescent dyes, such as fluorescein, absorb ultraviolet light and re-emit visible light. They are similar to the paints used in black-light posters. The dyes can be coupled chemically to antibody molecules, without altering the antibody's reactivity, to make a fluorescent antibody.

The **direct technique** is used to locate antigen. A fixed slide (usually a tissue section, but also perhaps a microbial culture or other potential source of antigen) is flooded with fluorescent antibody for a period. Uncombined antibody is then removed and the slide is viewed with the UV microscope. Where antigen has complexed with antibody, the fluorescent dye will emit visible light as the UV radiation hits it. The structure may be spotted or outlined with light, corresponding to the distribution of antigen in the specimen.

UV microscope; Chapter 3, p. 71

Box 13.1

Rapid Immunological Identification of Bacteria

By now, your laboratory experiences will have shown you that standard microbial growth, isolation, and identification procedures are slow. When it takes an efficient laboratory 48–72 hours to give the physician an exact identification of the bacterium attacking a seriously ill patient, the pressure to find faster methods becomes considerable.

In the past couple of years, many new "rapid identification" tests have been developed and marketed. These are all based on slide agglutination, using commercially prepared, standardized antibacterial antibodies. These may be bound either to dead staphylococci or latex beads. In the figure a positive agglutination test is confirmatory for *Neisseria gonorrhoeae*. Compare with the negative, nonagglutinated control (top).

On one day, the attending physician examines the hospitalized patient and, concerned about evidence of infection, orders a culture. The culture is promptly obtained, sent to the laboratory, and plated. The next morning (16–18 hours after plating), the laboratory technician sees suspect colonies on the primary isolation plate. A "rapid" procedure is directly applied, identifying the organism to strain within

Rapid diagnosis

minutes. The results are promptly recorded and transferred; the report is waiting in the patient's chart for the physician's daily rounds the following day.

So far, rapid tests are not available for many microbial pathogens, so this scenario remains more a dream than a reality in the laboratory diagnosis of the majority of infections. However, the area of rapid immunoidentification is expanding rapidly, in parallel with that of rapid biochemical identification. Within a very short time, overnight identification of most pathogens will undoubtedly be standard laboratory practice.

The **indirect technique** may be used to detect fixed antibody. The binding of antibody to tissue antigen occurs in many *in vivo* situations, especially immune complex disease. (See Chapter 14.) The binding is invisible under ordinary conditions. However, it may be visualized as follows. A rabbit or other animal is immunized with human globulin, a foreign protein. The test animal makes antibodies against human globulin—antihuman Ig. These antibodies are coupled to fluorescent dye and then applied to the specimen. The fluorescent antihuman antibodies bind to

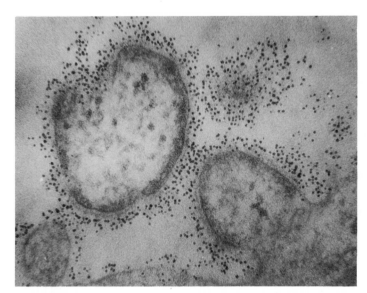

13.14
Immune Electron Microscopy Antibodies can be chemically conjugated to ferritin, an electron-dense, iron-containing substance. When a sample containing surface antigens is exposed to ferritin-conjugated antisera, the antibody binds to the specific surface sites. When later the sample is viewed with the electron microscope, the antigen–antibody combinations are clearly visible as dark dots. Here human cells in culture have been infected with the P-3 virus. Their surfaces appear covered with dark dots. These are the viral surface antigens exposed as the viruses leave their host cells by budding.

the human immunoglobulins which are in turn bound to the antigen—a three-layered sandwich.

Immune electron microscopy (IEM) uses visualizing techniques that label AG–AB complexes with substances that are electron-dense (Fig. 13.14). One approach is to complex specific antibody with the iron-containing marker **ferritin.** When ferritin-labeled antibody is immobilized on antigen-containing structures, the complexes show up as dark spots under the electron beam.

Human Blood Groups

An important application of immunology is in blood banking, the matching of donor and recipient in blood transfusion.

The ABO grouping   Red blood cells, like all of our cells, have genetically determined surface antigens. Certain of these are unique to red blood cells. The human RBC membrane may have the Type A or Type B polysaccharide antigen, both, or neither. The serum will always contain antibody to the antigens we do not have, probably induced by environmental exposure to a cross-reacting antigen. The antibodies clump RBCs and, in the presence of complement, lyse them. Thus incompatibility between donor and recipient blood leads to severe or fatal transfusion complications.

Bloods are first **typed** to establish group identity, then **cross-matched** to check that clumping due to ABO incompatibility or any other cause does not occur.

TABLE 13.4
The human ABO blood groups

BLOOD TYPE	RED BLOOD CELL ANTIGENS	SERUM ANTIBODIES	CAN RECEIVE BLOOD FROM
O	None	Anti-A and anti-B	Type O
A	A	Anti-B	Type A, Type O*
B	B	Anti-A	Type B, Type O*
AB	A and B	None	Type AB, Type A*, Type B*, Type O*

*In emergency situations *only*, bloods of a different type may be given. There will usually be some side effects due to donor antibodies hemolyzing recipient cells.

The **Coombs test** detects incomplete nonagglutinating antibody. In some cases, the antibodies formed, such as human anti-RhD antibody, can agglutinate red blood cells *in vivo*, but not *in vitro*. This is because the RBCs have negative surface charges, and their repulsive effect is stronger than the antibody-binding effect. After red blood cells from one party and serum from the other have been allowed to react, antiglobulin antibodies are added. These react with the incomplete antibodies bound to RBCs forming an agglutinating complex.

The Rh system   The Rhesus antigen (Rh) in several forms may be part of the RBC membrane also. About 85 percent of the population of the United States is Rh positive. In Rh negatives, anti-Rh antibody is not normally present. There is no incompatibility on **first** contact, because sensitivity is not yet present. Following contact, however, an immune response (sensitization) occurs in about 50 percent of individuals and anti-Rh antibodies are made. Destruction of Rh positive RBCs will result on subsequent contacts.

Sensitization occurs more commonly when RBCs from an Rh-positive fetus enter the circulation of an Rh-negative mother. This usually occurs during the birth process. The mother will then produce anti-Rh antibodies. Those of the IgG variety will cross the placenta of a future fetus and attack Rh-positive fetal RBCs, a condition called **erythroblastosis fetalis.** This rarely affects a first child, but may affect later children if they are Rh positive. Sensitization can be prevented by giving the mother a massive injection of anti-Rh antibodies just after delivery. These antibodies block development of an anti-Rh response by removing the antigen before it can sensitize. These transferred antibody molecules are short-lived and because they disappear within a few weeks, they do not pose a threat to subsequent pregnancies.

Maternal–fetal interaction; Chapter 21, pp. 636–644

Other RBC antigens   Other genetically determined antigen groups have been discovered, but they are of limited medical importance, either because they are quite rare or because they do not often create incompatibility problems in transfusion.

Determination of the complete spectrum of RBC antigens is useful in legal cases in the identification of an unknown person, in the determination of a source of a blood sample, or in proving a certain male innocent of siring a certain child.

Laboratory Tests to Detect Antigens

When infection occurs, the causative agent often releases significant amounts of its surface antigens into the body fluids. If these can be detected by their ability to react with specific diagnostic antibodies, it may be possible to pinpoint the exact pathogen present within a few minutes or hours rather than the one or two days needed for standard bacteriological culture.

Countercurrent immunoelectrophoresis (CIE)   CIE is a technique in which patient serum containing unknown antigens is allowed to diffuse through an electrical field and react with various specific antisera. Formation of precipitation zones in the presence of one antiserum shows that the patient sample contains that antigen. Among other applications, CIE can identify a number of viruses, including hepatitis B virus, influenza, the adenoviruses, dengue virus, and other arboviruses.

Latex agglutination tests (LATs)   LATs take advantage of the fact that latex beads can be made to bind specific antibodies. When latex beads are exposed to a sample containing the corresponding antigen, agglutination occurs. These are rapid slide tests that take only a few minutes. These tests have been adapted for identification of the streptococcal groups, diagnosis of rheumatoid arthritis, and determination of pregnancy, to name a few.

Staphylococcal coagglutination tests (SCTs)   SCTs take advantage of the fact that certain protein-rich strains of *Staphylococcus* bind to the Fc portion of some IgGs, leaving the Fab portion free. They thus become inert antibody carriers like the latex beads. When mixed with antigen-containing samples, such as spinal fluid, agglutination occurs. The tests, which require only a two-minute reaction time, may be done on a slide or in a capillary tube. SCT tests are available commercially for streptococcal grouping and identification of the gonorrhea agent.

CELL-MEDIATED IMMUNITY

In cell-mediated immunity (CMI) the target is a cell. It may be a bacterium or a cell harboring an intracellular parasite. It may be a host cell that has become apparently nonself because it bears new antigens of viral or tumor origin. It may be the cell of a tissue graft from an antigenically nonidentical donor. The antigen will be attacked by effector cells, either T-cells or macrophages activated by the immune lymphocytes.

When T-cells respond to antigen, they undergo division (blastogenesis) and differentiation just as B-cells do. In responding, they give rise to several functionally

TABLE 13.5
Subgroups of T-cells

SUBGROUP	CHARACTERISTICS
T-helper cells	Cooperate with B-cell in induction of antibody synthesis Cooperate with T-cell in induction of CMI
T-suppressor cells	Interact with B-cell to suppress development of anti-body-mediated immunity (AMI)
T-cytotoxic cells	Attach to specific antigenic markers on cellular target and lyse it by membrane disruption
T-delayed hyper-sensitivity cells	Attach to specific antigenic marker on cellular target, then release lymphokines that summon and activate macrophages

different lymphocyte subpopulations. These differ in the specific role they play as well as in surface markers. The subpopulations are microscopically identical and populate the same portions of the lymphoid tissues, such as the paracortical region of lymph nodes and marginal areas in spleen. In general, T-cell areas surround B-cell areas.

T-helper cells (Table 13.5) are regulatory cells that cooperate with B-cells in initiating an antibody-mediated response. Helpers may also be essential for development of the T-cells.

T-suppressor cells are also regulatory. They reduce the intensity of AMI responses. They can also moderate CMI by suppressing activity of killer cells. The role of the two regulatory T-cells, helpers and suppressors, is to balance the intensity of the simultaneously developing AMI and CMI responses.

T-cytotoxic cells are effectors also known as killer cells. They can attach directly to an antigen-bearing cell and cause irreversible (lethal) changes.

T-delayed hypersensitivity cells are effector lymphocytes and lymphokine producers. DHS cells contribute directly to macrophage activation. Delayed hypersensitivity reactions are side effects that usually accompany effective cell-mediated immunity.

Lymphokines

Thymus-derived lymphocytes produce soluble products called **lymphokines**. These have distinct physiological effects on surrounding cells. Lymphokines resemble the inflammatory mediators in that they act locally rather than circulating like the immunoglobulins. Lymphokines orchestrate the combined attack of T-effector cells and macrophages on the antigen.

T-cells responding specifically to a variety of antigens release apparently indistinguishable lymphokines. In contrast to antibodies, lymphokines are not directed against the antigen. They are nonspecific. Some of the lymphokines are

Chemotactic factor. This substance is chemotactic for macrophages, and summons them for the degradation and disposal of antigen.

Macrophage inhibition factor (MIF). This substance has two effects. It reduces the movement of macrophages and thus keeps them, once summoned, from moving away. It also increases their phagocytic ability.

Specific macrophage arming factor. This substance is believed to help direct macrophages against specific antigenic targets.

Interferon. This substance, described in Chapter 11, interferes with viral replication in host cells. It is produced by many types of cells in addition to lymphocytes. The immune interferons may have blocking or suppressing effects in immunity.

Interferon; Chapter 11, pp. 324–326

Armed and Activated Macrophages

Recall that the nonspecific phagocytes are effective only against certain types of microbial pathogen. When macrophages are **armed** by contact with lymphokine, their effectiveness is greatly increased. Arming adds new surface receptors for more efficient phagocyte–prey contact. Internally, the macrophage granules increase in number and size. The armed macrophage appears to be target-specific.

Armed macrophages, continuing the lymphokine-triggered differentiation, later become **activated** macrophages. These are even larger in size, highly motile, and packed with enzyme granules. At this point, the macrophage is no longer target-specific but shows enhanced activity against all cellular antigens. Thus in the later stages of a CMI response against tuberculosis, the patient may show enhanced, nonspecific resistance against other intracellular parasites such as systemic fungi.

Induction of Cell-Mediated Immunity

Evidence of response   Antigen entering the tissues or circulation is carried to and lodges in the nearest mass of lymphoid tissue. Contact with antigen stimulates selected T-lymphocytes. Blastogenesis occurs in the thymus-dependent regions, and effector T-cells appear in the circulation within three days, increasing in number up to one week. Later, if antigenic stimulus declines, it appears that the effector cells may be converted into memory cells that continue to circulate. Secondary responses follow a similar pattern and, like AMI, are more rapid and stronger than primary responses.

Cellular activities   T-cell to T-cell cooperation is required. It appears that the same pattern of two different lymphocytes contacting two different determinants on the antigen (the "second signal") is present.

Laboratory Tests of Cell-Mediated Immunity

In contrast to the variety of practical tests for antibodies, there are few tests for functional CMI. Most tests require a laboratory with cell-culture capabilities. Only the skin test is done in routine clinical settings.

Skin testing   If an individual has intracellular infection, cellular immunity develops. The existence of the committed lymphocyte populations can be revealed by intradermal injection of antigen extracted from the organism. An area of reddening and swelling, a **delayed hypersensitivity**, develops around the site after 24 hours. If the individual has not been immunized, there is no response. Skin tests using the appropriate purified antigens can be used in the diagnosis of any infection in which CMI development is an expected result. The tuberculin test is the most common example.

Delayed hypersensitivity; Chapter 14, pp. 417–420

MIF test   The lymphokine MIF (macrophage inhibition factor) is produced by T-effector cells when in contact with antigen. MIF will inhibit migration of macrophages from the mouth of a capillary tube (Fig. 13.15). By testing the ability of an individual's lymphocytes to produce this factor, a general assessment of their ability to mount a CMI response is possible. The mixed lymphocyte reaction (MLR) and cell-mediated lymphocytolysis test (CML), discussed in the next chapter, may also be used to measure the reactivity of an individual's T-lymphocyte population.

(a)

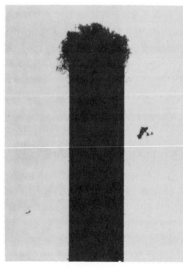

(b)

13.15
The MIF Test   Peritoneal exudate cells (lymphocytes and macrophages) are placed in glass capillary tubes and put into a nutrient medium. Normal uninhibited macrophages gradually migrate out of the mouth of tube to form a circle of cells (a). However, if antigen is present in the culture fluid, the lymphocytes there react by producing lymphokines, including migration inhibitory factor (MIF). This substance restricts the macrophages' movements (b). This test thus detects the presence of an antigen to which the donor's lymphocytes have been sensitized. It can be used to measure the effectiveness of T-lymphocyte responsiveness in an individual. The MIF test indicates in part how well the person can mount a cell-mediated immune response.

IMMUNITY AND PROTECTION FROM INFECTION

Interaction and cooperation between AMI and CMI work to eradicate infections and prevent their recurrence. We will now look at how these cells, substances, and structures actually work together to defend you against microbial challenge.

Acquiring Immunity

The immune state, permanent or transient, can arise by several means. The term **immunization** can be applied to any of these means.

Active immunization   This occurs when the body is presented with an antigen. The body responds by mounting AMI or CMI responses, or both. Because lymphocyte clones have been selected and stimulated, yielding both actively responding cells and memory cells, the response is self-perpetuating. In theory, an active immunity is lifelong. In practice, it may lapse after ten or more years.

    Naturally acquired active immunity results from an actual case of the disease, either clinical or unapparent.

Immunization; Chapter 27

    Artificially acquired active immunity develops following intentional administration of antigen, or **vaccination. A vaccine** is a manufactured product containing an altered natural antigen, such as attenuated or killed infective agent, or a toxoid. The effectiveness of vaccination depends on the amount of antigen given, how effectively it reaches lymphoid depots, and how similar the vaccine antigen is to the native antigen. Many "shots" provide as much protection as naturally acquired immunity; others do not.

Passive immunization   This is a temporary state. Immunity is transferred to a nonimmune recipient in the form of preformed antibodies or injections of lymphocytes. Preformed antibodies are naturally degraded in time; infused lymphocytes do not divide or colonize and eventually die out. Passive immunization is immediate—the recipient gets protection from the moment of transfer. However, the duration of its effectiveness is no more than a few weeks at best.

    The recipient is not stimulated to develop its own response to the antigen against which it is being protected. However, if the transferred material is perceived as foreign, it may elicit an immune response. Repeated passive immunization in the adult almost always provokes severe hypersensitivity reactions eventually.

    Natural passive immunization is critical to the survival of the newborn child. Fetuses and newborns have little immunological capacity of their own. Although the fetus may start IgM production in the last days of gestation, significant antibody does not appear until one to two months after birth. IgG levels may not reach full strength until age four. CMI functions are similarly slow to develop. Immunological protection of the young in the first months of life depends largely on the mother.

    The placental membranes are permeable to maternal preformed IgG. The baby is born with partial immunity to those diseases to which its mother is immune. After

Passive immunity in infancy; Chapter 21, pp. 642–643

TABLE 13.6
Some examples of immunization

TYPE	DISEASE	IMMUNIZING SUBSTANCE
Natural active	Chickenpox	Varicella virus
Artificial active	Pertussis	Killed bacteria
	Tetanus	Tetanus toxoid
	Polio	Live attenuated polio virus
	Pneumonia	Pneumococcal capsule antigen
Natural passive	Virus infection	IgG, IgA, specific against virus
	Diphtheria	IgG, IgA, specific against diphtheria toxin
	Polio	IgG, IgA, specific against virus
Artificial passive	Hepatitis B	Hyperimmune serum—specific IgG, IgM
	Rabies	Hyperimmune serum—specific IgG, IgM

birth, maternal IgG is reinforced by secretory IgA from breast milk, for the duration of the nursing period. In traditional societies in which lactation continues for up to two years, these antibodies bridge the gap between birth and the time when the child's own antibody production becomes adequate. If formula feeding is substituted for breast-feeding, the infant's immunological protection is at best marginal.

Artificial passive immunization can be achieved by injecting hyperimmune serum containing a high titer of preformed antibody. This method can protect high-risk individuals exposed to a serious infection. Children with immunodeficiency risk fatal complications if they develop chickenpox. Antichickenpox globulin given within 96 hours of exposure will abort the infective cycle of the virus. Rabies immune globulin is given to those bitten by a rabid animal. Clinical studies have also been done with passive transfer of CMI. A soluble product, **transfer factor,** extracted from lymphocytes can be used to control severe *Candida albicans* infections.

Participation of Different Kinds of Antibodies

Each class of antibody has its own role in the defense network. Active immunization usually produces antibodies of several classes.

IgA is produced by lymphoid tissues in gut, respiratory tract, and glandular tissue. Some IgA circulates, but most is secreted in saliva, mucus, milk, semen, and other body fluids. IgA protects the mucosa, the major route of entry into the body for pathogens. The saliva of an individual who is immune to diphtheria contains an

Box 13.2

Hybridomas—New Source of Antibodies to Order

The traditional sources of antibodies for research or diagnostic use have been the sera of immunized animals or humans. Such sera do indeed provide the desired antibodies, but they also contain a diverse, unpredictable mixture of antibodies to other antigens, some of which may cross-react, blurring the precision of the test results. Complex absorption and purification techniques are needed before using naturally obtained immune sera; these operations not only increase the cost but reduce the final yield of the desired antibody.

The ideal means of getting specific antibody production would be to collect continuously the immunoglobulin product of a single B-lymphocyte clone—*monoclonal* antibody. This can now be done using hybridoma cells in culture.

B-lymphocytes from the spleen can be isolated in agar and shown to produce specific antibody, but they cannot be continuously cultured. Myeloma cells derived from a bone marrow tumor do produce monoclonal antibodies continuously—the Bence-Jones proteins—but their specificity cannot be selected.

The hybridoma technique employs the controlled fusion of these two lines of cells—a myeloma line and a spleen cell line. First, normal mice are repeatedly immunized with the antigen for which the monoclonal antibody is desired. Then the mice's spleens are removed. The spleen cells, some of which are responder B-lymphocytes, are caused to fuse with the myeloma cells forming hybrid cells—the hybridomas. By selection procedures, those hybridomas that are producing the needed antibody are found and isolated as clones. The investigator can now establish a continuous cell culture that is committed to the production of the single antibody. Because all cells of the clone are genetically identical, their immunoglobulin product has one unique amino-acid sequence and three-dimensional structure.

The applications of the hybridoma technique are almost limitless. One recent report demonstrated the successful production of a monoclonal antibody that binds to a unique tumor antigen found on several types of human leukemia cells. Given the high level of specificity of the immunoglobulin, the authors suggest that this monoclonal antibody has valuable diagnostic potential. Furthermore, it could eventually prove to be directly valuable in immunotherapy of these leukemias.

Source: J. Ritz; J.M. Pesando; J. Notis-McConarty; H. Lazarus; and S.F. Schlossman, 1980, A monoclonal antibody to human acute lymphoblastic leukemia antigen, *Nature* **283**: 583–585.

IgA that neutralizes diphtheria toxin. The feces of those immune to cholera contain an IgA that blocks the bacillus's adherence. Anti-influenza IgA in respiratory tract secretions and antipolio IgA in the gut are far more important in actual protection than are circulating IgG or IgM because they inactivate the agent **before** infection can occur. Researchers are investigating the feasibility of immunizing against influenza by inhalation of vaccine virus that results in localized production of IgA, rather than by injection that produces circulating IgG and IgM.

TABLE 13.7
Biological activities of immunoglobulin

CHARACTERISTIC	IgG	IgA	IgM	IgD	IgE
Distribution	Serum Tissue fluid	Serum Mucous secretions Glandular secretions	Serum Tissue fluid	Serum Lymphocyte surface	Serum Mast cell surface
Complement fixation	+	−	+	−	−
Crosses placenta	+	−	−	−	−
Secreted in milk	−	+	−	−	−
Binds phagocytes	+	−	−	−	−
Roles	Neutralization of toxins, agents in body fluids Protects newborns	Defends routes of entry; preventive	First to appear; effective against organisms in blood	Lymphocyte receptor(?)	Controls parasites; hypersensitivity(?)

IgG is the predominant circulating antibody. It crosses the placenta and protects the fetus. It effectively combines with macromolecular antigens. IgG-antigen complexes are complement-fixing and thus promote complement-mediated immune adherence and phagocytosis.

IgM is the first antibody the infant produces. Because of its multiple combining sites, IgM effectively binds to cellular antigens promoting complement-mediated cell lysis. IgM and IgG have overlapping functions and probably work together to control most infections. However, IgM appears first on antigenic stimulus; it is perhaps more important in convalescence, while the later-appearing IgG prevents recurrence.

IgD's role is still unclear. There is some evidence that it is a component of the surface recognition systems of lymphocytes.

IgE is usually a very small fraction of serum globulins, but in some individuals it may be vastly increased. It is believed to aid in the expulsion of intestinal parasites. IgE antibodies are also responsible for certain types of hypersensitivity.

Participation of Different Kinds of CMI

The antimicrobial response   T-delayed hypersensitivity cells migrate to the antigen location at which they release lymphokines that summon and arm macrophages. These macrophages ingest and destroy the intruders. In the process, inflammatory substances are released; the mopping-up operation is thus surrounded by an area of inflammation. This **delayed hypersensitivity** response is a frequent side effect, as in tuberculosis, and a reliable indicator of effective antimicrobial CMI.

The anticellular response Populations of cytotoxic T-cells are the effectors; they attach to and kill the antigenic cell. Macrophages are also involved. Host tissue destruction is a frequent side effect. The cellular destruction around a viral skin rash is believed to be due to the killing of the host cells harboring virus. By means of the anticellular response, virus replication is halted, intracellular pathogens are exposed to AMI, and tumor growth is (occasionally) arrested. Grafted tissues erode at the margin and are rejected.

Protective Cooperation of the Two Types of Immunity

Protection against bacteria Extracellular bacterial infections (Table 12.3) are controlled almost entirely by AMI. Antibody neutralizes bacterial toxins, opsonizes bacteria for more effective phagocytosis, and activates complement by the classic pathway leading to lysis.

Intracellular bacterial parasites are controlled almost entirely by CMI because antibodies do not enter cells. The antimicrobial response controls acute infections, such as those by *Listeria* or *Brucella*, in which the agent is quickly eliminated. CMI also contains chronic infections such as tuberculosis in which the agent is never completely eradicated; its multiplication is held in check only so long as CMI levels remain high.

Protection against viruses Recall that some animal viruses are cytolytic. They are released from the host cell into body fluids, then penetrate the next cell. During the extracellular period they may be neutralized by antibody. Other viruses either are shed one at a time or move via intercytoplasmic bridges from one cell to another

TABLE 13.8
Immunological reactions that depend on cell-mediated immunity

REACTION	EXAMPLE
Cellular immunity	Eradication of measles virus
Delayed hypersensitivity	Positive skin reaction to tuberculin in tuberculosis cases
Graft rejection	Rejection of mismatched kidney, heart
Graft versus host reaction	Lethal result of transplant of mismatched bone marrow
Autoimmune disease	Lymphocytic destruction of host's own thyroid gland (Hashimoto's disease)

without ever leaving the protection of the membrane. In herpesviruses, measles, and poxviruses, antibody is of little use in arresting disease. Anticellular CMI is directed against the virus-infected tissue, which is destroyed along with the virus. Antibody is useful in preventing viral infections if it can neutralize the virus inoculum prior to its finding refuge in host cells. Persistent virus infections (Table 11.5) succeed in escaping immune surveillance by means as yet unknown.

Protection against fungi The superficial fungi that infect skin are extracellular; they are controlled by nonspecific means and by delayed hypersensitivity. The systemic fungi, such as *Histoplasma*, *Coccidioides*, and disseminated *Candida*, are controlled by CMI mechanisms. Antibodies may be present but they appear to play little role in the control process.

Systemic mycosis; Chapter 19, pp. 566–573

Protection against parasites After most protozoan or helminth (worm) parasites have gained entrance, they are rarely completely eliminated. Containment is the most that immune mechanisms can achieve. Antibodies may assist in killing extracellular parasites, such as trypanosomes. CMI is useful against *Toxoplasma*, which multiplies intracellularly. However, most parasite life cycles have highly resistant resting stages called **cysts.** These persist indefinitely despite high levels of immunity.

There is some evidence that IgE plays a role in the control of helminth infestations. Levels of this immunoglobulin are always elevated in parasitized individuals.

The two types of immune response, antibody-mediated and cell-mediated immunity, work together to prevent and control infection. The humoral branch is highly effective in preventing infection because the invader is met by antibody at all body portals and in all body fluids. Should the agent escape antibody-mediated destruction and gain entrance to host cells, the infected cells will become the target of a cell-mediated response. CMI will destroy the agent's source of nutrients—the host cell—and expose it to destruction.

SUMMARY

1. Specific immune defense depends on a two-limbed system.
2. Both limbs of the immune response need nonspecific defenses to complete the antigen's eradication. The limbs cooperate with and regulate each other. Antibody-mediated immunity develops when lymphocytes produce immunoglobulins that attach to antigen. Once antigen–antibody combination occurs, phagocytic cells and complement dispose of the antigen. AMI is useful against agents on mucosal surfaces, in body secretions, or in the circulation.
3. Cell-mediated immunity develops as other classes of lymphocyte attach to and recognize cellular antigens. These include host cells that harbor intracellular parasites, tumor cells, and noncompatible foreign tissue grafts. Lymphocyte-released lymphokines activate macrophages, and cytotoxic lymphocytes kill cells on contact.

4. Lymphocytes are the basic cell of lymphoid tissue. They originate from an ancestral stem cell and undergo differentiation during embryonic development. During that time, some pass through the thymus and become T-cells, which carry out the CMI response and regulate the AMI response. Others undergo their processing in another primary lymphoid organ, probably the bone marrow. These become B-cells, which produce the AMI response by giving rise to antibody-forming plasma cells. B-cells also appear to regulate CMI responses. Following processing, both types of lymphocyte populate secondary lymphoid tissues in separate but adjacent areas. Lymphocytes migrate continuously from one lymphoid area to another. Both B- and T-cells are in a resting, nondividing state prior to antigenic stimulation.

5. Antigens are substances lymphocytes recognize as foreign. Lymphocytes respond by rapid division and differentiation into enlarged, more active cell populations or clones. Antigens are large molecules containing an antigenic determinant and a carrier.

6. The human body distinguishes between "self" and "nonself." This very subtle distinction is "learned" by the fetus. The clonal selection theory states that, early in development, lymphoid tissues contain many genetically unique cells, each recognizing one antigenic determinant. During fetal life, contact of lymphocyte with its antigen results in its death. All lymphocytes that could attack self do encounter "their" antigen and are wiped out, giving permanent tolerance. After birth, something quite different occurs. Antigen contact has a positive selective influence, stimulating the lymphocyte to divide.

7. There are five classes of immunoglobulins produced by plasma cells. Each has two or more combining sites, variable amino-acid sequences folded into three-dimensional grooves that hold antigen. Other portions of the Ig molecule attach the antigen–antibody complex to phagocytes, activate complement, or assist in transporting the immunoglobulin across the membrane of the secretory cell.

8. Antibody-mediated immunity develops when B-cells are stimulated. B-cells always have a small amount of their specific antibody incorporated as a membrane receptor. This allows them to attach and bind an antigen. The antigen–antibody complex enters the cell and promotes blastogenesis, leading to the production of plasma cells and memory cells. The stimulated B-cells and their progeny form germinal follicles in the cortex of lymph nodes and in other lymphoid regions.

9. Plasma cells release immunoglobulin into serum and other body fluids. Many testing methods are available to detect and measure specific antibodies. We can also use preformed antibodies to identify antigens, as in blood banking.

10. Cell-mediated immunity is induced when antigenic attachment to a specific T-cell receptor leads to blastogenesis and development of several populations of cells. Cytotoxic T-cells kill foreign or antigenically labeled cells on contact. Delayed hypersensitivity T-cells liberate lymphokines on contact with antigen. Lymphokines in turn summon and activate macrophages.

11. Skin testing detects the delayed hypersensitivity side effects of CMI as evidence for the immune state. Other CMI tests detect the ability of lymphocytes to produce lymphokines.

12. AMI and CMI are both induced by active immunization. AMI can be passively acquired by transfer of immunoglobulins. Passive immunity protects the fetus and newborn until its capacity for immune response develops.

13. Prevention of infection depends on secretory antibodies at the body portals and circulating IgG and IgM.

14. Antibodies also control infections by extracellular parasites and viruses with an extracellular phase between rounds of replication. CMI is primarily responsible for dealing with systemic fungi, and those bacteria, viruses, and parasites that multiply intracellularly. AMI and CMI must both be operating at full efficiency and in balance with each other to provide functional defense.

Study Topics

1. It has been shown that the eight-week-old, breast-fed human infant to whom oral, live-virus, polio vaccine is given does not show an antibody response equal to the formula-fed infant. Why?

2. Administration of a killed-virus vaccine tends to result in development of antibody-mediated immunity, whereas an attenuated live-virus vaccine stimulates cell-mediated immunity. Why?

3. An individual genetically lacking in B-cell functions will be likely to suffer recurring infections by what class of agents? See Table 12.3.

4. Which portion of the immune response is compromised in the person lacking a normal complement pathway?

5. Explain how clonal selection can lead to both immune tolerance and immune response.

6. Be able to compare the pathways of lymphocyte differentiation leading to the fully responding B- and T-lymphocytes.

7. Compare the functions of lymphokines, inflammatory mediators, and antibodies.

8. How does interferon function as a lymphokine?

Bibliography

Bigley, Nancy J., 1975, *Immunological Fundamentals.* Chicago: Yearbook Medical Publishers.

Burnet, F. M. (ed.), 1975, *Immunology.* A collection of readings from the *Scientific American.* San Francisco: W. H. Freeman.

Relevant to this chapter:

Burnet, F. M., 1976, Tolerance and unresponsiveness, pp. 114–119.

Cooper, M. D., and A. R. Lawton, III, 1974, The development of the immune system, pp. 58–71.

Jerne, N. K., 1973, The immune system, pp. 49–57.

Golub, Edward, 1977, *The Cellular Basis of the Immune Response: An Approach to Immunobiology*, Sunderland, Mass.: Sinauer.

McCluskey, R. T., and S. Cohen (eds.), 1974, *Mechanisms of Cell-mediated Immunity.* New York: Wiley.

Roitt, Ivan, 1977, *Essential Immunology* (3rd ed.), Oxford: Blackwell Scientific Publications.

Rose, N. R., and H. Friedman (eds.), 1980, *Manual of Clinical Immunology* (2nd ed.) Washington, D.C.: American Society for Microbiology.

14

Dysfunctional Immunity

We have seen that the immune system is a complex, powerful weapon for defense against microorganisms that exert unremitting ecological pressure. On the rare occasion when primary and secondary defenses fail, survival hangs on the effectiveness of immunity. In its absence, we are helpless to prevent microbial colonization.

Unfortunately, our most essential defender can become our most traitorous attacker as well. The immune system may betray us simply by not reacting—resulting in immunodeficiency. It often fails, too, to rid us of tumors. Immune effectors are destructive factors; if they are present in excess, or misdirected, for example, at a life-saving, grafted kidney or at our own irreplaceable organs, they can cause great trouble.

IMMUNODEFICIENCY

Immunodeficiency is far from uncommon. For example, about one in 500 individuals of European descent do not produce secretory IgA. The lack of IgA usually does not cause major detrimental effects. The failure to produce IgA is a recessive genetic trait. Most genetic immunodeficiencies are fortunately much rarer. Yet everyday stresses are such that each of us, during life, has brief or extended periods of marginal immunological protection.

Clinical Evidence of Immunodeficiency

Recurring infections   Frequent or uncharacteristically severe infection is a primary sign of immunodeficiency. In infants, one infection crisis rapidly follows another. Illnesses that are normally mild are protracted and debilitating.

The infective agents are frequently not only pathogens (Table 12.2) but also opportunists. Pneumonia caused by *Streptococcus pneumoniae* or the protozoan *Pneumocystis carinii* occurs. Severe generalized *Herpes simplex* lesions or extensive colonization by the fungus *Candida albicans* are common. Appropriate antimicrobial drugs are only temporarily effective.

Opportunists; Chapter 12, p. 347

Increased rate of malignancy   Immunodeficient children develop malignancies at a rate 100 to 1,000 times higher than do normal children. Adults on immunosuppressive drug regimes share a similar risk. The tumors are characteristically those of the lymphoreticular tissues.

Primary Immunodeficiency Diseases

A primary immunodeficiency is one present at birth. Pedigree analysis usually suggests that such diseases follow the rules of Mendelian inheritance. The cause of a primary immunodeficiency is structural; the deficient function simply does not develop.

Types of deficiency   Anatomical study may reveal the complete or partial lack of a key lymphoid organ such as the thymus. Alternatively, all anatomical structures may be present but, because of incomplete differentiation of stem cells, certain tissues lack mature functional cells. Both organs and their cells may also be present and appear to be normal, but antigenic stimulus does not lead to a proliferative response. Unresponsiveness may also be associated with a specific metabolic defect.

There are many different primary immunodeficiencies, and more are being described all the time. The primary immunodeficiencies are manifested in inadequacies of B-cell function, T-cell function, or some combination of both. These states have often been referred to as "experiments of nature." They help researchers understand the normal immune system, just as auxotrophic mutants of bacteria helped investigators map out normal bacteria's metabolic pathways.

Auxotrophs; Chapter 9, p. 269

DiGeorge's syndrome, a T-cell deficiency   Abnormal embryological development produces rudimentary or absent thymus and parathyroid glands. The infants resulting from this abnormal development are deficient in all T-cell responses, including cell-mediated immunity and graft rejection. Without T-helper cells, they cannot

TABLE 14.1
Primary immunodeficiency diseases

DISEASE	CMI RESPONSE	AMI RESPONSE	CLINICAL FEATURES
DiGeorge's syndrome	Poor	Poor	Thymus does not develop; abnormal calcium metabolism due to parathyroid defect; recurrent viral, fungal, and intracellular bacterial infection
Bruton's agammaglobulinemia	Normal	Poor	Serum Ig low; recurrent bacterial infections, rheumatoid arthritis
IgA deficiency	Normal	Decreased IgA level	Respiratory and gastrointestinal infections; atopy and autoimmune disease
Severe combined immunodeficiency	Absent	Absent	Lymphoid tissues devoid of lymphocytes; severe, overwhelming infection
Common variable hypogammaglobulinemia	Variable	Poor	Plasma cells not produced; recurrent infection, intestinal malabsorption

mount a directed antibody response against most antigens even though B-cells are available. Immunoglobulins are present but are not protective. The infant suffers recurrent viral and fungal infections. The outlook is bleak, although transplantation of fetal thymus has been tried with some success.

Bruton's agammaglobulinemia, a B-cell deficiency   This disease, the result of a recessive sex-linked gene, occurs in males. Their immunoglobulin levels are extremely low and their lymph nodes lack germinal centers and plasma cells. These children suffer from bacterial infections caused by extracellular agents, but they deal quite well with intracellular agents including viruses. Survival is possible with the use of passive immunization and antibiotics as necessary. Active immunization, however, fails to produce an antibody response.

Severe combined immunodeficiency (SCID)   In this rare disease, both effector limbs are nonfunctional. The lymphoid stem cell line failed to differentiate. The few lymphocytes present show no immunological responsiveness and lymphoid organs are vestigial or absent. Some patients have a deficiency of the enzyme adenosine deaminase (ADA). Why this should cause the stem cell failure, if it does, is unknown. SCID patients are kept alive by residence in a sterile environment. Compatible bone marrow or fetal liver transplants have in some cases reconstituted their lymphoid tissues successfully.

Secondary Immunodeficiency

In Chapter 12 we discussed the stresses that compromise nonspecific defense. These stresses all affect the immune system too. However, the immune system is vulnerable to additional challenges.

Malnutrition   Protein starvation (kwashiorkor) reduces the amount of free amino acids available for protein synthesis; serum protein levels drop. Immunoglobulins, of course, are proteins, so antibodies fall below protective levels. This condition is seen in starving children and adults. It may also develop in elderly individuals, alcoholics, the mentally ill, or others who do not take care of themselves and lack an adequate diet.

Infection   The challenge of fighting an infection often depletes reserves of lymphocytes and granulocytes. Viruses such as measles actually have a cytotoxic effect on lymphoid cells. Syphilis, leprosy, and malaria depress cell-mediated immunity.

Proliferative disorders   Proliferative disease of lymphoid or reticuloendothelial cells ranges from a relatively benign disease, such as infectious mononucleosis, through the slowly progressive myelomas to the malignant leukemias and Hodgkin's disease. Immunodeficiency commonly results. Although lymphocytes or macrophages may be present in excess, they are not immunologically competent because transformation is associated with a loss of mature function.

Immunosuppressive therapy   **Immunosuppressive drugs** may be used to treat pathological overactivity of the immune system, or to suppress graft rejection. Such treatment is becoming increasingly common. Recent advances make more selective treatment possible. Some immune function is maintained during treatment, and it recovers after treatment is stopped. If immunosuppression is indefinitely prolonged, patients eventually succumb to acute infection. They also have many persistent infections and secondary malignancies.

HYPERSENSITIVITY

Hypersensitivities are abnormal physiological states. They result from combination of antigen with immune effector at the wrong time, in the wrong place, or in the wrong proportions. The damage may be either transitory or chronic.

The known hypersensitivities are placed in four categories on the basis of the mechanism responsible for symptoms (Table 14.2).

Types I, II, and III are called **immediate hypersensitivities** because symptoms appear within minutes after the sensitized individual contacts the antigen. The immune effectors are immunoglobulins, and sensitivity is therefore transferable by serum.

Type IV is **delayed hypersensitivity**, and it is cell-mediated. Symptoms appear 24 hours or more after contact with antigen. The symptoms are mediated by lymphokines from T-lymphocytes. Type IV hypersensitivity is passively transferable only by lymphocyte transfer.

The distinction between immunity and hypersensitivity is not clear. The mechanisms are much the same. For example, hypersensitivity develops only after the indi-

TABLE 14.2
The four types of hypersensitivity mechanism

	TYPE I	TYPE II	TYPE III	TYPE IV
Name	Classic	Cytotoxic	Immune complex	Delayed
Immune effectors	IgE	IgG, IgM	IgG, IgM	T-lymphocytes
Accessory factors	Mast cells, anaphylactoid mediators, eosinophils, eosinophil mediators	Complement	Complement, inflammatory mediators, eosinophils, neutrophils	Lymphokines, macrophages
Antigen	Soluble or particulate	Cell surface antigen	Soluble protein	Cell surface antigen
Maximum reaction time for response in sensitized individual	30 min	Variable	3–8 hr	24–48 hr
Symptoms relieved by antihistamine	In some cases	No	No	No
Symptoms relieved by corticosteroid	In some cases	Yes	Yes	Yes
Passive transfer by	Serum	Serum	Serum	Lymphocytes or transfer factor

vidual has had an initial immunizing or **sensitizing** contact with antigen to which he or she mounts a response. Subsequent contacts with antigen are **provocative**—the combination of immune effectors with antigen leads to adverse results. The term **allergen** is often used for an antigen that elicits hypersensitivity.

In many hypersensitivities, the allergen, although foreign, is **not** a serious threat to health. Thus we become hypersensitive to cat hair, pollen, poison ivy, or cosmetics. Human beings also vary widely in their tendency to become hypersensitive to these harmless substances, whereas most normal human beings respond positively to truly pathogenic antigens. The conventional usage is, very simply, to call a response a hypersensitivity if it is more harmful than helpful. If its protective effects outweigh its hazards, it is an immunity.

Type I: Anaphylaxis and Atopy

IgE; Chapter 13, pp. 381–383

Mechanism The mediators of Type I reactions are specific IgE antibodies. These are produced in significant amounts only by certain individuals. IgE attaches to the surface of the highly granular **mast cells.** Mast cells are found in connective tissue underlying epithelial surfaces in skin, respiratory tract, gastrointestinal tract, and elsewhere. The mast cell's ancestry is unclear. Some consider it to be a nonspecific

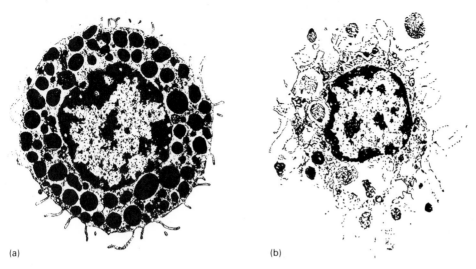

(a)   (b)

14.1
Mast Cell Degranulation (a) A mast cell. The cytoplasm is filled with large electron-dense granules. These contain histamine and other inflammatory substances. (b) A mast cell immediately after the allergen has bound to its surface by specific IgE. The granules are being dramatically released into the environment. Their contents will cause the symptoms of a Type I hypersensitivity.

basophil; others believe that it is a specific, distinctively developed form of lymphocyte. Its granules contain pharmacologically active mediators, including histamine, similar to the inflammatory substances. The mediators' effects are countered by specific "antidote" substances released by eosinophils.

When antigen touches the surface of an antibody-primed mast cell, it is bound by protruding regions of the IgE molecule. Binding induces the mast cell to **degranulate** (Fig. 14.1), releasing its mediators into tissue or the blood stream (Table 14.3). The mediators cause an immediate response. The effect does not last long because the active substances are rapidly degraded enzymatically or counteracted by antidote substances.

Anaphylaxis   Anaphylaxis, the most severe Type I hypersensitivity, occurs when antigen is introduced systematically into a sensitized host. A bee sting (a natural hypodermic injection), or injected medication may provoke anaphylaxis. Antigen circulates rapidly to many mast cell-containing areas, and a large mediator dose is suddenly released into body fluids. The first symptoms are apprehension, flushed skin, itching, and nausea. In human beings, capillaries dilate, causing a sharp drop in blood pressure. Simultaneously, a slow contraction of smooth (involuntary) muscle fibers occurs in the bronchioles and elsewhere. The filled lungs cannot be

TABLE 14.3
Clinical expressions of hypersensitivity

TYPE	RESPONSE
I	Anaphylactic response to drugs, foreign protein, insect stings
	Asthma—bronchial spasm, pneumonitis
	Atopic allergies—hay fever, food intolerance, urticaria
II	Transfusion reaction
	Erythroblastosis fetalis
	Hyperacute graft rejection
	Autoimmune disease, e.g., idiopathic thrombocytopenic purpura
	Drug reaction (if drug is bound to cell surface)
III	Arthus reaction to injected or inhaled antigen
	Serum sickness
	Chronic graft rejection
	Drug reaction (if drug is bound to serum protein)
	Autoimmune disease—rheumatoid arthritis
IV	Granulomatous response to infection
	Positive Mantoux test
	Acute graft rejection
	Autoimmune disease—Hashimoto's disease
	Contact dermatitis
	Tumor immunity

emptied because of bronchial constriction. Asphyxiation and cardiac failure follow. A severe episode may be survived if rapid administration of adrenalin and support are provided. The survivor returns to normal within a short time.

Penicillin injections and insect stings are the most common causes of death by anaphylaxis. An estimated one in six individuals is **potentially** hypersensitive to penicillin; however, only a small fraction of these become sensitized. Wise precautions include reviewing the individual's medical history and observing the individual after injection.

Atopic allergy   This is the localized expression of the Type I reaction. The common complaint "hay fever" is an example. The antigen, usually plant pollen, is inhaled and absorbed to the respiratory mucosa, causing degranulation of underlying mast cells. The results are excessive mucus production, localized epithelial swelling and, in **asthma**, bronchial smooth muscle spasm. Coughing, sneezing, and difficulty in breathing persist as long as antigen is present in the environment. Symptoms are relieved by **antihistamines**. Skin and digestive atopies follow a similar mechanism (Fig. 14.2). The only difference between atopic allergy and anaphylaxis is one of

Compare with inflammatory response; Chapter 12, pp. 368–369

TABLE 14.4
Interaction between mast cells and eosinophils in Type I hypersensitivity

MAST CELL		EOSINOPHIL	
SUBSTANCE	BIOLOGICAL EFFECT	SUBSTANCE	BIOLOGICAL EFFECT
Eosinophil chemotactic factor (ECF-A)	Summons eosinophils	—	Eosinophils migrate to source
Histamine	Vasodilation, increased permeability	Histaminase	Breaks down histamine
Slow-reacting substance (SRS-A)	Slow contraction of smooth muscle	Aryl sulphatase	Counteracts effect of SRS-A
Platelet-activating factor (PAF)	Causes platelets to release clotting factors	Phospholipase	Counteracts effect of PAF
Serotonin	Similar to histamine	—	
Heparin	Anticoagulant	—	
Prostaglandins	Role uncertain		

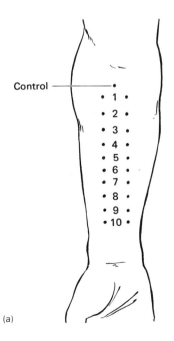

Control

• 1 •
• 2 •
• 3 •
• 4 •
• 5 •
• 6 •
• 7 •
• 8 •
• 9 •
• 10 •

(a)

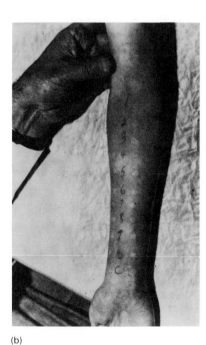

(b)

14.2
Skin Testing for Allergy (a) Droplets of solutions containing minute amounts of different pure allergenic substances are injected into the skin of the inside forearm according to a standard pattern. (b) If the patient is allergic to an antigen, a reddened area forms at the site of its injection. The responses are compared to a control spot where only sterile saline was injected.

degree. In the exquisitely hypersensitive individual, localized contact with antigen has on occasion provoked symptoms of anaphylactic severity.

Atopic individuals can sometimes be **desensitized.** They receive a series of injections containing small amounts of the antigen to which they are allergic. The procedure has variable success. The mechanisms of successful desensitization are not known. They may depend on induction of specific IgG as a blocking antibody, or development of suppressor T-cells.

Type II: Cytotoxic Response

Complement function;
Chapter 12, p. 364

Mechanism   Antigen-labeled cells are destroyed by combination with specific IgG and complement. *In vitro*, complement-mediated lysis is the visible outcome. *In vivo*, complement-mediated immune adherence produces phagocytic killing.

Examples of Type II reactions   Type II interaction destroys incompatible red blood cells following a mismatched transfusion. Released hemoglobin raises blood viscosity and places stress on the heart and kidneys, sometimes fatally. **Erythroblastosis fetalis,** the attack of maternal anti-Rh IgG on fetal cells, is another Type II reaction. Cytotoxic hypersensitivity is also responsible for hyperacute graft rejection and some types of autoimmune diseases, to be discussed shortly.

Type III: Immune Complex Disease

Mechanism   **Immune complexes** are combinations of soluble antigen, antibody, and complement. The complexes may circulate or be deposited in tissue. Tissue damage occurs wherever the complexes localize. Immune complex disease is an extension of normal protective antigen–antibody combination; it becomes abnormal when there are large antigen or antibody excesses. The complexes formed exceed the phagocytes' disposal powers. The antigen is usually a protein. The protein may be a carrier for a haptenic molecule such as a drug. The antibody will be IgG or IgM, both of which are complement-fixing.

When complement is fixed, the reaction cascade causes acute inflammation at the site at which the complexes are deposited. The reaction subsides within a few days, although the healing of resulting lesions may take much longer.

The Arthus reaction is a localized inflammation in which complexes have been deposited in small blood vessels. It often leads to PMN infiltration and extensive necrosis at the site of antigen immobilization. This may occur in muscle on repeated administration of tetanus toxoid. In **farmer's lung disease,** an Arthus reaction in lung tissue is provoked by repeated inhalation of antigenic bacterial and mold spores.

Serum sickness follows injection of serum products containing foreign protein, such as a series of passive immunizations. Although serum proteins from another species such as the horse sensitize readily, even human globulins may eventually sensitize.

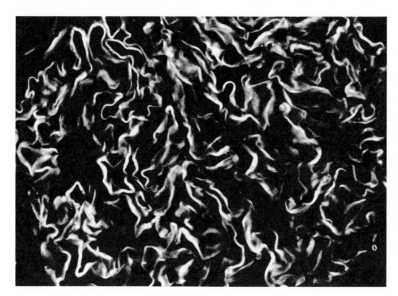

14.3
When antibody binds to antigen in the kidney, or when immune complexes are deposited in the kidney, they activate the complement system leading to localized cellular destruction. This is called glomerulonephritis. Indirect fluorescent antibody technique has been used to make this picture. Anti-kidney antibody bound to the glomerlar membrane has been stained with a fluorescent anti-human IgG antibody. This reveals the locations at which damage is occurring.

To be effective in their protective role, large doses of immune globulins must be given. They persist for a period long enough to sensitize. About 6–8 days after a first injection, sufficient antibody has been produced for significant immune complex formation. If the individual has been previously sensitized, however, the reaction will develop more quickly.

The antigen is present in large excess over available antibody. Circulating complexes form and drift. They cause inflammation that makes it possible for them to infiltrate the walls of blood vessels. Signs of inflammation—fever, swollen lymph nodes, rashes, and painful joints—appear. Complexes may infiltrate the glomerular bed of the kidney, causing immune complex **glomerulonephritis** (Fig. 14.3). This is a sequel to infections by *Streptococcus pyogenes* and persistent viruses. It is also a feature of certain immunodeficiency and autoimmune conditions.

Type IV Hypersensitivity–Delayed

Delayed hypersensitivity is cell-mediated and transferable only by lymphocytes. In some cases (the so-called allergy of infection), it is an unavoidable side effect of cell-mediated immunity's efforts to control the agent. Contact sensitivity represents an attempt to destroy cells to which chemicals such as poison ivy oil, heavy metals, or drugs such as penicillin have become coupled.

Mechanism   Tissues in which delayed hypersensitivity is occurring are infiltrated by T-lymphocytes. These release lymphokines that summon and activate macro-

Lymphokines; Chapter 13, pp. 396–397

phages. Antibody and inflammatory mediators are of reduced significance. The DHS response destroys the target cells. A surrounding **granuloma** containing lymphocytes, macrophages, and giant cells may form.

Delayed hypersensitivity in infection   An intracellular infective agent is eliminated by CMI attack on the harboring cell. Thus tissue destruction is inevitable, its severity determined by the persistence and distribution of the agent. Elimination of measles virus from skin produces a minor rash, quickly healed. By contrast, the primary tubercular lung lesion may develop into a cavity. Immune lymphocytes, macrophages, target cells, and "innocent bystander" cells are all destroyed in the area in which the agent has lodged and multiplied. In arrested cases, the cavity is walled off with calcium deposits without further damage, although the healed lesion persists for life. If arrest does not occur, lung destruction advances. This is not because of microbial products. Rather it is because of active but ineffective CMI. Large areas of lung tissue may be destroyed.

The skin test response   The **Mantoux** tuberculin test demonstrates reactivity to tuberculosis antigen. It uses an accessible site, skin, where cellular destruction will be limited and quickly reversible. Intradermal injection of antigen is followed within a few hours by the appearance of a reddened circle. Mononuclear cells infiltrate antigen-containing tissue creating an area of **induration** or hardening (Fig. 14.4). The reaction peaks after 24 to 48 hours, then gradually fades. The zone diameter may indicate the intensity of reactivity. In individuals actively immunized with the BCG tuberculosis vaccine, the hypersensitivity response may be strong enough to pro-

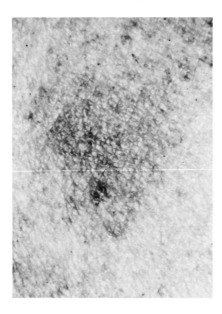

14.4
Mantoux test   This is a highly accurate test for delayed skin hypersensitivity to antigens of the tuberculosis organism. A purified protein derivative (PPD) from *Mycobacterium tuberculosis* is injected intradermally. A positive reaction is the formation of a large, reddened, hardened area at the site of injection. The size of the zone increases with increasing degrees of hypersensitivity.

TABLE 14.5
Common allergens

INGESTED	INHALED	INJECTED
Grains	Tree pollen	Bee venom
Nuts	Grass pollen	Wasp venom
Eggs	Weed pollen	Hornet venom
Fruits	Flower pollen	Yellow jacket venom
Milk products	Bark dust	Penicillin
Aspirin	Mold spores	Allergen extracts
Penicillin	House dust	ACTH
Hormone preparations	Insecticide	Heroin
Isoproterenol	Face powder	Insulin
	Cat dander	
	Dog dander	
	Horse dander	
	Feather dust	
	Microorganisms vaporized from humidifiers	
	Cocaine	
	Irritant smoke	

duce ulceration. A positive test can be interpreted to mean that the patient is presently harboring the tuberculosis organism or has experienced infection by the agent.

Skin tests may be used to demonstrate delayed hypersensitivity to any antigen derived from an intracellular infective agent. Other examples are skin tests to detect sensitivity to the agents of leprosy, histoplasmosis, or coccidioidomycosis.

Contact dermatitis   This term is used to describe delayed hypersensitivity to environmental antigens rubbed into the skin. A wide variety of materials may sensitize, usually after prolonged exposure. Workers in industry become sensitized to chemicals, bakers become sensitized to nutmeg or cinnamon, and photographers become sensitized to processing solutions. Most of us can become sensitized to the oils in the leaves and stems of poison ivy and oak.

When the sensitized individual gets the antigenic substance on skin, antigen becomes absorbed to epithelial cells. The area is invaded by immune T-lymphocytes, and reddening, induration, weeping of serous fluids, and tissue breakdown follow. Itching may be intense, but scratching is harmful because it further damages tissue, enhancing the inflammation.

Treatment   Anti-inflammatory drugs such as the corticosteroids control delayed hypersensitivity. Antihistamines are of little use. Steroid therapy is immunosuppressive; it should not be used when the hypersensitivity reflects an active CMI response that is fighting an ongoing infection. Steroids can control the cell-mediated aspects of "first-set" graft rejection and autoimmune disease.

Anti-inflammatory drugs; Chapter 12, p. 369

Immunity in Contrast to Hypersensitivity

In concluding this discussion of hypersensitivity, let us emphasize that the distinction between immunity and hypersensitivity is not based on the antigenic substance itself, but on the route of administration, the dosage, the timing, the individual, and all sorts of unknown factors.

Penicillin, for example, can be administered orally, by injection, or in the form of a topical ointment. For most people, contact (by any route) gives the desired therapeutic effect, with no immune manifestations. Some persons develop IgE antibodies. These can produce fatal anaphylaxis if penicillin is later injected. If the penicillin is taken orally, gastrointestinal atopy will result. In other individuals, it is IgG and IgM antibodies to penicillin that develop. In these persons, penicillin injection might produce Type II cytotoxicity or Type III "serum" sickness. Health personnel who handle penicillin preparations every day may develop contact dermatitis, a Type IV hypersensitivity. Nonprescription medications like aspirin and illegal substances like heroin may also produce a similar range of immunological states.

TRANSPLANTATION

The replacement of failing organs with healthy ones is a long-standing medical goal. As surgical techniques advance, more graft procedures are tried. Many fail, but not because the surgeon's hands lack skill. Rather, our bodies' efficient attack on nonself antigens, so vital in day-to-day survival, poses a major barrier to transplantation.

An **autograft** is tissue moved from one area of an individual's body to another. Skin grafts may be moved from one area of the body to replace burned tissue; bone transplants from the tibia may be used for facial reconstruction. An **isograft** is tissue transferred from one identical twin to another. Autografts and isografts are always immunologically compatible. They fail only because of technical problems.

Allografts are tissues transferred from one member of a species to another. Kidney transplantation is usually an allograft. Although it is not impossible that these transplanted organs will be antigenically identical, the probability is low unless the donor and recipient are related. Genetically unmatched grafts work only if the recipient's immune system is suppressed or absent. **Heterografts** are transfers of tissue between members of different species. Pigskin is used to form a temporary covering over burn wounds until autografting can be done.

Histocompatibility

In vertebrates, each species possesses a **major histocompatibility complex** (MHC). This is a cluster of genes concerned with the regulation of the immune response, synthesis of complement factors, and antigenic labeling of tissues. The human MHC, referred to as the HLA (histocompatibility locus antigen) region, contains four principal genetic loci. Of these, HLA-A and HLA-B control the synthesis of tissue antigens that evoke strong serological responses. HLA-C antigen evokes a weaker

14.5
(a) The HLA region in the human being. The four HLA loci are closely linked regions of chromosome b. They were lettered before their exact order was determined. The numbers are the designations of some of the alleles at each locus. On a given chromosome, only one allele would be present. The chromosomal region would bear four genes, one each of the possible alleles. (b) Possible genetic structure of a human male and female. Since human beings are diploid (that is, they have paired chromosomes), two HLA regions are shown for each. Each person received one chromosome from the mother and one from the father. Human beings are very diverse genetically; of the eight alleles, only one is shared (Dw_1) in this hypothetical example. (c) When this couple has children, each child will receive one maternal and one paternal chromosome. This produces four different possible combinations. Each child shares 50 percent of its HLA information with each parent, but parent and child will **not** be identical. (d) Each child has a 50 percent chance of sharing half its genetic material with another sibling, but a 25 percent chance of sharing no information. Each child also has a 25 percent chance of being HLA identical to an existing sibling because he or she gets the same chromosome combination.

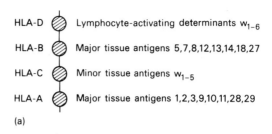

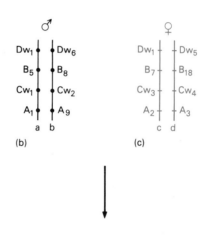

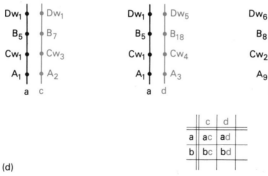

(d)

immune response. HLA-D governs lymphocyte activation; its roles are regulatory. In order to have a perfectly matched graft, one that stimulates *no* immune response in the recipient and is readily accepted, donor and recipient must match for all HLA loci.

The histocompatibility antigens   Each of the four genetic loci has several alternative genetic forms or alleles. Because the loci are closely linked, the four alleles present on a chromosome tend to be inherited together (Fig. 14.5). Each gene specifies a

TABLE 14.6
Currently used grafting procedures

GRAFTED MATERIAL	SOURCE	SUCCESS RATE
Heart	Allograft	Short-term survival; only ~10% >2 yr*
Kidney	Isograft	Almost always successful
	Allograft	Good success if live donor
		Related donor and/or closely matched to recipient
		Moderate success if cadaver kidney
Liver	Allograft	Short-term only in human beings; in pig very successful
Skin	Autograft or isograft	Excellent success
	Allograft	Moderate if exactly matched
Bone marrow	Isograft	Excellent success
	Allograft	Must have perfect match to avoid graft-versus-host response
Cornea Cartilage Tendon	Allograft	Immunologically privileged, excellent success
Bone Heart valve	Allograft Devitalized tissues	Excellent success
Aorta sections	Synthetic material	Excellent success; preferable to natural tissue

*Survival rates for all grafted tissues vary widely among medical centers.

different tissue-bound antigen. The antigens are immunoglobulin-like and express their genetic uniqueness as a unique amino-acid sequence. The antigens are inserted into the membrane of body cells. Some cells, such as lymphocytes, have many antigenic markers. Others, such as red blood cells, have very few markers. This is why a blood transfusion, which is really a transplant, is possible at all.

When donor tissue is introduced into a recipient, the recipient produces an immune response that destroys cells labeled with any marker not shared by the recipient, i.e., cells that are **nonself.**

Tissue matching   The antigens of proposed donor or recipient cells can be identified by exposing them to test antisera produced in experimental animals. There are regional and national organ banks, and registries of potential donors and recipients. If the HLA antigens can be matched, then two other tests may be done to verify compatibility.

The **mixed lymphocyte reaction** (MLR) mixes donor and recipient lymphocytes under cell structure conditions. It detects whether or not donor lymphocytes provide a strong enough antigenic stimulus to induce blastogenesis in recipient lymphocytes. If the recipient lymphocytes respond in the test tube, they will also respond in the body, and the tissue will be rejected.

The **cell-mediated lymphocytolysis test** (CML) mixes target (i.e., donor) cells (labeled with radioactive chromium isotope) with recipient lymphocytes. If they are antigenically stimulated, the recipient lymphocytes will kill target cells. Cytolysis releases the label, which can be measured, indicating the degree of cell injury. Again, target cell killing is a bad predictor for the future of such a graft.

It is rare indeed that donor tissue will pass all tests with a perfect score that indicates the tissue will be completely nonantigenic if surgically implanted. These tests give a useful indication of the degree of mismatch and thus of the severity of the rejection episodes to be expected.

Privileged sites   Certain tissues, such as the cornea of the eye, are not extremely vascularized. Thus an antigenic grafted tissue has no contact with lymphoid tissue and does not sensitize. Mismatched tissue can be implanted without any problem. An anatomical site protected from immunological surveillance is said to be privileged.

The Rejection Mechanisms

When implanted tissue is rejected, one or a combination of the four hypersensitivity mechanisms is at work (Table 14.7). There are several types of rejection episodes.

Acute rejection   Acute or first-set rejection develops after several weeks. The patient first becomes sensitized, then effector cells begin the attack by a Type IV mechanism. During the rejection episode, lymphocytes migrate to and accumulate in the graft, releasing lymphokines and promoting macrophage activity. Gradual

TABLE 14.7
Balance of factors influencing graft survival

POSITIVE INFLUENCES	NEGATIVE INFLUENCES
Development of tolerance to the graft	Cytotoxic T-cells
Production of blocking antibodies	Activited macrophages
Immunosuppressive therapy	Antibody-mediated Type II cytotoxicity
	Immune complex formation

Some or all of these factors may be present to varying degrees. The continued survival of a graft depends on the changing balance among them.

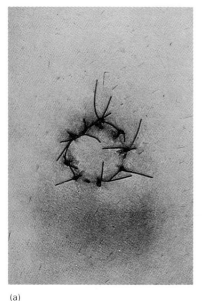 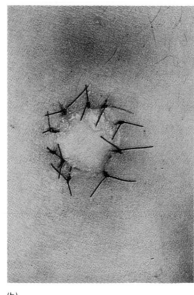 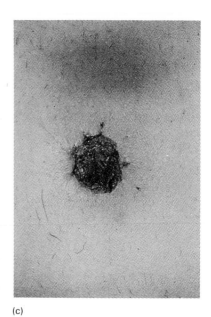

(a)   (b)   (c)

14.6
Skin Grafts   (a) The success or failure of a graft procedure depends largely on whether or not blood circulation is established between the graft and the graft bed. When this is successful, the graft rapidly attaches and blood flows from the graft bed into the grafted tissue. (b) If there is a rejection response, the immunological effects block the circulation. The grafted tissue appears pale due to the lack of blood flow. It will shortly lose viability and will be sloughed (c).

destruction of graft tissue occurs. Antibodies also develop, and although there may be some antibody-mediated cytotoxicity, antibody may also promote graft survival by blocking T-cells.

The graft may be irreversibly devitalized and slough off, or the response may subside so the tissue can survive (Fig. 14.6). There may be repeated episodes of acute rejection during the life of a graft.

Hyperacute rejection   This problem, called second-set rejection, occurs (rarely) if the recipient has preexisting antibodies to donor tissue. These antibodies may have been stimulated by blood transfusions or previous transplants, or they could be cross-reacting antibodies originally induced by some other stimulus. The mechanism is a Type II hypersensitivity involving the combination of IgG or IgM antibodies with the donor cells and complement. The implanted tissue is rapidly (in minutes to hours) infiltrated by PMNs and isolated from the circulation by fibrin clots. Once isolated, the grafted tissue rapidly dies.

Chronic rejection   Chronic rejection is a progressive deposition of immune complexes in the endothelial layers of the arteries that feed the graft. Eventually the blood supply fails. Chronic rejection is a Type III reaction and it threatens grafts that have survived acute rejection.

The recipient seldom develops true tolerance for the foreign antigens. Instead, the body keeps trying to get rid of the implanted material. The final outcome depends on the balance between those immunological factors that favor and those that work against retention.

Immunosuppression Immunosuppression treatment is usually essential for graft survival. Commonly used treatments are presented in Table 14.8.

TABLE 14.8
Immunosuppressive therapy

THERAPY	RESULT	LIMITATIONS/ SIDE EFFECTS	USE
Irradiation of bone marrow, lymphoid tissue	Destroys stem cells, reduces numbers of all blood cells	Very general	Not used in human beings
Irradiation of blood in tubing outside body	Kills circulating lymphocytes	Reaches only small percentage of immuno-competent cells	Rarely used in human beings
Azathioprine	Interferes with purine synthesis; cytotoxic; prevents lymphocyte responses		Renal transplants Autoimmune disease
Cyclophosphamide	Alkylates and cross-links DNA; antimitotic	Leucopenia	Rheumatoid arthritis SLE nephritis
	Prevents lymphocyte response	Bone marrow depression	Renal and bone marrow grafts
	Inhibits humoral response	Toxicity to other rapidly dividing tissues	
Methotrexate	Folic acid antagonist Antimitotic Lymphocytes inhibited		Occasional use in renal transplant
Prednisone	Anti-inflammatory; suppresses T-cell function Suppresses antibody formation	Hormone imbalance Electrolyte disturbance	Atopic allergy Types I, III, and IV hypersensitivity Renal transplant Most autoimmune states
Antilymphocyte globulin (ALG)	Cytotoxic for lymphocytes; major effect on CMI	Types I and II hyper-sensitivity May contain antibodies against RBCs, platelets, and kidneys	Renal and bone marrow transplants

Antilymphocyte globulin (ALG) is derived from the serum of horses immunized with human T-lymphocytes. ALG temporarily reduces the number and effectiveness of recipient lymphocytes. Steroids, especially **prednisone,** are anti-inflammatory. They show some lymphocyte toxicity *in vitro,* but this effect is of doubtful significance *in vivo.*

Antimitotic agents halt the growth of all rapidly dividing cells. The drugs listed in Table 14.8 are used not only as immunosuppressants but also for cancer chemotherapy. This explains the unavoidable connection of antitumor therapy with immunosuppression.

Graft-versus-host disease (GVH)   If unmatched immunocompetent lymphoid tissue such as bone marrow, postfetal thymus, or spleen is implanted into a host (recipient) who is not immunocompetent, the graft is not rejected. On the contrary, the graft recognizes the **recipient** as foreign. Graft (donor) T-lymphocytes proliferate and attack the recipient's antigen-labeled tissues. Wasting and death are the usual results. GVH is the major barrier to immunological "reconstruction" of persons with primary immunodeficiency disease.

The Maternal–fetal Interaction

When a mother is carrying a fetus, she is always in a sense the recipient of a partly unmatched "graft." For nine months, she maintains the graft without any evidence of a developing rejection process. This is despite the fact that maternal lymphocytes do penetrate the **trophoblast,** the fetal layer of placenta, the point of embryonic contact with maternal cells (Fig. 21.12). Fetal lymphocytes also circulate in the mother

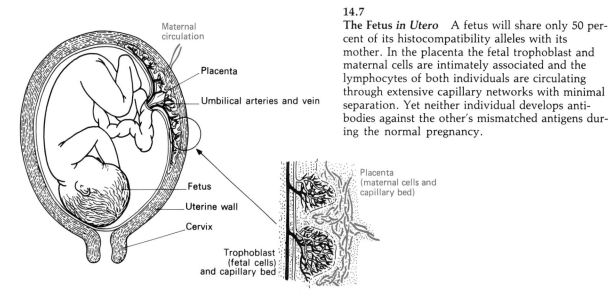

14.7

The Fetus *in Utero*   A fetus will share only 50 percent of its histocompatibility alleles with its mother. In the placenta the fetal trophoblast and maternal cells are intimately associated and the lymphocytes of both individuals are circulating through extensive capillary networks with minimal separation. Yet neither individual develops antibodies against the other's mismatched antigens during the normal pregnancy.

during pregnancy and after birth. There is no universally accepted explanation for this special case of immune tolerance. Many feel it is due to a thick, mucous layer on the trophoblast (Fig. 14.7) preventing effective contact.

What was the pressure that led to the evolution of an HLA system? Since evolution certainly did not anticipate tissue transplantation, we must conclude that the rejection mechanism is part of a system needed to meet some other need or challenge. The mother–child interaction provides us with a viewpoint.

In very rare cases, some of the rapidly proliferating embryonic cells actually invade the maternal uterus. Then they stop differentiating, continue rapid division, and become a malignant tumor. This is probably so rare because HLA incompatibility normally prevents it. However, the fact that it does occur shows us how, without the histocompatibility barrier, our bodies could readily become cell culture vessels for any introduced rapidly dividing cell. Malignancy would be a truly contagious disease. As it is, a malignant cell introduced by an unmatched "donor" during sexual or other contact will be eliminated by the histocompatibility mechanism.

Our knowledge of the genetics of immune regulation is growing rapidly at this time. Researchers are quite sure that the genes of the MHC are the crucial regulatory genes for all immune responses. The biological importance of the regulatory genes is much larger than previously suspected.

TUMOR IMMUNOLOGY

A naturally arising tumor is basically "self," that is, it shares the HLA antigens of the present normal tissue. At the same time, it becomes to some extent "nonself," acquiring new tumor antigens. It is a mosaic—in other words, it gives the immune system mixed signals. Both the AMI and the CMI limbs respond to tumor antigens, but the response is as likely to be enhancing to tumor growth as it is to be suppressing. Immunotherapy could actually cure malignancy if we could master the complexities and contradictions of the immune response to tumors. *In vitro* and *in vivo* observations in experimental animals and human beings have generated an enormous mass of results producing very few hard conclusions. Let us look for some useful generalizations about this complex interaction.

Antigenic Uniqueness of Tumors

Tumor antigens All tumors possess unique surface antigens. This is equally true of tumors induced by carcinogenic chemicals, by viral infection, or by unknown causes. Many **tumor-specific antigens** (TSAs), or **tumor-specific transplantation antigens** (TSTAs), have been purified and characterized. They are sufficiently antigenic to produce rejection of a tumor isograft when the recipient has been preimmunized with tumor extract. They are not antigenic enough to lead to rejection if the recipient has not been preimmunized.

Transformation; Chapter 11, pp. 329–330

Oncofetal antigens   Many tumors bear antigens characteristic only of fetal tissues and not found on normal adult cells. This may represent a derepression of genes normally shut off early in embryonic development. These **oncofetal markers** are the **α-fetoprotein** produced in hepatic cancers and the **carcinoembryonic antigen** (CEA) that appears in intestinal tract cancers. Detection of these antigens has been explored as a means of early diagnosis.

Tumor antigens can stimulate both antibody production and the development of immune T-lymphocytes. Thus there is no doubt that tumors can alert the immune system and stimulate a response. The unanswered questions arise when we try to evaluate the net protective effect of such responses in the intact host.

Immune surveillance   Until recently, the **immune surveillance** hypothesis was the accepted view of how the immune system prevented malignancy. This hypothesis proposes that newly arising malignant cells, or developing small clones, are continually detected and eradicated by immune mechanisms. An error or a temporary lapse permits the occasional tumor clone to escape and grow too large to be dealt with. Although this proposal still sounds sensible, most evidence shows that the immune system is really not particularly good at eliminating tumor cells. This suggests that surveillance is not the immune system's primary contribution to tumor prevention. Later, we will examine another concept that is currently more fashionable.

Mechanisms of Tumor Establishment

Small initial population   Initially, of course, all tumors start from a single cell. Even with rapid division, the tumor will still be microscopic for days or weeks. During this time it may not make contact with the immune system. Alternatively, it may induce tolerance. The tumor, then, is said to "sneak through" the defensive network.

Immunosuppressive effects of tumors   Some degree of immunosuppression is seen in most clinical cancer. Immunosuppression is most pronounced with malignancies of the lymphoid tissues themselves, such as myeloma or Hodgkin's disease. Once the tumor is established, it may cripple further attacks by the immune system.

Blocking effectors   The serum of some cancer patients contains **blocking antibodies** that block T-lymphocyte attack against their own cancer cells when these are mixed together *in vitro*. Blocking antibodies levels seem to go up during periods in which the tumor is progressing and seem to go down during periods of remission. Antitumor T-lymphocytes can also be shown to block a cell-mediated rejection response in mice. Blocking leads to **immunological enhancement** of tumor growth.

Tumors in immunodeficient patients   Tumors arise with a greater frequency in immunodeficient individuals. Such tumors are mainly lymphoreticular malignancies. In both animals and human beings, the lymphoreticular neoplasms have been closely linked with oncogenic virus infection. In animals, immunodeficiency

TABLE 14.9
Balance of factors affecting tumor development

SOURCE	POSITIVE FACTORS FAVORING REMISSION	NEGATIVE FACTORS ENHANCING GROWTH
Intrinsic	Active cell-mediated immunity, nonspecific stimulation, specific stimulation, Cytotoxic T-cells Activated macrophages	Blocking antibodies Tumor-induced immunosuppression Tumor-induced tolerance
Resulting from therapy	Tumor irradiation Surgical removal or reduction of antigenic mass Chemotherapeutic drug killing of tumor	Therapy-induced immunosuppression

Alternations in the balance of factors may produce a cyclical pattern of growth and remission.

promotes multiplication of oncogenic viruses. These observations suggest another hypothesis on the role of the immune system in contrast to malignancy.

Oncogenic viruses; Chapter 11, pp. 328–333

Perhaps the immune system, designed to prevent infection of all sorts, similarly protects against cancer by eliminating oncogenic viruses. Its primary role, then, is preventive. However, if it fails in this respect, the immune system has not yet evolved foolproof means of coping with the cancerous cell that results.

Immunotherapy

It is highly probable that, appropriately directed, the patient's immune system could become the major factor in curing cancer. In actual clinical experience, immunity is not a consistently beneficial influence. In addition, the standard therapeutic approaches—radiation and chemotherapy—reduce immunocompetence. We are clearly not using the full potential of the immune system yet. Some exploratory advances in immunotherapy have, however, been made.

Vaccination   If a malignancy is caused by a known virus, transmitted horizontally by an infectious process, vaccination is feasible. These conditions are met by some of the DNA oncoviruses. An example is Marek's disease in chickens, completely controllable since vaccines became available. An Epstein–Barr virus vaccine, now under development, could conceivably eliminate Burkitt's lymphoma. Unfortunately, the causative virus (if any) has not been found for other human malignancies. In animals, and probably also in human beings, RNA-containing retroviruses pass vertically from mother to offspring so infection induces fetal immune tolerance. This would effectively prevent immunization.

Nonspecific stimulation of CMI   The experimental evidence indicates that CMI is the major factor in destruction of tumor tissue. Tumor patients almost always have low CMI activity; tumor growth coincides with decreasing CMI; tumor regression, with increasing CMI.

BCG vaccine and killed suspensions of *Corynebacterium parvum* have a nonspecific CMI stimulatory effect. Such substances are called **adjuvants**; they are sometimes added to vaccines to boost the vaccine's immunogenicity. Adjuvant injection into or around accessible tumors, such as melanomas or breast tumors, often promotes regression. Side effects may be extensive, however. These treatments are less effective for less accessible tumors. Clinical refinement of this approach continues.

Specific stimulation   Tumor cells or their extracts have been added to adjuvants in an effort to raise a specific CMI response. It is still too early to evaluate the effectiveness of this approach.

AUTOIMMUNE DISEASE

Chronic, progressive disease is often a slow degeneration of one or more types of tissue. The actual causes can be extremely difficult to ascertain. Some degenerative diseases have an underlying immune mechanism; the tissue destruction results from hypersensitivity. The immunity is directed against self; thus it is called **autoimmunity.**

The Nature of Autoimmunity

Autoantibodies   The clonal selection hypothesis predicts that we will be completely and permanently made tolerant to all self-antigens during fetal development. It should be impossible for autoimmunity to develop, but it does. At least 50 percent of persons over 70 years of age have circulating antibody against one or more body components. Antinuclear and antithyroid antibody are commonly encountered examples.

There is an important distinction between autoimmunity and autoimmune disease. Only when the autoimmune factors attack tissue does disease result.

Accurate figures for the prevalence of autoimmune disease in various age groups do not exist, largely because of disagreement about which degenerative diseases have an autoimmune etiology.

Autoimmune **disease** develops much more commonly in females than males. It is rare in the young. There is some evidence for genetic predisposition.

Diseases Affecting Blood Cells

The autoimmune nature of an illness is particularly clear in the cases of two blood diseases in which cell populations are consistently depleted. In **autoimmune hemo-**

TABLE 14.10
Some characteristic autoimmune diseases

DISEASE	AUTOANTIBODIES TO	CLINICAL PICTURE
Autoimmune hemolytic anemia	Red blood cell antigens	Hemolysis; anemia
Idiopathic thrombocytopenic purpura	Platelets	Hemorrhages in skin and mucous membranes
Systemic lupus erythematosis	DNA, nuclear antigens	Facial rash, lesions of blood vessels, heart, and kidneys
Rheumatoid arthritis	Immunoglobulin	Inflammation and deterioration of joints and connective tissue
Goodpasture's syndrome	Glomerular basement membrane	Pulmonary hemorrhage and kidney failure
Sjogren's syndrome	Salway duct cells	A form of rheumatoid arthritis with failure of salivation and tear secretion
Pernicious anemia	Intrinsic factor	B_{12} absorption from intestine prevented, severe anemia
Myasthenia gravis	Acetylcholine receptor on muscle	Progressive neuromuscular weakness
Addison's disease	Adrenal gland	Weakness, skin pigmentation, weight loss, electrolyte imbalance
Hashimoto's thyroiditis	Thyroid antigens	Goiter, abnormal changes in thyroid gland
Graves' disease	Thyroid antigens	Stimulation of thryoid secretion; eye and skin infiltration

lytic anemia, RBCs are lysed by low affinity or incomplete IgG antibodies reacting with diverse cellular combining sites. In some cases, hemolysis is aggravated by chilling of the extremities. All cases reveal anti-RBC antibodies by the Coombs test.

Idiopathic (cause unknown) **thrombocytopenic purpura** (ITP) is a syndrome in which platelets are much depleted. As a result, purpura (tiny hemorrhages in the skin and tissue) occur. Antiplatelet IgG is present. The IgG mainly originates from splenic clones, so splenectomy is often helpful.

Generalized Diseases

In some cases, immunological attack affects many areas of the body. **Systemic lupus erythematosis** (SLE) is marked by such symptoms as a facial rash, fever, intestinal complications, and progressive kidney failure. It is much more common in females. Antibodies against many different autoantigens can be demonstrated. The most typical are antibodies against nuclear substances such as native and denatured DNA, histones, and nucleoprotein complexes. Antiliver and antikidney reactivity may also be seen.

Box 14.1

Autoimmunity—The Value of Multiple Hypotheses

How can autoimmunity occur so frequently? Why is it apparently a time-dependent process, since the incidence increases dramatically with age? There is no single generally accepted hypothesis, but let us pose five possible ways of thinking about this paradox.

(1) One hypothesis suggests that autoimmunity could arise through the proliferation of a "forbidden clone" that for some reason resisted eradication during the fetal period and survived. (2) Another hypothesis proposes that if some self antigen, absent or not "exposed" during the fetal period, suddenly became exposed, its responder clones would not have been made tolerant and could still react. (3) A third hypothesis proposes that perhaps in situations in which B–T or T–T cell cooperation was required, only one of the two clones was destroyed. The other could later bypass the fail–safe mechanism by getting its second signal from a nonspecific source. (4) Another hypothesis proposes that exposure to an environmental nonself antigen simply resulted in antibodies that cross-reacted with a self-antigen. (5) A fifth hypothesis proposes that the body might react to a virus whose envelope was derived from cell membrane and contained cellular antigens. The cell could be tricked into mounting an immune response against the host cell itself.

As you can see, investigators can become very ingenious in proposing hypotheses to explain phenomena they do not understand. These examples are presented in order to make a point about the value of hypotheses in advancing knowledge. Most theories presented to you, in this and other textbooks, seem rather self-evident. They seem self-evident because they have been around for a long time and are amply bolstered by evidence. It is well to remember that when Watson and Crick presented their hypothesis of DNA structure, or when F. M. Burnet presented his hypothesis of clonal selection theory, that each hypothesis was just one among many provisional theories to explain unresolved questions. However, the competing theories have now been eliminated by experimental activity, and those of Watson, Crick, and Burnet are accepted as true.

Each of the autoimmunity hypotheses above is now being experimentally challenged. A few years from now, there may be just one generally accepted idea left. The successful hypothesis may be one of the above, a combination of them, or something completely new. In the meantime, each hypothesis will have stimulated worthwhile experiments. Some of the greatest breakthroughs come in attempting to verify a hypothesis that turns out to be false!

A confirmed diagnosis depends on a positive **LE test.** Blood from an SLE patient is drawn and allowed to stand an hour. The serum contains IgG antinuclear antibody. During the waiting period the antibody causes swelling and extrusion of nuclei from the white blood cells. Stained preparations will show LE cells (Fig. 14.8), which are heavily distorted PMNs stuffed with nuclei they have ingested. The fatal

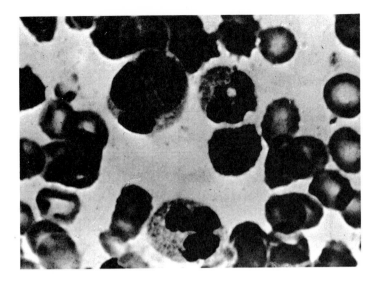

14.8
LE Cells   In systemic lupus erythematosis, one diagnostic sign is the formation of LE cells. An LE cell is a neutrophil that has phagocytosed a nuclear body. The nuclear body is a degenerate nucleus that has been extruded from a lymphocyte disordered by antinuclear autoantibodies.

outcome of SLE is due to a Type III immune complex attack on kidney glomeruli, leading to renal failure.

Rheumatoid arthritis (RA) is characterized by progressive joint degeneration, with thickening of the soft tissues and erosion of bony surfaces. There may also be spleen enlargement; low, white blood cell count; and cardiac, pulmonary, or renal complications. Diagnosis rests on the demonstration of **rheumatoid factor** (RF), an autoantibody that reacts with normal or partly denatured IgG. Thus an immunoglobulin is the **antigen** in this disease. In involved joints and organs, aggregates of immunoglobulin accumulate and provoke inflammatory changes. Many feel that RA is not a primary autoimmune disease but rather that it begins as a result of some exogenous cause such as an infection. The autoimmune pathology arises as a misdirected secondary response.

Localized Disease

There are many examples of organ destruction with a clear autoimmune component. Hashimoto and Graves' diseases involve the thyroid gland, stimulating or depressing hormone secretion. In a given individual, there are often autoantibodies to tissues that have not become targets in that specific patient, but may become so in other autoimmune diseases. It is as if a fail–safe mechanism has been bypassed, and all sorts of self antigens have the green light to stimulate immune responses.

Management of Autoimmune Disease

Immunosuppressive therapy is widely used, but it is only palliative and does not cure. Surgical removal of the affected tissue is rarely possible. If new kidneys are im-

TABLE 14.11
Conditions in which the contribution of autoimmunity is not clear

CONDITION	EVIDENCE OF AUTOIMMUNITY
SSPE	Very high levels of antimeasles antibody in spinal fluid
Slow virus disease	No direct evidence; tissue destruction resembles auto-immune disease
Syphilis	Tertiary stage CNS destruction associated with few microorganisms; antitissue antibodies present
Infertility in females	Antiegg antibodies produced by some females
Infertility in males	Antisperm antibodies (IgA) in semen of some infertile males
Diabetes	Presence of antipancreatic antibodies (rare)
Ulcerative colitis	Immune complexes in area of lesions
Chronic hepatitis	High levels of smooth muscle antibody

planted to replace those that have failed, the new kidneys will surely be destroyed by the same mechanism that damaged the original organs. Reasoned immunological approaches are sorely needed. We need to know how to turn the process off. Without a clear understanding of why these conditions arise and continue, we cannot do this.

SUMMARY

1. The immune system can cause diseases. The mechanisms of functional and dysfunctional immunity are basically similar. They differ in the nature of some effectors, the massiveness of the response, and the target.

2. Immunodeficiency is primary if a genetic defect is the underlying cause. It is secondary if it results from stress to the host. The primary immunodeficiencies are rare—failures of B-cell or T-cell differentiation. They are cured by transplanting immunocompetent cells from a donor. Secondary immunodeficiencies result from malnutrition, infection, or lymphoreticular neoplasms. They can be induced by radiation, steroids, and cancer chemotherapy.

3. Hypersensitivity is excessive or misplaced response to an antigen. Types I, II, and III are mediated by immunoglobulin, and reactivity can be passively transferred in serum. The responses are immediate and subside quickly when the antigen is removed.

4. Type I hypersensitivities occur when antigen combines with IgE-labeled mast cells. The mast cells release active mediators. These cause atopic allergy when lo-

cally released in small amounts or cause anaphylaxis when systemically released in larger amounts.

5. Type II reactions are cytotoxic. IgG or IgM antibody and complement attach to antigenic cells and destroy them, as in mismatched transfusions.

6. Type III immune complex diseases result from soluble antigen–antibody–complement complexes forming near the site of antigen entry (Arthus reaction) or circulating and lodging in lymphoid filters and kidneys (serum sickness, glomerulonephritis).

7. Type IV hypersensitivity is cell-mediated, and onset after antigen contact is delayed. Manifestations are granuloma formation around the site of an infection, induration at the site of antigen injection, contact dermatitis, and cell-mediated destruction in autoimmunity and graft rejection.

8. Successful transplantation of grafted tissue from donor to recipient requires a close match between the histocompatibility antigens of the paired tissues. Perfect matches occur only between genetically identical individuals. There are several tests to evaluate the acceptability of donor tissue to recipient lymphocytes. Imperfectly matched tissue is always a target for immunological rejection, which may be partially controlled by immunosuppressive therapy. Fetal tissue, always unmatched, is not rejected by the mother, but fetal cells are prevented from invading the mother's tissue.

9. Naturally arising tumors evade destruction and may indeed induce toleration. The contribution of immunity to antitumor defense is less than might be desired. The immune system deals more effectively with oncogenic agents than with the tumor cells themselves. Generally accepted treatments for malignancy are immunosuppressive, as is tumor growth itself. Efforts to stimulate immune system aggression against tumors are still in their infancy.

10. Autoimmune disease occurs when the individual responds to self-antigens, and destroys target tissues or organs. The process can be controlled to some extent by immunosuppression. Autoimmunity is paradoxical in view of the clonal selection process. Many hypotheses have been put forth to explain autoimmune disease. They are guideposts for those searching for cures for the immunopathological diseases.

Study Topics

1. Compare the results of (a) implanting a graft of fetal lymphoid tissue into a child with DiGeorge's syndrome and (b) implanting a graft of adult lymphoid tissue into a normal, unrelated child.

2. Can you think of some other hypotheses for the origin of autoimmunity?

3. Explain the relationship between delayed hypersensitivity and cell-mediated immunity.

4. Immunosuppressive therapy produces secondary immunodeficiency. Prepare a list of states in which immunosuppressive therapy is advisable and inadvisable.

5. What steps could be taken to prevent instances of anaphylaxis?

6. Ten years ago, any person who received a serious flesh wound was given an injection of tetanus toxoid if longer than six months had elapsed since the last tetanus immunization. Now the injection is not recommended unless it has been five years or more since the previous immunization. What would you expect the reason for this change might be?

7. It is observed that cancer patients are usually immunodeficient. One might say that they are immunodeficient because they have cancer, or they have cancer because they are immunodeficient. What is the evidence for either position?

Bibliography

Review articles

Burnet, F. M. (ed.), 1975, *Immunology*. A collection of readings from the *Scientific American*, San Francisco: W. H. Freeman.

Burnet, F. M., 1976, Immunodeficiency: investigations since 1957, pp. 173–176.

Reisfeld, R. A., and B. D. Kahan, 1972, Markers of biological individuality, pp. 203–211.

Books

Bigley, Nancy J., 1975, *Immunologic Fundamentals*, Chicago: Yearbook Medical Publishers.

Burnet, F. M., 1972, *Autoimmunity and Autoimmune Disease*, Philadelphia: F. A. Davis.

Roitt, Ivan, 1977, *Essential Immunology* (3rd ed.), Oxford: Blackwell Scientific Publications.

Thaler, M. S.; R. D. Klausner; and H. J. Cohen, 1977, *Medical Immunology*, Philadelphia: J. B. Lippincott.

15

Disease
and Populations

In becoming acquainted with infectious disease, we looked first at the biological traits of the invading organism. Then we analyzed the contribution of nonspecific and immune defenses to prevention of and recovery from disease. We focused on the relationship between an individual host and a specific parasite population—a single case of infectious disease. Now we will turn our attention to the principles of infectious disease transmission within populations. This type of knowledge is crucial for control or eradication of communicable diseases.

EPIDEMIOLOGY

A Definition of Epidemiology

Epidemiology is the quantitative study of the occurrence of disease in populations. The discipline initially developed through study of communicable disease outbreaks. However, its methods were quickly adapted for investigating noncommunicable diseases as well. Today's epidemiologist may be equally concerned with outbreaks of bacterial food poisoning—a communicable condition—and lead poisoning—an environmental disease. He or she may also search for factors contrib-

437

uting to diseases such as diabetes or stroke, for which the underlying cause is not known.

The epidemiologist's viewpoint on disease differs from that of other health-care professionals. A doctor or nurse is directly involved with individuals, but the epidemiologist is concerned with groups. Hospital personnel see ill people and carry out therapeutic activities. Epidemiologists rarely render personal health services. They are concerned with prevention; they aim to minimize new cases of disease and, wherever possible, to eliminate the disease from the population.

Epidemiologists work closely with clinical practitioners, however. Doctors provide data to the epidemiologists by reporting cases of disease as they are encountered. The epidemiologist, in turn, communicates the findings to practitioners and the public to assist their personal and public decisions about health issues.

Questions an Epidemiologist Can Answer

In most communicable diseases the causative agent is known, and the diseases are generally amenable to control or cure. Examples of some useful epidemiological questions include:

Which antibiotic gives the most dependable treatment for typhoid fever, based on the outcome of many treated cases?

In a sizable clinical trial, is a new vaccine actually effective in preventing measles?

What is the reservoir of the Legionnaire's Disease organism?

How effective is swamp drainage in reducing cases of mosquito-borne diseases such as malaria?

In noncommunicable diseases such as heart attack and lung cancer, there are usually several contributing factors rather than a single clear cause. Cigarette smoking is clearly **associated** with a vastly increased risk of lung cancer. But people who have breathed sparklingly clean air all their lives and have never been near a cigarette or another smoker occasionally develop the disease. So smoke is not the **causative factor;** it is one of several **predisposing factors** that augment the effect of the (unknown) causative agent. Some typical questions applicable to noncommunicable disease include:

What effect will different types of exercise programs have on reducing the risk of heart attack?

Is the long-term use of medication to relieve hypertension justified by a significant reduction in circulatory disease?

What types of public education are most effective in persuading smokers to quit?

Does occupational exposure to asbestos seem to be associated with respiratory disease?

The epidemiologist may also be a detective, the role that most often attracts public attention. When a disease outbreak develops, the epidemiologist is responsible for finding the cause, providing the key control information, and bringing health workers together in a cooperative effort to break the chain of transmission.

Public Health

Epidemiologists must have an overview of existing conditions. Although the scope of a problem may be as small as a single hospital nursery, it also may be statewide, national, or international. A network of public agencies coordinates the collection, analysis, and dissemination of epidemiological information.

Within hospitals, epidemiology of hospital-acquired (nosocomial) disease is carried out by an **infection control team,** with coordination among the hospital services. All data on infectious diseases, as well as other required information, are entered in hospital records.

Infection control; Chapter 28

Some or all of these data are transmitted to city or state bureaus or departments of health, routinely or on request. Community health is also monitored by those agencies that assume the responsibility for dealing with local or regional outbreaks.

The United States Public Health Service collects data from cities, states, and territories. At the Center for Disease Control (CDC) in Atlanta, Georgia, data are analyzed. A weekly report is prepared for dissemination to the health professions and the public. The CDC also operates educational programs, research projects, and reference laboratories to assist local and state laboratories.

Diseases do not respect national boundaries. The international surveillance and control activities of the World Health Organization (WHO) have become increasingly important. With the steady growth of high-speed transportation and mass dislocations of populations, the risk of global outbreaks is magnified. The WHO, with the cooperating governments, has carried out successful multinational vaccination campaigns against yellow fever and cholera. In one of the great achievements of twentieth-century medicine, smallpox has been eradicated through the efforts of WHO (Fig. 15.1). The WHO information network gives advance warning of the approach of new influenza strains.

15.1
Smallpox Vaccination   On market day in a small village in Ethiopia, a World Health Organization vaccination team administers vaccine to people of all ages. Simultaneously, they gather information about possible cases of smallpox. Since 1977, no new cases of smallpox have been reported in this area, the last endemic area for smallpox in the world.

Public health is often a matter of public policy. Public-health agencies frequently provide practical recommendations to the policymakers and legislators. Our lives are made safer by water and air safety regulations, food purity standards, and quality control criteria for pharmaceuticals.

COMMUNICABLE DISEASE TRANSMISSION

The Transmission Cycle

Each new case of a communicable disease is the end result of a chain of events (Fig. 15.2). It is also potentially the first step in a subsequent chain that will produce yet more cases. If one step can be prevented, communicable disease transmission is arrested.

The organism's transmission efficiency in a particular place and time determines the **prevalence** of the disease, or actual proportion of the population that is infected at any given time. **Endemic** diseases are those that infect a relatively constant, small percentage of a population. **Sporadic** diseases are those in which there are small, localized, unpredictable outbreaks. Typhoid fever in the United States follows this pattern. An **outbreak** becomes an **epidemic** when the number of cases rises significantly above the expected "background" level. Although these two terms mean much the same thing, the word "epidemic" has much more frightening connotations. It is seldom used unless there is a major threat to the health of a great many people. Many acute infectious diseases can cause **pandemics**—epidemics threatening several nations. Influenza outbreaks are always pandemic in nature; a new strain arising in the USSR in 1977, for instance, became established in epidemic form throughout

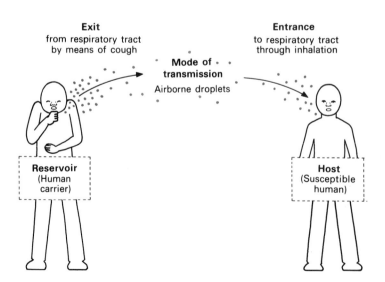

Exit
from respiratory tract
by means of cough

Mode of transmission
Airborne droplets

Entrance
to respiratory tract
through inhalation

Reservoir
(Human carrier)

Host
(Susceptible human)

15.2
Transmission of Infectious Disease   The disease agent passes from a reservoir, in which it multiplies and/or maintains viability. There must be a means of exit and a mode of transmission. In this case the human reservoir coughs droplets of infectious respiratory secretions containing the agent into the air. Air currents transmit the droplets to the vicinity of another person. The second individual inhales these droplets unknowingly. Should he or she be susceptible to the agent in question, a new active case of the disease may result.

TABLE 15.1
Sexually transmitted diseases

DISEASE	AGENT	CLASSIFICATION
Candidiasis*	*Candida albicans*	Fungus
Chancroid	*Hemophilus ducreyi*	Bacteria
Conjunctivitis, inclusion	*Chlamydia trachomatis*	Bacteria
Gonorrhea	*Neisseria gonorrhoeae*	Bacteria
Granuloma inguinale	*Calymmatobacterium granulomatis*	Bacteria
Hepatitis B*	*Hepatitis* B Agent	Virus
Genital herpes	*Herpes simplex* Type 2	Virus
Lymphogranuloma venereum	*Chlamydia trachomatis*	Bacteria
Nongonococcal urethritis	*Chlamydia trachomatis* (and other agents)	Bacteria
Pediculosis (pubic lice)*	*Phthisus pubis*	Insect
Scabies*	*Sarcoptes scabiei*	Insect
Syphilis	*Treponema pallidum*	Bacteria
Trichomonas*	*Trichomonas vaginalis*	Protozoa

*Also transmitted by nonsexual means.

Western Europe and North America within a few weeks. The great historic pandemics of smallpox, plague, and measles periodically decimated the populations of Europe.

Whether an agent's distribution is endemic, sporadic, epidemic, or pandemic depends on a complex, multifactorial interaction among agents, people, and events.

Living Reservoirs

Human beings   The most important reservoir for human infectious diseases is other human beings. Most of our disease agents are best adapted to survival and multiplication in human bodies. For diseases such as smallpox and measles, there is no other significant reservoir. A reservoir is said to be **infective** when it is actively releasing viable pathogenic agents.

A human being will often be a reservoir when acutely ill, but he or she may be equally infective while a **carrier.** Healthy carriers are those with inapparent infections, convalescent carriers are those who no longer have symptoms but are still carrying and shedding the agent, and incubatory carriers are persons excreting significant numbers of the agent before their own symptoms appear. The relationship of the period during which a person is infective to the period of illness varies from one disease to another.

TABLE 15.2
The major groups of vector insects and arachnids. Some diseases are also carried by genera in addition to those listed

CLASS	GROUP	GENERA	DISEASE	AGENT
Arachnida (eight-legged)	Ticks	*Americanum*	Q fever	*Coxiella burnetii*
		Ornithodoros	Relapsing fever	*Borrelia* sp.
		Amblyoma	Queensland typhus	*Rickettsia conori*
		Dermacentor	Rocky Mountain spotted fever	*Rickettsia rickettsii*
		Haemaphysalis	Tularemia	*Francisella tularensis*
	Mites	*Leptotrombidium*	Scrub typhus	*Rickettsia tsutsugamushi*
		Allodermanyssus	Rickettsial pox	*Rickettsia akari*
Insecta (six-legged, wingless)	Lice	*Pediculus*	Epidemic typhus	*Rickettsia prowazekii*
			Relapsing fever	*Borrelia* sp.
			Cholera	*Vibrio cholerae*
			Impetigo	*Streptococcus pyogenes*
			Trachoma	*Chlamydia trachomatis*
Insecta (six-legged, winged)	True bugs	*Panstrongylus* *Rhodnius* *Triatoma*	American trypanosomiasis	*Trypanosoma cruzi*
	Fleas	*Xenopsylla*	Plague	*Yersinia pestis*
		Nosopsyllus	Endemic typhus	*Rickettsia typhi*
			Tapeworm	Several genera
	Mosquitos	*Anopheles*	Malaria	*Plasmodium* sp. Arboviruses
		Culex	Encephalitis	
		Aedes	Dengue	Dengue agent
			Tularemia	*Francisella tularensis*
		Haemogogus	Yellow fever	*Arbovirus*
	Flies	*Chrysops*	Tularemia	*Francisella tularensis*
		Phlebotomus	Leishmaniasis	*Leishmania donovani*
		Glossina	African trypanosomiasis	*Trypanosoma* sp.
		Simulium	River blindness	*Onchocerca volvulus*
	Cockroaches	*Blatella*	Amebic dysentery	*Entamoeba histolytica*
		Periplanata	Hepatitis A	Type A virus
			Salmonellosis	*Salmonella*

A human being's normal flora may also serve as the reservoir of an opportunistic agent. This agent may be "transmitted" to another area of the body and establish an **endogenous** infection.

Animals **Zoonoses** (singular, zoonosis) are animal diseases transmissible to human beings. The major reservoir of the plague organism is wild rodents. Reservoirs for the equine encephalitis virus are horses, wild birds, and possibly snakes. In a zoonosis, the disease is much more prevalent in the animal population than in human beings, who are accidental hosts.

Arthropods Insects and arachnids play a dual role in disease transmission. They are reservoirs when the agent reproduces within them, yielding an infective stage. The malaria parasite completes the sexual stages of its life cycle in the mosquito and the asexual stages in human beings. Ticks, mites, and fleas are reservoirs of the rickettsial diseases. Insects and arachnids are also often responsible as **vectors** for the mechanical transmission of the agent from reservoir to susceptible individual.

Nonliving Reservoirs

Soil Soil is a primary reservoir when infectious agents survive and multiply there. Most pathogenic fungi live in the soil and produce copious spores that are infective if inhaled. Soil is a secondary reservoir when it is contaminated with infective human or animal waste products or remains. The agents in these contaminants do not multiply, and their survival is usually short. However, while they remain viable, the soil can be a threat. Fertilizing fields with fresh human excreta may make soil a reservoir of infection.

Water Few human pathogens are true water organisms although *Pseudomonas aeruginosa* and *Flavobacterium meningosepticum* (important hospital opportunists) multiply well in water. However, civilized society contaminates large volumes of available water supplies with human and industrial waste. Essentially all **surface waters** (ponds, lakes, and rivers) in the United States are potentially infective, with the possible exception of those in wilderness areas. Fecal organisms may survive for long periods in fresh water; in salt water they lose viability much faster. Water collectors in household humidifiers, nebulizers, and air conditioners, as well as in large institutional or industrial climate-control devices, frequently become colonized with microorganisms. These agents are sprayed into the air in droplets. A cloud of tiny droplets is an **aerosol.**

Air Air does not support microbial multiplication. Most organisms originally enter the air in droplets or dust clouds, and many remain viable in the air for long periods. Spore-formers and some viruses are well adapted for airborne survival without special protection.

15.3
A Restaurant Kitchen   Foods are prepared and stored for large numbers of persons in restaurants and cafeterias. If poor techniques are used, a large outbreak of infectious disease may occur. Personnel must pay attention to cleanliness of the facility, personal cleanliness, and adequate cooking and refrigeration.

Other agents are shed from the body in droplets of body fluids such as respiratory or rectal mucus. As the water evaporates from these, the organisms remain, protected by a dried mucoprotein film. The residual **droplet nuclei**, very much lighter than the original droplets, remain airborne and drift almost indefinitely.

Dusts and danders may be derived from skin flakes or shed hairs. These can serve as a stimulus for atopic allergy attacks or may carry viable microorganisms.

Food   Fresh, raw food is with few exceptions perfectly safe. It is a primary reservoir only when an animal from which the meat or milk derived was infected with a zoonotic disease. Pork or bear meat may be a source of trichinosis, and unpasteurized dairy products a source of brucellosis. Much raw poultry and the shells of eggs are contaminated with *Salmonella* derived from the intestinal contents of the birds.

Salmonellosis; Chapter 20, pp. 558–560

The majority of dangerous food has been made hazardous by mishandling. Clean foods are inoculated with pathogens by dirty hands, dirty utensils, or contaminated water. Often foods are then allowed to stand at incubation temperatures that allow bacterial multiplication. Almost any food may become a reservoir by these means (Fig. 15.3).

Unpasteurized milk may be a potential primary reservoir for the agents of tuberculosis, brucellosis, and Q fever. It is also a favorable medium for the growth of secondary contaminants.

Exit from the Reservoir

The usual means of exit from a living reservoir is body fluids that have contacted the infected area. Saliva, tears, respiratory secretions, exudates, blood, urine, feces, genital secretions, and skin flakes or crusts may provide means of exit. Some infections are not transmitted while no exit is available. The viable tuberculosis bacilli in a healed lung lesion are physically trapped as long as the lesion remains inactive. Intestinal flora cannot cause peritonitis or septicemia as long as the gut permeability barriers remain intact. A malaria patient who does not live where the *Anopheles* mosquito breeds does not transmit by natural means, although the blood of such a person may be infective if removed by a syringe needle.

Mode of Transmission

Transmission, to be successful from the parasite's point of view, must be linked to the agent's ability to survive in the outside world.

Direct transmission

This occurs when the susceptible person touches, inhales, ingests, or is bitten by the reservoir. The pathogenic agent is not exposed to a hostile environment for any significant time.

A disease is **horizontally** transmitted when it is spread among the members of a group from person to person. It is **vertically** transmitted when it is spread from parent to child during reproduction (with egg or sperm cell), during fetal development, or during the birth process.

STDs; Chapter 21, pp. 624–636

The sexually transmitted diseases provide examples. Syphilis is spread horizontally, almost exclusively by direct contact of epithelial surfaces during sexual activity. Herpes virus may be transmitted vertically to the fetus as it comes in contact with infected cervical surfaces during childbirth. Rubella virus is transmitted vertically to the fetus, resulting in congenital malformations.

Trichinosis is transmitted only by the ingestion of inadequately cooked animal flesh containing the parasite's cysts.

Insects and arachnids are important in the infectious disease cycle as transmitters. They are usually called **vectors. Arthropod-borne** diseases are those transmitted primarily or exclusively by insects or arachnids. If the insect or arachnid is the reservoir too, it is a **biological** vector. The bite injects saliva or regurgitated blood stored from a previous blood meal. If these liquids contain infectious agents, direct transmission has occurred. Arthropod-borne diseases, because of the requirement for the biological involvement of the vector in the cycle, can be transmitted by only one or a very small number of arthropod species. The distribution pattern of the disease coincides with the ecological range of the vectors.

Indirect transmission

This occurs when the susceptible person contacts food, drink, clothing, or inanimate objects previously contaminated by contact with the reservoir. Some indefinite length of time elapses between the time the **fomite**, or object bearing the agent, was originally contaminated and the moment at which the susceptible individual touches, wears, or eats it. During that time the agent is not multiplying and is exposed to an alien environment. Indirect transmission occurs when you eat an apple that, because it has been washed in fecally contaminated water, carries a gastrointestinal agent on its skin. The so-called **oral-fecal** route of transmission of gastrointestinal diseases requires this type of ingestion of fe-cally contaminated material. Similarly, you can acquire an upper respiratory tract infection by picking up a used tissue, transferring the agent to your hands, and then touching your mouth.

Insects carry out indirect transmission as **mechanical** vectors. The housefly and the cockroach walk on garbage or feces, contaminate their feet, then walk on uncov-ered food and inoculate it with the organisms.

This consideration of direct and indirect means of transmission is useful because it encourages us to think about how many ways an infective agent may move or be moved through a population. It is sobering to reflect, in addition, that if **you** were the susceptible host in any of the examples given, you would see and feel nothing to warn you of your risk. Covering food, washing your hands after blowing your nose or going to the bathroom, even if you think your fingers remained clean, and cook-ing your pork thoroughly even though you know the farmer who raised it, are always sensible.

Entrance to the Susceptible Host

Unavoidable normal activities such as breathing, eating, and socializing constantly admit organisms to our bodies. Routes of entrance include nose, mouth, eye, ear, skin, anus, urethra, and vagina. New artificial entrance routes may be opened by trauma, wounds, or surgery. Tetanus, gangrene, and other wound infections cannot occur without the prior creation of an abnormal entrance.

The ability to take advantage of a normal or abnormal entrance, it should be re-membered, depends to a certain extent on the infecting organism's virulence.

Population Susceptibility

An infective agent may have a different impact on two different populations because of variations in susceptibility.

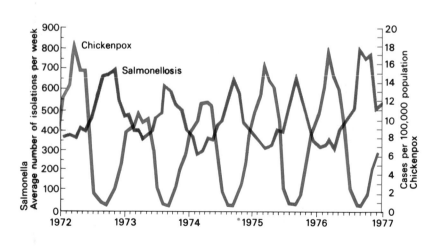

15.4
Seasonal Variations in Diseases This graph compares the monthly rates of chickenpox and salmonellosis in the United States over a period of four years. Chickenpox, like most res-piratory infections, is most com-mon in the late winter months. Salmonellosis, like many gas-trointestinal infections, is most common in the summer months. From graphs like these, epidemi-ologists can anticipate probable outbreaks and issue warnings or public information statements.

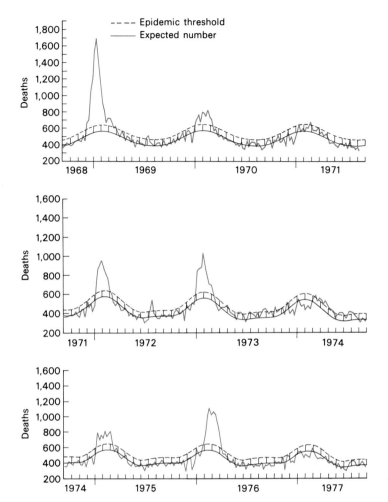

15.5
Influenza Patterns Annually, the number of influenza-related deaths in the United States peaks around February. On the basis of past experience, an "expected number" can be calculated. A specified increase over this number occurs in certain years. When the deaths exceed the epidemic threshold, as they do in about two out of three years, an epidemic can be said to be occurring.

Physical and social conditions You will recall that malnutrition, underlying diseases, or emotional stress predispose the individual toward disease. When such conditions are widespread in a population, the group too becomes much more susceptible. The physical hardships of inadequate or overcrowded housing, the stresses of displacement and loss experienced in wartorn areas, and the acute challenges of natural calamities such as earthquakes or hurricanes invariably amplify infectious disease. Endemic disease flares into epidemic outbreaks.

Factors predisposing toward disease; Chapter 12, pp. 357–358

Many infectious diseases show regular seasonal variations in incidence. These may be due either to physical or social factors or both (Fig. 15.4). When the peak incidence exceeds a certain level, an epidemic is said to be present (Fig. 15.5).

Box 15.1

The World's Refugees

Human history, from Exodus to today, is marked by wars of conquest followed by the mass expulsion of entire populations from their homelands. Some of these events, like the eviction of the Indochinese "boat people" from Vietnam, are well publicized. Others, such as the recurrent local wars in Africa, receive little international attention. Where will the next wave of displaced persons originate and arrive? No one knows. But it is certain that in our troubled times, there will continue to be a long succession of these episodes of human misery. It is equally certain that infectious disease will occur in epidemic proportions to add to the sufferings of the refugees.

International health agencies, such as the World Health Organization and International Red Cross, work with private philanthropies, such as Oxfam and Save the Children, wherever disasters of any nature require relief efforts. Epidemics are always to be expected, fanned by the malnutrition and poor or nonexistent housing and sanitation endured by the refugees. It is not easy to predict in advance what the dominant infectious challenge may be.

Kampuchean (Cambodian) refugees fleeing to Sakaeo in Thailand during 1979 and 1980 entered refugee camps at the rate of two to eight thousand per day. There were few persons under five or over 55. Most individuals in those age groups had not survived the protracted and arduous flight to the Thai border. The refugees' dominant health problems were malnutrition, malaria, and upper respiratory disease.

Indochinese refugees leaving Vietnam by boat were carried by the prevailing ocean currents to Hong Kong or Malaysia. Up to 14,000 persons monthly are accepted for resettlement in the United States, after their pressing health problems are treated. In the refugee camps, measles, tuberculosis, hepatitis B, and intestinal parasites have been the most severe problems. Efficient action on the part of the relief teams has been able to bring these diseases under control very rapidly after the refugees have reached a safe haven.

Age distribution   Each disease affects one age group more than others (Fig. 15.6). Mumps and chickenpox are most common in the young, and pneumonia is more common in the elderly. Populations have different proportions of their members in each age group. As these change, disease incidence may change. In the United States and Europe, a declining birthrate, in conjunction with increased life expectancy, have reduced the proportion of young persons and increased the proportion of the elderly. Changes in disease prevalence are predictable.

Herd immunity   This rather unflattering term is used to designate the proportion of a population functionally immune to an agent and able to resist the challenge of exposure. Unless there is a high density of susceptible individuals in a group, an

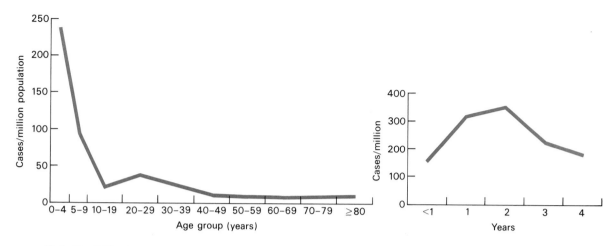

*Excluding California

15.6

Rates of Shigellosis by Age Group   Shigellosis reaches its peak incidence between ages 1 and 2 (insert), then drops off rapidly.

agent will spread in an endemic fashion, with few and isolated cases. If there is a very low density of susceptibles, the cycle of transmission is very rarely completed (Fig. 15.7).

For example, in a city of 300,000 people with a standard age distribution, between 2,500 and 5,000 new cases of measles a year must occur in order for the virus to perpetuate itself. If these new young susceptibles can be made resistant by vaccination, the virus will die out.

It is not essential to immunize every single child, but it is desirable. Theoretically, if most but not all children were vaccinated, and only 1,000 susceptible (unvaccinated) children were added per year, the virus would gradually disappear. However, from the point of view of individual human beings, this would not be satisfactory. In the interim, many of these unprotected children would be ill, and one or two might die or suffer permanent disability. Thus public health programs can control an infectious disease if they can achieve a reasonable herd immunity level estimated at about 70 percent for some diseases. All cases can be prevented only if the immunity level approaches 100 percent.

Social organization   Diseases spread by direct contact, such as smallpox, measles, and the common cold, require not only a susceptible population but also an urbanized society. These diseases probably did not exist before *Homo sapiens* evolved from a tribal hunting and gathering species to a settled village and city species several thousand years ago. It has been suggested that at that point certain animal

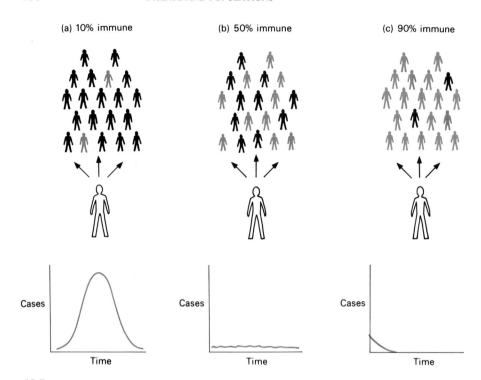

15.7

Herd Immunity   Epidemics spread most rapidly in populations such as (a) in which only a
small percentage of the people have resistance to the infective agent. The epidemic develops rapidly, and rapidly subsides after all the susceptibles have experienced the disease
and become immune. (b) When about half of the group is immune, the infective agent
spreads more slowly to the remaining susceptibles. The incidence of disease is sporadic,
and the infectious agent persists at a low level over a long period. (c) When a high percentage of the group is immune, an introduced case of the infectious disease is rarely transmitted at all, and the agent disappears from the population as soon as the original cases
have recovered.

viruses gained virulence for human beings as a result of repeated human-to-human
passage, leading to "new" infectious diseases.

The kinetics of an epidemic   When the numbers of new cases are graphed against
elapsed time, epidemics have very defined shapes (Fig. 15.8). No epidemic goes on
forever—it more or less quickly runs out of new victims. The rapidity of the epidemic's development depends on the density of susceptibles; the denser the susceptible population density, the more explosive the outbreak.

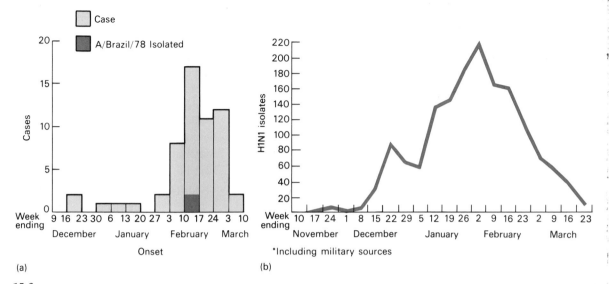

15.8

Kinetics of Influenza Outbreaks (a) This graph represents the pattern of influenza cases among nursing students in a small hospital. Note that clinical illness was present at a low level for more than a month before the real outbreak developed. Once the necessary conditions for transmission were present in February, the agent quickly infected all susceptible students. Note that isolation of the influenza virus occurred in only two cases. (b) Total virus isolations (influenza H1N1) reported for one winter season in the United States. Isolations follow the same type of pattern as do cases. However, at present, there are many times as many cases as successful laboratory isolations of the virus.

TRACING AND ARRESTING EPIDEMICS

The process of tracking down and dealing with an epidemic follows certain steps. The chase always poses novel challenges and difficulties to the investigator.

Reporting Cases

Certain infectious diseases are by law designated as **notifiable diseases.** Health practitioners are required to report any cases of these diseases they treat to local or state public health officials. Notification gives a more or less accurate estimate of the prevalence of a disease. Of course, not all cases are seen by physicians, and not all treated cases are reported, so most diseases are underreported. The notification system is a mechanism to assemble the needed data on disease prevalence in a central office, making it possible to spot an unexpected number of cases or sporadic outbreaks. A decision may then be made to investigate.

TABLE 15.3
Water-borne diseases. All are transmitted by fecally contaminated water unless otherwise noted

DISEASE	AGENT	CLASSIFICATION
Amebic dysentery	*Entamoeba histolytica*	Protozoa
Amebic meningoencephalitis	*Naegleria fowleri*	Protozoa*
	Acanthamoeba sp.	Bacteria
Bacillary dysentery	*Shigella* sp.	Bacteria
Cholera	*Vibrio cholerae*	Bacteria
Acute gastroenteritis	Many	Bacteria, virus
Giardiasis	*Giardia lamblia*	Protozoa
Hepatitis A	Hepatitis A agent	Virus
Leptospirosis	*Leptospira* sp.	Bacteria†
Bilharzia	*Schistosoma* sp.	Metazoa
Typhoid fever	*Salmonella typhi*	Bacteria

*Free-living organisms inhabiting warm fresh water; not a human organism.
†Transmitted by urine-contaminated water.

Identifying the Infectious Agent

The causative agent must be obtained, from each patient if possible, to establish whether all the cases are really part of a single outbreak. It would be possible to have an outbreak of 52 diarrheal illnesses, for example, in which one case could be a coincidental food allergy and another an undetected ulcerative colitis. Such cases should be carefully eliminated because they will confuse the issue.

Once the agent is isolated, it will be subjected to the most precise possible identification methods, such as bacteriophage typing, serotyping, antibiograms, and the complete spectrum of metabolic tests. If all isolates coincide, there is strong evidence that all cases had a single source. Isolation of an identical organism from the suspected source will confirm the attribution.

Searching for the Reservoir

The identity of the infectious agent provides immediate clues to the type of reservoir to be expected. *Salmonella* gastrointestinal illness suggests fecally contaminated foods; staphylococcal upset suggests foods inoculated from sores on the body of a food handler. Respiratory illnesses, transmitted by air currents, originate in secretions expelled by a carrier or, less commonly, in vapor sprayed from an air-conditioning device.

In some epidemics the source is never found. Following the first outbreak of Legionnaire's disease in Philadelphia in 1975, many sources were eliminated but the real one was not pinpointed. Three years passed, diagnosis of this condition became more common, and improved techniques for isolation of the agent were developed. In the meantime, the hotel in which the outbreak centered was totally remodeled, so the source was presumably completely removed. It has since become clear that one reservoir of *Legionella* is large-scale cooling towers and climate-control systems. This finding helped to explain why many cases occur in the summer months and cluster around large buildings, such as hotels and hospitals, in which this type of equipment is used.

Breaking the Chain of Transmission

Swift, directed action can often arrest an epidemic before it achieves its full destructive potential. This action may involve one or a combination of approaches.

When there is a single apparent source, it will become the focus of control. Infective persons can be treated, trained in personal cleanliness, or switched to dif-

TABLE 15.4
Airborne diseases*

SOURCE OF AIRBORNE CONTAMINANT	DISEASE	AGENT	CLASSIFICATION
Spores from soil	Anthrax	*Bacillus anthracis*	Bacteria
	Aspergillosis	*Aspergillus* sp.	Fungus
	Blastomycosis	*Blastomyces dermatiditis*	Fungus
	Coccidioidomycosis	*Coccidioides immitis*	Fungus
	Cryptococcosis	*Cryptococcus neoformans*	Fungus
	Histoplasmosis	*Histoplasma capsulatum*	Fungus
Aerosols	Chickenpox	Varicella agent	Virus
Droplets or droplet nuclei	Common cold	Many	Virus
	Diphtheria	*Corynebacterium diphtheriae*	Bacteria
	Influenza	Influenza agent	Bacteria
	Legionellosis	*Legionella pneumophila*	Bacteria
	Measles	Measles agent	Virus
	Meningitis	*Neisseria meningitidis*	Bacteria
	Mumps	Mumps agent	Virus
	Rubella	Rubella virus	Virus
	Smallpox	Variola agent	Virus
	Tuberculosis	*Mycobacterium tuberculosis*	Bacteria
	Whooping cough	*Bordetella pertussis*	Bacteria

*Most are also spread by direct contact and/or contaminated fomites as well.

ferent occupations. Leaking water or sewer mains can be fixed, herds of infected dairy cows can be destroyed or milk from the cows kept off the market, inadequately processed batches of canned goods can be recalled, and contaminated shellfish beds can be closed to harvesting.

Where there is a large reservoir, such as a group of carriers, it is more effective to interrupt transmission. During community influenza outbreaks, nonessential personnel and visitors are frequently barred from hospitals. Schools are closed to arrest measles epidemics. Isolation and segregation procedures are also useful, especially in hospitals.

Animals and arthropods serving as reservoirs and/or vectors may be eliminated or kept from contact with human beings.

There may be a campaign to seek out and actively immunize all susceptibles in epidemic zones and surrounding areas. The effectiveness of this approach depends on the length of the incubation period. Because influenza has an incubation period of from one to three days, no immediate protection can be gained from immunization after an epidemic has begun. A reduction in the number of cases will occur starting two weeks or so after the immunization campaign, by which time most susceptibles will have already experienced the disease. For good results, influenza vaccination must occur well in advance of possible exposures or epidemics. Measles outbreaks, however, can be arrested by crash vaccination campaigns in a threatened community in conjunction with the exclusion of unvaccinated children from school. Rabies has an incubation period measured in weeks or months. When rabies in wild animals reaches epidemic proportions in an area, public health officials will initiate a campaign to get all pet dogs and cats vaccinated, while dog and game officers will attempt to destroy any dogs or cats that have reverted to the wild state.

Passive immunization, if available, may be used to temporarily protect abnormally susceptible individuals. During chickenpox outbreaks, there is an increased need for passive protection of immunosuppressed children.

PUBLIC HEALTH AND DISEASE PREVENTION

The community's health is best protected by a cooperative mixture of governmental surveillance and regulation, informed and dedicated medical practice, and educated self-protection by individuals. Routine public health measures are generally adopted throughout the developed world; they are seen as offering the best possible chance of preventing epidemics.

Safe Water Supplies

An uncontaminated source of water for drinking, washing, and food preparation is essential to avoid outbreaks of gastrointestinal disease. All public water supplies are subject to regulation. The treated water they supply is regularly analyzed for chemical and microbial contaminants.

Box 15.2

A Classic Epidemiologic Investigation

An outbreak of shigellosis occurred May 17–30, 1979, among hospital employees in a children's hospital in Pennsylvania. Thirty-two percent of employees reported being ill. Two hundred eighty employees and visitors with complaints of vomiting and/or diarrhea presented to the employee health service and were cultured; 142 (51 percent) had positive stool cultures for *Shigella sonnei.* Staffing problems during the outbreak were severe, and the hospital was closed to new admissions for a three-day period.

Questionnaires were sent to 1,700 employees to determine the symptoms of disease and places where these persons had eaten from May 16–21; a food-specific history was obtained from those who had eaten in the hospital cafeteria. One thousand ninety-three questionnaires (64 percent) were returned. Analysis showed a strong association between illness and eating in the hospital cafeteria ($p < .0001$). Based on 78 culture-confirmed cases and 150 well controls, significant associations were found between illness and consumption of tuna salad ($p \le .0001$) and eating food from the salad bar ($p \le .0001$). A negative association between illness and consumption of hot foods was also found.

One cafeteria employee had diarrhea on May 17, the first day of the outbreak. She had

been exposed at home to a child with severe diarrhea before onset of her illness. This employee was found to be culture positive for *S. sonnei.* She had worked on May 17 and May 21 and was responsible for preparing all salads and sandwiches in the employee cafeteria, where visitors also sometimes ate. The two peaks in the outbreak were on May 19 and May 23—consistent with the one- to two-day incubation period of foodborne shigellosis.

The organism identified from culture-positive individuals was resistant to ampicillin and tetracycline. and sensitive to trimethoprim-sulfamethoxazole. All symptomatic individuals were treated with a five-day course of the latter drug, or with furazolidone, if they were sulfa sensitive. For cafeteria employees, three negative rectal cultures—taken at one-day intervals at least 48 hours after antibiotic therapy had ended—were required before a culture-positive individual could return to work. Other culture-positive hospital employees were permitted to return to work after 48 hours of therapy. No hospitalized patients became culture-positive for *Shigella* as a result of the outbreak.

Source: *Morbidity and Mortality Weekly Report* **28:** (42) October 26, 1979, p. 498.

One essential test is the quantitative determination of the number of **coliform bacteria.** Coliforms are lactose-fermenting, Gram-negative bacilli, including *Escherichia coli* and closely related organisms. These organisms occur naturally only as normal intestinal flora of human beings and higher animals. Their presence in a water sample indicates that it contains fecal material. Thus coliforms are referred to

15.9

Water Treatment   When water is supplied to the consumer for drinking, household, and industrial uses, it must be free of pathogenic microorganisms and have relatively few, harmless microorganisms. The desired quality is obtained first by adding aluminum compounds to the water. These compounds form a gel-like coagulation that traps many particulate impurities, including microorganisms. The coagulated material is allowed to settle out. Then the water is passed through a sand or other type of filter and chlorinated to whatever level is required by the remaining organic impurities in the water. It is then ready for use.

as **indicator** organisms. The indicator organisms are not themselves usually pathogenic, but it is a valid assumption that if they are present, there is fecal contamination. The water is apt to be unsafe, not to mention unappealing.

Almost all municipal water supplies derive from contaminated surface water. Therefore the water must be purified by settling, filtration, chlorination, or a combination of these methods. The treatment (Fig. 15.9) needed may vary on a seasonal basis, since heavy rainfall brings heavy loads of contaminants. The dose of chlorine is adjusted to the level of contamination in order to obtain a final concentration sufficient to kill or inactivate all potential pathogens. Water-borne outbreaks do happen periodically. These may be caused by leaky water mains through which contaminated ground water mingles with the clean water, breakdowns in chlorinating equipment, or overloading of an old, inadequate system. Many cities' underground water mains and central pumping equipment, approaching one hundred years of age, are showing progressive deterioration. This means that failures are becoming increasingly common. Because the vast amounts of money needed to replace these systems are not available, malfunctioning water systems will remain a major urban health concern for the foreseeable future.

Individual home water supplies are usually safe if local guidelines about distance of wells from sewage disposal facilities are observed. You can have your water checked by public health laboratories, conveniently and for a minimal charge, whenever you wish.

Sewage Disposal

Sewage consists of three components, derived from different sources. The average human being excretes urine of low bacterial count, and feces that contain billions of microorganisms per gram. Water that has been used for toilets is called **black water;** it is heavily contaminated with organisms of human origin, and is presumed hazardous. We also wash ourselves, our clothing, and our dishes, rinsing away millions of organisms of both human and environmental origin. Contaminated wash water is called **gray water;** its microbial load and potential threat is very much less than that of black water. **Storm runoff** from gutters, storm drains, and culverts may be heavily loaded with microorganisms, but these are predominantly soil flora and few are hazardous to human beings.

When microorganisms of human origin are dumped into water or soil, they do not usually survive very long. In rural areas of low population density, custom rarely calls for sophisticated means of sewage disposal. The septic tank or even the outhouse is adequate because the microbial load dies off quickly. In larger communities, sewage accumulation poses an overwhelming threat to public health, and municipal sewage treatment is mandated.

Sewage treatment facilities were developed in parallel with water supplies in most communities. Together, water and sewage treatment have almost eliminated water-borne disease in this country. Only in underdeveloped areas without these facilities do such great killers as typhoid and cholera still cause epidemics.

Sewage treatment may involve up to three stages of treatment. **Primary treatment** consists of the collection of sewage through a network of pipelines, screening to remove large debris (which is removed and burned or buried) and sedimentation in tanks. The sedimentation process separates sewage into two components, sediment or sludge (the solid waste matter) and **effluent** (the liquid wastes). Both are heavily contaminated with microorganisms. A large plant may produce many millions of gallons of effluent and tons of sediment or sludge per day. If only primary treatment is used, the effluent will be disinfected (usually chlorinated) and discharged into a nearby body of water. Chlorination renders the effluent noninfectious, but it still contains noxious substances. The sludge may be burned or dumped; it is also still hazardous because many human pathogens survive primary treatment.

Secondary treatment follows primary in more up-to-date systems (Fig. 15.10). Secondary treatment encourages microbial decomposition of organic wastes and favors the elimination of pathogens by ecological competition. Effluent and sludge are mixed and aerated. Microbial action reduces the bulk and offensiveness of the sludge, and also reduces the organic and microbial load of the effluent. Chlorination is still required before effluent is released. Controversy persists over the use of the treated sludge as a fertilizer. In some communities, it is considered safe for agricultural use, and is hauled away by farmers. In other areas, its use on food crops is banned, and most of it is burned or dumped.

Tertiary treatment is a relatively new development. It incorporates additional chemical treatments to precipitate out contaminants, or the growth and harvest of algae to remove contaminating plant nutrients such as nitrate or phosphate ion. All the water can be recovered in entirely drinkable condition if the system is working properly. It can be recycled to the community or discharged to the environment without causing ecological disruptions.

Individual home sewage has traditionally been delivered to a **septic tank**, in which sedimentation occurs. The effluent is passed through a drainage field into the surrounding soil in which microbial contaminants lose viability and chemical impurities decompose. The sediment undergoes slow, anaerobic breakdown in the tank. Ecological concerns have spawned many alternative systems utilizing chemical

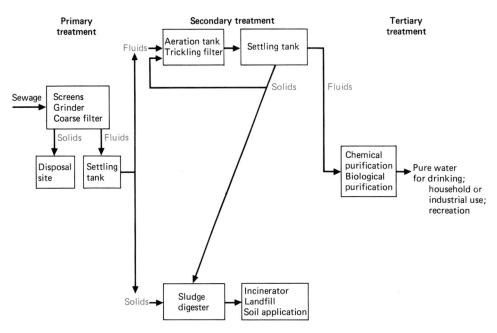

15.10

Advanced Sewage Treatment   By using three levels of treatment, sewage can be converted
back to clean water. This is an extremely expensive process, however, and many communities carry out only primary treatment, or a combination of primary and secondary
treatment. Primary treatment removes some portion of the solids from the sewage.
Secondary treatment reduces the microbial content of the fluids and reduces the bulk of
the solids. During both of these phases, pathogenic microorganisms decline rapidly due to
intense microbial competition. A much reduced bulk of sludge may be disposed of at this
stage. The fluid effluent still contains much soluble material that makes it unsuitable for
reuse. It can, however, be discharged into large bodies of water without too much environmental disruption. Tertiary treatment removes essentially all remaining organic contaminants. It may also be designed to remove inorganic compounds such as nitrates and
phosphates, as well as toxic heavy metal ions.

disinfection, incineration, or contained composting of human domestic wastes.
Most of these systems are still in the testing and evaluation stage.

Food Safety

Meat can be hazardous if it comes from a diseased animal or if it is improperly refrigerated. Inspection programs operate in all commercial meat-packing facilities to
detect diseased animals. Rigid standards of dressing, cutting, grading, and refrigeration are enforced. Special tests may be carried out to detect antibiotic or pesticide
residues in meat, *Salmonella* or other bacterial contaminants in chicken, or encysted
parasites in beef or pork. Governmental inspection does not necessarily ensure that
the meat or poultry are free of all pathogenic microorganisms, however. Most meat-
caused disease outbreaks now can be traced to home-raised or wild animals.

Refrigeration and handling of foods in supermarkets and restaurants is also subject to regulation and inspection.

Milk is classified by its raw bacterial count, which indicates both the cleanliness of the producer's dairy facility and the length of storage. Milk must be pasteurized prior to commercial sale. The milk is heated to a temperature at which all pathogens are killed. Pasteurization also significantly reduces the total bacterial count. Pasteurized milk is not sterile, but it is free from pathogens and keeps much longer than nonpasteurized milk. Processed milk is periodically analyzed to make sure that pasteurization has been adequate.

Much of the food consumed in the United States is imported. In order to be imported, foreign canned foods, dairy products, sausages, and frozen foods must be prepared by methods as stringent as those required for domestic products.

Canning, properly carried out, is one of the safest ways of preserving food. Properly canned foods are sterile. However, inadequately processed foods may support growth and toxin production by *Clostridium botulinum*, and be potentially lethal. Over 20 billion cans of food are commercially prepared in the United States per year. Rigid testing schedules are designed to prevent release of inadequately processed cans. Each batch passed through the cannery must be tested microbiologically. Should a batch of contaminated food reach the market, resources of the public health agencies are mobilized to recall every can in warehouses or on store shelves. The public is alerted by mass media announcements and instructed to return, or throw out, unopened cans that have already been purchased. Botulism is rare—only about 70 cases per year are reported in the United States—and it almost always results from the consumption of products that have been carelessly canned in the home.

Botulism; Chapter 20, pp. 609–610

Controlling Human Reservoirs

In certain diseases, active surveillance of known carriers and screening to detect unsuspected carriers can pay large dividends.

Some fraction of recovered hepatitis patients continue to harbor infectious virus indefinitely. A small percentage of those who recover from typhoid·fever (Typhoid Mary was a notorious example) retain the bacterium in gall bladders and excrete live pathogens in their feces. Systematic case follow-up of hepatitis and typhoid patients will identify these carriers. They can receive treatment and/or instruction in hygienic measures to avoid infecting their contacts.

Several of the venereal diseases characteristically produce inapparent infections in the female. She may be infective for months before symptoms develop that cause her to seek treatment. Routine examinations that include *Neisseria gonorrhoeae* culture pick up a substantial number of positives. These women and their sexual contacts may then be treated before serious clinical complications occur.

Reducing Susceptibility

Immunization of infants and children can prevent epidemic childhood diseases such as measles and polio, and can initiate a lifelong immunity against age-independent

Immunization; Chapter 27

diseases like smallpox and tetanus. Remarkable decreases in infant mortality have been achieved. One sad result of the success story of vaccination is public apathy about these supposedly vanquished diseases. Compliance with immunization recommendations is steadily decreasing. Herd immunity levels for polio have dropped to what many authorities believe is the danger level. At the moment, public resources are being heavily committed to reversing this trend.

Travelers expose themselves to new diseases when they leave their country of origin. International vaccination recommendations are prepared and continually updated by the World Health Organization.

Military personnel are exposed to similar risks; in a combat situation, they will also have to endure physical hardships that increase their susceptibility. Recruits are routinely immunized against a wide range of potential pathogens.

Medical and veterinary personnel have unusual occupational hazards. They may find it advisable to obtain immunization against agents for which the population at large does not need routine protection. In addition, they need to be much more alert to detect incipient disease.

Controlling Animal Diseases

Dairy herds are regularly tested for tuberculosis and brucellosis; infected animals are destroyed. The reservoir of rabies is wild animals. People are rarely exposed by contacts with the wild reservoir. Recently there have been many rabies exposures because families have adopted wild animals, such as de-scented baby skunks, that later develop rabies. People are often exposed by a two-stage process. A dog or other pet is bitten by a rabid wild animal, contracts the disease, then bites its owner. The almost universal requirement that evidence of rabies vaccination be presented before a dog may be licensed minimizes this risk. However, there are many unvaccinated and unlicensed dogs, often running loose, and they are responsible for most human rabies exposures.

Rodents and their insect parasites transmit specific infections such as the bubonic plague. They also increase infection rates by the nonspecific spread of garbage and filth. Rat control is constantly necessary in urban areas.

Controlling Insects

Where arthropod-borne diseases are endemic, there is usually a large reservoir of infected individuals. Effective immunization is rarely possible, but aggressive insect control may yield good results (Fig. 15.11). Mosquitos can be controlled by draining their stagnant water breeding areas, oiling the surface to kill larvae, or applying pesticides to kill adults. Spraying habitations can eliminate biting bugs. Improving the opportunities for personal hygiene will eliminate lice. If insects are uncontrolled, protective clothing, effective repellents, and screened sleeping quarters may be effective.

As epidemiology has developed, the prevalence of preventable infectious disease has plummeted. Epidemics used to be common, evoking a feeling of terror and helplessness. They are now rare (and newsworthy) and actually touch very few of

15.11
Control of Insect Vectors   Killing mosquito larvae by
spraying pesticides, to control the spread of malaria. This
swampy area below the citadel of Kabul in Afghanistan is a
prime breeding ground for this vector. Preventing the maturation of the larvae will stop the transmission of the
malaria parasite.

our lives. More and more diseases are being added to the list of preventable diseases.
We are being effectively protected against many avoidable hazards. We must remember, though, that this is an ongoing campaign. Although the disease becomes
uncommon, the disease agent is usually still around. Unrelenting preventive efforts
will continue to be needed for the foreseeable future. Every person has an important
stake in disease prevention.

SUMMARY

1. Epidemiology is the study of the occurrence and spread of disease in populations. Its goals are the control and eventual elimination of communicable and
noncommunicable disease, and the institution and maintenance of public health
safeguards.

2. Communicable disease agents pass from a reservoir of infection via an exit
route, carried by one or several modes of transmission. They then make entrance
to a susceptible host. Living reservoirs include human beings, animals, and
arthropods; nonliving reservoirs include soil, water, air, and food. Direct means
of transmission such as contact, inhalation, ingestion, and arthropod bite transmit organisms that have little capacity to survive in a hostile environment. Indirect transmission, by contact with contaminated objects, food, or water, trans-

mits hardier microorganisms. The act of transmission is almost always undetected.

3. Population susceptibilities vary widely. The introduction of an agent may or may not cause epidemics; outbreaks may vary in their severity. Adverse physical and social conditions, certain patterns of age distribution, and low levels of herd immunity all increase population susceptibility.

4. Epidemiologists investigate outbreaks. An epidemic becomes evident when notifiable diseases or other unusual states are reported to epidemiological officials. These officials carry out field investigations, attempt to isolate an infective agent, and to get the most complete possible identification. They search for the reservoir(s) and the means of transmission responsible for the epidemic. The outbreak can then be arrested by whatever means seem most practical.

5. Public health measures include provision of safe water supplies and systematic collection and treatment of sewage. Techniques in these areas are continually improving, as are the stresses on the systems imposed by an ever-increasing demand for fresh water, most of which is promptly turned into an ever-larger volume of sewage.

6. Food safety is improved by meat inspection, restaurant and processor inspection, pasteurization of dairy products, testing of dairy herds, and strict controls on cannery operation.

7. Human reservoirs are detected by case followup and routine screening. Universal immunization for certain serious diseases is recommended, and special immunizations are provided for travelers, military personnel, and medical and veterinary workers.

8. The effect of animal diseases on the human population is minimized by vaccination or elimination of infected animals. Arthropod control is important in reducing arthropod-borne diseases.

Study Topics

1. List the communicable diseases that are endemic, sporadic, and epidemic in your community. How might these differ from those found in an equatorial community or a research colony in Antartica?

2. What methods of treatment are used for your household water supply and your household sewage? What are their strengths and weaknesses?

3. What arguments can be advanced for and against 100 percent vaccination programs?

4. Suppose you were investigating an outbreak of gastrointestinal disease that had struck a large number of individuals following attendance at a banquet. What steps would you take?

Bibliography

Books

Austin, D. F., and S. B. Werner, 1974 *Epidemiology for the Health Sciences*, Springfield, Ill.: Charles C Thomas.

Benenson, Abram S. (ed), 1975, *Control of Communicable Diseases in Man* (12th ed.), Washington, D.C.: American Public Health Association.

Friedman, Gary D., 1974, *Primer of Epidemiology*, New York: McGraw-Hill.

Standard Methods for the Examination of Water and Wastewater (13th ed.), 1971, Washington, D.C.: American Public Health Association.

Periodicals

The following periodicals are available free to any health professional or student on request. They contain the latest information on notifiable and other infectious diseases, case histories of epidemics, and updated immunization recommendations.

Morbidity and Mortality Weekly Report. Published weekly by Center for Disease Control, Atlanta, Georgia 30333.

State and City Epidemiology Newsletters.

III

INFECTIOUS DISEASES

Corncob Plaque

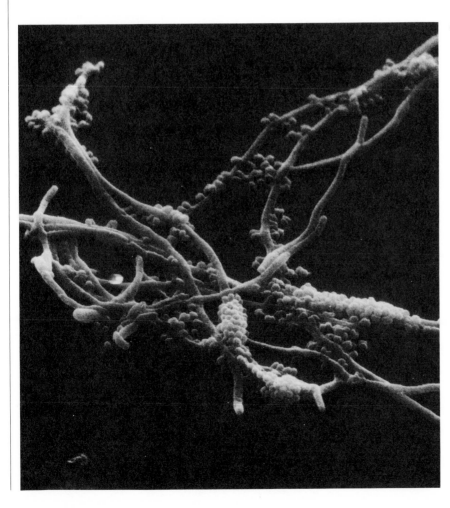

16

Oral
Disease

We begin our systematic study of the microbial diseases of human beings with a survey of oral disease and also with the most universal kind of oral disease—tooth decay and tooth loss. Most persons are completely unaware that dental caries (tooth decay) is an infectious disease process. It is thus completely controllable, at least in theory, by the same methods we use to deal with other microbial disease. Much current dental practice is therapeutic, devoted to remedying the results of past neglect and microbial destruction. However, we have right now the necessary information and techniques to bring the rampant epidemic of dental disease under control. All that is needed is the active participation of every individual in the population at risk, which means everyone who has natural teeth. Against no other infectious disease do we have quite such a good chance to protect ourselves. Furthermore, no one else can really do it for us. An understanding of the microbial processes that produce dental disease should help create the necessary motivation to protect against it. In no other area of health care does the responsibility for maintaining health rest so clearly on the individual.

467

Box 16.1

The Benefits of Civilization

Most primitive peoples enjoyed very good dental health, although they may have suffered severely from other infectious diseases. The coming of civilization has often reversed this picture, and as tribes were relieved of the burdens of smallpox, cholera, or other plagues, they developed epidemic oral disease in exchange.

This pattern was documented by dental officers of the Royal Navy on periodic visits to the remote, inhospitable island group of Tristan da Cunha, located in the middle of the South Atlantic. In 1932, the population of 162 inhabitants subsisted on a Spartan but balanced diet of potatoes, fish, eggs, and a little milk and vegetables. Only when a ship called (usually less than once a year) did the inhabitants have the chance to obtain white flour or sugar. After 1932, ships docked more and more fre-

quently; eventually the Tristan Development Company set up a trading operation. By 1955, although the staple foods had not changed much, the inhabitants were supplementing their diets with 4.5 lb of white flour and 2 lb of sugar per week per person. The effect on dental health is starkly illustrated by the following figures.

DATE	PERCENTAGE OF PERSONS WITH ALL TEETH FREE FROM CARIES	PERCENTAGE OF PERSONS WITH HEALTHY GUMS
1932	83	96
1937	50	69
1952	22	58
1955	12	48

Source: Adapted from J.C. Drummond, 1959, *The Englishman's Food*, London: Readers Union, pp. 161–164.

ANATOMY AND PHYSIOLOGY OF THE ORAL CAVITY

The portion of the oral cavity with which we will be concerned contains the structures designed for the intake and chewing of foods, specifically tongue, salivary glands, and the teeth and their supporting structures.

Anatomy of the Mouth

Figure 16.1 shows the major structures of the tooth and gum. Note the great variety of tissue types that provide a diversity of microenvironments for microbial growth.

Hard versus soft surfaces   The inner cheek, gum surfaces, and tongue are all soft, and the tongue surface is very porous. The oral epithelium is not reinforced by the protective structural protein, **keratin**, except on the gingiva (gum) nearest to the tooth, where there is extensive wear and tear.

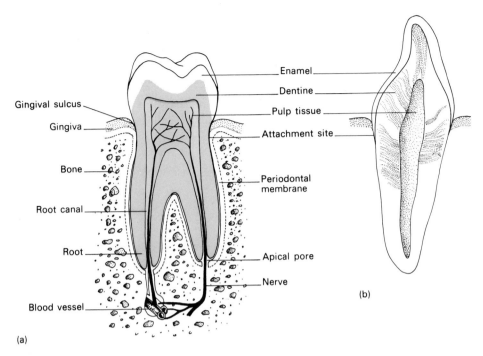

16.1
(a) Diagram cross section of a molar or grinding tooth. (b) Cross section of an incisor or cutting tooth, ground to show internal structure. Note the raylike appearance of the dentine caused by the dentine tubules.

Intact tooth surfaces are impermeable. Normal tooth enamel is the hardest substance in the body. When perfectly clean, it resists microbial adhesion and chemical attack.

Renewable contrasted with nonrenewable surfaces   The soft epithelial surfaces are readily colonized by microorganisms, but they are also renewable, so the surface layers slough off regularly. This creates an enforced turnover of the adherent microbial population and places an upper limit on its numbers.

Tooth surfaces are nonrenewable. If microorganisms successfully attach themselves via organic films, they and their descendants tend to remain and increase in number, unless they are mechanically removed by eating coarse food, brushing, or dental prophylaxis. Furthermore, damage is not naturally repaired.

Tooth structure   A tooth is a layered structure made up of several different materials. The **enamel** is a crystalline substance made up of various salts of calcium and phosphate plus an organic matrix of protein. The underlying **dentin** is similar in chemical composition, but less hard. Dentin is arranged in tubules that render it porous, so introduced substances and microorganisms can percolate slowly through it. Normal dentin is opaque and gives the tooth its white color. Neither enamel nor dentin contains living cells. The **pulp** is soft and cellular, with a blood supply and nerve endings that enter through the tiny **apical pore** at the end of the tooth root.

Tooth attachment   The tooth is attached to the underlying **alveolar bone** by a system of **ligaments** made up of calcified and uncalcified collagen fibers. These tie the bone to the **cementum** lining the tooth socket. Collagen fibers also reinforce the attachment of the gingiva to the cementum and to the tooth surface at the point at which the enamel coat ends. The **gingival sulcus** is the pocket in which gum overlaps tooth.

Physiology of the Mouth

The mouth is an extremely hospitable environment for microorganisms. Its temperature is warm and it is uniformly moist. Because of the periodic presence of foods, nutrients are plentiful. Different physicochemical conditions prevail in various areas of the mouth, establishing a series of distinct microenvironments.

Saliva   **Saliva**, the secretion of the salivary and mucosal glands, is the main conditioning feature of the oral environment. About 1,500 mL of saliva daily pour through the mouth and are swallowed. The saliva normally has a pH of about 6.8 and contains high concentrations of calcium ions and dissolved oxygen. In addition, it bears significant amounts of secretory IgA and traces of other immunoglobulins. Microbial metabolism tends to alter the saliva's pH and to consume the dissolved oxygen, but the continuous supply of saliva maintains homeostasis in those areas it penetrates freely.

pH   The pH of oral secretions is raised by addition of the weak base **urea**, a saliva constituent. The ingestion of acid foods such as coffee or citrus juices lower the pH. Microorganisms also produce acid fermentation products from carbohydrates. Because fermentation is an anaerobic activity, acid production is affected not only by carbohydrate intake but also by oxygen availability.

Eh; Table 8.2   Eh   The redox (oxidation–reduction) potential or Eh varies greatly among oral microenvironments. The available supply of oxygen is provided in the form of dissolved oxygen in the saliva, and consumed by aerobic microbial metabolism. Thus the effective Eh of an area reflects the balance between saliva access and the biomass of metabolizing aerobic microorganisms. Expansion of anaerobic areas coincides with the appearance of pathological signs.

Oral Microenvironments

The groove between the side of the tooth and the fold of gum that extends over it is normally no more that two millimeters deep (Fig. 16.2). It tends to

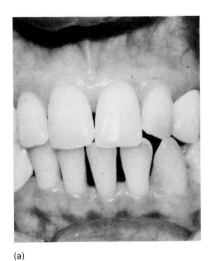

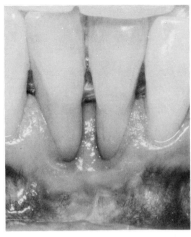

16.2
(a) Normal, clean healthy teeth. The gums are smooth, not swollen. There is no accumulated material on the tooth surface. (b) A diseased region. The plaque is prominent at the gum line and between the teeth. It has caused inflammation and the gingiva and interdental papillae are swollen and malformed. The neglect that caused this state has been going on for a long time; periodontitis has caused the retreat of the gum and exposed the roots of these teeth.

(a)   (b)

be an anaerobic environment that readily collects food debris and microbial agents. The microbes metabolizing the food reduce available oxygen levels to an Eh value that permits the growth of strict anaerobes. The sulcus is flushed by immunoglobulin-bearing tissue fluid that helps in clearing debris.

Gingival surfaces   These bear a relatively tough, keratinized epithelium usually colonized by aerobic organisms.

Inner cheek surface   This is soft, nonkeratinized tissue, replaced about every four days. Like the gingival surface, it is constantly washed by saliva that minimizes accumulations.

Tongue   The tongue surface is soft and highly porous because of the pits containing taste buds. Microorganisms accumulate forming a film or coating, especially during sleep when cleansing mouth movements are at a minimum. The flora is aerobic.

Tooth surface   The normal tooth surface is extremely smooth and shiny, and foods do not adhere. However scratches, cracks, pits, and fissures are acquired with age and hard use. These provide improved opportunities for microbial adherence and growth. Microbial growth is aerobic on exposed surfaces, but may become anaerobic in pits or fissures or within adhering accumulations of organic matter and bacteria.

NORMAL FLORA OF THE MOUTH

All major microbial groups except algae may colonize the normal human mouth (Table 16.1). Organisms are always present in large numbers and in complex mixtures. A carefully analyzed sample from one tiny spot may reveal 40 or more bacte-

TABLE 16.1
Distribution of normal oral flora

ENVIRONMENT	NUMBERS	COUNTING METHOD	PREDOMINANT ORGANISMS
Saliva	40 billion/mL 110 billion/mL	Aerobic culture Anaerobic culture	*Streptococcus salivarius* *S. mitis* strains *Veillonella* sp.
Inner cheek epithelium	5–25 bacteria/cell	Microscopic count	*S. mitis* *S. salivarius* *S. sanguis*
Tongue surface	100 bacteria/cell	Microscopic count	*S. salivarius* Gram-positive filaments *S. sanguis*
Gingival crevice debris	16 billion/g 40 billion/g	Aerobic culture Anaerobic culture	Gram-positive filaments *Veillonella* sp. *S. mitis*
Plaque	25 billion/g 46 billion/g	Aerobic culture Anaerobic culture	Gram-positive filaments *S. mitis* *S. sanguis* *S. mutans**

*Variable, from 0 to 50 percent.

rial species. Thus a disease process can rarely be ascribed unequivocally to a single type. Furthermore, the same organisms may be found in both the normal and the diseased mouth. The onset of dental disease is not caused by the introduction of a pathogen. On the contrary, it is an endogenous process, an extension of the normal flora's activities to tissue-destroying levels.

Important Bacterial Groups

Classification of bacteria; Chapter 4, pp. 101–127

Oral bacteria are numerous and directly involved in oral disease. Eight types have major significance.

Streptococci—Gram-positive, chain-forming organisms—are extremely numerous. They are facultative anaerobes and occupy all oral microenvironments to varying degrees. Important metabolic traits are the production of lactic and other acids from carbohydrate, and the manufacture of long-chain, sticky polymers such as **dextran** and **levan** from sugars. These polymers strongly adhere to tooth enamel and become a major component of dental **plaque,** the coating that forms on unclean teeth. *Streptococcus mutans* is presently considered by most investigators to be the major contributor to dental caries.

Lactobacilli are predominantly anaerobic, Gram-positive rods. They are strongly fermentative and highly acid-tolerant. They produce copious acid and survive at pHs low enough to dissolve tooth enamel. Lactobacilli are always found in carious lesions and probably play a role in the decay process.

Gram-negative diplococci such as the aerobic *Neisseria* and *Branhamella* and the anaerobic *Veillonella* are common in the mouth. Their contributions to the oral environment are usually harmless.

Actinomycetes are Gram-variable, anaerobic, often branching rods that clump together to form a mycelial mat. Actinomycetes are found in the gingival sulcus and in deeper levels of surface plaque. They may cause the deposition of calcium salts in plaque, forming **calculus,** a hard, irritating material. Calculus wounds the gum and leads to **periodontal disease** and progressive deepening of the gingival sulcus. *Actinomyces israelii* may cause serious oral and pulmonary disease.

Bacteroides are Gram-negative, anaerobic rods. In the mouth they are restricted largely to the gingival sulcus. Their numbers are greatly increased in gum disease. It is believed that *Bacteroides* produce endotoxin, leading to a destructive and progressive inflammatory response. They also produce proteolytic enzymes that erode the fibrous tooth supports.

Capnocytophaga, an anaerobic, gliding bacterium, is being increasingly studied in relation to its normal and abnormal roles in the anaerobic environments of the mouth.

Fusiforms such as *Fusobacterium* species are filamentous and Gram-variable; they grow in anaerobic areas. Single cells are large and sometimes tapered at both ends. These organisms proliferate in periodontal disease.

Spirochetes such as *Borrelia, Treponema,* and many unnamed spirochetes occur in anaerobic environments of the mouth. *Borrelia,* in association with fusiform bacilli, is found in increased numbers in cases of **acute necrotizing ulcerative gingivitis** (ANUG), more commonly known as trench mouth or Vincent's angina. Their role in producing this state is unclear.

Corynebacteria of the group called **diphtheroids** are universal oral inhabitants. They are found in areas of plaque, and might contribute to periodontal disease, but are usually assessed as nonpathogenic.

Acquisition of the Normal Flora

During and after birth, a baby begins to collect organisms from its environment. Probably every known species of microorganism eventually finds its way into the baby's mouth. Only those microorganisms that are well-adapted to oral conditions multiply sufficiently fast to avoid being eliminated by salivary wash-out. Retention and multiplication depend on the microorganisms' forming a stable attachment to a surface. Some species specifically attach by means of pili or fimbriae. Prior to the eruption of the first tooth, all surfaces are soft and smooth; anaerobic environments are at a minimum, so the flora is largely aerobic. When teeth appear, anaerobic environments are formed. The first or **deciduous** teeth are usually widely spaced, providing few tight spots to trap food.

Subsequent development of the flora will depend somewhat on dietary habits. Firm or rough foods promote cleansing and removal of aggregates, whereas soft or sticky foods tend to adhere. Frequent ingestion of carbohydrates, especially refined sugars, strongly accelerates plaque accumulation while providing a substrate for fer-

TABLE 16.2
Important microbial members of the oral flora

GROUP	SPECIES	O$_2$	RELEVANT PHYSIOLOGICAL TRAITS	ASSOCIATION WITH DISEASE
Streptococci	*Streptococcus salivarius*	FA	Acid, levan production	Infectious endocarditis
	S. mitis	FA	Acid production	Pulpitis, infectious endocarditis
	S. sanguis	FA	Acid, dextran production	Stomatitis
	S. mutans	FA	Acid, dextran, levan production	Caries, infectious endocarditis
Lactobacilli	*Lactobacillus acidophilus*	FA	Acid production, acid tolerant	Caries
	L. casei	FA	Acid production, acid tolerant	Caries
Gram-negative diplococci	*Neisseria sicca*	AE	Weak acid production	Plaque formation
	Branhamella catarralis	AE	Nonfermentative	Caries, tongue lesion
	Veillonella parvula	AN	Nonfermentative	Plaque formation, mixed infections
Actinomycetes	*Actinomyces israelii*	AN	Proteolytic, acid production	Actinomycosis
	A. naeslundii	AN	Mineralization, acid production	Periodontitis, caries, actinomycosis
	A. viscosus	AN	Acid production	Periodontitis, root caries, actinomycosis
	Bacterionema matruchotii	FA	Mineralization	Periodontitis
	Rothia dentocariosa	AE	Produces lactose from glucose	Caries
Bacteroides	*Bacteroides oralis*	AN	Proteolytic	No pathogenicity
	B. melanogenicus	AN	Proteolytic, produces collagenase, endotoxin	Periodontitis, ANUG
Fusiforms	*Leptotrichia buccalis*	AN	Acid production	No pathogenicity
	Fusobacterium necrophorum	AN	Acid production	ANUG
Spirochetes	*Treponema macrodentium*	AN	Nutritionally fastidious	ANUG
	T. vincentii	AN	Nutritionally fastidious	ANUG
Diphtheroids	*Corynebacterium* sp.	V	Heterogeneous	Pulpitis, root canal infection
	Arthrobacter crystallopoietes	AE	Normally found in soil	Root surface caries

FA = facultative anaerobic; AE = aerobic; V = species vary; AN = anaerobic.

mentative acid production. Sugar is also a selective influence favoring the growth of caries-producing streptococci and lactobacilli. The proportion of these two cariogenic groups drops dramatically when a completely sucrose-free diet is adopted (Table 16.2).

Microbial numbers also fluctuate with oral hygiene. Brushing, flossing, and professional cleaning not only reduce total microbial mass but also shift the species distribution away from cariogenic types. The removal of plaque restricts the extent of the anaerobic environments in which disease processes start and progress. The healthy mouth contains a predominantly aerobic or facultatively anaerobic, acidogenic flora. The diseased mouth's flora shifts toward anaerobic, proteolytic species.

Natural Defenses of the Mouth

Any effect that tends to reduce the number of oral bacteria, or control harmful activities such as acid production, could be called defensive. The oral defenses are numerous, and appear to have been quite effective until the recent increase in sucrose consumption launched a potent dietary assault. The shedding of epithelium to regularly rid surfaces of colonizers, the flow of saliva, and the partial control of oral pH through salivary buffering have already been noted.

Saliva contains many microbial inhibitors. These include **lactoferrin,** an iron-binding substance; the enzyme **lysozyme,** which kills Gram-positive bacteria by destroying their cell walls; and hydrogen peroxide and thiocyanate, which are bactericidal when enzymatically activated. **Dextranase,** an enzyme that depolymerizes dextran, is produced by up to 14 percent of oral flora and appears to reduce colonization of *S. mutans.* Secretory IgA (sIgA) is present in all oral secretions, and IgG and IgM appear in the tissue fluid that leaks into the gingival sulcus. sIgA coats and prevents attachment of bacteria to epithelium and teeth. However, there is little evidence that specific antibody can totally control normal oral flora, especially in the presence of overwhelming growth.

Leukocytes are found in many areas of the oral cavity, and the inflammatory response occurs when conditions are abnormal. For example, when plaque accumulates in the gingival sulcus, the adjacent gum becomes inflamed. When caries invades the pulp, inflammation (pulpitis) will occur in an effort to eliminate microorganisms. When invaders pass through the apical pore into the underlying tissues, abscess and granuloma formation occur there. At each stage, inflammatory defenses attempt to prevent further spread, but they may not readily eliminate the cause of the irritation. Thus the inflammation tends to become chronic. Damage done is often not reversible by healing, but requires dental reconstruction.

PLAQUE

Plaque is a coating formed of microorganisms plus organic matrix. It forms on all tooth surfaces that are not frequently and adequately cleaned. It is not formed solely from food, as plaque accumulation continues in the mouths of persons fed only by gastric tube for extended periods. However, frequent ingestion of sugar accelerates plaque accumulation, the necessary first step in the development of all dental diseases. If we could completely prevent plaque buildup, most dental disease as we know it would cease to exist.

Plaque Formation

There is essentially no such thing as a completely clean tooth. When your teeth are carefully cleaned and polished by the dentist, that desirable state lasts only for about five minutes before the sequence of events that produces plaque begins again. Routine home care does not prevent plaque deposit, but does periodically remove part or all of the accumulation.

Acquired pellicle   The clean enamel surface is so smooth that few organisms can adhere to it directly. However, the enamel has a slight negative surface charge, and therefore is coated by cationic salivary proteins immediately after it is cleaned. This **acquired pellicle** is delicate, and it is readily removed by mechanical rubbing. However, the pellicle is soon replaced.

True plaque   Dome-shaped deposits of aerobic cocci and filamentous bacteria attached by one end rapidly appear on the pellicle (Fig. 16.3). Because bacterial surfaces are negatively charged, the attraction is mainly electrostatic. The microorganisms come from pits in the enamel, saliva, and the other oral surfaces. In ecologists' terms, this is the **primary community**. These early colonists of the community are aerobic.

16.3
Plaque   A scanning electron micrograph of corncob plaque reveals its microbial structure. Filamentous bacteria adhering at one end to the tooth or plaque surface are surrounded by a coating of cocci.

Microbial growth, plus the precipitation of more salivary protein, leads to buildup of plaque biomass. Plaque resembles a connective tissue, with living microbial cells embedded in a large amount of organic matrix. As plaque matures, layers nearest the tooth surface undergo **stagnation** as they become more isolated from salivary dissolved oxygen, pH buffering, and calcium ions. The flora in these regions shifts to predominantly anaerobic forms. This is a **secondary community**, dependent on environmental conditions created by the primary community.

Dextran is formed from the glucose moiety of the sucrose molecule when this is hydrolyzed by certain organisms, most notably *Streptococcus mutans*. Dextran is believed to stabilize the initial attachment of microorganisms to teeth as well as contributing to a rapid increase in plaque biomass. **Dextranase,** an enzyme produced by other microbial species, may have a protective role in minimizing these effects.

Microbial Activities in Plaque

Microorganisms are present and active at all depths in the plaque. Their effect depends greatly on diet. Soluble molecules such as sugars, acids, and oxygen diffuse slowly into and out of the plaque.

Acid production   If microelectrodes that detect pH are placed under the plaque next to the tooth surface, changes in pH following exposure to different foods can be detected. When sugars are ingested, they diffuse into the plaque. Then pH rapidly drops as a result of microbial fermentation (Fig. 16.4). Acid production goes on after the sugary food has been removed from the mouth because some sugar is trapped in the plaque; the acid end products are also trapped at the tooth surface. When between-meal consumption of sugar occurs frequently, pH levels rarely return to normal, and the enamel is exposed to a continuous acid assault.

Toxin production   Microorganisms in plaque at or below the gumline produce inflammatory endotoxins. They also release exoenzymes such as collagenase and chrondroitin sulfatase.

16.4

Effects of Exposure to a 10 Percent Glucose Solution   A microelectrode placed in the plaque measures the sharp fall of pH as lactic acid bacteria in the plaque convert the sugar to lactic acid. A pH minimum is reached. The critical pH, which varies slightly among individuals, is that level at which calcium salts on the tooth surface begin to dissolve. The critical level is usually at around pH 5.0. Note that diffusion of acid away from plaque, shown by rising pH, is slower than pH drop. The thicker the plaque or more prolonged the sugar contact, the longer the pH will stay at or near the critical point with resulting damage to the tooth.

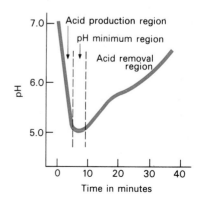

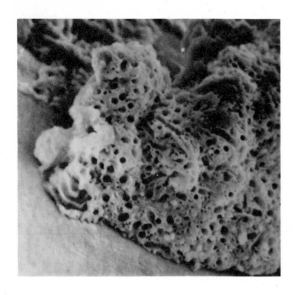

16.5
Calculus   This SEM of calculus looks like a coral growth. The resemblance reflects structural similarities. The structure of calculus is an accretion of calcium salts. The pores are holes in which bacterial cells were embedded before the specimen was prepared for the microscope.

Calculus formation   Soft plaque that forms at or under the gingival margin tends to calcify with time, forming **calculus** (Fig. 16.5). This material differs from soft plaque because it resists mechanical removal; it cannot be brushed off. In calcification, calcium and phosphate ions deposited from saliva come to constitute up to 80 percent of the mass. Some investigators believe this is a defensive reaction rather like the walling off of a tuberculous lesion, and that the host tissues are mainly responsible.

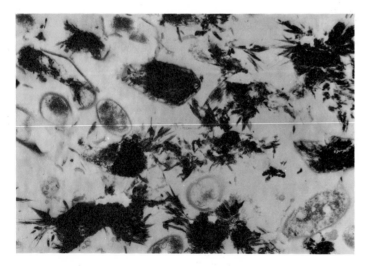

16.6
Bacterionema matruchotii   When *B. matruchotii* strain 14266 is incubated *in vitro* in a calcifying medium for four weeks, large numbers of dark, pointed calcium phosphate crystals form both intracellularly and in the medium. It is believed that this organism contributes to the creation of calculus in the oral cavity.

However, it seems more likely that microorganisms cause the condition. *Actinomyces naeslundii* and *Bacterionema matruchotii* (Fig. 16.6) in particular can cause calcium phosphate precipitation.

DENTAL CARIES

On exposed tooth surfaces, the end result of microbial metabolism within plaque is caries. This process is extremely widespread in developed nations. Of persons presently reading this page, only one out of twenty does not have a developing lesion! Of teeth requiring extraction from 15- to 24-year-olds, almost 80 percent are lost to rampant caries.

Acid Decalcification

Caries is a chemical attack on an extremely hard, resistant material. Enamel is a mixture of calcium salts and proteins, both of which can be dissolved at acid pHs. Recall that when an ionic compound dissolves, its ions dissociate one from another. In the presence of many free ions, an equilibrium will exist in which dissociation is equalized by the tendency of some ions to reassociate to form the parent compound. We have already seen that in some areas of plaque, calcium salts accumulate. In other areas of plaque, in the presence of acid, they will dissociate and the enamel will decalcify.

The pH at which the saliva buffer fails to protect the enamel from dissolution is the **critical pH**. This varies somewhat with tooth structure and local conditions. In general, when the pH falls to about 5.0 (Fig. 16.7), material is lost from the enamel surface and a white spot forms. This may be an indicator of developing caries. The rate of progression of the lesion depends mainly on the extent of acid exposure the tooth endures.

The calcium ions in saliva inhibit acid decalcification by shifting the equilibrium to favor reassociation of the inorganic matrix. Fluoride ions in drinking water, foods, or dentrifices stabilize calcium salts and reduce dissociation at low pH.

Caries cannot occur without plaque. This is because plaque provides a lodging point for acid-forming bacteria on the tooth surface, an anaerobic region for fermentation, a trap to hold and concentrate the produced acids, and a barrier to stabilizing salivary calcium ions.

Progression of Caries

The original lesion enlarges slightly every time the local pH drops to the critical point. Over a period of weeks or months, the cavity extends into the dentin, where microorganisms and their products spread more rapidly through the dentin tubules,

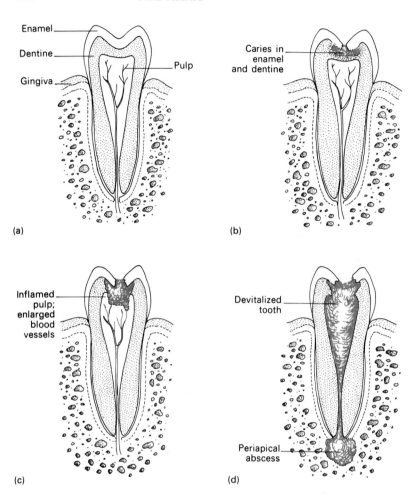

- Enamel
- Dentine
- Gingiva
- Pulp

(a)

- Caries in enamel and dentine

(b)

- Inflamed pulp; enlarged blood vessels

(c)

- Devitalized tooth
- Periapical abscess

(d)

16.7
The Progression of Untreated Caries
(a) Normal tooth. (b) A carious lesion developing on the crown of a molar. This is an area in which plaque readily accumulates. The enamel has eroded and the lesion now involves the top layers of the dentine. It can be readily treated at this stage. (c) The dentine has been penetrated, and microbial invaders have reached the pulp. The inflammatory response here would be causing acute pain. This tooth can be saved with extensive reconstruction. (d) An abscess has formed in the bone. At this point, the individual's general health is also threatened.

and soften it (Fig. 16.8). Eventually, the lesion penetrates the pulp cavity, where an inflammatory response to microbial invaders causes acute pain. The live cells of the pulp may be killed as a result and the tooth **devitalized.** An infection that spreads beyond the tooth may cause **periapical abscesses** in the underlying bone or at the gum surface (gum boils). In some cases, spreading infections of the bone (**osteomyelitis**) may be the outcome.

Dentists treat caries of the enamel and dentin by removing affected material and replacing it with a hard, reconstruction material such as amalgam—the common **filling.** Caries may reactivate under the filling or around its borders if the cavity was inadequately sterilized or sealed. In addition, the margins of the filling are a prime location for new plaque development. If the pulp has been affected, the pulp cavity and root canals must be entirely emptied of cellular material, any adjacent abscesses

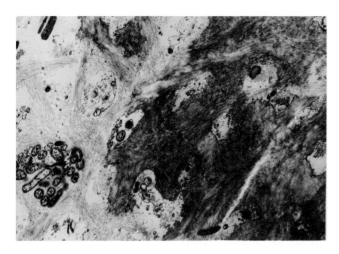

16.8
Serratia marcesens cells (dark ovals) invading dentinal tubules from the pulp canal (light area on left). Once bacteria gain entrance to the dentine layer, they may penetrate rapidly to the pulp cavity or to other areas of the dentine.

drained, and the cavity sterilized and filled. Such devitalized filled teeth may well last indefinitely.

PERIODONTAL DISEASE

Periodontal (around the tooth) **disease** is the inflammatory involvement and potential destruction of the soft tissues supporting the tooth and its roots. It is prevalent in older age groups; as much as 97 percent of the population is affected to some extent by age 45. From that age on, about 80 percent of tooth loss is caused by this condition. Like dental caries, periodontal disease is almost completely preventable with conscientious tooth care.

There are three clinically recognizable common forms of periodontal disease. All are endogenous and develop because of uncontrolled proliferation of microbes normally present in the gingival sulcus.

Gingivitis

Gingivitis is the most common periodontal abnormality. It is an inflammation of the gingiva, which become swollen, deeply red in color, and bleed easily. There may be excessive gum growth over the tooth surface. However, the side of the tooth does not separate from its attachments.

The underlying cause of gingivitis is accumulated microbial biomass and debris (plaque and calculus) in the gingival crevice. Material collected from the diseased sulcus may contain four times the viable microbial count of material from the normal crevice.

A combination of factors, including wounding by calculus, bacterial endotoxins, and antibody-mediated complement activation, cause a classic inflammatory response. In addition, hormonal factors such as pregnancy and psychological stress increase susceptibility.

The physical condition of the gingiva is profoundly affected by hormonal changes in the body. Pregnant women often notice this. Hormonal changes during pregnancy make the gums much more susceptible to irritation and bleeding. Inflammation subsides when the provoking microbial aggregates and calculus are removed.

Periodontitis

Periodontitis follows untreated gingivitis. If severe inflammation persists anywhere in the body, we have seen that it can involve and kill adjacent deeper tissues. This is also true in the periodontal region.

The attachment of gingival epithelium to enamel retreats, expanding the debris-containing pocket. The pocket becomes profoundly oxygen-depleted, selecting for strict anaerobes that produce hydrolytic enzymes. Hyaluronidase depolymerizes the intracellular glue of epithelial tissue, destroying its normal architecture. Chrondroitin sulfatase digests the viscous ground substance of underlying connective tissue. Collagenase breaks down fibers binding the tooth. The development of a **pocket** exposing the neck and roots of the tooth is diagnostic of periodontitis. The detached tissue surfaces are usually ulcerated, with pus formation (pyorrhea).

There is evidence that a CMI response involving T-lymphocytes reactive to the pocket flora is important in the later aspects of this state. Lymphokines are released. One of the lymphokines is an **osteoclast-activating factor.** This factor causes osteoclast bone cells to remove calcium salts from bone. Bone resorption, in concert with fiber destruction, loosens the tooth in its socket; it eventually falls out. In addition, the abnormally exposed roots become prone to **root caries.**

Acute Necrotizing Ulcerative Gingivitis (ANUG)

ANUG is a specific clinical state rarely found in individuals under 13 or over 30 years of age. It was once believed to be communicable, but more recent analysis suggests that it occurs in individuals of the same age who share the same stresses, not necessarily the same silverware.

ANUG develops suddenly, with necrosis of the interdental epithelium of the gum. Pain, bleeding, and a whitish coating or membrane will appear. In deep lesions, large numbers of *Capnocytophaga* (Fig. 16.9), *Eikenella* (Fig 16.10), spirochetes, and fusiforms proliferate.

ANUG responds promptly to antibiotic administration and local cleaning. If ANUG is not quickly treated, residual tissue deformity may result. Scarred areas are susceptible to other subsequent periodontal disease.

ANUG patients usally have a history of acute anxiety or stress. The condition is common in military recruits or in students during examination periods. Elevated levels of certain steroids can be demonstrated. Thus ANUG appears to be precipitated

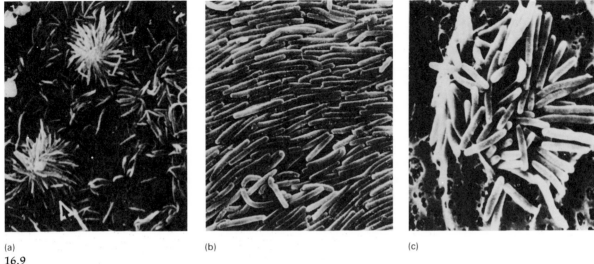

(a)   (b)   (c)

16.9
Capnocytophaga (a) *C. gingivalis,* (b) *C. ochraceae,* and (c) *C. sputigena.*

primarily by altered host defenses permitting shifts in the ecological balance of the normal flora.

Treatment

Therapy for periodontal disease focuses on removal of the offending masses of microorganisms. If deep pockets have already been formed, surgical reconstruction can reduce the depth of the pocket. The patient can then successfully keep the area clean by brushing.

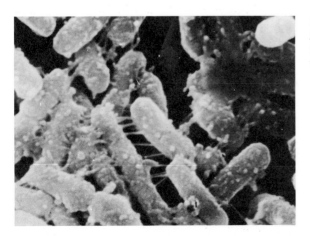

16.10
Eikenella corrodons   This Gram-negative bacillus is often found in the lesions of acute necrotizing ulcerative gingivitis.

OTHER ORAL INFECTIONS

Disease foci in the teeth or supporting tissues may cause disseminated infection. In addition, many microbial diseases of other body systems show oral manifestations (Table 16.3).

Infections of Bone

Osteomyelitis   Bone infection may occur wherever microorganisms enter spongy bone or red marrow. The agents may also migrate under the **periosteum** (the membrane covering the bone). Tooth pulp infections caused by advanced caries, septic reconstruction, or trauma may spread from the tooth to the jawbone via the apical pore. Extraction or severe periodontitis may also introduce organisms into bone.

TABLE 16.3
Bacterial diseases with oral manifestations

DISEASE	AGENT	PRIMARY INVOLVEMENT— SYSTEM OR ORGAN
Actinomycosis	*Actinomyces israelii;* other *Actinomyces* sp.	Skin, lung
Anthrax	*Bacillus anthracis*	Skin, lung
Diphtheria	*Corynebacterium diphtheriae*	Nasopharynx
Gonorrhea	*Neisseria gonorrhoeae*	Urogenital tract
Leprosy	*Mycobacterium leprae*	Skin, peripheral nerves
Meningitis	*Neisseria meningitidis*	Central nervous system
Pneumonia	*Streptococcus pneumoniae Klebsiella pneumoniae*	Lower respiratory tract
Scarlet fever	*Streptococcus pyogenes*	Upper respiratory tract
Pharyngitis Staphylococcosis	*Staphylococcus aureus*	Skin, any body area
Syphilis	*Treponema pallidum*	Urogenital tract
Tetanus	*Clostridium tetani*	Neuromuscular
Tuberculosis	*Mycobacterium tuberculosis*	Lung
Tularemia	*Francisella tularensis*	Lymphoid
Whooping cough	*Bordetella pertussis*	Lower respiratory tract

*Transmission to susceptible dental personnel or clients.

Abscess formation usually occurs first, and with adequate body defenses the organisms are walled off before they reach the marrow area. However, on occasion microbial spread does occur. This results in osteomyelitis accompanied by intense pain, high fever and chills, and necrosis of the affected bony areas. Common causative organisms include α-hemolytic streptococci, *Staphylococcus aureus*, Gram-negative rods such as *Klebsiella pneumoniae*, and *Streptococcus pyogenes*. Antibiotics, sometimes supplemented by surgical drainage, usually arrest the disease.

Dry socket   Following tooth extraction, a clot normally forms over the socket to form the protective meshwork through which healing will occur. In from 1 to 4 percent of cases the clot becomes dislodged, either through mechanical damage or infection. Tissues in the region become devitalized and osteomyelitis may occur. Healing is retarded unless the lesion is recleansed and antibiotics administered.

ORAL MANIFESTATION	POTENTIAL HAZARD OF TRANSMISSION*	INFECTIVE LESIONS OR SECRETIONS	CHAPTER REFERENCE
Cervicofacial swelling, osteomyelitis	Low	Yes	18
Pustules on lip, face, pharynx	High	Yes	17
Lesions of oral mucosa, paralysis of palate	Moderate	Yes	18
Stomatitis, gingivitis	High	Yes	20
Leproma of soft tissues, especially palate	Low	Yes	22
Petechiae of oral mucosa	High	Yes	22
Stomatitis, parotitis, osteomyelitis	Moderate	Yes	19
Rash on soft palate, raspberry tongue	High	Yes	18
Osteomyelitis, parotitis, periapical abscesses	Moderate	Yes	17,23
Primary chancre may be oral, secondary mucous patches, gummae	High	Yes	21
Paralysis of jaw muscles	None	No	22
Ulcers of tongue and other mucosa	High	Yes	19
Stomatitis	High	Yes	24
Oral hemorrhage, tongue trauma	Moderate	Yes	19

Focal Infections

A focal infection is a localized infection which acts as a source of microorganisms that migrate to other areas and "seed" new infections. Microbial spread is usually bloodborne, or **hematogenous**. Such migrating organisms usually colonize only already damaged tissues, or areas of scar tissue.

Bacteremia   Oral microorganisms, either from healthy or diseased mouths, are readily introduced into the blood stream, and **bacteremia** results. Normally, such intruders are promptly removed by the reticuloendothelial network without significant effect on the individual.

Tooth brushing, the use of a toothpick, or chewing rough foods may all cause slight bleeding and bacteremia, particularly if some degree of gingival inflammation is present. With active periodontal disease, about 25 percent of people develop bacteremia routinely on brushing. More extensive introductions occur during dental

TABLE 16.4
Viral diseases with oral manifestations or transmission hazards

DISEASE	AGENT	PRIMARY INVOLVEMENT —SYSTEM OR ORGAN
Chickenpox	Varicella-zoster virus	Skin
Fever blisters	Herpes simplex Type I	Skin, oral cavity
Hand-foot-and-mouth disease	Coxsackie A	Upper respiratory tract, skin
Hepatitis B	Hepatitis B virus	Liver
Herpangina	Coxsackie A	Upper respiratory tract
Burkitt's lymphoma	Epstein-Barr virus	Lymphoid tissue
Influenza	Influenza virus	Lower respiratory tract
Measles	Rubeola virus	Skin, upper respiratory tract
Mumps	Mumps virus	Salivary gland
Polio	Poliovirus	Nervous system
Salivary gland virus disease	Cytomegalovirus	Gastrointestinal, salivary gland, congenital
Smallpox	Variola virus	Skin, upper respiratory tract

manipulations such as scraping of calculus, fillings, and extractions. In 85 percent of extractions **transitory** bacteremia (lasting 5–10 minutes) may be detected with careful laboratory technique. Bacteremia also originates from infected pulp, periodontitis, and jaw fracture.

Infectious endocarditis   This is a bacterial attack on the heart valves. It is a constant threat for the person with heart abnormalities due to rheumatic fever, congenital malformations, or cardiac surgery. Microorganisms may lodge and multiply on the altered surface, and the resulting heart damage may be quite severe. Mortality, even in the antibiotic era, is still around 20 percent. The pathogenesis of the disease is covered fully in Chapter 24.

Infective endocarditis; Chapter 24, pp. 693–696

The patient's medical history reveals the risk of endocarditis. For at-risk individuals, good oral health and hygiene are especially important. These measures minimize risk by reducing the bleeding on brushing and the number of dental manipulations needed. When dental treatment, particularly surgery, is planned, antibiotics

ORAL MANIFESTATION	POTENTIAL HAZARD OF TRANSMISSION*	INFECTIVE LESIONS OR SECRETIONS	CHAPTER REFERENCE
Vesicles on gingiva and mucosa	High	Yes	17
Primary acute stomatitis, recurrent ulcerations of lip and oral mucosa	High	Yes	17
Stomatitis	High	Yes	18
None	High	Yes	20
Oropharyngitis, lesions of tongue and soft palate	High	Yes	18
Tumors of jaw and salivary glands	Low	Yes	24
Inflammation or ulceration of oral mucosa	Moderate	Yes	20
Koplik's spots, inflamed painful mucosa	Moderate	Yes	17
Parotitis, stomatitis, painful chewing	Moderate	Yes	20
Paralysis of jaw	High	Yes	22
None; saliva infective	High	Yes	21
Vesicles on oral mucosa, especially soft palate	High	Yes	17

*Transmission to susceptible dental personnel or clients.

may be given both immediately before and for a short time after the treatment. The highest standards of asepsis are called for in the treatment of these and all dental patients.

Ludwig's Angina

This is a rare mixed infection of the connective tissue in the floor of the mouth. Swelling pushes the tongue upward; pain, fever, and difficulty in swallowing are present. Death due to respiratory obstruction may occur. The source is usually a periapical abscess that perforates the jaw under the tongue. From that site the organisms spread into the soft tissues. Antibiotics and tracheotomy largely prevent fatalities.

TABLE 16.5
Fungus and protozoan diseases with oral manifestations

DISEASE	AGENT	PRIMARY SITE
CAUSED BY FUNGI		
Aspergillosis	*Aspergillus* sp.	Lung, any damaged tissue
Blastomycosis; North American	*Blastomyces dermatitidis*	Lung
Blastomycosis; South American	*Paracoccidioides brasiliensis*	Lung, skin, viscera
Candidiasis	*Candida albicans*	Skin, upper respiratory tract, urogenital, systemic
Coccidioidomycosis	*Coccidioides immitis*	Lung
Cryptococcosis	*Cryptococcus neoformans*	Lung, central nervous system
Geotrichosis	*Geotrichium* sp.	Gastrointestinal, lower respiratory tract
Histoplasmosis	*Histoplasma capsulatum*	Lung
Sporotrichosis	*Sporothrix schenckii*	Skin, lymphoid tissues
CAUSED BY PROTOZOA		
Leishmaniasis	*Leishmania* sp.	Skin, mucocutaneous or visceral lesions (several forms)

Oral Manifestations of Nondental Diseases

Many diseases not directly affecting the teeth and related structures affect the oral tissues. These are covered in other chapters in which the infective agent is considered in connection with its more typical clinical presentation. For ease of reference, however, the oral manifestations of these diseases are presented in Tables 16.4 and 16.5.

Infectious diseases in which oral lesions occur are an important occupational hazard to dental practitioners. Contaminated instruments in the dental office may function as a mode of transmission of disease. High-speed drills, cooled and lubricated by water sprays, spread aerosols as effectively as coughs or sneezes. Microorganisms may be spread over large areas and remain airborne for hours.

ORAL MANIFESTATION	POTENTIAL HAZARD*	INFECTIVE LESIONS OR SECRETIONS	CHAPTER REFERENCE
Inflammatory granuloma of soft palate or epiglottis	None	No	19
Lesions of tongue, palate, and gingiva	None	No	
Lesions of oropharynx, gingiva	None	No	
Thrush (infants) denture stomatitis, root canal infections	Low	Yes	18
Ulcers of lips, oral cavity	None	No	19
Lesions of all soft tissues	None	No	
Stomatitis	None	No	19
Ulcers of tongue	Low	Yes	19
Ulcers of oral soft tissues	Low	Yes	23
Lesion of oral mucosa in mucocutaneous form	Low	Yes	17,20

*Transmission to susceptible dental personnel or clients. Most systemic fungal infections show no evidence of person-to-person transmission; in others, it is rare.

PREVENTION OF ORAL INFECTION

With present techniques one can prevent both dental caries and periodontitis, with constant attention to oral hygiene and correct dietary habits. Only highly motivated individuals actually carry through on such a program. In our society it has become almost a tradition to ignore the personal responsibility for one's own physical welfare. We prefer to turn that responsibility over to the health-care professionals. Since human nature is not likely to change in the near future, practical-minded researchers are searching for methods of achieving extended plaque prevention that will work despite casual oral hygiene and unwise dietary habits. Effective prevention must take into consideration the many interactive factors that produce caries (Fig. 16.11).

Plaque Prevention

Plaque buildup is the predisposing state without which dental disease would not exist. Plaque accumulates in each mouth at a different rate. The present preventive goal is to train persons in effective brushing habits that will frequently and completely remove freshly formed plaque. This has the dual advantage of getting rid of the plaque while it is still soft, before calculus forms, and avoiding thick layers in which anaerobic conditions occur. Brushing also encourages keratinization of the gingiva and stimulates its blood supply. Tongue brushing significantly reduces microbial counts. Flossing removes plaque from interdental surfaces and cleans the gingival crevice.

One useful aid in learning to brush and use dental floss not just vigorously but effectively is the **disclosing stain.** Plaque itself is not readily visible, but if a tablet of nontoxic, tasteless, vegetable dye is chewed after brushing, it will brightly color any plaque that brushing missed. A second brushing can be done to remove the plaque. After using disclosing stain a few times, the person learns which areas need extra attention.

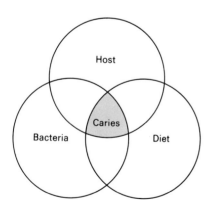

16.11
The Multifactorial Concept of Caries   This diagram, first constructed by P. H. Keyes in 1962, points out that caries develops as a result of the interaction of three factors. There must be certain types of acidogenic and plaque-synthesizing bacteria, the diet must supply ample carbohydrate, and the host must fail either in innate defenses, toothbrushing, or both. If one of these factors can be corrected, caries will be arrested.

The chemical **chlorhexidine** inhibits microbial growth in the oral cavity. In the form of a mouthwash, used daily, the substance binds avidly to the tooth surface. It is released slowly causing great reduction in the numbers of viable oral bacteria. Plaque production is accordingly minimized too. However, chlorhexidine has an undesirable taste and some side effects. Other antimicrobial chemicals and a variety of antimicrobial drugs have also been tried in the form of mouthwashes, dentifrices, and topical applications. None have had unqualified success. Dextranase, the microbial enzyme that breaks down the dextran component of plaque, shows some promise in plaque reduction.

Caries Prevention

In addition to being prevented by plaque control, caries is prevented by other specific activities.

Using fluoride   The fluoride ion, when incorporated into the crystalline lattice of developing enamel, or absorbed to the surface of the fully formed tooth, lowers the pH at which the tooth will start to erode. Thus weaker acid concentrations have little bad effect. The fluoride-deficient tooth will suffer erosion under much less severe stresses. Fluoride is somewhat bacteriolytic; it also inhibits anaerobic metabolism in bacteria. Best effects are obtained when the fluoride is ingested by the pregnant woman during periods of fetal tooth development, and by children in early and later childhood. Topical applications and/or fluoride-containing dentifrices are also highly effective in reducing caries.

Controlling diet   Carbohydrate is a potential substrate for acid-forming bacteria. The **cariogenicity** of the diet varies with the type of carbohydrate ingested and the frequency of ingestion. Sucrose is by far the most damaging carbohydrate because it contributes both to plaque formation and pH drop. Many individuals in the United States get as much as 30 percent of their daily calories from sucrose, often consumed in a dozen or more snacks. This practice sets up the precise oral environment that most favors caries development. Plaque formation is continual, as is the acid insult.

Rational eating habits for caries prevention include minimizing sucrose intake, avoiding frequent or prolonged contact with sugar snacks, and careful mouth rinsing after sugar ingestion if brushing is not practical.

Replacement sweeteners are potentially useful for control of weight, diabetes, and dental caries. The sugar alcohol **xylitol** has many desirable characteristics for decay control. It is not acted upon by the oral flora, but it is regularly metabolized by the human cell.

Vaccination   If vaccination against the cariogenic oral microorganisms became possible, then a single immunization could theoretically protect you against caries for a long period. Vaccines containing whole cells or antigens of *Streptococcus mu-*

Box 16.2

Alternate Sweeteners

Most human beings (and some other animals) undeniably have a "sweet tooth." We love the taste of sweet foods—thus our search for alternates for sucrose. Three groups of people should avoid sugars—diabetics, persons concerned about their teeth, and those who are weight-watchers.

In terms of dental health, the most desirable replacement sweetener is one that bacteria cannot ferment or use as a substrate for dextran production. Sugar alcohols fulfill these criteria. For a diabetic, the sugar should be one that is not metabolized like glucose. Xylitol, a five-carbon sugar alcohol, fulfills this criterion.

Although, unlike saccharin, it is metabolized by normal routes, those routes do not include glycolysis and the Krebs cycle. Thus, xylitol does not contribute to the diabetic's metabolic imbalance.

At the moment, xylitol is the best substitute for sugar we have. It is palatable, it is noncariogenic, and it is tolerated by diabetics. It may cause shifts in the normal bowel flora. Only one group of people would be disappointed in xylitol—those who want to avoid calories. Because it is a harmless, natural food substance, it does what a food ought to do—it provides energy.

SWEETENER	CLASSIFICATION	SOURCE	SWEETNESS COMPARED TO SUCROSE	FERMENTATION BY ORAL BACTERIA	USE AS SUBSTRATE FOR PLAQUE FORMATION	METABOLISM BY HUMAN BEINGS
Sucrose	Disaccharide	Fruits, vegetables	Equal	Rapid	Strong	Rapid
Glucose	Simple sugar	Fruits, vegetables	Less	Rapid	Moderate	Rapid
Fructose	Simple sugar	Fruits, honey	More	Rapid	Moderate	Rapid
Mannitol	Sugar alcohol	Plants, marine algae	Less	Slow	Slight	Slight
Sorbitol	Sugar alcohol	Fruits, vegetables, animal tissue	Less	Slow	Slight	Slight
Xylitol	Five-carbon sugar alcohol	Fruits, vegetables, animal tissue	Equal	None	None	Rapid
Saccharin	Complex	Synthetic	More	None	None	None

TABLE 16.6
Effect of human foods on caries development in rats

	CARIES SCORE, MEAN OF EIGHT RATS	
FOOD MATERIAL TESTED	ADDED TO NONCARIOGENIC DIET	ADDED TO CARIOGENIC DIET
Basic diet only: control group	0.0	11.3
Popcorn	0.0	1.8*
Whole milk	0.0	4.2*
Cabbage	0.0	7.0*
White bread and butter	4.9	22.1
White bread and raspberry jam	10.2	33.0
Rye bread	12.7	29.0
Chewing gum	14.0	27.2
Caramels	16.0	50.4
Baby cookies	18.5	22.0
Bananas	21.0	73.7
Chocolate sandwich cookies	23.8	92.2
Raisins	30.9	61.6
Soft drink	29.6	40.2
Milk chocolate	34.1	41.2
Sucrose	62.1†	51.9

*These foods are anticariogenic because they reduced the caries score below that of the control group.
†Sucrose addition converts a noncariogenic diet to a strongly cariogenic one.
Source: R.M. Stephan, 1966, Effects of different types of human foods on dental health in experimental animals, *J. Dental Res.* **45:** 1551.

The food material was added to a basic noncariogenic or cariogenic diet and fed to groups of eight rats over a prolonged time. Then the rats' teeth were scored for percentage of carious lesions.

tans have been tested in laboratory animals with some effect. There is no consensus that these vaccines would be reliable, effective, and without side effects, in human beings. However, anticaries vaccination remains a future goal.

SUMMARY

1. Deterioration and disease of the teeth and associated structures affect almost 100 percent of the population. Tooth decay and periodontal disease are a result

of the growth and metabolism of endogenous microorganisms. Poor oral hygiene and high-sugar diet tip the microbe–host balance in favor of enhanced bacterial growth.

2. The mouth contains both hard and soft surfaces. It provides both aerobic and anaerobic environments for microorganisms. The moist, warm, nutrient-rich oral cavity is bathed, oxygenated, and buffered by the saliva.

3. The normal flora of the mouth is huge in numbers and diverse in species. The streptococci (acid-formers—some of which produce dextran) and the lactobacilli are implicated in the creation of low pH environments in which tooth enamel will erode. *Actinomyces* and *Bacteroides* are anaerobes active in pits and fissures on the tooth surface and in the gingival sulcus. They contribute to periodontal disease when they multiply excessively. Fusiform bacilli and spirochetes are increased in the lesions of ANUG. All oral disease is related to overgrowth of normally harmless oral microflora.

4. The mouth is defended by the mechanical cleansing effects of the tongue and cheek musculature, salivary bathing, bacterial inhibitors, secretory IgA, and the inflammatory response.

5. Plaque is a mixture of living and dead microorganisms, salivary proteins, and the polysaccharides dextran and levan. Ingested sucrose is used by some bacteria to make dextran. As plaque builds up, the deeper layers become anaerobic, and calcium deposits turn soft plaque into calculus.

6. Within plaque, the microbial mass forms organic acids, toxins, and hydrolytic exoenzymes destructive to the teeth or supporting structures. Dental caries develops when prolonged acid exposure demineralizes the enamel. Calcium salts and acid-soluble protein matrix are both lost. The carious lesion can eventually invade the pulp and devitalize the tooth.

7. Periodontal disease is an inflammation of the gingiva and supporting structures of the tooth, caused by irritating calculus at or below the gumline. Advanced cases—periodontitis—require radical measures. ANUG is an acute, severe form of gingivitis that appears to be brought on by stress.

8. Oral disease prevention clearly calls for microbiological control. Plaque formation can be reduced by minimizing dietary sucrose. Plaque buildup can be controlled by conscientious oral hygiene. Fluoride significantly reduces caries development. Antimicrobial compounds and vaccination are still in the research stage but bear promise.

Study Topics

1. Calcium salts undergo a variety of reactions in the mouth that lead to both dissociation and deposition. Outline these and the factors that favor or inhibit them.

2. If secretory IgA in the mouth is not directed against the oral flora, what might its role be?

3. What is the proposed role of *Streptococcus mutans* in production of caries?

4. Frequent sucrose ingestion contributes to caries production in three ways. What are they?

5. What are the microbial activities in the periodontal pocket?

6. How are teeth loosened by microbial activities?

Bibliography

Bowen, W. H.; R. J. Genco; and T. C. O'Brien (eds.), 1976, *Immunological Aspects of Dental Caries*. Washington, D.C.: Information Retrieval, Inc.

Bowen, W. H., et al., 1975, Immunization against dental caries, *British Dental Journal* **139**: 45–58.

Burnett, George W.; H. W. Scherp; and G. S. Schuster, 1976, *Oral Microbiology and Infectious Disease* (4th ed.) Baltimore: Williams and Wilkins.

Gibbons, R. J., and J. van Houte, 1975, Bacterial adherence in oral microbial ecology, *Ann. Rev. Microbiol.* **29**: 19–44.

Nolte, William A., (ed.), 1977, *Oral Microbiology* (3rd ed.), Saint Louis: C. V. Mosby.

Silverstone, Leon M , 1978, *Preventative Dentistry*, Fort Lee, N.J.: Update Books.

Socransky, S. S., 1970, Relationship of bacteria to the etiology of periodontal disease, *J. Dental Res.* **49** (Supplement 2): 203–222.

Stephan, R. M., 1966, Effects of different types of human foods on dental health in experimental animals, *J. Dental Res.* **45**: 1551.

Stiles, H. M.; W. J. Loesche; and T. C. O'Brien (eds.), 1976, *Microbial Aspects of Dental Caries*, Washington, D.C.: Information Retrieval, Inc.

17

Skin Infections

Skin covers all of our exposed surfaces, forming our major point of contact with the microbe-filled, outer world. Skin is structurally and physiologically adapted for the role of microbial barrier. Only exceptional microorganisms can infect or pass through the undamaged skin.

However, because of its exposed position, the skin is constantly at risk of trauma, and invasion may follow trauma. Skin infections, although usually effectively localized, have the potential to invade deeper tissues. They may spread to other parts of the body via blood or lymphatic circulation. Similarly, infective agents, particularly viruses, may be brought to the skin from systemic multiplication sites.

ANATOMY AND PHYSIOLOGY OF THE SKIN

Skin is an organ. It is composed primarily of epithelial and connective tissue, and also contains glandular, circulatory, nervous, and muscular elements. Its environment is hostile to most microorganisms. Its extensive normal flora is composed only of certain specially adapted types.

Anatomy of the Skin

Epidermis   The outermost layer of the skin is the **epidermis** (Fig. 17.1). It is non-vascular epithelial tissue of variable thickness. Epithelial tissue is characterized by regularly arranged, tightly bound cells cemented together by an intercellular muco-polysaccharide called **hyaluronic acid.** This tissue is an effective barrier to most microorganisms. A deep layer of mitotic cells replenishes shed or damaged cells, providing steady replacements. They progressively flatten, dehydrate, and acquire a strong, flexible protein called **keratin.** The surface layers are composed of dead, hardened, and protein-rich cells. These layers are continuously flaked off and shed.

Dermis   Beneath the epidermis is a somewhat thicker layer of connective tissue, the **dermis.** This is composed of mucoprotein matrix containing scattered fibers of the proteins elastin and collagen. There are also isolated cells, including fibroblasts, mast cells, lymphocytes, and macrophages. Blood capillaries and lymph sinuses penetrate throughout. Sensory nerve endings of all types are present.

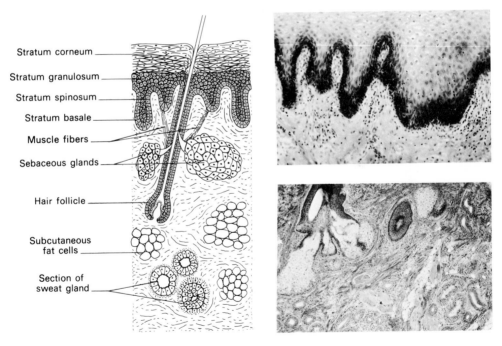

Stratum corneum
Stratum granulosum
Stratum spinosum
Stratum basale
Muscle fibers
Sebaceous glands
Hair follicle
Subcutaneous fat cells
Section of sweat gland

17.1
The Skin   Skin is composed of two layers. The outer layer, the epidermis, is composed almost entirely of dead cells. This hardened surface protects the deeper tissue from microbial invasion and water loss. The underlying dermis is living. It contains pilosebaceous units, capillaries, nerves, and protective white cells.

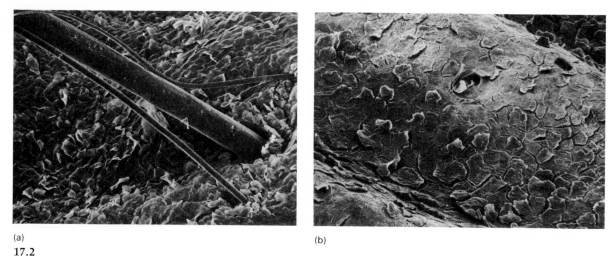

(a)

(b)

17.2

Scanning Electron Microscopy of the Skin (a) The surface of the scalp. Three hairs of different diameter are seen issuing from their follicles. Deep in the follicles, pockets of anaerobic bacteria live and multiply. (b) The surface of the big toe. The opening of one sweat gland at a pore is visible. This surface looks as hostile to life as the surface of the moon. This area requires special adaptations for survival. Microorganisms that live on the skin surface must tolerate salty, often dry conditions.

The dermis contains glandular units. The **sweat glands** are networks of coiled tubules with ducts that pass through the epidermis to the skin surface. They are particularly numerous in the **axilla** (armpit), on the palm, and on the soles of the feet. **Pilosebaceous units** (Fig. 17.2) are complexes containing a **hair follicle** and one or more **sebaceous glands** that secrete **sebum** (oil) to the skin along the hair shaft. Their concentration varies widely over the body.

Subcutaneous tissue   The connective tissue beneath the dermis is not properly part of the skin. It anchors the skin to underlying structures. In some areas, fat cell deposits provide padding, insulation, and reserve energy supplies.

Physiology of the Skin

The physiological state of the epidermis is quite specialized. It is conditioned by the chemical nature of the glandular secretions that bathe it. Sebum is composed principally of fats and oils that coat the skin. Sweat, which serves an excretory and cooling function, contains many salts. As it evaporates, the surface liquids become very concentrated ionic solutions. The pH of the skin secretions is moderately acid. Surface skin layers are, of course, freely exposed to oxygen. Only in the pilosebaceous regions are moderately anaerobic conditions found.

Nutrients to feed the cells of the dermis and lower levels of epidermis arrive via the capillary bed; tissue fluids are drained away via the lymphatics. This network

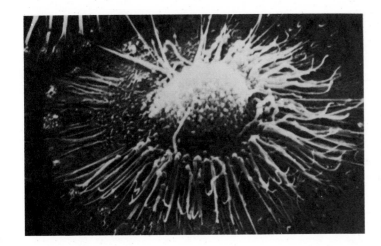

17.3
A Macrophage The biological activity of macrophages, numerous in the dermal layer of skin, is very evident in this SEM of a particularly active cell in culture. The numerous pseudopodia are immunochemical sensory devices, assessing everything they adhere to for self and nonself antigens.

may readily become a vehicle for hematogenous spread of microorganisms that enter the skin. It is also a passageway for defensive macrophages (Fig 17.3).

Nutrients for the growth of skin flora include the fats and oils of sebum, amino acids provided by the sweat, and whatever materials can be scavenged from the dehydrated and generally impermeable surface epithelium. Many skin fungi or **dermatophytes** are able to use keratin readily.

Although moisture is continuously supplied by sweat, its loss through evaporation means that most skin areas are relatively dry. In these environments water for solubilization of nutrients and wastes is not freely available to microorganisms. In general, the skin is a poor environment for the growth of most microorganisms, especially pathogens. Only specially adapted groups are able to survive.

NORMAL FLORA OF THE SKIN

Distribution

Microorganisms, in particular bacteria and fungi, colonize the skin in tremendous numbers. They adhere to all exposed layers of epithelium. Rigorous scrubbing, which removes the loose outer layers, simply exposes the organisms resident on the deeper, vital layers. In the dermis, organisms normally reside only in the pilosebaceous units.

Exact description of the normal flora is made difficult by the fact that many common forms have not as yet been adequately classified. There is also much individual variation in the composition of one's personal normal flora. It has even been suggested that a person's flora, left behind like a fingerprint on anything touched, is sufficiently unique so that its analysis could serve as fingerprinting in criminal identification.

Both numbers and species distribution vary greatly from one area of the body to another (from 10^3 to 10^8 per cm^2), and with climate, age, state of health, and degree of personal cleanliness.

Important Groups

Propionibacteria *Propionibacterium acnes,* an anaerobic nonsporeforming Gram-positive bacillus, is found in most skin samples. It multiplies in the pilosebaceous units. The distribution parallels the density of these units. *P. acnes* utilizes the fats and oils for growth and is carried to the surface with the secretion. A closely related species, *P. granulosum,* is usually present in lower numbers. Because the skin condition known as acne involves pilosebaceous blockage and these organisms are almost always present, it was assumed that *P. acnes* was causative and so it was named. However, it is now felt that acne is a multifactorial condition in which the role of microorganisms is unclear.

Staphylococci *Staphylococcus epidermidis* is a normally nonpathogenic, facultatively aerobic, Gram-positive coccus found in more than 85 percent of samples. It is highly salt-tolerant and colonizes the skin surface and an occasional pilosebaceous unit. The closely related potential pathogen *S. aureus* is found in 5 to 25 percent of samples, particularly from moist areas such as the nose, perineum, and axilla.

Micrococci *Micrococcus luteus* is a strictly aerobic, Gram-positive coccus that produces a bright yellow pigment. It is salt-tolerant, widely distributed, but seldom numerous.

Aerobic diphtheroids These organisms are poorly differentiated, pleomorphic, Gram-positive rods. Species such as *Corynebacterium xerosis* are most numerous in moist areas such as the axilla, and they are found in about half the samples from this site. Like most of the Gram-positive skin flora, they actively digest lipids. The fatty acid breakdown products they release help maintain ecological balance on the skin surface, as they are directly inhibitory to nonadapted species.

The by-products of fatty acid metabolism are highly odorous, the main contributors to body odor, and a primary target of cosmetic design. Deodorants and antimicrobial soaps are formulated to inhibit growth of Gram-positive organisms. The use of deodorants and antimicrobial soaps usually results in a shift to Gram-negative, nonlipid-digesting species. These species, however, have a slightly higher potential for overstepping the thin line that divides the "normal" from the pathogenic. It appears that, whereas you cannot have too clean a mouth, it may be unwise to have a too well-scrubbed skin. However, the systematic routine of scrubbing hands in the hospital is the single, most important step in preventing disease transmission.

Types of fungi; Chapter 5, pp. 137–149 **Yeasts** *Pityrosporum ovale* is a small budding yeast, actively lipophilic, that is generally found on the scalp and oily skin areas. It has been implicated in causing dandruff. *P. orbiculare,* which also has a mycelial phase, is more common on the trunk. It may become the causative agent of **tinea versicolor,** a superficial infection in which it develops mycelia among the epithelial cells.

Candida albicans is a dimorphic fungus, usually found in the yeast phase. It is sometimes seen in normal skin, but is also noted as an opportunistic pathogen. C. parapsilopsis is a more common (1 to 15 percent of samples) nonpathogenic relative.

Other organisms   The enteric streptococci, *Pseudomonas aeruginosa,* and the Gram-negative enteric bacilli such as *Escherichia coli,* are uncommon except in areas exposed to fecal contamination. The acid-fast bacillus *Mycobacterium smegmatis* is sometimes encountered in oily, moist areas.

Interactions of Bacteria on the Skin

The skin and its flora establish a restrictive or selective environment that limits colonization by the less well-adapted but more pathogenic species of microorganism. The resident flora rapidly and efficiently consumes available nutrients, providing unbeatable competition to the less-adapted and therefore slower-growing potential invader. Furthermore, metabolic end products, such as short-chain fatty acids, of resident species are actively inhibitory to other bacteria, not only limiting the growth of normal residents but also preventing establishment of opportunistic pathogens.

Bacterial interference is the process whereby host colonization by one species is prevented by the presence of a previously established colonist. It has been considered as a clinical tool. Newborns are particularly vulnerable to colonization by *Staphylococcus aureus.* If the colonizing strain is highly pathogenic, the results may be disastrous. Staphylococcal infections spread in epidemic fashion among the infants in the hospital nursery. Experimental implantation in the noses of newborns of nonvirulent *S. aureus* strains such as 502A may result in successful colonization. Undesirable strains may be excluded long enough to get the newborn past the high-risk period. This concept of biological control—using one bacterium to limit another—will undoubtedly undergo further development.

Staphylococcal infections of newborn; Chapter 21, pp. 644–645

BACTERIAL INFECTIONS

This chapter considers organisms that can establish disease states in essentially normal skin that has neither suffered major trauma nor has areas of devitalized tissue. Those disease states that follow burns, bites, or other wounds are discussed in Chapter 23.

Infections Caused by Staphylococcus aureus

Causative agent   *Staphylococcus aureus* is a Gram-positive coccus which shows a microscopic pattern of grapelike clusters. Colonies on agar, particularly enriched media, may have pale to golden yellow pigment. On blood agar, *S. aureus* usually

TABLE 17.1
Clinical terminology of skin diseases

	TERM	DEFINITION
Conditions	Acne	An inflammatory eruption involving the pilosebaceous apparatus; many types, not always infective
	Dermatitis	General inflammation of the skin; many causes, not always infective
	Impetigo	A superficial pyoderma marked by vesicle formation and crusts
	Pyoderma	Any spreading, pus-forming infection of the skin
	Rash	A cutaneous eruption, localized or generalized, usually described by type of lesion as vesicular rash; not always infective
	Ringworm	A superficial fungus infection of the skin and appendages; many types; also called **tinea**
Lesions	Furuncle	A localized, pus-forming infection originating in a hair follicle
	Carbuncle	An abscess of skin and subcutaneous tissues, an extension of a furuncle invading multiple follicles
	Macule	A small, discolored patch or spot on skin, not elevated above the general surface
	Papule	A small, circumscribed solid elevation on the skin
	Pustule	A small, circumscribed elevation on the skin containing pus
	Ulcer	A lesion on skin or mucous surfaces marked by erosion and loss of tissue
	Vesicle	A small, circumscribed elevation on the skin containing serum.

produces β-hemolysis due to excretion of the membrane-damaging **alphatoxin.** It is differentiated from *S. epidermidis* on the basis of pigmentation, β-hemolysis, and because it ferments mannitol. An important confirmatory step is the **coagulase** test. This detects an *S. aureus* isolate's ability to produce an enzyme that coagulates normal plasma. DNAase production is another confirmatory reaction.

Strain identification of *S. aureus* isolates for epidemiological purposes depends on bacteriophage typing.

Pathogenesis   Infection by various strains of *S. aureus* in various areas of the skin and under different conditions causes different disease states. Staphylococcal infections are usually well localized by host defenses. The human adult normally resists systemic invasion by this agent.

Impetigo is a superficial infection of the skin found almost exclusively in small children; often it is caused by combined infection of *S. aureus* with *Streptococcus pyogenes.* The agent probably invades a small lesion such as a scratch or insect bite. *S. aureus* phage type 71 is a common agent. This strain produces a large amount of hyaluronidase, an enzyme that digests hyaluronic acid, an epithelial, cementing substance.

The enzyme permits the lateral enlargement of the lesion. The lesions are covered by thin crusts; they usually heal without scarring. The organism is readily spread by contaminated fingers to other areas of the body and other children.

Furuncles are individual, localized abscesses (boils) that affect adults when hair follicles or sebaceous glands become obstructed. Multiplication of the organism creates inflammation and local cell death. Pus forms; eventually the abscess wall weakens, and the furuncle drains. The contents of facial furuncles may drain or be squeezed backwards into the cranial sinuses, with life-threatening results. Healed furuncles may leave scars, particularly in the axilla or on the face.

Carbuncles are abscesses on the back of the neck and upper back. This particularly thick and inelastic area of skin tends to prevent normal drainage of the abscess. It develops laterally and into underlying connective tissue, attaining great size. Surgical drainage may be required, plus antibiotics to control bacteremia.

The newborn has very little resistance to staphylococcal infection, which may take one of two courses. Some strains form multiple individual lesions causing **pemphigus neonatorum,** a severe form of impetigo which spreads rapidly over the entire body. Infections by phage type 2 yield the **staphylococcal scalded-skin syndrome** (SSSS). Here the action of a virulence factor called **exfoliatin** causes a rapidly spreading separation and peeling of the epidermis. This condition may progress to bacteremia, systemic involvement, and death within 36 hours.

Therapy and immunity   Community-acquired *S. aureus* usually responds well to many antimicrobial drugs such as the penicillins. However, there is a high incidence of genetically acquired resistance. Susceptibility testing of each target strain is necessary to guarantee selection of an effective drug. Recently, concern over increasing frequency of multiple resistance in hospital-acquired strains has led to a changed therapeutic strategy. Topical disinfectants are now recommended, rather than antibiotics, for the treatment of accessible skin infections.

As first mentioned, most adults have a high level of resistance to *S. aureus* infections primarily because of nonspecific phagocytosis. High levels of circulating antibody are also usually seen; these may serve in complement-mediated opsonization. These protective mechanisms are not adequate in newborns, and may fail in adults with diabetes, immunosuppression, or deficient reticuloendothelial function. Experimental immunization has not helped these people.

Epidemiology   The nasal carrier is the reservoir of *S. aureus* most frequently implicated in transmitting disease. From 10 to 30 percent of individuals are temporary or permanent carriers. They disseminate the organism directly and indirectly, via nasal secretions, contaminated fomites, hands, air, or food.

When the importance of the nasal carrier in transmission was first recognized, systematic nasal culture surveillance was recommended for nursery and surgical staffs. However, it has since become clear that an individual's carrier status may change from day to day, making monthly surveillance fruitless. Preventive efforts are now being directed to training personnel in correct precautionary behavior. This includes strict aseptic procedures in dealing with high-risk populations, and appropriate methods of handling contaminated materials.

Infections Caused by *Streptococcus pyogenes*

Causative agent *S. pyogenes* is a Gram-positive coccus that often, especially in liquid culture, grows in chains. It is facultatively anaerobic, fermentative, and catalase-negative. It is metabolically fastidious and requires enriched media for successful isolation and maintenance. On blood agar it forms tiny colonies surrounded by a zone of β-hemolysis. Hemolysins are produced; streptolysin S acts aerobically, while streptolysin O, which is inactivated by oxygen, acts anaerobically. On primary isolation, *S. pyogenes* can be distinguished from other small β-hemolytic colonies because it is inhibited by the antimicrobial drug bacitracin, provided in a filter paper "A-disc" pressed onto the streaked area.

Bacitracin; Chapter 26, p. 777, Tables 26.2 and 26.4

The streptococci are classified by serological typing. The Lancefield typing scheme detects carbohydrates located just exterior to the rigid structural layers. From 18 to 20 lettered groups are recognized. Group A, to which *S. pyogenes* belongs, is the most important in human disease.

The Group A streptococci also bear, exterior to the carbohydrate layer, a layer of **M-proteins** organized in projections, or **fimbriae.** M-proteins inhibit phagocytosis, and appear to provide the bacterium's primary mechanism for attachment to the tissue it will parasitize. Serological typing of M-proteins is used to pinpoint the source of *S. pyogenes* strains in outbreaks. T- and R-proteins of cell wall are also used. The M-, T-, and R-proteins together form a second comprehensive typing scheme.

Some strains of Group A streptococci also possess, as their outermost layer, a hyaluronic acid capsule. Because this substance is identical to epithelial cementing material, it protects the bacterium from phagocytosis and interferes with antibody production.

Immunological tests; Chapter 13, 385–395

A streptococcal isolate can be presumptively identified as *S. pyogenes* on the basis of β-hemolysis and bacitracin susceptibility. It will then be grouped—identified as Group A—by fluorescent antibody techniques, latex agglutination, staphylococcal coagglutination, or other immunological means.

Pathogenic *S. pyogenes* can cause impetigo or pyoderma, alone or in mixed infections with *S. aureus.* The lesions are large and have a thick, brownish crust. Healing is usually prompt and complete, but because of the high level of pathogenicity of the organism, the possibility of developing complications is significant. In

TABLE 17.2
A comparison of *Staphylococcus aureus* and *Streptococcus pyogenes*

FEATURE	*Staphylococcus aureus*	*Streptococcus pyogenes*
Microscopic morphology	Gram-positive coccus in grapelike clusters	Gram-positive coccus in pairs and chains
Species identification	Yellow pigment, aerobic β-hemolysis, salt-tolerant, mannitol-fermenting, catalase-positive, coagulase-positive	Nonpigmented, β-hemolysis; aerobic and anaerobic, catalase-negative, bacitracin-sensitive, Group A carbohydrate
Strain identification	Bacteriophage typing	Serological identification of M- and T-proteins
Extracellular products	Hemolysins Hyaluronidase DNAase Penicillinase Exfoliatin	Hemolysins DNAase Hyaluronidase Nephritogenic factors Rheumatogenic factors
Pathogenicity	Low for normal adults, high for the newborn and the compromised	Highly pathogenic for all individuals
Types of infection	Impetigo Furuncles and carbuncles Neonatal infections Wound infections Postsurgical infections Secondary pneumonias	Impetigo Erysipelas Wound infections Puerperal fever
Chronic or delayed states	Chronic disseminated disease, abscesses, osteomyelitis	Glomerulonephritis Rheumatic fever

mixed infections, penicillin treatment against the streptococci is usually sufficient to produce a cure.

Erysipelas is an uncommon, more severe form of pyoderma in which the invading organism moves deeper into the skin and spreads laterally via the underlying lymphatics. It develops most commonly on the face or neck, following minor trauma or surgery, but lesions may occur anywhere. Because septicemia often follows, this infection had a high mortality before antibiotics were developed; it is still regarded as very serious.

Skin infections with certain M-types such as 49, 55, and 57 carry the risk of poststreptococcal kidney complications. These strains secrete a soluble nephritogenic antigen that attaches to glomerular tissue in the kidney. Antibody synthesis leads to formation of destructive immune complexes. **Glomerulonephritis** ensues, usually about ten days after the primary infection. Blood and protein in the urine, edema, and hypertension are the features of this condition. Recovery is usually com-

Immune complex disease; Chapter 14, pp. 416–417

plete and recurrences are rare. Less commonly, skin infection is followed by rheumatic fever.

Therapy and immunity   A mild *S. pyogenes* infection is self-limiting. However, because of the risk of serious poststreptococcal complications, active diagnosis and therapy are advisable. The disease should not be allowed to run its course. *S. pyogenes* responds promptly to penicillin and many other antimicrobial drugs.

Circulating protective antibody to each individual M-type develops following exposure. However, because there is no cross-protection for other strains, numerous *S. pyogenes* infections may afflict one individual. Clinically useful vaccines have not been developed.

Epidemiology   The reservoir for the *S. pyogenes* organism is the human nasopharynx and, to a lesser extent, other orifices. Many persons are asymptomatic carriers. The organism is spread in contaminated secretions, in aerosols, and on hands. Open lesions are also highly infective. This organism is both virulent and readily communicable; thus thorough precautions must be used to prevent its spread.

Other Bacterial Infections

Anthrax   Anthrax is a disease of cattle, sheep, and other animals caused by the spore-forming, Gram-positive *Bacillus anthracis*, a facultative anaerobe. It is now very rare in human beings, occurring by means of occupational contact with diseased animal products. A skin lesion develops at the site of bacterial entry. If left untreated, pneumonia and/or septicemia may develop leading to death. Antibiotics usually produce cure for the septicemic form but not for the pneumonic form. Anthrax is prevented by stringent surveillance for infected animals and disposal of contaminated products. A vaccine has become available for high-risk individuals, such as persons working in tanneries or handling wool fleeces or swine bristles.

Pseudomonas infections   *P. aeruginosa* can infect the nail beds and toe webs. Dishwashers, or others whose hands are wet most of the time, and persons with disturbances of the normal flora or underlying debilitative disease are susceptible.

FUNGAL AND PROTOZOAN INFECTIONS

The skin is a favored location for the development of fungal parasites, most of which require aerobic environments, tolerate a hostile environment, and have limited nutritional requirements. Such fungi are called dermatophytes.

Dermatomycosis   This is a general term for mycotic (fungal) parasitism of the skin. It may involve infiltration of superficial epithelium, of the dermis, or of the subcutaneous layers.

Box 17.1

Rash Associated with Whirlpool Baths

Twenty-seven members of a new racketball club in South Paris, Maine, developed a rash illness in the period December 9, 1978 to January 6, 1979. In a follow-up investigation, the club's two whirlpools were found to be statistically associated with illness.

The rash began as a single crop of discrete, maculopapular lesions, a few millimeters in diameter, that soon developed either a vesicle or a pustule on the apex. The lesions crusted over in a few days, and by the seventh day they were disappearing without treatment. Most of the lesions were on the trunk or proximal extremities. The lesions were predominant around the axillae and pelvis.

Although the majority of patients—who included 16 men and 11 women—had no symptoms, eight had painful lymph nodes, seven reported headache, and five noted muscle aches. Three patients had chills and low grade fever. Five of the patients, three men and two women, had painful breasts.

The cause of the rash was not initially apparent, but a survey of physicians and school nurses in the area indicated that only members of the racquetball club were affected. Results of a questionnaire demonstrated a significant association between using the club on December 10, a day of unusually heavy use, and becoming ill during the next two days. A significant association was found between using the men's or women's whirlpool on December 10 and developing rash within the next two days. No

association was found between rash and the use of any other facility at the club.

Pseudomonas aeruginosa was isolated from the skin lesions of two of the patients on December 12, and December 19, respectively. One of these isolates was sent to the Center for Disease Control for serotyping and found to be serotype 0–11. A culture of water from the men's whirlpool, taken on December 19, grew *P. aeruginosa*, serotype 0–11.

Investigation revealed that the two implicated whirlpools had been chlorinated by hand each morning. Peak levels of free residual chlorine, measured on the morning of December 19, were 0.7 parts per million (ppm) in the men's whirlpool and 1.2 ppm in the women's whirlpool.

Once the statistical association between illness and use of the whirlpools had been demonstrated, the whirlpools were closed from December 19 to December 29. During this period the filters were changed, and the whirlpools were drained and acid-washed.

No more cases occurred until after January 1, when the whirlpools were reopened. Three women who had used the women's whirlpool on January 3 developed a rash January 5–6. Automatic chlorinators that maintain a free chlorine residual level of more than 1 ppm were installed on January 6. No subsequent cases have been reported.

Source: *Morbidity and Mortality Weekly Reports* **28** (182), 1979.

One dermatomycosis that many people experience at one time or another is **tinea pedis,** or athlete's foot. We will use it as an example. Others are summarized in Table 17.3.

TABLE 17.3
The superficial and cutaneous dermatomycoses

CONDITION	AGENTS	LOCATION	PATHOGENESIS	TREATMENT*
Tinea barbae	*Trichophyton mentagrophytes* *Microsporum canis*	Bearded areas	Localized inflammation, may become ulcerated	Griseofulvin
Tinea capitis	*Microsporum canis* *M. audouini* *Trichophyton* sp.	Scalp	Several forms; hair loss, inflammation, scaling, temporary baldness	Oral griseofulvin Topical antifungal ointments
Tinea corporis	*Microsporum* sp. *Trichophyton* sp.	All body areas	Flat, spreading ring-shaped lesions	Oral griseofulvin Topical antifungal ointment
Tinea cruris	*Candida albicans* *Epidermophyton floccosum* *Trichophyton* sp.	Groin	Flat, spreading lesions	Antifungal ointments Gentian violet solution
Tinea pedis and manuum	*Trichophyton rubrum* *T. mentagrophytes* *Microsporum* sp.	Hands and feet	Scaling, cracking and inflammation between toes or fingers; blisters	Oral griseofulvin Topical ointments or dusts
Tinea nigra	*Cladosporium werneckii*	Palms	Dark brown or black discoloration	Topical disinfectants including sodium thiosulfate tolnaftate
Tinea unguium	*Trichophyton* sp. *Epidermophyton floccosum* *Candida albicans*	Nails	Thickening, deformation of nail; nail may disintegrate	Oral griseofulvin
Tinea versicolor	*Pityrosporum orbiculare*†	Body surfaces	Scaly discolorations without inflammation	Topical disinfectants as above
Black piedra	*Piedraia hortai*	Surface of hair shaft	Nodules of dark hyphae and spores surround hair	Topical disinfectants as above
White piedra	*Trichophyton beigelii*	Surface of hair shaft	Nodules of white hyphae surround hair	Topical disinfectants as above

*May vary with location and severity.
†*Malassezia furfur* is the traditional name; appears to be identical to *P. orbiculare.*

Causative agents Several members of the genera *Trichophyton, Microsporum,* and *Epidermophyton* may be found. In the United States, *T. rubrum* is most common. Human parasitic fungi are deuteromycetes. These do not reproduce sexually, but may form spores. Dermatophytes grow slowly in the dead skin layers, utilizing the keratin for growth. They thus may also invade hair and nails.

Identification is by microscopic examination of tissue from the affected area looking for characteristically shaped **microconidia** or **macroconidia** (Fig. 17.4) and mycelial growth. The fungi can be isolated on differential selective media containing generous amounts of glucose and antibiotics to suppress bacteria. They develop slowly, often requiring two to three weeks. The resulting growth can be identified by colonial and microscopic morphology.

Fungal reproductive structures; Chapter 5, pp. 145–147

Pathogenesis Tinea pedis is a type of **ringworm.** Itching and burning between and around the toes is followed by the appearance of vesicles. In chronic cases, peeling and ulceration may occur, especially on the soles. The actual symptoms seem to be due not only to mycotic growth, but also to a delayed hypersensitivity response to fungal antigens in the affected tissue. Other cutaneous mycoses create similar conditions on other body areas, and sometimes cause hair loss.

Therapy and immunity Treatment may involve antifungal drugs such as griseofulvin given orally. Ointments or powders containing clotrimazole, miconazole, or tolnaftate are effective. No protective immunity develops.

Antifungal drugs; Chapter 26, pp. 778–779

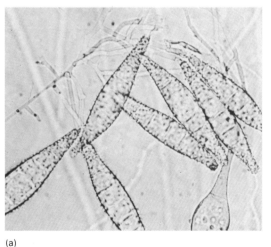

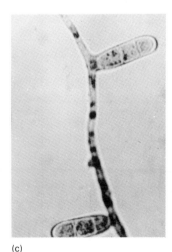

(a) (b) (c)

17.4
Dermatophytes The macroconidia are sufficiently distinctive to aid in identification of these fungi. (a) Macroconidia of *Microsporum canis,* the most common cause of ringworm in the United States. (b) *Microsporum gypseum* is also a cause of ringworm of the scalp and body. (c) *Epidermophyton floccosum* is one cause of "athlete's foot."

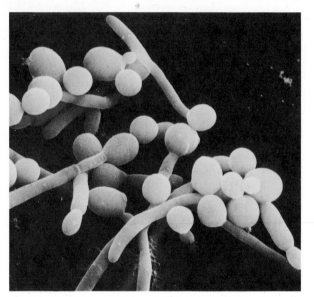
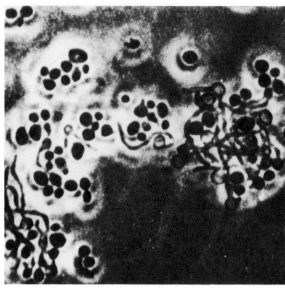

(a)

(b)

17.5

Candida albicans (a) When *Candida* is cultivated on agar, it rapidly (in two to three days) forms creamy colonies of yeast cells. At the edge of such colonies pseudohyphae may protrude into the media. Blastospores form at the tips of these structures. (b) Another view of pseudohyphae formation. One of the primary ways of identifying *Candida* is by observing the formation of germ tubes by yeast cells inoculated in serum.

Epidemiology   Fungal spores contaminate floors and footwear, from which they are picked up. Active multiplication, to the point where symptoms appear, usually is associated with the wearing of hot, moist, unventilated footwear. Spread is controlled by disinfection of bathing areas, wearing protective coverings, and other commonsense measures.

Candidiasis

Causative agent   *Candida albicans* and, to a lesser extent, several other species of *Candida* are opportunists, able to cause infections when there are predisposing environmental and constitutional factors. The fungus is dimorphic. It grows as a budding yeast under most laboratory cultural conditions, but develops **pseudomycelia** (Fig. 17.5), chains of elongated cells, under semianaerobic conditions or in tissue. When yeast cells from a colony are incubated in serum, characteristic **germ tubes** develop. Identification is by cultivation on differential media and microscopic examination. The organisms are facultative anaerobes. They ferment sugars, and have few nutritional requirements.

Pathogenesis *Candida albicans* infects most body systems. On skin it causes **intertrigo,** an infection of areas in which two layers of skin rub together, such as between fingers, under the breast, and between the buttocks. Vesicles appear followed by scaling, and the lesions are difficult to heal. It is experienced by physicians. dentists, and others who repeatedly immerse their hands in water. Debilitating factors such as diabetes, obesity, and alcoholism also predispose individuals to *Candida* infections. Chronic infections of the nail (paronychia) may develop. **Perleché** is a related cracking or fissuring of the corners of the mouth.

Mucocutaneous candidiasis is a condition of immunodeficient individuals, most commonly children. Lesions proliferate over mucous membrane and skin, eventually involving the whole body. The underlying problem appears to be related to depressed CMI. Antibiotics cannot eradicate the organism in the absence of directed macrophage activity. Transfer factor, a soluble extract of immune T-lymphocytes, has been given experimentally with good, if temporary, effect.

Therapy and immunity Treatment with nystatin ointment, Vioform (iodochlorhydroxyquin) or other antifungal agents is effective for superficial infection. Other drugs are employed for systemic infection. It appears that the normal adult has effective CMI against *C. albicans.* Immunization is not practicable.

Epidemiology The organism often inhabits the mouth, urogenital tract, and gastrointestinal tract of human beings. About half of the healthy population carries *C. albicans* at one or more of these sites. Transmission is frequent and occurs through a wide variety of means. The important issue in the development of candidiasis is not infection (acquiring the agent), but possessing the predisposing factors that allow it to produce disease.

Leishmaniasis

Cutaneous leishmaniasis is a disease of skin and and mucous membranes caused by the flagellate protozoans *Leishmania brasiliensis* or *L. tropica.* Lesions develop around the bite of the arthropod vector, the sandfly. *L. tropica* lesions tend to heal spontaneously with scarring, but *L. brasiliensis* infections often show a two-phase pattern in which healing of the initial sore is followed after a latent period by eruption of ulcers on the mucous membranes of the nose, mouth, respiratory tract, or vagina (Fig. 17.6). Extensive tissue destruction, disfigurement, and death may ensue if the disease remains untreated. Antimony compounds such as Pentostam, and antimalarial drugs are used. This infectious disease occurs in subtropical and tropical climates.

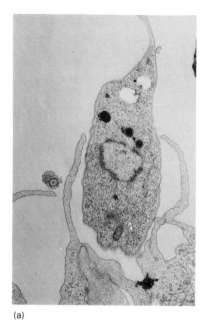

(a)

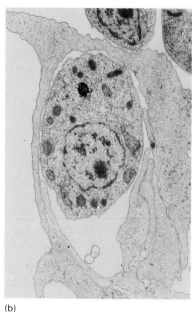

(b)

17.6

Leishmania donovani **Undergoing Phagocytosis** (a) A hamster macrophage extending slender protoplasmic extensions around the protozoan parasite. The ingestion process has been studied and shown to be a cooperative process. The protozoan parasite has adhesion factors on its surface. Its flagella tangle with the macrophage pseudopodia, aiding prolonged contact. The macrophage, in turn, ingests the parasite. (b) The parasite has been completely encircled by not one but many layers of pseudopodia. Unfortunately, this does not signal the impending destruction of the invader but rather the first, essential step in establishing disease. *Leishmania donovani* multiplies preferentially inside macrophages.

VIRAL INFECTIONS

Viruses multiply avidly in epithelial tissues, including the skin (Table 17.4) and mucous membranes. We include here virus diseases whose most prominent symptom, a rash or eruption, appears on the skin, although the viruses may be found in other tissues.

TABLE 17.4
Epidemiology of skin infections

DISEASE	RESERVOIR	TRANSMISSION	INCUBATION PERIOD, DAYS
Anthrax	None: source is spores	Contact with infected animal products, soil	2–5
Pseudomonas infections	Water, soil, human beings	Various: airborne, water droplets, fecal contamination of hands	Short

Cold Sores or Fever Blisters

Causative agent   The herpes simplex virus is a large, envelope-bearing DNA virus. Serotype I predominates in oral and skin lesions, whereas Type 2 is more common in genital tract infections. The virus can be identified by fluorescent antibody staining of material from lesions. It can also be readily isolated and identified in cell culture. Herpes multiplication *in vivo* is characterized by periods of active viral replication alternating with periods of latency. The virus is also called Herpesvirus hominus (HVH).

Pathogenesis   Exposure to herpes virus early in life is very common. Primary infection is usually completely asymptomatic, but in 5–10 percent of persons it gives rise to fever, malaise, and an acute stomatitis or encephalitis. The characteristic recurrent fever blister is the secondary phase of the disease. These lesions occur on the lip, gum, or tongue. A period of tingling or burning is followed by the appearance of one or more vesicles. These later ulcerate, then heal without scarring. Weeks or months may pass before the next recurrence.

Therapy and immunity   Specific therapy is not usually prescribed for fever blisters. Herpes infection prompts the development of AMI, but the antibodies are not protective. CMI helps in the containment and elimination of replicating virus in each episode of cold sores. In the neonate or the immunosuppressed adult, herpes replication may become generalized and life-threatening.

Epidemiology   The human being is the only reservoir of the herpes simplex virus. Exposure to the virus is so common that 70–90 percent of adults possess antibodies. Transmission occurs primarily by saliva and hands. Persons with active herpes lesions should avoid newborns or other highly susceptible individuals.

PERIOD DURING WHICH COMMUNICABLE	PATIENT CONTROL MEASURES*	EPIDEMIC CONTROL MEASURES
Spores infective indefinitely, no person-to-person transfer	Isolate wound or patient, disinfect soiled materials	Case reporting Animal surveillance Importation regulations
While symptoms persist	Disinfect soiled materials	Search for source; institute improved technique

See also Isolation section, Chapter 25.

(continued)

(Table 17.4 continued)

DISEASE	RESERVOIR	TRANSMISSION	INCUBATION PERIOD, DAYS
Staphylococcal infections	Human beings	By hand or fomite; from nasal carrier or lesion	4–10
Streptococcal infections	Human beings	Direct contact with patient, carrier; also milk or foodborne	1–3
Tinea pedis	Human beings	Spores on floors, washrooms, clothing	Unknown
Candidiasis	Human beings	By contact with secretions, contaminated objects	2–5; variable
Leishmaniasis	Domestic animals	Bite of sandfly	Days to months
Fever blister	Human beings	Contact with saliva, lesions; by contaminated hands; bite	To 14
Varicella-zoster	Human beings	By contact or inhalation of respiratory secretions or skin lesions	13–17
Measles	Human beings	By contact with respiratory secretions and urine	8–13
Rubella	Human beings	By contact with respiratory secretions and perhaps urine and feces	14–21
Smallpox	Human beings	Respiratory discharges, lesions, and contaminated materials	7–17

Source: Adapted from Abram S. Benenson, 1975, *Control of Communicable Diseases in Man* (12th ed.).

PERIOD DURING WHICH COMMUNICABLE	PATIENT CONTROL MEASURES*	EPIDEMIC CONTROL MEASURES
While lesions are present or carrier state persists	Disinfect dressings. Avoid contact with infants and debilitated persons	Search for source, treat with antibiotic and segregate
While discharge persists; until 24 hr after penicillin therapy	Isolate until noncommunicable. Disinfect discharges and soiled material	Determine source and manner of spread
While lesions persist and viable spores are present	Disinfect socks and shoes	Cleansing of floors, showers
While lesions persist	Disinfect secretions and soiled materials	Not important in skin infection
While lesions contain parasites	Protect patient from flies	Insect destruction Mass diagnosis and treatment
While lesions are present; saliva infectious even while asymptomatic	Those with lesions should avoid newborns, children with skin trauma, immunosuppressed	Not applicable
From five days before eruption to six days after last lesions; while lesions are present	Exclude child from school, disinfect soiled articles	None
From onset of symptoms to four days after rash	Isolation only in hospital. Disinfection: none	Reporting; community vaccination; universal vaccination desirable
From one week before to four days after onset of rash. Affected neonates shed virus indefinitely	Isolate only from susceptible pregnant women. Disinfection: none	Reporting; community vaccination; universal vaccination desirable
From development to complete healing of lesions	Isolate; strict precautions, disinfect soiled articles	No longer needed

Chickenpox and Shingles

Causative agent   The varicella–zoster or herpes zoster virus causes both chicken-pox (varicella) and shingles (zoster). It is structurally similar to other herpes viruses (Fig. 17.7a). Laboratory identification is made by detection of characteristic inclusion bodies in material from lesions, cell culture isolation, or detection of specific antibodies in convalescent serum.

Pathogenesis   Varicella is a disease of children, in whom it is usually mild. Adults without immunity may contract it and may have very severe cases. Following exposure, an incubation period occurs, during which the virus multiplies in respiratory epithelium. Then the virus is disseminated via the blood to the skin of the trunk and face. The virus then replicates in the skin. A vesicular rash, fever, and malaise appear at this time, marking the onset of clinical illness. In more prolonged cases, successive crops of vesicles may appear at intervals. The vesicles crust over and heal, with occasional scarring. Complications such as secondary pneumonia and encephalitis are very rare in children, but more common in adults.

Shingles occurs principally in older adults. It is a reactivation of the varicella-zoster virus that has lain latent in the body since recovery from a childhood case of chickenpox. Crops of vesicles appear on the skin along peripheral nerves of the face and trunk. They are accompanied by intense burning pain and fever. Healing may take several weeks and may be followed by transient paralysis or neuralgia. Zoster appears frequently in persons receiving cancer treatment and in those with immunodeficiency.

Antiviral drugs;
Chapter 26, pp.
781–783

Therapy and immunity   Specific therapy is not frequently used; however, ara-A has been used successfully in zoster. Chickenpox usually stimulates development of a lifelong antibody response that prevents subsequent attacks of varicella. This clearly does not eliminate virus, probably because most latent virus is in a region inaccessible to antibody, underneath the myelin sheath of the nerve cell. However, the immune system must generally be able to keep the virus in latency because immunosuppression has such unfortunate effects in promoting zoster. There is no vaccine for varicella–zoster virus. High-risk infants with documented exposure may receive temporary protection via injections of specific immune globulin.

Epidemiology   The human being is the only host. Each person who has recovered from childhood infection must be considered a lifelong carrier. However, the virus is probably transmitted only by droplets of respiratory secretions from an active case, or contact with the vesicles or fomites contaminated with such discharges. Zoster lesions contain infectious virus that may cause chickenpox in young contacts. Sick children should be kept from school and contact with other susceptible children for one week or until six days after the last group of vesicles appears. A child with chickenpox may pose a serious threat to an immunosuppressed contact, especially in the hospital.

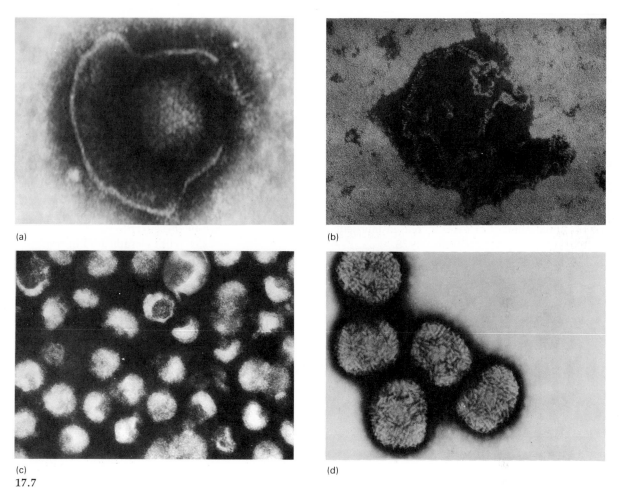

(a)

(b)

(c)

(d)

17.7
The Viruses of Skin   (a) Varicella-zoster virus. This is a large DNA virus that causes chickenpox and shingles. (b) Measles virus, the causative agent of measles, an RNA virus. (c) Rubella virus, the causative agent of German measles, an RNA virus. (d) Vaccinia virus. The surface of this oval virus, which is among the largest known, is covered with tubular ridges. That is a DNA virus, used in the manufacture of smallpox vaccine.

Measles

Causative agent   The **rubeola virus,** a paramyxovirus, is an envelope-bearing, RNA virus (Fig. 17.7b). "Wild" virus is **monotypic,** that is, only one type is distributed in nature. The virus can be isolated from patient samples and identified in cell culture by its cytopathic effect, the formation of multinucleated giant cells. Giant cells can also be identified by indirect fluorescent antibody microscopy in slides of pharyngeal epithelium. The virus multiplies in epithelial or neural tissues.

Box 17.2

The Eradication of Measles

Given the fact that we have a highly effective, safe, readily available measles vaccine, there should be no more measles. Yet over the period from 1974–1978, more than ten years after this vaccine was licensed for general use, we were still averaging around 25,000 cases of measles per year in the United States, and a small number of measles deaths. More worrisome still, the annual number of cases increased markedly each year in 1976, 1977, and 1978.

In response to the challenge offered by this discouraging situation, the then Secretary of the Department of Health, Education and Welfare announced on October 1, 1978, that the complete eradication of measles would become an official goal of United States health policy. The techniques to be employed include increased emphasis on universal infant vaccination, requiring proof of vaccination for school admission, and epidemiological surveillance. Through the use of these techniques, it is hoped that measles will be eradicated in the United States within ten years.

Pathogenesis   After an incubation period of from 9 to 11 days, during which the virus multiplies asymptomatically in the respiratory mucosa, the virus is disseminated to the entire body, localizing in the skin and mucosa. In over 90 percent of cases, the first overt symptom is the appearance of a rash of **Koplik's spots** on the oral mucosa. One or two days later the rash emerges over the rest of the body, while fever, malaise, sore throat, photophobia, and other symptoms develop. Rapid improvement normally occurs within three days, and the virus is completely eliminated during the convalescent period. In one case out of 1,000, encephalitis develops. This may produce neurological damage or death. Secondary bacterial pneumonias also occur. Between the two, the mortality from measles in the United States is about one death per 1,000 cases. In areas of poverty and malnutrition, the mortality rate may be 100 times greater. Subacute sclerosing panencephalitis (SSPE) is a very rare, slow-virus, degenerative disease caused by altered rubeola virus. It is noncommunicable. Rubeola virus can cause fetal damage.

Slow virus disease; Chapter 11, p. 328, Chapter 22, pp. 667–668

Therapy and immunity   In uncomplicated cases, treatment is symptomatic. Antimicrobial drugs may be given to deal with secondary bacterial pneumonias.

Natural active immunity follows clinical or asymptomatic illness. A live-virus measles vaccine, prepared in chick embryo cell culture, is available and recommended for all children at age 15 months. If given prior to 12 months, effective protection may not develop. The vaccine is also recommended for any adults who lack a history of either effective vaccination or physician-diagnosed measles. An earlier killed-virus vaccine was proved ineffective. It has been discontinued, and persons who received it should be revaccinated with the live-virus vaccine.

Epidemiology   Measles is endemic throughout the world as a disease of small children. In rural areas of Africa, where vaccination has not been adopted, measles causes 21 percent of all deaths in children from birth to four years of age. The case fatality rate is 5 percent, about 50 times higher than it is in the United States. Measles incidence and distribution in developed nations has been significantly modified by vaccination programs. Since the measles vaccine was introduced in 1963, the reported incidence has dropped 90 percent, with the annual saving of about 400 lives. Most cases now occur in children ten years of age and older, in junior or senior high school. This age group contains the most susceptibles now for three reasons. Members of this age group either were not vaccinated in early childhood, or they were vaccinated by earlier, less effective methods, or individual members lost their immune status.

Measles vaccination programs; Chapter 27, p. 797 and Box 27.2

Rubella (German Measles)

Causative agent   The **rubella virus** (Fig. 17.7c) is a relatively large, enveloped RNA virus. It can be isolated from throat swabs, feces, and urine, and be identified by its biological effects in cell culture. Confirmation of a clinical diagnosis is obtained by demonstrating an increase in hemagglutination-inhibiting antibodies in convalescent sera.

Pathogenesis   This is typically a disease of small children, although susceptible adults also contract it. After an incubation period of two to three weeks, a mild rash spreads rapidly over the body. Mild fever, sore throat, swollen lymph nodes, and coldlike symptoms may be present although they are rarely severe. Almost 30 percent of cases are completely asymptomatic, however. The disease usually disappears in two or three days. Its importance lies in the fact that if rubella is contracted by a woman in the first trimester (three months) of pregnancy, the virus multiplies in the fetus. In over 30 percent of such cases, the virus produces severe fetal malformation or fetal death. Infants surviving to term are often born with the **congenital rubella syndrome** including abnormalities of eye, ear, central nervous system, genitalia, and heart. Such infants shed live rubella virus for years.

Therapy and immunity   No specific therapy is indicated. Natural infection confers lifelong immunity. Live-virus rubella vaccine is available, prepared either in duck embryo or human diploid cell culture lines. It is recommended for use for all normal infants over 12 months of age and any unvaccinated child up to the age of puberty. It is usually administered in a trivalent formula called MMR (measles, mumps, and rubella) at 15 months. Hospital employees should have current rubella vaccination. Any woman of reproductive age without evidence of rubella immunity should be immunized, providing she is not pregnant and avoids pregnancy for three months after vaccination.

Epidemiology   The incidence of rubella dropped radically, following the introduction of vaccine in 1969. Many cases still occur, unfortunately, principally in those 15 years of age or older. The fact that the remaining pool of susceptibles is in the reproductive age group explains the continuing birth of about 50 affected infants per year in the United States. At present, public health efforts are being expanded to obtain active immune status not only in small children, but also in adolescent females.

Other Viral Infections

Smallpox   In ten years smallpox (Fig. 17.7d) went from being the most serious of skin diseases to one that can be included, for historical reasons only, in an "other" category. The eradication of this human bane is surely one of the great triumphs of twentieth-century medicine. In 1967, the disease was endemic in 33 countries, and there were 10 to 15 million cases a year. The World Health Organization established an eradication program in 1967. It was based on surveillance and containment of the disease. New cases were actively sought (a substantial bounty was given to persons reporting each new case) and an "immunological barricade" was created around each case by isolation and vaccination of all contacts. Smallpox was eradicated first in the Far East, Indonesia, India, and Pakistan. Then efforts shifted to the African continent. The last naturally occurring case was reported in Somalia, East Africa, in 1977. Because there is no animal reservoir in nature, no asymptomatic human carriers, and no remaining endemic region, smallpox now could only reappear if laboratory stocks of virus were released. Steps to reduce the number of laboratories holding cultures of smallpox virus and to reinforce security are underway.

Warts   Human warts are caused by the papilloma virus, a small DNA virus. It is spread by direct contact or by contaminated floors and clothing. Probably a small lesion is necessary for access. Common sites are hands and soles of the feet. The wart is a type of benign tumor. It is usually self-limiting and may spontaneously and suddenly regress. This feature has encouraged the vast repertoire of folk-medicine cures for warts, all of which "work" some of the time.

EYE INFECTIONS

The eye is a sensory organ, anatomically a part of the nervous system. Its diseases are included here because the parts of the eye—eyelids, conjunctiva, and corneal surface—that usually develop infections are all composed of epithelium and modified connective tissue, similar to the skin in many of their features, and continuous with it.

Anatomy and Physiology of the Eye

Anatomy   The eye is a complex sensory organ (Fig. 17.8). The visual structures are enclosed in an impermeable, multilayered coat. Protection is so effective that micro-

17.8
Eye structures Tears are secreted by the lacrimal gland and pass through the many tear ducts to flush the conjunctiva and cornea from all sides. The tears then are drained away into the nasopharynx. The conjunctiva lining the inside of the eyelids is continuous with the cornea covering the exterior surface of the eyeball.

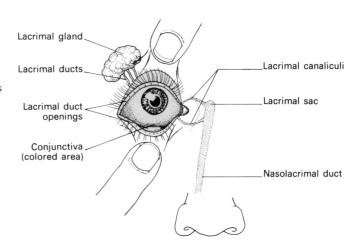

Lacrimal gland
Lacrimal ducts
Lacrimal duct openings
Conjunctiva (colored area)
Lacrimal canaliculi
Lacrimal sac
Nasolacrimal duct

organisms rarely enter the eyeball proper. This occurs only by hematogenous spread (e.g., in toxoplasmosis) or gross trauma.

Among the surfaces exposed to the environment are the eyelids, thin muscular structures continuous with the facial skin. They are lined with a thin mucous epithelium, the **conjunctiva,** that is fused to the surface margin of the cornea, forming a closed **conjunctival sac.** The cornea that forms the outer covering of the exposed portion of the eye is composed of connective tissue and has no blood supply; it is thin, tough, and transparent. **Lacrimal** glands supply tears via a network of lacrimal ducts, and the tears are carried away by the **nasolacrimal duct** and pass into the nasopharynx. Organisms on the eye surface are confined to the conjunctival sac and the lacrimal structures.

Physiology The secretions that wash and lubricate the eye surface are composed of mucus, sebum, and tears. True tears prevent drying and minimize mechanical damage by dust; they are thus essential for maintaining the integrity of tissue. Constant tear secretion and drainage prevents microbial colonization because most microorganisms cannot attach firmly enough to avoid washout. In addition, tears are a source of lysozyme that damages Gram-positive bacterial cell walls. Yet the scanty normal flora of the conjunctiva, which includes staphylococci, streptococci, diphtheroids, and some anaerobes, is mainly Gram-positive. Microorganisms that cause eye infection (Table 17.5) are either intracellular parasites, protected from washout, or have superior powers of attachment, like the gonorrhea organism.

Bacterial Infections

Gonococcal conjunctivitis *Neisseria gonorrhoeae,* best known as a venereal parasite of the genitourinary tract, may cause serious conjunctivitis in the infant infected during the birth process or the adult infected by inoculation with contaminated geni-

TABLE 17.5
Epidemiology of eye infections

DISEASE	RESERVOIR	TRANSMISSION	INCUBATION PERIOD, DAYS
Gonococcal conjunctivitis	Human beings	By secretions of infected genitourinary tracts; direct contact	1–2
Pinkeye	Human beings	Contact with discharges of eye or upper respiratory tract; insects	1–3
Trachoma	Human beings	Direct contact with ocular discharges, other discharges, soiled articles, flies	5–12
Inclusion conjunctivitis	Human beings	Infected genital discharges; direct contact	5–12

tal secretions. Severe damage and blindness may result in the newborn. Administration of silver nitrate or penicillin drops in the eyes of all neonates is required by law to prevent this condition.

Pinkeye   This highly contagious disease, also called **acute bacterial conjunctivitis,** is caused by the small, Gram-negative bacillus *Hemophilus aegyptius.* The disease is a conjunctivitis with profuse watery discharge and swelling.

Other infections   Staphylococci can cause inflammation of the eyelids and abscesses at the eyelid margin. Streptococci may cause lacrimal sac abscesses, cellulitis, and acute conjunctivitis. *Streptococcus pneumoniae* causes several severe eye problems, including **keratitis** (corneal inflammation), corneal ulcer, and chronic inflammation of the tear ducts. *Moraxella lacunata* causes a chronic inflammation of the conjunctiva and lids at the corners of the eye, particularly in debilitated persons and in hot, dusty environments.

Chlamydial Infections

Trachoma   This is a highly communicable keratoconjunctivitis caused by the intracellular parasite *Chlamydia trachomatis.* The agent causes inflammation of the conjunctiva, invades the cornea, and may cause gross deformity of the eyelids, loss of

PERIOD WHEN COMMUNICABLE	PATIENT CONTROL MEASURES	EPIDEMIC CONTROL MEASURES
While discharges are present or until 24 hours following specific therapy	Isolation for 24 hr following antibiotic, disinfect soiled articles	None: control venereal gonorrhea
During active infection	Keep children from school; disinfect soiled articles	Adequate treatment of patients and contacts Insect control
While active lesions are present	Disinfection of eye discharges and contaminated articles	Mass treatment Hygiene education
While genital infection or eye infection persists	For 48 hr after beginning treatment isolation; aseptic precautions	Investigation of venereal transmission Chlorination of swimming pools

vision, and possible blindness. In the endemic areas, including the Middle East, certain regions of the United States, and parts of North Africa and South America, it is a major cause of preventable blindness. The agent is spread by direct contact with infected eyes or their secretions. Tetracycline is used therapeutically. There is no vaccination.

Inclusion conjunctivitis   This is a related, much less severe condition caused by different strains of *C. trachomatis.* Inclusion conjunctivitis is a disease of newborns and adults; the symptoms are acute conjunctivitis but only superficial corneal involvement. It usually disappears spontaneously. Chlamydia can also cause neonatal pneumonia.

The source of the organism is actually the adult urogenital tract. *C. trachomatis* causes about 50 percent of cases of **nongonococcal urethritis,** one of the most common venereal diseases. The agent can be spread to the eye by contaminated hands, and during the birth process. Contaminated swimming pools have also been implicated.

Nongonococcal urethritis (NGU); Chapter 21, pp. 633–634

Virus Infections

Adenovirus infections   Keratoconjunctivitis caused by adenoviruses occurs following surface trauma by dirt irritation or clinical procedures involving the eye.

Herpes infections   Herpes simplex Type I can cause several types of eye lesions. Recurrent corneal ulceration can cause blindness. It is treated with idoxuridine.

Smallpox   In active cases of smallpox, the vesicular rash involves the eye surfaces in about 20 percent of cases, and blindness due to residual scarring is a very common aftereffect.

SUMMARY

1. The skin is our most important barrier to infection, covering all of the exposed parts of the body except its orifices. Its anatomy features a constantly renewed exterior layer of hardened, dead epithelium, with tight intercellular boundaries to block microbial entrance. The physicochemical environment of the skin surface is hostile; available moisture is low, and salts provide a high osmotic pressure. The nutrients, including keratin, fats, and small amounts of amino acids, do not support most human pathogens.

2. The skin hosts a large microbial population composed of types adapted to survival under the selective environmental conditions. The flora is mainly Gram-positive, salt-tolerant, and able to metabolize fats secreted in sebum. By-products of fat metabolism aid in microbial population control. Several anaerobic species inhabit the deeper dermal regions.

3. Infection of the skin follows stress. Trauma and incompletely developed or temporarily compromised host defenses lead to skin infection. Disturbance of the normal flora or alteration of the physiological state of skin may provide a foothold for the pathogen.

4. Two species of bacteria, *Staphylococcus aureus* and *Streptococcus pyogenes*, cause most bacterial skin conditions. *S. aureus* primarily causes abscesses of varying degrees of severity. Certain strains with specialized extracellular products can cause spreading infections. *S. pyogenes* may cause impetigo or erysipelas. If caused by certain M-types, there is a risk of delayed complications, particularly glomerulonephritis.

5. Certain dermatophytic fungi and dimorphic yeasts can invade superficial or deeper layers of skin, causing the various forms of ringworm. *Candida albicans* is an opportunistic pathogen that invades newborns or compromised individuals.

6. Several human viral diseases show a skin eruption or rash as the major sign. Within the spots, viruses are present and multiplying; they are the target of a CMI response that gives rise to the localized tissue destruction. Other symptoms, such as fever, malaise, and prostration, are effects of the inflammatory response and viral multiplication in other areas such as the respiratory mucosa. This pattern, with various degrees of severity, describes chickenpox, measles, rubella, and smallpox. In both measles and chickenpox, virus may persist after recovery, in the latent form, causing later serious disease. Smallpox has now been eradicated; it is hoped that measles will be eradicated next.

7. The surface structures of the eye, including eyelids, conjunctiva, lacrimal network, and corneal surface are occasionally subject to infection. The eye is protected principally by the constant, effective mechanical cleansing by the tears, which contain antibacterial substances.

8. The eye is a secondary site of infection for two venereal disease organisms—*N. gonorrhoeae* and *C. trachomatis*. Infections occur in adults and children due to transfer of contaminated secretions to the eye. Newborns are infected in the birth process. Trachoma is the most serious of the eye infections. It is endemic in hot, dry areas of the world causing a vast toll of blindness, mostly in children.

Study Topics

1. Compare the roles of sweat, sebum, and tears as microbial control factors.

2. How does the skin participate in the benefits of immunological protection?

3. If an infective agent does colonize the skin, it will remain localized so long as it does not penetrate to which dermal structures?

4. Why would it be extremely difficult to eradicate the varicella-zoster virus from the human population over a ten-year period as smallpox was eradicated? Why is measles a more suitable candidate for eradication, as well as a more important target?

5. What role does personal hygiene play in the effectiveness of the skin as an infection barrier?

6. Describe how dermatophytes cause the symptoms of ringworm.

Bibliography

Cohen, J.O. (ed.), 1972, *The Staphylococci*, New York: Wiley.

Maibach, H.I., and G. Hildrich-Smith (eds.), 1965, *Skin Bacteria and Their Role in Infection*, New York: McGraw-Hill.

Marples, Mary J., 1974, The normal microbial flora of the skin. In F.A. Skinner and J.B. Carr (eds.), *The Normal Microbial Flora of Man.* Society for Applied Bacteriology Symposium, Series 3, New York: Academic Press.

Read, Stanley E., and John B. Zabriskie (eds.), 1980, *Streptococcal Diseases and the Immune Response*, New York: Academic Press.

Shinefeld, H.R., et al., 1974, Bacterial interference between strains of *Staphylococcus aureus*. *Ann. N.Y. Acad. Sci.* **236**: 444–455.

Skinner, F.A., and L.B. Quesnel (eds.), 1978, *The Streptococci*, New York: Academic Press.

Tachibana, Dora K., 1976, Microbiology of the foot, *Ann. Rev. Microbiol.* **30**: 351–376.

Youmans, G.P., et al., 1980, *The Biological and Clinical Basis of Infectious Diseases*, 2nd ed., Philadelphia: W.B. Saunders, Chapters 36 and 37.

18

Upper Respiratory Tract Infections

The upper respiratory tract (URT) is composed of passageways that connect adjacent organs to the outside world. It supports a large, complex, and primarily protective normal flora, and is highly susceptible to infection.

With a few important exceptions, the URT infections are caused by viruses. URT infections are generally mild, self-limiting, and loosely defined. The common cold may be caused by any one of nearly one hundred different viral agents. Exact identification of the causative agent is so time-consuming that the results often become available only after the disease has mended.

The bacterial diseases, on the other hand, although they make up only 5 to 10 percent of URT infections, are clearly defined entities. Prompt diagnosis is important because specific treatment can be used and may prevent the development of serious complications.

ANATOMY AND PHYSIOLOGY

The Anatomy of the Upper Respiratory Tract

The upper respiratory tract is composed of ducts, passages, and vestibules. The airway is the paired **nasal passages.** Air is inspired through a convoluted pathway that begins with the **nares** (nostrils). Air currents are directed against baffles in which they pass over moist mucous membranes.

Larger airborne particles, over 10 μm in diameter, are trapped on hairs and sticky mucous surfaces. The moistened, warmed, and cleansed air then passes through the **pharynx,** a central chamber, and is drawn downward into the lower respiratory tract (LRT), which begins at the **larynx.**

The term **nasopharynx** is often used to designate the airway (Fig. 18.1a), as distinct from the food passageway, which is the **oropharynx.** Although the passageways merge, differences in their physiological states make them distinct microbial

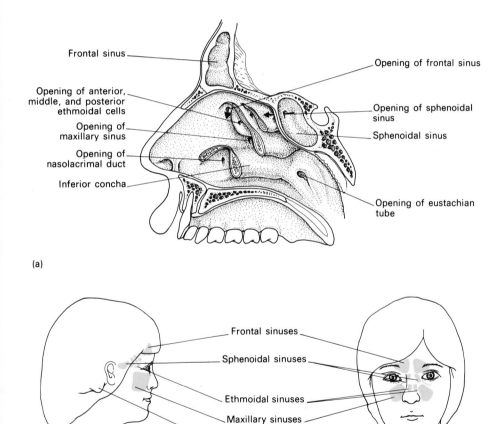

(a)

(b)

18.1
Structures of the Upper Respiratory Tract (a) The nasopharynx. When air is inhaled, it must pass over a number of surfaces before it enters the pharynx. The openings of the sinuses, the lacrimal ducts, and the eustachian tubes are all found in the nasopharynx. (b) The sinuses are membrane-lined air spaces in the bones of the skull.

environments. The point at which the gastrointestinal tract and the lower respiratory tracts leave the pharynx is marked by the **epiglottis,** a small flap of tissue that closes over the airway during swallowing.

The **lacrimal ducts** drain from the inner margin of each eye into the nasopharynx; the tears carry microorganisms, debris, and lysozyme.

The **middle ear** is connected to the pharynx by means of the slender **eustachian tube.** This tube is a device for equalizing atmospheric pressure on both sides of the eardrum. The duct can be readily compressed or collapsed, especially in small children.

Lymphoid tissues; Chapter 13, pp. 374–378

Tonsils are masses of lymphoid tissue. There are several groups of these in the nasopharynx. The **palatine** tonsils, embedded in the lateral wall of the pharynx, are the ones whose swelling is called **tonsillitis.** Swelling of the **pharyngeal** tonsils, on the posterior wall, is the condition called **adenoid disease.** Swelling of the **tubal** tonsils may obstruct the eustachian tubes and thereby contribute to middle-ear infection (otitis media).

The cranial bones underlying the face contain numerous **sinuses** (Fig. 18.1b). Sinuses are cavities lined with mucous membrane. They are air-filled, and continuous with the nasal passages. Their architecture varies greatly from individual to individual. Inflammation of the paranasal sinuses is **sinusitis.** The mastoid process, an extension of the temporal bone beneath the ear, is a spongy bone filled with numerous tiny cavities, the **mastoid cells.** Penetration of infection into these bony cavities causes **mastoiditis.**

Epithelium, amply supplied with mucous glands, lines all parts of the respiratory tract. Most of the tissue is **ciliated epithelium,** with a blanket of cilia on the exterior surface of each cell. The ciliary motions are coordinated, moving a continuous layer of mucus. The **mucociliary elevator,** continuous down the LRT to the ends of the smallest bronchioles, brings mucus up. Mucociliary action in the other passages also brings mucous secretions to the mouth, where they are swallowed, and the microorganisms trapped in the mucus are consigned to the stomach acid for disposal.

Physiology of the Upper Respiratory Tract

In the nasopharynx, gradual evaporation of the secretions creates an environment that is like the skin, rich in dissolved salts. The pH of the secretions is normally about 6.5. The intensely vascular lining of the nasal passage responds to changes in temperature and humidity by swelling, causing congestion. Such changes also seem to increase the tissue's susceptibility to infection.

In areas other than the nasal passages, physiological conditions are suitable for fastidious microorganisms. The bacterial flora derives its nourishment from secretions. Respiratory tract infection causes inflammation, increasing the volume of secretions. The secretions may be plentiful and thin (runny nose) or thick and viscous.

The temperature of the surfaces and secretions of the URT is usually two or more degrees lower than the interior body temperature. This has a selective effect on the types of microorganisms that are present.

The main defenses of the respiratory epithelium include secretory IgA and variable amounts of other immunoglobulins. Phagocytic PMNs and macrophages readily pass from the circulation into the mucous coat.

NORMAL UPPER RESPIRATORY TRACT FLORA

The nasal passage, nasopharynx, and oropharynx harbor different groups of microorganisms. All are closely adapted to their environment and have limited potential for survival outside their habitat. There is a number of genera represented. Nonpathogenic members of these genera, the normal flora, are present in large numbers in healthy human beings. Each genus also contains one or more members with similar ecological adaptations that are either frank pathogens or opportunists. These harmful species find it easy to persist in the carrier state. Respiratory tract secretions readily transmit disease; healthy carriers are a key factor in the spread of respiratory infections.

The Nasopharyngeal Flora

The flora of the anterior portion of the nasal passage is similar to and derived from the external skin flora, with Gram-positive, salt-tolerant organisms such as *Staphylococcus aureus, S. epidermidis,* and the aerobic diphtheroids predominating. This flora can be conveniently sampled with a sterile cotton swab. In various population groups, from 10 to 30 percent or more of the individuals will be shown to carry *S. aureus.* The carrier state may develop or clear on a day-to-day basis, and many of the isolates are potential hazards.

Skin flora; Chapter 17, pp. 449–501

The flora of the posterior nasopharynx gradually becomes similar to and merges with the oropharyngeal flora. This area is difficult to sample. A sterile *nasal swab* must be introduced on a slender, flexible wire. The procedure may be used for isolating the diphtheria organism from carriers.

The Oropharyngeal Flora

The oropharyngeal flora is composed of organisms associated with the teeth, gums, tongue, and saliva, and organisms that preferentially multiply when attached to the pharyngeal epithelium. Since one prerequisite for pathogenesis is attachment to the target tissue, potential pathogens are in this second group.

In obtaining a specimen for laboratory study, it is imperative that the **throat swab** be properly collected. The sterile swab should be rubbed firmly over the posterior pharynx and tonsillar area in which the potential pathogens may be found.

Contact with the oral surfaces and saliva should be avoided because the organisms found there are not clinically significant. They are also so numerous that they may overgrow the culture and prevent isolation of the true pathogens.

In addition to the organisms of the nasal passage, the oropharynx harbors several species of α-hemolytic streptococci, including the potential pathogen *S. pneumoniae*. The Gram-negative diplococci, most importantly *Branhamella catarrhalis*, *Neisseria sicca*, and other neisseriae, and the tiny, fastidious Gram-negative bacilli of the genus *Hemophilus* are also common.

Gram-negative enterics are quite uncommon, although *Klebsiella pneumoniae*, a potential pathogen, may be found. Hospitalized patients readily develop some Gram-negative oral flora. Appearance of significant numbers of enteric organisms or *Pseudomonas* indicates altered body defenses and is a matter for some concern.

Anaerobic organisms such as *Bacteroides*, the anaerobic streptococci, and diplococci such as *Veillonella* are a minor component.

Laboratory Analysis

Samples taken from the URT inevitably contain a large, mixed flora. Separation of normal flora from pathogens is a delicate, skilled procedure. The first essential is a properly collected sample, taken from the area that it claims to represent, and collected without contamination from other areas. It should be cultured without delay on appropriate media. Sheepblood agar plates are streaked by a technique designed to yield isolated colonies and reveal both aerobic and anaerobic hemolysis. The medical technologist learns to recognize significant colony morphologies and to subculture colonies of presumed pathogens for completed identification. There is always a problem in interpretation when a pathogen is found in very low numbers. Accurate reporting of the observed prevalence of the pathogen in the sample may be of value to the physician in forming a clinical judgement.

Streak plate; Chapter 7, pp. 189–190

The Carrier State

Carriers; Chapter 15, pp. 441–442

Carriers as reservoirs of infection were discussed in Chapter 15. Carriers spread respiratory infections by direct contact, via airborne droplets, or by contaminating fomites. During the cold winter months or in crowded habitations, such as dormitories and military barracks, carrier rates may rise alarmingly. *N. meningitidis*, causative agent of meningococcal meningitis, has been known to reach carrier rates approaching 90 percent in small populations. The usual carrier rate appears to be about 15–20 percent. Persons immune to diphtheria can be healthy carriers. The carrier rate for diphtheria is sporadically high in crowded urban populations and in the rural South and Southwest. For these reasons, the respiratory secretions of any individual, sick or well, must be considered potentially hazardous. Responsible hygienic measures to prevent the dissemination of respiratory secretions should become habitual.

TABLE 18.1
Bacterial flora of the upper respiratory tract

NORMAL FLORA	PREVALENCE IN NASOPHARYNX	PREVALENCE IN OROPHARYNX	SPECIES WITH PATHOGENIC POTENTIAL
Staphylococcus epidermidis	*	†	
S. aureus	†	‡	*S. aureus*
Corynebacterium xerosis, C. pseudodiphtheriticum (diphtheroids)	†	†	*C. diphtheriae*
Alpha and nonhemolytic *Streptococcus* sp.	§	*	*S. pyogenes*, Group A
S. pneumoniae	§	‡	*S. pneumoniae*
Neisseria sicca and *N. subflava*	§	†	*N. meningitidis* *N. gonorrhoeae*
Branhamella catarrhalis	§	†	
Hemophilus influenzae and *H. parainfluenzae*	§	‡	*H. influenzae*
Enterobacteria	§	§	*Klebsiella pneumoniae*

*Found in almost all samples.
†Found in more than half of the samples.
‡Found in less than half of the samples.
§Found in only an occasional sample.

BACTERIAL DISEASES

The serious bacterial diseases of the URT, streptococcal pharyngitis and diphtheria, comprise a very small fraction of the total cases of URT infection. These diseases will be covered in detail, however, because of their potential severity and the value of differential identification in successful therapy.

Streptococcal Pharyngitis

Acute pharyngitis (sore throat) is a common complaint, particularly in children and adolescents, but it happens to us all periodically throughout life. The symptoms are pain and reddening and inflammation of the pharyngeal epithelium. Lymphoid masses may swell and become coated with exudates, and fever or malaise may also be present. Almost all cases are self-limiting, and the symptoms disappear after a few days without treatment. From 90 to 95 percent of all cases of pharyngitis are nonbacterial, caused by viruses of many types.

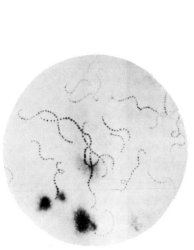

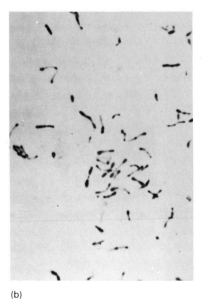

18.2
Bacterial Pathogens of the Upper Respiratory Tract (a) *Streptococcus pyogenes*, a Gram-positive organism, is often seen in short chains and pairs. It causes the most important form of bacterial pharyngitis. (b) *Corynebacterium diphtheriae* is a Gram-positive non-sporulating rod. The cells are often highly granular and may be club-shaped. Toxigenic strains cause diphtheria, involving the nasal or oral pharynx and occasionally the larynx and trachea as well.

(a)   (b)

In 5 to 10 percent of all pharyngitis cases, the agent is *Streptococcus pyogenes*, Group A (Fig. 18.2a). Streptococcal sore throats are not much different clinically from viral sore throats although exudate production is more likely in the former. Differential diagnosis rests on recovery of the organism from throat culture. *S. pyogenes* pharyngitis is common during the winter months among individuals 5–14 years of age.

Presumptive laboratory identification may be available in less than 24 hours. Antimicrobial therapy with penicillin or erythromycin, instituted promptly, will relieve symptoms and prevent the development of sequelae such as otitis media, mastoiditis, and abscesses.

Another pressing reason to identify and treat sore throats caused by *S. pyogenes* is that, if untreated, about 3 percent of these patients will develop **rheumatic fever.** This is a serious, often recurrent, hypersensitivity that damages heart valves and causes arthritis. Rheumatic fever is discussed further in Chapter 24.

Rheumatic fever; Chapter 24, p. 702

Scarlet fever is a form of streptococcal pharyngitis in which the infecting strain produces an **erythrogenic toxin.** The toxin causes a diffuse rash, probably a hypersensitivity response. Toxigenic strains, uncommon in the United States for the past few years, often have enhanced virulence.

Diphtheria

Causative agent   Diphtheria is caused by *Corynebacterium diphtheriae*, a non-sporulating, Gram-positive rod. Its microscopic appearance is unusual. The cells often look beaded as a result of numerous intracellular **granules.** Also, because of the way in which dividing cells separate, they tend to lie together in parallel arrays called **palisades.** They are often very irregular in shape (Fig. 18.2b).

The organism can be isolated from throat or nasopharyngeal swabs by incubation on Loefflers medium, a blood agar medium containing **potassium tellurite.** This substance strongly inhibits normal flora but not *C. diphtheriae*, allowing the pathogen to develop without being overgrown by other species. Corynebacterial colonies develop a gray to black color on tellurite medium, which aids in presumptive identification.

A confirmed identification requires demonstration that the isolate produces **diphtheria toxin,** the major virulence factor of this species. This may be done by injecting culture filtrates into guinea pigs, or with *in vitro* immunological techniques.

Diphtheria toxin is an exotoxin, produced only by strains of *C. diphtheriae* lysogenized by the β-phage. Phage-free strains are nontoxigenic. The toxin is a protein, 62,000 daltons, that crosses the membrane of the susceptible host cell and then combines enzymatically with NAD. In this form it inhibits an enzyme needed for elongation of peptide chains in protein synthesis. The blocked cell eventually dies. The heart, kidney, and peripheral nervous system are the most severely affected. The organs undergo degenerative changes; **demyelination** (loss of the myelin sheath) incapacitates nerve cells.

Lysogeny; Chapter 10, pp. 295–297

Pathogenesis   *C. diphtheriae* initially infects the oropharynx, or less commonly, the nasopharynx. Multiplication of the bacilli attached to the surface epithelium causes inflammation, and this in turn leads to leakage of serum proteins and white blood cells onto the pharyngeal surface. This exudate coagulates to form a tough, whitish **diphtheritic membrane,** one major diagnostic feature of the disease. The membrane is composed of fibrin, living and dead bacteria, PMNs and macrophages. Although unpleasant, the membrane itself is not life-threatening unless the infection extends into the trachea and the resulting membrane growth obstructs airflow.

Systemic pathogenesis is caused by toxin produced in the pharynx, absorbed into the bloodstream, and distributed to the body. The toxin's effects are delayed. Cardiac abnormalities progressing to myocarditis and congestive heart failure may appear after a week. Neurologic effects—muscle weakness and variable paralysis—develop after a month or more. The mortality of treated or untreated diphtheria has remained at about 10 percent for the past 50 years. This figure reflects the fact that once toxin has been absorbed by the target cells, specific therapy cannot reverse its effects.

Therapy and immunity   One form of specific therapy for diphtheria is the administration of antitoxin, hyperimmune horse serum. Horse serum requires preliminary testing to rule out previous sensitization to horse antigens; one should be prepared to deal with anaphylactic shock should it occur. This approach is used very rarely now. Antibiotics—penicillin and erythromycin—can reduce the microbial popula-

Box 18.1

Diphtheria—California

On May 10, 1979, a 68-year-old man with no history of diphtheria immunization was seen at a Los Angeles County Health Department clinic with a sore throat. When clinicians noted a pharyngeal membrane, they took a throat culture, and the patient was referred to a hospital with a diagnosis of diphtheria. A physician in the ear, nose, and throat clinic of the hospital stripped the membrane and sent it for culture, made a diagnosis of pharyngitis, and sent the patient home on oral penicillin. The next day, the county health department again referred the patient to the hospital because of the suspicion of diphtheria, but he was sent home with no change in the diagnosis or treatment.

Corynebacterium diphtheriae was cultured from the throat membrane fragments, and guinea pigs were inoculated to test for toxigenicity. When the test was found to be positive on June 1, the county health department was informed of the culture results.

The patient, whose throat had healed, was hospitalized on June 4. An electrocardiogram showed a right bundle branch block that had not changed from the pattern a year earlier. The patient was discharged without parenteral penicillin or antitoxin treatment. Two throat cultures taken during admission were reported as negative on June 12 and 13. On June 25, the patient returned to the hospital with trouble swallowing and a nasal quality to his voice. He was admitted with a diagnosis of diphtheritic bulbar neuropathy and was put on tube feeding for the dysphagia. At present, he has a neurologic deficit involving the muscles of the soft palate.

On June 4, the county health department cultured the throats of 20 family members, all of whom had been well. The culture of a 9-year-old granddaughter, who had been previously immunized, was positive for *C. diphtheriae;* that culture was also toxigenic. The toxoid was given. The immunizations of all family members have been brought up to date, and all members remain well. Because the index patient had made a trip to Albuquerque, New Mexico, four weeks before the onset of illness, the New Mexico State Health and Environment Department was notified about the case. All family members in New Mexico who were visited by the patient were followed up for immunizations; all members have remained well.

This case illustrates the importance of keeping adults up to date with recommended immunizations and of considering the diagnosis of "childhood" diseases in unimmunized adults.

Persons with suspected or proven diphtheria should receive both parenteral penicillin and diphtheria antitoxin after their skin has been tested for hypersensitivity.

This case also illustrates a growing tendency in American health-care delivery. The patient was referred back and forth among different agencies. The responsibility for his case was divided. This division resulted in a lack of concerted action on his behalf.

Source: *Morbidity and Mortality, Weekly Reports* **28**, 1979.

tion in the pharynx, minimizing further toxin production and promoting healing of the lesions under the membrane.

Diphtheric toxin is a potent antigen. Therefore recovery from the disease confers permanent antitoxic immunity. A person will not experience the acute form of diphtheria twice. However, the antibody response does not prevent pharyngeal colonization. Recovered persons may be pharyngeal carriers.

Diphtheria toxin is produced commercially, treated with formalin and alum, and converted to a **toxoid.** In this form it is immunogenic but not toxic. The toxoid is recommended for all children, and it is usually given with other materials in the trivalent **DPT vaccine.** Adults may be reimmunized if they are exposed, but a much reduced dose of toxoid is used to avoid the risk of a hypersensitivity reaction.

The **Schick** skin test has been used to measure circulating antitoxin and evaluate immunity and/or hypersensitivity to diphtheria toxin and toxoid. One can assess the susceptibility of exposed persons and determine whether revaccination is necessary and safe. Minute amounts of toxin and/or toxoid are injected into the skin of the forearm; the site is observed for the extent and duration of reddening.

Epidemiology   The incidence of diphtheria has dropped greatly since immunization became generally available in 1923. Only sporadic outbreaks occur now, primarily in the winter months in urban areas and the southern and southwestern United States. There are about 60 cases per year. Diphtheria was originally a disease of the first five years of life. Now it is more common in older individuals who are long past their childhood immunization or who had never been immunized.

One reservoir is the human pharynx. A small number of healthy carriers harbor the organism. Another source, especially in warmer climates, is **cutaneous diphtheria.** Skin lesions such as insect bites readily become infected with C. *diphtheriae* and may become highly infective. The mode of transmission is droplets spread from the pharyngeal reservoir or direct contact from skin lesions.

Control follows universal immunization. However, because vaccinated persons continue to serve as reservoirs, diphtheria cannot be eradicated. Immunization will remain necessary for the foreseeable future.

Otitis Media

Causative agents   **Otitis media** is infection of the middle ear. It is estimated that about 70 percent of cases of otitis media are bacterial in origin (Fig. 18.3). The most common causative agents are listed in Table 18.2. The remaining cases are probably all caused by viruses, although these are not readily demonstrated.

Pathogenesis   Otitis media almost exclusively affects small children. Acute cases are marked by pain and reddening and bulging of the **tympanum,** or eardrum. The symptoms are caused by accumulated fluid in the enclosed middle ear, retained by

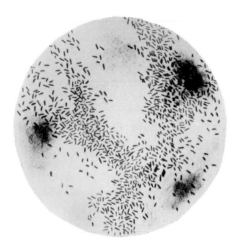

18.3
Hemophilus influenzae, a tiny pleomorphic Gram-negative rod, may be entirely normal in the URT. However, under certain conditions pathogenic strains, especially Type B, may cause serious illness including meningitis, otitis media, pneumonia, epiglottitis, and septicemia.

obstruction of the eustachian tube (Fig. 18.4). In chronic cases, exudate is present in the middle ear, but other symptoms may be absent.

The middle ear is readily colonized by the normal oropharyngeal flora. When these multiply to large numbers in the obstructed area, they cause inflammation and

TABLE 18.2
Causative agents of otitis media in children

ORGANISM	PERCENTAGE OF PREVALENCE	COMMENTS
Streptococcus pneumoniae	25–50	Serotypes 1,3,6,14,18,19,23
Hemophilus influenzae	15–25	Rarely Type b, many untypable
Streptococcus pyogenes	Less than 6	
Branhamella catarrhalis	2–9	
Staphylococcus epidermidis	Less than 10	
Staphylococcus aureus	Rare	In neonates
Enterobacteria	Rare	In neonates
All bacteria	About 70	
All viruses	About 30	

Source: G.S. Geibick and P.G. Quie,,1978, Otitis media: the spectrum of middle ear infection, *Ann. Rev. Med.* **29**: 285.

18.4
The Ear   The middle ear is an enclosed space, connecting to the exterior for pressure equalization by the eustachian tube. When the masses of tonsillar material around this tube become swollen, the eustachian tube may be pressed shut so that the middle ear does not drain. Overgrowth of flora results in infection, bulging of the tympanic membrane, and sometimes damage to the delicate auditory structures.

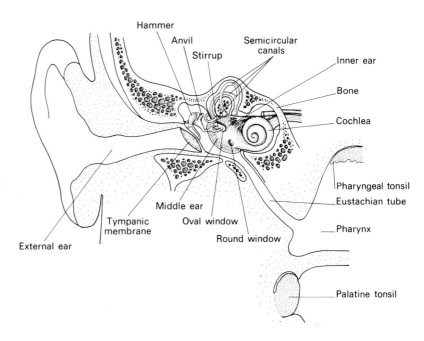

sometimes perforation and scarring of the eardrum. Repeated or prolonged cases may result in hearing loss. Mastoiditis is another potential complication.

Treatment   It is not usually practical to isolate and identify the organism, because isolation requires aspiration of fluid from the middle ear. Antimicrobial agents effective against the most probable organisms are prescribed. A combination of erythromycin and sulfa, or ampicillin, has the best chance of success.

Recurrent or chronic cases are frequently treated surgically by the installation of tympanotomy tubes to provide a continual drain. This procedure gives improvement in about 65 percent of cases. Surgical removal of tonsils and adenoids is no longer routine in view of evidence that indicates that it is not usually effective.

Epidemiology   Otitis media is a surprisingly common condition. Studies of Eskimo children in Alaska and of children in South Carolina and Alabama showed that from 76 to 84 percent of the children experienced at least one episode of acute otitis media. If the initial episode occurred in the first year, the chance of recurrence was high. Up to 20 percent of school-age children have chronic asymptomatic otitis media.

Preventive measures are not well worked out mainly because the predisposing features are unclear. Vaccination against key strains of *S. pneumoniae* and *H. influenzae*, still under development, could reduce the incidence of otitis media by 50 percent.

Other Bacterial Infections

S. aureus, S. pneumoniae, and other opportunistic members of the normal flora rarely cause infections of upper respiratory tract structures. *Neisseria gonorrhoeae* and *N. meningitidis* are occasional causes of pharyngitis.

VIRAL INFECTIONS

About 90 to 95 percent of URT infections are caused by viruses. Our knowledge of these viral infections lags far behind our knowledge of the bacterial infections. One of the main reasons is that the viral disease patterns are overlapping. There is usually no reliable symptomatic difference among "common colds," caused by rhinoviruses, coronaviruses, or echoviruses. When you consider that over half of all known human viruses infect the upper respiratory tract, usually causing about the same symptoms, you will understand the potential confusion. Recently, large-scale epidemiological studies, which sought identification of the causative agent in each case of URT illness, have detected some patterns.

Acute Rhinitis

Rhinitis is an inflammation of the nasal mucous membrane, resulting in a profuse watery secretion, sneezing, and lacrimation. This is the all-too-familiar common

TABLE 18.3
Frequency of infection with respiratory agents

AGENT	FREQUENCY OF ISOLATION OF AGENT	
	ALL ILLNESSES	THROUGH MEDICAL CONSULTATION
Rhinovirus	38.5	23.3
Parainfluenza virus	16.9	22.2
Group A streptococci	13.3	22.2
Influenza virus	11.9	15.0
Respiratory syncytial virus	5.9	5.6
Adenovirus	4.5	3.9
Enterovirus	4.3	3.3
Other*	4.7	4.5

Source: A.S. Monto and B.M. Ullman, 1974, Acute respiratory illness in an American community, *JAMA* **227**: 164.

*Coronaviruses were not included among the viruses the investigators looked for.

These data are derived from a six-year study of the entire population of Tecumseh, Michigan.

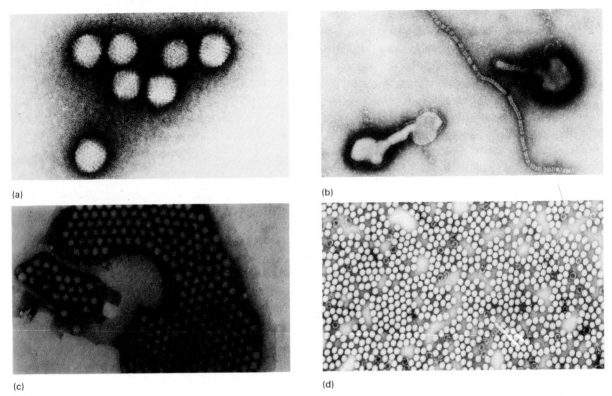

(a)

(b)

(c)

(d)

18.5
Viruses of the Upper Respiratory Tract (a) The rhinovirus of the common cold, a tiny DNA virus. (b) A para-myxovirus, the parainfluenza virus. Note irregular shape, visible envelope. One particle has been ruptured, and the nucleic acid, an RNA-protein helix, is visible in its extended form. (c) Mumps virus. This is a paramyxovirus containing DNA, approximately 90 nm in diameter. Its envelope, which contains hemagglutinin and neuraminidase units, is clearly seen. (d) A closely related paramyxovirus obtained from an infected chick embryo. It has been ruptured, and the nucleic acid, an RNA-protein helix, is clearly visible leaking out.

cold. It is so common that it is estimated that on any given winter day in the United States, 10 percent of the population—that is, in excess of 20 million people—have cold symptoms! Children under five average six colds per year; children from 5–14 have five colds a year, and those over 14 have about three colds a year. There is a great difference, largely unexplained, in individual susceptibility.

Causative agents   The rhinoviruses (Fig. 18.5a) are RNA viruses. They replicate best in cell cultures incubated at 33°C rather than 37°C, and seem to be restricted in their infectivity to the cooler areas of the respiratory tract. There are at least 56 strains proven to infect human beings, and many additional strains of unestablished pathogenesis.

The coronaviruses, a recently discovered group of envelope-bearing, RNA viruses, appear to be a major cause of colds in adults. Respiratory syncytial viruses cause colds principally in small children.

Pathogenesis The common cold is normally mild and self-limiting. Respiratory epithelium becomes inflamed and hypersecretive, but there is no appreciable cellular destruction. The watery secretion carries interferon and immunoglobulins, and probably helps bring the infection under control by washing away virus, reducing viral replication and, once an immune response appears, neutralizing new crops of virus. Colds can become serious only in small children whose incompletely developed immunity is unable to prevent viruses from advancing into the LRT. Colds, like other primary respiratory tract infections, appear to increase susceptibility to secondary LRT infections.

Therapy and immunity There is no specific therapy. Symptomatic medications such as decongestants and analgesics do not alter the course of the disease.

Each cold confers some immunity to the agent responsible. However, there are so many different agents that if each person had three colds a year, becoming immune to one agent each time, it would take 30 or more years to acquire a full spectrum of protection. This multiplicity of agents has so far baffled researchers attempting to formulate vaccines.

Epidemiology The cold is a disease of modern urbanized life. Colds confined to small geographically isolated populations quickly "burn themselves out." With the

TABLE 18.4
Role of viral agents in URT illness

AGENT	TYPICAL URT CONDITIONS	LRT INFECTION
Rhinovirus	Rhinitis	No
Coxsackie A, various strains	Rhinitis, herpangina, hand-foot-and-mouth disease	Yes
Influenza, Type A	Rhinitis is the only symptom in some cases	Yes
Respiratory syncytial	Rhinitis, pharyngitis in older children and adults	Yes
Parainfluenza Types 1,3	Rhinitis	Yes
Adenoviruses	Pharyngitis, rhinitis, ARD	Yes
Echoviruses	Rhinitis with fever (rare)	Yes
Coronaviruses	Rhinitis in adults	Yes

advent of rapid transportation, cold viruses are widely disseminated. Furthermore, we live with central heating that exposes the nasal mucosa to hot, dry air in the wintertime and thereby places a new stress on the target tissue.

Nonbacterial Pharyngitis

Sore throat, fever, and reddening of the pharynx (acute respiratory disease or ARD) may be caused by viral agents. The adenoviruses, large DNA viruses (Fig. 18.5b) were first isolated from adenoid biopsies. There are 31 antigenic types that may cause colds, ARD, and many other diseases. Adenovirus ARD is most commonly found in military installations, summer camps, and in other dense populations.

The Coxsackie viruses may cause an uncomplicated pharyngitis in young children. Certain strains cause **herpangina,** a more severe condition marked by high fever, gastrointestinal symptoms, and the development of vesicular lesions on the tonsils and palate. With Coxsackie virus strain A-16, the lesions erupt on hands, feet, and occasionally on buttocks as well. This is called **hand-foot-and-mouth** disease.

Mumps (Infectious Parotitis)

Causative agent   The mumps virus is a paramyxovirus (Fig. 18.5c and d), a large envelope-bearing, RNA virus. It has hemagglutinin and neuraminidase units incorporated into the outer layer.

Viral envelope; Chapter 11, pp. 306–307

Pathogenesis   Mumps is characteristically a disease of small children, but it may infect susceptibles of all ages. The primary site of viral replication is the parotid salivary gland. In typical cases one or both parotids become greatly swollen. Pain on talking or eating may be pronounced. The mumps virus may attack other glandular tissues, and the pancreas and reproductive organs may have swelling too. Orchitis (testicular swelling) is seen in about 20 percent of postpubertal males. Mumps virus can also invade the nervous system, and transient meningoencephalitis may be present in about 10 percent of cases. Nerve deafness, fortunately usually unilateral, occurs rarely. The symptoms are usually transient and disappear within a defined period. Severe complications or death are rare.

Therapy and immunity   There is no specific therapy. The disease, either active or asymptomatic, confers permanent immunity. There is no truth to the tale that you are not immune if you "had it only on one side."

There is a highly effective, live-virus vaccine available, recommended for all infants and other susceptibles, especially males. It is usually given in the form of the trivalent MMR (measles, mumps, rubella) vaccine.

Epidemiology The virus is spread through direct contact with saliva, respiratory secretions, or fomites (Fig. 18.6). The actively ill—those in the incubation period and those with asymptomatic cases—may all be infectious. The chances for control depend largely on the universal adoption of vaccination. There are still about 13,000 cases per year in the United States.

FUNGAL INFECTIONS

Candida albicans;
Chapter 17, pp.
510–511

The major significant fungal infection of the upper respiratory tract is **thrush**, caused by *Candida albicans*, affecting the tongue, oral mucosa, and oropharynx. The mucosa become acutely inflamed and the reddened patches become covered with a grayish-white exudate that may mimic the diphtheritic membrane. Thrush occurs in about 5 percent of newborns, developing within the first few days of life. It may also

TABLE 18.5
Epidemiology of URT infections

DISEASE	RESERVOIR	TRANSMISSION	INCUBATION PERIOD IN DAYS
Streptococcal pharyngitis	Human beings	Direct contact, milk, water, food	1–3
Diphtheria	Human beings	Direct contact, airborne, rarely fomites, raw milk	2–5
Herpangina, hand-foot-and-mouth disease	Human beings	Direct contact with nose and throat discharges, feces, droplets	3–5
Common cold	Human beings	Direct oral contact, droplets, soiled articles, hands	1–3
Acute respiratory disease (ARD) due to adenovirus	Human beings	Direct oral contact, droplets, soiled articles, hands	Few days to week or more
Mumps	Human beings	Droplets, direct contact with saliva	12–26

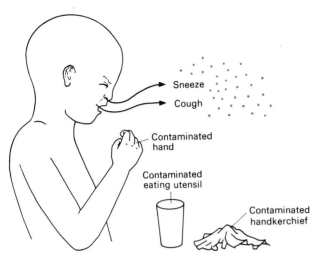

Sneeze

Cough

Contaminated hand

Contaminated eating utensil

Contaminated handkerchief

18.6
Transmission of Respiratory Disease   The victim's respiratory secretions are infectious from a period about one day before symptoms appear and during the first five days of symptoms. Another person catches the cold when he or she inhales airborne droplets from an uncovered sneeze or cough, touches or uses contaminated eating utensils or tissues, or has direct contact such as kissing.

PERIOD WHEN COMMUNICABLE	PATIENT CONTROL MEASURES	EPIDEMIC CONTROL MEASURES
Untreated, while discharge present	Disinfect discharges, soiled materials. Especial attention to manual contamination risk.	Trace carriers; determine source of food and/or milk contamination, if any
Two to four weeks, while organism is present	Isolation, minimum 14 days, disinfection of all patient-owned articles	Immunization of exposed group, especially infants and children; quarantine contacts; initiate search for carriers
During acute stage; while virus is present in stools	Disinfect discharges, feces, and soiled articles	Notice to physicians. Isolation of cases and febrile children pending diagnosis
From one day before onset to five days after	Stifle sneezes, dispose of tissues in sealed bag, keep hands clean	None
For duration of active disease; virus may have latent period	Stifle sneezes, dispose of tissues in sealed bag, keep hands clean	Avoid crowding
From six days before symptoms to nine days later. Inapparent cases may be communicable	Isolation for up to nine days after swelling develops, disinfection of soiled articles	Routine immunization of all infants

be expected to occur in as many as 10 percent of elderly patients. Thrush often develops in the immunodeficient patient, especially if previous antibiotic therapy has substantially altered the normal flora of the oropharynx. Oral *Candida* infections are treated by the administration of nystatin solution.

SUMMARY

1. The upper respiratory tract is the site of our most common infections. In order for pathogens to establish themselves, they must be able to attach to the epithelial lining and compete effectively with the normal flora. Secretory immunoglobulins in mucus may prevent infection in the resistant individual by preventing attachment.

2. Almost all upper respiratory tract infections are mild and self-limiting, and multiplication of the pathogenic agent causes little or no tissue destruction. This is rather generally true of the virus infections that compose 90 percent or more of the caseload. Bacterial infections, while less common, are potentially far more serious.

3. Streptococcal pharyngitis caused by *Streptococcus pyogenes* is an acute, self-limiting disease. Since it carries a significant risk of developing rheumatic fever, prompt antibiotic treatment is needed.

4. Diphtheria caused by toxigenic strains of *Corynebacterium diphtheriae* is always a severe illness with a significant mortality. It can be completely prevented by vaccination with diphtheria toxoid.

5. Otitis media, in severe or recurrent cases, leads to loss of hearing. It is caused by one of several bacterial and viral agents, and is most common in children under five.

6. Nonbacterial pharyngitis, the common cold, and other viral conditions are seldom serious. Their major importance is in their huge social cost—lost time and misery. They may also predispose the convalescent patient to lower respiratory tract disease, which is inherently more severe. The causative viruses are numerous and ubiquitous, and there are no practical therapeutic or control measures at present.

Study Topics

1. When you go to your doctor complaining of a sore throat, what considerations might the doctor keep in mind in diagnosing and treating your case?

2. Why is it not wise to expect the doctor to prescribe antimicrobial drugs for every case of sore throat?

3. Describe the sequential steps in four hypothetical examples of rhinovirus transmission.

4. In what circumstances do sporadic outbreaks of diphtheria occur nowadays?

5. What is the difference between controlling an infectious disease and eradicating it, epidemiologically speaking?

6. Because diphtheria is so rare, physicians may practice for an entire career without seeing a case, and they do not ordinarily suspect it when seeing "sore throat" patients. Explain how this situation relates to the fact that the mortality rate is higher among the first cases in an outbreak than among the later cases.

7. Why has there been hesitation over the advisability of developing live adenovirus vaccines? (*Hint:* See Chapter 11.)

Bibliography

Books

Fenner, Frank J., and David O. White, 1976, *Medical Virology* (2nd ed.), New York: Academic Press.

Hoeprich, Paul D. (ed.), 1977, *Infectious Diseases* (2nd ed.), New York: Harper & Row.

Knight, V. (ed.), 1973, *Viral and Mycoplasmal Infections of the Respiratory Tract*, Philadelphia: Lea and Febiger.

Youmans, Guy P.; Philip Y. Paterson; and Herbert M. Sommers, 1980, *The Biological and Clinical Basis of Infectious Diseases* (2nd ed.), Philadelphia: W.B. Saunders.

Articles

Giebnick, G. Scott, and Paul G. Quie, 1978, Otitis media: the spectrum of middle ear infection. *Ann. Rev. Med.* **29:** 285–306.

Monto, A.S., and B.M. Ullman, 1974, Acute respiratory illness in an American community. The Tecumseh Study, *JAMA* **227:** 164–175.

19

Lower
Respiratory Tract
Infections

The lower respiratory tract (LRT), including the respiratory airway and the lungs, is a critical organ system. Its function—gas exchange—cannot be seriously compromised without threatening the life of the individual. LRT infection is thus inherently more serious than URT infection. One disease, pneumonia, is the most important microbial cause of mortality in the United States.

Most of the LRT infections are diseases of lowered host resistance. Others become serious only when host defenses fail to contain them. Severe LRT infection is found principally among the young, the elderly or immobilized, and the immunodeficient.

ANATOMY AND PHYSIOLOGY

Anatomy of the LRT

The airway   The **larynx** is the organ of speech (Fig. 19.1). It is composed of a number of cartilages and contains voluntary muscles including the vocal cords. The **epiglottis** is a flap of tissue on the top of the larynx that folds to seal off the airway while food is swallowed.

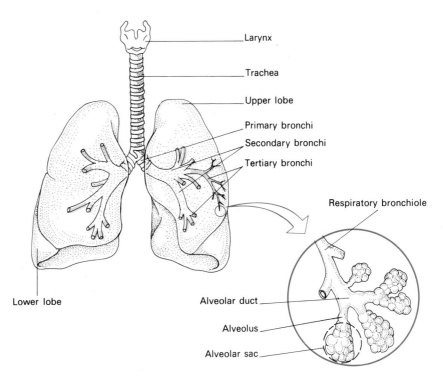

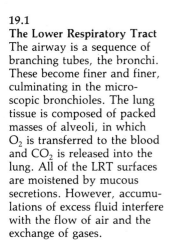

19.1
The Lower Respiratory Tract
The airway is a sequence of branching tubes, the bronchi. These become finer and finer, culminating in the microscopic bronchioles. The lung tissue is composed of packed masses of alveoli, in which O_2 is transferred to the blood and CO_2 is released into the lung. All of the LRT surfaces are moistened by mucous secretions. However, accumulations of excess fluid interfere with the flow of air and the exchange of gases.

The **trachea** is a thin-walled, straight tube supported in the open position by U-shaped rings of cartilage, open toward the back. Smooth muscle braces these rings.

The **bronchi** are the extensions of the trachea that enter the lobes of the lungs, continually branching into smaller tubules that eventually have a combined diameter almost 100 times that of the trachea. The smallest terminal members of the bronchial tree are the **bronchioles.** As the tubules grow smaller, the amount of cartilage in their walls decreases and the amount of smooth muscle increases. In the bronchioles, each tubule is wrapped in two antiparallel spiral bands of smooth muscle fibers that provide it great contractility. This contractility permits great variation in the diameter of the bronchioles.

The airway is lined by ciliated epithelial tissue (Fig. 19.2) down to the ends of the bronchioles. The epithelium also contains mucus-producing cells, loose aggregations of lymphoid tissue, and elastic fibers.

Lungs   The **lungs** proper are organized in lobes, two for the left and three for the right, and further subdivided into lobules. Because each subdivision is served by a particular branch of the bronchial tree, infectious tend, at least at first, to be localized to one lobe.

The bronchioles end in **alveoli,** thin-walled sacs that make up the bulk of lung tissue. Alveolar chambers are one cell thick and richly netted over by capillaries.

19.2
The Respiratory Mucosa   The surface, as shown in this scanning electron micrograph, bears clumps of cilia emerging from ciliated epithelium (CE) and the mucus-secreting goblet cells (GC).

Thoracic cavity   The **thoracic cavity,** in which the heart and lungs lie, is lined with two independent layers of membrane, the **pleura.** These are moist and glide easily over each other during the breathing movements. The space between the layers, the **pleural cavity,** may become filled with fluid during infection.

Physiology of the LRT

The airway secretions   Mucus production is a normal function of the airway lining tissue. The mucociliary elevator operates as a continuous cleansing system bringing trapped particles from the base of the bronchioles up to the pharynx. Certain factors increase mucus production and cause the material to become thicker, more tenacious, and difficult to move. These changes may be due to irritants, such as smoke or dust, and inflammation due to an infectious process. The thickened secretion is called **sputum.** The person who develops sputum will normally attempt to expel it by coughing. However, the cough may be ineffective, especially if the individual lacks muscular strength or if the sputum is highly viscous. Accumulation of sputum that cannot be expelled is a serious condition because the sputum prevents air flow and gas exchange.

Bronchial constriction   The bronchi can rapidly constrict to a small fraction of their normal diameter, obstructing air passage and causing acute respiratory distress. Stimulation of the smooth muscle fibers by the sympathetic nervous system is one cause. In **asthma,** the contraction may be caused by psychogenic, allergic, or infectious mechanisms. When overt infection occurs, inflammatory swelling of the epithelial lining may also create obstruction.

Bronchial dilation   Bronchodilation is caused normally by neurological stimulation of the smooth muscle; dilation increases air flow. Drugs such as epinephrine, aminophylline, isoproterenol, and corticosteroids also have this effect.

The lung secretions and defenses   Alveolar surfaces are designed for gas exchange. Oxygen and carbon dioxide cross the thin tissue barrier by dissolving in the aqueous film on the alveolar surface. This fluid contains a **surfactant** that keeps the lung surfaces from sticking together when they expand. Absence of surfactant in premature infants may lead to **hyaline membrane disease** and predispose the infants to acute respiratory infection. Accumulation of fluid in the lung due to circulatory malfunction or inflammatory processes severely reduces gas exchange.

The lung is policed by **alveolar macrophages.** These ingest foreign particles that escape the filtering mechanism of the URT. In general, only particles smaller than 10 μm reach the lungs; most particles smaller than 0.2 μm are simply exhaled. Most bacterial cells and small droplet nuclei are in the size range (from 0.2 to 10 μm) of particles that reach the lung and remain there. The alveolar macrophages are highly efficient scavengers, however. Only microorganisms that resist phagocytic digestion are successful LRT pathogens.

Macrophages; Chapter 12, p. 367

Physiological response to infection   The presence of irritants in the lung stimulates inpouring of fibrin-loaded fluid from the blood. The fibrin forms a gelled **consolidation** that blocks gas exchange in the affected area, but helps contain the infection. Immunoglobulins, especially IgA, are concentrated in the area of inflammation. The local macrophage population is rapidly augmented by huge numbers of circulating mononuclear cells and PMNs migrating through the capillary walls. If the irritant is microbial, these defenses will usually deal with the problem, and the inflammatory effects will subside. However, if the inhaled material is resistant to degradation (asbestos fibers, smoke particles, peanuts), the continued futile attempts of the host defenses to remove it may result in extensive, irreparable damage to the lung.

Normal Flora

The lower respiratory tract has no normal flora in the usual sense. Although microorganisms are regularly carried down from the upper tract, they are usually quickly removed. There is no resident, multiplying population. For this reason, the invading pathogen meets no ecological resistance in the form of microbial competitors. It must cope only with host defenses in order to establish itself.

BACTERIAL INFECTIONS

LRT infections can be caused both by bacteria and viruses, but bacteria play a much larger role than in the URT (causing up to 50 percent of infections). There are relatively few pathogenic species, and most are susceptible to antibiotics. The viral

Box 19.1

Dragon's Breath—A Microbiological Fairy Tale

The well-known fire-breathing of dragons indicates their nutrition as cellulose fermentors producing methane, supplementary to their common diet of fair maidens and daring knights. Considering the average daily methane production of a recent ruminant (the cow), the quantities of combustible gas available to a sizable monster for flame-throwing purposes must have been substantial indeed. The method of ignition may well have been connected to the occasionally reported "gnashing of teeth," with the dental deposition of heavy metals acting as flint.

A highly possible reason for the extinction of the fire-breathing species could be bursts of oxygen by catalase reactions—during periods of evolutionary trials and errors—that were apt to cause disastrous internal explosions. It could well have been part of a successful dragon-slayer's strategy (most likely the heroes kept such tricks to themselves) to excite their foe into an irregular breathing pattern that resulted in the fatal backfiring.

Source: Adapted with the kind permission of Dr. Holger W. Jannasch from "Breathing Fire," *ASM News* **42**: 601–602, 1976.

infections do not respond to chemotherapy, but tend to be less severe and have low mortality rates. Fungal infections are uncommon and restricted to certain geographical and occupational exposure patterns.

The LRT infections are often described clinically in terms that bear no reference to the infective agent. A collection of commonly used terms is supplied in Table 19.1 for reference.

There are some general patterns for transmission of LRT infections. Inhalation of infective agents is by far the most common means of transmission. The agent may become airborne in the form of aerosolized respiratory secretions, dried droplet nuclei, or spores in dust clouds. Quite often it appears that an agent will first establish itself in the URT in a carrier state. It will then periodically be inhaled into the LRT as air is inspired—a form of self-inoculation. Only if host defenses fail will this self-inoculation "take" and clinical disease result. A less common route by which microorganisms reach the lung is hematogenous spread from infectious foci elsewhere in the body. Bloodborne agents are frequently anaerobic.

Inhalation pneumonia follows exposure to irritating or corrosive chemical vapors or anesthesia. Aspiration pneumonia is a result of the inhalation of vomitus, occurring in unconscious accident victims and alcoholics.

It is often difficult to obtain laboratory identification of the causative agent. Sputum cultures are the usual first step. The patient is instructed to produce sputum

TABLE 19.1
Clinical terminology of LRT infection

AREA	TERM	USAGE
Airway	Laryngitis	Inflammation of the mucous membrane of the larynx
	Tracheitis	Inflammation of the mucous membrane of the trachea
	Bronchitis	Inflammation of the mucous membrane of the bronchial tubes
	Croup	Inflammation of airway in small children
	Bronchiolitis	Inflammation of smallest bronchial tubes
	Influenza	Acute viral respiratory disease caused by myxovirus
Lung	Pneumonia	Inflammation of the lungs
	Lobar pneumonia	With consolidation of part or all of one or more lobes
	Interstitial pneumonia	Chronic inflammation without consolidation
	Primary pneumonia	Occurs in host without preexisting respiratory disease
	Secondary pneumonia	Occurs as a complication of some primary infectious disease
Pleura	Pleurisy	Inflammation of the pleura
	Empyema	Pus in the pleural cavity

Note also combined terms (e.g., laryngeotracheitis, laryngeotracheobronchitis, bronchopneumonia) whose meanings will be evident.

deep and prolonged coughing, delivering the thick adhesive material into a sterile container. Simple expectoration of saliva is not adequate in that the sample contains URT organisms. Even correctly obtained sputum samples frequently fail to yield the causative agent on culture because of overgrowth by URT contaminants. To circumvent this problem, samples may be obtained by bronchial brushing or transtrachial needle aspiration. However, these difficult, invasive procedures are not often used. Most LRT pathogens cause bacteremia, so the agent can often be recovered by taking blood cultures. All cultural methods require one or more days.

Recently, interest has focused on diagnosis not by culture of the organism but by identification of antigens produced by the offender in body fluids such as serum, sputum, or urine. Countercurrent immunoelectrophoresis (CIE), staphylococcal coagglutination tests (SCT), and latex agglutination tests (LAT) are in use or under development. These specifically identify the antigens of each major bacterial LRT pathogen where present. Such methods are generally sensitive, specific, and fast.

Immunological techniques; Chapter 13, pp. 391–393

Whooping Cough

Causative agent   Whooping cough or **pertussis** is usually caused by the small, Gram-negative bacillus *Bordetella pertussis*. *B. parapertussis* and adenovirus may also cause this condition. *B. pertussis* is ovoid or coccobacillary; when multiplying in the host or freshly isolated, it is encapsulated. The organism is aerobic and slow-growing. It can be isolated on a glycerin–potato agar to which blood is added. Penicillin, which does not affect this organism, is often incorporated into isolation media to suppress competing normal flora. Samples are effectively and simply collected by allowing the patient to cough directly on the open sterile plate. Fluorescent antibody staining of secretions may provide a rapid identification.

Pathogenesis   When *B. pertussis* multiplies in the airway, it causes irritation by producing exotoxins and endotoxins. Thick, mucilaginous fluid is secreted forming a meshlike tangle of cilia and bacteria on the walls of the trachea and bronchi. During a catarrhal period of about two weeks, the symptoms are mild and resemble those caused by the common cold. Then the paroxysmal stage begins, with fits of rapid, violent, and uncontrollable coughing several times daily. These fits usually fail to expel the offending material, leave the patient exhausted and anoxic, and may induce vomiting or small hemorrhages. This stage also lasts about two weeks. It is followed by a long convalescence during which the symptoms gradually disappear. Secondary infections may occur during this stage.

Therapy and immunity   Antimicrobial therapy using erythromycin, tetracyline, or chloramphenicol stops the agent's growth but does not shorten the illness once the respiratory tract lesion is established. Supportive measures to provide oxygen, control electrolytes, and maintain adequate nutritional intake are needed until the paroxysmal phase ends.

Active disease confers lifelong immunity, but circulating antibody levels fall off rapidly. For this reason, even immune mothers give little or no passive immunization to their infants. Early vaccination with the DPT vaccine is essential. The P-component is killed *B. pertussis* or crude bacterial extracts, and it contains much endotoxin. This material is responsible for the common side effects (fever and soreness) associated with these immunizations. Reimmunization later in life may be indicated, as protection is not permanent.

Epidemiology   The young infant and child is most susceptible to whooping cough, although cases occur in the elderly. The winter months have the peak incidence, but cases occur throughout the year. Transmission occurs when droplets shed from the respiratory tract of individuals in the catarrhal stage are inhaled by susceptible persons.

Epiglottitis

Epiglottitis is one form of **croup**, a generalized syndrome in which the larynx or adjacent structures is obstructed. Coughing, respiratory distress, and complete respiratory tract obstruction may occur. Laryngitis, laryngotracheitis, or laryngo-tracheobronchitis are other clinical varients of croup. The usual bacterial agent is *Hemophilus influenzae* Type b, a tiny, fastidious, Gram-negative rod. Croup may also be caused by a variety of viruses. *H. influenzae* is facultatively anaerobic and requires an enriched medium containing two blood components, the X-factor (hematin) and the V-factor (NAD). It is isolated on chocolate agar (agar containing heated, denatured blood) to which NAD has been added.

Hemophilus strains have an antigenic polysaccharide capsule. There are six capsular types, designated a-f, but only Type b is commonly pathogenic. Isolates may be typed by the **Quellung** reaction, in which bacterial cells are mixed with specific typing serum. The polysaccharide capsule swells in the presence of the matched serum. Countercurrent immunoelectrophoresis (CIE) is a newer technique to identify capsular antigens.

Pathogenesis Susceptible children (and occasionally adults) experience coldlike symptoms. Then, suddenly, the infected epiglottis swells to many times its normal size. Complete airway obstruction requiring tracheostomy may occur.

H. influenzae can also cause pneumonia, meningitis, and septicemia. Its role in secondary, lethal pneumonia during the 1918 influenza epidemic is the source of its name. The name is misleading, however, since the bacterium does **not** cause influenza.

Therapy and immunity Until recently, *H. influenzae* was regularly treated with ampicillin. However, the spread of resistant strains has made it necessary to use chloramphenicol and sulfa compounds in some areas of the country.

Contact with capsular antigens stimulates a protective immunity. Normal early childhood exposure to the relatively small number of antigenic types is sufficiently immunizing so that *Hemophilus* infections are not common past early childhood except in the immunodeficient. Vaccines, particularly against the virulent Type b, are under development.

Epidemiology *H. influenzae* is widely distributed in the normal pharyngeal flora of both adults and children. There is constant exposure from aerosols, direct contact, and contaminated objects. Control of the sporadically occurring cases is not possible at present.

Bacterial Pneumonia—General Features

The clinical entity called pneumonia is marked by the accumulation of fluid in the lungs, usually visible as an opaque area in the chest X-ray. Fever, cough, respiratory distress, and acute pain are usual. The bronchial tree may or may not be involved. The characteristic pathogenesis involves damage to one or more lung lobes. Fluid accumulation is localized or **lobar;** the microorganisms multiply extracellularly in the exuded fluids, which make a fine culture media. The extent of alveolar tissue damage varies with the causative agent.

Pneumococcal Pneumonia

Causative agent About 90 percent of bacterial pneumonia cases are caused by *Streptococcus pneumoniae*. This is a Gram-positive coccus that characteristically forms pairs or short chains of pairs. The cells are enclosed in a smooth, thick polysaccharide capsule essential for virulence.

The organism is identified following primary isolation on blood agar, by observation of tiny α-hemolytic colonies inhibited by filter paper discs impregnated with **optochin** (ethylhydrocupreine hydrochloride). *S. pneumoniae* is bile soluble, and its cells lyse when suspended in solutions of the bile salt, sodium desoxycholate. It is further distinguished from other α-hemolytic streptococci by its ability to ferment inulin.

There are more than 80 serotypes of *S. pneumoniae*. The most important antigens are the capsular polysaccharides, but type-specific M-proteins can also be identified. Type 3 is the most virulent.

Pathogenesis Pneumococcal pneumonia begins with a severe, shaking chill, followed by fever. **Pleurisy**—inflammation of the pleura—is usually present, creating intense pain. The pleural cavity may fill with fluid and may, in extreme cases, become infected (empyema). Within the lung, multiplication of the bacteria in a lobe causes inflammation. Fibrin-bearing serous fluid forms a circumscribed consolidation that prevents expansion and contraction of the lobe and blocks gas exchange. Bacteremia is very common; one reliable way of establishing that *S. pneumoniae* is present is via blood cultures.

Convalescence, even when antibiotics are used, depends on development of an effective AMI response against the capsular antigen of the invading strain. Once antibodies are present, opsonization promotes effective phagocytosis. As microbial multiplication is arrested, gradual resolution and reabsorption of the fluid occurs, and the lung tissue returns to normal. Remarkably, there is almost no tissue destruction or residual loss of alveolar function.

Therapy and immunity Penicillin and its derivations effectively arrest the disease. Mortality remains significant, however, because many persons contracting pneumococcal pneumonia were already severely ill from other causes. Removal of residual microorganisms depends on phagocytic action. If body defenses are unable to respond, this essential follow-up step may not occur.

Protective immunity follows contact with pneumococcal polysaccharide, but it is type specific and not very long-lasting. A pneumococcal polysaccharide vaccine, containing antigens derived from the 14 most common pathogenic serotypes, was released in 1977. It is recommended for the elderly and those with chronic respiratory and cardiac disease or other predisposing conditions.

Epidemiology Pneumococcal pneumonia is a disease spread by carriers. From 20 to 70 percent of individuals carry pneumococci in the nasopharynx some of the time. Epidemic outbreaks occur in institutional settings. Single-serotype outbreaks happen when a single carrier has introduced a new strain to a susceptible population. Often, however, there are multiple strains, and it seems more likely that the outbreak is caused primarily by a dramatic increase in susceptibility in the group.

Sporadic cases occur most frequently in children under five and in older persons, particularly during the winter. Pneumococcal pneumonia is frequently the critical event that ends the life of a terminally ill patient.

Other Bacterial Agents of Pneumonia

Hemophilus influenzae   Pneumonia caused by *Hemophilus influenzae* occurs most commonly in children, but may also occur in adults in whom protective immunity to capsular antigen is absent or resistance has been lowered by a recent viral infection. The clinical manifestations are similar to pneumococcal pneumonia, but chills and pain are usually absent. Bacteremia is common, and meningitis is a frequent complication.

Klebsiella pneumoniae   This is a facultatively anaerobic, Gram-negative rod of the fermentative enteric group. It is an important cause of pneumonia in alcoholics, diabetics, and other adults with metabolic disturbances or chronic cardiopulmonary disease. Clinically, symptoms may include repeated and prolonged chills. The sputum is extremely viscous and reddish due to blood leaking from damaged alveoli. There may be extensive tissue destruction leaving scar tissue on recovery. The mortality rate has been quite high even with treatment, since septicemia is readily established. Cephalothin or gentamycin, singly or together, are the recommended antibiotics. *K. pneumoniae* is a normal inhabitant of the human gastrointestinal tract, and sometimes is a URT resident. Infection is correlated with altered defenses, and there is no specific prevention.

Mycoplasma pneumoniae   The mycoplasmas are the smallest, free-living cellular organisms. Although they are considered to be bacteria, they differ substantially from them as they lack the peptidoglycan-containing cell wall, include sterols in their cytoplasmic membrane, and have an unusual division mechanism. *M. pneumoniae* and *Ureaplasma urealyticum* are pathogenic for human beings.

Mycoplasma species are difficult to isolate and cultivate, requiring high concentrations of serum or other sterol sources and prolonged incubation periods. They are readily obtained by cell culture, in which they grow not in but intimately attached to the membrane of the living cell (Fig. 19.3). Mycoplasmas are said to be cell-adapted; that is, they need close association with a host cell for best growth. This is because they use the host cell membrane of a source of materials for their own exterior layer. As a result of this "borrowing," there is a high degree of antigenic similarity between host and parasite that confounds the normal process of immunological stimulation. Mycoplasmas are not effectively removed by immune mechanisms.

In the LRT, mycoplasmas cause **primary atypical pneumonia**, atypical because the agent escapes isolation by standard bacteriological tests. The pneumonia is

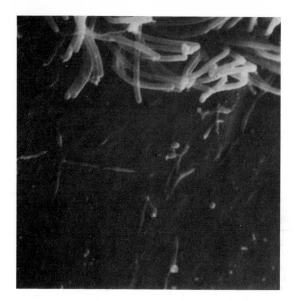

19.3
Mycoplasma pneumoniae organisms can be seen attached to the surface of a ciliated epithelial cell. Note the background, which is the outer surface of the cell membrane. Attached to this surface are many irregularly shaped *Mycoplasmas*, some spherical, some elongated. A tuft of cilia is seen at the upper right.

marked by slight sputum production, little consolidation, and no tissue damage; the mortality, understandably, is very low. Other clinical manifestations include URT illness, pharyngitis, bronchitis, and eye or ear infection.

Presumptive diagnosis hinges on demonstration of **cold agglutinin.** This substance agglutinates human type O red blood cells at 4°C but not at 37°C. Complement fixation tests can confirm specific antimycoplasmal antibodies in convalescent sera. Tetracycline is the drug of choice. There is no vaccine and no specific preventive measures.

Pseudomonas aeruginosa   Mucoid strains of this ubiquitous opportunist cause pneumonia in the compromised, particularly in children with cystic fibrosis. The organism is in general multiply resistant to clinical useful antibiotics; nosocomial (contracted in the hospital) cases may result from coinfection by two or more *P. aeruginosa* strains with different antibiograms, making effective therapy very difficult.

Legionella   Although it sometimes seems that we are familiar with all the microbial agents of infectious disease, nature has periodic surprises in store. In July, 1976, a large outbreak of severe pneumonia-like disease occurred at an American Legion convention in Philadelphia. Despite the coordinated efforts of the nation's top epidemiological experts, the causative agent eluded isolation for many months. When it became possible to detect the agent in tissue, and titer specific antibodies against it in patient sera, it became clear that this was not a "new" pathogen but rather one that had been around a long time. On serological evidence, it was implicated in unsolved epidemics going back to 1965. Legionnaire's disease, as it was immediately named, aroused much public interest and concern. It was seen as a potential new "plague." However, the actual incidence of legionellosis is low and it is not primarily an epidemic disease.

The causative agents are the three species *Legionella pneumophila, L. bozemanii*, and *L. micdadei*. (Fig. 19.4). They are Gram-negative bacilli normal in structure but lacking some endotoxin components and having unusual concentrations of certain branched-chain lipids. Genetic analysis using DNA homology indicates that they are not closely related to any of the other human Gram-negative pathogens or microflora. The agent can be isolated and maintained on media containing cysteine and iron salts, such as F-G and charcoal–yeast extract agars. Growth at 35°C in candle jars is slow, and colonies become visible at 4–5 days. *L. pneumophila* produces a β-lactamase enzyme that degrades both penicillin and cephalosporin-type antibiotics. The bacilli can be identified in pleural fluid, sputum, or lung tissue by direct fluorescent antibody microscopy. ELISA tests may be used for antigen detection. There is also a much milder form of the disease, called **Pontiac fever,** in which the incubation period is shorter, the symptoms are confined to the respiratory tract, and there is no mortality.

ELISA; Chapter 13, pp. 390–391 and Figure 13.13

Erythromycin, rifampin, and the tetracyclines have been used with success in therapy. Convalescent patients develop antibody against the organism. Detection of specific antibodies by indirect fluorescent antibody tests, microagglutination or

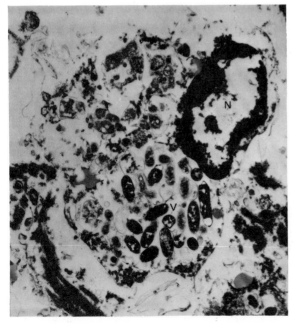

19.4
Legionella pneumophila An alveolar macrophage from a clinical specimen contains the rod-shaped organisms. This species, widely distributed in soil and water, is only rarely detected in the human lung. However, when infection does occur, it may be a serious multisystem condition. Many bacilli are in a phagocytic vacuole (V) adjacent to the nucleus (N).

ELISA techniques confirms the clinical diagnosis. Tests for immunological detection of antigen in body fluids, to aid in early diagnosis, are also under development.

Epidemiology of the disease has remained unclear. There is no direct evidence of person-to-person spread in clusters of cases, but rather a common source is suspected. Cases also occur singly. The typical patient is a middle-aged male, a smoker with some chronic condition. Summer months are peak, especially for clustered outbreaks. Isolation of the causative agent from soil, streams, and climate control systems has led to the hypothesis that the reservoir is soil or water. *L. pneumophila* can be shown to be part of the normal microbial flora of soils, where it acts as a decomposer species. The agent becomes airborne in vapor or dust clouds arising from construction sites, and is inhaled. There are no preventive measures.

Tuberculosis

Tuberculosis, which used to be called the "white plague," is progressing toward effective control. However, more than 20,000 new cases a year still develop in the United States. Attack rates in impoverished areas of the world are very much higher. Methods for diagnosis, treatment, and prevention are all well developed, but most require intensive, long-term, or costly medical care and patient cooperation.

Causative agent   Tuberculosis is traditionally defined as the chronic progressive pulmonary disease caused by *Mycobacterium tuberculosis.* However, very similar illness can also be caused by several other **atypical mycobacteria** species. For purposes of clarity, we will focus on *M. tuberculosis* as the primary agent since it is by far the one most commonly involved in serious disease.

Acid-fast stain method;
Table 3.2

The genus *Mycobacterium* is composed of slender, rod-shaped organisms. Their cell walls contain up to 60 percent lipids. Their Gram-stain reaction is variable and uninformative. At the same time, the cell-wall lipid renders mycobacteria unusually resistant to aqueous chemical solutions such as dyes, acids, bases, and most disinfectants. They can be stained well if the smear is steamed with the dye; once the dye is driven in, it is resistant to acid extraction (acid-fast). This property is the basis for the valuable differential staining technique.

Mycobacteria are strictly aerobic, metabolically versatile, but extremely slow-growing with generation times of 18–24 hours. To isolate the organism, sputum must be chemically digested with a basic solution containing mucolytic agents. This digestion step liquefies the sputum materials and kills most contaminants other than the mycobacteria. The sample is then incubated for up to eight weeks on Lowenstein–Jensen or Middlebrook media in an atmosphere resembling that of alveolar air, with 5–10 percent CO_2. Colonies develop extremely slowly; they are typically dry, wrinkled, and pale yellow. A variety of cultural tests may be used to differentiate *M. tuberculosis* from other species of mycobacteria. Perhaps its most useful determinative trait is its ability to produce niacin, an ability not shared by atypical species that could otherwise be confused with it.

M. tuberculosis manufactures no clearly defined virulence factors, although a lipid called **cord factor,** which makes the cells grow in coiled arrangements, is almost always found in isolates from active disease. Tissue destruction appears to be related entirely to induced delayed hypersensitivity. The DHS response may be either helpful in controlling the disease or instrumental in destroying adjacent tissue and accelerating the agent's spread.

Pathogenesis There is a distinction between tuberculosis **infection** and tuberculosis **disease.** Infection used to be an almost universal occurrence, but overt progressive disease was and still is much less common. The course of the disease is quite variable among individuals of different age groups and even at different points in one person's life.

The causative organism is characteristically inhaled in the form of droplet nuclei, penetrates to the alveoli, and is phagocytosed by the alveolar macrophages. **Primary infection** begins. Nonspecific macrophages cannot enzymatically inactivate ingested mycobacteria, and they actively multiply. Irritation of lung tissue causes an influx of fluid and PMNs. Shortly thereafter the organism enters the lymphatic system and lodges in lymph nodes and other organs throughout the body. There is little actual tissue destruction. During this early **exudative phase,** the lymphoid tissues receive the antigenic stimulus that causes the development of specific CMI, apparent as a positive **tuberculin test** from three to four weeks postinfection.

As CMI develops, specific informed lymphocytes and macrophages arrive at the infective site. These ingest and destroy the bacteria. Dissemination of the bacteria via lymphatics stops. A **granuloma** develops as the defensive cells surround the lesion; some cells fuse to form Langhans giant cells. This is the **productive phase.** The destructive effects of hypersensitivity appear with **caseation,** a breakdown of host and bacterial cells to an acellular, cheesy mass at the center of the infective focus. The lesion becomes surrounded by a fibrous layer that blocks exodus of organisms and materials. The enclosing process is **tubercle** formation. If the process is successful, the infection is arrested or resolved. Anaerobic conditions within the tubercle are extremely unfavorable to the mycobacteria; few survive, and they do not multiply. In most cases, the foregoing events have been entirely asymptomatic, and there will be no further tubercular activity during the individual's life span. The tubercle will gradually calcify. If it is large enough, it may show up on a chest X-ray. The individual's positive tuberculin test will persist indefinitely.

Progressive tuberculosis occurs either when the tubercle is not adequately walled off or when the lesion opens up again. The host has lost the battle to control the infection via CMI. Now only the tissue-destroying effects of the hypersensitivity are seen. An open **cavity** appears and rapidly enlarges. Bacterial multiplication is rapid and unchecked in the aerobic environment. Symptoms of fever, fatigue, weight loss, and cough appear. Disease can be said to be present. The mycobacteria may be further disseminated via the bronchial network, leading to new foci in the distant lung tissue, in the larynx, and occasionally in the pharynx. The disease may

Development of CMI; Chapter 13, pp. 395–398

still be arrested if immune defenses are improved by rest and diet and if antimicrobial drugs are systematically administered.

Secondary or **reinfection/reactivation tuberculosis** occurs in some individuals later in life when a healed primary lesion reactivates. The disease has the characteristics of progressive tuberculosis, with much tissue damage. Onset is usually insidious. Such individuals are the major reservoir for new primary infections in susceptible infants and children.

Mycobacteria seeded throughout the body and uncontrolled may cause multiple lesions, a condition called **miliary** tuberculosis. Resulting clinical states may mimic almost any other disease. Miliary disease occurs following ingestion of milk from cows infected with *M. bovis.* With modern pasteurization, this disease has become very rare.

Therapy and immunity   A therapeutic plan starts with efforts to improve the patient's general living conditions. This may require hospitalization, but more often can be achieved at home. Chemotherapy is essential. There are more than ten useful antimicrobial drugs available for the clinician to choose from. Some drugs have undesirable side effects and are used only if less toxic drugs fail. *M. tuberculosis* rapidly develops resistance to any drug used singly, so combinations of two or preferably three drugs are prescribed. **Isoniazid, streptomycin, ethambutol,** and **rifampin** are presently most favored. Drugs must be taken with great regularity over periods from 12 to 36 months. Patient noncompliance often leads to reactivation of drug-resistant strains. This then necessitates retreatment with substances that are even more toxic.

Immunological control by the host is the determining factor in whether the disease advances or is contained. CMI is essential, and AMI appears to be unimportant. Tuberculin hypersensitivity is a readily assessed indicator of active antimycobacterial CMI. **Tuberculin** is an extract of the organism. A form called **PPD (purified protein derivative)** is used in internationally standardized dosages. The substance

Figure 14.4   may be administered in several ways. The **Mantoux** method, an intradermal injection, is most accurate. The **tine** test is much quicker and requires less technical skill, but is less accurate. A positive response is a reddened, indurated (hard) area at the site of administration. Details of interpretation vary with the different testing procedures and dosages of antigen. False positives may occur as a result of cross-reactions with other antigens; BCG vaccination; or contamination of the tuberculin, equipment, or infection site. False negatives occasionally occur in individuals with advanced tuberculosis and/or a very poor CMI response.

Protective vaccination has been developed. The **BCG vaccine** (Bacille Calmette-Guerin) contains a live, attenuated strain of *M. bovis.* BCG was initially developed in Europe and has gained more acceptance there than in the United States where it is little used. BCG may be used only before primary infection takes place. Thus its use is restricted to young children or to those with negative tuberculin tests with a high risk of exposure. The protective effect of BCG is debatable. Clinical trials conducted

TABLE 19.2
Characteristics of atypical mycobacteria

SPECIES	GROUP	CULTURAL FEATURES	SOURCE	DISEASE IN HUMAN BEINGS
Mycobacterium kansasii	I	Colonies have yellow pigment produced on exposure to light	Raw milk, soil	Chronic pulmonary disease, especially in older white males
M. marinum	I	Yellow pigment in light, best growth at 31–32°C	Fish, warm water	Cutaneous lesions in fish handlers
M. ulcerans	I	Growth at 30–33°C	Soil, river mud	Granulomatous skin ulcer
M. scrofulaceum	II	Yellow pigment in dark, lack cord factor	Soil, water	Suppurative cervical adenitis in children, chronic pulmonary disease in adults
M. avium	III	Nonpigmented to buff, best growth at 40°C	Birds	Chronic pulmonary disease; rare
M. intracellulare	III	Colonies thin, translucent	Soil	Chronic pulmonary disease, older white males
M. fortuitum	IV	Nonpigmented to buff, rapid growth	Soil, water, numerous animal species	Cutaneous and ocular lesions; chronic pulmonary disease
M. bovis	—	Short, fat rods, very small colonies on standard media	Cows, other animals, birds	Miliary tuberculosis

in the 1950s gave protection rates of from 15 to 80 percent. Recently completed studies in India showed that the vaccines used there had little or no protective effect.

Epidemiology   The morbidity and mortality rates for tuberculosis dropped rapidly in the early part of the century, due almost entirely to improving standards of living. Even today, sociology is perhaps as important as medicine in determining who is at risk of tuberculosis and who is not.

The disease is transmitted almost exclusively by the inhalation of droplet nuclei recently expelled by a person with active cavitary tuberculosis. Although the bacillus remains viable for long periods in dust and on fomites, it probably becomes noninfective. Transmission of *M. bovis* via dairy products has been entirely arrested wherever tuberculin testing of dairy herds and pasteurization of milk is practiced.

When antimicrobial therapy is begun promptly, patients become noninfective within a few weeks. Isolation is generally impractical. Most patients are sent home with instructions in how to control their respiratory secretions—covering the nose and mouth while coughing and careful disposal of soiled materials.

Preventing new cases requires tuberculin testing of the active case's known contacts. Any whose tuberculin test converts from negative to positive should receive prophylactic antimicrobial therapy for one year. Inactive cases (persons with healed lesions) should receive periodic chest X-rays to detect reactivation if it occurs. This is of particular importance for those whose occupations involve close contact with children, such as teachers. The chest X-ray is no longer used in routine screening of healthy persons because of the radiation exposure it delivers unnecessarily. Preventive immunization with BCG vaccine is used selectively, such as for infants of tubercular mothers or for personnel of hospital tuberculosis units. Outside the United States, it is used much more widely.

Actinomycosis

Oral flora; Chapter 15, pp. 471–475

Actinomycosis is a chronic disease caused by the strictly anaerobic, filamentous bacterium *Actinomyces israelii* and other species. The actinomycetes normally inhabit the gingival crevice and other anaerobic microenvironments of the mouth. *Actinomyces* may invade tissues of the face and neck, and lung and pleura, the lower gastrointestinal tract, and bone.

The lesions formed are granulomatous. Infection becomes established only in damaged tissue with extremely low oxygen levels. As *Actinomyces* multiples, yellowish filaments adhere and become cemented together by an extracellular polysaccharide, forming **sulfur granules.** The granules are readily seen on microscopic examination of infected tissues, and are diagnostic. The actinomycetes can be identified in tissue by fluorescent antibody techniques. The lesions spread through connective tissue eventually to the body surface, forming a draining sinus on the face or chest wall, for example. Actinomycosis responds to long-term penicillin treatment. There is no vaccination or specific preventive measure.

VIRAL INFECTIONS

Influenza

Influenza is the most common identifiable virus infection of the LRT. It is an epidemic disease, recurring yearly during the colder months throughout the world. Extreme genetic variability in influenza virus leads to periodic emergence of new strains for which there is no preexisting population immunity.

Causative agent   The influenza viruses are **myxoviruses.** Their nucleic acid is RNA, and the genome consists of seven separate pieces. The envelope bears several thousand spikes of antigenically distinctive hemagglutinins (H) and neuraminidases (N). These proteins are inserted into the host-cell membrane at the time of virus maturation. When the virion buds out through the cell membrane into the external environment, this modified membrane becomes its envelope (Fig. 19.5). Different virus

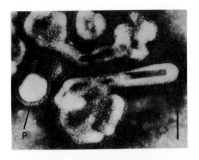
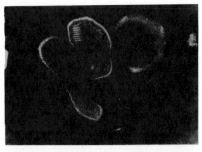

19.5
The Influenza Virus Influenza virus is an enveloped RNA virus. Its shape is highly variable. The surface projections (P) are the peplomers on its envelope. These peplomers undergo frequent chemical changes due to changes in the viral genes, in Type A Influenza. The peplomer proteins are important in the virus's adhesion to the respiratory epithelium, and its ability to hydrolyze mucus and parasitize tissue.

strains vary in their genetic information and thus instruct the host cell to make different H and N antigens.

Typing of internal capsid antigens determines the A, B, or C classification. Types A and B are pathogenic; Type C causes illness only in infants. The A, B, or C classification is genetically stable. Influenza isolates are also serotyped by their envelope H- and N-antigens, and designated by the place and date they first appeared. Thus we have strain designations such as A/Texas/1/77 (H_3N_2) or A/Brazil/11/78 (H_1N_1). The envelope antigens of Type A are subject to frequent change, leading to strains with new epidemic potential. Major antigenic shifts may reflect mutation or recombination of genetic elements among human strains, or with strains found in nonhuman reservoirs such as swine.

Pathogenesis In the usual form of the infection, a sudden onset of fever, chills, and muscle aches may be accompanied by pharyngitis or gastrointestinal upset. A 3- to 5-day period of illness is common. Recovery is slow, and relapse may follow overexertion.

The LRT becomes involved if the airway is inflamed. Coughing, sputum production, and chest pain may be present. Recovery is much slower. Primary influenza pneumonia may develop (rarely) in the elderly or those with underlying cardiopulmonary disease. Its progression is rapid and cannot be halted with antimicrobial agents. Mortality is high.

Major complications may interfere with recovery from influenza. The most important complication is **secondary bacterial pneumonia** (Fig. 19.6). Since mortality

TABLE 19.3
Role of viral agents in LRT illness

VIRAL AGENT	TYPICAL LRT INVOLVEMENT
Coxsackie	Influenza-like illness
Influenza Type A	Tracheobronchitis, croup, occasional primary pneumonia
Influenza Type B	Tracheobronchitis
Influenza Type C	Mild respiratory tract disease in children
Respiratory syncytial	Mild to severe bronchiolitis and bronchopneumonia in infants
Parainfluenza Types 1 and 2	Common cause of croup in young children
Parainfluenza Type 3	Uncommon but serious pneumonia in infants
Adenovirus	Fatal pneumonia in infants (Types 1, 2, 3, 7, and 7a) Atypical pneumonia in military recruits (Types 2, 4, and 7)
Rubeola	Giant cell pneumonia in immunodeficient children
Varicella	Interstitial pneumonia in adults; complication of normal disease
Cytomegalovirus	Pneumonia in immunodeficient adult; associated with use of renal dialysis or open-heart surgery

19.6
Viral Infections Predispose to Secondary Bacterial Infections In this experiment, human epithelial cells in culture were infected with P_3 virus. The viral infection causes them to fuse, forming syncytia, the background material of this SEM. When these virus-infected cells are then exposed to bacterial cells, the bacteria (here *Streptococcus pneumoniae*) adhere actively to the surface of the infected cells. This suggests a mechanism for the bacterial complications of viral respiratory infections.

from influenza by itself is uncommon, the actual mortality rate is more accurately reflected by the report of **pneumonia–influenza** (P and I) deaths.

Reye syndrome is a condition found mainly in children. A virus infection, most commonly influenza, is followed by sudden loss of consciousness, central nervous system abnormalities, fatty infiltration, and potential failure of the liver and kidneys. Mortality is about 30 percent. The actual relation of the viral infection to the development of Reye syndrome is not clarified.

Therapy and immunity   **Amantidine,** an antiviral drug, reduces the severity of symptoms in some persons if it is given during the incubation period and early disease. It is not yet generally used. Prophylactic vaccination is widely practiced but not completely successful. Since influenza viruses undergo frequent antigenic change, new vaccines must be repeatedly developed, distributed, and administered in advance of rapidly approaching epidemics. Sometimes, as with the swine flu vaccination program of 1976, an epidemic is anticipated, vaccine is prepared and distributed, and the expected outbreak does not materialize. Annual vaccination against current influenza strains is recommended for high-risk individuals—the elderly and those with underlying cardiopulmonary conditions predisposing to pneumonia.

A change in the route of vaccine administration is being investigated. A vaccine inhaled as an aerosol stimulates respiratory tract lymphoid masses to produce specific secretory IgA. These antibodies are present at the site of viral attack and prevent infection. Injected antigen, on the other hand, stimulates circulating antibodies, primarily useful to restrict the spread of an already established infection.

Guillain-Barré disease, an unexplained syndrome characterized by muscular pain and weakness and neurological abnormalities, has occurred sporadically over the years. Many cases were associated with receipt of the swine flu vaccine in 1976. This, along with the nonappearance of swine flu, became a primary reason for terminating the vaccination campaign. The causal relationship between Guillain-Barré disease and the vaccine, which contained no live virus, is very unclear.

Epidemiology   Influenza is spread primarily by droplets of respiratory secretions. The incubation period is short (1–3 days), so epidemics spread explosively among susceptibles. Epidemics are international, and new influenza strains spread within weeks from one hemisphere to another. Crowds and assemblies should be avoided during epidemic periods. Universal vaccination probably is not justifiable. However, vaccination campaigns aimed selectively at high-risk individuals can save an estimated 30,000 lives in severe epidemic years.

Viral Pneumonia

Viral pneumonia is characterized by a gradual onset, less pronounced fever, and little chilling. There is less sputum production than there is with bacterial pneumonia. The lung tissue is diffusely affected and the agent, multiplying intracellularly, affects the interstitial tissue. Viral pneumonia is less severe than bacterial pneumonia; however, no specific antimicrobial therapy has been found to treat it.

FUNGAL INFECTIONS

The lung is a prime site for fungal pathogenesis because of the abundant oxygen provided by inspired air. Most fungi are aerobes and, being eucaryotic, are also characterized by slow growth. Because they multiply slowly, they rarely establish infections in areas with a large normal flora, such as the URT, because they compete poorly. However, since the LRT lacks a resident flora, fungi do well once they are established. Deep infections, called **systemic mycoses,** have a slow onset and development. When deep tissue and organs are affected, fungal infection provokes strong delayed hypersensitivity. Local tissue response includes inflammation and granuloma formation. Resolution or expansion of the lesion depends on the balance between multiplication of the agent and the effectiveness of the host response. When the infection is not controlled, after a time the fungus becomes generally disseminated to other areas of the body, often with fatal results.

Histoplasmosis

This widely distributed mycosis is endemic in at least 30 states of the United States and in more than 60 other countries. In its pathogenesis, immunological picture, and epidemiology it provides a typical example of the systematic mycoses.

Causative agent   *Histoplasma capsulatum* is a dimorphic fungus of the deutero-mycete group. In nature, growing in soil enriched by the droppings of bats or birds, such as starlings and chickens, or on laboratory media, it develops in the hyphal form (Fig. 19.7). The white to brown mycelial mold forms numbers of microconidia and macroconidia. The conidial spores are infectious when inhaled. *H. capsulatum* preferentially parasitizes the reticuloendothelial system. In tissue it develops as a budding oval yeast. Pseudohyphi develop in necrotic areas. The agent may be isolated from any affected body fluid with modified Sabouraud's medium or other formulas incubated at 30°C. Identification is based on specific colonial and microscopic morphology.

Pathogenesis   Most infections are entirely asymptomatic and give no signs of overt disease. Three forms of histoplasmosis can be distinguished. The **primary acute** form is the most common. After the spores are inhaled, they germinate and the fungi enter macrophages to multiply. The developing lesion resembles that of primary tuberculosis. Symptoms range from none (asymptomatic infection) to fever, respiratory distress, cough, weight loss, and joint pains. The agent disseminates to all other parts of the body. A CMI response develops that brings the agent under control. The pulmonary lesion may heal completely, or there may be residual scarring or calcification.

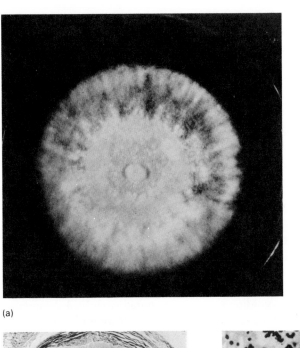

(a)

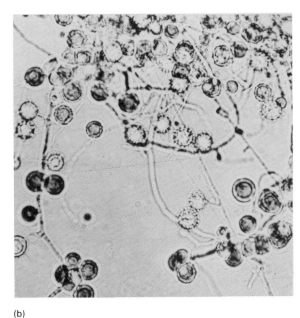

(b)

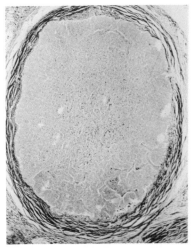

(c)

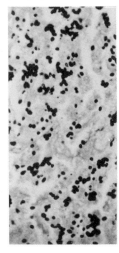

(d)

19.7

Histoplasma capsulatum (a) A colony on Sabouraud's agar after 30 days' incubation. Note that the cottony, white filamentous growth fills most of the plate. (b) In the mycelial form, macroconidia are produced. These are infectious if inhaled. (c) In the lung, *Histoplasma* produces primary lesions that are usually healed or resolved. This is a healed lesion. The outer layer is fibrous and calcified. The caseous interior contains debris and some *H. capsulatum* cells in the yeast phase (tiny black dots). (d) In a higher magnification oval yeast cells are obvious. Should the lesion reactivate, these cells, which are still viable, could give rise to progressive disseminated disease.

In a very few cases, principally in infants and children, histoplasmosis progresses to a **severe disseminated** form. The CMI response is too weak to control fungal multiplication. Lung lesions are less prominant, but lesions may develop in reticuloendothelial tissues of the liver, lymph nodes, adrenal glands, and the gastrointestinal tract. Without prompt treatment, mortality is certain.

A **chronic, cavitary** form, again similar to progressive tuberculosis, may develop in adults. It advances over a period of years.

Therapy and immunity   Therapy is not usually indicated in mild primary histoplasmosis. The antimicrobial drug, **amphotericin B,** markedly improves survival in severe cases. Surgical removal of diseased tissue may be needed.

A CMI response to the fungus produces a delayed hypersensitivity to **histoplasmin,** an antigen derived from cells of the fungus. Effective CMI is essential to control the invading organism successfully; progressive or chronic disease indicates an underlying CMI deficiency. Histoplasmin reactivity is assessed by skin test. The histoplasmin test is useful in diagnosis, but there is a high frequency of false positive reactions. Complement-fixation tests on paired sera are far more reliable. A rising titer indicates an active disease process; in the disseminated form it implies a poor prognosis. There is no vaccination available.

Epidemiology   In the endemic area, particularly in the Ohio and Mississippi river valleys, more than 80 percent of the adult population have positive histoplasmin tests. This indicates the pervasiveness of inapparent or mild infection in these areas. There are about 40–50 deaths per year. The organism clearly has a relatively low virulence, since the frequency of disease is slight compared to the number of total infections.

Preventive measures include avoiding soil contaminated with bird droppings. Bird roosting sites, chicken coops with old accumulations of droppings, bat-infested

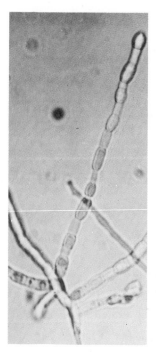
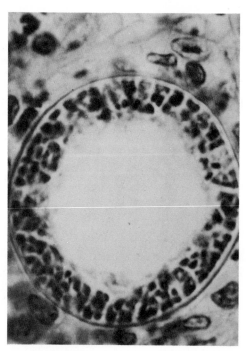

19.8
Coccidioides immitis (a) Growth on agar or in soil is mycelial. Arthrospores (chains of irregular oval bodies) are formed as reproductive structures. The arthrospores are scattered when mature. They are infective when inhaled. (b) A partly matured spherule in lung tissue. In a warm-blooded host, *Coccidioides* forms spherules. These become filled with multiple cells. When the spherule ruptures, these daughter cells can reproduce again in other areas of tissue, extending the affected area. However, they are not infective to other human beings. Human infection is largely accidental; it has no natural place in the life cycle of the fungus.

(a)   (b)

Box 19.2

An Airborne Outbreak

The San Joaquin Valley in California is an important endemic area for coccidioidomycosis.

A violent windstorm in the San Joaquin Valley on December 20 and 21, 1977, created extensive dust clouds that spread to many areas of California. State and local health officials became concerned that dust bearing the arthrospores of *Coccidioides immitis* would expose people outside the regions endemic for coccidioidomycosis to the disease. During the first 24 days of January 1978, 11 percent of 656 sera obtained from persons with suspected coccidioidomycosis and submitted to the Kern County Health Department for tube precipitin tests were positive for *C. immitis*; by comparison, 2 percent of samples submitted in January 1977, 9 pecent of samples submitted in January 1976, and 6 percent of samples submitted in January 1975 were positive. During the same period, the University of California at Davis reported that 18 percent of samples submitted were positive compared with 4 percent of samples tested in January 1977. Several of these patients lived outside endemic regions of the state. Persons traveling through the San Joaquin Valley during the storm, or those exposed to dust clouds from the area, experienced a transitory increased risk of coccidioidomycosis.

Source: Adapted from *Morbidity and Mortality Weekly Report* 27: 55, February 17, 1978.

areas such as caves and attics, or dust from these areas, contain infective conidia. Masks should be worn when contact is unavoidable. Soil decontamination is not practicable.

Transmission almost always occurs by inhalation, and occasionally by ingestion, of conidial spores produced by the organism growing in its mycelial form. Laboratory cultures may be infective. The pathogenic budding form produced in host tissues produces no spores and is not infectious. Person-to-person transmission does not occur.

Other Mycotic Infections of the Lung

Coccidioidomycosis, caused by *Coccidioides immitis*, occurs commonly in desert areas in the southwestern United States, Mexico, and South America. As much as 80 percent of the population may show immunological evidence of infection in these areas. Visitors to these areas may also become infected, so isolated cases appear frequently in remote, nonendemic areas. The dimorphic fungus, multiplying in soil or on laboratory media, produces **arthrospores.** These are infectious when inhaled.

In the host, the arthrospore develops into a **spherule** (Fig. 19.8). Within the spherule, nuclear division yields thousands of **sporangiospores.** In two to three days the spherule ruptures, releasing mature spores for further dissemination. Sporangio-

TABLE 19.4
Epidemiology of LRT infections

DISEASE	RESERVOIR	TRANSMISSION	INCUBATION PERIOD IN DAYS
Whooping cough	Human beings	Aerosols, droplet nuclei, soiled articles	10 (7–21)
Pneumococcal pneumonia	Human beings	Aerosols, droplet nuclei	1–3
Other bacterial pneumonias	Human beings, water, soil	Aerosols, droplet nuclei, dust	1–3 Variable
Mycoplasmal pneumonia	Human beings	Droplets, oral contact	14–21
Tuberculosis	Human beings, cattle	Droplet nuclei, milk	4–12 wk
Actinomycosis	Human beings	Endogenous	Days or months after trauma
Influenza	Human beings, animals (Type A)	Direct contact, droplets, soiled articles	1–3
Histoplasmosis	Soils with high organic content	Inhalation of airborne spores	5–18
Coccidioidomycosis	Soil	Inhalation of arthrospores	7–28
Aspergillosis	Decaying vegetation	Inhalation of spores	Days to weeks
North American blastomycosis	Soil	Inhalation of spores	Weeks to months

Source: Adapted from A.S. Benenson, 1975, *Communicable Diseases of Man.*

PERIOD WHEN COMMUNICABLE	PATIENT CONTROL MEASURES	EPIDEMIC CONTROL MEASURES
For seven days after start of therapy	Isolation; disinfection of discharges and soiled articles	Exclude nonimmune children from school, active immunization of all infants with DPT
Until discharges are noninfective; three days after penicillin therapy begins	Precautions for disposal of secretions	Immunization of high-risk individuals
While infectious agent is present in discharge	Isolation for *S. aureus* and *S. pyogenes*; secretion, precaution for others	Not generally applicable
Less than ten days	Disinfection of nasal and throat discharges	None available
While tubercle bacilli are discharged and open lesion is present	Isolation usually not necessary; sound routine hygiene, disposal of secretions, rapid specific drug therapy	Prophylactic drugs and/or immunization of contacts; follow-up of all tuberculin-positive contacts
Not applicable	None	None
Up to three days after clinical onset	No isolation No special precautions	Surveillance and reporting; vaccination of high-risk persons
Noncommunicable	Disinfection of discharges	Cleanup of suspected sites
Noncommunicable (cultures infective)	Disinfection of soiled articles; organism may grow on casts or dressings in infective spore-forming phase	Dust control-oiling, grass planting
Noncommunicable	Ordinary cleanliness	None
Noncommunicable	Disinfection of sputum, discharges, soiled articles	None

19.9

Endemic Areas of Systemic Mycosis Histoplasmosis (color) is endemic in many states of the eastern central United States. Coccidioidomycosis (gray) is endemic in dry or desert areas from Texas westward. However, because people now move frequently from one area of residence to another, and visit distant regions for business or pleasure, it is not unexpected to encounter cases of either disease far from the endemic area. Careful questioning will usually establish a recent visit to the endemic area.

spores are not infectious to human beings. Because arthrospores are not formed in tissue, person-to-person transmission does not occur.

The pathogenesis of coccidioidomycosis resembles that of histoplasmosis. The common form is benign and acute, occasionally giving severe symptoms for up to four weeks, but tending to be self-limited. However, less common progressive disseminated and chronic cavitary forms are seen. Pregnant females, and males of highly pigmented races are particularly affected in the endemic areas (Fig. 19.9).

Skin tests using the antigen **coccidioidin**, and serological tests of the complement fixation type are used in diagnosis. These are not perfectly reliable because of cross-reactions with other fungal pathogens. *Coccidioides* can be isolated on suitable media. However, laboratory cultures form plentiful arthrospores and are exceedingly hazardous to laboratory personnel. Treatment with amphotericin is essential in the more severe forms of the illness. Recovery gives permanent protective immunity. There is no specific preventative action.

Aspergillosis *Aspergillus fumigatus* and other related species cause a lung disease marked by development of **fungus balls**, granulomas, necrosis, and cavitation. Aspergillosis is almost always secondary to immunological impairment. *A. fumigatus* is widely established in nature, especially in decaying vegetation such as spoiled hay. The mode of transmission is spore inhalation.

Blastomycosis **North American Blastomycosis**, or Gilchrist's disease, is caused by *Blastomyces dermatitidis.* Beginning as a respiratory infection, it often becomes progressive with osteomyelitis and cutaneous lesions; lesions may occur in any part of the body. Treatment with amphotericin may arrest systematic blastomycosis. The organism appears to be acquired from soil, but the conditions influencing distribution are unknown. The disease is rare and causes about two deaths a year.

Geotrichosis   This is a bronchial, pulmonary, or oral infection usually secondary to an immunosuppressive condition. The fungus, *Geotrichium candidum*, is widespread in soil, on fruits, and in dairy products, and thus is ingested with great frequency. The pulmonary form of geotrichosis resembles a chronic bronchitis or mild pneumonia; the oral disease is much like thrush. Geotrichosis tends to clear up as the primary disease is successfully controlled.

SUMMARY

1. Lower respiratory tract infections can become severe and life-threatening because they cause inflammation leading to secretion and accumulation of tissue fluids forming sputum. Sputum may be viscous or semisolid, difficult to expel, and obstructive to the proper flow of air through the airway and exchange of gases at the alveolar surface.

2. Mycobacteria, viruses, and fungi also cause cell and tissue destruction. The agent may be cytolytic, or the delayed hypersensitivity response that accompanies cellular immunity may destroy cells harboring the antigenic target.

3. Lower respiratory infections are most common and severe in the weakened or immunologically compromised host. Risk increases with age, underlying cardiopulmonary disease, or a history of smoking, diabetes, or alcoholism.

4. In the healthy individual, a mild viral infection of either URT or LRT may temporarily reduce resistance, allowing a secondary pneumonia to develop.

5. Tuberculosis and the systemic fungal infections are usually asymptomatic, benign, and rapidly resolved. Progressive or chronic infections, which are serious and life-threatening, correlate with lowered resistance due to underlying disease, malnutrition, or other compromising factors.

6. Bacterial infections of the LRT can be effectively treated with suitable antimicrobial drugs if the patient is not otherwise overwhelmingly ill. Active tuberculosis requires combined drug therapy for an extended period. Viral infections receive general supportive therapy. Systemic fungal infections are treated with amphotericin, but mortality is high if treatment is not begun soon enough.

7. Vaccination is available for selected LRT infections, including pertussis, pneumococcal pneumonia, tuberculosis, and influenza. Pertussis vaccination is essential for all infants. Vaccination for the other diseases is usully recommended only for high-risk groups or individuals.

Study Topics

1. Why does the lower respiratory tract not support a normal flora?

2. Compare the clinical features of the different bacterial pneumonias.

3. What would be the major element of a campaign to reduce the incidence of tuberculosis in a rundown inner-city area? How might the problems differ from those encountered in a remote tribal village?

4. Review the techniques for laboratory identification of the bacterial pneumonia agents.

5. Compare the antimicrobial drugs used to treat each type of LRT infection.

6. In histoplasmosis of the disseminated type, a rising titer of circulating antibody is said to indicate a poor prognosis. Yet increasing antibody levels are usually considered to be protective. What is the resolution of the apparent paradox?

Bibliography

Articles

Austwick, P. K. C., 1972, The pathogenicity of fungi. In *Microbial Pathogenicity in Man and Animals*, Symposium 22, Society for General Microbiology, Cambridge: Cambridge University Press.

Stanbridge, Eric J., 1976, A reevaluation of the role of mycoplasmas in human disease, *Ann. Rev. Microbiol.* **30:** 169–177.

Books

Conant, N. F., et al., 1974. *Manual of Clinical Mycology.* Philadelphia: W. B. Saunders.

Emmons, C. W., et al., 1977, *Medical Mycology* (3rd ed.), Philadelphia: Lea and Febiger.

Jones, Gilda, and G. Ann Hebert (eds.), 1979, *"Legionnaire's"—the Disease, and Bacterium, and Methodology,* Atlanta: Center for Disease Control.

See also books listed at the end of Chapter 18.

20

Gastrointestinal Tract Infections

The gastrointestinal tract carries out a complex sequence of mechanical, chemical, and absorptive operations on the food we eat. When all goes well, it gives us a feeling of great well-being. However, when pathogenic microorganisms or their toxic products enter with our dinners and upset the GI tract's function, we become acutely miserable. Few of the gastrointestinal infections and food intoxications are life-threatening, but they are a very common source of minor illness and discomfort. Most are spread by the oral–fecal route via water, food, or fingers.

ANATOMY AND PHYSIOLOGY

The GI tract is a continuous, hollow tube running through our bodies. The hollow part or **lumen** contains food in the process of being digested, the digestive secretions, and a complex microbial flora. This digestive mixture is actually external to our bodies if the physical structure of the tract is intact. Few pathogens can pass through the intestinal wall and cause systemic infection.

Anatomy of the GI Tract

We have already discussed the mouth and oropharynx in Chapters 16 and 18. We will be concerned here with the alimentary tract, from the epiglottis to the anus, and with its accessory glands, the liver and pancreas (Fig. 20.1).

Digestive organs   The **esophagus** is a straight tube, about 25 cm long in the adult. It is flexible and collapsible, and joins with the stomach at the **fundus.** This entrance is regulated by the **cardiac sphincter.** This sphincter closes to prevent stomach contractions from squeezing gastric contents back up into the esophagus.

The **stomach** is an expandable, muscular reservoir that collects food during a meal, and changes shape as food is added or emptied out. Processed food leaves the stomach via the **pyloric sphincter.** About 1 mL of gastric contents is expelled with each stomach contraction.

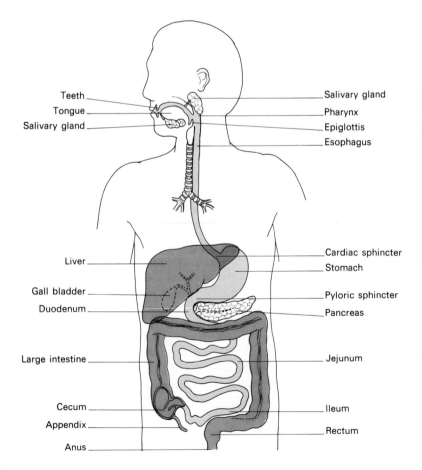

Teeth
Tongue
Salivary gland

Salivary gland
Pharynx
Epiglottis
Esophagus

Liver

Gall bladder
Duodenum

Large intestine

Cecum
Appendix
Anus

Cardiac sphincter
Stomach

Pyloric sphincter
Pancreas

Jejunum

Ileum

Rectum

20.1

The Gastrointestinal System and Associated Glands The GI tract is a hollow tube, continuous from mouth to anus. Food passes through, being digested by enzymatic secretions contributed by the digestive glands. Only small molecules leave the tract to enter the body proper. The huge microbial population of the lower tract cannot normally move into surrounding areas.

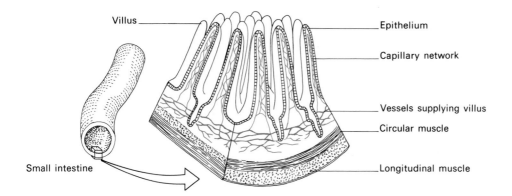

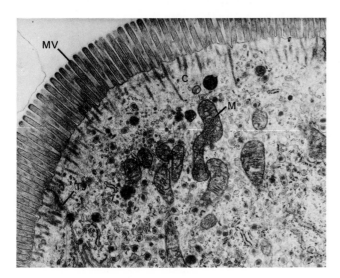

20.2
Intestinal Epithelium This freeze-etched electron micrograph shows three cells and a fraction of a fourth. They are bordered with *microvilli* (MV) increasing the absorptive capacity. Tight junctions (TJ) occur at the intercellular borders. Mitochondria (M) and cytoplasm (C) are also indicated.

The stomach lining is studded with simple **tubular glands** that produce hydrochloric acid, digestive enzymes, mucus, and serotonin. These substances compose the highly acidic **gastric juice.**

The **small intestine** is a long (about 610 cm), much-folded tubule. It is made up of three continuous regions, the **duodenum,** the **jejunum,** and the **ileum.** Stomach contents empty into the duodenum, as do bile and pancreatic secretions. Enzymatic hydrolysis of food macromolecules occurs throughout the small intestine. Muscular movements move the fluid mixture along, mix it, and bring it into contact with the intestinal lining for absorption.

The lining of the small intestine is arranged in mobile finger-like projections called **villi,** specialized for absorption (Fig. 13.3). The deep **crypts of Leiberkuhn** lie between. Secretory **goblet cells** and other specialized cells (Fig. 20.2) release mucus, enzymes, and hormones. Loose aggregates of lymphoid tissue called **Peyer's patches** are abundant in the intestinal mucosa.

The **large intestine** has a greater diameter than the small intestine and is thick-walled and muscular. It is specialized for the reabsorption of water from the residue left after the food has been digested. The ileum joins the large intestine not directly, but at a right angle. The main portion of the bowel beyond the junction is the **colon;** the blind pouch below the junction is the **cecum.** A slim, twisted, and coiled tube, the **appendix,** is at the end of the cecum. The cecum and appendix are partially iso-lated structures, potential sites of obstruction and infection.

The **rectum** ends the colon, and connects with the **anal canal.** Wastes pass out the anus when three sets of sphincters are voluntarily relaxed, or when propelled by powerful, unregulated peristalsis.

Liver   The liver is a large four-lobed structure lying on the right side of the body immediately under the diaphragm. It has many functions but its role in digestion is to supply bile to assist in the emulsification and digestion of fats. The liver's internal structure is a mass of **lobules.** In a lobule, blood flows in one direction through spongelike sinuses and is cleansed of the bile components. Bile flows in the opposite direction in canaliculi to bile ducts and eventually to a collecting sac, the **gall-bladder.** The gallbladder, a blind sac, is also a frequent site of infection.

Pancreas   This lobular organ produces a majority of the digestive enzymes, as well as hormones such as insulin and glucagon. The enzymatic secretion is mixed with food in the small intestine.

Peritoneal cavity   The organs of digestion, with the exception of the esophagus, lie in the peritoneal or abdominal cavity. Like the thoracic cavity, this hollow space is lined with one layer of serous membrane, the **parietal peritoneum,** and the organs are covered by folds of another layer, the **visceral peritoneum.** A membrane fold, a **mesentery** or **omentum,** suspends and attaches each loop of intestine.

Physiology of the GI Tract

A complete discussion of the physiology of digestion is unnecessary for our pur-poses. We will limit our overview to those aspects of normal function most fre-quently altered by microbial action.

Motility   The GI tract contents are propelled through the system by rhythmic waves of involuntary, smooth muscle contraction, **peristalsis.** About three contrac-tions per minute advance the contents about 1 cm. The normal rate may be tempo-rarily altered by stimuli such as urination, infection, emotional stress, diet, or numerous other factors. Retardation (constipation) or acceleration (diarrhea) may occur. Microbial products affect intestinal movements directly by irritating or in-flaming the bowel or indirectly by interacting with the central nervous system.

TABLE 20.1
Volume and functions of GI secretions

SECRETION	APPROXIMATE VOLUME IN ML/DAY	DIGESTIVE ROLE	MICROBIOLOGICAL ROLE
Saliva	1,200	Dissolves, lubricates food Enzymatic attack on starch	Cleansing role essential to control size of oral microbial flora
Gastric juice	2,000	Lubrication Acid hydrolysis Enzymatic attack on protein	Acidic pH limits flora of stomach and may inactivate some pathogens and toxins Acid also breaks down some drugs and vaccines so they cannot be given orally Acid activates botulinum toxin
Pancreatic juices	1,200	Neutralizes acidity of gastric juice Source of enzymes to initiate food hydrolysis	Renders pH of contents suitable for microbial growth
Bile	800	Digestion and absorption of fat, cholesterol Vitamin D promotes peristalsis	Selective for bile-tolerant microflora
Intestinal juice	3,000	Source of enzymes to complete digestion of proteins, starch, fat, sugars	Keeps mixture dilute and moving, limits numbers in small bowel
Total	8,200		

Role in water balance Fluid secretions of the various glands enter the GI tract in large amounts. About eight liters of water per day are transferred from the internal fluid reserve of the body to the lumen of the GI tract. Almost all of this fluid is recovered by reabsorption in the large intestine if the intestinal contents are moving through at the normal, slow rate. In diarrhea, rapid passage of highly fluid wastes allows for little reabsorption, and there is a significant water loss. In moderate cases, **dehydration** occurs. As fluid loss continues, **hypovolemia** (abnormally low blood volume) develops. This may culminate in shock and death. The normal adult blood volume is about 10 liters, so the uncompensated loss of several liters of water per day in severe diarrhea cannot be sustained for long.

Electrolyte balance   The secretion and reabsorption of water is accompanied by the movement of key electrolytes such as Na^+, K^+, Cl^-, and HCO_3^-. Electrolytes are essential for the function of nerve and muscle tissue and regulate the internal pH of body fluids. Massive electrolyte loss associated with GI infection thus compromises all body functions. Diarrhea, for example, promotes loss of HCO_3^- ion, leading to acidosis. Persistent vomiting, however, with loss of hydrochloric acid, may lead to alkalosis. To maintain homeostasis, complex shifts and stresses result as other organs—lungs and kidneys—attempt to readjust the pH.

Nutritional role   The soluble building block molecules released by enzymatic hydrolysis of food must be adequately absorbed in order for the individual to be well nourished. Chronic diarrhea often contributes to malnutrition. In a sort of vicious circle, malnutrition also increases the rapidity of water loss from the serum to the bowel in diarrhea.

Immunological role   Many antigens make their first contact with the immune system via the aggregates of lymphoid tissue in the mucosa of the GI tract. These **Peyer's patches** produce large amounts of IgA **coproantibodies,** which effectively prevent pathogen attachment to mucosa.

NORMAL FLORA OF THE GI TRACT

While the normal human body is composed of about 10^{13} cells, it supports around 10^{14} microbial colonists. Thus "they" outnumber "us" by ten to one! Most of these microorganisms inhabit the GI tract, predominantly the mouth and the large intestine. In thinking of the distribution of microorganisms in the GI tract (Table 20.2), it is helpful to visualize a tube containing a flowing stream. The rate of flow is fast in some regions, such as the small intestine, and slow in others, such as the large bowel. The chemistry of the stream varies with the secretions that are added to it at various points, the diet, and the individual's physiological state. Microbial colonization probably requires attachment to the surface epithelium in all fast-moving areas. This is not true in the colon, where the flow rate is slow.

There is a terrific variety of available habitats, matched by an equal variety in the microbial species that can be isolated. Over 300 species of microorganisms have been recovered by various sampling methods. Some of these undoubtedly were not true colonists, but were just passing through at the moment of sampling.

Organisms Found in Different Organs

Esophagus and stomach   The esophagus probably has no appreciable resident flora because it is simply a passageway. Some bacteria may attach to epithelial surfaces; certainly, damaged surfaces have the potential of becoming colonized.

TABLE 20.2
Normal flora of regions of the GI tract

GROUPING	STOMACH	SMALL INTESTINE	LARGE INTESTINE
Anaerobes			
Bacteroides sp.		†	*
Fusobacterium sp.			*
Lactobacillus sp.	*	*	†
Clostridium sp.	†	†	†
Other anaerobes	†	†	*
Facultative anaerobes			
Staphylococcus		†	†
Streptococcus	†	*	†
Enterobacteriaceae (coliforms)	†	*	†
Yeasts (general)	*	†	†
Candida	†	†	†
Other yeasts	†	†	†

* = Dominant members of microflora.

† = Also present in the majority of samples in significant numbers.

The stomach has traditionally been described as sterile when empty because the fasting pH drops to about 2. However, some investigators have demonstrated a limited acid-tolerant flora attached to the gastric mucosa. In any case, microbial **activity** in the normal stomach is not significant.

The role of stomach acidity in inactivating intestinal pathogens is unpredictable. This protective mechanism may be effective in the empty stomach, but in the presence of a meal—which is, after all, the vehicle by which many pathogens enter—the pH of the stomach contents may not be sufficiently low.

Small intestine Rapid movement of contents through the duodenum and jejunum prevents microbial buildup in the lumen. However, microorganisms do survive attached to the mucosa. These have been studied in animal biopsies, but are poorly understood in the normal human being. Biopsies recovered at surgery from ill human beings are probably unrepresentative. Thus the contribution of normal microorganisms is unclear.

The number of organisms increases rapidly in the ileum because the flow rate slows down.

Diverticula (blind sacs), ulcerated areas, and regions of immobility or obstruction promote extensive microbial growth. Bad effects may include vitamin depletion, gas production leading to distention, and tissue breakdown.

Large intestine   After the fluid mixture enters the colon, its movement slows down greatly, and microbial numbers soar. Feces, generally representative of the large bowel contents, contain about 10^{12} bacteria per gram, about 40 percent by weight of the total mass. These bacteria are predominantly obligate anaerobes, living off food residues not successfully attacked by human enzymes. They also utilize shed epithelial cells, mucus, and host enzymes secreted in the upper tract that have outlived their usefulness except as microbial nutrients.

Role of the Normal Flora

Compare rumen micro-flora; Chapter 6, pp. 176–177

Bacterial enzymes do not appear to help make otherwise indigestible foods available to us, in contrast to their essential role in the nutrition of the cow. Gut bacteria do produce some toxic end products (such as ammonia) that pass into the blood and are detoxified by the liver and or excreted by the kidneys. Gut bacteria may actually produce carcinogenic compounds, such as nitrosomines from nitrites.

The intestinal microflora plays a significant role in competitive exclusion of pathogens. Many enterobacteria produce **bacteriocins**, substances lethal to closely related bacterial strains. Some also produce antibiotic-like substances. Nutritional competition may also play a role. Toxic metabolites are produced: H_2S produced by the anaerobes represses the growth of *E. coli*, and short-chain fatty acids inhibit *Salmonella*, *Shigella*, *Pseudomonas*, and *Klebsiella*.

Facultative anaerobes immediately utilize any free oxygen that diffuses into the gut from the mucosal surface. They establish a very low redox potential, excluding oxygen-requiring organisms, but simultaneously creating a niche for the growth of strict anaerobes.

Intensive antimicrobial therapy disturbs the normal bowel flora. This leads to unchecked multiplication of the antibiotic-resistant remnant of the resident flora. *Candida albicans* or *Staphylococcus aureus*, normally very minor members of the microflora, can become predominant. The ecological disturbance is normally temporary, and the system returns to its normal population base when antibiotic therapy stops.

Key Groups of Microorganisms

Anaerobic infections; Chapter 23, pp. 672–675

Strict anaerobes   Up until a few years ago, if you had asked any microbiologist to name the dominant gut bacterial species, the answer would undoubtedly have been *Escherichia coli*. Improved methods of anaerobic sampling and cultivation have since showed us that *E. coli* and other facultatives were just a minor component, the tip of the iceberg. The strict anaerobes—*Bacteroides*, *Fusobacterium*, *Eubacterium*, *Peptococcus*, and others—outnumber the facultative enterobacteria by perhaps 1,000 to 1.

Anaerobic attack on food residues is fermentative. It yields a mixture of gases, including hydrogen, carbon dioxide, methane, and hydrogen sulfide. These gases are partially reabsorbed, but the remainder is passed as **flatus.** Short-chain fatty acids, decarboxylated amino acids (amines), and odoriferous by-products such as indol and skatol are also produced. In concentration, these are potentially toxic not only to the bacteria but also to the individual. When constipation causes prolonged retention of feces, systemic toxicity is added to the general discomfort.

Strict anaerobes do not appear to cause pathogenic changes in the intact, normal intestine.In the presence of structural abnormalities, they may cause problems due to excessive multiplication. When an ulcer perforates, anaerobes may escape into adjacent areas and cause peritonitis and abscesses.

Lactic acid bacteria   Fermentative organisms producing lactic acid from carbohydrate include the genera *Lactobacillus, Streptococcus,* and *Bifidobacterium.* These Gram-positive rods and cocci are fastidious in their growth requirements. In the newborn they are the first microbial colonists of the GI tract. Lactic acid bacteria are the dominant organisms of the stomach and upper sections of the small intestine in adults. In the ileum and colon, they are still a significant factor, but their numbers become relatively less. These organisms are of low pathogenicity.

The enterobacteriaciae   *Bergey's Manual of Determinative Bacteriology* places the families Enterobacteriaceae and Vibrionaceae in Part 8. These two groups of Gram-negative facultatively anaerobic rods contain many very closely related species liberally represented among the normal gut flora. They also include almost all of the bacterial GI pathogens. It is frequently difficult to tell at what point a normal contributor has ceased to be normal and has established a pathogenic relationship with the host.

Much systematic effort is being expended to refine the classification and means of identification of these bacteria. At present, the family is separated into five tribes, each containing one or more genera and species. The relationships as presently agreed on are diagrammed in Fig. 20.3. Be aware that changes are proposed so frequently that this diagram may need some revision by the time you use it.

Species differentiation is based largely on biochemical testing. All enterobacteria attack glucose fermentatively and are oxidase negative. Certain genera, those usually considered normal, ferment lactose. The serious pathogens such as *Salmonella* and *Shigella* do not. The tests in the IMViC series, urea hydrolysis, deaminases and decarboxylases, and carbohydrate fermentations, are useful determinative traits. Commercially developed rapid identification methods are widely used in clinical laboratories, replacing the older, more tedious, test-tube methods. Test-tube methods, however, are preferred by reference laboratories to identify elusive atypical strains.

The enterobacteriaceae are significant in the lower small intestine and large bowel, but comprise less than 1 percent of the total flora. Their metabolic contributions are similar to those of the previous two groups. Plasmids, often carrying genes

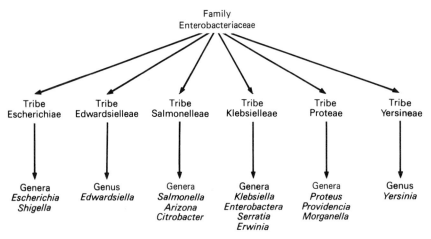

20.3
Classification of the Enterobacteria The scheme of Edwards and Ewing (1972) separates the family Enterobacteriaceae into six tribes, groups that generally share key biochemical and genetic traits. These are further subdivided into genera, and the genera into species. Biochemical tests such as lactose fermentation, H_2S production, and urease are important in identification.

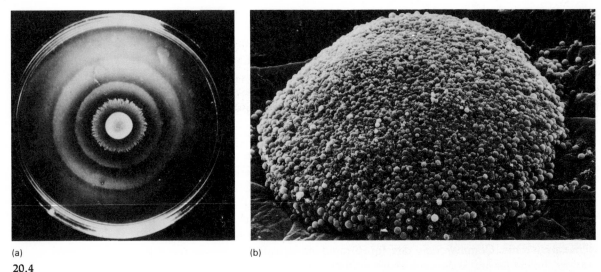

(a)   (b)

20.4
Proteus Colonial Forms (a) Certain *Proteus* species have active motility that periodically leads to swarming. When swarming, the organisms swim rapidly over the water film on the agar surface, forming a diffuse growth. Swarming periods are followed by periods of less active migration, giving the colony a banded appearance. When active swarmers are present in the mixed flora from a primary isolate, they may make it impossible to recover isolated colonies of other organisms. (b) This SEM photograph shows a colony of a nonmotile, wall-deficient strain of *Proteus mirabilis*. The picture gives a clear view of how colonies of nonmotile bacteria develop.

Box 20.1

Crisis under the Streets

In the older major cities of the United States, such as New York, Boston, and Chicago, water and sewerage systems were installed a long time ago. Some portions of these systems are more than 100 years old, and they are deteriorating rapidly.

Let us use New York City, for an example, with the understanding that the situation is equally as bad in a number of other cities. New York City has 6,200 miles of water mains of varying sizes and materials. Lying under the streets, they are subjected to a continuous stress from traffic vibration and soil compaction. Cast iron and concrete mains crack, and steel mains erode. Water main breaks in New York are a daily occurrence. They are also increasing—from 415 in 1969 to almost 600 in 1979.

Because water is supplied at high pressure, these breaks have so far not led to mixing of drinking water with contaminated ground water or sewage. Thus they are more an inconvenience than a health threat. However, another class of plumbing accident is more hazardous. Cross-connections are the illegal plumbing connections between water lines and cooling or waste-water lines. These continue to

occur in large numbers because there is insufficient labor to enforce plumbing codes and to make rigorous inspections. A cross-connection often leads to waste water siphoning back into the drinking water source. In 1975, poisonous chromates from an air conditioner siphoned into a drinking water source and hospitalized 20 people. In 1978, public concern over Legionnaire's disease in the Murray Hill area led to a systematic plumbing inspection of all the buildings in the area. Although *Legionella* was not found, plenty of other problems were. The area drinking water was found to be heavily contaminated with *E. coli*, indicating sewerage leakage. Over 50 illegal cross-connections were discovered within a two-block area. One wonders what would be found if this detailed inspection could be extended to other areas of the city.

Deteriorating equipment, broken mains, and illegal plumbing connections are all correctable problems, but unfortunately, correcting them would require huge sums of money. There is no foreseeable source for these sums in today's hard-pressed urban budgets. Even maintaining the status quo of the existing sewerage is next to impossible.

for multiple-drug resistance, are freely exchanged among the members of the enterobacterial family. Antibiotic therapy may actually encourage the growth of resistant strains by suppressing their competitors, the much more susceptible anaerobes and lactobacilli. Enterobacteria also produce bacteriocins. Some enteropathogenic strains produce **enterotoxins** that induce a massive diarrhea.

Practically all enterobacterial genera have been associated with GI disease in rare or not-so-rare cases. However, isolation of enterobacteria from feces is not, in the absence of symptoms, indicative of disease. The exceptions are the genera *Salmonella* and *Shigella*, whose presence is always abnormal. *Proteus* species may cause gastrointestinal disease or urinary tract infection (Fig. 20.4).

Coliform organisms (lactose-fermenting enterobacteria, especially *E. coli*) are, as we have seen, universally found in human feces. Since coliforms are readily isolated and identified, they are used as **indicator** organisms to prove the presence of fecal contamination of food, water, or other samples.

BACTERIAL INFECTIONS

Bacterial infections of the GI tract have a wide range of symptoms and severity. These reflect the invasiveness and toxigenicity of the infecting organism as well as the susceptibility of the person infected.

Typhoid Fever

Typhoid fever is the most severe of the epidemic GI infections. Long a major killer, it is now rare in the developed nations. It remains a serious problem in less-developed areas.

TABLE 20.3
Clinical terminology of GI infection

TERM	DEFINITION
Appendicitis	Inflammation of the vermiform appendix
Colitis	Inflammation of the colon; not always infective
Diarrhea	Discharge of frequent fluid bowel movements
Diverticulitis	Inflammation of a small pocket; especially in the colon
Dysentery	Diarrhea with frequent watery stools, with pain, and passage of blood and mucus
Enteric fever	Severe enteric infection with fever; typhoid fever
Enteritis	Inflammation of the intestine; marked by diarrhea
Enterocolitis	Inflammation of the mucous membrane of small and large bowel; blood and mucus may be passed
Food poisoning	Inexact term for disease that can be traced to food eaten—may be due to pathogenic bacteria or toxins
Gastroenteritis	Inflammation of both stomach and intestine; often marked by both vomiting and diarrhea
Peritonitis	Inflammation of the peritoneal membranes
Travelers' diarrhea	Diarrhea of sudden onset and short duration affecting travelers in all parts of the world

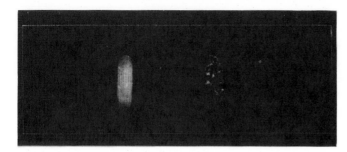

20.5
Slide Agglutination Test for *Salmonella* When a small amount of the pure culture is mixed with the homologous antiserum, (right) visible agglutination occurs, confirming the organism's serotype. A control containing *Salmonella* cells and serum without the type-specific antibody remains evenly cloudy (left), without visible clumping.

Causative agent   The genus *Salmonella* contains three species, *typhi, enteritidis,* and *choleraesuis.* Salmonellae are motile and can survive for limited periods outside the body in moist environments, such as food, water, and soil. *Salmonella* can be isolated from the mixed flora of the gut by plating on highly selective media. Recovery is improved by a preliminary incubation of the fecal sample in selenite or tetrathionite broth to eliminate lactose-positive enterobacteria. *Salmonella* is typically lactose-negative and H_2S-positive. After presumptive identification, serological tests are essential to determine which of the more than 2,100 serotypes of *Salmonella* has been recovered (Fig. 20.5). Typhoid fever is always caused by *S. typhi,* although severe infections (**paratyphoid**) caused by other strains may occasionally be confused with it.

Selective media; Chapter 7, p. 192

Pathogenesis   The bacilli pass through the mucosal lining of the upper small intestine into the underlying lymphoid tissue. They are phagocytosed and disseminated. Intracellular multiplication occurs during the incubation period. Then the organisms emerge and multiply extracellularly, and symptoms begin. The average duration of the disease is four weeks. During the first week there is bacteremia, with fever, headache, and general discomfort largely attributable to bacterial endotoxin. As the disease continues into the second week, organisms become widespread in tissues. *S. typhi* can be recovered from the blood during the first week of the disease, then from the stool for a variable period.

Bacilli multiplying in the gallbladder reinfect the intestine, and erosive lesions of the mucosa and Peyer's patches develop. Rosy spots develop on the skin of the abdomen; they contain viable *S. typhi.* Fever and prostration become more severe. In the second or third week, as the patient's condition worsens, intestinal hemorrhage and perforation, or pneumonia, may appear as complications. Fecal and urine cultures are usually positive for the agent. Convalescence begins in the fourth week, as effective CMI develops. About 3 percent of patients develop chronic infections of the gallbladder or urinary bladder. Antibiotics may eradicate the carrier state, but surgical removal of the gallbladder is occasionally required.

Therapy and immunity   Chloramphenicol is the drug of choice, its occasional toxicity outweighted by its *in vivo* effectiveness. There are some resistant strains.

Chloramphenicol; Chapter 26, pp. 748–777

Steroids and extensive supportive therapy may be essential in more severe or advanced cases.

Recovery usually confers a lifelong immunity. Although circulating antibody to O (somatic) and H (flagellar) antigen is produced, it appears that cellular immunity is the actual protective factor. Increase in the titer of specific antibody during the course of the disease (the **Widal** reaction) is a useful confirmation of the diagnosis. Protective killed-organism vaccines have been available for a number of years. Due to a high concentration of endotoxin in the vaccine, injections produce unusually

TABLE 20.4
Epidemiology of bacterial GI infections

DISEASE	RESERVOIR	TRANSMISSION; VEHICLE	INCUBATION PERIOD IN DAYS
Typhoid fever	Human beings	Oral–fecal; contaminated food and water. Urine may be infective	7–21
Salmonellosis	Poultry, other animals	Oral–fecal, contaminated meat, poultry, and eggs	½–1½
Bacillary dysentery	Human beings, primates	Oral–fecal, direct contact, water, food, and flies	1–7
Cholera	Human beings	Oral–fecal by water, flies, and hands	2–3
Vibrio parahemolyticus	Salt water and silt	Contaminated fish and shellfish, ingested	½–1
EEC disease	Human beings	Oral–fecal, during birth, by direct contact or fomites	½–3
Yersiniosis	Human beings, cats, and dogs	Oral–fecal	3–7

pronounced discomfort of systemic illness. Sadly, the protective effect is relatively weak and of short duration. In consequence, vaccination is used only by persons that anticipate immediate exposure. Improved immunization methods are sorely needed.

Epidemiology   *S. typhi* is found only in human beings. It is spread by the oral–fecal route. The reservoir is the convalescent or chronic carrier, and the vehicle is water, food, or soil contaminated with untreated sewage. Up to 10^7 organisms must be

PERIOD WHEN COMMUNICABLE	PATIENT CONTROL MEASURES	EPIDEMIC CONTROL MEASURES
As long as agent is in excreta. Ten percent carry three months; 2–5 percent are permanent carriers	Isolate; disinfect wastes and soiled articles	Intensive search for case or carrier and vehicle
Variable—days to weeks, temporary carriers are common	Remove from food handling, patient, or child care. Enteric precautions	Report. Investigate food sources and handlers
During acute illness or during short carrier state	Isolate; rigid personal and enteric precautions	Report. Investigate food sources and handlers
While stool is positive, usually a few days, rarely months or years	Isolation not necessary if staff practices strict cleanliness, disinfection of wastes and soiled articles	Boil or treat water, supervise food, investigate transmission
Noncommunicable	No special precautions	Report. Determine source of exposure.
Not known; during fecal colonization	Isolate infected infants and suspects. Disinfect discharges and soiled articles	Admit no more babies to nursery. Provide separate staff for exposed babies. Investigate source of outbreak
As long as symptoms persist. Some chronic carriers	Remove infected person from food preparation and patient or child care. Enteric precautions	Report to local health authority

Adapted from: A. S. Benenson, 1975, *Communicable Diseases of Man.*

ingested in order for disease to develop; smaller inocula are probably effectively neutralized by local intestinal defenses. The severity of the subsequent infection appears to increase with larger inocula. Case incidence increases where there is poor sanitation associated with untreated sewage, privies in close proximity to wells, and the use of frankly polluted drinking water. Improved sanitation has been the major means of reducing the toll taken by this disease. In the United States, sporadic cases (about 400 a year) are either acquired abroad or traced to contact with chronic carriers, often elderly persons who have been carriers for decades. Systematic follow-up measures are now taken to ensure that all recuperating patients stop excreting the bacilli. Typhoid fever is gradually disappearing entirely in the United States, except for imported cases.

Salmonellosis

About 40 serotypes of *Salmonella* in the species *enteritidis* and *choleraesuis* cause human infection of varying intensities. In some classifications, these receive separate species names. These organisms, plus numerous nonhuman serotypes, are primarily animal pathogens. Their transmission to humans is accidental. Gastroenteritis marked by nausea and diarrhea, or enterocolitis with diarrhea and varying degrees of mucosal damage may result. The infections are usually self-limiting; symptoms disappear after a few days and antibiotic therapy is not usually needed. *S. choleraesuis* is occasionally associated with extraintestinal infection such as osteomyelitis, empyema, and infection of grafts or prostheses following surgery. Concurrent infection by bloodborne parasites such as *Plasmodium* or *Bartonella* and RBC abnormalities such as sickle-cell anemia also predispose to *Salmonella* infection.

The natural history of foodborne gastroenteric *Salmonella* is of great interest. The recent rapid increase in incidence can be directly related to changes in modern food production and consumption patterns, the mass production of eggs and meat, and the increasing trend towards "fast" and convenience foods. The chicken and other domestic fowl are the main reservoir of *Salmonella* strains, and a large proportion of chicken meat is inoculated during butchering with organisms accidentally released from the viscera. Laying hens also may contaminate eggs. Although *Salmonella* is killed by cooking, surfaces and utensils that have been in contact with the raw food may inoculate other food that will not be cooked.

The market feed cycle for domestic meat animals utilizes rations that contain the reprocessed viscera or wastes of other animals. Inadequately decontaminated fish meal is fed to chickens. Chicken or pig wastes may be fed to healthy beef animals. The recipients become *Salmonella* carriers. Thus the agent spreads from chickens to other domestic animals.

Recent outbreaks indicate the role of convenience foods in *Salmonella* transmission. Our dietary habits have changed so that, on the average, more than 50 percent of our meals are eaten away from home. More mass consumption "fast" foods are being offered and consumed. Precooked turkeys and rare roasts of beef are sold for banquets and at delicatessen counters. They are consumed cold without further cooking. These meats have been the source of many *Salmonella* outbreaks. Incomplete heating of the interior of the meat during initial preparation is at fault. Gravies held on steam tables for prolonged periods have infected customers of fried chicken take-out chains. An outbreak traced to a restaurant may be due either to contaminated food sources or to sloppy sanitary practices on the part of kitchen help, some or all of whom may be *Salmonella* carriers. When an outbreak originates at a restaurant, there are usually many more cases than if a similar situation arises in the home. Despite intensive surveillance, the incidence of salmonellosis has increased by 50 percent in the last ten years.

Bacillary Dysentery (Shigellosis)

There are four serogroups in the genus *Shigella*, A–D, and all are associated with disease. Initial isolation of this agent can be accomplished on a differential medium such as MacConkey's or a mildly selective formula such as SS or Hektoin agar. Any lactose and H_2S-negative, nonmotile organisms recovered can be further differentiated by biochemical patterns and the identification confirmed serologically. *Shigella flexneri* (B) and *S. sonnei* (D) in the United States and the classic *S. dysenteriae* (A) in other parts of the world are the primary agents of shigellosis.

Pathogenesis   The characteristic disease is an enterocolitis. *Shigella* penetrates the mucosal cells of the large intestine, causing inflammation and local microabscesses. The organism is not invasive, so the lesions remain shallow, but may cover a large area. Symptoms include fever, pain, and diarrhea with blood and mucus. In the healthy adult, shigellosis resolves spontaneously in about a week. In debilitated adults or small children, fluid and electrolyte loss may become life-threatening. Mortality is about 0.5 percent when adequate medical care is provided.

Therapy and immunity   Tetracycline, chloramphenicol, and ampicillin can all be used on susceptible strains. However, *Shigella* is the organism in which **resistance transfer factors** (RTFs) were first demonstrated. RTFs have been responsible for a worldwide spread of resistance to sulfa drugs and tetracycline in *Shigella* species.

Resistance transfer factors; Chapter 9, p. 275

Recovery confers partial protective immunity. Vaccine is not available for general use; an oral attenuated live vaccine is being tested in high-risk populations.

Epidemiology   Human beings are the only reservoir, and the organism is shed by patients and convalescents for one to three months. Prolonged carriage is rare. The organism persists for up to three days in sea water and for 30 days in milk, eggs, and seafoods. The infective dose may be as small as 200 bacilli; accidental infections of laboratory and health workers are regrettably common. Children under five have the highest case rates. Shigellosis is often seen in homosexual males. Enclosed populations in mental institutions, homes for exceptional children, and internment camps have very high rates of infection, and very severe manifestations.

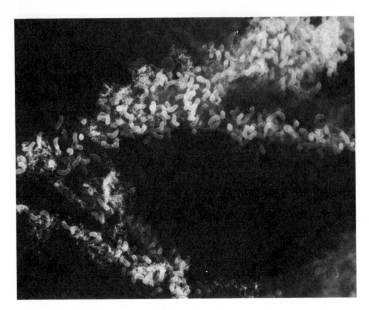

20.6
Vibrio cholerae Large masses of curved vibrios are seen attached to strands of intestinal mucus. This motile Gram-negative bacillus causes cholera. Cholera is relatively common in the Far East. Long thought to be eradicated in the United States, it has recently reactivated in the coastal regions around the Gulf of Mexico and the Mississippi delta.

Cholera

Causative agent *Vibrio cholerae* is a Gram-negative, curved rod (Fig. 20.6). Its family, Vibrionaceae, is closely associated with the enterobacteria. Vibrios have an epidemiologically important ability to survive outside the body in cool water. The biotype El Tor, serotypes Inaba or Ogawa, are the most common forms in the western hemisphere; they cause a high percentage of mild or inapparent infections. The classic strains, more pathogenic, are prevalent in the Far East.

 V. cholerae can be isolated from stool specimens on selective media containing tellurite and/or with a high pH. Alkaline peptone broth (pH 8.4), Monsur (TTG), or TCBS agars may be used. Further identification depends on microscopic observation of the typical C-shaped or S-shaped cells, biochemical differentiation, and serological typing.

Pathogenesis Cholera is a disease caused by a toxin. The vibrio colonizes the small intestine and attaches to the mucosa but does not penetrate or damage its surface. It then produces an **enterotoxin** called **choleragin** that attaches to the host-cell membrane and activates the cell to produce the enzyme **adenyl cyclase.** This enzyme in turn catalyzes a great increase in cellular cAMP, a regulatory nucleotide that promotes the secretion of fluid and electrolytes from the cell into the lumen of the gut (Fig. 20.7). The end result of toxin production is voluminous diarrhea. The outcome of disease is directly dependent on the degree of fluid-electrolyte imbalance and the ability of the patient to stand the stress. If the patient is strong, the process subsides naturally in a few days. Otherwise, hypovolemia leads to apathy, shock, and death in 50–70 percent of patients ill enough to require hospitalization if they do not receive adequate therapy.

Cyclic AMP; Chapter 2, pp. 56–57

Therapy and immunity Fluid-electrolyte replacement is all that is needed to reduce the mortality rate sharply. In Western countries we take for granted the instant availability of sterile intravenous fluids and of personnel to properly administer them. However, in remote areas of the world these may be totally unavailable, especially in amounts sufficient to deal with an epidemic of thousands of cases. In the cholera outbreak which followed the India–Pakistan war of 1971, international health agencies were able for the first time to move in massive support teams; they were able to save most patients they could get to. There has been encouraging success with the use of oral or nasogastric fluids, which need not be sterile and can be mixed on the spot. Antimicrobial drugs are of slight value.

Natural immunity is short-lived, as is that derived from current killed-bacteria vaccines. Several different new vaccines are being developed and tested. It appears that specific IgA secreted by the intestinal mucosa is the key preventive factor.

Epidemiology The asymptomatic carrier is of major importance. Human beings are the only reservoir and spread the agent via fecal contamination of water and (to a lesser extent) food. Recent cases in the United States have been traced to contaminated crayfish and shellfish. Sanitary methods are the practical means of prevention and control.

Other Vibrios

Marine vibrios Salt-tolerant *Vibrio* species are found in coastal waters throughout the temperate zones, and collect in large numbers on fish and shellfish. *Vibrio*

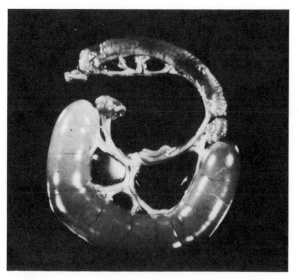

20.7
Effect of Cholera Toxin The rabbit ileal loop test demonstrates that cholera toxin increases the secretion of fluid into the lumen of the bowel. Segments of the ileum have been tied off. The upper segment (control) was perfused with saline solution. The lower segment was exposed to cholera toxin. It is completely distended with secreted fluid. In the human being, of course, this secreted fluid would give rise to an acute diarrhea.

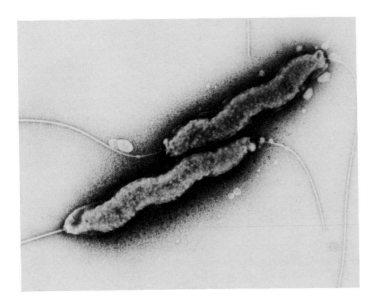

20.8
Campylobacter fetus var. *jejuni* This is a
Gram-negative spirillum that causes a significant percentage of gastroenteritis cases. It is
microaerophilic, and requires special media
and incubation conditions for isolation.

parahaemolyticus was first recognized in severe enteritis in Japan, where raw fish is
a favored delicacy. More recently, vibrio enteritis has been seen in the United States
in association with shrimps, crabs, and oysters, either raw or inadequately cooked.
The symptoms are very similar to bacillary dysentery. Rarely, *V. parahaemolyticus*
causes wound infection and/or septicemia in fishers or others exposed to seawater.

Vibrio alginolyticus has been recovered from external ear infections, wound infections and septicemias, but not GI infections. Lactose-fermenting *Vibrio* isolates
are increasingly recognized in severe enterocolitis followed by septicemia. Although
these infections are rare, the agent appears to be extremely invasive; underlying
alcoholism, diabetes, or some other debilitating condition is usually, although not
always, seen.

Microaerophilic vibrios   *Campylobacter fetus* var. *jejuni* (Fig. 20.8) (previously
Vibrio fetus) has been considered a zoonotic agent causing enteritis or abortion in
cattle, sheep, goats, dogs, pigs, and fowl. It is now also implicated in enteritis, particularly of children. Bacteremia and disseminated infection occur occasionally. In
one study with infants under eight months, *C. fetus* was implicated as the agent in 31
percent of gastroenteritis cases.

Some authorities feel that *C. fetus* var. *jejuni* is more important than *Salmonella* or *Shigella* in gastrointestinal disease. Transmission of *C. fetus* var. *jejuni* through
the town water supply prostrated 20 percent of the population of a small city; raw
milk from an infected cow spread the disease among many members of a farm community. The agent is also frequently foodborne.

Enteropathogenic *Escherichia coli* (EEC)

Causative agent Certain serotypes of *E. coli* have been implicated in two types of GI infection, one a diarrhea syndrome similar to cholera and the other an invasive enterocolitis resembling bacillary dysentery. The conditions are produced by quite separate mechanisms. The diarrhea syndrome appears in infants, and also in adults as travelers' diarrhea or **turista.**

Infant diarrhea has long been a scourge of hospital nurseries; the role of *E. coli* was noted only about 15 years ago. The reason for the delay is that *E. coli* is a member of the normal bowel flora, so its appearance on clinical cultures traditionally resulted in a report of "normal flora." To show that a strain is enteropathogenic requires demonstration of the **heat-labile enterotoxin** (LT) in culture filtrates. Methods involving whole animals or cell culture are being replaced by immunological techniques including ELISA and SCT assays. The gene for toxin production is transmitted by an episome. Newly developed tests search for a **colonization factor** that aids in virulence, or for adherence factors (Fig.20.9).

SCT assay; Chapter 13, p. 395

Pathogenesis The heat-labile toxin interacts with small bowel mucosa, in much the same way as cholera toxin interacts, although LT has a much lower potency. Outpouring of fluid and electrolytes leads to a diarrhea–dehydration condition that can be rapidly fatal to newborn infants in whom the case fatality rate is about 16 percent.

Dysentery-like cases involve *E. coli* strains with unusual invasiveness. The factor responsible is not known.

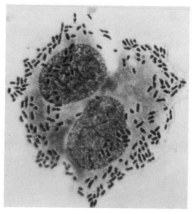

(a)

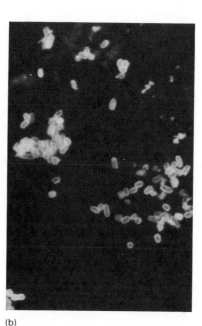

(b)

20.9
Escherichia coli (a) *Escherichia coli,* like many pathogens, may exhibit a strong, specific adherence to target cells. In this laboratory assay for adhesiveness, active binding of this isolate to human fetal intestinal cells on a glass cover slip is observed. Adhesion is a virulence factor in *E. coli,* contributing to its ability to cause kidney infection and enterocolitis. (b) *E. coli* stained with fluorescent antibody. This sample was obtained from a fecal smear of an infant with diarrhea. The technique has been used for rapid laboratory identification of the agent.

Therapy and immunity   Fluid replacement is essential in all severe diarrhea. Antimicrobial drugs are of limited value, since multiple-drug resistance is widespread.

Postinfection immunity probably occurs, but its extent is unclear. There does appear to be a cross-reaction among the numerous antigenic types of enterotoxin. The search is on for a broad-spectrum enterotoxoid vaccine that could be useful for infants under two years of age and for travelers to remote areas.

Epidemiology   Infant diarrhea is common in hospital-delivered babies and hospitalized infants, but less common in home-delivered children. In underdeveloped countries, however, the latter are very much at risk, too. It appears that the newborn's sterile GI tract rapidly becomes colonized with *E. coli* strains from its attendants; these may be harmless to the adult carrier but dangerous to the newborn. Breast-feeding confers some protection. Travelers' diarrhea, or turista, is often contracted by Westerners visiting other countries. Scattered epidemiological studies show that enteropathogenic *E. coli* strains are the primary villians.

Yersiniosis

Yersinia enterocolitica and *Y. pseudotuberculosis* are enterobacteria that produce severe enterocolitis, appendicitis-like symptoms, and arthritis-like pain. Although they have been most common in Scandinavia and Western Europe, foodborne outbreaks have recently been seen in the United States. Healthy swine, drinking water, and rivers also serve as sources.

Mixed Infections

Many infections of the peritoneal cavity, the bowel, and the intestinal glands are characteristically caused by two or more organisms working synergistically. The infection may develop spontaneously, postoperatively, or by extension of an intestinal infection that resulted in perforation. These are endogenous infections in which a structural abnormality is complicated by overgrowth of of normal flora. The strict anaerobes are usually a key part of the troublesome population.

Appendicitis   Acute appendicitis occurs in more than 200,000 persons a year in the United States. The opening of the appendix becomes obstructed, leading to microbial overgrowth, inflammation, abscess, and rupture. Pain, constipation, and elevated white blood cell count are common symptoms; surgical removal is the usual therapy.

Diverticulitis   A diverticulum may be present for many years without causing trouble; most older persons have one or more. Occasionally, the sac becomes obstructed and inflamed. Hemorrhage or perforation may occur. Diverticulitis often responds to conservative treatment with antibiotics.

Peritonitis   When irritating or infectious agents enter the peritoneal cavity, they induce the inflammatory state of peritonitis. Peritonitis is usually localized at first. Symptoms may be mild or may mimic a number of other conditions. Antimicrobial therapy is followed by surgical correction of the precipitating problem.

VIRAL INFECTIONS

The intestinal tract is a common portal of entrance of infective viruses. Most move to other tissues before establishing a disease state, and have been dealt with elsewhere in this book. The viruses that cause gastroenteritis and hepatitis are discussed here because they actually infect GI tissues and organs.

Polio; Chapter 22, pp. 664–667

Viral Gastroenteritis

Almost everyone has had the "24-hour flu," "intestinal flu," "green death," or whatever the local nickname may be. These terms are misleading; the causative agent has nothing to do with the influenza virus.

Viral gastroenteritis is characterized by rapid onset, with repeated vomiting and/or diarrhea; it is usually over within 12 to 24 hours. Moderate reversible changes in the small intestine lining are seen. The infant under two years may have more prolonged symptoms and severe dehydration. Recent advances in virus identification have pinpointed certain of the responsible agents.

Infantile disease   In infants the **rotavirus** (Fig. 20.10) is the most common cause of winter diarrhea, and it is an important cause of mortality in less-developed societies.

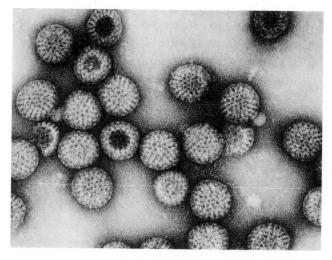

20.10
The rotavirus is a common cause of infant diarrhea.

The rotavirus is a relatively large RNA virus. The incubation period of rotavirus infection is 48–72 hours, vomiting is the main symptom, and illness lasts two to seven days. Rotaviruses may be carried by adults, but rarely cause illness in persons older than two years. The virus is detected by immune electron microscopy, CIE, ELISA, or radioimmunoassay of virus in stool samples. Recovery stimulates immunity, and immune nursing mothers play a major role in protection of infants. Incidence in the breast-fed child is much lower than in the formula-fed child, probably due to the secretion of antirotaviral IgA in milk.

Immunological techniques; Chapter 13, pp. 391–393

The **Norwalk agent** is the best studied and perhaps the commonest gastrointestinal virus affecting adults. It is a **parvovirus**, a tiny DNA virus. The incubation period of the disease is 24–36 hours, nausea is the major symptom, and the duration is about 24 hours. Recovered individuals may shed virus in the stool for one month. There is a rather short-term, type-specific immunity. Adenoviruses, coxsackieviruses, polioviruses, and echoviruses have also been implicated in adult viral gastroenteritis.

Viral Hepatitis

Hepatitis is an inflammation of the liver, marked in most cases by necrosis of hepatic cells. Reduced liver function is apparent as **jaundice**—retention of bile giving a yellow tinge to skin and eyes. Some hepatitis cases are caused by noninfectious processes such as chemical exposure, but most are of viral origin. There are two separate well-studied viral agents, Types A and B, and information suggestive of at least two other "non-A, non-B" agents. None of the viral agents has been successfully cultivated in the laboratory, so little is known about them as yet. We will discuss the types simultaneously, to facilitate comparison.

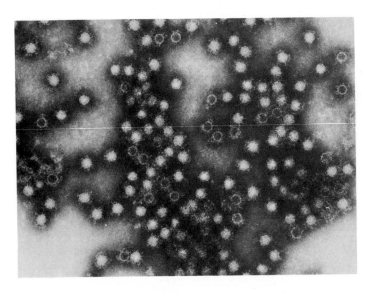

20.11
The first electron micrograph of the virus-like particles associated with non-A, non-B hepatitis. They are spherical, and measure about 27 nm in diameter.

TABLE 20.5
Comparison of types of hepatitis

CHARACTER	TYPE A	TYPE B	NON-A NON-B
Clinical name	Infectious hepatitis	Serum hepatitis	None
Principal sources	Food and water	Blood	Probably blood
Most common transmission	Oral–fecal	Injection	Probably injection
Clinical onset	Acute	Gradual	Probably gradual
Incubation period	15–50 days	50–180 days	Unknown
Chronic cases	Rare	Rather common	Unknown
Prevention by immune serum	Effective	Ineffective	Unknown
Viral particle	20 nm and 40 nm	20 nm and 20 nm tubular body 40 nm Dane particle	27 nm particle
Serological diagnosis	IEM of virus in stool. Anti-HAV detected by RIA	Hb_s-AG or anti-Hb_s antibody detected by many means	Not yet. Suggested when A and B tests are negative

Causative agents **Infectious hepatitis** is the common description of the disease caused by the hepatitis A virus (HAV). The agent is believed to be a small RNA virus, not yet fully identified, but similar to an enterovirus. It is exceptionally resistant to heat and cold inactivation. Identification is by immunological means (see below).

Serum hepatitis is the traditional if inaccurate term for the disease caused by hepatitis B virus (HBV). The causative agent, a DNA virus, has not been adequately characterized. Several types of viral particles appear in patient sera. The largest of these, the 27-nm **Dane particle,** may be the infective virus. Very little is known about the causative agents of non-A and non-B hepatitis (Fig. 20.11).

Pathogenesis The incubation period of infectious hepatitis is about 30 days. It is followed by a **preicteric** (before jaundice) phase of about a week, marked by fever, abdominal tenderness, nausea, and loss of appetite. The **icteric phase** with actual jaundice is marked by increasing liver disease, as indicated by rising levels of liver enzymes in the serum. The clinical symptoms usually disappear fairly rapidly, and

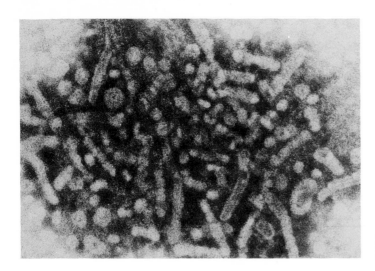

20.12
Hepatitis B virus.

recovery is complete after about three months. Liver regeneration may take up to a year and occasional mild relapses are common. Mortality is about 2 percent.

Serum hepatitis has a much longer incubation period, averaging 60 days, and a gradual onset. The preicteric and icteric symptoms are generally as above, but the disease is apt to be more severe. In individuals with poor cell-mediated immunity, the disease may be progressive and fatal. In addition, some individuals never completely recover, have chronic symptoms, and become permanent HBV carriers.

Non-A, non-B hepatitis (NANB) is clinically indistinguishable from Type B. However, analysis of antibody response pinpoints the existence of at least two distinct diseases, one with a short and one with a long incubation period. It is currently believed that the NANB virus(es) are the primary cause of transfusion-mediated hepatitis.

Therapy and immunity   There are no specific therapeutic measures for hepatitis; bed rest is no longer rigidly enforced. Experimentally, interferon therapy shows promise.

Serum antibodies to the hepatitis viruses appear during the course of the illness, and their differential detection is the usual basis for a diagnosis. Anti-HAV is diagnosed by a competitive binding radioimmunoassay test. The antigens of the HAV particle have not yet been studied. HBV particles (Fig. 20.12) have several antigens to detect. The surface antigen HB_s-A_g, or **Australia antigen,** appears on the surface of particles found in serum. It is detected in serum by RIA. Antibody to HB_s may also be specifically tested. There is also a core antigen and a corresponding antibody. Non-A, non-B hepatitis is a diagnosis of exclusion—clinical hepatitis without increase in either anti-HAV or anti-HBV immunoglobulins.

In general, it appears that recovery from one form of viral hepatitis confers complete protection against repeats of that type, but not against any other. Apparent repeats noted in the past may have been due to the very recently detected non-A, non-B viruses. Sequences of four cases have been documented in drug addicts, each presumably caused by a serologically distinct virus. It is unclear how many more agents remain to be discovered.

Immune serum globulin (ISG) has a definite role in immunoprophylaxis of exposed persons. If HAV exposure has occurred, nonspecific ISG should be given as soon as possible. If the ISG is given within two weeks of exposure, 80–90 percent of the cases may be prevented or converted to inapparent infections. Following HBV exposure, ISG or specific hepatitis B immunoglobulin (HBIG) should be given within one week. There are no licensed vaccines for viral hepatitis, although an experimental HBV immunization is under development.

Epidemiology Spread of these viruses occurs by separate but overlapping patterns. In general, hepatitis viruses may be transmitted by the oral–fecal route, accidental or intentional administration of blood or blood products, saliva, other direct contact, or sexual intercourse. The oral–fecal is the most common for HAV, with contaminated water or food as the vehicle. Shellfish from sewage-flooded beds carry an HAV risk. Precautionary measures depend on high hygienic standards.

Although oral–fecal transmission can occur, the serum inoculation method is more characteristic of HBV. Contaminated blood products, intravenous equipment, and syringes are at fault. Thus hepatitis is a particular problem in the hospital environment. Medical personnel may be infected by a minor accidental needle scratch. Groups with especially high risks are small children in day-care or nursery school situations in which careful handwashing is difficult to enforce, residents of institutions for the mentally retarded, homosexuals, and drug abusers who share equipment.

The natural history of HAV in undeveloped areas contrasts sharply with its natural history in industrialized nations. Hepatitis in children is usually inapparent or mild, whereas in adults it is more likely to be severe. Where modern water and sewage systems are unknown, mild childhood disease is common, and there are few severe adult cases. In more developed areas, most adults are susceptible. Although there are fewer total cases, a larger proportion of cases are serious. A sort of "spare the child, afflict the adult" pattern emerges, similar to that seen with the other childhood diseases and polio. Western tourists have a high hepatitis risk when they travel off the usual tourist routes. They should first obtain immune serum prophylaxis.

HBV and NANB hepatitis, spread primarily by injection, are almost exclusively diseases of civilization. Although the incidence of HAV is decreasing in the United States, HBV and NANB seem to be increasing.

PROTOZOAN INFECTIONS

Protozoans can invade and colonize the intestine. They are frequent members of the normal flora of nonhuman animals, but in human beings their benign presence is questioned. In diseases such as amebic dysentery, they are associated with ab-

TABLE 20.6
Epidemiology of viral and protozoan GI infections

DISEASE	RESERVOIR	TRANSMISSION, VEHICLE	INCUBATION PERIOD IN DAYS
Viral gastroenteritis, infant	Human beings, animals ?	Probably oral–fecal, water, food	1–2
Viral gastroenteritis, adult	Human beings	Probably oral–fecal	1–2
Hepatitis A	Human beings, primates	Oral–fecal, direct contact, injection	15–50
Hepatitis B	Human beings, primates	Parenteral inocula-tion of blood, personal contact	50–180
Amebiasis	Human beings	Water contaminated with cysts, hand-to-mouth	Variable 2–4 wk
Giardiasis	Human beings, animals (?), beaver	Oral–fecal; water and hand-to-mouth	6–22
Balantidiasis	Human beings, pig	Oral–fecal; ingestion of cysts. Water, raw food, flies, soiled hands	Unknown

Adapted from: A. S. Benenson, 1975, *Communicable Diseases of Man.*

normal bowel conditions and are believed by all except the most difficult to satisfy to be the causative agents.

Amebic Dysentery

Causative agent   *Entamoeba histolytica* is in the Class Sarcodina, and moves via pseudopodia. It has a two-stage life cycle. The resting stage, or **cyst** form, is en-

PERIOD WHEN COMMUNICABLE	PATIENT CONTROL MEASURES	EPIDEMIC CONTROL MEASURES
Up to eighth day after onset	Isolate infants; enteric precautions	Search for vehicle and source
During acute stage, possibly shortly after	Enteric precautions, no disinfection	Search for vehicle and source
Maximum infectivity last half of incubation period, into icteric phase	Enteric precautions first two weeks, disinfect blood, wastes, and contaminated objects	Determine source, provide ISG for exposed contacts
Blood infective long before symptoms, through acute and chronic phases. Chronic state may persist for years	Enteric precautions, disinfect materials contaminated with blood	Search for cases among patients attending some clinic or service, strict control of blood products
While cysts are passed, perhaps for years	Carriers excluded from food handling, sanitary disposal of feces	Determine source and vehicle
Entire period of infection	Sanitary disposal of feces	Determine source and vehicle
During period of symptoms	Sanitary disposal of feces	Determine source

closed in a resistant exterior coating. The cyst coat allows it to survive in the exterior environment and, on ingestion, to pass through the acid stomach without destruction. Once in the intestine, the cysts germinate and *E. histolytica* multiplies in the **trophozoite** form (Fig. 20.13), a typical shapeless amoeba containing a single nuclear body. Trophozoites continue to be produced while the organisms multiply within the intestinal wall, but in the lumen of the gut, cysts are formed. Both may be shed in the stool. Although the organism can be cultivated *in vitro*, diagnostic identifica-

Protozoa, classification; Chapter 5, pp. 149–153

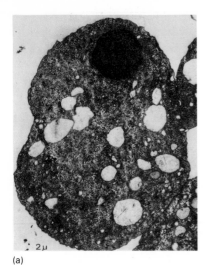

(a)

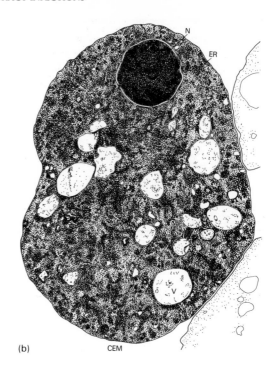

(b)

20.13
(a) *Entamoeba histolytica.* (b) A drawing of this cell's nucleus (N), vacuole (V), cell membrane (CEM), and endoplasmic reticulum (ER) are shown.

tion is based on demonstrating the organism in specially prepared stool specimens. *E. histolytica* must be differentiated from other occasionally seen amoebas that appear to be without clinical significance.

Pathogenesis   The trophozoites can invade the intestinal mucosa. Inflammation and diarrhea, often sporadic or recurring, are the usual symptoms. Sometimes there is pronounced invasiveness and ulcers develop. Ulceration is probably a synergistic bacterial attack on issues weakened by the amoebas. Bleeding and intestinal perforation develop in severe, untreated cases. Hematogenous spread may lead to liver abscesses or, less commonly, to lung or brain involvement.

Antiprotozoan drugs; Chapter 26, p. 783   **Therapy and immunity**   The drug of choice is metronidazole (Flagyl) given by mouth. Surgical treatment of abscesses is needed if drug therapy alone is not effective.

Amebic infection usually leads to an antibody response, useful in diagnosis but without apparent protective value. Repeat infections occur, and there is no vaccination.

Epidemiology   The reservoir is human beings, and the agent is spread via oral–fecal transmission. Water, foods, and flies are often implicated. Cysts are infective whereas trophozoites probably are not. Thus the chronic carrier, who sheds mainly cysts, is potentially more hazardous than the acute case who sheds many trophozoites but few cysts.

The disease is endemic throughout the world, with case incidence inversely correlating with sanitary standards. Recent studies indicate that about 4 percent are infected in selected urban industrialized areas and from 5–30 percent in rural areas. In the United States a major survey in 1976 found the agent in 0.6 percent of stool specimens examined.

Giardiasis

Diarrheal infection due to *Giardia lamblia* (Fig. 20.14) is the most common intestinal parasitism in the United States, affecting an estimated 3.8 percent of the population. The agent is a member of the Class Mastigophora; its trophozoites bear several flagella. The oval, motile stage contains two nuclei positioned so that the cell appears to have two large, owlish eyes. The cyst stage, by which *Giardia* is transmitted, is oval and is marked by a central, rod-shaped axostyle.

Giardiasis, the active disease state, affects the small intestine. The disease has an acute onset, with symptoms of cramps, diarrhea, bloating, flatulence and, occasionally, constipation. Complications are rare and recovery occurs within a few days. Giardiasis may affect persons of any age. Metronidazole is the usual therapeutic drug. *(Flagyl)*

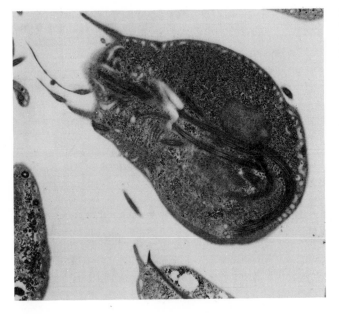

20.14
Giardia lamblia   This is a transmission electron micrograph of a thin section through the trophozoite form of the parasite.

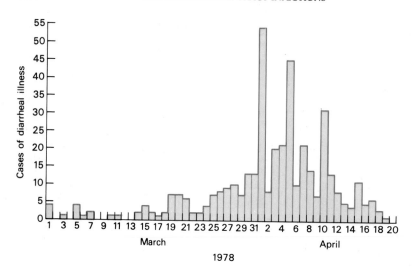

20.15
Giardiasis outbreak, Vail, Colorado This epidemic, totaling about 55 cases, was caused by a sewerline obstruction that allowed sewage to leak into the town's water supply. After the problem was found and corrected, the outbreak gradually ended.

The cysts are spread by hand-to-mouth inoculation, water, and (rarely) food. It has traditionally been associated with sewage contamination (Fig. 20.15). However, recent surveys found that its incidence in the United States was highest in the sparsely populated states of Maine, Minnesota, and Colorado. Many cases have developed following wilderness camping, where sewage contamination was extremely unlikely. It is becoming clear that there must be a **sylvatic** (wild animal) reservoir, and the beaver is the most likely candidate.

Balantidiasis

Balantidium coli, a ciliate, is the probable agent of an ulcerative condition of the lower bowel, usually so mild as to be symptomless. Invasive disease is usually seen only in the debilitated individual. The pig is an important reservoir and source of human infection.

FOOD INTOXICATION

Food can be an important vehicle in almost all diseases of infectious origin that involve the GI tract. It is important to distinguish between two types of foodborne disease. In a **foodborne infection,** such as salmonellosis, the agent itself is ingested and multiplies in the gut, causing disease. In a **food intoxication,** the agent has synthesized a toxin in the food prior to the food's ingestion. It is the ingested toxin, usually without any significant further microbial growth in the gut, that causes the disease state. Disease may be present without infection. Food intoxications are almost without exception preventable with good food handling and preservation practices (Fig. 20.16).

(a)

(b)

(c)

20.16

Food Safety   (a) Raw fish and shellfish may carry many infectious agents if they are harvested from contaminated waters. Shellfish beds are monitored for fecal contamination by local health laboratories to establish guidelines for safe harvest. Both raw and cooked fish and shellfish spoil rapidly. They must carefully be held under refrigeration to retard the process. (b) Raw chicken meat may frequently be contaminated with *Salmonella* and *Campylobacter*. Careful attention to the cleaning and plucking of the chicken, followed by controlled refrigeration, minimizes this problem. (c) Prepared foods, especially those made with extensive chopping and mixing, may be the vehicles for outbreaks of *Staphylococcus aureus* food poisoning. The cooks should have scrupulously clean hands and cover any lesions with dressings or gloves before preparing food. Once prepared, the foods should be stored under refrigeration until immediately before use.

Intoxications with Gastroenteric Symptoms

Certain of the toxins upset gut function, via brain centers. Their ingestion results in gastroenteritis clinically indistinguishable from many of the conditions we have previously discussed. Although unpleasant, these toxicities are rarely serious.

Some strains of *Staphylococcus aureus* make an enterotoxin. When absorbed from food, this neurotoxin affects the brain center controlling the vomiting reflex. Two to eight hours after ingestion of toxin, the person begins to vomit forcefully and repeatedly. Diarrhea occasionally develops as well. The symptoms disappear within 24 hours or less.

The problem often originates with a food handler who is a nasal carrier or has open skin lesions bearing enterotoxigenic strains of *S. aureus*. Less commonly, the

Box 20.2

Recipe for an Epidemic

In November, 1978, a Thanksgiving dinner shared by approximately 350 students, faculty, and guests at a college in Florida resulted in an outbreak of staphylococcal food poisoning. At least 54 individuals developed an illness characterized by the abrupt onset of nausea and vomiting, followed by diarrhea. None reported fever. The onset of the majority of cases occurred four to five hours after eating; the range was two to eight hours. Two persons were hospitalized.

Food histories obtained from 64 persons who ate the meal incriminated ham as the vehicle of transmission. Bacterial cultures of leftover food items confirmed the epidemiologic findings; $>10^7$ *Staphylococcus aureus* colonies per gram were isolated from the ham and turkey. Phage typing was not done, and no specimens from patients or food handlers were cultured.

Because of the holiday, the dinner was prepared by a large number of students instead of the usual kitchen personnel. Interviews with these students indicated that hams and turkeys had been partially thawed at room temperature and that the temperature controls on the ovens used for cooking were not functioning properly. The sliced ham and turkey had been stored at temperatures favorable for the growth of *S. aureus* (less than 60°C; or 140°F) for as long as eight hours before serving.

Improper thawing of meat and storing of prepared foods provided the conditions necessary for this outbreak to occur. This incident serves as a reminder of the importance of having trained, supervised personnel involved in the preparation of food for the public.

Source: Adapted from *Morbidity and Mortality Weekly Report* **28**: 153, April 6, 1979.

vehicle may be milk from a cow with staphylococcal mastitis. Foods are accidentally inoculated during preparation with large numbers of *S. aureus*. Then the food is held for some hours at temperatures sufficiently warm for bacterial multiplication and toxin production. Certain foods are frequently incriminated because (1) they are good staphylococcal media and (2) they are habitually mishandled. These foods include cream-filled baked goods, salads with mayonnaise, cured meats such as ham, fish dishes, and dairy products.

Clostridium perfringens Type A causes an intoxication very similar to staphylococcal food poisoning. The toxin is a neurotoxin, and the symptoms differ only in the longer (~18 hours) incubation period. Rarely, necrotic enteritis is associated with Type F. *C. perfringens* is a spore-former, exceptionally widely distributed in soil and the intestines of warm-blooded animals. Spores can be expected to be found on the surface of most fruits and vegetables. The types of food that are potential problems can be deduced from the fact that *C. perfringens* is both anaerobic and thermophilic. Pots of soups or stews are cooked at 100°C, then often held in a warm condition on the back of the stove, or reheated briefly for subsequent meals. Turkeys may be stuffed hours in advance of cooking, then cooked by slow methods so their interiors reach only moderately warm temperatures. Crock-pots or other slow cookers cook foods over long periods at temperatures well below boiling. All of these provide conditions ideal for *C. perfringens.* Prevention is based on the always sensible practice of keeping cold foods cold and hot foods hot. No moist food should ever be allowed to sit for a long period at temperatures above 4°C or below 100°C.

Certain variants of *Bacillus cereus* may also be enterotoxin producers.

Intoxications with Systemic Effects

Certain ingested bacterial toxins affect the nervous or circulatory systems, and thus cause very general and potentially life-threatening disruptions. The most important of these is botulism.

Causative agent   *Clostridium botulinum* is an anaerobic, Gram-positive, spore-forming rod widely distributed in soil. The spores are a natural contaminant of many foodstuffs. Although spores are frequently ingested, on germination they do not survive in the human gut. With the exception of infants under one year, human beings cannot apparently develop gastrointestinal disease by ingesting spores or cells of *C. botulinum.* Infant botulism is discussed in Chapter 21.

Infant botulism; Chapter 21, p. 647

The vegetative cells survive only in highly anaerobic environments. Growth is inhibited by low pH, but not by mild salting. In the course of growth, the bacilli excrete toxin. The preformed toxin is activated, not destroyed, by stomach acid. **Botulin,** the active product, blocks transmission of nerve impulses by inhibiting the release of the neurotransmitter, acetylcholine.

Pathogenesis   Preformed botulin is absorbed from the intestine, and enters nerve fibers. The first neurological functions to show the effect are vision and balance. Double vision, dizziness, lack of coordination, nausea, and vomiting are followed by inability to swallow, a flaccid paralysis of voluntary muscles, and respiratory failure.

Therapy and immunity   The primary treatment is immediate immunotherapy. A trivalent antitoxin containing antibodies to Types A,B, and E toxins is administered. Supportive measures to assist respiration are usually essential. Recovery is slow, and appears to depend on removal or neutralization of the toxin within the relatively inaccessible nerve tissue. Vaccination, although possible, is not practiced because the actual risk is very low.

Epidemiology   Canned foods are the usual sources of toxin because the interior of a sealed can provides superior anaerobic growth conditions. Commercial canning processes are designed to assure that the interior of each processed can reaches temperatures sufficient to kill the spores of *C. botulinum.* Inspection and quality-control checks are constant, and failures are extremely rare. Safe canning techniques are essentially similar to autoclaving, using both elevated heat and pressure. Unfortunately, this degree of care is not equaled by all home canners. Sodium nitrite and nitrate are added to commercially prepared meats. These chemicals raise the redox potential, inhibiting clostridial growth.

Processing instructions given in cookbooks and instruction manuals should be followed with meticulous care. Oddly enough, we often hear of grandmothers who have been feeding their families inadequately canned foods (boiled, not pressurized) for many years without incident. At these times, it is clear that some people are just born lucky.

Fewer than 50 cases of foodborne botulism are reported per year in the United States, and the mortality rate is less than 10 percent. Almost all cases are traced to home-canned foods. Recent outbreaks have been traced to home-processed olives, beans, or fish. Uncooked foods such as potato salad, cured or processed meats, and sausage can be sources too.

Toxins of fungal origin   Fungi are important food spoilage agents. Grains, nuts, or beans are relatively stable foods that are usually stored dry. Fungal growth may occur if these foods are incompletely dried or stored in a damp place. In some cases, the food is obviously inedible and the only loss is economic. In others, the fungally contaminated food may remain edible enough to be consumed by hungry people or to be superficially cleaned up and marketed wholesale. Illness and death may follow. *Aspergillus flavis*, growing on peanuts and stored animal feeds, causes abortion and death in sheep and cattle. Its toxin, the **aflatoxin**, is one of the most potent hepatic carcinogens known. Aflatoxin may find its way into the human diet in peanut butter and perhaps on homegrown or poorly stored peanuts and grains. Its significance as a human poison remains unclear.

Ergotism is a destructive condition. It develops in persons who have consumed grains on which the plant pathogen *Claviceps purpurea* has multiplied, producing ergot, a complex mixture of toxins. Ergotism results in a hemorrhagic breakdown of capillaries, gangrene of the extremities, and death. One of the active compounds in ergot was once used as an abortion-inducing drug.

Algal toxins   The algae rarely find their way into a discussion of medical microbiology. However, certain algal species may poison fish, livestock, and poultry as well as human beings. Certain marine algal blooms, called **red tides**, are composed of the dinoflagellate *Gonyaulax tamarensis* (Fig. 20.17). This algae produces a **paralytic shellfish poison** (PSP). As clams, mussels, and oysters feed, their digestive organs become packed with algae. People then eat the shellfish and, if a sufficient dose of toxin is ingested, symptoms similar to botulism develop. PSP or *saxitoxin* is a neurotoxin, somewhat less potent than botulin but the most potent algal toxin known. Monitoring programs along coastlines where red tides occur warn the public during periods when harvesting shellfish is unsafe.

Dinoflagellates; Chapter 5, pp. 134–135

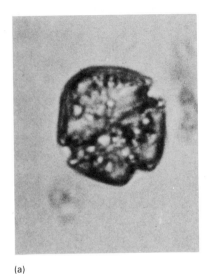

(a)

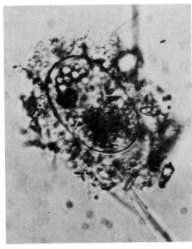

(b)

20.17
Gonyaulax tamarensis, a dinoflagellate   One of the few algae
with pathogenic effects, this organism synthesizes saxitoxin, a neurotoxin. When the algae are ingested
by human beings via shellfish, PSP
(paralytic shellfish poisoning) can
occur. This may be a threat during
algal blooms off the New England,
Gulf, and Pacific coasts. (a) The
motile form, seen in warm waters
and during the summer; (b) the resting cyst, seen in cold winter waters.

SUMMARY

1. The gastrointestinal tract is readily disturbed by microbial activity. This may
be endogenous (overgrowth of the normal flora) or exogenous (the introduction
of a pathogenic organism or its toxic products).

2. Infectious disease may disturb normal motility, produce excessive fluid loss,
lead to ulcerative attack on the intestinal mucosa, or progress to extraintestinal
invasion. Short-term, nonspecific gastroenteritis may be caused by bacterial parasitism, bacterial toxins in food, or viruses. On the basis of symptoms alone it is
usually not easy to distinguish among these causes.

3. Typhoid fever, bacillary dysentery, and cholera are the most serious bacterial infections. Typhoid and cholera have been controlled by the installation of
modern water- and sewage-treatment systems. The diseases are now prevalent
only in underdeveloped areas.

4. Salmonellosis and shigellosis continue to be major and increasing problems
in advanced nations.

5. Hepatitis is caused by at least two and probably four or more different
viruses. None has been successfully cultured *in vitro*, so our knowledge about
them remains inadequate.

6. Food toxicities are due to ingestion of performed bacterial, fungal, and algal
toxins. Staphylococcal food poisoning is common but mild. Other forms are less
common.

7. The careful regulation of the food producing and food processing industries
attempts to minimize foodborne outbreaks and toxicities traceable to commercially provided foods. Home cooks and canners need to also pay more attention
to preventive measures.

Study Topics

1. In what ways are microorganisms excluded from the body by being kept in the GI tract? What barriers must be passed in order for an organism to become disseminated?

2. Give a number of reasons why the lower bowel supports a different and larger flora than the small intestine.

3. There are two quite different bacterial products that have, unfortunately, been given the same name, enterotoxin. Distinguish between them on the basis of producer organism, target tissue, and result.

4. Many of the infectious agents discussed in this chapter have only recently been identified. List some of them. Why, do you suppose, have these agents, some of which are extremely common, escaped detection so long?

5. As you look back over the chapter, you will note that there is an almost complete lack of effective vaccination against these diseases. Speculate on some possible causes.

6. If you experienced a short, acute bout of nausea, vomiting, and diarrhea, from which you recovered after one or two days, name three possible microbial causes.

Bibliography

Collins, M., 1978, Algal toxins, *Microbiological Revs.* **42**: 725–746.

Craig, J. P., 1972, The enteropathic enteropathies. In *Microbial Pathogenicity in Man and Animals.* Society for General Microbiology, Cambridge: Cambridge University Press, pp. 129–155.

Hoeprich, Paul D. (ed.), 1977, *Infectious Diseases* (2nd ed.), New York: Harper & Row.

Mosley, James W., et al., 1977, Multiple hepatitis viruses in multiple attacks of acute viral hepatitis, *New England Journal of Medicine* **296**: 75–78.

Richards, Karen L., and Steven D. Douglas, 1978. Pathophysiological effects of *Vibrio cholerae* and enteropathogenic *Escherichia coli* and their exotoxins on eucaryotic cells. *Microbiological Revs.* **42**: 592–613.

Savage, Dwayne C., 1977, Microbial ecology of the gastrointestinal tract. *Ann. Rev. Microbiol.* **31**: 107–133.

Zuckerman, Arie J., and Colin R. Howard, 1979, *Hepatitis Viruses of Man*, New York, Academic Press.

21

Urinary and Reproductive Infections

In human beings, the urinary and reproductive systems are anatomically closely associated. Both contain a normal flora in their outer sections. Opening to the outside, both may become infected by ascending microorganisms. The sexually transmitted diseases (STDs) are increasingly common conditions of sexually active adults. We will also examine the infections of mother and child both before and immediately after birth.

ANATOMY AND PHYSIOLOGY OF THE URINARY TRACT

Anatomy of the Urinary Tract

The kidneys   The kidneys are paired organs about 10–12 cm long, located behind the peritoneum. They are composed of an external **capsule,** under which lies the **cortex.** Beneath the cortical layer is the **medulla** and central to the kidney is the collecting network of **calices** and the **renal pelvis.** Collected kidney filtrate is pooled in the kidney pelvis, then passed down the **ureters** to the **urinary bladder** (Fig. 21.1).

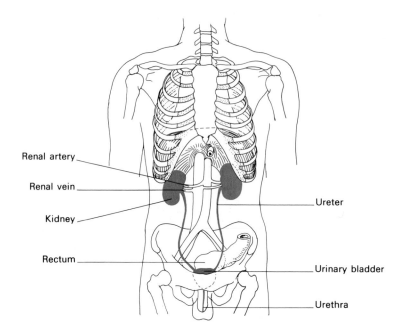

Renal artery

Renal vein

Kidney

Rectum

Ureter

Urinary bladder

Urethra

21.1
The Urinary Tract Urine is produced in the kidney, flows down the ureters, and is collected in the bladder from which it is voided via the urethra. Ascending infections readily colonize the bladder, especially in females.

Within the kidney tissue are packed many individual filtering units or **nephrons,** composed of a **corpuscle** and a **tubular** network. Most of the corpuscular portions of the nephrons lie in the kidney cortex and most of the tubular portions lie in the medulla, although there is some overlap (Fig. 21.2). The corpuscle is the actual filtering unit, containing a **glomerulus** (a tuft of capillaries) and a **Bowman's capsule,** (a tiny collecting vessel). Strong hydrostatic pressures force fluid filtrate through the membrane barriers from the circulation into the receiving vessel. As the filtrate passes along the tubular network, some water and electrolytes are reabsorbed. Kidney diseases such as **glomerulonephritis,** that involve the cortical area, impair filtration. Infections of the medulla and collecting network—**pyelonephritis**—interfere with reabsorption and drainage of urine from the kidney.

The ureters The ureters are slender tubules about 26–28 cm long. Their walls contain smooth muscles that produce a type of peristalsis. This action moves urine along even when you are lying down, and also helps to oppose backflow or **refluxing** of urine from the bladder. The ureters enter the **bladder** from the rear through slitlike openings.

The bladder The bladder is a highly elastic receptacle surrounded by a strong, muscular coat. The mucosal lining is deeply folded when the organ is empty, but becomes thin when it is full. Urine is voided via the **urethra,** controlled by an internal sphincter at the base of the bladder.

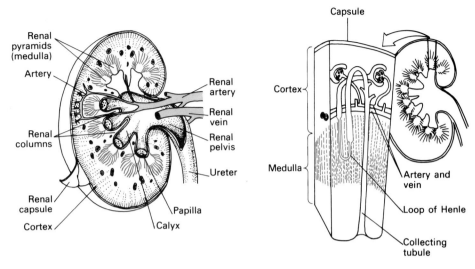

21.2
The Kidney   In the kidney the circulatory system and the excretory system interact. The cortex contains packed glomeruli, which are tufts of capillaries. Fluid bearing soluble wastes is forced out through these into the Bowman's capsules. The tubular loops of the nephron are also surrounded by capillaries. As urine proceeds down through the medulla, much of the water is reabsorbed, concentrating the urine. Urine collects in the pelvis of the kidney before passing down the ureters.

The urethra   In females the urethra is short, from 2.5 to 2.8 cm, and fused to the anterior wall of the vagina. It exits between the clitoris and the vaginal opening, and is closed by the external sphincter. The urethral opening is close to both anal and vaginal openings. There is a high probability of transfer of flora. In the male, the urethra is much longer, and serves both excretory and genital functions. Its opening at the end of the penis is distant from the anal opening, so incidental contamination with GI flora is unlikely.

Physiology of the Urinary Tract

The urine   The physiology of urine production is extremely complex. In an average day, about 150 liters of filtrate are produced; all but 1.5 L is reabsorbed. The resulting urine contains about 60 g of solids, including wastes that serve as adequate nutrients for many species of bacteria. Urine, therefore, which is sterile when secreted in the normal individual, is a potential culture medium both in the bladder and after voiding.

Both concentration and pH of urine are highly variable because the kidney performs homeostatic adjustments to correct for daily water, salt, and protein intakes.

TABLE 21.1
Abnormal substances in urine

SUBSTANCE	ABNORMALITY SUGGESTED BY ITS PRESENCE
Pus (white blood cells)	Infection of urinary tract—bladder or kidneys
Blood (red and white blood cells)	Bladder infection with inflammatory attack on mucosa, nephritis
Albumin	Glomerulonephritis or nephrosis
Ketones	Diabetes
Sugar	Diabetes
Indol and skatol	Bacterial putrefaction in large intestine, constipation
Bacteria	Cystitis, pyelonephritis

Acid urine is somewhat more bacteriostatic than neutral or basic urine. The mechanical flushing action of urine is a valuable anticolonization force.

Certain substances are abnormal in urine, and their presence indicates ongoing disease (Table 21.1).

Defenses of the urinary system   The bladder mucosa produces secretory IgA, some IgG, and possibly IgM. Lysozyme is a component of normal urine; its secretion increases during infection. The cells of the bladder mucosa are phagocytic.

Immunoglobulins are also produced in the kidney. They may be involved in an immune complex hypersensitivity that damages glomeruli. In the medulla, high concentrations of solute and ammonia partially inactivate complement, so immune complex formation does not occur. However, bacterial killing by the classic complement pathway is also rendered less efficient.

Immune complex disease; Chapter 14, pp. 416–417

ANATOMY AND PHYSIOLOGY OF THE REPRODUCTIVE TRACTS

Anatomy of the Reproductive System

The male   The reproductive organs in the male are the **testes**. These produce **sperm,** collected in the **epidydimis** and conducted via ducts across the exterior bladder surface (Fig. 21.3). The sperm is mixed with **seminal fluid,** the secretion of the **seminal vesicles, prostate,** and **Cowper's glands.** Semen enters the urethra within the tissue of the prostate, then is delivered to the outside through the **penis.** The urethra, the prostate and, less commonly, higher structures may become infected.

The female   The reproductive organs in the female are the **ovaries**. These produce **ova** (eggs) during the fertile years. The ova pass out of the ovary into the peritoneal

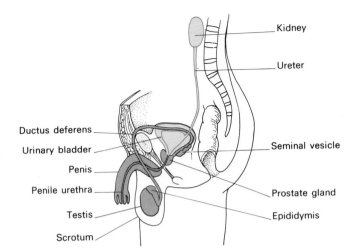

21.3
The Male Reproductive System   Semen is produced when sperm from the testes is mixed with the secretions of the accessory glands. The urethra serves as a common duct for both semen and urine. Because of the length of the urethra, urinary tract infection is uncommon in males. Genital infections are usually limited to the urethra and prostate gland.

Labels: Kidney, Ureter, Ductus deferens, Urinary bladder, Penis, Penile urethra, Testis, Scrotum, Seminal vesicle, Prostate gland, Epididymis

cavity, then are swept up by the finger-like projections at the end of the **fallopian tubes** and carried down to the **uterus** (Fig. 21.4). Because the ends of the fallopian tubes are open to the peritoneal cavity, ascending infection may extend itself from the reproductive tract into other abdominal regions.

The **uterus** is a thick-walled, muscular organ specialized for support of a fetus. Its lining, the **endometrium,** varies in thickness with the phases of the menstrual cycle. When conception occurs, the endometrium receives the ovum, which invades the top layers. These layers become modified as the maternal half of the **placenta.**

The **cervix** is the muscular cuff at the opening of the uterus. Its surface epithelium is readily colonized by certain pathogens; the **Papaniculaou** smear, usually performed to detect cancerous cells, will also show tissue changes due to the presence of viruses or chlamydia.

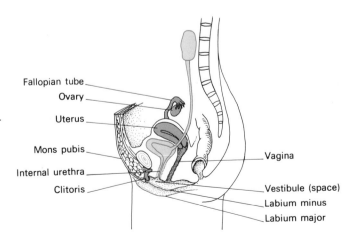

21.4
The Female Reproductive System   In females the reproductive tracts are completely separate from, but adjacent to, the urinary tract. The vagina has a large normal flora that may ascend to infect the uterus and ultimately migrate to the peritoneal cavity via the open ends of the fallopian tubes.

Labels: Fallopian tube, Ovary, Uterus, Mons pubis, Internal urethra, Clitoris, Vagina, Vestibule (space), Labium minus, Labium major

The **vagina** is the elastic, ribbed tube that connects the uterus to the exterior. Its opening is immediately adjacent to the anal opening. The vagina is also subjected to minor mechanical trauma during coitus, and infections may be readily established. Because of the lack of pain-sensitive nerve endings within the vagina, inflammation may not be perceived and infection may be undetected.

The **vulva** comprises the external genitalia, including the **labia, clitoris,** and adjacent tissues. The vulva is often infected simultaneously with the vagina, especially in the prepubertal female.

Physiology of the Reproductive Tracts

The male   Portions shared with the urinary tract are partially protected by urine flow. However, this defense does not extend to the accessory glands. Although infections do not readily reach the prostate gland or seminal vesicles, once there they are difficult to eradicate. Chronic prostatic infection increases in incidence with age. The resulting swelling and inflammation can exert pressure on the urethra and make urination difficult or impossible.

The female   Vaginal secretions vary in chemical composition with age. At puberty, as sexual activity with its increased potential for infection begins, the vaginal secretions become more protectively acidic. Frequent epithelial replacement places an upper limit on colonization by normal flora. The uterine macrophages deal with organisms that pass through the cervical opening. Immunoglobulins and lysozyme are secreted.

The menstrual cycle periodically changes the metabolic activity and growth rate of the surface tissues. These normal changes, or the use of female hormones for birth control or menopausal problems, may increase the susceptibility of the vagina to infection.

Certain personal habits, such as wearing very tight underwear, and overuse of perfumed soaps, douches, and vaginal deodorants, tend to cause local irritation of the vulva and vagina. Moderate bathing habits preserve healthy tissue surfaces better than excessive pursuit of cleanliness.

NORMAL FLORA

The external urethra possesses a normal flora, which dwindles rapidly beyond the first centimeter. The vagina, however, possesses a large microflora throughout.

Flora of the Urethra

Many organisms may be cultured from the external opening of the urethra. These include organisms derived from adjacent skin flora and normal fecal flora. *Staphylococcus epidermidis* is usually most prominent. Both aerobes and anaerobes are

TABLE 21.2
Normal flora of urogenital tracts

GROUP	ANTERIOR URETHRA, BOTH SEXES	VAGINA OF ADULT	PERCENTAGE OF INDIVIDUALS
Bacteria	*Staphylococcus epidermidis*	*Staphylococcus aureus*	(5–15)
	Anaerobic micrococci	α-hemolytic streptococci	(47–50)
	Group D streptococci Nonpathogenic *Neisseria*	Enterococci or Group D streptococci	(30–90)
	Lactobacillus sp.	*Streptococcus pyogenes*	(5–20)
		Anaerobic streptococci	(30–60)
	Aerobic diphtheroids	Nonpathogenic *Neisseria*	
	Escherichia coli	Lactobacillus	(50–75)
	Klebsiella sp.	Aerobic diphtheroids	(45–75)
	Proteus sp.	*Clostridium* sp.	(15–30)
	Pseudomonas sp.	*Actinomyces* sp.	(25–75)
		Enterobactereaceae	(15–40)
		Hemophilus sp.	
Fungi	Yeasts (irregularly)	*Candida albicans*	(30–50)
Protozoa	*Trichomonas vaginalis*	*Trichomonas vaginalis*	(10–25)

Source: Adapted from Paul D. Hoeprich (ed.), 1977, *Infectious Diseases*, New York: Harper & Row.

Species found in clinically normal individuals. Figures in parentheses are percentages of positive samples in different surveys, given where available.

represented. It is believed that the flushing and bacteriostatic action of urine is the major barrier to ascending microbial activity. The role of flagellar action in upward migration is unclear; however, the great majority of bacteria that reach the bladder and establish themselves are motile.

There is no indication that the urethral normal flora protects by competitive pressure or production of inhibitory substances.

Microorganisms are not normally found in the upper urethra, bladder, ureters, or kidneys.

Flora of the Vagina

The vaginal flora is a mixture of lactobacilli, anaerobic *Bacteroides* and *Clostridium*, aerobic diphtheroids, staphylococci, and yeasts. The composition of the flora changes with hormonally induced changes in the physiology of the vaginal mucosa. Bacterial metabolism in turn conditions the pH of the vaginal secretions. The flora of the prepubertal female is not strongly acidogenic, as lactobacilli are a minor component. With puberty, the lactobacilli become dominant. They ferment sugars supplied by the vaginal secretions, lowering the pH to 3.5–4.5. This low pH inhibits growth of many invaders, including the enteric flora.

The normal flora does not cause problems in the absence of trauma or physiological imbalance. Antibiotics may upset the flora and change microbial distribution, since the lactobacilli are highly susceptible to most antimicrobial agents. Overgrowth with *Candida albicans* may result. *Streptococcus, Clostridium,* and *Bacteroides* introduced into the uterus during abortion or childbirth can cause a fatal infection. It is interesting that normal hormonal changes in pregnancy suppress growth of these genera, lessening the risk somewhat.

TABLE 21.3
Clinical terminology of urogenital infections

TERM	DEFINITION
Cervicitis	Inflammation of the mucous membrane or deeper structures of the cervix
Cystitis	Inflammation of the mucous membrane of the urinary bladder
Dysuria	Difficulty or pain in urination
Endometritis	Inflammation of the uterine lining
Glomerulonephritis	Renal disease with inflammatory changes in glomeruli, not the result of kidney infection
Nephritis	Inflammation of the kidney; many forms and causes
Nephrosis	Renal failure, usually with destruction of tubular epithelium; many possible causes
Oophoritis	Inflammation of the ovary
Orchitis	Inflammation of the testes
Pelvic inflammatory disease	Acute or chronic inflammation in the pelvic cavity, especially salpingitis
Prostatitis	Acute or chronic inflammation of the prostate gland
Pyelonephritis	Inflammation of the renal pelvis due to local bacterial infection
Salpingitis	Inflammation of the fallopian tubes
Urethritis	Inflammation of the urethral mucosa
Vaginitis	Inflammation of the vaginal mucosa
Vulvovaginitis	Inflammation of both vulva and vagina, or of the vulvovaginal glands

URINARY TRACT INFECTION

Primary infections of the urinary system are almost always bacterial. Viral or fungal infections may spread to the urinary system secondarily from foci on other areas.

Cystitis

Bacteriuria   **Cystitis,** or bladder infection, is a very common condition. Almost every woman has at least one episode of clinical disease during her lifetime. The condition is marked by **bacteriuria**—bacteria in the urine. Accurate laboratory diagnosis requires quantitative determination of the number of bacteria present. A clean urine sample—a "clean catch"—should be obtained. The patient first cleanses the external urogenital area with soap or a disinfectant solution and rinses carefully. Then he or she partially empties the bladder, then collects a **midstream sample** in a sterile container.

If the urine cannot be cultured by an appropriate method within one hour, it should be refrigerated at 4°C until laboratory services are available. If not, organisms present will multiply, giving rise to misleading results. To get around this problem, some urine collection vessels are now supplied containing a stabilizing agent. This chemical inhibits bacterial multiplication during the period between collection and laboratory analysis.

Following plating and incubation, counts of less than 10,000 bacteria/mL are considered negative, since this many organisms could be introduced into sterile urine simply by passage through the urethra. Counts between 10,000 and 100,000/mL are suspicious, and repeat cultures may be ordered. Counts greater than 100,000/mL clearly indicate infection in a properly handled sample.

TABLE 21.4
Agents causing cystitis in different groups of patients

COMMUNITY-ACQUIRED		HOSPITAL-ACQUIRED		SPINAL CORD INJURED	
ORGANISM	PERCENT	ORGANISM	PERCENT	ORGANISM	PERCENT
Escherichia coli	57	*Escherichia coli*	39	*Proteus rettgeri*	30
Klebsiella pneumoniae	13	*Proteus mirabilis*	18	*Pseudomonas aeruginosa*	23
Proteus mirabilis	8	*Pseudomonas aeruginosa*	14	*Proteus mirabilis*	13
Enterococcus	8	*Enterococcus*	11	*Escherichia coli*	8
Pseudomonas aeruginosa	7	*Klebsiella pneumoniae*	10	*Proteus vulgaris*	6
Proteus rettgeri	5	*Serratia marcescens*	8	*Klebsiella pneumoniae*	6
Other	2			Other	14

Source: G.P. Youmans et al., 1975, *Biological and Clinical Basis of Infectious Diseases.*

Infection in contrast to disease   Surveys of healthy persons show that bladder infection is quite common. About 1–2 percent of school-age girls, 5–6 percent of young, pregnant women, and up to 10 percent of elderly women have bacteriuria. Affected females outnumber males by about 30:1 in each age group, except for newborns and for elderly men with prostate obstruction. It is a good idea to review again the distinction between **infection** and **disease** in this context. Although many are infected (and apparently for prolonged periods), episodes of disease are rare and sporadic.

Clinical disease occurs when there is invasion and inflammation of the bladder wall and urethra. There is pain on urination; because the bladder loses elasticity, the patient needs to urinate frequently. Pus and blood may be seen in the urine. Left untreated, cystitis usually resolves spontaneously. Antimicrobials are usually given because they relieve the symptoms and reduce the risk of ascending kidney disease. Ampicillin and the sulfa drugs are common choices.

Predisposing factors   Overt cystitis in women often follows the pressure and trauma of sexual intercourse and pregnancy. Local tumors, **calculi** (stones) in the bladder, prolonged bed rest, spinal cord injuries, diabetes, and insertion of catheters or instruments are predisposing in both sexes.

The list of causative agents is large. Different species predominate according to whether the infection was community-acquired (by a normal person in the course of daily life), hospital-acquired, or associated with spinal cord injury.

In the normal person, the agent is almost always derived from the intestinal flora. About 90 percent of cases are due to *Escherichia coli* or other related enterobacteria such as *Proteus mirabilis*. *Proteus* has an active urease, and breaks down urea to ammonia and CO_2. These products make the urine pH more alkaline, decreasing the solubility of calcium salts in the urine and possibly inducing stone formation. *Streptococcus faecalis* causes most of the remaining 10 percent of cases.

In the hospitalized patient a larger variety of agents may be found, perhaps related to GI-tract colonization by hospital strains of bacteria. Like all hospital-acquired pathogens, antimicrobial resistance factors in these agents are common. *Serratia marcescens* and *Klebsiella pneumoniae* may be especially troublesome.

Pyelonephritis

Origin   Bacteria may reach the pelvis and collecting tubules of the kidney by either the ascending or the hematogenous route. Ascending infections are more common. At least 50 percent of women with acute clinical cystitis also have infection of one or both kidneys, often unsuspected. Ascent may be promoted by endotoxin released by the Gram-negative organisms in the urine. Endotoxin acts on smooth muscle and weakens the slit-like closures of the ureters allowing urine to flow back; it may also reduce ureter peristalsis. Bacterial pili allow them to adhere firmly to the ureter epithelium and prevent them from being flushed out.

Endotoxin effects; Chapter 12, pp. 355–356

TABLE 21.5
Most common agents of pelvic inflammatory disease

TYPE	NAME
Anaerobes	*Bacteroides fragilis*
	Peptostreptococcus
	Other *Bacteroides* sp.
Aerobes	α-hemolytic streptococci
	Diphtheroids
	Neisseria gonorrhoeae
	Staphylococcus epidermidis

Each group ordered by approximate decreasing frequency.

Staphylococci, Group A streptococci, *Mycobacterium tuberculosis,* and systemic fungi may infect the kidney, brought by the blood from distant foci.

Manifestations Acute pyelonephritis features high fever, chills, severe pain, and spasm of the small intestine. Prompt treatment with an appropriate antimicrobial drug can usually bring the pathogen under control so that residual kidney damage will be minimal. Chronic pyelonephritis, in which the symptoms are not pronounced, may not receive early attention. Obstruction and fibrosis of kidney ducts develops; eventually the kidney becomes nonfunctional and shrinks.

Other Kidney Conditions

Glomerulonephritis This is predominantly, an immunological disease. Immune complexes adhere to glomerular membranes, with subsequent complement-mediated cell damage. Glomerulonephritis may follow skin infections with *Streptococcus pyogenes.* The condition is also an aspect of autoimmune disease such as systemic lupus erythematosis.

S. pyogenes; Chapter 17, pp. 503–506

Microorganisms brought to the glomeruli by the blood rarely infect, due to the efficient local phagocytosis.

Perinephric abscesses Bloodborne bacteria can establish infection under the renal capsule. *Staphylococcus aureus* abscesses may be seen in disseminated staphylococcosis in the diabetic.

21.5
Leptospira interrogans   This species of spirochetes can cause leptospirosis in dogs, cattle, swine, rodents, and human beings. The agent is transmitted via urine or direct contact with infected animals or carcasses. There are many different antigenic variants causing different types of disease.

Spirochetes; Chapter 4, pp. 116–117

Leptospirosis   Spirochetes of the genus *Leptospira* (Fig. 21.5) cause a variety of infections in domestic animals, rodents, and human beings. These affect but are not limited to the urinary tract. Infectious urine is a primary means of spreading leptospirosis.

GENITAL TRACT INFECTION

Tables 12.2 and 12.3

Infections of the genital tract may be caused by many groups of microorganisms (Table 21.6). Some are transmitted predominantly by sexual intercourse, and these are called **sexually transmitted diseases** (STDs), a term that is gradually replacing the term **venereal** disease. However, there is no absolute distinction between STDs and non-STDs, because most genital diseases can be transmitted in several ways. Many conditions formerly thought to be transmitted by nonsexual means are now known to be commonly spread by sex as well. In consequence, the list of STDs has grown rapidly. It is perhaps correct to assume that any genital infection, and most GI infections, **may** be spread sexually.

Box 21.1

Current Trends in STDs

Several recent national surveys have attempted to establish the true incidence of the various STDs and determine changes. Traditionally, five historically defined venereal diseases—syphilis, gonorrhea, chancroid, granuloma inguinale, and lymphogranuloma venereum—have been considered. However, other STDs, more recently recognized, have been shown to be more prevalent in both men and women. The accompanying table shows incidence in a recent six-month study involving six STD clinics. The results do not necessarily reflect the incidence in the whole population.

Gonorrhea is still the most common single STD, constituting about 37 percent of STD diagnosis in men and about 33 percent in women. A large culture testing program evaluated close to nine million cultures obtained at random from women undergoing physical examinations for a variety of reasons. Of these women, a total of 4.7 were culture positive; the majority were not seeking care for venereal disease and were unaware of their condition. Women with inapparent infection such as these serve as the reservoir of the agent.

After ten years of rapid increase from 1965–1975, gonorrhea incidence has remained about the same over the period 1975–1979. This is a great improvement, of course, but the public health officials would like very much to see the incidence begin to decline. It seems unlikely that this will be accomplished until some form of vaccination becomes available.

Nongonococcal urethritis is now marginally the most prevalent STD in men, while nonspecific vaginitis in women is a close second to gonorrhea. Although these figures of

	CASES PER 100 VISITS TO STD CLINICS	
DIAGNOSIS	MEN	WOMEN
Gonorrhea	24.0	15.7
Nongonococcal urethritis	24.8	—
Genital herpes	3.4	1.5
Venereal warts	4.3	3.0
Syphilis	1.7	1.4
Trichomonal vaginitis	—	10.4
*Candida vaginitis	—	6.1
Nonspecific vaginitis	—	11.3
Scabies	1.3	0.4
Pediculosis pubis	2.9	1.6
Molluscum contagiosum	1.0	0.3
Chancroid	0.1	0.1
Lymphogranuloma venereum	<0.1	0.0
Granuloma inguinale	0.0	0.0
Total positive diagnoses	63.6	51.7

*Transmitted primarily by nonsexual means.

persons seeking medical attention for known or suspected STDs do not show it, genital herpes (often inapparent) is a common and rapidly increasing STD problem.

It is now being suggested that an expanded number of STDs be sought with an expanded spectrum of testing services, and reported. In reality, the old standby STDs, gonorrhea and syphilis, are only the tip of the iceberg.

Source: *Morbidity and Mortality Weekly Reports* **29**, 1979.

TABLE 21.6
Epidemiology of sexually transmitted diseases

DISEASE	RESERVOIR	TRANSMISSION	INCUBATION PERIOD IN DAYS
Chancroid	Human beings	Direct sexual contact, other direct contacts	1–14
Gonorrhea	Human beings	Direct sexual contact or contact with fresh infectious exudates	2–5
Granuloma inguinale	Human beings	Direct sexual contact	8–80(?)
Lymphogranuloma venereum	Human beings	Direct sexual contact, indirect contact with soiled clothing or bed linen	5–30
Pediculosis (pubic lice)	Human beings	Intimate contact, usually sexual	7–14
Syphilis	Human beings	Direct sexual contact, contact with contaminated blood, transplacental	21
Nongonococcal urethritis and vulvovaginitis	Human beings, other animals(?)	Direct sexual contact (others?)	5–7
Herpes	Human beings	Direct sexual contact	Variable

Source: Adapted from A. S. Benenson, 1975, *Control of Communicable Diseases in Man*, (12th ed.).

PERIOD WHEN COMMUNICABLE	PATIENT CONTROL MEASURES	EPIDEMIC CONTROL MEASURES
As long as agent is present in lesion or discharging lymph nodes, usually weeks	Avoid sexual contact until all lesions healed	Search for contacts of two weeks before, after onset
Months or years, especially in untreated female. Ends within hours of specific therapy	Avoid sexual contact until noncommunicable. Disinfect soiled articles	Search for contacts of ten days before onset, treat
Unknown, probably for duration of open lesions	Avoid close personal contact until lesions heal	Search for contacts, especially homosexual
Variable, while active lesions are present (weeks to years)	Avoid sexual contact until lesions heal, dispose of soiled articles carefully	Search for contacts
As long as lice are present	Apply effective insecticides, disinfect clothing, bedding, and other vehicles	All household and direct personal contacts should be examined and treated, if necessary
Variable; during stages 1 and 2 and in relapses; possibly in latency. Becomes noncommunicable 24 hr after specific therapy	Avoid sexual contact until noncommunicable; disinfect articles soiled with exudate	Search for contacts of three months prior to appearance of first stage, six mos. prior to second stage, one year prior to latent stage; treat
Unknown	Appropriate drug therapy if agent identified; avoid sexual contact until symptoms clear	Concurrent treatment of spouses, sexual contacts; active contact search not required
As long as lesions are present; virus may be excreted even when visible lesions appear healed	Avoid sexual contact; disinfect soiled articles	None

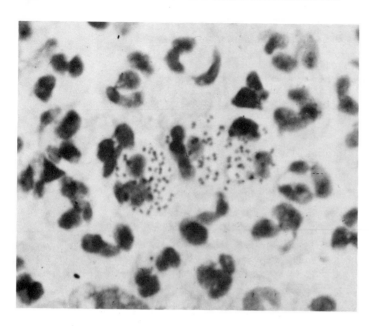

21.6
Intracellular *Neisseria gonorrhoeae* The exudate from the urethra of a male gonorrhea patient contains many neutrophils, indicating an inflammatory response. These PMNs ingest the Gram-negative diplococci. A Gram-stain of the exudate shows this diagnostic sign.

Gonorrhea

Causative agent Gonorrhea, the most common STD worldwide, is caused by the Gram-negative aerobic diplococcus *Neisseria gonorrhoeae* (Fig. 21.6). This species is characterized by its ability to grow only in an atmosphere containing elevated CO_2. Biochemical tests of importance are the oxidase test, which is positive for all *Neisserias*, and inability to ferment lactose, sucrose, and maltose. The organism is usually cultured on chocolate agar, which contains hemolyzed blood. Thayer–Martin medium containing antibiotics is usually used. One use for this medium is in **transport** cultures, sealable bottles or plates designed to be inoculated in the doctor's office or clinic and shipped to a laboratory. These containers provide a gas mixture rich in CO_2. Since *N. gonorrhoeae* survives poorly outside the body, especially on a dry swab, appropriate collection measures are essential for its recovery. Individual colonies are spot-tested for oxidase positivity. Species identity can be confirmed with fluorescent antibody methods and sugar fermentations.

Pili; Chapter 3, p. 91

The gonococci bear many pili on their cell envelopes. Pili are a key to pathogenesis because they facilitate attachment to the mucosal surface. Gonococci also possess plasmids, not thought to be involved in antimicrobial resistance.

N. gonorrhoeae strains have become progressively more resistant to penicillin since its introduction. At first, the resistance was relative, due to undetermined mechanisms. All gonococcus strains continued to respond to higher doses. In 1976, new strains appeared that produced a penicillinase enzyme and were absolutely resistant. These have spread worldwide, but not at as rapid a rate as was first feared.

Pathogenesis   In the male, gonococci invade the submucosal tissues of the urethra, and less commonly the rectum and pharynx. After an incubation period of 2–7 days, pain and burning on urination appear, along with a purulent discharge. Therapy is promptly sought by most individuals, but some either are asymptomatic or choose to ignore the discomfort. In untreated gonorrhea, the symptoms will subside after about eight weeks, but chronic epidydimitis or urethral stricture are common sequels. There is 1–2 percent chance of developing disseminated gonorrhea.

In the female, the gonococcus invades the cervical lining, and less commonly the urethra, rectum, pharynx, and associated glands. Cervical gonorrhea is largely asymtomatic, the only clue being the appearance of a purulent discharge. In both male and female, examination of a Gram-stained smear of the discharge will show the typical Gram-negative diplococci within PMNs. This finding is diagnostic. In untreated gonorrhea in the female, about 20 percent of cases will progress to **pelvic inflammatory disease** (PID) with involvement of the fallopian tubes and ovaries. In some cases it will cause peritonitis. This risk is multiplied from 4–7 times if an intrauterine device (IUD) is in place. Chronic PID and sterility are the major risks of untreated gonorrhea in women. About 3 percent of women experience bacteremia and disseminated infection.

Disseminated gonorrhea is marked by fever, arthritic pain in many joints, and skin lesions. Myocarditis, hepatitis, and meningitis are less common.

In children, infection of the eye during birth (neonatal gonorrhea) was once a major cause of infant blindness. It is now prevented by the obligatory use of silver nitrate or penicillin eye drops at birth. Older children may develop gonorrhea as a result of sexual molestation.

Therapy and immunity   Injected penicillin has long been used as the drug of choice, in combination with probenecid, a substance that prolongs the penicillin's action. The increasing resistance of the organism has dictated escalation in the dose. Where penicillinase-positive strains are likely, ampicillin-probenicid combinations given by mouth, or spectinomycin, are used. Follow-up cultures are essential to confirm cure. In addition, removal of the gonococcus often (up to 50 percent of cases) does not completely alleviate the symptoms. This is because a secondary STD is also present, calling for a different form of treatment.

Probenicid; Chapter 26, p. 764

Microbial resistance to drugs; Chapter 26, pp. 755–756

There is much debate about the status of the immune response in gonorrhea. However, documented reinfection with an identical strain has been shown, so it seems clear that immunity is not protective. Efforts to develop gonococcal preparations immunogenic enough to have vaccine potential have not yet paid off.

Epidemiology   Spread of gonorrhea by other than sexual means is very rare. It is usually communicated by asymptomatic males or females. Not every contact results in disease, but repeated exposure to infected sex partners will eventually do so. The normal vaginal flora (Fig. 21.7) may play a role in controlling some exposures that do not result in disease.

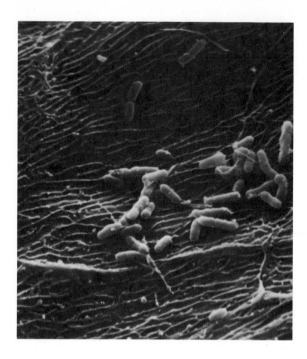

21.7
Normal Human Vaginal Flora A segment of vaginal epithelium is shown with numerous adherent bacilli.

Control hinges on the identification and treatment of exposed **contacts.** This includes all sexual partners of the diagnosed case within ten days prior to onset of symptoms. A person receiving treatment becomes noninfective within 24 hours after the antimicrobial drug is administered. The short incubation period of gonorrhea makes the identification and treatment of sex contacts, before they have the opportunity to spread the disease further, very difficult. Contact tracing is formally the responsibility of public health officials, but it is also a personal responsibility of the infected individual. The etiquette of such obligations has not yet been handled by Emily Post.

On the average, about one million cases of gonorrhea are reported each year in the United States. The actual number of cases is probably significantly higher.

Syphilis

Causative agent   *Treponema pallidum* is a spirochete. It is virtually impossible to visualize with common Gram-stain techniques. It takes the stain poorly and its extreme slenderness places it beneath the resolving power of ordinary light microscopes. Silver stains, which coat the cell with a metallic layer, are useful. Live spirochetes in the fluid from lesions are visible in dark-field because of their undulating movements. *T. pallidum* has never been grown *in vitro*. Therefore, little is known of

Dark-field microscopy;
Chapter 3, pp. 70–71

its metabolism. The organisms are rapidly killed by heat, drying, soap and water, and a large variety of disinfectants and antimicrobials.

Pathogenesis The spirochetes enter the body through minute abrasions on the genitals or any other mucosal surface that touches a lesion borne by the sexual partner. Once in the body, the agents migrate immediately to the lymphatics and bloodstream, from which they infect all tissues. However, the development of symptoms is slow and proceeds in stages, depending on the localization of treponemal growth at each time.

The **primary stage** occurs about three weeks after infection, and appears as a **hard chancre** at the site of spirochete entry. The lesion is typically an open painless ulcer with a hard base and some serous exudate. In the male, the penis is the usual site; but in the female the chancre is most often in the vagina or on the cervix, where it is not readily detected. The exudate from the lesion is swarming with spirochetes and extremely infective. Dark-field examinations of the fluid will show the spirochetes for a positive diagnosis. The chancre heals spontaneously in 3–6 weeks, leaving little scarring.

About 6–8 weeks later, the **secondary stage** develops, in which the widely distributed spirochetes cause various symptoms. This stage may mimic other infectious states. There are usually skin lesions and mucous patches; these are all highly infective. Hair loss, swollen lymph nodes, and such general symptoms as fever, malaise, and sore throat are also seen. This phase also fades away spontaneously.

The **latent** phase follows, and may be months or years in duration. In the first four years, relapses to infective stage 2 may occur. In the later years, the disease does not relapse and is infective only by blood transfer.

Tertiary syphilis will eventually appear in about 35–40 percent of untreated cases who do not die first of other causes. There are three forms, all reflecting destructive changes caused by hypersensitivity. The benign form features large ulcers called **gummas** that may occur in any organ. These are destructive but localized. The cardiovascular form consists of an erosion of the aorta. Neurological forms (tabes dorsalis or paresis) may show psychosis, paralysis, and loss of sensory functions.

Therapy and immunity Syphilis is customarily treated with penicillin, but several other drugs appear to be equally effective. Cure is possible in all stages but tertiary, where the infection can be arrested but the tissue destruction cannot be reversed.

There is no protective immunity and no vaccination for syphilis. The role of the immune system appears to be equivocal; most investigators feel that the tissue destruction of the third stage is due principally to hypersensitivity mechanisms.

Circulating antibody to the syphilis organism appears early in infection. These antibodies also cross-react with a substance called **reagin,** a lipid extracted from heart muscle. A bewildering variety of serological tests for syphilis have been developed. The earliest were **nontreponemal** tests. They used reagin, not the organism, as the test antigen because of its ready availability and low cost. Complement-fixa-

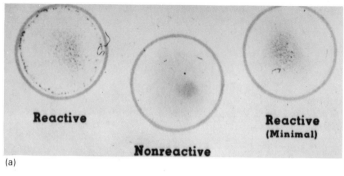

(a)

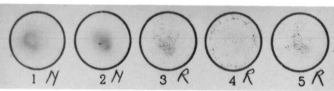

(b)

21.8

A Positive Syphilis Test The control card (a) shows the effect of using standard sera. A positive or reactive serum yields large peripheral antigen–antibody clusters. A nonreactive test remains evenly suspended, and a minimum reactive shows small clumps. The five test sera (b) shows two negatives and three reactives of varying degrees of strength.

tion tests such as the Wasserman are now rarely done. Rapid slide tests such as the VDRL and RPR (Fig. 21.8) are still in general use as screening tests. Their usefulness is limited by the fact that they may react positively in other treponemal infections and give false positives in other diseases. Some may be adapted to quantitative titration.

Treponemal tests are more sensitive and specific, but they are expensive because of the limited supply of *T. pallidum* antigen, which must be obtained from animal tissues. Most effective is the TPI (*Treponema pallidum* immobilization) test, used in Canada. Fluorescent antibody methods (the FTA-Abs test) are more generally used. Treponemal antigen tests are usually reserved for persons whose serum gave a positive result on a rapid slide test, or those with a suspicious clinical picture.

Epidemiology The sole reservoir of *T. pallidum* is human beings. It is spread largely by sexual contact, but may also be transmitted by mother to fetus, or by infective blood. The second and third methods are very rare, and syphilis is as close as any disease gets to being an absolute STD. Any open lesion is infective, and the probability of contracting disease on exposure is about 10 percent. Given the long incubation period, contacts can usually be traced and incubatory cases cured before their primary lesion appears and they become infective. For this reason, case-finding methods are very much more successful in the control of syphilis than in gonorrhea. The active public fear of syphilis helps too. Antibiotic therapy arrests the infection and renders the person noninfective within 24 hours. Sexual contacts of primary syphilitics within the past three months and of secondary syphilitics within the past six months should be traced and receive penicillin therapy.

Nongonococcal Urethritis (NGU)

This is a catchall title for cases of urethritis (male) in which the gonococcus is not present. NGU is a sexually transmitted condition and, by all measures, is more widespread in the United States than either gonorrhea or syphilis, particularly in college students and upper income groups. Several agents have been implicated.

Chlamydia trachomatis appears to cause some 30–50 percent of NGU cases in males. This ubiquitous agent is also responsible for inclusion conjunctivitis, another venereal disease called **lymphogranuloma venereum,** as well as **trachoma,** and **infantile pneumonitis.** Strain variations appear to dictate the direction of pathogenesis. *C. trachomatis* is an obligate intracellular parasite that forms characteristic inclusion bodies in the host cell. It is effectively treated with sulfa or tetracycline. Although most infections do not produce significant complications or aftereffects, some cases in females lead to pelvic inflammatory disease. Cultural and serological methods of diagnosis have been developed. *Mycoplasma* species are frequently associated with urogenital infection. Mycoplasms are cell-adapted, lack a cell wall, and parasitize by means of a firm attachment to the host cell membrane (Fig. 21.9). *Urea-*

Chlamydia; Chapter 4, pp. 125–126

Mycoplasma; Chapter 19, p. 555

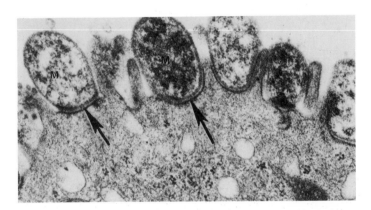

(a)

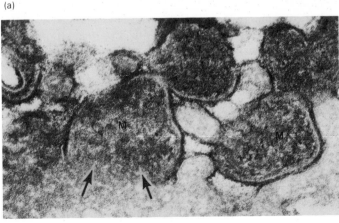

(b)

21.9
Mycoplasma–Host Interaction   (a) *Mycoplasma* (M) lacks a cell wall. When it associates with the host cells, a gradual process of membrane modification (arrows) occurs. (b) In chronic infection, the membrane between host and parasite cell visibly breaks down (arrows).

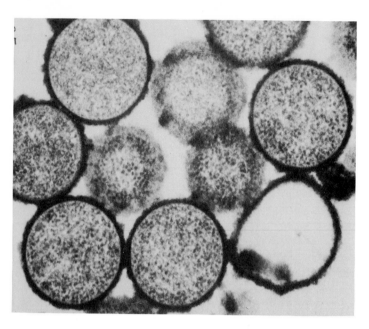

21.10
Ureaplasma urealyticum   This mycoplasma is a major cause of nonspecific sexually transmitted diseases (STDs) in both sexes.

plasma urealyticum, the so-called T-strain mycoplasma (Fig. 21.10), is frequently isolated from the inflamed urethra. This species has an active urease, and this raises the possibility that its irritating effect on the urethral lining may be due to the release of ammonia. It has also been suggested that the elevation of urine pH may contribute to stone formation. *Ureaplasma* may be routinely identified by culture. Both spectinomycin and tetracycline are effective drugs; because tetracycline also controls *Chlamydia,* it is most often used. No agent is recovered in from 10–20 percent of NGU cases.

Vulvovaginitis

Vulvovaginitis and **nonspecific vaginitis** are catchall terms for inflammations of the vulva, vagina, and cervix. These conditions are primarily STDs, but certain of the agents (e.g., *Candida* and *Trichomonas*) are resistant to cold and drying, and are readily transmitted nonsexually by fomites. *Candida* infections may also be endogenous.

Chlamydia trachomatis and *Ureaplasma* are frequently involved. Another significant agent is *Gardnerella vaginalis.* This small, Gram-negative organism, also called *Hemophilus vaginalis,* is found in normal vaginal flora, but also has been associated with vaginitis. It responds to ampicillin therapy.

Candida albicans, a normal vaginal resident, will cause vulvovaginitis if overgrowth occurs. The yeast can readily be seen in stained smears; it is treated with topical nystatin.

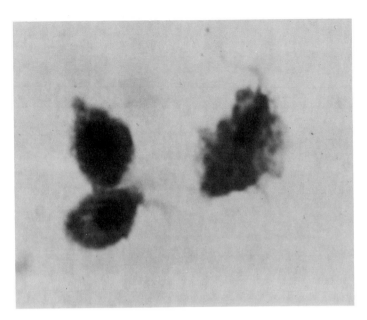

21.11
Trichomonas vaginalis This protozoan, a flagellate, is a very common causative agent of nonspecific vulvovaginitis in women. It may be transmitted both sexually and by indirect means.

Trichomonas vaginalis is found in 30–40 percent of cases. This is a flagellate protozoan that can be seen in freshly prepared, unstained, wet mounts of the vaginal discharge (Fig. 21.11). It has been treated with metronidazole (Flagyl) or povidone–iodine creams.

Herpes simplex Type II, and less commonly Type I, may cause either vulvovaginitis or cervicitis. The vesicular lesions appear in crops, alternating with latent periods. The true incidence of herpes genital infection is not known, but it is extremely widespread and appears to be second only to gonorrhea. Like all viral infections, herpetic disease is difficult to treat. Idoxuridine creams have been used. Genital herpes, although not immediately more serious than similar infections, has two extremely worrisome features. It may infect the children of the herpetic woman, both via intrauterine infection and during the birth process. Also, chronic infection of the cervix has been associated with increased rates of cervical cancer. Males also have genital herpes.

Herpes simplex; Chapter 17, p. 513

Nonspecific Genital Infections

Pelvic inflammatory disease PID has already been mentioned as one of the serious complications of gonorrhea. However, it may also be an ascending infection caused by the vaginal flora. The anaerobes *Bacteroides fragilis* and *Peptostreptococcus*, the enterobacteria, and the α-hemolytic streptococci and diphtheroids are the most common agents. PID may develop spontaneously, but often follows surgery or abortion. The symptoms resemble acute appendicitis. Peritonitis and septicemia may follow, or it may result in sterility. Thus although PID usually subsides without treatment, antimicrobials are always given to prevent more serious trouble. This condition is very common.

Endometritis   Infection of the uterine lining may follow sexual transmission but more commonly follows abortion, surgery, caesarian section, or septic vaginal deliveries, in which case it is called **puerperal fever.** The causative agents are usually the normal vaginal species. Its most threatening feature is the high rate of septicemia that follows. This outcome was horribly common in one period of obstetrical practice, after the invention of obstetrical forceps and before the institution of aseptic delivery techniques. During that time as many as one in eight women having a hospital ("modern") delivery died of septicemia. Today, such deaths may result from illegal, nonprofessional abortions.

Prostatitis   Acute or chronic prostatitis is a common complaint of older males. Prostatitis is usually caused by *E. coli*, although other enterobacteria, enterococci, and (rarely) the sexually transmitted organisms are found. In acute prostatitis there is swelling and intense pain, as well as fever and urinary obstruction. The condition usually responds promptly to antimicrobials effective against Gram-negative bacteria. Chronic bacterial prostatitis, on the other hand, may have few symptoms. However, the infection serves as a focus to initiate repeated bouts of cystitis, otherwise usually rare in the male. The symptoms may be lower back pain or urinary frequency. The urine may or may not show bacteriuria. Treatment is complicated by an unusual diffusion barrier between the plasma and the prostatic fluid where drugs are needed. Trimethoprim-sulfa combinations have shown promise, some chronic prostatis is nonbacterial. It is suspected that chlamydias, mycoplasmas, or viruses may be the causative agents.

INFECTIONS OF THE PRENATAL PERIOD

The fetus *in utero* is protected, fed, and insulated from most traumas of independent life. However, it is not totally isolated from infection. Many agents, predominantly viruses, cross the placenta. Some of these can kill or deform the fetus.

Anatomy and Physiology

During the developmental period, it is anatomically correct to regard the placenta and the fetus as temporary maternal organs. The structure and function of these "organs" should be examined in order to make a discussion of congenital infection clearer.

Fetal membranes and placenta   In the early days of the fertilized egg's development, it acquires an outer **trophoblast** layer. At seven or eight days this layer becomes

TABLE 21.7
Terminology of fetal and infant development

TERM	TIME PERIOD
Ovum	Zero to seven days or until implantation
Embryo	Eight days to six weeks
Transition	Six weeks to two months
Fetus	Two months until birth
Neonate	Birth to four weeks
Infant	Four weeks until walking begins

known as the **chorion**, and it is covered with finger-like villi. On reaching the uterus, the egg starts to bury itself in the uterine wall. The underlying maternal mucosa is destroyed by embryonic enzymes, liberating nutrients that temporarily nourish the embryo. The **placenta** then develops (Fig. 21.12). The fetal circulation is connected to its half of the placenta by the **umbilical** vessels. The placenta is well formed by the third week of pregnancy, and by the fifth month it occupies one half of the uterine surface.

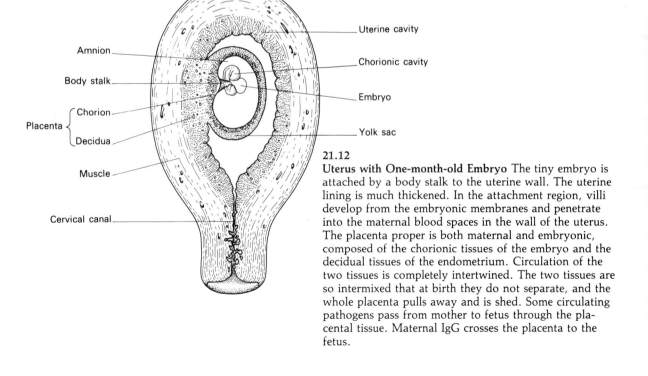

21.12
Uterus with One-month-old Embryo The tiny embryo is attached by a body stalk to the uterine wall. The uterine lining is much thickened. In the attachment region, villi develop from the embryonic membranes and penetrate into the maternal blood spaces in the wall of the uterus. The placenta proper is both maternal and embryonic, composed of the chorionic tissues of the embryo and the decidual tissues of the endometrium. Circulation of the two tissues is completely intertwined. The two tissues are so intermixed that at birth they do not separate, and the whole placenta pulls away and is shed. Some circulating pathogens pass from mother to fetus through the placental tissue. Maternal IgG crosses the placenta to the fetus.

Transfer across the placenta   The fetus has its entire chemical contact with the world via the placenta. Almost all plasma components pass the placenta, including water, electrolytes, and serum proteins such as immunoglobulins. Wastes and acquired chemical substances such as alcohol, anesthetics, antimicrobial agents, and other drugs move across too. Blood cells cannot pass the placenta, so the fetus is dependent on its own developing red and white blood cells. Oddly, the fetus's white blood cells do enter the maternal circulation, perhaps inducing toleration to prevent immune rejection.

Intrauterine infection   Virus particles, many times smaller than blood cells, can enter the fetal circulation when the mother has viremia. A few cellular bloodborne pathogens do also. The result is an **intrauterine infection.** Some agents cause developmental disturbances and thus create birth defects. These are called **teratogens.** Infections in the first trimester are most likely to produce malformation because the basic differentiation of tissues and construction of organ systems is occurring then. Later infections compromise the function of already developed organs.

Spontaneous abortion is one result of intrauterine infection. Alternatively, the fetus may survive to term, but be critically ill at birth with sepsis and multiple organ involvement.

When intrauterine infection has occurred, the fetus may respond by beginning immunoglobulin synthesis early. If an infant shows a high level of IgM antibody at birth, this is strong evidence of prenatal infection.

Fetal Infections

Cytomegalovirus (CMV) infection   This is the most common intrauterine infection. CMV may be found in about 1 percent of all live-born neonates. Most of these infants display no apparent ill effects, but at a later time some 10 percent of these reveal some mental retardation. These figures imply that intrauterine CMV infection is an important cause of mental retardation in children. A smaller number of infants will have clinical disease with liver and spleen enlargement, jaundice, and other neurological defects. Few of this group of children survive.

The epidemiology of CMV is complex. The virus is a herpesvirus and therefore has the potential for establishing a latent state. Its exact prevalence is uncertain, but because CMV infection appears in from 50–90 percent of immunosuppressed organ transplant recipients, presumably by reactivation, it is clearly common. The virus, found in the vaginal secretions of 10 percent of normal women, can be transmitted either across the placenta or at birth. The potential for an effective vaccine is slight because serious questions remain about the persistence and potential oncogenicity of the viral genome, and methods to determine attenuation are not available.

Rubella; Chapter 17, pp. 519–520   Congenital rubella syndrome   When the mother experiences rubella during the first trimester of fetal life, the virus is transmitted to the fetus and causes diminished tissue growth, tissue necrosis, and chromosomal damage. The most severely affected

TABLE 21.8
Agents that cause intrauterine infection

GROUP	ORGANISM
Viruses	*Rubella *Herpes simplex Types I and II *Cytomegalovirus Varicella–zoster Poliovirus Coxsackievirus Echovirus Influenza viruses Hepatitis Measles Variola–vaccinia Mumps
Bacteria	*Treponema pallidum* *Listeria monocytogenes* *Mycobacterium tuberculosis* *Mycoplasma* sp.
Fungi	*Candida albicans*
Protozoa	*Toxoplasma gondii* *Plasmodium* sp. *Trypanosoma* sp.

*Most significant.

fetuses die and are aborted. Damage in survivors shows up most strongly in the cardiovascular and nervous systems, especially the sense organs. Mental retardation is common. The virus persists for several months following birth, sometimes with continuing damage. Due to the widespread use of live rubella virus vaccine, the incidence of congenital rubella syndrome is now very low. The vaccine virus is generally agreed to be harmless if inadvertantly given to a pregnant woman, but intentional administration is strictly avoided.

Toxoplasmosis *Toxoplasma gondii* (Fig. 21.13) is an obligate intracellular protozoan parasite with a worldwide distribution. It infects many domestic and wild animals and human beings. It carries out sexual reproduction in the intestine of members of the cat family that shed **oocysts** in the feces. These may infect other animals and possibly human beings. In the infected nonfeline animal, cysts give rise to active **trophozoites** that invade all tissues. Once a sufficient immune response develops, the multiplication of trophozoites is curtailed, and the remainder forms **tissue cysts** that persist for the life of the individual. Such tissue cysts are found in 10 percent of lamb or beef samples and 25 percent of pork. Ingestion of such meat, inadequately cooked, transmits the infection to human beings.

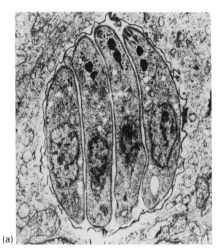

(a)

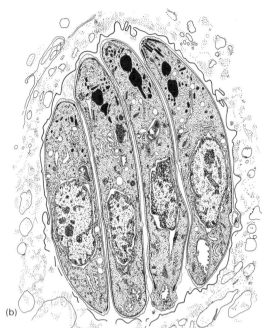

(b)

21.13
Toxoplasma gondii (a) Merozoites of *T. gondii* embedded in the intestinal wall. (b) A drawing showing details of the structures.

Human infection is usually inapparent, occasionally shows mild lymphoid tissue involvement, and rarely gives rise to overwhelming disease. This last condition is usually related to immunodeficiency, however. Positive serological evidence of past infection is found in from 2–93 percent of persons of childbearing age, depending on geographical area. In the United States, the range is about 25–45 percent.

The woman who has had toxoplasmosis earlier in life and brought it under control will not transmit it to a fetus when she becomes pregnant. However, if the disease is contracted during pregnancy, particularly in the last trimester, the fetus will be infected and tissue cysts will form. Most infected newborns are asymptomatic at birth. However, cysts may later reactivate. Then trophozoites will be released, and can cause severe permanent destruction in the retina of the eye or the central nervous system. Immune tolerance may be responsible for the congenitally infected child's inability to arrest the disease. Infants showing active disease at birth may have multiorgan involvement, and their rate of mental retardation is about 80 percent.

Active disease may be detected in serological tests of pregnant women. The TORCH series, combining serological tests for antibody to *Toxoplasma*, *Rubella*, *Cytomegalovirus* and *Herpes*, is widely used. If the Toxoplasma test is positive, pyrimethamine plus sulfonamide can then be given. This treatment will reduce the chance of symptomatic neonatal toxoplasmosis. The same combination is used to treat affected neonates. No available drug inactivates the tissue cysts. Thus treatment is never curative.

Congenital syphilis When *Treponema pallidum* is present in the mother's serum, as it is throughout infection up to the later latent phase, it can cross the placenta after the sixteenth week of pregnancy. Twenty-five percent of affected fetuses die before birth and another 30 percent die shortly after. Signs include lesions of the skin and mucous membranes, bone deformity, and liver and kidney disease. Bone deformities, deafness, and joint deformity may not become apparent until later in life.

If positive serological diagnosis of maternal syphilis is obtained early in pregnancy, penicillin administered before the sixteenth week will not only cure the mother but will also prevent intrauterine infection of her fetus.

Chorioamnionitis Ascending infection late in pregnancy can result in bacterial multiplication within the amniotic sac. There is a great risk if the fetal membranes rupture, and then delivery is delayed for 24 hours or longer. The bacteria threaten the fetus, as they may be inhaled and cause overwhelming lung infection. They are a risk to the mother, as they may invade her circulation and cause septicemia. Furthermore, the presence of a huge bacterial "culture" makes it difficult to perform a caesarian section without serious contamination.

Maternal Infections

In general, the pregnant woman is not overly susceptible to infectious disease, but there are some notable exceptions. The mortality rate for influenza contracted in the last third of pregnancy is much higher than for nonpregnant women. The risk of paralytic forms of polio increases. The urinary system undergoes both mechanical stresses (due to compression) and physiological stresses due to the need to excrete for a fetus. These increase the chance that cystitis will develop or that inapparent conditions such as chronic pyelonephritis will flare up.

As the rate of maternal infections rises, so also does the rate of newborn infections. Infants born to mothers with uncontrolled urinary tract infection have a six times normal risk of neonatal infection, a pressing argument for thorough prenatal care.

INFECTIONS OF THE NEONATAL PERIOD

At birth the child leaves a sterile world and enters one filled with microorganisms. This stress could possibly overwhelm the incompletely developed immune defenses, at a time when the baby is also assuming the functions of respiration, digestion, and excretion. The neonate's immunological defenses are augmented by immunoglobulins passively transferred from the mother. The mother's lymphoid system is an accessory immunological organ for the baby. The separation begun at birth is not complete until the infant is no longer receiving maternal antibodies and becomes immunocompetent.

Neonatal Immunity

The infant's own defenses   A fully developed term infant possesses the same lymphoid cells, tissues, and organs as the adult, and has a larger and more active thymus. However, the lymphocytes are not completely programmed, and some have not totally completed their differentiation.

Neonatal macrophages are normal except for reduced chemotactic responsiveness. Inflammatory responses are below adult levels. Although CMI is functional, as shown by the ability of lymphocytes to show a mitogenic response *in vitro*, the *in vivo* responses such as delayed hypersensitivity are low. The serum contains immunoglobulins of all types, but normally most, if not all, are derived from the mother. The fetus is able to make some IgG and IgM as early as two months before birth. Usually, however, this potential has not been used. The exception is intrauterine infection that stimulates fetal antibody production of elevated IgM.

Complement levels are only half those of the infant's mother, and there are low levels of properdin. Thus opsonizing activity is weak, and phagocytosis is only partially effective. Serum lysozyme and interferon levels, on the other hand, are elevated.

In summary, the baby is able to commence antibody synthesis when stimulated. Its phagocytic capacities, on the other hand, are rather minimal.

The infant's innate immune capacities are stimulated by the normal flora that colonize the gut and by the challenge of environmental microbes. It is one of the ironies of nature that these contacts are, on the one hand, a great threat to the baby's survival and, on the other, an essential initiation to life without which the capacity to defend itself will not develop.

Role of IgG; Chapter 13, pp. 381–383

Transplacental passive immunization   During the later months of pregnancy, large amounts of maternal IgG enter the fetal circulation. These antibodies confer variable protection, after birth, against agents to which the mother is immune. Infants receiving only this immunological gift will have infections, but they will be milder. Protection continues for as long as the antibodies last, which is no more than six months.

There are two major limitations to this protection. The neonate's antibody titer depends on how high the maternal antibody titer is. Because adults retain very low titers of pertussis antibody, effective levels are rarely transferred, even though the mother was once fully immunized. Furthermore, IgG's role is to control established infections, not to prevent infection. Transferred antibody level may also be extremely low when the mother's serum proteins are depleted by malnutrition.

Enteromammary passive immunization   In recent years, breast-feeding is becoming reestablished in its natural place in the human life-cycle. Researchers have taken a serious look at the immunological properties of human milk. It is now clear that human milk contains a wide range of immune factors.

The immunological donations of milk supplement the infant's immature defense capabilities. In most societies, it used to be traditional to nurse an infant for about two years, in some cases longer. This two-year nursing period extends a sort of

Box 21.2

The Immunological Gift

Human milk is far more than a food. Its nutrient role can be adequately performed by a number of formulas based on cow's milk, soybean hydrolysates, and other substances. These, however, have no immunological benefits.

Human milk contains a variety of immunoglobulins. Secretory IgA is found in abundance in the colostrum, the first thin, lipid-free fluid produced by the breast. Secretory IgA is continually secreted in milk throughout the nursing period, along with lesser amounts of IgG and IgM. The source of the sIgA is the mother's gut lymphoid tissue; sIgA is transported to the mammary gland, then concentrated in the milk. Recently, it has been shown that when a new microorganism colonizes the infant, it normally colonizes the mother too. She is stimulated to produce specific sIgA antibodies. These are transferred to the baby to assist it in controlling the new threat. IgA, by preventing microbial attachment, protects the neonate from microbial infection.

Milk antibodies also appear to protect the infant from hypersensitivities. The infant gut lining is readily penetrated by potential allergens in the diet. Milk antibodies combine with these antigens and appear to prevent allergic sensitization. Food allergies are thus very uncommon in the breast-fed child.

Human milk also contains as many white blood cells per milliliter as are found in the mother's blood. She donates about 25 percent of her daily monocyte production to her infant. These milk monocytes become large, active macrophages once in the baby's gut. There they phagocytize ingested pathogenic organisms. They also augment the infant's deficient phagocytic potential by secreting complement. Other milk leucocytes include B- and T-lymphocytes. Once in the gut, these may secrete antibodies. It is thought that maternal lymphocytes may play a key role in informing the infant gut lymphoid tissue about its function, and stimulating the lymphoid tissue to become operational.

Although cow's milk also contains these factors, antibodies against bovine diseases have no apparent value for human infants. The cow's white cells do not survive pasteurization and refrigeration; if they did, they probably could not interact with human lymphoid tissue to stimulate it.

immunological umbrella over the child until the time, between two and three years, when its AMI and CMI become adequate.

It has long been clear that infants fed human milk rather than cow's milk had a reduced incidence of infectious diseases, food allergies, and intestinal malfunction such as colic. Formulas gained acceptance in the face of this knowledge because of their convenience. Formulas were also thought to be nutritionally superior, since children fed formula put on more weight. Recent clarification of human milk's immunological benefits, more accurate evaluation of the neonate's nutritional requirements, and a new appreciation of the psychological and sociological role played by breast-feeding led to a 1979 resolution of the American Academy of Pe-

TABLE 21.9
Epidemiology of neonatal infections

DISEASE	RESERVOIR	TRANSMISSION	INCUBATION PERIOD IN DAYS
Diarrhea, enteropathogenic *Escherichia coli*	Human beings	Oral–fecal; water, food, weighing scales, dressing tables, contaminated hands	1–3
Neonatal inclusion conjunctivitis	Human beings	Contact with infected vaginal discharges during birth	5–12
Neonatal gonorrheal ophthalmia	Human beings	Contact with infected birth canal	1–2
Group B streptococcosis	Human beings (?)	Contact with adult carriers, usually maternal vaginal secretions at birth	1–3
Neonatal staphylococcosis	Human beings	Spread by hands of hospital staff; airborne fomites	4–10
Thrush	Human beings, others	Contact with secretions of mouth, skin, vagina, and feces of patients and carriers	2–5

Source: Adapted from A.S. Benenson, 1975, *Control of Communicable Diseases in Man*, (12th ed.).

diatrics to urge all mothers to breast-feed their infants, unless medical reasons preclude this.

Infections of the Neonate

During the first month of life, the infant is highly susceptible to infections, often contracted in the hospital nursery. This situation is not solely due to carelessness or negligent care, although they are the occasional cause. Rather, it is because the first few days of life are the riskiest, and the nursery happens to be where the infant is at that time. The highest standards of care, constant observation, and rapid, effective

PERIOD WHEN COMMUNICABLE	PATIENT CONTROL MEASURES	EPIDEMIC CONTROL MEASURES
For duration of fecal colonization, more than one month	Isolate infected infants and suspects; disinfect discharges and soiled articles	Suspend maternity service unless separate, clean nursery and staff available
While exudate is produced; while genital infection persists in female	Isolate for 48 hours; aseptic techniques prevent nursery transmission	Search for sexual contacts; treat maternal infection
For 24 hours following specific treatment or until discharges cease	Isolate for 24 hours; conjunctival discharges carefully disposed of	Routine administration of silver nitrate eye drops at birth prevents disease
While infection is present	Isolate; disinfect all discharges and soiled articles	None
As long as purulent lesions continue to drain or carrier state persists	Isolate all cases and suspects; careful disposal of dressings and soiled articles	Determine characteristics of epidemic strain; search for source. Group isolation and quarantine
For duration of lesions	Segregate afflicted infants, disinfect secretions and contaminated articles	Segregation; increased emphasis on cleanliness

evaluation of any adverse signs can minimize the danger and help keep minor problems from escalating into major ones.

Staphylococcal infections   In the newborn, staphylococcal colonization occurs rapidly. If the colonizing strain is highly virulent, one of two syndromes may result. **Pemphigus neonatorum** is an impetigo-like skin infection, with a profuse crop of vesicular lesions over large areas of the body. The staphylococcal **scalded-skin** syndrome is a spreading skin infection in which the superficial layers of the skin separate and slough off. Both syndromes carry a risk of fatal invasion of the lungs or of septicemia.

Staphylococcus aureus; Chapter 17, pp. 501–504

Lancefield grouping;
Chapter 23, p. 694

Group B streptococci   *Streptococcus agalactiae,* Lancefield Group B, is now recognized as a leading cause of neonatal sepsis. These organisms, found in the vaginal flora of 12.5 percent of normal mothers in one study, cause septicemia or meningitis in neonates and children up to several weeks old. There are two types, early and late onset. The initial symptoms may be mild irritability and failure to nurse, without fever. If treatment is delayed or ineffective, it can progress rapidly to septicemia and death. Alert observation, leading to early intervention, may be of critical importance in the outcome.

Bacterial meningitis   The risk of meningitis in the newborn is great. Predisposing factors include traumatic labor and delivery, prematurity or low birth weight, primary immune deficiency, and early antimicrobial therapy for other infections.

    Listeria monocytogenes, a small, motile, facultatively anaerobic, Gram-positive rod, is less frequently diagnosed as the cause of a meningitis appearing from 1–4 weeks after birth. The agent is widespread in nature, in nonhuman animals and in human beings. *Listeria* may be acquired at birth, just prior to birth, or transplacentally. The earlier exposures almost always lead to abortion or stillbirth, but neonatal infection can be treated successfully with erythromycin if diagnosed in time.

    Gram-negative bacilli such as *Escherichia coli,* strain Kl, and other enteric organisms are important meningitis agents in the infant under two months of age. Because the newborn receives little IgM from mothers, its ability to opsonize Gram-negatives is low. The mortality rate is from 40–80 percent. *Flavobacterium meningosepticum,* found in many samples of clean tapwater, has been seen as the cause of some nursery outbreaks of meningitis.

Pneumonia   The newborn's first gasp draws in the liquids in its mouth, including amniotic fluid and vaginal secretions. If these are heavily contaminated, pneumonia may result. The risk of pneumonia is greatest in the premature infant whose lungs may not be fully developed.

EEC disease; Table 20.4

Other infections   Gastroenteritis due to enteropathogenic strains of *E. coli* and other enterobacteria has already been mentioned. GI upset due to viruses is uncommon, until a few months have passed and passive immunity has waned.

    The newborn may become infected with herpes simplex virus if the mother had open herpetic lesions in the birth canal at the time of delivery. The virus may be inhaled or swallowed. Early in life, herpes multiplies readily in nerve tissue, causing encephalitis. Mortality is high, even with newly introduced drugs such as ara-A (Vidarabine).

Candida albicans;
Chapter 17, pp.
510–511

    Thrush, an oropharyngeal infection of *Candida albicans,* is especially common in neonates. It is contracted by exposure to a maternal vaginal infection; it spreads readily in nurseries via direct contact with secretions or contaminated fomites. Thrush is seen as a reddening of the oral membranes, with gray-white patches of exudate. Thrush is not a severe condition, and usually disappears shortly, as the neonate's defenses improve and the competitive normal flora establishes itself. The discomfort, however, may cause the baby to feed poorly, sleep fitfully, and gain weight slowly.

Tetanus may occur in the newborn when the umbilicus has been cut with grossly contaminated instruments or soiled with dirt.

Pneumocystis carinii is a small protozoan; it causes pneumonia in infants with primary immunodeficiency.

One of the most mystifying causes of infant mortality, usually after the neonatal period, from about three months to three years, is the **sudden infant death syndrome** (SIDS). An apparently normal infant inexplicably suffers respiratory arrest and dies in its crib. Dozens of explanations have been advanced; probably all fit some cases and none fit all. One recent explanation is microbiological, and deserves our notice.

In 1977, in the United States and England, infants were brought to doctors suffering from "floppiness"—lack of muscle tone, inability to hold up the head, weak or absent suckling motions. It was found that the "floppy" infants had *Clostridium botulinum* Type B in their intestinal flora. The botulinum toxin was being produced and absorbed, and was causing muscle weakness. Antimicrobial therapy and supportive measures will cure infant botulism.

Clostridium botulinum; Chapter 20, pp. 609–610

Since one effect of botulinum toxin is respiratory arrest, a connection has been proposed between infant botulism and SIDS. It is proposed that rapidly accumulating toxin may in some cases lead to SIDS before other symptoms (floppiness) become apparent. It has been strengthened by the finding of *C. botulinum* in the bowel flora of some SIDS victims.

Many samples of honey contain *C. botulinum* spores, and honey has been implicated as the source in about 25 percent of the documented cases of infant botulism. Some sources now recommend that honey not be fed to infants under one year of age. Since *C. botulinum* cannot establish itself in the gut of persons older than one year, the rest of us may continue to enjoy honey without fear.

POSTPARTUM INFECTIONS OF THE MOTHER

Puerperal sepsis   At birth, the placenta separates from the uterine wall, leaving much of the uterine lining traumatized and bleeding. This exposed tissue is a prime site for microbial invasion if bacteria are introduced to the uterus. The resulting overwhelming endometritis is called puerperal sepsis or childbed fever. The pathogen may migrate to the bloodstream almost unhindered, producing septicemia. Prior to the antibiotic era, childbed fever was almost inevitably fatal. Now, prompt therapy saves most victims. With modern aseptic techniques, in uncomplicated deliveries, this disease should be only a historical curiosity. But, sadly, techniques are used by fallible human beings, and cases (and deaths) do still occur.

Mastitis   The nursing mother may develop infection within the mammary glands, or **mastitis.** *Staphylococcus aureus* is the usual pathogen. Swelling, reddening, and pain of the affected area of mammary tissue develops. Milk flow from that portion of the breast may be arrested. In extreme cases, abscess formation may occur.

SUMMARY

1. The urinary and reproductive systems in the male use the urethra as a common passage to the exterior. Infections of the male urethra may involve both systems. Infections of the bladder and kidneys are far less common in males than in females.

2. In the female, the two systems are anatomically separate, yet their proximity leads to the high rates of urinary tract infection in women.

3. Both the urethra and the vagina have normal flora. The vaginal flora illustrates the protective effect of resident organisms. In the woman of reproductive years, the flora creates an acidic physiological environment hostile to the most likely invaders, the bowel flora.

4. Cystitis is usually caused by bowel enterobacteria. It is often a symptomatic episode in a long history of inapparent infection.

5. Bladder infections may cause ascending infections that affect the renal pelvis, collecting tubules, and medulla. Acute pyelonephritis is readily detected and cured; thus it has little residual effect. Chronic pyelonephritis may be present for months or years before being detected, with cumulative, irreversible damage.

6. Male genital infections primarily involve the penis and urethra. Later, in untreated disease, the accessory glands may be attacked. In the female, such infections involve the vulva, vagina, and cervix, progressing if untreated to the endometrium of the uterus, the fallopian tubes and ovaries, and potentially the peritoneal cavity.

7. The sexually transmitted diseases are those transmitted primarily by sexual acts. They are regarded as serious if they lead to sterility or life-threatening complications. Gonorrhea and syphilis fall in this category. In less serious STDs, such as lymphogranuloma venereum and chancroid, the destruction is limited to the genital organs. Still other STDs, such as nongonococcal urethritis, appear to be self-limiting and nondestructive. Syphilis, herpes, and perhaps other diseases can be transmitted to the unborn child.

8. Nonspecific genital infections, such as pelvic inflammatory disease, endometritis, and prostatitis, are not usually considered STDs. Although the agent is sometimes a venereal organism, it more commonly is not. The infection can be associated only rarely with any specific sexual contact. These conditions may also develop in persons who are habitually celibate.

9. Infections can be transmitted across the placenta to the fetus when the agent is bloodborne in the mother. Infection of the fetus in early pregnancy may lead to spontaneous abortion or severe malformation. Later in pregnancy, fetuses may survive and be born with acute systemic infection. Many intrauterine infections cause mental retardation that is not apparent until later in childhood.

10. Immediately after birth and for its first month, the newborn is especially susceptible to normal pathogens and opportunists. It is protected by maternal IgG passed through the placenta, that slowly wanes in effectiveness. A wide range of maternal immune factors are continuously supplied in the milk as long as nursing continues. These minimize but do not prevent infection. Some infective agents commonly seen in newborns but not in mature individuals include the Group B streptococci, *Listeria* and *Clostridium botulinum* Type B.

11. During pregnancy, maternal risks in later months include increased incidence of pulmonary and urinary infection. After birth, introduction of organisms into the uterus may lead to puerperal fever. The mammary glands may become infected and suffer mastitis.

Study Topics

1. Trace the movement of an ascending urinary tract infection from its source to the kidney. Compare the patterns seen in males and females.

2. How do female hormonal changes vary the pH of vaginal secretions? What is the effect on infection susceptibility?

3. Why would a person with spinal cord injury have an increased susceptibility to cystitis and different types of causative agents?

4. It is a common statement that increased sexual freedom has contributed to the increase in STDs. This is certainly true, but there is more than one reason why. List some.

5. Intrauterine infections have different results depending on the period of pregnancy in which they occur. Why?

6. By reference to Chapter 13, if necessary, prepare a list of specific immunological functions to which breast milk might contribute in the newborn.

Bibliography

Fiumara, Nicholas J., 1978, The sexually transmissible diseases. *Disease-a-month* **25** (3).

Hoeprich, Paul D., (ed.), 1977, *Infectious Diseases* (2nd ed.), New York: Harper & Row.

Marano, Haro, 1979, Breast-feeding. *Medical World News,* February 5:62–78.

Sanford, Jay P., 1975, Urinary tract symptoms and infections *Ann. Rev. Med.* **26**: 485–498.

South, Mary Ann, and Charles A. Alford, Jr., Congenital intrauterine infections. In E. Richard Stiehm et al. (eds.), 1980, *Immunological Disorders in Infants and Children,* (2nd ed.) Philadelphia: W. B. Saunders.

22

Nervous System Infections

The nervous system is the most complex of body systems. Because of extensive anatomical protection, it is relatively invulnerable to infection. The nervous system is normally sterile; there is no normal flora. But infection, when it does occur, may be devastating. Irreversible loss of central nervous system function spells death. In nonfatal infections, death of nerve cells may cause permanent neurological impairment.

ANATOMY AND PHYSIOLOGY

Anatomy of the Nervous System

Cells of the nervous system   The basic conducting cell is a **neuron** (Fig. 22.1a). It is composed of a **cell body,** containing the nucleus and organelles. One or more nerve fibers or **processes** extend from it, in some cases for long distances. Processes that conduct impulses towards the cell body are **dendrites;** those that conduct impulses away from the cell body and establish communication with receptor cells are called **axons.** Many nerve cell processes are **myelinated,** that is, covered by a multilayered

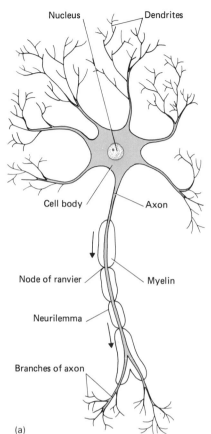

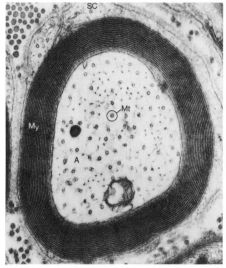

(b)

22.1

The Neuron   (a) The neuron, basic conducting cell of the nerve tissue, is composed of a cell body and two processes. A nerve impulse is received by the dendrite, passes through the cell body, and moves down the axon to the next receptor. (b) The axons of many nerves are myelinated. The myelin sheath (My) is made up of special complex lipids. It is secreted by the Schwann cell (SC). Within the sheath is the axon (A) containing many mitochondria (Mt) that provide the ATP for stimuli conduction.

insulating sheath of myelin, a complex lipoprotein (Fig. 22.1b). Demyelination causes loss of function or death of the neuron.

A signal passes between the axon of one neuron and the dendrite of the next by triggering the release of neurochemicals at the **synapse**, a point at which the two nerve fibers come very close to each other but do not actually touch. Other chemicals are then released to break down the first neurotransmitter and thus terminate the signal. Some microbial toxins (e.g., botulin and tetanospasmin) chemically mimic the normal synaptic neurotransmitters, causing aberrant neurological function.

Central Nervous System

The central nervous system (CNS) is composed of the brain, the spinal cord, and their coverings, the **meninges** (Fig. 22.2).

These structures are completely enclosed by the axial skeleton. This skeleton has openings for passage of blood vessels, the spinal cord, and the cranial nerves. These openings, and the very thin, porous bone underlying the cranial sinuses, are

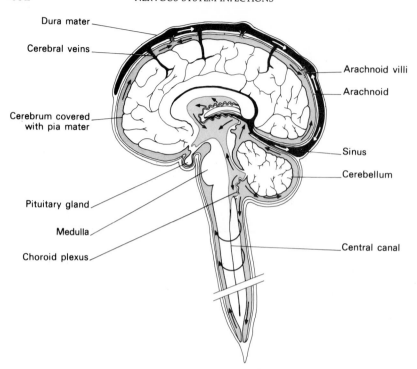

Dura mater

Cerebral veins

Cerebrum covered
with pia mater

Pituitary gland

Medulla

Choroid plexus

Arachnoid villi

Arachnoid

Sinus

Cerebellum

Central canal

22.2
The Brain   The brain is composed of three major areas, the cerebrum, cerebellum, and medulla. The organ is covered and protected by three membranous layers, the dura mater, arachnoid, and pia mater. The cerebrospinal fluid originating in the choroid plexus circulates around the brain and its coverings (arrows).

routes by which infective agents may enter. Similarly, the spinal column is penetrated by openings through which the peripheral nerve trunks enter (Fig. 22.3).

The anatomy of the human brain is exceedingly complex. For our purposes it will be sufficient to review the main areas and their functions. The **cerebrum** is concerned with sensory perception, voluntary action and its coordination, and abstract thought. The **cerebellum** is responsible for motor coordination. The **medulla** and **pons,** directly connected to the spinal cord, regulate involuntary vital functions such as respiration, peristaltic action, and heartbeat.

The spinal cord receives and coordinates nerve impulses from the peripheral spinal nerves. **Afferent** (in-coming) signals typically enter via the **dorsal root,** and are referred to other nerve cells within the **horns** of the structure. The **anterior horn** contains the cell bodies of those nerve cells that control muscle contraction. This tissue is frequently attacked by the poliovirus.

The **gray matter** of the brain and spinal cord contains cell bodies and unmyelinated nerve fibers. **White matter** is composed of bundles of myelinated nerve fibers.

In the CNS, the neurons are interspersed with **neuroglia,** composed of four types of nonconducting, supportive cells. In the brain, these outnumber neurons by 100 to 1. Some form myelin, others are potentially phagocytic.

The meninges are the connective tissue coverings of the CNS. The outer layer, or **dura mater,** is very tough. Beneath is the thinner **arachnoid membrane,** separated

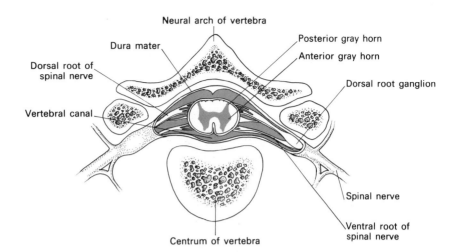

**22.3
Cross Section of the Spinal
Cord** Incoming (afferent)
nerve impulses enter the
spinal cord and are transmitted to other neurons that
relay them to appropriate
brain centers and/or to out-
going (efferent) neurons that
transmit the response to the
tissues.

Labels in figure: Neural arch of vertebra; Dura mater; Posterior gray horn; Anterior gray horn; Dorsal root of spinal nerve; Dorsal root ganglion; Vertebral canal; Spinal nerve; Ventral root of spinal nerve; Centrum of vertebra

from the innermost **pia mater** by the subarachnoid space. All three layers follow the
shape of the nervous tissue more or less closely and extend out through the skeletal
pores with the cranial and spinal nerve roots.

Peripheral nervous system   Neurons exterior to the CNS constitute the peripherial
nervous system (PNS) and all make contact with it. **Sensory** neurons have their dendrites in the skin, sense organs, viscera, and muscle. Their cell bodies lie in **ganglia**
(enlargements of nerves) at some distance, and from there their axons extend to the
CNS. **Motor** neurons or other effectors usually have their dendrites and cell bodies
within the central nervous system, and only their long axons extend to the muscle
fiber to activate it.

A **nerve** serving a particular area will contain many neuron processes, both
dendrites of sensory neurons and myelinated axons of motor neurons. A nerve contains no cell bodies. The neuron processes are arranged in bundles with several layers of connective tissue coating. These coatings are impermeable to blood components, such as antibodies, as well as antimicrobial agents. A microbial agent within a
peripheral nerve may be effectively shielded from inactivation.

There are 31 pairs of spinal nerves, originating from the spinal cord, and twelve
pairs of cranial nerves, originating from the brain and exiting through openings in
the skull.

Physiology of the Nervous System

We will look only at those functions directly affected by microbial invaders.

Cerebrospinal fluid   The cerebrospinal fluid (CSF) is a clear, watery secretion that
fills the spaces between the brain and spinal cord and the skeletal enclosures. The

fluid also circulates slowly between the meningeal layers and into the hollow areas of the brain. It is formed by filtration through capillary walls, and seeps away by osmosis. The quantity is regulated to give a constant, slight pressure. Any increase in the intracranial fluid pressure is abnormal. CSF functions as a hydrostatic cushion and, to some extent, mediates nutrient exchange. Cells are not normally found in CSF. The presence of white blood cells indicates CNS inflammation and infection.

Circulatory supply   The brain demands a large proportion of the body's nutrient and oxygen supply, which is delivered by capillary networks in both gray and white matter. There is no lymphatic circulation. An important physiological offset of circulatory interaction with brain tissue is the so-called **blood–brain barrier.** Chemical exchanges between capillary and tissue always depend on diffusion, in any area of the body. In the brain, the local tissue environment is such that there is no barrier to diffusion of natural nutrients and oxygen. However, many drugs are not able to diffuse efficiently from the blood into the brain tissue or the CSF. The concentration of many antimicrobials in normal brain tissue may be only 1/20 of the serum concentration. It can be difficult to achieve therapeutically useful concentrations of drug in the brain tissue or CSF. The barrier effect is the result of the microanatomy of brain capillaries, which have an unusually thick basement membrane surrounding them.

Drugs that cross blood–brain barrier Chapter 26, p. 777

Fortunately, inflammation increases the permeability of capillaries, reducing the barrier effect. This allows some diffusion of drugs during infection when they are most needed. Even so, extraordinarily high dosages of antimicrobials are usually required to control CNS infections.

The blood–brain barrier also partially protects against microbial invasion. CNS infections do occasionally arise from hematogenous spread. However, this occurs much less often than microbial spread to other organ systems. This confirms the antimicrobial effect of the barrier.

Protective mechanisms   The nervous system is protected primarily by its skeletal and membranous coverings. If these fail, it is defended by opsonizing activity. This requires accumulation of phagocytic cells and serum factors from the blood, which occurs only in case of inflammation/infection.

Routes of infection   A prime route of invasion is by extension of an upper respiratory infection or a skin infection in the head area. Penetration through the cranial sinuses, middle ear, or mastoid bone is possible. Such events are uncommon except in neonates. Hematogenous spread of agents resident in the oropharynx, skin, lungs, or GI tract also occurs. Ascending infections are those that become established first in peripheral nerve endings, then migrate to the CNS under the nerve sheath. Trauma such as a skull fracture or surgery is yet another pathway for invasion.

BACTERIAL INFECTIONS

Bacteria may infect the meninges and, less commonly, the underlying brain tissue (Table 22.1). Because of the diffusion barrier, these infections are difficult to treat.

TABLE 22.1
Clinical terminology of nervous system infections

TERM	DEFINITION
Meningitis, also called leptomeningitis	Inflammation of the meninges
Meningoencephalitis	Inflammation of brain and its coverings
Poliomyelitis	Inflammation of gray matter of spinal cord
Encephalitis	Inflammation of brain tissue, many types

The rapidity with which bacterial meningitis progresses makes it essential to start antimicrobial therapy immediately. Broad-spectrum antibiotics or combined therapies are often initiated, designed to cope with the most likely organisms. These choices can be readjusted later if necessary when laboratory results identifying the causative agent and its antimicrobial susceptibilities become available.

Meningococcal Meningitis

Causative agent Epidemic and sporadic meningitis are most commonly caused by *Neisseria meningitidis*. This fastidious, Gram-negative diplococcus has cultural requirements and biochemical behavior similar to those of the gonococcus. It may be

Compare *N. gonorrhoeae;* Chapter 21, pp. 628–630

TABLE 22.2
Biochemical differentiation of some *Neisseria* and associated species

| CHARACTERISTIC | NEISSERIA | | | | | BRANHAMELLA |
	FLAVESCENS	GONORRHOEAE	MENINGITIDIS	LACTAMICA	SICCA	CATARRHALIS
Growth on nutrient agar, 25°C	+	−	−	+	+	+
Growth on nutrient agar, 35°C	+	−	−	−	+	+
Growth on Thayer–Martin medium	−	+	+	+	−	−
Glucose fermentation	−	+	+	+	+	−
Lactose fermentation	−	−	−	+	−	−
Sucrose fermentation	−	−	−	−	+	−
Maltose fermentation	−	−	+	+	+	−
Fructose fermentation	−	−	−	−	+	−

isolated on Thayer–Martin or other chocolate agars in a 5 percent CO_2 atmosphere. Species identification depends heavily on sugar fermentation tests. The organism's polysaccharide capsule is important in pathogenesis because it inhibits phagocytosis. Serological typing of the capsular substance identifies at least seven major groups (A, B, C, D, X, Y, and Z) and numbered strains. Endotoxin, shed from the cell wall during all phases of growth, is also of great importance in the organism's pathogenesis. Endotoxin provokes hemorrhages, intravascular coagulation, and systemic shock.

Pathogenesis   Infection may or may not develop following oropharyngeal colonization. The outcome depends on immune response. Within 7–10 days after colonization occurs, immunocompetent individuals have formed specific anticapsular antibodies. These, in serum, prevent further invasion, leading to a healthy carrier state. However, if the immune response is inadequate, bacteremia develops. The organisms may then multiply in the blood to cause **meningococcemia,** or may invade the meninges to cause **meningitis.** In the worst cases, both aspects are combined. Development of the disease is often explosively rapid. Unless treatment is both prompt and appropriate, mortality is high.

In meningococcal meningitis, the primary signs are extreme headache, immovably stiff neck, and many PMNs in the cerebrospinal fluid. The organism is not always visualized or successfully cultured from the CSF. **Petechiae** (small skin hemorrhages) are often present. In the septicemic form, large hemorrhagic lesions develop from which the causative organism can be demonstrated in a direct Gram-stained smear of exudate. Circulatory failure is rapid and often irreversible. The cause of death is endotoxin shock.

Therapy and immunity   Penicillin, ampicillin, or penicillin plus sulfonamide are given in massive, intravenous doses. Chloramphenicol may be substituted in hypersensitive individuals. Oral rifampin is used for prophylaxis of exposed individuals.

Immunity develops on exposure to Types A and C capsular material, but Type B is not immunogenic. Monovalent vaccines for Groups A and C, and a bivalent A–C vaccine, have been licensed for use in high-risk populations. Type C vaccine is given routinely to United States military recruits. Vaccine may also be of benefit during epidemics, for travelers, and for exposed contacts of cases.

Meningococcal vaccine; Table 27.3

Epidemiology   Meningococcal meningitis occurs principally in late winter and early spring. It affects children, young and older adults. Military recruits have always had a high risk of epidemic outbreaks. The healthy carrier is the primary reservoir, and the agent is spread by droplets and direct contact. Since the meningococcus is readily inactivated outside the body, indirect spread is unlikely. The carrier rate ranges widely, depending on the population surveyed and the season. There is a periodic change in the predominance of the various strains. At present, Groups B, C, and Y are the most widespread.

TABLE 22.3
Bacterial meningitis agents

DISEASE	NUMBER OF CASES	PERCENT	INCIDENCE*	CASE FATALITY RATE
Hemophilus influenzae	1,885	46	1.24	7.1%
Neisseria meningitidis†	1,095	27	0.72	13.5%
Streptococcus pneumoniae	456	11	0.30	28.2%
Group B *Streptococcus*	130	3	0.09	22.4%
Listeria monocytogenes	68	2	0.04	29.5%
Other	235	6	0.15	36.6%
Unknown	212	5	0.14	16.7%
Total bacterial meningitis	4,081	100	2.69	13.6%
Meningococcemia	289		0.19	25.1%

*Cases per 10^5 population, estimated July 1978 for the 38 reporting states.
†Excludes cases of meningococcemia alone.
Source: *Morbidity and Mortality Weekly Reports.* **28**: 277, 1979.

Other Bacterial Meningitis Agents

In different age groups, other pathogens are major causes of meningitis (Table 22.3).

Streptococcus pneumoniae may initiate meningitis by extension of middle-ear infection. Thus it is seen frequently in small children, in whom otitis media is common. Also, *S. pneumoniae* pneumonia usually has a bacteremic phase, and meningitis may follow pneumonia in the older patient. Penicillin is the usual therapeutic agent, and mortality is high.

Streptococcus pneumoniae; Chapter 19, p. 554

Hemophilus influenzae may be a meningitis agent in children under five, but not in adults unless they are immunodeficient. Overall, this bacterium may be the most frequent causative agent of meningitis. The middle ear is a frequent source. Mortality is not as high as with other agents, but there is an increased chance of learning disability. Ampicillin was formerly the drug of choice, but at least 15 percent of *H. influenzae* strains are now resistant. Combined chloramphenicol–ampicillin therapy is recommended until a laboratory report on drug susceptibilities is obtained.

Hemophilus influenzae; Chapter 19, p. 553

In the neonate, *Escherichia coli, Streptococcus agalactiae* (Group B) and *Listeria monocytogenes*, are prominent problems. These agents and other opportunists occasionally appear in the severely compromised adult.

Neonatal infections; Chapter 21, pp. 641–647

Abscesses and Focal Infections

Bloodborne organisms occasionally establish abscesses beneath the meninges, or in brain tissue. *Staphylococcus aureus* is most common, but others, including anaerobes, are seen. Abscesses may form in the subdural or subarachnoid spaces, or at the junction between gray and white matter. Therapy requires intensive antimicrobial chemotherapy and often surgical drainage. Mortality is high and damaged tissue is not replaced.

Miliary tuberculosis; Chapter 19, p. 560

Progressive pulmonary tuberculosis may give rise to many distant foci. Tubercular brain lesions are granulomatous and may be multiple.

Leprosy

Leprosy, a slowly progressive infection caused by *Mycobacterium leprae* (Fig. 22.4), initially involves the peripheral nerve endings in the skin. The bacilli multiply within Schwann cells—the cells that produce myelin sheath material. The tissue supplied by the affected nerve becomes insensitive and is often inadvertently damaged, leading to disfiguration. Involved tissues often become much swollen and enlarged.

In **tuberculoid** leprosy, the milder form of the disease, host CMI defenses restrict the spread of the lesion. It is self-limited, although there is severe local damage. The agent cannot penetrate the central nervous system. A positive skin test with antigen is diagnostic.

Lepromatous leprosy develops if CMI is inadequate. The lesions are multiple and eventually the entire skin surface may be affected. Bacteremia develops, and the organism is seeded to internal organs. Blindness and severe deformity result; death is due to secondary infection or other conditions. Lepromatous leprosy is particularly common in individuals with parasitic infections such as onchocerciasis, in whom the CMI responses are compromised. The lepromin test may be negative if CMI is sufficiently depressed.

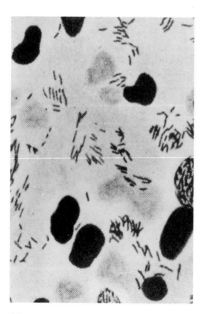

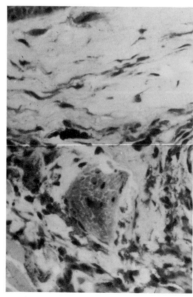

22.4
Mycobacterium leprae (a) The causative agent of leprosy is an acid-fast organism. It has not yet been cultivated in cell-free media. It is seen here as bacilli lying among numerous body cells. (b) Lesion caused by *M. leprae* in tissue.

(a)

(b)

Leprosy remains a mysterious disease in several ways. The agent is not cultured in bacteriologic or cell culture media; thus little is known about it. The bacillus can be cultivated in the armadillo. Further, although it is now clear that leprosy is not particularly communicable, its exact means of transmission is not known.

Dapsone or acedapsone are the antimicrobial drugs of choice. Rifampin is also becoming well accepted. Three to six months of therapy renders active patients convalescent and noninfective. Therapy is usually continued for several years.

In 1975 the World Health Organization reported about 3.6 million registered leprosy cases. The proportion of lepromatous cases tended to be low in African countries, but higher in the Americas. The percentage of these patients receiving regular treatment ranged from 41 to 74.

FUNGAL INFECTIONS

Systemic pulmonary mycoses occasionally also form granulomatous lesions in the brain. In **cryptococcosis,** the neurological form of the disease is clinically the most important, although skin and skeletal forms are also seen.

Cryptococcosis is caused by the yeastlike organism, *Cryptococcus neoformans,* long considered a deuteromycete. However, the sexual reproductive phase of the fungus has recently been discovered, revealing that it is a basidiomycete similar to the rusts and smuts. *Cryptococcus* multiplies in avian feces worldwide, most significantly for human beings in droppings of the common pigeon. Transmission probably involves inhalation of dust particles derived from dried bird droppings. The initial infection is pulmonary, and the vast majority of cases remain inapparent. In this, it closely resembles coccidioidomycosis and histoplasmosis. Dissemination from the lungs is rare, and associated with immune impairment. The CNS is the preferred target of disseminated disease, and deep lesions occur in both gray and white matter. Exudates accumulate on the brain surface; meningitis is also present. Antimicrobial drugs may be used but the prognosis is very poor. Diagnosis is based on culture of the organism from cerebrospinal fluid and use of latex agglutination tests to identify fungal antigen in body fluids.

Compare with Histoplasmosis; Chapter 19, pp.

VIRAL INFECTIONS

From 30–40 percent of meningitis cases are caused by viruses. Because these are not recovered on routine bacteriological examination of the cerebrospinal fluid sample, the term **aseptic meningitis** is used, even though there is an infective agent present. It may be caused by the mumps virus, coxsackie viruses of Types A and B, and echovirus.

The disease usually has a gradual onset, a short duration, and is relatively benign with few late complications. It is most common in children and young adults. A few cells (mononuclear white cells) are found in the CSF. The agents enter by their typical respiratory or intestinal routes, and meningitis follows viremia. Treatment is supportive, and recovery is usually complete.

Arthropod-borne Encephalitis

Causative agents   More than 350 viruses are classified as **arboviruses**—arthropod-borne agents (Fig. 22.5a). By definition, an arbovirus can multiply in both arthropod and vertebrate tissues. The arthropod is both host and vector for the virus, which it transmits by its bite. The arboviruses are RNA viruses, currently placed in the togavirus group, and separated into three subgroups, alpha viruses, flaviviruses, and bunyaviruses. A limited number of these viruses is of importance in the North American continental area. Some important characteristics are summarized in Table 22.4.

Pathogenesis   Characteristically, the arthropod bite gives rise to a viremia, allowing the agent to enter neurons and supporting tissues in the CNS. Early signs are headache and fever; then sensory disturbances, confusion, muscular weakness, and convulsions may occur. Swelling and hemorrhage of the brain may be present. The severity of the infections varies considerably among the different agents, depending largely on the degree of neuronal necrosis and quantity of destroyed tissue. Only the supporting tissue can be repaired, and permanent mental impairment typically follows severe cases.

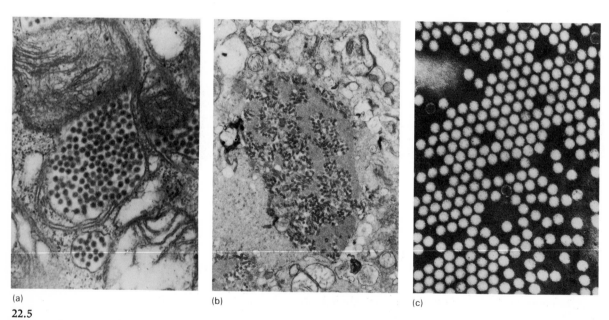

(a)   (b)   (c)

22.5

Viruses That Attack Nerve Tissue   (a) The EEE (Eastern Equine Encephalitis) virus is the most virulent of the arthropod-borne encephalitides. It is transmitted by mosquitos from infected horses to human beings. Viruses are seen here in brain tissue. (b) Rabies virus multiplies in brain tissue causing encephalitis. The intracellular maturing virus forms Negri bodies, which confirm evidence of rabies infection. (c) The poliovirus attacks preferentially the cell bodies of motor neurons. By destroying these, the virus makes it impossible for nerve impulses to activate the affected muscle, and paralysis results.

TABLE 22.4
Characteristics of arthropod-borne encephalitis in North America

DISEASE	GEOGRAPHICAL RANGE	VECTOR	ANIMAL HOST	AGE GROUP	MORTALITY RATE PERCENTAGE	SEQUELAE
Eastern equine encephalitis	New Hampshire to Texas, Eastern United States	*Culiseta Aedes Culex*	Birds, horses	Children	60–75	90% of survivors
Western equine encephalitis	All states west of Mississippi River	*Culex*	Birds, horses, snakes (?)	Infants adults <50	5–15	Moderate
Venezuelan equine encephalitis	Northern South America, Central America, Florida, and Texas	*Psorophora Aedes*	Horses, rodents	All ages	0.6	Low
Saint Louis encephalitis	West Central and Southern United States, Panama, Trinidad	*Culex*	Birds	Adults <50	2–11	Low
California encephalitis	North Central and Southwest United States	*Culex Aedes*	Rabbits, squirrels, field mice	Children	1	Low
Colorado tick fever	North Central and Northwest United States	*Dermocentor*	Ground squirrel	Children, adult males	Low	Low

Source: Modified from G.P. Youmans et al., 1975, *Biologic and Clinical Bases of Infectious Diseases,* Philadelphia: W.B. Saunders.

These diseases are not, so far as is known, transmitted from human being to human being.

Therapy and immunity   There are no specific antiviral drugs for the RNA viruses, so therapy is supportive. Recovery leads to protective immunity.

Killed-virus vaccines are available but in limited amounts. Laboratory workers may receive appropriate experimental vaccines. The vaccine for Japanese B encephalitis, a very severe disease common in the Far East, is manufactured in Japan and widely used to control epidemics in India.

Epidemiology   Mosquito control is most important. Periodic outbreaks of encephalitis may be associated with seasonal flooding or heavy rains. Water-holding receptacles around dwellings, such as discarded tires or metal cans, provide breeding sites for mosquitos. Removing breeding areas and taking due precautions to minimize mosquito contact are wise preventive measures.

Since the horse may be a major host, vaccination of equines in endemic areas is advised or required. Proof of vaccination for Venezuelan equine encephalitis (VEE),

Eastern equine encephalitis (EEE), and/or Western equine encephalitis (WEE) may be required in order to ship horses from one area to another. Because of the rarity of the disease, human vaccination is limited to the applications given above.

Rabies

Only two individuals have ever survived a documented case of rabies, once symptoms developed. The public has a great horror of rabies, a horror that is useful in that it motivates essentially all exposed persons (several thousand a year) to seek prophylactic treatment. The treatment is clearly effective, since in the United States there are usually fewer than five human cases per year.

Causative agent   The rabies virus is a large, complex, envelope-bearing, RNA virus, readily inactivated by sunlight, drying, or chemical agents (Fig. 22.5b). It can be cultivated in a wide range of cell culture systems and in duck eggs. Laboratory diagnosis is based on direct microscopic study of brain tissue from rabid animals or deceased patients. In brain tissue, the virus forms characteristic inclusion bodies called **Negri bodies.** These are unequivocally identified by a fluorescent rabies antibody (FRA) test on brain tissue.

Pathogenesis   The bite of the rabid animal implants saliva containing infective virus particles. The virus rapidly adsorbs to the nearest peripheral nerve endings. It enters the regional nerve trunk and moves by unknown means toward the CNS. The length of the incubation period appears to depend on the area of implantation, which may vary both in terms of the numbers of nerve endings and distance to the CNS. Thus bites on the face, where the tissue is richly supplied with nerve endings in close proximity to the brain, are considered to have the greatest risk.

Viral multiplication occurs in numerous tissues, including salivary gland, kidney, pancreas, and the cornea. However, the clinical manifestations of the disease are caused by destruction of gray matter. Lesions develop in the cerebellum, hippocampus, cerebral ganglia, pons, and medulla of the brain.

Once viral multiplication in the CNS becomes developed, the first stage of the disease is one of excitability and the characteristic **hydrophobia.** This is a painful spasm caused by an uncoordinated swallowing reflex. Later, a paralytic and comatose state develops. In some cases, the excitability does not develop, and rabies may not be suspected or diagnosed until autopsy.

Therapy and immunity   Therapy is supportive. In 1970, a young boy was successfully carried through a rabies episode, but this success cannot be reliably repeated. A number of persons have been maintained for some weeks, but have eventually succumbed.

Rabies vaccination was pioneered by Pasteur, using crude extracts of brain tissue containing an attenuated virus. Other nerve vaccines followed, but they were strongly allergenic. The prolonged course of vaccinations gave plenty of time to

TABLE 22.5
Postexposure prophylaxis in rabies

| | STATUS OF ANIMAL | | |
NATURE OF EXPOSURE	AT TIME OF EXPOSURE	DURING TEN DAYS	RECOMMENDED TREATMENT
Contact, no lesions; indirect contact	Rabid	Rabid or dead	None
Licks of skin; scratches or abrasions; minor bites (covered areas or arms, legs, trunk)	Suspected*	Healthy	Start vaccine. Stop if animal is healthy for five days or proved negative.
	Rabid; animal unavailable for observation		Start serum and vaccine.†
Licks of mucosa; major bites (multiple or on face, head, neck, or finger)	Suspect or unavailable	Healthy	Start serum and vaccine. Stop if animal remains healthy,
	Rabid		Complete course of vaccination.

*All unprovoked bites are suspected until animal is proved negative by fluorescent antibody test.
†Health personnel who have had inadvertent, unprotected contact with a rabid patient, and had mucosal or skin lesions at the time of exposure, should receive this treatment.

develop severe, sometimes fatal, hypersensitivities. At present, two vaccines are licensed, neither of which is produced in brain tissue. The duck embryo vaccine (DEV) is generally used and the human diploid cell vaccine (HDCV) is at present restricted to persons with confirmed exposure who are known to be allergic to egg products. Increasing availability of HDCV is expected, and it will probably eventually replace DEV.

Severely exposed individuals are given rabies vaccine and simultaneously receive rabies immune globulin (RIG). The recommendations for prophylactic treatment are quite complex. They are summarized in Table 22.5; for more complete information, a current reference should be sought. No treatment is required unless the skin is broken or a mucosal surface has been contaminated by the animal or human patient's saliva.

Epidemiology   Both domestic pets and wild animals can serve as reservoirs and transmit the disease to human beings. Where pet vaccination is widely used (the United States, Western Europe, Canada), wild animals have replaced dogs as the most significant source of rabies. The skunk and fox account for the majority of documented animal cases (about 3,000 a year) in the United States. Rabies is also seen in all other mammals (it is very rare in rodents). Other than by bite, transmission has been documented by inhalation of dust in caves inhabited by rabid bat colonies, and by two corneal transplants. Since saliva is infective, persons who have had direct, unprotected contact with a patient may be considered exposed, especially if cuts, scratches, or lesions were present on their skin or mucous membranes.

Certain preventive measures are of great importance. Immunize all domestic pets, especially domesticated wild animals such as pet skunks and raccoons. Avoid wild animals that, because they are ill and sluggish, seem unusually "friendly"—putting out of your mind all those cute Disney creatures. If you are bitten, be sure the wound is immediately, thoroughly washed and swabbed with soap and water for at least 20 minutes. This first-aid measure gives a good chance of removing the virus before it enters the nerve tissue. Seek medical attention promptly.

Other Viral Agents of Encephalitis

The agents of viral encephalitis are identified in only 30 percent of cases. Primary infection occurs with mumps and herpesviruses. Encephalitis develops (rarely) as a late manifestation of chickenpox, rubeola, rubella, and vaccinal infection. The pathogenesis is as described before, with swelling and hemorrhage followed by necrosis of brain tissue. Specific therapy is available for treatment of the DNA viruses, in particular the herpes and pox groups. Idoxuridine, cytosine arabinoside (ara-C) and now adenine arabinoside (vidarabine or ara-A) have been used with a significant reduction of mortality.

Poliomyelitis

Causative agent   The polioviruses are small RNA viruses belonging to the enterovirus group (Fig. 22.5c). There are three types, I, II, and III, present in nature both as wild-type pathogenic strains and as attenuated vaccine strains. The viruses can be propagated in monkeys and in cell cultures of monkey and other tissue. They are readily identified antigenically.

Pathogenesis   Infection is frequently inapparent if modified by the presence of partial immunity. In the unimmunized person, subclinical cases often occur, but the severe paralytic form becomes more common.

The virus is ingested or possibly inhaled. It moves from the original site via the lymphoid tissue to the blood, and then across the capillary wall to the neurons of the CNS. In particular, there is a high affinity for the cell bodies of the motor neurons of the spinal cord and brain stem. An encephalitic destruction may occur. Meningitis is also present; this will be the predominant feature of nonparalytic polio. In paralytic forms, loss of function of muscles activated by the destroyed neurons occurs. In fatal cases, irreplaceable functions of the medulla and cerebellum are lost. Paralytic effects, once developed, are usually irreversible, although in some cases reeducation can develop compensatory movement patterns.

Box 22.1

An International Polio Outbreak

Since the widespread adoption of polio vaccination, outbreaks of polio have been rare in the industrialized nations because outbreaks cannot occur unless there is a large number of susceptible individuals in a relatively dense cluster. These conditions were fulfilled recently, resulting in a significant outbreak among members of the Amish community, a group that for religious reasons has traditionally refused vaccination.

The originating virus appears to be a Type I strain that was isolated from polio patients in Kuwait in 1977. By some means, the virus was transported to the Netherlands. The outbreak started there in April, 1978. The first patient was a 14-year-old girl from a village near Utrecht. She attended a large regional school attended by many Amish children from neighboring communities. A rapid development of the outbreak followed, all cases being in unvaccinated persons who were members of the Amish communities.

During this summer, Amish visitors from the Netherlands traveled to Canada. Six Canadian Amish, in British Columbia, Alberta, and Ontario, developed polio, with the same Type I wild-type viral strain. Each of these individuals had had contact with the visitors from the Netherlands.

As of October 12, 1978, there had been 110 cases, 80 of which were paralytic. United States health officials were saying cautiously that there was no evidence that the virus had spread to the United States.

However, in late summer, 1978, an Amish family from Ontario moved to a town in Pennsylvania. In January, 1979, the first United States polio case appeared in that town. After a three-month lag, other cases, totaling 16, appeared. These occurred in four states and Canada, and all except two were Amish persons.

A rapid immunization program was developed to serve the Amish communities in the United States and Canada, and the majority of nonvaccinated persons accepted the vaccine. It is now believed that the outbreak is finally over.

Note that the persons who transmitted the virus from one area to another were all perfectly healthy. The majority of persons acquiring poliovirus experience asymptomatic disease Only 1 in 100 at the most develop paralytic disease. By these figures, the roughly 100 paralytic cases would suggest up to 10,000 asymptomatic cases over the same period. Passage of the virus among successive asymptomatic persons accounts for the long lag between the summer, 1978, and spring, 1979, phases of the epidemic. This sequence of events also illustrates quite forcefully why there is a continuing need for polio vaccination.

Therapy and immunity   Therapy is supportive, designed to minimize chances of secondary infection, and prepare for possible rehabilitation.

Recovery gives rise to a lifelong protective immunity. Both inactivated polio vaccine (IPV) and oral live-virus vaccine (OPV) are available. OPV is generally recommended in the United States. It should be given to all normal infants; the

TABLE 22.6
Epidemiology of nervous system infections

DISEASE	RESERVOIR	TRANSMISSION	INCUBATION PERIOD
Meningococcal meningitis	Human beings	Direct contact and droplets from carriers, occasionally cases	2–10 days
Leprosy	Human beings	Not established; probably via skin lesion or respiratory tract	Average 3–5 yr
Cryptococcosis	Soil, bird excreta	Presumably by inhalation	Unknown
Rabies	Many wild and domestic mammals (rarely rodents)	Saliva to bite, scratch or mucosa; airborne; tissue graft	Usually 2–8 wk variable
Poliomyelitis	Human beings	Direct contact with pharyngeal secretions or feces; milk	7–12 days

Source: Adapted from A. S. Benenson, 1975, *Control of Communicable Diseases in Man* (12th ed).

breast-fed infant will not respond as fully because of maternal antibody and should be reimmunized in childhood.

There is a small risk associated with the use of OPV. The live virus is a mutant strain with greatly reduced affinity for nerve cells. It does multiply in the recipient's intestine, and is shed in the stool for a period. Immunodeficient persons may acquire full-blown paralytic polio from vaccine virus. There is also a slight possibility of the virus's reverting to a virulent form. Most of the ten or so cases of polio each year in the United States can be shown to be caused by vaccine virus.

In some European countries, IPV is used exclusively. It is a less effective antigenic stimulus because there is no live virus to multiply in gut lymphoid tissue. However, IPV is safer for persons whose immunological capacities might not be able to control multiplication of even the weakened vaccine virus, and reversion is not an issue.

Epidemiology   Both wild and vaccine viruses are spread primarily by the oral–fecal route. Contaminated swimming areas and shellfish have been of particular impor-

PERIOD COMMUNICABLE	PATIENT CONTROL MEASURES	EPIDEMIC CONTROL MEASURES
Until bacteria are no longer present in naso-pharyngeal discharges; until 24 hr after therapy is initiated	Isolate until 24 hr after start of chemotherapy; disinfect discharges and soiled articles	Early diagnosis; relieve crowding of target group; prophylaxis with sulfonamide or rifampin
While bacilli are present in skin; until three month's dapsone therapy completed	If patient is on dapsone, isolation not required; disinfect nasal discharges and dressings	Search for possible family or contact cases
Not transmitted from person to person	Disinfect discharges and dressings	None
For 3–5 days before onset of symptoms and during course of disease	Isolate for duration of illness; disinfect saliva and materials soiled with it. Attendants wear protective gown and rubber gloves	Control disease by pet vaccination; control stray dog population
As early as 36 hr after infection; virus in feces for 3–6 weeks or longer. Most infections 7–10 days before onset and after onset	Isolate for not more than seven days if hospitalized; disinfect throat, fecal discharges, and articles of clothing	Mass vaccination with oral vaccine at earliest sign of outbreak; search for sick persons among contacts

tance. Before the introduction of vaccination, the peak incidence of polio was during the hot summer months. Vaccination is the primary control measure, responsible for a decline of paralytic polio from more than 18,000 cases in 1954 to only eight in 1976. At present, polio is still prevalent in countries without vaccination or adequate sanitation. However, infant exposure during the period of passive maternal immunization is almost universal in such areas. This leads to a very mild infection; only a small percentage of cases are symptomatic and few of those cause paralysis. On the other hand, in westernized society, outbreaks may lead to paralysis in 70 percent or more of cases that tend to occur in older children and adults. Recently, epidemics occurred in religious communities that do not practice vaccination.

Slow Virus Diseases

Irreversible neurological degeneration is the hallmark of several slow virus diseases, such as Creutzfeld–Jacob disease, kuru, progressive multifocal leukencephalopathy

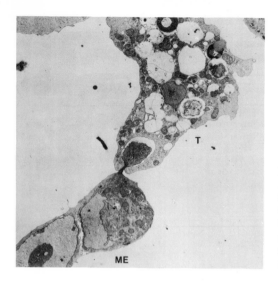

22.6

Amebic Meningoencephalitis   *Naegleria fowleri*
trophozoite (T) snipping off a fragment of a mouse
epithelial (ME) cell. When this amoeba gains access
to the brain, it systematically consumes brain tissue. Drugs have not generally been effective in
stopping its relentless feeding. Only one human
case has survived this fortunately rare disease.

Slow virus diseases;
Chapter 11, p. 328

(PML), and subacute sclerosing panencephalitis (SSPE). The biological basis of slow
virus disease was discussed in Chapter 11. Viral agents have been recovered from
brain tissue in several of these conditions; other similar diseases are under study.
Long incubation periods and slow degenerative change, associated with a hypersen-
sitivity-mediated autoimmune attack, are characteristic.

PROTOZOAN INFECTIONS

Primary Amebic Meningoencephalitis (PAM)

In 1965, a new neurological infection, caused by free-living, aquatic amoebae, was
described. About ten cases of PAM per year are documented in the United States,
almost all fatal. Most cases are caused by *Naegleria fowleri,* a small amoeba found
in brackish or stagnant warm water. Other cases have been traced to *Acanthamoeba*
and *Hartmanella* species. Amoebae have been demonstrated living in hot springs,
puddles, and cracks in the walls of concrete swimming pools.

   It seems likely that the motile trophozoite is inhaled during swimming, passes
the nasal mucosa and the cribiform plate, and gains access to the brain. Such penetration must be due to an unlikely combination of events because cases are so rare.
Once present, the amoeba attacks and consumes the brain tissue (Fig. 22.6). Diagnosis is often made by viewing the motile amoebae in microscopic examination of the
spinal fluid.

SUMMARY

   1. By contrast with respiratory or gastrointestinal infections, nervous system infections are uncommon, affecting only a small percentage of the population.

They are usually very severe. Most have high mortality rates and tend to leave residual handicaps.

2. The major protections of the nervous system are anatomic. The CNS is sheathed in the meninges and encased in the axial skeleton. Peripheral nerve fibers are individually insulated by a myelin sheath. Bundles of fibers (nerves) have impermeable connective tissue coverings. Diffusion between the circulation and brain tissue is restricted. This reduces the migration of microorganisms and nonlipid soluble drugs to brain tissue.

3. Meningitis is a bacterial infection of the meningeal space. It is most common in neonates and infants, but is found in all age groups. Different agents, however, predominate in different groups of people. Meningococcal meningitis is the typical, very severe form in adults.

4. Cryptococcosis is a rare, often fatal systematic mycosis in adults. Other disseminated fungi and tuberculosis may produce granulomatous lesions of the brain.

5. Viruses have a particular affinity for nervous tissue, as they do for epithelial and glandular tissue. Viral multiplication in the cell bodies of neurons may destroy them. Neuron loss is not repaired, and leads to permanent neurological defects. Encephalitis and poliomyelitis are two examples. Partially effective drug therapy is available against the DNA viruses, but there is none for the RNA-containing rabies, polio, and togavirus groups.

6. Surveillance and vaccination have led to the almost complete control of rabies and polio in the United States. However, there appears to be no possibility of eradicating either.

Study Topics

1. What is the difference between a meningitis and an encephalitis in terms of the types of tissue attacked? What does this difference mean in terms of the patient's making a complete recovery?

2. Compare the relative advantages of inactivated and live-virus polio vaccination.

3. Several viruses studied in earlier chapters seem able to establish latency in peripheral nerves. Review these and discuss the mechanisms that allow such long-term survival.

4. Poliomyelitis and primary amebic meningoencephalitis have a similar case incidence of about ten cases a year. However, one is treated as a serious public health threat while the other is considered an unfortunate natural accident. Why?

Bibliography

Hoeprich, Paul D. (ed.), 1977, *Infectious Diseases* (2nd ed.), New York: Harper & Row.

Plotkin, Stanley A., and Tadeusz Wiktor, 1978, Rabies vaccination, *Ann. Rev. Med.* **29:** 583–591.

Underman, A. E., et al., 1978, Bacterial meningitis, *Disease-a-month* **24** (5).

Youmans, G. P., et al., 1980, *Biologic and Clinical Bases of Infectious Diseases* (2nd ed.), Philadelphia: W. B. Saunders.

23

Wound
Infections

Whenever a wound occurs, whether minor or severe, infection is a possible sequel. The body's natural defenses normally deal rapidly with microbial invaders in small wounds. However, large wounds require special treatment, i.e., cleansing to remove organisms introduced at the moment of trauma, and dressing to exclude them during the healing period.

PATHOLOGY OF WOUNDS

Because a wound is by definition abnormal, it cannot be said to have any normal anatomy or physiology. However, predictable pathological changes occur in the structure and function of wounded tissues. These alterations predispose to infection.

Structural Changes

In any wound, cells die. Areas of devitalized necrotic tissue develop. A portion of the capillary bed is destroyed. If the wound penetrates the skin, respiratory, gastro-intestinal, or urogenital surface (and there are very few wounds that do not), the underlying tissue is exposed, and surface flora have direct access to underlying tissue.

Physiological Changes

Inflammation   Traumatized tissue has a reduced blood supply. One result of tissue destruction is the release of inflammatory substances. These cause local vascular changes (Chapter 12) that work cumulatively to isolate the damaged area. Intravascular coagulation of blood may obstruct capillary flow. Capillary permeability also increases; and large amounts of serous fluid may exude. This fluid loss is very pronounced in a burn.

Anaerobiosis   Normal tissue is oxygenated by its blood supply. The Eh (oxidation–reduction potential) of normal tissue ranges between $+0.126$ and $+0.246$, depending on the body area.

Eh; Table 8.2

Damaged tissue, by contrast, becomes anaerobic. When a wound occurs, capillary obstruction reduces the tissue's oxygen supply. The adjacent tissues quickly exhaust this oxygen, thus lowering the Eh. In addition, the metabolism of damaged tissue may become fermentative, lowering the pH. This also contributes to lowered Eh. Oxygen-deprived tissues may die, thus expanding the area of necrosis. If the Eh drops to negative values, from -0.11 to -0.25, anaerobic bacteria are able to multiply. These anaerobes cause the majority of the most serious wound infections.

Foreign materials   Wounds often contain foreign objects, such as bits of metal, wood, sand or gravel, shreds of fabric, or bullets. Surgical wounds contain sutures or clips. These closures, although sterile, are still foreign irritants. Foreign materials may provoke continuing inflammation. If large enough, they may greatly prolong or prevent healing. It is essential that foreign objects be quickly and completely removed from wounds.

Surgical closures such as sutures and clips have been designed to be minimally irritating. Most either dissolve gradually or are removed as soon as healing is well advanced. Even so, the presence of sutures clearly predisposes to staphylococcal wound infections. Whereas more than 10^7 cells of *Staphylococcus aureus* are required to initiate infection in an open wound, as few as 10^4 cells can initiate infection adjacent to a stitch (Table 23.1).

FLORA OF WOUNDS

Both bacteria and fungi colonize wounds, but bacteria are by far the more common. Viruses rarely colonize traumatized tissue because they need healthy, vital host cells in order to survive. A wound, such as a bite, may provide viruses a portal of entry. But the rabies virus, which is introduced in a bite, migrates to other, undamaged tissue to replicate.

Rabies; Chapter 22, pp. 662–664

TABLE 23.1
Some clinical terminology applied to wounds and wound infections

CONDITION	DESCRIPTION
Cellulitis:	An inflammation of cellular or connective tissue
Debridement:	Excision of devitalized tissue from surface and area of wound or burn
Gangrene:	Necrosis due to obstruction of blood supply
Gas gangrene:	Gangrene due to infection with anaerobic bacteria, especially gas-forming *Clostridium* species
Ischemia:	Reduced blood supply due to mechanical obstruction of vascular system
Myositis:	Inflammation or infection of muscle tissue
Necrosis:	Pathogenic death of cells or tissue resulting from irreversible damage
Osteomyelitis:	Inflammation of the bone marrow and adjacent bone and cartilage

Endogenous Distinguished from Exogenous Infection

As we know, there is a normal flora on all external body surfaces and the mucosa that lines body orifices. Normally, skin and mucosa are intact barriers, and the flora cannot penetrate. In most accidental traumas, the endogenous human flora introduced is potentially more dangerous than most of the exogenous environmental organisms from soil and water. Endogenous infection is the rule in GI trauma; perforating the viscera releases the bowel flora into the peritoneal cavity. Similarly, infections following septic abortion or delivery usually derive from the normal vaginal flora. On the other hand, surgical infections are very frequently exogenous. They can often be traced to pathogens carried by a member of the surgical team.

Theoretically, wound infections are entirely preventable. A wound is an obvious risk, and methods for cleansing and disinfecting are well developed and readily available. In practice, some injuries are overlooked or ignored by the recipient, and so do not get adequate attention. Overwhelming injuries, such as some burns, are so long in healing that keeping them uninfected requires skill, perseverance, and luck.

Anaerobic Infections

Wounds are a prime site for anaerobic microbial pathogenesis. *Bacteroides*, *Fusobacterium*, *Clostridium*, and other anaerobic genera are prominent. Although anerobes are well represented in the normal flora, they do not seem able to cause disease unless they encounter damaged tissue. Remember that in the mouth, the development of plaque in the gingival sulcus leads to mechanical tissue damage, prolifera-

TABLE 23.2
Types of wounds and their vulnerability to infections

WOUND	CHARACTERISTICS	TENDENCY TO INFECTION
Bite	Tissue is crushed, torn, and punctured to varying degrees	Very high. Oral flora introduced, anaerobic environment with devitalized tissue may be present
Burn, 1st degree	Skin reddened, no blisters	No significant risk
Burn, 2nd degree	Reddening and blistering	Some risk of infection as blisters break. Quick healing minimizes problem
Burn, 3rd degree	Full thickness burn; epidermis destroyed	Very high; damaged skin does not regenerate. Patient's defenses compromised. Healing takes an indefinitely long period; extreme care needed
Burn, 4th degree	Full thickness burn; underlying structures are destroyed	
Contusion	Bruise; coagulated blood in tissue	None, unless skin or mucous membrane broken; if so, obstruction to blood flow may contribute to anaerobic conditions
Crushing	Combined bruising and laceration may be extensive; tissue death	Serious if internal organ ruptured, skin broken. May lead to gangrene, osteomyelitis, peritonitis. Devitalized tissue should be removed
Fracture, closed	Broken bone; tissue adjacent also traumatized. No skin break	Not significant
Fracture, open	Bone fragment breaks through skin or mucosa; compound fracture	High; surface flora may be introduced into bone marrow, joint. Osteomyelitis possible
Laceration	A tear in surface tissues	Significant if skin or mucous membrane broken; seriousness depends on location and extent
Puncture	Tissue is devitalized, often in form of inverted cone. Extent of damage may be invisible from exterior	If anaerobes introduced, and lips of wound allowed to close, severe infection may result. Must be encouraged to heal from inside out

tion of anaerobes, then periodontitis. Without the wounding effect of calculus, the disease process does not begin.

Periodontitis, Chapter 16, p. 482

Oxygen tolerance   The anerobic genera vary greatly in their tolerance for oxygen. Moderate anaerobes, sometimes described as **aerotolerant,** may possess the enzymes superoxide dismutase and/or catalase. These enzymes can remove free oxygen from the environment even though the bacteria cannot use it for respiration. Such species

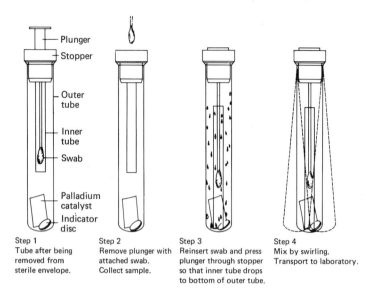

Plunger
Stopper
Outer tube
Inner tube
Swab
Palladium catalyst
Indicator disc

Step 1
Tube after being removed from sterile envelope.

Step 2
Remove plunger with attached swab. Collect sample.

Step 3
Reinsert swab and press plunger through stopper so that inner tube drops to bottom of outer tube.

Step 4
Mix by swirling. Transport to laboratory.

23.1
Anaerobic Culture Collection and Transport Device   This is a simple mechanism for providing anaerobic collection and transport conditions. The sample can be collected directly from the patient at bedside into the container that will be used for transport. Insertion of the specimen swab activates the catalytic removal of oxygen and provides a CO_2/H_2 atmosphere.

tolerate surprisingly high O_2 concentrations (although less than that found in air or normal tissue). Strict anaerobes lack oxygen-disposal enzymes and are killed by minimal O_2 exposure. Because the aerotolerant group survives temporary aerobic conditions during sampling, they were the first to be discovered and studied. Strict anaerobes have been recognized only since adequate collection, transportation, and incubation methods have been adopted.

Collection and cultivation   The sample originates in an area isolated from the air, and must be collected and transported in a way that keeps air exposure to a minimum (Fig. 23.1). Specialized swab-and-tube systems are commercially available. In these, a small amount of oxygen-free medium is provided; the receiving tube is filled with nitrogen or another inert gas. Carefully sealed, these transport systems will maintain the viability of anaerobes until they reach the laboratory.

Oxygen-free media must be used for strict anaerobes. This, in practice, means freshly prepared media; although autoclaving drives dissolved oxygen out of media, O_2 readily redissolves during storage. **Thioglycollate** is a nontoxic reducing agent. When added to liquid media, it reacts with the dissolved oxygen and removes it.

Neomycin or kanamycin added to solid media select against facultative anaerobic overgrowth and aid in the recovery of strict anaerobes.

Incubation   Incubation containers must be free of oxygen during anerobic growth. Many labs utilize a sealed-jar system. Introduction of the cultures into the jar admits oxygen. After loading and sealing, a catalyst is activated. This promotes a chemical

23.2
Anaerobic Chamber Samples to be inoculated are introduced through the left-hand door into an interchange chamber. The air is removed from this chamber and replaced with gases. Then the samples can be removed from the interchange chamber to the inner chamber. There the worker may carry out all necessary procedures using the two glove ports. Specimens to be incubated are placed in the box above.

reaction that utilizes oxygen as one of the reactants and thus consumes it. An indicator solution of methylene blue dye is placed in a visible position in the vessel. This redox indicator turns from blue to white when the Eh falls to suitable levels for anaerobic growth. In large-scale applications, anaerobic cabinets supplied with an artificial atmosphere are practical (Fig. 23.2). Individual sealed-bag systems may be used for individual cultures.

Mixed Infections

Wound infections are usually **polymicrobic;** that is, two or more bacterial strains are present. These may form a structured microecosystem. Typically, both facultative and true anaerobes are present. The facultative organisms utilize whatever oxygen is available, thus lowering the Eh of the area. Their early development in the wound further provokes inflammation that, as previously seen, also promotes oxygen shortage. If the Eh drop continues, moderate and then strict anaerobes start to proliferate. The anaerobes produce hydrolytic enzymes. The enzymes break down tissue components such as glycogen and proteins, releasing free sugars and amino acids. These nutrients support further growth of the whole symbiotic microbial community. The situation tends to escalate unless direct intervention occurs.

Clinical laboratories are increasingly alert to the need to search for more than one isolate from wound samples. Previously, those reported were often the organisms most readily cultured and not necessarily the most important. Pathogenic isolates must be separately evaluated for antibiotic susceptibility. The only effective therapy will be a drug or drug combination to which all respond.

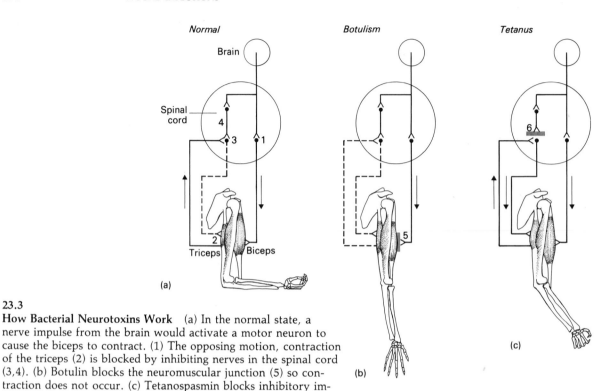

Normal　　　*Botulism*　　　*Tetanus*

Brain

Spinal cord

4

3　1

Triceps　Biceps

2

(a)

5

(b)

6

(c)

23.3

How Bacterial Neurotoxins Work　(a) In the normal state, a nerve impulse from the brain would activate a motor neuron to cause the biceps to contract. (1) The opposing motion, contraction of the triceps (2) is blocked by inhibiting nerves in the spinal cord (3,4). (b) Botulin blocks the neuromuscular junction (5) so contraction does not occur. (c) Tetanospasmin blocks inhibitory impulses (6). Thus both muscles contract, causing a spasm.

TABLE 23.3

Properties of some important species of *Clostridium*

SPECIES	MOTILITY	FERMENTATION OF			
		GLUCOSE	MALTOSE	LACTOSE	SUCROSE
Clostridium botulinum	+	+	+	−	+ or −
C. tetani	+	−	−	−	−
C. perfringens	−	+	+	+	+
C. histolyticum	+	−	−	−	−
C. novyi	+	+	+	+	−
C. septicum	+	+	+	−	−
C. difficile					

ACCIDENTAL TRAUMA

There are many types of accidental trauma. They differ, of course, in their seriousness depending on location and extent. Most important, each provides a different type of opportunity, such as a newly created microenvironment, for microbial proliferation.

Microenvironments; Chapter 6

In emergency medicine, immediate wound cleansing is the most important single determinant of whether or not infection will develop. However, cleansing may be incomplete, for very human reasons. If the patient is not breathing, or the heart has stopped, he or she will be dead within minutes. Obviously these problems take priority. However, after the crisis is passed, the time must be found to carry out meticulous cleansing and disinfection.

Tetanus

Much feared, but now entirely preventable, tetanus is a very uncommon disease in developed nations. However, because the spores of the agent are ubiquitous, it remains a constant threat.

Causative agent   *Clostridium tetani* is a Gram-positive, strictly anaerobic spore-forming bacillus. Its natural habitat is the intestinal tract of warm-blooded animals. It is spread by feces to soil in which the spores remain viable indefinitely. Via soil and dust, spores may be transferred to almost any surface or object.

Spores; Chapter 3, pp. 89–90

C. tetani is difficult to isolate, and isolation efforts often fail. If isolation is successful, the organism can be characterized by inability to ferment sugars, indol production, and pronounced swarming motility. Confirmed identification depends on the demonstration of its major pathogenic factor, the neurotoxin **tetanospasmin**. This extremely potent exotoxin causes muscles to go into prolonged rigid contractions or spasms (Fig. 23.3). This happens because tetanospasmin blocks neurological

NITRATE REDUCTION	INDOL PRODUCTION	TOXINS	DISEASES
−	−	Botulin	Botulism; wound infection
−	+	Tetanospasmin	Tetanus
+	−	Alphatoxin collagenase, others	Gas gangrene; food poisoning
−	−		Gas gangrene; food poisoning
+	−	Alphatoxin	Gas gangrene; food poisoning
−	−	Alphatoxin, others	Gas gangrene; food poisoning
		Cytotoxin	Antibiotic-associated colitis

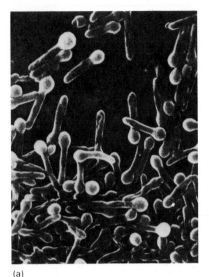

(a)

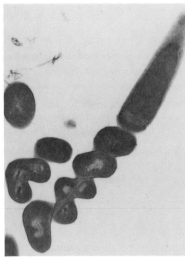

(b)

23.4
Clostridium **Species**   (a) *C. tetani* is the agent of tetanus. Note the swollen polar spores. These spores, numerous in the feces of warm-blooded animals, are also common in soil. The spores are introduced by a puncture wound and germinate if the wound environment is sufficiently anaerobic. (b) *C. perfringens* is one agent of gas gangrene. Wounds are contaminated by spores. On germination, the bacterium actively ferments muscle sugars, producing progressive tissue destruction and large amounts of gas.

inhibiting impulses. These impulses normally prevent random muscular contractions and terminate completed actions. Muscles innervated by toxin-saturated nerve tissue exhibit a gradually increasing tendency to contract uncontrollably.

In the lab, toxin in culture filtrates may be demonstrated by injection into mice, in cell culture, or by *in vitro* immunological methods.

Pathogenesis   The spores of *C. tetani* (Fig. 23.4) may be introduced into any wound with soil or fecal contamination. However, they germinate only under very anaerobic conditions. Thus tetanus most characteristically follows a puncture wound. This wound may be a very small one—a thorn prick—and it may completely heal without evidence of infection. The spores also do not always germinate immediately. Therefore, the incubation period is highly variable.

Clinical tetanus occurs after the spores germinate, when growth of *C. tetani* gives rise to toxin. The toxin is absorbed by peripheral nerves. The nerves that control the jaw muscles are often affected first; the spasm called lockjaw may be the first clear symptom. As toxin levels increase, other voluntary muscles become progressively affected, and death may come due to respiratory failure and exhaustion. Aspiration pneumonia, due to uncontrolled swallowing or regurgitation of food, is a common complication.

Therapy and immunity   First, certain steps are taken to prevent the situation from worsening. Toxoid (see below) should be administered immediately to stimulate production of antibodies if the patient has a history of previous tetanus vaccination. Antitoxin may be given to any patient who has never been immunized. Antimicrobials, usually penicillin, are given to kill the vegetative bacteria in the wound and thereby to interrupt the toxin production.

However, toxin already present and fixed to nerve cells will not be inactivated by these measures. It will continue to have its effect until it is gradually metabolized. Other therapeutic measures are designed to support the patient through this period, which may be two to six weeks long. Muscle relaxants, sedatives, and intensive artificial supports may be required. In special treatment centers, the mortality rate may be held to 10 percent, but overall it remains around 60 percent. Death is most likely in persons over 50.

The toxin is a potent antigen. It is converted to a nontoxic but still antigenic toxoid by formalin treatment and alum precipitation. Administration of toxoid, as given to infants in the DPT series, or in the form of monovalent injections, leads to a protective immunity lasting from 10–20 years in normal individuals. Toxoid booster shots used to be given routinely to persons receiving wound care, but it became clear that an Arthus-type hypersensitivity could develop if too frequent antigen stimulation occurred. At present, toxoid is generally given as part of wound treatment only if more than ten years has elapsed since the last toxoid administration. The use of antitoxin preparations such as hyperimmune horse serum has been discontinued. Antitoxin of human origin is available for those few individuals for whom it is indicated.

Arthus reaction; Chapter 14, pp. 416

Epidemiology   In the United States, a common manifestation of tetanus is in a male over 50, who develops the disease during the summertime in association with outdoor farming, gardening, or construction activities. He is susceptible because many years have elapsed since his childhood vaccinations. Neonates also get tetanus as a result of gross contamination of the umbilical stump or circumscision scar. It is obvious that both neonatal and adult tetanus are fully preventable.

Gas Gangrene

Gas gangrene is an infectious disease. It occurs when devitalized tissue is invaded by anaerobic bacteria that are potent gas-formers. Because the organisms liberate a mixture of potent toxins to the blood stream, the disease has systemic effects. Microbial multiplication constantly expands the necrotic area. The disease may advance inexorably unless radical surgery is carried out.

Many species of clostridia may be seen in gas gangrene, but *Clostridium perfringens* (Fig. 23.4) is the most common and has the most severe effects. It is only moderately anaerobic and can multiply in areas such as surface wounds containing devitalized tissue in which oxygen is reduced but not completely absent. The bacillus actively ferments glucose liberated from muscle glycogen reserves, creating pockets of gas that distort and devitalize adjacent tissue. It also manufactures an extensive collection of hydrolytic enzymes that liquefy tissue, break down cell membranes, and cause hemolysis. *C. septicum* and *C. novyi* are also encountered. Rarely, wound botulism due to infection with *C. botulinum* has been documented. *Bacteroides* species and anaerobic streptococci may cause a similar condition.

The infection may take several forms. Cellulitis, relatively less severe, occurs when underlying connective tissue is invaded. True gas gangrene or myositis occurs when muscle itself is affected. Because the organisms grow very rapidly at the expense of muscle sugar, it is more feared. Life-threatening puerperal infection occurs when invasion occurs following a septic childbirth.

Within about three days, evidence of gas distension in the tissues is seen. The patient shows personality changes due to systemic toxicity, becoming disoriented and obstreperous. Prompt treatment with penicillin or related antibiotics, and surgical removal of affected tissue and adjacent regions (debridement) are indicated. These measures can hold the mortality to about 20 percent. In some cases, exposure to hyperbaric oxygen (at elevated pressures) is used. Gas gangrene now develops in less than one in a thousand trauma patients, largely because of effective debridement and wound cleansing.

Bites

Oral flora; Chapter 16, pp. 471–475

A bite is a very likely site for a serious polymicrobic infection. The bite introduces a large sample of the oral flora of the biting individual. This flora is preselected for growth within the biter's body. It is thus automatically more likely to infect than is a bunch of soil organisms rubbed into a cut or laceration.

Human bites Oddly enough, we tend to take animal bites much more seriously than human bites, although our attitude should be exactly the opposite. The anaerobic flora of the human mouth include such genera as *Actinomyces* and *Bacteroides* as normal flora. A significant number of persons may also carry the true pathogen *Streptococcus pyogenes.* If a human bite is not adequately cleaned, or is prematurely closed by suture or tight bandaging, infection is inevitable. Such wounds should be treated as extremely serious threats. Human bites are demonstrably more dangerous than those of nonrabid animals.

Animal bites The oral flora of a normal animal is adapted to that species. It is not quite as likely to thrive when transferred to human tissue. There are, of course, exceptions, and animal bites should receive immediate medical attention.

Pasteurella multocida is the most common pathogen in animal bites. These Gram-negative bacilli, closely related to the plague and tularemia agents, can cause severe infections in human beings. This organism responds to therapy with antimicrobials.

There are two distinct forms of **rat-bite** fever. In the United States, *Streptobacillus moniliformis* is (rarely) seen to cause a disease characterized by fever, headache, chills, swollen lymph nodes, arthritis, and a rash on palms and soles. It responds promptly to penicillin or tetracycline. The organism can be acquired from bites of mouse, weasel, cat, and squirrel. In the Far East, but occasionally in the West too, *Spirillum minus*, a spiral organism, is seen in a milder disease lacking arthritis. Penicillin or streptomycin is useful.

Monkey or other primate bites are more hazardous to human beings than are those of less closely related animals. Persons who work closely with caged monkeys have been exposed to a number of diseases, some of them, such as Marburg disease, being extremely pathogenic to human beings.

Skin Ulcers

Bedsores or **decubitus ulcers** are only partly a microbiological problem. They originate because of localized tissue breakdown that results from reduced circulation due to prolonged pressure on an area, in the bedridden patient. Common sites are areas over bony crests such as hipbones, ankles, or elbows. The devitalized tissue becomes colonized by a mixed microbial flora whose contribution to a worsening problem is unclear. In any case, healing bedsores depends primarily on restoring circulation to the area and secondarily on controlling the flora. Topical disinfectants are more useful than are antibiotics in this situation.

Fungal Infections

Traumatic contact with wood splinters or thorns may introduce a variety of soil fungi. Several of these cause chronic, slowly progressive, and usually benign conditions. **Sporotrichosis** is an invasion of cutaneous and lymphatic tissues by *Sporothrix schenkii*. Found largely in warm climates, sporotrichosis is common in field-workers, nursery-workers, and others of similar occupation.

Following accidental inoculation, for example by a blackberry thorn, a small lesion develops at the site. Regional lymph nodes draining the area may become enlarged. Rarely, pulmonary or disseminated disease has developed. Sporotrichosis may be self-limiting; it responds rapidly to potassium iodide or amphotericin treatment.

BURNS

In terms of their potential for fatal infection, severe, full-thickness burns are the most challenging of all wounds. The patient undergoes not only the visible tissue destruction but profound metabolical alterations as well. The burn site differs from other types of wound in that anaerobiosis in not a major feature. Recovery is slow and may be extended over a prolonged period by the need for grafts to reconstruct the damaged area. Infection most commonly begins 2–4 weeks after injury. However, the burn site presents an unparalleled opportunity for microbial invasion throughout the recovery period.

Control of Infection in Burns

Seriously burned patients are customarily stabilized, then transferred to special burn-care centers. There their special metabolic, pain control, and psychological needs will be effectively met while medical and surgical care is supplied.

Protective isolation is frequently used. That is, the patient is placed in an environment in which he or she is protected from exposure to environmental flora or the normal flora of the staff. Prophylactic penicillin or nafcillin is administered for a period. The burn is cleansed initially and daily with silver sulfadiazine. These methods have effectively reduced the risk of infection with such common human pathogens as *Staphylococcus aureus* and *Streptococcus pyogenes*. Now, the vast majority of burn infections are caused by the highly resistant opportunist *Pseudomonas aeruginosa*. Occasional epidemics are caused by *Klebsiella pneumoniae*, *Enterobacter cloacae*, *Providencia stuartii*, *Serratia marcescens*, and *Flavobacterium meningosepticum*. Burn pathogens are usually aerobic organisms.

One major function of burn care is frequent observation of the appearance of the burn margins or **eschar** (covering that forms over the burn) to detect the early, possibly curable stages of infection. In some centers, frequent routine biopsy and quantitative culture of the burned area and of blood are performed. If infection occurs, appropriate antimicrobials can be administered. However, bacteria readily move from the burned tissue into the circulation; chances are poor that the severely burned patient will survive after septicemia develops.

Septicemia; Chapter 24, p. 697

Pseudomonas aeruginosa Infection

Causative agent   *P. aeruginosa* is a small, highly motile, Gram-negative rod. It is aerobic, nonfermentative, and extremely adaptable in its nutritional and environmental tolerances. Its normal habitat is wide-ranging. It is found not only in the human GI tract but also in aqueous solutions of many types.

The organism is presumptively identified on the basis of its blue-green pigment, **pyocyanin**, its fluorescent yellow pigment, **fluorescein**, its very strongly positive oxidase test, and its distinctive grapelike aroma. Confirmatory identification is based on serotyping, phage typing, and other specialized tests.

Pathogenic strains of *P. aeruginosa* produce at least three virulence factors. **Exotoxin A** is an inhibitor of mammalian protein synthesis. Its mode of action is like that of diphtheria toxin. **Protease** and **elastase** exoenzymes directly attack tissue components and contribute to invasiveness. These three factors are found more commonly in strains isolated from severely ill patients than they are in strains originating from well persons or nonclinical environments.

Pathogenesis   On entry to the burned area, the organism multiplies rapidly in the tissue fluid leaked from the wound. Its proteases contribute to penetration of deeper layers and loss of eschar. The exotoxin contributes to failure of skin grafts and scarring. Since both PMN function and CMI are depressed in the burned patient, the agent is frequently able to initiate a septicemia with systemic distribution of the toxic products.

Therapy and immunity   *P. aeruginosa* is resistant to a wide range of the better-tolerated antimicrobial agents. It may respond to such agents as gentamicin, amikacin, or carbenicillin, alone or in combination. The percentage of strains resistant to gentamicin is rapidly increasing. These antimicrobials must be used with great caution because of renal and neurological toxicities. Protease inhibitors infused into the burn site are being tried experimentally with some success.

There has always been interest in the possibility of immunological controls for *Pseudomonas.* A killed-bacterial vaccine has been administered with variable success to some burn patients in the early phases of their treatment. Since the importance of the exotoxin has become obvious, antitoxins and now formalinized toxoids have been developed for experimental trial. It is to be hoped that these will prove clinically useful.

Epidemiology   *Pseudomonas aeruginosa,* an exceptionally hardy organism, not only survives but multiplies in some commonly used hospital disinfectants, liquid   Table 28.7

Box 23.1

Surgical Infections Are Decreasing

Over the period 1975–1978 a steady decrease has been noted in most categories of surgical wound infections. The improvements have been most significant in university-teaching hospitals, and smallest in the small community hospital.

The improvements are due to a general acceptance of the newest advances in aseptic techniques, plus the intensive surveillance of hospital infection control programs and the growing influence of professional peer review.

Some of the changes require expensive new technology, such as high-efficiency particulate air filters and elaborate unidirectional air flow systems. These contribute to the much-publicized increase in medical care costs. However, by far the majority of changes are relatively in-

SERVICE	1975 RATE*	1978 RATE
Surgery	170.1	144.6
Obstetrics	75.2	95.1
Gynecology	125.7	104.4
All services	86.0	72.1

*Rates = infections per 10,000 discharged patients.

expensive because they are simply alterations in human behavior. Adoption of one-way traffic flow, use of satisfactory scrub-and-preparation techniques, and alterations in the timing of pre-surgical antimicrobials are not at all costly.

The only increased rate is that for obstetric services. This is proposed to be related to the much increased rate of caesarian sections.

soaps, and water reservoirs for respiratory therapy apparatus. Although new lurking places are repeatedly discovered and cleaned up, it is probably literally impossible to exclude the organism from the burn unit unless we also exclude human beings.

SURGICAL WOUNDS

Surgical asepsis;
Chapter 25, pp.
744–745

Data on the incidence of post-surgical infection have in the past been hard to obtain. They varied greatly among medical centers, among surgeons, and among procedures. With the general introduction of hospital infection-control programs, such data are now being routinely collected. This activity is very helpful in finding and removing preventable causes of surgical sepsis.

From the point of view of maintaining surgical asepsis, the most difficult problems occur when the surgery involves the GI, GU, or respiratory tracts in which there are large resident floras. These surfaces are difficult if not impossible to disinfect. Many of the flora are suitably adapted to survive when introduced into other body skin areas. Skin, by contrast, can be effectively disinfected (not sterilized); if skin organisms are transferred to deep tissues, few have the capacity to survive.

A few years ago, when we had a great deal of misplaced confidence in the potency of antimicrobial drugs, massive drug therapy preceded intestinal surgery,

TABLE 23.4
Agents of surgical infection

AGENT	PERCENT OF WOUND INFECTIONS IN WHICH ISOLATED
Staphylococcus aureus	15.4
Escherichia coli	15.2
Group D streptococci	10.3
Proteus spp.*	7.2
Klebsiella spp.	5.4
Staphylococcus epidermidis	4.8
Pseudomonas aeruginosa	4.6
Enterobacter spp.	4.1
Nongroup D streptococci	5.4
Bacteroides fragilis	2.9
Other anaerobes	6.7
No pathogens	4.7
No culture	6.5

*The letters spp. indicate that several species are involved.

with its goal being the sterilization of the bowel. Analysis of the clinical data showed clearly that this approach was of no use, and may have actually increased the rate of postsurgical complications. The routine "prophylactic" administration of antimicrobials is now a matter of active debate. If they are given, they should be administered 30 minutes prior to and during the surgical procedure.

Human pathogens may be brought to the surgical suite by the staff. Most postsurgical infections at this time are caused by either *Staphylococcus aureus* or *Streptococcus pyogenes.* Careful followup frequently traces these to a carrier in the operating room. Symptoms usually appear several days after surgery. They may be localized to the area of suture and incision or may reflect a more general deepseated process. Treatment may involve reopening and cleansing the site and/or treatment with carefully selected antimicrobials.

Antimicrobials penetrate abscessed areas poorly. Abscess formation is characteristic of the postsurgical infection. Drainage of the material leads to reduced inflammation, better antimicrobial drug access, and recovery.

Microbiological goals of postsurgical nursing care are to minimize pathogens' access to the healing incision and to provide surveillance for the earliest signs of trouble.

SUMMARY

1. The process of wounding initiates a series of changes in the affected tissue. When a large area is affected or the wound is inaccessible or the healing time is prolonged, the infection risk is considerable.

2. Wounds sometimes become infected by the standard bacterial pathogens such as *Staphylococcus* or *Streptococcus.* However, the infecting agents are more often part of the normal flora of skin or mucosa. Wounds are usually colonized by a mixed population with diverse metabolic activities. A symbiotic relationship among the species promotes their survival in the damaged tissue.

3. Anaerobiosis is a factor in most wound environments. Anaerobic bacteria, derived from soil, feces, or the normal body flora, are a major factor in wound infections. Anaerobic members of polymicrobic mixtures are the most significant in pathogenesis. Specialized laboratory techniques are used to isolate and cultivate anaerobes.

4. Tetanus and gas gangrene are specific infections of anaerobic wounds. Once common causes of death, following occupational and battlefield casualties, they are now rare. Tetanus is completely preventable by prophylactic immunization with toxoid.

5. Burns present special problems in wound management. There is a general alteration of the patient's defenses, and healing may be extended over many months. The opportunist *Pseudomonas aeruginosa* is the major threat.

6. Surgery may lead to infection, especially if surgery involves areas with a highly developed normal flora. Escape of bacteria into underlying tissues can lead

to abscess formation if the agents are not controlled by the host defenses. Other surgical infections occur when pathogens are transferred in the operating theater from a carrier to the surgical field. The inflammatory response to sutures seems to assist the pathogen to gain a foothold.

Study Topics

1. Outline the sequence of events that leads to the development of anaerobic conditions within a wound.

2. What specific steps can be taken to minimize the development of anaerobiosis in a wound?

3. Describe the characteristics of *Pseudomonas aeruginosa* that render it well adapted to cause burn infections.

4. Referring to Chapter 14 if necessary, describe precisely why frequent administration of tetanus toxoid can be harmful.

5. What are the differences among aerotolerant anaerobes, facultative anaerobes, and strict anaerobes?

Bibliography

Balows, A., et al. (eds.), 1974, *Anaerobic Bacteria: Role in Disease*, Springfield, Ill.: Charles C Thomas.

Hoeprich, P. D., (ed.), 1977, *Infectious Diseases* (2nd ed.), New York: Harper & Row.

Smith, L. de S., 1975, *The Pathogenic Anaerobic Bacteria* (2nd ed.), Springfield, Ill.: Charles C Thomas.

24

Circulatory and Lymphoreticular Infections

The phagocytic network, circulating antibodies, and lymphoid cells—the body's major defenses—are focused on the blood and lymphatic circulations and their associated organs. For an agent to infect here, in the body's central stronghold, requires the agent to have specialized mechanisms. There are, however, many serious circulatory pathogens. Each has a mechanism by which it is protected—such organisms "hide out" by residence in a sheltered microenvironment.

ANATOMY AND PHYSIOLOGY

The circulatory system comprises the heart, blood vessels, and lymphatic vessels (Fig. 24.1). Their normal anatomy and physiology is complex, and you might find reference to a good, general anatomy text useful. We will restrict our coverage to those structures and functions commonly affected by microbial pathogenesis.

Lymphoid organs;
Chapter 13, p. 375

The structure and function of the lymphoid organs were presented and discussed in Chapter 13.

687

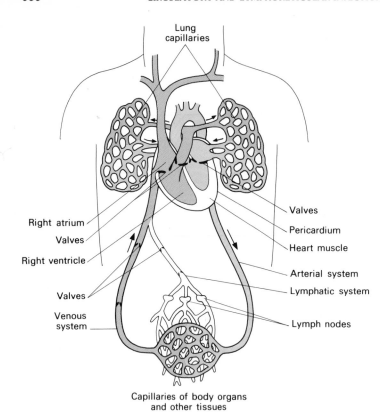

24.1
The Human Circulatory System Both blood and lymphatic vessels are shown.

Anatomy of the Circulatory System

The heart   The heart, a compact muscular organ (Fig. 24.2), is specialized for efficient blood pumping. The heart is suspended in the thoracic cavity and covered by the pleural layers forming the **pericardium.** The heart itself is a mass of cardiac muscle, the **myocardium,** and it is lined inside, as is the entire circulation, with a delicate, smooth network of **endothelium.** The heart contains four chambers. The flow of blood into, through, and out of these chambers is unidirectional. Backflow is prevented by a system of **valves.** Valves are leaflets or folds of endothelium reinforced with dense fibrous connective tissue. Swelling, erosion, or deformation of these valves reduces the heart's efficiency.

The vascular network   **Arteries** conduct blood from the heart. They are characteristically rigid, with thick, muscle-reinforced walls (Fig. 24.3). They are lined with endothelium, lying over a subendothelial, connective layer. The intact arterial wall is almost completely impermeable to living cells or molecular substances.

Arteries subdivide into smaller branches when they reach the tissues in which the final division into **capillaries** takes place. A capillary wall is a single, endothelial

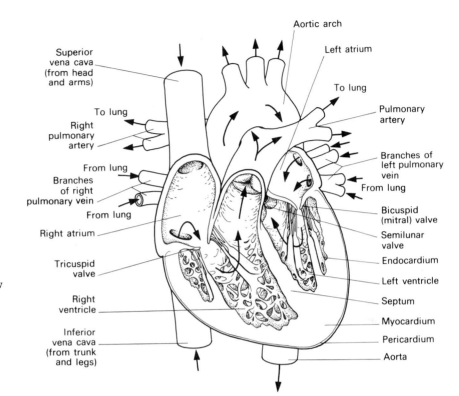

24.2
The Human Heart Blood flow through the heart is regulated by the cardiac valves that prevent backflow under the pressure of contracting heart muscle. The heart is lined with smooth endocardium.

Superior vena cava (from head and arms)

Aortic arch

Left atrium

To lung

To lung

Right pulmonary artery

Pulmonary artery

From lung

Branches of right pulmonary vein

Branches of left pulmonary vein

From lung

From lung

Right atrium

Bicuspid (mitral) valve

Tricuspid valve

Semilunar valve

Endocardium

Right ventricle

Left ventricle

Septum

Inferior vena cava (from trunk and legs)

Myocardium

Pericardium

Aorta

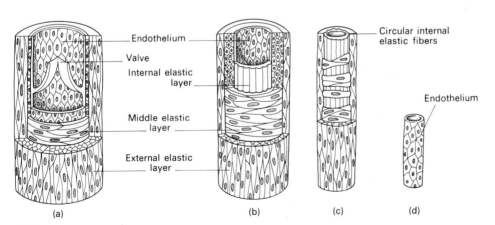

Endothelium

Valve

Internal elastic layer

Middle elastic layer

External elastic layer

Circular internal elastic fibers

Endothelium

(a)   (b)   (c)   (d)

24.3
Blood Vessels (a) Cross section of a vein. Veins are thinner-walled than arteries. (b) Cross section of an elastic artery. Arteries have thick, muscular walls. The inner lining is impermeable to the passage of blood cells or substances. (c) An arteriole. (d) A capillary.

cell thick. It is enclosd by an acellular **basement membrane.** Solutes readily enter and leave, but migration of cells across this barrier is partially restricted. Lymphocytes and PMNs can move from capillaries to tissues. This movement is much increased in inflammation. RBCs do not leave intact capillaries.

In spongy organs, such as the spleen, liver, and certain endocrine glands, the circulation divides into large, irregular passageways called **sinusoids** rather than into capillaries. Circulating blood, phagocytic RE cells, and lymphoid tissue are closely associated in these organs.

Veins, thinner-walled than arteries, collect blood and return it. The returning blood is under little pressure because of its distance from the heart. Backflow is prevented by action of valves in the vein wall. These valves, and the vein endothelium, are more susceptible to inflammation than is the arterial lining.

The lymphatics   Lymph is fluid that has been forced out of capillaries into tissues under the pressure of the heart's pumping action. Lymphatic vessels, composed of endothelium and connective tissue, collect lymph and carry it toward the heart, merging with the blood circulation at the **vena cava. Lymph capillaries** are composed only of endothelium, lacking a basement membrane. They are readily entered by infective agents that are then transported in the lymph. As you will recall, lymph passes through aggregates of filtering tissue, the **lymph nodes,** in which susceptible agents are promptly destroyed by RE cells. Resistant agents, however, may survive and multiply in the lymph node.

Physiology of the Circulatory System

Integrity of the vascular network   Given effective heart action, the most important factor affecting fluid flow is the condition of the blood vessel walls. A perfectly smooth lining offers minimal resistance to blood flow and no foci for clots to form. Irritations, erosions, or surgical procedures can create rough spots, areas of turbulent flow or scar lines ("rapids"). These provide niches in which bloodborne bacteria can attach and escape phagocytosis. Turbulence also promotes clotting. An abnormal surface may collect clotted blood that further impedes blood flow.

Permeability   Capillary permeability is significantly increased by inflammatory changes. Abnormal permeability may follow endothelial multiplication of organisms such as the rickettsiae or exposure to bacterial endotoxins. When capillaries rupture, petechiae form on the skin, gums, and on other tissues throughout the body. Hemorrhage is the result of the rupture of a larger vessel.

Coagulation   Blood's ability to clot is a primary life-saving mechanism. Clots not only arrest blood loss but also wall off areas of potential infection. However, clotting can also be initiated by inappropriate or unrelated stimuli. One such stimulus is an irregularity in a vessel lining. Another is bacterial products. Anaerobic bacteria in particular produce extensive intravascular coagulation near an infection focus.

Thrombi—clots or fragments of clots—may break away and then lodge downstream. An **embolus** is such a clot that has completely occluded a small vessel. When the nutrient and oxygen supply is cut off by an embolus, the tissue normally served by the blocked vessel suffers ischemia and may die. Thrombosis is a grave matter in brain, lungs, or kidneys.

Physiological changes in lymphoid tissue   The mitotic activity of lymphoid tissue depends on the degree of antigenic stimulation it is receiving. When infective agents enter the lymphatic circulation and are filtered out by a draining lymph node, the node will swell. This **lymphadenitis** may reflect both recruitment of additional lymphocytes from the circulation and increased cell division. The swelling will usually persist as long as the pathogen does.

Lymphoid activity; Chapter 13, pp. 378–380

A cellular immune response may develop around the lymphoid tissue containing the pathogen. Granuloma and ulceration may result.

Routes of Infection

The lymphatic route   It is easier for microorganisms to invade the lymphatic system than to invade the blood circulation because the lymphatic system's capillaries are thinner-walled. Many infectious agents entering the lungs (e.g., *M. tuberculosis*) or the gut (e.g., *S. typhi*) move promptly into the lymphatics and are disseminated. This pattern is also seen in systematic viral infections such as measles.

Typhoid fever; Chapter 20, pp. 586–590

Recall, though, that microbial contact with lymphoid tissue has its beneficial aspects too. It stimulates effective immune responses. In some infections, brief residence of a pathogen in the lymphoid tissue evokes sufficient immunity to suppress the infection. We are concerned here with diseases in which the organism is **not** promptly controlled. It either multiples destructively in the lymphoid tissue itself or enters the blood circulation. There seems little doubt that, in the absence of obvious inoculation, the usual route to circulatory infection is from surface tissue to lymphatics to the blood.

Wounds   Trauma lets microorganisms enter the bloodstream directly. In the normal individual, highly virulent species such as *Streptococcus pyogenes* can persist and infect. Low-virulence organisms escape phagocytosis only if they encounter an area of devitalized tissue or an abnormal surface as they are carried along. When this happens, they may survive and cause a local surface infection.

Bacteremia—bacteria present in the blood—may become **septicemia**—bacteria multiplying in the blood. This will occur if the agent is an extracellular pathogen of high virulence, if the host's defenses are severely depressed or compromised, or if there is an attack site such as a damaged heart valve.

Insect vectors   Many insects feed on vertebrate blood. They may withdraw blood containing an infective agent from one host and regurgitate a portion to infect their next host. Or, alternately, the infective agent multiplies in the insect's organs. It

TABLE 24.1
Clinical terminology of circulatory infection

TERM	DEFINITION
Embolism	Obstruction or occlusion of a vessel by a transported clot or vegetation, a mass of bacteria, or other foreign material
Endocarditis	Inflammation of the endocardium or lining membrane of the heart, involving the valves or the heart walls
Lymphadenitis	Inflammation of a lymph node or nodes, often seen as a swelling
Lymphangitis	Inflammation of the lymph vessels, often seen as a red streak under the skin
Myocarditis	Inflammation of the muscular walls of the heart
Pericarditis	Inflammation of the pericardium
Phlebitis	Inflammation of a vein
Rheumatic	An indefinite term applied to various conditions with pain or other symptoms of the joints and musculoskeletal system. One of these is rheumatic fever
Rheumatoid	Resembling a rheumatic condition
Thrombosis	The formation of a thrombus or clot or its presence
Thrombophlebitis	Inflammation of a vein with secondary thrombus formation

may supply a continuing inoculum in its saliva. As an insect bites, it also commonly defecates, and the feces may be loaded with the pathogenic organism. The bite itches, and the host scratches, thereby inoculating the puncture site with the insect's feces.

The females of certain insects incorporate the pathogen into their eggs. This **transovarial**, vertical transmission can maintain a disease in an area for years, independent of animal hosts.

Blood Culture Techniques

In most circulatory infections, correct diagnosis and treatment depend heavily on isolating the agent from a blood sample. Blood culture has proved to be a most challenging task. Various commercially developed systems have been marketed for collection, incubation, and detection of microorganisms from the blood (Fig. 24.4).

Literally, almost every known microorganism has been isolated from the blood at some time. Collection procedures must attempt to prevent introduction of misleading skin flora: multiple cultures from separate sites are advisable. Initial samples should be obtained before antimicrobial therapy is begun.

Even in full-scale circulatory infections, the bacterial count of blood will probably be less than 50 cells/mL. Three 10-mL samples of intravenous blood collected

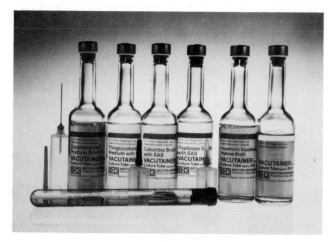

24.4
Blood Culture Media   A wide variety of media is available such as supplemented thioglycollate broth for anaerobic culture and hypertonic supplemented broth for the cultivation of fragile bacteria and L-forms, and regular media for the cultivation of facultative anaerobes. Anticoagulants (SPS, SAS) are necessary. The vacuum system shown here provides a continous sterile pathway from patient to container and delivers a 2, 5, or 10 mL sample as needed. Vents may be introduced to provide desired gas exchange.

one or two hours apart from different sites are recommended. Complement in the collected sample may be bactericidal, so the blood must be neutralized by five-fold dilution into 40 mL of sterile fluid media.

Culture methods must be flexible enough to provide for the possible detection of aerobes, anaerobes, and microaerophils. Enriched complex media are most frequently used. Normally pairs of culture bottles are used, one with a 10-percent CO_2 atmosphere, and one with an oxygen-free atmosphere (Fig. 24.4). Hypertonic media may be used to allow growth of wall-deficient bacteria.

Growth may be detected microscopically by the Gram stain or by staining smears with acridine orange. This dye causes any bacteria present to fluoresce. Periodic blind subculture to fresh plated media is essential to allow isolation and identification of the offending organism. Growth can also be observed by detecting release of radioactive CO_2 from labeled substrates provided in the medium.

There are several reasons for false negative results. A fastidious or slow-growing organism may not be detected, the agent may be an intracellular parasite not cultivable by these means, or antimicrobial therapy started before the samples were obtained has rendered the organisms incapable of growth in culture. Sadly, this does not always mean that they are equally inhibited within the patient.

BACTERIAL INFECTIONS OF THE CIRCULATORY SYSTEM

Infective Endocarditis

Many agents cause infective endocarditis, which is an inflammation of the endocardium, particularly in the area of the heart valves. There is a spectrum of severity of which the **acute** and the **subacute** forms are the extremes (Table 24.2). In its acute form, endocarditis is a rapidly progressive, frequently fatal process in which normal heart valves are colonized and destroyed. *Staphylococcus aureus, Streptococcus*

TABLE 24.2
Comparison of acute and subacute forms of infective endocarditis

FEATURE	ACUTE FORM	SUBACUTE FORM
Important agents	Pathogenic bacteria include:	Relatively nonvirulent bacteria include:
	Staphylococcus aureus *Streptococcus pyogenes* *S. pneumoniae*	Viridans streptococci Group E streptococci Enterococci Enterobacteria
Underlying condition	Normal heart or implanted prosthesis	Congenital or acquired heart abnormality
Provoking incident	Trauma; surgical infection	Minor trauma, tooth-brushing, dental work, gyne-cological procedures, other surgery or catheterization
Onset	Moderately rapid	Insidious
Duration	Less than six weeks to death if untreated	More than six weeks
Pathogenesis to heart valve	Vegetation, destruction of underlying tissues	Vegetation, little destruction
Potential for secondary or general infection	Usually progresses to general infection or septicemia	Little potential for extension to other normal tissues

pyogenes, or other true pathogens are the agents. The infection readily spreads to involve large areas of the vascular system. It occurs following mechanical invasion of blood vessels and fortunately is quite rare.

The subacute form is more slowly progressive, because the agent is of low virulence. It requires preexisting abnormality of the heart valve. Spread to other, normal tissues is rare. Invariably fatal if untreated, it can almost always be cured if the organism(s) are isolated and an appropriate antimicrobial therapy designed. Although there are many possible agents, we will discuss the disease in reference to the non-Group A streptococci, which cause 60–70 percent of all cases, to familiarize you with this important collection of bacteria.

Group A *Streptococcus;* Chapter 17, p. 504

Group B *Streptococcus;* Chapter 21, p. 646

Causative agents   The Lancefield classification is based on the serological detection of surface antigens. Serology is the most effective way of identifying these organisms because their morphology and biochemistry are not highly distinctive. Latex agglutination or coagglutination techniques are rapid, inexpensive, and easier to perform than is the traditional Lancefield technique.

TABLE 24.3
The non-Group A streptococci

LANCEFIELD GROUP	REPRESENTATIVE SPECIES	HEMOLYSIS	BIOCHEMICAL DIFFERENTIATION	DISEASES CAUSED
B	*Streptococcus agalactiae*	Beta	Hydrolysis of sodium hippurate	Neonatal meningitis Neonatal septicemia Suppurative arthritis
C	*S. equi* *S. equisimilis* *S. dysgalactiae*	Beta		Respiratory infection Infective endocarditis (IE) and bacteremia
D	Enterococcus group: *S. faecalis* *S. faecium* *S. durans*	None (or variable)	Growth at 10°C, 45°C Growth at pH 9.6 Growth in 6.5% NaCl Growth in bile-esculin Penicillin-resistant	Abdominal abscesses IE and bacteremia Urinary tract infection
	Nonenterococcus group *S. bovis* *S. equinus*	None (or variable)	Growth in bile-esculin not in 6.5% NaCl	IE and bacteremia
F	*S. anginosius*	Variable		Abscesses Sinusitis rare Meningitis
H	*S. sanguis*	Alpha		IE and bacteremia
K	*S. salivarius*	Alpha		IE and bacteremia
UNGROUPED				
Aerobic	*S. mitis*	Alpha		IE, bacteremia
	S. mutans	Alpha	Polymerizes sucrose	Dental caries, IE
	S. pneumoniae	Alpha	Ferments inulin, bile soluble	Pneumonia IE
Anaerobic or microaerophilic	*Peptococcus magnus saccharolyticus prevotii* *Peptostreptococcus micros anaerobius intermedius*	Variable		Necrotizing pneumonia Abscesses Wound infections Gynecologic infections

Some characteristics of the non-Group A streptococci are summarized in Table 24.3. *S. mutans, S. sanguis,* and the enterococci are the most prominent in infective endocarditis.

Pathogenesis   Subacute bacterial endocarditis (SBE) can develop only if there is a preexisting heart valve lesion. The most common cause of such a lesion is a hypersensitivity, such as that produced in rheumatic fever. Hypersensitivities can produce valve distortion and scarring. Tertiary stage syphilitic lesions may also destroy a valve. Congenital heart deformities are common with Down's syndrome (mongolism). Valve prostheses implanted by cardiac surgery and the lipid-containing plaques that develop in atherosclerosis may also provide a focus.

SBE commences when normal flora organisms of low virulence enter the blood, usually from the mucosa. Dental and gynecological procedures are a common source, as the oral and vaginal floras are rich in streptococci. The bacteria attach to the valve surface as the blood passes through the heart. They persist, and a **vegetation** forms at the site. The vegetation is a loose, irregular aggregate of fibrin containing microcolonies of bacteria, over a necrotic mass of granuloma tissue and connective tissue fibers. The mass enlarges slowly, in a downstream direction. Periodically, fragments break off and are carried along to lodge in smaller blood vessels.

Classic symptoms of a severe general infection, such as fever and toxicity, are accompanied by heart murmurs, fatigue, and signs of impaired cardiac function. Serious or lethal developments follow lodging of emboli in the spleen, liver, kidney, or cerebral circulation.

Therapy and immunity   There is no cure without antimicrobial therapy. However, choice of the drug(s) and dosages to use may be extraordinarily complex. The vegetation protects the bacteria so that neither antimicrobials nor immune factors penetrate it effectively. The patient's ability to clear the bacteria without aid has obviously failed, so the drug(s) used must be bactericidal. After the minimum lethal concentration (MLC) of the drug is determined, a sufficient dose is given to provide about ten times that concentration in the serum. Therapy must be extended over many weeks, and follow-up blood cultures done to ensure complete clearance of the agent. Synergistic combinations of drugs have been extensively tried.

Immunity has limited effect on the progress of the disease. The development of immunization has not been considered worthwhile.

Epidemiology   Certain individuals have a clearly identifiable risk because of their medical history. Before they undergo any procedure likely to cause bacteremia, a brief course of prophylactic penicillin should be given. This measure generally prevents valve colonization, at least in the healthy person. In hospitalized patients, penicillin-resistant organisms are significantly more frequent, so this technique is not as effective.

Dental health procedures; Chapter 16, pp. 490–493

Thrombophlebitis

Thrombophlebitis accounts for about 10 percent of hospital-acquired infections. It is an irritation of the venous endothelium, developing at or near the site of a surface lesion. It may develop under an infected abrasion or decubitus ulcer. Most often, it appears at the site of an improperly maintained intravenous catheter. Inserting a needle through the skin layers always provides a potential for infection. Irritation is caused by the continued presence of the needle, movement, and mechanical stress. The risk increases proportionate to the length of time the needle stays in place. Daily care of the site is crucial, and rotation of the intravenous equipment to a new site within 72 hours is recommended. Because of the risk of skin microbes working their way up the tubing to contaminate the fluid reservoir, all parts of the unit except the catheter itself require daily changing.

Septicemia

Septicemia is rarely a primary disease. It may appear as the ultimate development of almost any infection. You have already read about it a number of times. At this point, a few observations should be added to expand your understanding of the real effects of microbial growth in the blood.

Uncontrolled multiplication of bacteria in blood seems to occur only when phagocytic potential is much lowered. The patient may have used up reserves of phagocytic cells or complement in an unsuccessful attempt to subdue a local infection elsewhere. Septicemia also is characteristic of the immunosuppressed, debilitated, or moribund person.

When bacteria multiply in blood, toxins may be released. Some exotoxins of the Gram-positive organisms are destructive to blood cells. Streptokinase, coagulase, and the fibrinolytic enzymes probably play a role in intravascular coagulation.

Endotoxin is released by dying Gram-negative organisms. In large amounts it produces a characteristic syndrome called **endotoxin shock** (sometimes called **septic shock**). Endotoxin causes adverse changes in vascular function. These include a loss of smooth muscle contractility in the arterial walls, weakening of the cardiac muscle, and capillary breakdown.

Endotoxin; Chapter 12, pp. 355–356

When Gram-negative organisms initiate septicemia, for a brief period most cells are in log phase, and endotoxin release is minimal. Then complement-mediated phagocytosis or lysis and random bacterial death begins to increase the free endotoxin. Symptoms of cardiovascular collapse develop. A cruel paradox binds the physician who needs to reverse the situation. Antimicrobials given in Gram-negative septicemia may kill the organisms, but will by doing so cause a large endotoxin surge that could cause the heart to fail. Without antimicrobials, the patient will surely die. The only way to beat this double bind is to monitor the patient's condition for the earliest indicators of septicemia. Antimicrobials may then be used while the bacterial population is still small.

Rocky Mountain Spotted Fever (RMSF)

The obligate intracellular parasites called rickettsiae cause a number of serious cardiovascular infections. These agents are currently divided into the spotted fever group, the typhus group, and the scrub typhus group. Q fever is a pneumonia caused by the related agent *Coxiella burnetii*.

TABLE 24.4
Characteristics of the human rickettsial diseases

GROUP	DISEASE	AGENT	GEOGRAPHY
Spotted fever	Rocky Mountain spotted fever	*Rickettsia rickettsii*	Western hemisphere
	Tick typhus	*R. australis*	Australia
	Rickettsialpox	*R. akari*	North America, Europe
	Boutonneuse	*R. conorii*	Europe, Africa, Middle East, India
Typhus	Epidemic typhus	*R. prowazekii*	Worldwide
	Brill–Zinsser disease	*R. prowazekii* latent in tissues, reactivates	South America, Europe
	Endemic typhus	*R. typhi* *R. mooseri*	Worldwide
Scrub typhus	Scrub typhus	*R. tsutsugamushi*	Asia, Australia, Pacific Ocean
Q fever	Q fever	*Coxiella burnetii*	Worldwide
Trench fever	Trench fever	*Rochalimaea quintana*	Europe, Africa, North America

The rickettsiae multiply in the vascular endothelium, causing a peripheral vasculitis. In the United States, the most common and serious is Rocky Mountain spotted fever. We will use this disease as a model for discussing rickettsial infection.

Causative agent   *Rickettsia rickettsii* and its antigenic varients cause the spotted fever group of diseases. This group also includes typhus, Boutonneuse, and rickettsialpox. Rickettsiae are typical bacteria in structure, and possess a Gram-negative type of cell wall. They divide by binary fission. These agents require an intracellular location for growth (Fig. 24.5). These deficiencies also ensure that if the agent leaves the host cell for a prolonged period, it will die of starvation. Thus a living vector is required for transmission of all of this group.

Rickettsiae may be cultivated in the yolk sac of eggs or in tissue culture. They can be readily harvested and are converted to standard antigen preparations. These

VECTOR	RESERVOIR	POTENTIAL SEVERITY	RASH	WEIL–FELIX REACTION
Tick	Tick, wild rodent, dog	Severe	Yes	+
Tick	Marsupial wild rodent	Moderate	Yes	+
Mite	Rodent	Mild	Yes	−
Tick	Rodents, dogs	Mild	Yes	+
Body louse	Human being	Severe	Yes	+
Body louse	Human being	Moderate	Yes or no	+
Flea	Rodents	Moderate	Yes	+
Mite	Rodents	Severe	Yes	+
Ticks; nonvector transmission	Domestic livestock	Mild	No	−
Body louse	Human beings	Mild	Yes	−

are used to identify specific antibodies in patient sera. Complement fixation, agglutination, and fluorescent antibody techniques are most useful in laboratory diagnosis of rickettsial disease. Antibodies formed during rickettsial disease also cross-react with antigens found in three strains of *Proteus vulgaris*, OX-19, OX-2, or OX-K. Mixture of patient sera with these *Proteus* antigens gives a positive agglutination, called the **Weil–Felix** reaction.

Pathogenesis   The Rocky Mountain spotted fever agent, *Rickettsia rickettsii*, is introduced by a tick-bite. After about 2–6 days, symptoms begin. There is a very high fever (up to 105–106°), extreme muscle tenderness, and a severe headache that does not respond to analgesics such as aspirin. The rash appears on the second or third day, on the ankles and wrists. This rash localization is highly unusual and is a valuable clue in diagnosis. There may be neurological abnormalities. If untreated, up to

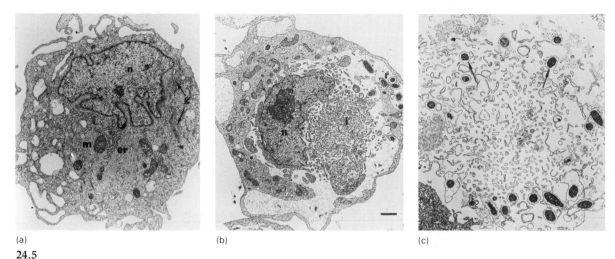

(a)   (b)   (c)

24.5
Rickettsial Infection   (a) Normal uninfected chick embryo cell in monolayer culture (n), nucleus; (m), mitochondrion; (er), endoplasmic reticulum; (g), Golgi apparatus. (b) Ninety-six hours after infection with *Rickettsia rickettsii*. Note empty spaces in cytoplasm, rickettsia (r) surrounded with membrane layer, and breakdown of nuclear membrane. There are many abnormal cytoplasmic islets (i). (c) After 120 hours, the remains of a disintegrating fibroblast contain membrane-covered rickettsiae. No other host cell structures are identifiable.

20 percent die as a result of vascular collapse, kidney failure, and other effects of widespread intravascular coagulation.

Therapy and immunity   Prompt therapy with chloramphenicol or tetracycline is life-saving. Deaths are frequently the result of missed diagnosis or delayed therapy. The same drugs are also generally used in treating the other ricksettial diseases.

Specific antibodies appear between the eighth and twelfth days and may aid in recovery. These antibodies protect against reinfection. A killed-bacterial vaccine is available, but it is recommended only for persons, such as lab workers, with high occupational risks.

Epidemiology   The RMSF agent is carried by wood ticks, dog ticks, and the Lone Star tick. The agent multiplies in the salivary gland of the tick (Fig. 24.6) and is transmitted to a mammalian host in the course of a prolonged blood meal. Many different kinds of mammals can be infected. Cycles involving alternate animal–tick transmission are important, as well as transovarial transmission within the tick vector species. Often a small geographical area will contain a highly infected tick population. When a person without protective clothing enters one of these regions, he or she may be bitten and thus exposed. Only a few miles away, the ticks may not be infected. Sensible persons choose clothing that prevents tick attachment. The skin and scalp should be examined frequently; a tick that is removed quickly is unlikely to infect you. Domestic pets should also be de-ticked, as they too may contract RMSF. The hands should be well washed afterwards, as all parts of the tick may be contaminated.

Most RMSF in the United States is found in the eastern states, particularly along the Atlantic Coast and in the mountains. It is uncommon in the Rockies at this time, although a few years ago it was a serious concern there.

Other Rickettsial Diseases

The typhus group of diseases is represented by three conditions.

Epidemic typhus fever *Rickettsia prowazekii* causes this disease, once widespread throughout the world, but now found mainly in regions of Asia and Africa. It is characterized by abrupt onset, extremely high temperature, severe headache, and a rash appearing first on the upper trunk. In severe cases, spleen and kidneys become involved, and there may be myocarditis or neurological changes. In some persons, the agent persists in an inactive form in the host's RE system after recovery. Therapy is similar to that described for RMSF. The Cox killed vaccine and a live rickettsial vaccine are available and recommended for travelers to the endemic areas.

Typhus vaccines; Table 27.3

Typhus is transmitted by the human body louse, usually found only under conditions of severe poverty, intense crowding, or dislocation. Insecticides are widely used to destroy the vector.

Brill–Zinsser disease This is a reappearance of typhus, in a mild form, in someone who has had typhus years before. There is often no rash. A vector is not needed to

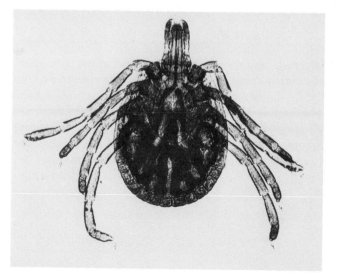

24.6
A Tick Wood ticks, dog ticks, and sheep ticks are important vectors of circulatory infection. When the tick bites (painlessly), it fastens its mouth parts into the tissue, injects a natural anticoagulant, and proceeds to ingest blood for number of hours. Pathogens can be introduced to the host during the feeding process.

initiate the disease; if lice are present, however, the person may serve as a reservoir to infect the lice and thus cause new cases.

Endemic typhus   Also called murine typhus, this mild infection is caused by *Rickettsia typhi*. It infects rats and mice and is transmitted to human beings by the rat flea's feces rubbed into a bite. In the United States, murine typhus is occasionally seen on the Atlantic and Gulf coasts.

Several forms of **scrub typhus** compose the third group of classic rickettsial diseases. The agent is *Rickettsia tsutsugamushi,* transmitted by mites. It is endemic in areas of the Middle and Far East and around the Indian Ocean. The human disease varies in severity. Symptoms include an eschar at the point of the mite bite, fever, headache, and rash. Mortality rates may be high.

Q fever, caused by *Coxiella burnetii,* is an atypical rickettsial disease. It is quite widespread, affecting many domestic animals as well as human beings. It is transmitted not only by tick bite but also by indirect means such as drinking raw milk, contact with infected animals, or inhalation. The agent survives for prolonged periods outside the host. *Coxiella* usually causes a type of atypical pneumonia, but also may produce infective endocarditis. The flash method of pasteurization kills the agents in milk.

Pasteurization; Chapter 25, p. 730

Rheumatic Fever

Streptococcal pharyngitis; Chapter 18, pp. 531–532

Rheumatic fever is not an infection. It is a tissue-destroying hypersensitivity, potentially developing in about 3 percent of individuals following Group A streptococcal pharyngitis. Apparently, in these persons, the initial streptococcal exposure induces antibodies that cross-react with normal tissue components, particularly in the heart, joints, and skin.

When such an infection occurs, within one to two weeks a group of generalized symptoms, called rheumatic fever, will appear. Carditis occurs in 40–50 percent of first attacks, and about two-thirds of these suffer residual valve damage. Other symptoms may include arthritis, rash, and chorea, but these do not cause permanent damage. Most persons experiencing a rheumatic fever attack have elevated serum levels of an antibody called ASO—**antistreptolysin O**—used as a diagnostic clue.

Each time streptococcal pharyngitis develops, there will be heightened antibody production and another attack of rheumatic fever will follow. The pattern of symptoms seen on the initial attack tends to be repeated. Thus damaged heart valves are often damaged further. Penicillin therapy given after the rheumatic lesions are present does not cause them to improve. This confirms the fact that the bacterium, which is penicillin-sensitive, is not actually present in the lesion. Anti-inflammatory drugs are useful.

Penicillin or long-acting penicillin analogs are usually given prophylactically to rheumatic fever patients. This may prevent subsequent pharyngeal colonizations by the allergenic bacterium.

VIRAL INFECTIONS OF THE CIRCULATORY SYSTEM

Dengue

Dengue, endemic in many tropical regions, is caused by a single-stranded RNA toga-virus. There are four serotypes and no cross-immunity, so persons may experience the disease several times. It is restricted to the latitudes 25° north to 25° south, and has not been endemic in the United States for some years. However, it does occur on the Caribbean Islands and has been brought back by returning tourists. The virus is spread by the mosquito *Aedes aegypti* and, because of the distribution of the vector, is predominantly urban. The reservoir is human beings and possibly monkeys.

The virus attacks the vascular endothelium. In classic dengue there is fever, muscle ache, rash, and headache, but the course is benign and recovery is prompt. In hemorrhagic dengue, found principally in small children, there is a change for the worse about two to six days after the onset of symptoms. There is generalized hemorrhage that produces bleeding gums, hypotension, and intravascular coagulation, and the mortality is high. There is no specific therapy, only supportive measures.

There is no vaccine for dengue at present. In fact, a strange observation has been made that raises questions about the wisdom of vaccination. Both laboratory and epidemiological studies have shown that when a high level of maternal antibody is passed to an infant, both the risk and severity of dengue infection in the child is **increased**. There are no clear explanations for this at present.

Yellow Fever

Yellow fever, caused by a flavivirus, is endemic in much of tropical South America and sub-Saharan Africa. It is seen in middle-latitude zones only as an import. In endemic regions, wild monkeys serve as a reservoir, and the virus is transmitted by mosquitos. Epidemic (or urban) yellow fever is transmitted by the insect from person to person. Endemic (or jungle) yellow fever is transmitted to people when they enter forests in which infected monkeys dwell, by a monkey–mosquito–person sequence.

The disease may have several manifestations. In its most severe form, there is massive hemorrhage, with bloody vomit and urine. The liver is affected, causing the jaundice that yellows the skin and gives the disease its name. In most cases, however, it is mild or asymptomatic. Live-virus vaccination with the 17D strain is recommended for persons anticipating exposure to yellow fever.

Yellow fever vaccine; Chapter 27, p. 805

Epidemic Hemorrhagic Fevers

A number of acute viral hemorrhagic diseases has been described in remote, rural, geographic areas. They all share the symptoms of vascular infection—high fever, muscle ache, hemorrhage, and coagulation. Several, such as Lassa fever, are notorious for their high mortality rates and have aroused global concern. Yet all these epidemics have remained contained. There is little systematic information yet available about this disease group. Rodent reservoirs and arthropod transmission are involved in several of these diseases.

Pericarditis

Viruses are the usual agents causing inflammation of the pericardium. The condition is marked by chest pain and accumulation of fluid in the pericardial sac. In advanced cases, the pressure exerted by this fluid severely compresses the heart. Viral cases usually clear spontaneously, but bacterial cases may be extremely refractory to treatment.

Myocarditis

Myocarditis is a term that covers a variety of conditions in which the contractility of heart muscle is reduced. Viruses are the most notable primary offenders. The coxsackieviruses were early recognized as a major myocarditis agent, and echoviruses, polioviruses, and influenzaviruses have since been added to the list.

Myocarditis also occurs as a complication of diphtheria, due to the action of diphtheria toxin on the heart, the cause of most fatalities. Endotoxins released in meningococcal meningitis and other severe, Gram-negative infections also weaken the myocardium. Myocarditis develops in advanced stages of tuberculosis, syphilis, and many parasitic diseases.

Many patients with viral myocarditis are but mildly affected. They exhibit easy fatigability and malaise, but recover spontaneously. With more severe cardiac symptoms, particularly toxicity, the patient may or may not recover full heart function.

BACTERIAL LYMPHORETICULAR INFECTIONS

Certain bacteria survive and replicate in lymphoid and reticuloendothelial tissues. In doing so they provoke a highly injurious, inflammatory process.

Plague

The term plague was historically applied to any serious epidemic disease. At the present time, this name is reserved for the disease caused by *Yersinia pestis.* Yersiniosis is synonymous with the historical **bubonic** plague—there is a swollen, ulcerated lymph node called a **bubo**—and the **black death**—surface hemorrhages may cause the skin and extremities to turn black. It is hoped that large-scale epidemics are a thing of the past. Persistence of a wild reservoir of the agent, as well as the rat–flea transmission system, prevents its eradication.

Causative agent   *Yersinia pestis* is a small, Gram-negative rod, presently classed with the enterobacteriaceae. It is slow-growing but not fastidious and can readily be cultivated on many laboratory media. In the host, it can multiply extracellularly in plasma or tissue fluid, but prefers the intracellular environment of a macrophage phagosome. *Y. pestis* has numerous virulence factors. A glycoprotein capsule protects it against phagocytic digestion. Exceptionally high levels of catalase protect against peroxide-mediated killing in the phagosome. The organism releases liberal amounts of endotoxin with catastrophic effects on peripheral circulation.

Pathogenesis If introduced via a flea bite, *Y. pestis* migrates to the regional lymph nodes, commonly in armpit or groin. It multiplies while there, and an extremely painful swelling and ulceration develops. Fever develops followed by the symptoms of endotoxin poisoning. Peripheral circulation is restricted, and capillary breakdown in the extremities may produce blackened and necrotic fingers and toes. If the yersiniae enter the blood, septicemia and generalized septic shock follow.

A pneumonic form of plague follows inhalation of airborne bacteria in contaminated secretions of a patient. This form progresses inevitably to septicemia and has a mortality approaching 100 percent if untreated.

Therapy and immunity Streptomycin and tetracycline are the drugs of choice, and they are successful in arresting most bubonic cases if given soon enough. The disease is now so rare, though, that the diagnosis may be missed. A positive identification of the agent may be made by applying fluorescent antibody to material aspirated from a bubo, and by cultural identification of the bacterium.

A heat-killed bacterial vaccine is available for persons at high risk.

Epidemiology Rodents such as the ground squirrel and prairie dog in the Southwestern United States, and native rodent species of the arid mountainous portions of the Middle East and other regions, are a permanent reservoir. Fleas transmit the disease among animals. Human beings can contract the disease either by capturing and skinning infected animals or by being bitten by an infected flea. Cats and dogs, by temporarily harboring infected fleas, can also bring the disease to human beings. Because the endemic areas are geographically remote, transfer to urban rodents (which might produce a massive human epidemic) does not currently occur.

Brucellosis

Several members of the genus *Brucella*, including *B. melitensis*, *B. suis*, *B. abortus*, and *B. canis*, cause this disease. The brucellae are slow-growing, fastidious, Gram-negative rods, requiring 10 percent CO_2 for incubation. They infect a wide range of mammals, both domestic and wild, causing various diseases. Spontaneous abortion is the most common manifestation in domestic herd animals. The placenta, fetus, GU secretions including semen, and the meat or milk derived from the infected animal may all be infectious. Farmers, veterinarians, and meat packers are the persons most commonly infected. The agent remains viable for 21 days in refrigerated meat;

it is destroyed by pasteurization of milk. Routine testing of dairy and meat animals is practiced in areas in which the disease is prevalent.

Brucella multiplies intracellularly in RE cells, producing granuloma and abscess of liver, spleen, bone marrow, and lymph nodes. Brucellosis has vague, undefined symptoms, including aches, chills, and sweating episodes. The fever goes up and down—thus its common name, **undulant fever.** Mortality is low if tetracyclines or trimethoprim-sulfa are given, although there is a risk of infectious endocarditis or osteomyelitis as complications.

Recovery confers some immunity against reinfection. A live, attenuated vaccine is administered to calves.

Tularemia

Tularemia may well be the most infectious of bacterial diseases because fewer than 50 bacilli can cause disease, gaining entrance through apparently unbroken skin. The agent *Francisella tularensis* is closely related and similar to the *Brucella* genus. It infects more than 100 animals and many insects and arachnoids, and may be transmitted to people by tick bite, contamination of the skin or conjunctiva, inhalation, or ingestion. It is surprising and fortunate that, with this great virulence and host versatility, human exposure remains quite limited.

Type A is the most serious form for human beings to contract. It is commonly acquired during rabbit-hunting season while hunters are skinning an infected animal. A lesion forms at the site of entry, and buboes develop in the draining lymph nodes. Should the agent enter the blood, the disease may resemble typhoid fever. Diagnosis is difficult because the symptoms are highly variable. The organism is difficult to cultivate, and exceptionally hazardous to laboratory personnel, so making attempts to culture it is usually unwise. Treatment with streptomycin will arrest the disease. Recovery confers a protective immunity. There is an experimental live, attenuated vaccine. Sensible preventive measures involve avoiding tick or deerfly bites and not handling wild rabbits that seem ill.

VIRAL LYMPHORETICULAR INFECTIONS

Viral agents frequently carry out preliminary replication in lymphoid tissue during the incubation period of an illness. Swelling of lymph nodes and tonsils in children is a well-known warning that an infectious disease is about to erupt. Sometimes the swelling recedes and no further symptoms are noted because the body's defense mechanisms have been successful. At other times viruses seem to establish a prolonged latency. Multiplication of the retroviruses (Chapter 11) in lymphoid and blood-forming tissues is a factor in at least some leukemias and lymphomas. A great deal more remains to be learned about viral interaction with the lymphoreticular tissues and its pathogenic effects.

Box 24.1

Tularemia, a Multifaceted Disease

Although tularemia is relatively rare (fewer than 200 cases a year in the United States), it can be acquired in a surprising variety of circumstances. For example:

☐ While hunting, a 19-year-old Washington state man found a dead rabbit. With his bare hands, bruised and scratched from his work as an auto mechanic, he cut off the front paws for good luck charms and gave them to a friend. The hunter developed a form of tularemia. His friend fortunately threw the rabbit paws away, and remained well.

☐ A 30-year-old Washington state man shot, killed, and skinned an apparently healthy black bear, cutting his left hand in the process. Tularemia developed within two days.

☐ Eleven cases of tularemia developed on an Indian reservation in Montana during early summer, the height of the tick season. The reservation dogs almost all had antibodies to tularemia, and eight of fourteen lots of ticks removed from them contained the bacterium.

☐ Four sheep shearers in Colorado developed tularemia after shearing sheep that appeared ill and covered with wood ticks.

The initial lesions appeared on their left hands. This is because shearers part the fleece with the bare left hand, while shearing with the right. Engorged ticks are often destroyed in the process, and the bacteria-containing gut contents contaminate the left hands.

☐ A 49-year-old Alaska man became ill after dressing a rabbit killed by his dog. A number of dead rabbits had recently been observed in the area, suggesting an active outbreak of rabbit tularemia.

☐ Two young Georgia boys became ill after handling a dead rabbit they had found.

☐ A laboratory technician in Colorado developed severe tularemia after working with a laboratory isolate of *Francisella tularensis.* Inhalation of aerosolized culture was suspected.

These cases range widely in mode of transmission, animal species involved, and geographical area. They also were highly variable in clinical symptoms and severity. All patients, however, recovered with antimicrobial therapy.

Source: Adapted from *Morbidity and Mortality Weekly Reports* **27** (12), **28** (44), 1979, and **29** (5), 1980.

Infectious Mononucleosis (IM)

This disease is one of at least three manifestations of infection by the Epstein–Barr virus (EBV).

Epstein–Barr virus
(EBV); Chapter 11, pp.
332–333; Tables 11.5
and 11.7

Causative agent   EBV, a member of the herpesvirus group, is a DNA virus. It can be demonstrated in B-lymphocytes during the acute phase of the disease, and probably multiplies in all the lymphoreticular tissues. The virus is identical to that isolated from tumor tissues in African Burkitt's Lymphoma (ABL); antibodies produced in ABL disease and in nasopharyngeal carcinoma (NPC), are identical to those found in IM.

Pathogenesis   Multiplication of the virus occurs during a long incubation period; then relatively mild fever, sore throat, and lymphadenopathy develop. Moderate enlargement of the spleen is common, and liver function tests are abnormal. Rashes or jaundice are less common symptoms.

The number of circulating lymphocytes and monocytes is much increased, and many of the lymphocytes are atypical. Another useful diagnostic observation is the presence of **heterophil** antibodies. These are antibodies, produced by a nonspecific response, that agglutinate specially treated sheep, horse, or ox red blood cells. Recently, methods have become available to test directly for anti-EBV antibodies.

IM is usually benign, with mild symptoms clearing in two to three weeks. Complications develop in less than 5 percent, probably associated with the failure of the lymphoid tissues to restrict viral multiplication.

Therapy and immunity   There is no specific therapy. Bed rest and analgesics for the sore throat may be advised.

Recovery confers immunity, not only against IM, but also apparently against the other EBV diseases.

Epidemiology   The patient excretes virus in the saliva. Droplets of airborne saliva, or saliva exchanges during kissing, are the most likely means of transmission. This disease affects only human beings, although closely related viruses have been found in the Old World monkeys. The distribution of IM is largely limited to the United States, Canada, Europe, and Scandinavia. It is quite rare in tropical or underdeveloped nations. ABL and NPC are predominantly found in Africa and Southern China, respectively. There is no accepted explanation for this curious distribution pattern.

Because of the potential carcinogenicity of the virus, the production of vaccines has been pursued with caution. A purified capsular protein preparation may have potential for safe and effective use, but this line of research is still in its earliest stages.

PROTOZOAN INFECTIONS

Two protozoan infections of the blood and tissues are among the world's greatest unsolved microbiological problems. Malaria, endemic through much of the semitropical and tropical zones, may infect in excess of 100 million persons per year. American trypanosomiasis (Chagas' disease) infects about 15 million persons,

24.7

The Life Cycle of *Plasmodium* in Human Beings The parasite completes two asexual phases in human beings. The first, or pre-erythrocytic phase, is the multiplication of merozoites in the liver. The second, or erythrocytic phase, is the replication of the parasite in red blood cells. Some of the parasites produced in red blood cells continue to induce new erythrocytic replication cycles. Others mature into male and female gametocytes.

When a mosquito bites such an individual, it will ingest these gametocytes. The malaria parasite will then be able to complete its life cycle by sexual replication in the mosquito.

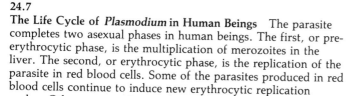
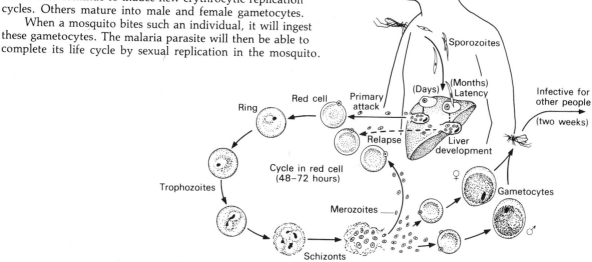

largely in South America. The total number of cases has already been greatly reduced by insect-control measures, but major problems still remain both in the areas of effective therapy and vaccination.

Malaria

Causative agent Four species of the sporozoan parasite *Plasmodium* cause malarias of differing degrees of severity. These organisms have a complex life cycle that for completion requires two hosts—the female *Anopheles* mosquito and the human being—and three developmental cycles (Fig. 24.7).

The first cycle, sexual **sporogony,** consists of the union of male and female gametes, acquired from a meal of human blood, in the stomach of the mosquito. The zygote then multiplies actively over a period of several days to produce many **sporozoites.**

Sporozoites migrate to the insect's salivary glands, and are injected into the human host at the insect's next feeding. The second or **pre-erythrocytic** cycle then occurs. The injected sporozoites enter liver cells and for 10 to 28 days undergo further division, yielding large numbers of **merozoites.** The merozoites are released into the blood to start the third or **erythrocytic** cycle.

TABLE 24.5
Comparison of human malarial diseases

FEATURE	Plasmodium falciparum	Plasmodium vivax	Plasmodium ovale	Plasmodium malariae
Incidence	Common	Common	Common	Common
Incubation period	12 days	14 days	14 days	30 days
Duration of pre-erythrocytic cycle	5.5 to 7 days	6–8 days	9 days	13–16 days
Persistence in liver	No	Yes	Yes	No
Duration of erythrocytic cycle	48 hours	48 hours	48 hours	72 hours
Number of new merozoites/RBC	8–24	12–24	6–16	6–12
Mortality	High if untreated	Low	Low	Low

Merozoites are highly invasive. They enter erythrocytes by inducing the cells to take them in by endocytosis. Once they are in the RBC, their rapid multiplication produces from 6 to 24 daughter merozoites, depending on the species of parasite. When these are mature, in 48 or 72 hours, the RBC ruptures and releases them. Some merozoites, instead of replicating asexually, commence a meiotic division in RBCs, leading to male **microgametocytes** and female **macrogametocytes**. These have no future in the human host, but if taken in by a mosquito will initiate sporogony again.

Natural infection in human beings is comprised of the second and third cycles, together referred to as **schizogony**. However, infection can also follow transfusion of human blood bearing malarial merozoites in the erythrocytic phase. When this happens, the recipient will immediately develop erythrocytic-stage malaria without a pre-erythrocytic phase.

In *P. vivax* and *P. ovale* malaria, inactive parasites may persist indefinitely in the liver, reactivating years after the initial episode. They not only cause the individual to relapse but also serve as a source of the parasite to infect others.

The four *Plasmodium* species are differentiated primarily by careful examination of blood films stained with Wright's or Giemsa's stains. A highly trained microscopist can identify the different species of *Plasmodium* by the appearance of various forms in the erythrocytic stage. Neither culture nor animal inoculation is useful. Indirect malarial fluorescent antibody techniques have received some use in the more sophisticated laboratories.

Pathogenesis The symptoms of malaria result largely from wholesale destruction of red blood cells. Successive crops of merozoites tend to be released with hemolysis at about the same time. Their release causes a sudden onset of fever and chills that usually last for no more than six hours. Anemia develops, and results in tissue anoxia. Hemoglobin is excreted in the urine; where this feature is pronounced, it earns the name blackwater fever. Even unparasitized RBCs become fragile and rupture. This appears to be caused by the host's developing a complement-mediated immune response directed against RBCs in general. In malaria caused by *P. falciparum* (the most severe type) intravascular coagulation, liver and kidney necrosis, and brain congestion may lead to death in about 10 percent, predominantly young children. Deaths in the other three types are uncommon.

Either chloroquine or Fansidar should be taken by anyone residing in a malarial area. On leaving the area, one should take primaquine to clear the liver of parasites. These precautions have been generally adopted by the armed services. These measures are largely responsible for the limited amount of malaria in Korean and Vietnam war veterans.

Box 24.2

Malaria Control in California

The Sutter and Yuba counties of California exhibit a set of unique circumstances that present a serious threat of mosquito-transmitted malaria. This rich agricultural area contains 120,000 acres of rice paddies. These provide a prime breeding ground for *Anopheles freeborni*, a mosquito species that serves as a vector for *Plasmodium vivax* malaria. The climate, which resembles that of the Punjab region of India, has attracted more than 5,000 Punjabi immigrants. There have been many imported cases of vivax malaria (over 70 in 1979).

The challenge is to detect infected persons and to treat them to eradicate the parasite in both its pre-erythrocytic and erythrocytic phases, while preventing the transmission of the parasite to the mosquito population. Should the parasite be passed from the imported human cases to the local mosquito population, malaria might become firmly established as an endemic disease in the area.

Control measures include adequate drainage of unused paddies and stocking fish in flooded paddies to eat the mosquito larvae. All newly arrived Punjabis are blood-tested. Any persons with positive evidence of vivax malaria receive antimalarial chemotherapy. Insecticides are applied as a fog over a one-half mile radius around the household and work place of any malaria patient. This process begins within 12 hours of the detection of any new case and is repeated weekly. These measures so far have been completely effective in preventing transmission of malaria to local residents.

Source: *Morbidity and Mortality Weekly Reports* **29** (5), 1980.

Therapy and immunity   Primary chemotherapy is directed against the erythrocytic merozoites. Chloroquine is the drug of choice because it has the greatest effectiveness and least toxicity. Unfortunately, strains of *P. falciparum* in many areas of the world have become resistant to this drug. In such areas, Fansidar, a combination of pyrimethamine and sulfonamide, is recommended. Drugs should be used for a period of several weeks.

Cure requires the eradication of the hepatic parasites in *P. vivax* and *P. ovale* malaria. Primaquine is used for this purpose.

Immunity follows recovery only after repeated attacks, and it is of short duration. As mentioned earlier, certain of the antibodies seem actually to increase the degree of hemolysis and anemia. No vaccine is available, although preparations of membrane components from killed parasites are being tested in laboratory animals. The creation of an effective antimalarial vaccine is one of the most urgent medical research priorities.

Epidemiology   Interaction of susceptible mosquito and human populations is crucial to the transmission of the disease. Persons with sickle-cell trait or disease have a reduced susceptibility to malaria. Large-scale mosquito control with DDT was introduced after World War II, with a dramatic reduction of malaria incidence. However, insects resistant to the pesticide soon emerged, and these gains were partially wiped out. Drainage and oil-filming of standing water are widely used to eradicate breeding places (Fig. 24.8). Household screening, protective clothing, and repellents keep the insects away.

Malaria ceased to be endemic to the United States a number of years ago, but it is frequently reintroduced by persons returning from endemic areas. These remain

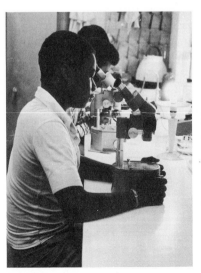

24.8
Mosquito Control in Nigeria   (a) Dissecting mosquitos to look for developing maleria parasites. This allows health authorities to find out what percentage of mosquitos in a given area carry the various *Plasmodium* species. (b) Eradication of the parasite is dependent on destroying the insect vector around homes.

(a)   (b)

"one-case epidemics" unless they become blood donors or share syringes in the injection of drugs. The risk of imported malaria can be removed by systematic use of prophylactic drugs.

Trypanosomiasis

The parasites *Trypanosoma cruzi* and *T. rangeli* cause American trypanosomiasis (Fig. 24.9). It is seen from Texas to Argentina and is a leading cause of cardiovascular death in males in endemic areas. Trypanosomiasis is transmitted by reduvid or triatomine bugs that infest open, wood-frame, and mud-floor dwellings. The related agents *T. rhodesiense* and *T. gambiense* cause African sleeping sickness, which is also a devastating disease.

The agent may multiply intracellularly in many tissues, in a nonmotile **amastigote** form. Multiplication and maturation yield a motile **trypomastigote** form, released into the blood when host cells rupture.

The biting insect passes the trypomastigote in its saliva or its feces, and these are rubbed into the itching bite wound. After an incubation period, the acute phase develops. The motile forms are numerous in the blood plasma; nonmotile organisms are multiplying in the myocardium, smooth and skeletal muscles, and sometimes the central nervous system. Generalized toxic symptoms are present. About 10 percent of cases die during this phase, usually of heart failure. The rest recover spontaneously. If significant tissue damage has been sustained, the disease passes into a chronic phase marked by cardiac disease and/or enlargement of the esopha-

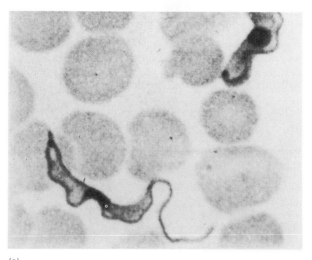

(a)

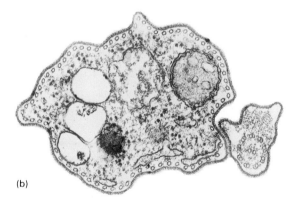

(b)

24.9
***Trypanosoma* and Its Vector**   (a) *Trypanosoma gambiense*, the agent of African sleeping sickness, in a stained blood smear. (b) *Trypanosoma vivax* trypomastigote stage, TEM, thin section.

TABLE 24.6
Epidemiology of circulatory and lymphoreticular infections

DISEASE	RESERVOIR	TRANSMISSION	INCUBATION IN DAYS
Rocky Mountain spotted fever	Ticks, dogs, rodents	Tick bite, tick feces	3–10
Typhus	Human beings	Body louse feces	12
Dengue	Human beings, monkeys, mosquitos	Mosquito bite	5–6
Yellow fever	Human beings, monkeys, mosquitos	Mosquito bite	3–6
Plague	Wild rodents, other animals	Flea bite, handling diseased animals, droplet inhalation	2–6
Brucellosis	Many domestic and wild animals	Contact with tissues, discharges, infected meat or dairy products	Variable
Tularemia	Many domestic and wild animals, especially rabbits	Almost any means, commonly handling diseased animals	3
Malaria	Human beings, (monkeys)	Mosquito bite, transfused blood, contaminated syringe	Variable
American trypanosomiasis	Human beings, domestic and wild animals	Feces of triatome bug	5–14
Infectious mononucleosis	Human beings	Person-to-person by pharyngeal secretions	2–6 wk

*In vector-borne diseases, the disease may be communicated to the **vector** at any time when the viable agent is present in blood or tissue fluid.

Source: Adapted from A.S. Benenson (ed.), 1975, *Control of Communicable Diseases in Man* (12th ed.).

PERIOD WHEN COMMUNICABLE	PATIENT CONTROL MEASURES	EPIDEMIC CONTROL MEASURES
Not directly communicable* from person to person	Remove all ticks from patient	Identify ticks: clear people from high-risk areas
Not directly communicable	Apply insecticide to delouse patient, especially hair, clothing, and bedding	Apply insecticide to all contacts
Not directly communicable	Patient housed in screened quarters	Mosquito destruction
Not directly communicable	Patient housed in screened quarters	All houses in vicinity should be sprayed, vaccinate susceptibles
Not directly communicable **except** when patient has pneumonic form	Rid patient and effects of fleas; strict isolation until 48 hr after antibiotic therapy begins; disinfect discharges	Intensive case-finding; careful management of contacts; flea and rodent destruction
Has evidence of communicability from person to person	Disinfect discharges	Search for vehicle of infection such as infected dairy herd
Not directly communicable	Disinfect discharges	Search for source of infection
Not directly communicable	Patient housed in screened quarters; disinfect materials contaminated with blood	Mass spraying, case-finding, chemoprophylaxis
Not directly communicable	None	Spray houses, improve dwellings
Unknown, presumably long	Disinfect articles soiled with nasal or throat discharges	None

(a)

(b)

24.10
Control of the Tsetse Fly (a)
Pathfinders capturing tsetse fly
specimens in Botswana. (b) Laboratory technician examining flies to
determine the type of trypanosome
present (Nigeria).

gus and colon due to smooth muscle destruction. These conditions may eventually
prove fatal. There is no effective specific therapy. The nitrofuran drugs have been
used with variable or incomplete success. Reliable protective immunity does not develop on infection; there is no vaccination. Epidemiological control is based on improving the housing conditions of the target population and spraying to eliminate
the vector (Fig. 24.10).

SUMMARY

1. The circulatory and lymphoreticular diseases develop following a specific sequence of events for the introduction of the agent. There are also mechanisms
that promote the agent's continued survival in the face of the host defense
mechanisms. Some agents accumulate a protective layer of clotted blood, while
others multiply intracellularly, in the endothelium, lymphocytes, or red blood
cells. Unchecked multiplication of an unprotected extracellular organism (septicemia) is an indicator that host defenses are paralyzed or absent.

2. The symptoms of the circulatory diseases reflect the modes of pathogenesis.
The circulation can distribute bacterial toxins and inflammatory substances.
These lead to fever, malaise, and chills. When the circulation is blocked by intravascular coagulation or when oxygen-carrying RBCs are destroyed, tissues become oxygen-starved. Muscle aches and liver, kidney, and neurological damage

result. Fragility of RBCs or capillary hemorrhage produce petechiae, bloody vomit, or darkened urine. Hyperactivity of lymphoid tissues provoked by bacterial parasitism leads to swelling and ulceration.

3. Insect vectors transmit most of these infections. An insect bite is the most effective way for the organism to be introduced to the circulation. The bite mechanism allows the organism to bypass the surface defenses of skin and mucosa, and directly enter the connective tissues in which capillaries abound.

4. The endemic area of a vector-transmitted disease corresponds to the habitat of the vector species. When pesticides first became available, it was hoped they would eradicate malaria and other vector-borne tropical diseases. However, vectors have become increasingly resistant. Pesticide use is being supplemented with biological means of reducing vector populations and protecting persons from their bites.

5. The antibody-mediated immune response is not particularly helpful in combating circulatory infection. Most of the agents survive in microenvironments in which they are not exposed to antibody. We do not have a clear understanding of how the body actually does bring these infections under control in convalescence. Thus traditional vaccination, which produces AMI, has been largely ineffective.

Study Topics

1. What preexisting conditions place a person at risk of acquiring infective endocarditis? What types of experience are likely to introduce the necessary inoculum? Prepare lists as complete as you can.

2. Continuous penicillin prophylaxis is recommended for persons with rheumatic fever, but not for persons with heart valve abnormalities. What reasons can you advance for this difference?

3. By review of previous chapters, prepare a list of infectious diseases that can culminate in septicemia.

4. Discuss personal means the individual can take to minimize the chance of being infected by (a) tickborne, and (b) mosquito-borne agents.

Bibliography

Hoeprich, Paul D. (ed.), 1977, *Infectious Diseases* (2nd ed.), New York: Harper & Row.

Rahimtoola, S. N. (ed.), 1977, *Infective Endocarditis*, New York: Academic Press.

Schlessinger, David (ed.), 1979, *Microbiology 1979* Part II, Mechanisms of Microbial Virulence.

Skinner, F. A., and L. B. Quesnel, 1978, *The Streptococci*, New York: Academic Press.

Trager, William, Erythrocyte-malaria parasite interactions, pp. 166–199.

Sherwan, Irwin W., *Plasmodium:* Biochemical consequences of life in a red cell, pp. 124–129.

Brubaker, Robert R., Expression of virulence in yersiniae, pp. 168–171.

Yekutiel, P., 1979, *The Eradication of Infectious Diseases: A Critical Study*, New York: S. Karger. Chapters on malaria and yellow fever eradication.

IV

CONTROL OF MICROORGANISMS

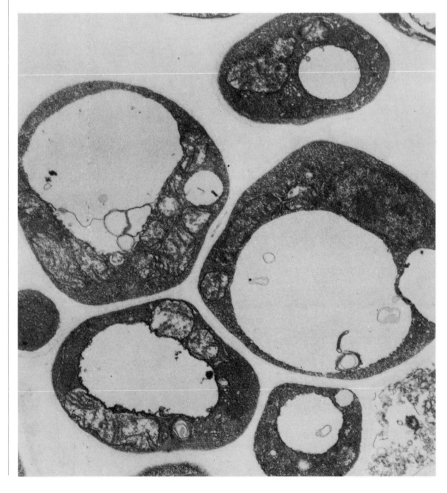

Cells of *Histoplasma capsulatum* Showing the Destructive Effects of Amphotericin B

25

Disinfection
and Sterilization

Disinfection and sterilization are techniques widely used for rendering foods, medications, and the environment less likely to transmit infection. In hospitals and other clinical agencies, vast amounts of time and money are spent on these functions. In our daily lives, we are deluged with advertisements for personal and home products that promise to render us healthier, safer, and more attractive. We need to understand these products' usefulness and limitations.

In this chapter, we will consider the basic concepts of disinfection and sterilization, stressing situations in which they are useful and pointing out circumstances that compromise their effectiveness. We will not give detailed instructions for the use of equipment and commercial products. These are best obtained from the manufacturers, and should be carefully followed. Each institution has a tendency to adapt its sanitation procedures to suit local conditions; the procedures may be learned as needed.

BASIC CONCEPTS OF MICROBIAL CONTROL

The field of disinfection and sterilization is an applied science. It draws heavily on basic concepts from two areas already covered. The effects of environmental variables on microbial survival, and the kinetics of microbial growth, discussed in Chapter 7, come together with the analysis of modes of transmission of infectious disease, discussed in Chapter 15. Taken together these define the occasions in which microbial control measures can and should be applied. You may find it useful to spend a few minutes looking back through this material before proceeding.

Some Terminology

A number of terms are commonly used (and not uncommonly misused), in designating the various levels of microbial control.

Sterilization   To **sterilize** is to treat an object so that there is a vanishingly small probability that any viable microorganisms will be present. An object that has been sterilized, practically speaking, is free of all life forms. No organism will be able to grow on a sterilized object even in the most favorable circumstances. In commercial canned foods, for example, the heat process should reduce the possibility of *Clostridium botulinum* contamination to one surviving spore in one of 10^{12} (one million million) containers. The term "sterilize" cannot correctly be applied to skin cleansing; it is not possible to sterilize skin without killing the tissue.

Nonviable microbial cells or structures will usually still be present after sterilization. They may be stained and examined microscopically, and usually appear to be severely distorted. Their chemical composition is in some respects often little altered. Microorganisms that have been killed to use in a vaccine are still antigenic, and microbial toxic products may be present and unchanged. Gram-negative bacteria produce fever-inducing substances, **pyrogens.** If large numbers of pyrogenic organisms contaminate an injectable medication or solution, subsequent sterilization may render it noninfective, but it will still be toxic to the recipient.

Pyrogens; Chapter 12, p. 352, Tables 12.4 and 12.5

Disinfection   **Disinfection** is the removal or destruction of pathogens. It also reduces the numbers of viable organisms in the material. Nonpathogenic environmental microorganisms may not be completely eliminated. The term "disinfectant" is correctly used to describe a treatment used on an inanimate object—linen, scalpel, air, or floor. **Antisepsis** is correctly used for a treatment applied to living skin or tissue. Often the same chemical preparation may be used in concentrated solution as a disinfectant and in a dilute solution as an antiseptic. Other disinfectants are not suitable for use on living tissue at all. **Decontamination** is used when not only living pathogens but also their potential toxic products are to be removed.

Other terms refer to procedures used to reduce the numbers of organisms. **Degerming** is the correct term for careful tissue cleansing, for example of skin. An effective antiseptic is the substance used. It may reduce the total bacterial population a thousandfold. **Sanitizing** is the systematic cleansing of inanimate objects to reduce the microbial count to a safe level. Examples are the usual procedures for cleaning lavatories and toilets.

Technical Approaches to Microbial Control

Sepsis is the uncontrolled presence of infectious agents or their toxins in the blood or tissues. Early and primitive medicine used septic techniques. The mental picture of a surgeon sharpening his dull scalpel on his boot heel before operating is a sufficient example.

When the role of microbes in disease became apparent, septic techniques were supplanted by **antiseptic** ones. This approach assumed that the presence of pathogens was inevitable but that they could be chemically killed. Surgeons operated in a cloud of disinfectant spray designed to prevent the development of infection. This technique was a significant improvement, but still far from perfect.

It has been replaced by **aseptic** technique. The goal of aseptic technique is exclusion of potential infective agents. This approach attempts to establish a barrier to infection by maintaining scrupulous sterility or cleanliness of all materials coming in contact with the uninfected patient. Asepsis applied to an infected patient, on the other hand, confines the infective agent to its site and prevents its spread. Physical and chemical disinfection and sterilization are essential to achieve these goals.

Killing in Contrast to Inhibition

A treatment that is lethal to an organism causes irreversible changes. Even when the organism is later placed in the most favorable conditions, it cannot initiate growth. For microorganisms, the inability to reproduce is the functional definition of death. Killing treatments are designated by the suffix -**cidal,** with a prefix that indicates the type of organism affected, e.g., **bacterio**cidal, **fungi**cidal.

Other treatments halt microbial growth. As long as the inhibitory substance or state is present, no growth occurs. However, the inhibition is reversible. If the inhibitory condition is removed, growth begins again. Thus spoilage may begin as soon as a frozen food is defrosted or the pickling brine is washed out of a salted food. When the effective concentration of a bacteriostatic disinfectant is reduced by dilution, microorganisms may be able to grow, sometimes right in the stock solution. Inhibitory treatments are designated by the suffix -**static,** e.g., bacteriostatic, fungistatic.

For many disinfectants, and also for some antimicrobial drugs, concentration makes the difference between -cidal and -static results. Use of the full recommended concentrations is of great importance.

Kinetics of Microbial Death

When a large population of microorganisms experiences a lethal treatment, in any given period of time a certain proportion dies. In the next equal time period, the same proportion dies. When plotted on a graph, a **logarithmic death curve** (Fig. 25.1) results. Practically, in sterilization this means that treatment time must be pro-

Growth curve; Figure 7.12

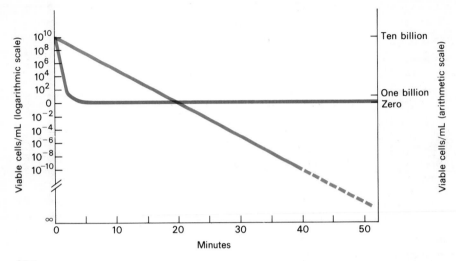

25.1

Logarithmic Death Curve   Microbial killing is exponential. Shown here is a hypothetical situation in which 90% of the viable population dies every two minutes. On the logarithmic scale (black) the starting concentration is 10^{10} viable cell/mL. At 20 minutes, it has dropped to 10^0 or one viable cell/mL. It contains to drop at the same rate, so at 40 minutes there is 10^{-10} viable cell/mL which translates to one viable cell in 10 billion mL. Another way of expressing this would be to say that of 10 billion one mL samples, all except one would probably be sterile. This might be considered an acceptable risk. The arithmetic scale (color) does not show negative numbers. Thus it looks like complete death has been achieved at about six minutes. This, however, is not strictly true because of the prolonged possibility of a few survivors.

longed until the **probability** of encountering a single survivor is extremely slight. Shortening the exposure time may leave a number of survivors.

The time required to sterilize depends on the initial microbial load. The larger the contaminating population, the longer the sterilization time. If objects to be sterilized or disinfected are thoroughly sanitized first, the microbial load will be reduced. Thus treatment begins farther down on the death curve and will have greater effectiveness.

Modes of Microbial Killing

The cell-killing effects of sterilants and disinfectants are largely nonspecific. That is, they damage all forms of life, although not necessarily all to the same extent. This is why disinfectants are toxic if taken internally. They must be used as directed to avoid adverse effects. Most cannot be allowed to contaminate foods, for example.

Sites of action   The proteins of a microorganism are vulnerable to attack by a number of means. **Denaturation** (irreversible change in the secondary and tertiary structure) results from contact with moist heat (which coagulates cellular protein) or alcohol (which precipitates proteins from the cellular materials). **Alkylation** (the replacement of hydrogen atoms by other groups) is induced by ethylene oxide gas (EtO) and the aldehydes.

Cell membranes are vulnerable to agents that interfere with the hydrophilic and hydrophobic interactions that hold the membranes together. Detergents and other surface active agents do this and so do alcohols and phenolic compounds.

Radiation attacks nucleic acids and proteins. Ultraviolet light alters DNA bases. Other radiations (X-rays, gamma rays) are less specific, causing ionization of many cellular constituents.

To kill a microorganism, the treatment must be applied at a sufficient strength for a long enough period so the microbial cell cannot repair the damage and recover its function.

Specificity   The destructive action of sterilizing and disinfecting agents or treatments is not specific at all. Any biochemical, whether inside or outside a cell, bacterial or human, may be attacked. This places numerous restrictions on use of sterilizing and disinfecting agents.

Effects of Environmental Variables

A disinfecting treatment that may be completely effective in one environment may be relatively ineffective in another. A number of conditions can interfere with the proper operation of a selected procedure.

Organic matter   Many materials to be treated contain or are coated by nonmicrobial organic matter. Saliva, blood, or excretions are present on many objects submitted for disinfection in the hospital. Many disinfectants are readily inactivated by such organic matter. Since disinfectants bind nonspecifically, the nonmicrobial organic matter absorbs the chemical too, reducing the amount available for antimicrobial action. One can either reduce the amount of organic matter by prewashing or increase the concentration of disinfectant to compensate for losses due to nonspecific binding. In chlorinating drinking water, the amount of active agent is adjusted to the amount of organic matter in the water source. Cities that draw their water supplies from rivers with a heavy organic load must use a great deal of chlorine. Their treated drinking water may end up safe but rather unpalatable.

Chlorination; Chapter 15, p. 457

pH   Most disinfectants operate most effectively at a neutral or slightly alkaline pH.

Temperature   The effectiveness of most disinfectants increases with elevated temperature because they depend on a chemical reaction with the target substance. Thus treatment should be carried out at room temperature or above, or be prolonged if cold temperatures are unavoidable. The time of exposure can often be shortened at elevated temperature, as in EtO sterilizers.

Oxygen   The presence of oxygen increases the lethal effect of radiation sterilization. Oxidizing agents may be effective in maintaining an elevated Eh that prevents the growth of anaerobes.

Inactivating ions   Certain disinfectants are ionic or charged and depend for their action on an electrostatic interaction with the microbial surface. Cationic detergents, for example, bind to negatively charged groups in the bacterial membrane. If other anions such as those found in soaps are present, they compete with the cell surface for the cation, and render the disinfectant ineffective.

Free access   The sterilizing or disinfecting agent must have free access to all contaminated surfaces and areas. Improper packaging, failure to separate parts before immersing equipment, or residual coatings of grease or wax may prevent effective action.

TESTING AND EVALUATION

The claims made for sterilizing agents and disinfectants and the recommendations for their use arise from considerable testing by the manufacturer. Testing methods are standardized. There are over 8,000 registered disinfectants. Those used on inanimate objects are regulated by the United States Department of Agriculture, and those used in foods or on human tissue are controlled by the Food and Drug Administration. The Association of Official Analytical Chemists (AOAC) procedures manual is the standard for evaluation procedures.

Initial Tests

Not all microorganisms are equally susceptible to killing by any technique. Certain forms have been pinpointed as particularly resistant. These include the tuberculosis organism, viruses (especially hepatitis B), and spores. Bacterial endospores are in general the most highly resistant of all life forms. Disinfectants or sterilants are tested against numerous different pure cultures to define their range of usefulness. Testing that uses pure cultures *in vitro* is considered **initial testing.**

In an antimicrobial treatment, there are two independent variables, the intensity of exposure and the time of exposure. Killing can be accomplished with different combinations. A high temperature for a short time may have the same effect as a lower temperature for a longer time. The choice of conditions depends on time available, stability of materials, and economic considerations.

Intensity of exposure   If the time of exposure is fixed, for example, to ten minutes, then one can measure the intensity of exposure (disinfectant concentration, tempera-

Box 25.1

Laboratory-associated Infections

In a modern microbiology laboratory, aseptic technique, disinfection, and containment procedures are practiced to a high level of sophistication. This, however, is a relatively recent development. Today, for example, samples for isolation and identification of *M. tuberculosis* are handled with extreme care. Workers operate within a restricted area, using biological cabinets and wearing complete protective clothing. They are regularly tuberculin-tested to detect infection. However, many of us can recall a very few years ago that tuberculosis work was done right out on the bench in the main laboratory. We used no special protections whatever and had all sorts of people milling around us. Few of us were ever tuberculin-tested either, so there are no data on the outcome of this sloppy procedure.

Several surveys have attempted to document the clearly underreported incidence of laboratory-associated infection. As of 1978, there were 4,079 clinically apparent incidents. This is but the tip of the iceberg, as subclinical infections generally outnumber recognized infections.

Diagnostic laboratory work is definitely hazardous. This is probably because the samples submitted to a clinical lab may contain any agent. The technologist does not know if the sputum just received contains normal flora or plague bacilli.

There have been at least 173 deaths from laboratory-associated infection. The vast majority of these occurred before 1950. In recent years, viruses have caused several deaths. In several instances, deaths have accompanied the first laboratory isolation of a newly discovered viral pathogen such as Lassa

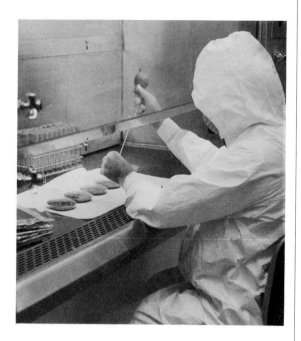

fever and Marburg disease. The most recent smallpox death (1978) occurred in a laboratory photographer.

The greatest technical hazard is the inhalation of an infectious aerosol. Aerosols may be produced by many actions such as blowing out pipettes, flaming heavily loaded loops, and popping off test-tube caps. Mouth pipetting and careless use of needles and syringes are clear, avoidable hazards. General acceptance of protective techniques and cautious procedures could eliminate much of the risk associated with the microbiologist's work.

Source: Data from R. M. Pike, 1979, *Ann. Rev. Microbiol.* **33**: 41–66.

ture, radiation intensity, etc.) needed to attain the desirable lethality. The **Z-value** is the temperature needed to obtain 90 percent killing in a fixed time period. The **thermal death point** (TDP) may also practically be measured on vegetative bacterial cells. This is the temperature that kills **all** members of a 24-hour broth culture of a given species of bacteria.

Duration of exposure   The intensity of exposure may be fixed, and the time left variable. Under these limits, the **D-value** (decimal reduction time) or time needed to kill 90 percent of the organisms, is determined. The **thermal death time** (TDT) is the time needed to kill **all** organisms in a 24-hour broth culture at a given temperature.

Z-values and D-values are used to compare control measures against each other. In practice, one uses Z-values to find the most effective treatment to do a job when only a certain amount of time can be allocated. D-values are used when the intensity of exposure may be limited, for example, by the instability of materials such as rubber or plastics, but the time of exposure can be extended within reasonable limits.

Chemical standards   Chemical disinfectants are frequently evaluated by comparing their effectiveness to that of phenol. A ratio called the **phenol coefficient** *(PC)* results. Modern disinfectants are generally much more effective than phenol, and have *PC*-values much greater than one. In evaluating survivors after exposure to a disinfectant, it is very important that the chemical be removed or neutralized before the treated culture is incubated. Otherwise, continuing inhibition may be mistaken for a killing effect. Since the *PC*-value is particularly dependent on environmental variables (species of microorganisms, growth phase, temperature, pH, etc.), any agents being considered for adoption should be tested **in use,** under the actual conditions in which they will be employed on a careful simulation.

In-use Tests

New disinfectants are routinely tested in conditions that simulate actual use. Mixed cultures are applied to simulated surfaces, then treated. The disinfectant ability is then evaluated. Later, the disinfectant may receive field trials in actual clinical settings. Such tests are essential before any honest claims can be made for the products.

Many large clinical facilities maintain routine microbiological testing schedules to monitor the effectiveness of their control programs and housekeeping. This permits prompt detection of equipment failure, ineffective lots or dilutions of disinfectant, and human error.

PHYSICAL AGENTS IN COMMON USE IN STERILIZATION

Heat

Moist heat   When hot air contains moisture, its effect is to coagulate protein. **Steam** (water in the gaseous state) at 100°C has more heat energy than liquid water at

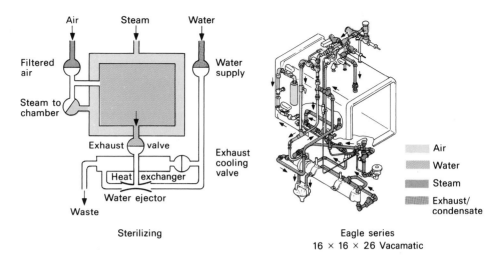

Sterilizing

Eagle series
16 × 16 × 26 Vacamatic

25.2
High-speed Vacuum Autoclave   A vacuum is used to draw air from the chamber and to draw steam into it. This speeds up the exchange process and provides effective removal of air.

100°C. This is because on conversion to steam the water molecule acquires additional energy, the **heat of vaporization.** When steam touches the surface of a cooler object, it condenses giving up its heat to the object. The temperature of steam can vary, depending on the atmospheric pressure. At sea level, in an unsealed vessel, steam can be no hotter than 100°C. In a pressurized vessel, steam may be heated to much higher temperatures, depending on the pressure imposed. The steam in turn will heat objects it touches to the elevated temperature.

Unpressurized boiling kills most vegetative microorganisms, but it is not reliable against hepatitis virus, and is completely useless against most bacterial spores. Flowing steam has similar effects, although its heating is more rapid. Neither can be relied on for sterilization.

The **autoclave** (Fig. 25.2) is a pressurized steam sterilizer. Materials to be sterilized are placed in a chamber, which is then sealed. Steam flows in from a generator and pushes out the room-temperature air. Once all air has been displaced, temperature-sensitive valves close, and steam continues to fill the chamber until the pressure is typically 15 pounds per square inch (psi). If all air was displaced, the temperature of the pressurized steam will reach 121°C. If some air was retained, the desired pressure may be obtained, but the temperature will not be high enough and sterilization may not occur. Timing of the process should begin only when a temperature gauge indicates a chamber reading of 121°C.

To kill, steam must penetrate throughout the load, touching all surfaces. Packing and wrapping should facilitate steam movement. For example, steam enters pipettes or other long, thin containers well only if they are horizontal. Containers must not be sealed completely. Screw-cap tops should be loosened. Wrappings impermeable to steam should not be used. Syringes should be disassembled.

It may take longer than 15 minutes for the interior of packages or the contents of large vessels to reach the necessary temperature. True sterilization begins only when *all* the contents of the autoclave have reached 121°C. It may be necessary to prolong the exposure time to 30 minutes or an hour in order to guarantee sterility in large volumes of liquid or dense objects.

Tapes containing heat-sensitive dyes may be used to package materials to be autoclaved. On completion of a sterilizing run, the dyes should have turned color in all areas of the load.

The spores of *Bacillus stearothermophilus*, the most heat-resistant organism known, are killed after 15 minutes at 121°C. Indicator cultures composed of a vial of *B. stearothermophilus* spores and a sterile vial of nutrient broth may be processed with each load, preferably at the center. Following cooling, the broth is aseptically added to the spores. After incubation for seven days, the sterilization procedure is judged adequate if no growth appears.

Heat stability of spores; Chapter 3, p. 90

Liquid and dry materials should be autoclaved separately because they require different final steps in the process. If the load contains liquids, the steam pressure must be released gradually after autoclaving so the liquids will not boil violently, causing breakage and loss. If dry materials are processed, the steam must be rapidly evacuated so they will dry off. A soggy sterile wrap is completely useless because it is rapidly permeated by microorganisms once removed from the chamber. Wrapped materials should not be removed until they have dried.

Autoclaving adversely affects some materials. Sharp-edged instruments may be rusted or dulled; thermolabile substances, oils, or greases may be evaporated; powders may be wetted; and equipment with tight-fitting surfaces cannot be penetrated completely. These items are all suitable for dry-heat sterilization, however.

Pasteurization is disinfection by heat. It is designed to kill pathogens and reduce total microbial count, while many harmless microorganisms survive. The process is applied not only to milk but also to other beverages produced commercially. The **batch method** or low-temperature–long-time (LTLT) method holds milk at 61.7°C (145°F) for 30 minutes. The **flash method** or high-temperature–short-time (HTST) method passes milk over heated plates. The milk is heated to 71.7°C (161°F) for 15 seconds, then quickly cooled. The flash method is generally preferred because it is more efficient for bulk milk handling, and also because it kills the Q fever agent *Coxiella burnetii*. This rickettsia is not killed by the lower temperature of the batch method.

Tyndallization or **intermittent sterilization** has its uses because it may be carried out with the most minimal equipment. A solution or food preparation in a closed container is first boiled one hour to kill all vegetative cells, then incubated overnight to encourage spore germination. It is then boiled one hour again, and again reincubated. Following boiling one hour a third time, it is presumed sterile. An alternative schedule specifies heat treatments of 60°C for one hour on five successive days.

Dry heat   Materials such as dry empty glassware may be sterilized by placing them in an oven at 180°C for one to three hours. The effectiveness is increased if the oven is provided with a circulating fan. Ovens may be used for all nonvolatile materials. Some types of glassware must be allowed to cool gradually to prevent cracking.

A home oven accurately set at 330°F may be used for sterilization. Materials should be heated for three hours.

Cold

The effect of cold is largely bacteriostatic; few organisms are actually killed by it. Refrigeration at 4°C slows spoilage because very few bacteria are able to grow at that temperature.

Psychrophiles; Chapter 7, p. 194

Slow freezing causes some microbial death due to intracellular ice-crystal formation. Flash freezing has a less lethal effect. Freezing to −10°C arrests microbial growth and completely kills some food pathogens including the agents of trichinosis and toxoplasmosis, but it does not kill the tularemia bacillus.

Superfast freezing in liquid nitrogen (−196°C) is an effective way to preserve viable cultures, and so is lyophilizing (freeze-drying).

When a refrigerated or frozen material is returned to warm temperatures, microbial growth may immediately begin again.

Radiation

Because the lethality of radiation is well known, it has been much investigated as a microbial control mechanism. Its applications remain limited.

Ionizing radiation   Radiation of wavelengths shorter than ultraviolet (see Fig. 3.1) and beams of subatomic particles cause ionization in matter they strike. The reactive ions and free radicals nonspecifically alter cellular biochemicals. Their effect on proteins and nucleic acids is most pronounced.

Spectrum; Figure 3.1

The energy level of radiation drops sharply with increasing distance from the source. Material to be irradiated must be positioned very close to the energy source. Some radiations such as alpha particles and UV have little penetrating power. Only very high-energy radiations such as gamma rays have enough penetrating power to sterilize the interior of large objects. Gamma rays from a cobalt-60 source are used industrially to sterilize mass-produced disposables such as syringes, petri dishes, and dressing materials. The great expense of installing and operating radiation equipment limits its use.

There has been much interest in the possibility of using radiation to sterilize and preserve foods. Foods irradiated with gamma rays at an intensity that kills selected pathogens such as *Salmonella* do not lose appreciable nutritional value and appeal. However, when **sterilizing** doses of gamma rays are used, foods become unpalatable and suffer reduced nutritional value.

Ultraviolet rays   Mercury-vapor lamps emit ultraviolet radiation in the 230–280 nm range. This radiation is strongly absorbed by nucleic acids and proteins. UV exposure may cause the formation of thymine dimers. These abnormal covalent structures prevent normal gene expression and DNA replication. Repair enzymes may reverse the lethal effects of UV, especially if the irradiated organisms are exposed to visible light shortly afterward. UV lamps are used to kill organisms in air or within relatively small, closed cabinets. The lethal effect occurs only at close range, in clean, dry air. UV is screened out by glass, plastic, and opaque coverings. The light is hazardous to human skin and eyes.

Visible light   When intense visible light such as sunlight is absorbed by some microorganisms, atmospheric oxygen is activated. The resulting **photo–oxidation** has a lethal effect on some microorganisms.

Ultrasound   Probes emitting ultrasonic vibrations are used to rupture microbial cells under controlled conditions. Ultrasound produces tiny bubbles in liquids. These bubbles collapse, creating tiny holes—**cavitation**—and the associated pressure changes cause cellular rupture. Ultrasound is not a reliable sterilant, however. Ultrasonic devices are more commonly used to cleanse intricate equipment, dentures, and jewelry because their scrubbing action penetrates microscopic fissures. Such objects, once cleansed, can be rapidly and effectively sterilized by other conventional means.

Filtration

Filtration; Chapter 2, p. 60

Rather than killing organisms, it is sometimes more desirable to simply remove them from air or a solution by filtration.

Depth filters   The early depth filters were designed like sponges, as a thick layer of porous material. Liquid passes through many tortuous channels, in which microorganisms will eventually become stuck before the liquid emerges into a sterile vessel. Filter materials include porcelain, asbestos, or porous glass. Depth filtration is slow, and because the particles enter the filter, it eventually clogs and requires a long and tedious cleaning. The filter may also absorb too much of the fluid, retain important substances such as enzymes, or introduce chemical contaminants.

Sieve filters   The **membrane** filter has almost entirely replaced the depth filter. Membrane filters function like sieves. They are made of synthetic polymers such as cellulose acetate. These polymers are precipitated from solution to form thin sheets with very tiny, uniformly sized pores. By using different precipitation conditions, any desired pore size can be produced. Membrane filters are sterilizable, inexpensive, and disposable.

A sterile filter is placed on a porous support, and the fluid to be filtered is passed through under vacuum. If membranes of a sufficiently small pore size are used, all bacteria are retained. No filter reliably removes the tiny, highly flexible mycoplasmas, chlamydias, and rickettsias, or the viruses.

Air Control Systems

A controlled flow of air, by directing the movement of airborne microorganisms, can be used as a barrier to infection. Air control systems may be used in several ways.

Positive pressure In a surgical unit, the operating room may be provided with a constant flow of sterilized air to maintain an atmospheric pressure slightly higher than the exterior environment. All air flows are from the inside (clean) towards the

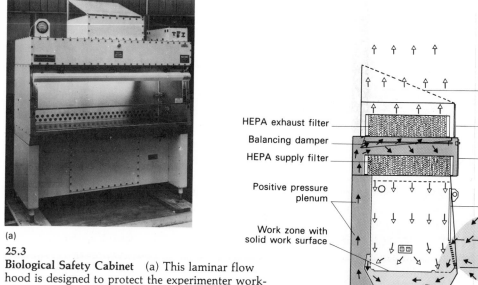

(a)

25.3

Biological Safety Cabinet (a) This laminar flow hood is designed to protect the experimenter working on low to moderate risk materials. It provides Class II containment. (b) Room air (dirty) flows in at the front opening where the worker's hands enter. It flows under and in back of the work zone (white area). A supply of HEPA-filtered clean air flows down over the work area, preventing contamination of cultures. All air vented to the outside (at top) has been sterilized by passage through a HEPA filter.

Expanded metal exhaust cover

HEPA exhaust filter

Exhaust filter access

Balancing damper

HEPA supply filter

Supply filter access

Positive pressure plenum

View screen

Work zone with solid work surface

Work opening

Negative pressure plenum

Blower

Dirty air →
negative pressure

Dirty air →
positive pressure

→ Clean air →

(b)

outside (unclean). Unsterilized air cannot enter because the interior pressure is higher. Random airborne contamination is thus effectively excluded.

High-efficiency particulate air filters (HEPA) are finding increasing use in air control. They are used to clean the air supplied to surgeries, to patients in total isolation units and to the "clean rooms" in which sensitive electronic equipment is assembled. HEPA filters also may be used to remove pathogens in contaminated air. The exhaust from special biological containment units used to study extremely hazardous microorganisms is vented through HEPA filtration.

Negative pressure   This technique may be used to isolate patients with highly communicable diseases. A negative pressure within the unit causes all air flows to move inward, preventing dissemination of the patient's pathogen.

Laminar flow transfer hoods are found in many laboratories. They supply an unidirectional layer of air over the work area. The negative pressure type is used for transferring *M. tuberculosis* cultures and other biohazards (Fig. 25.3). A constant inward flow of air protects the worker sitting outside. Positive pressure types may be used for sterile transfers, cell culture work, or in pharmacies in the formulation of sterile intravenous fluids.

CHEMICAL AGENTS IN COMMON USE IN STERILIZATION

Few chemicals are able to sterilize, but many can disinfect. None are suitable for all applications. Different families of chemical disinfectant have been developed, each with its own strengths and weaknesses.

Desirable Qualities

In the olden days, any disinfectant or antiseptic worth its salt had to smell bad and sting like fire. Newer classes of disinfectants do not have these disadvantages.

A perfect disinfectant, which unfortunately does not yet exist, would destroy all pathogenic microorganisms effectively in a reasonable period of time. Many current candidates are unable to kill one or more of the problem types, such as spores or acid-fast bacilli.

The chemical should be readily soluble and an active wetting agent that coats surfaces and penetrates into cracks and fissures. It should not evaporate too rapidly. These conditions are met by most disinfectants, especially when they are used in appropriate formulations.

The chemical should be nontoxic. It should not irritate or blister, kill living tissue, or cause allergic responses. Newer antiseptics generally meet these criteria, when they are used as directed.

Similarly, the disinfectant, if used on inanimate materials, should not corrode metals, dull blades, degrade rubber or plastics, or leave toxic residues.

The practical acceptance of the chemical is increased if it is inexpensive and if it does not stain or have an objectionable smell.

The disinfectant should not be readily inactivated by dilution with fluids to be treated or by combination with organic matter or soil.

Surfactants

Surfactant compounds lower the surface tension of water. As a result, the solution more readily wets all available surfaces. Such wetting agents are also called **surface active.** Although few surfactants have significant antimicrobial effects by themselves, they are frequently incorporated into mixtures with more active disinfectants. This gives the product much improved penetrating or wetting powers, and thus augments the access of the active ingredient to the microorganism.

Natural **soaps,** the sodium salts of the long-chain fatty acids, were the earliest surfactants. Their active moiety, the fatty acid, is an anion so they are said to be anionic. They have little -cidal activity, but are useful in routine degerming.

The term **detergent** may be generally used for any surfactant used for cleansing. However, it usually indicates a varied group of synthetic cleansers that have largely replaced soap. Home laundry and dishwashing detergents are anionic, more strongly charged than soaps. Thus they have some cell-membrane lytic effect.

Cationic detergents contain a positively charged ion, usually a pentavalent nitrogen, to which is attached one or more large hydrophobic chains. This group of chemicals are quaternary ammonium compounds, or **quats.** Such molecules share the dual nature of membrane phospholipids, having both hydrophobic and hydrophilic regions. The molecules bind avidly to membranes and viral envelopes, and disrupt them. Cationic detergents are probably useful only as sanitizers. They are -cidal only against some vegetative organisms, not spores or acid-fast bacilli. Weak solutions support growth of *Pseudomonas aeruginosa* and some enterobacteria. Mixture with anionic detergents or soap completely inactivates them. Zephiran, Ceepryn, and Roccal are commonly encountered brand names.

Halogens

Chlorine, bromine, iodine, and their compounds have a long history in microbial control.

Chlorine gas, chloramines, and compounds such as hypochlorites that yield the ion $HOCl^-$ in water, are almost universally used to sanitize drinking water supplies and swimming facilities. It is believed that chlorine oxidizes microbial protein. Chlorine compounds are inexpensive and effective against a complete range of organisms. However, they are corrosive to certain materials and irritating to tissues; because their gases are volatile, their effect is of short duration. When large amounts of organic matter are present, the amount of chlorine used must be increased to compensate for it. More chlorine is added than will be immediately consumed to react with the organic matter. A margin of **residual chlorine** is required for complete safety.

A home preparation of two ounces of household bleach per gallon of water provides an effective disinfectant.

Iodine was once widely used in an alcoholic solution or **tincture.** The tincture stung, stained, and evaporated so readily that it often became highly concentrated and caused serious burns. It does not consistently kill spores.

Iodine can be chemically bound to a surfactant carrier that releases it gradually. These **iodophors,** such as povidone, are water-soluble and stable. Povidone-iodine

Cationic detergents or quats

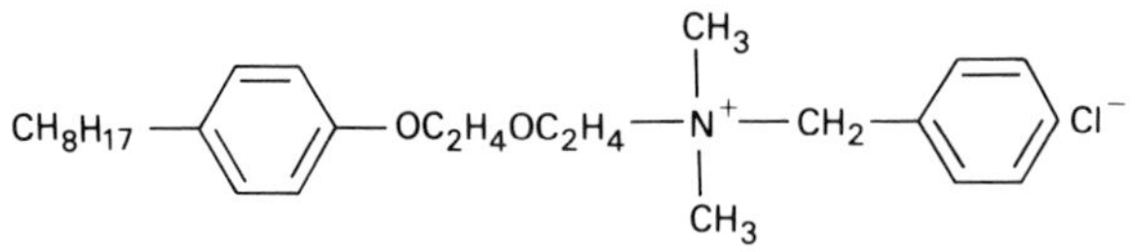

CH_8H_{17} —〇— $OC_2H_4OC_2H_4$ — $\overset{\overset{CH_3}{|}}{\underset{\underset{CH_3}{|}}{N^+}}$ — CH_2 —〇 Cl^-

Benzethonium chloride

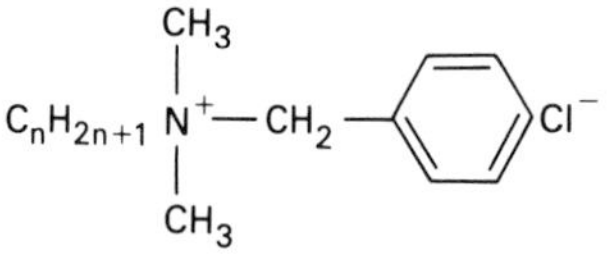

C_nH_{2n+1} $\overset{\overset{CH_3}{|}}{\underset{\underset{CH_3}{|}}{N^+}}$ — CH_2 —〇 Cl^-

Benzalkonium chloride

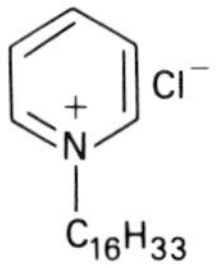

〇 Cl^-
$\underset{C_{16}H_{33}}{N^+}$

Cetylpyridinium chloride

Phenolics

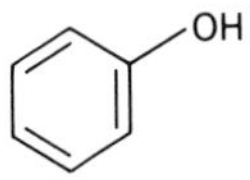

Phenol

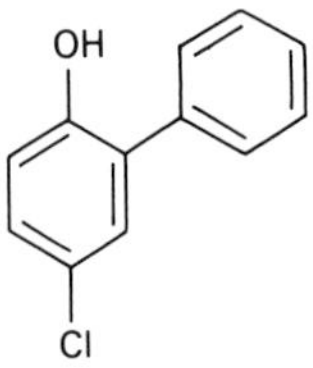

Chlorophene

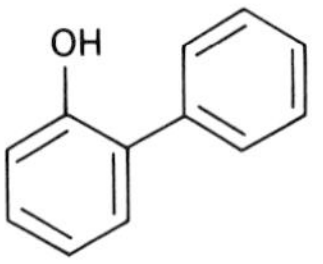

Orthophenylphenol

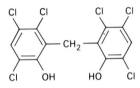

Hexachlorophene

Alcohols or aldehydes

CH_3 — CH_2 — OH

Ethyl alcohol

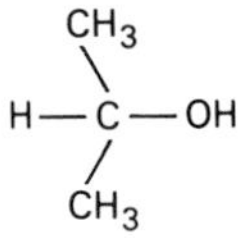

H — $\overset{\overset{CH_3}{|}}{\underset{\underset{CH_3}{|}}{C}}$ — OH

Isopropyl
alcohol

$\overset{H}{\underset{H}{}}C{=}O$

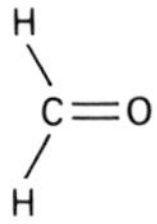

Formaldehyde

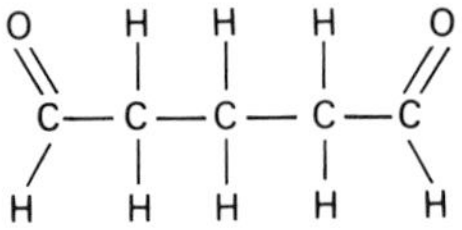

Glutaraldehyde

Gases

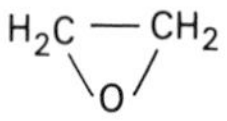

H_2C — CH_2
  \\ /
   O

Ethylene oxide

CH_2 — CH_2
 | |
 O — C${=}$O

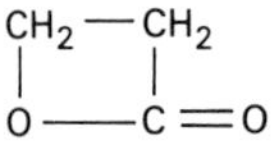

Betapropriolactone

25.4
Structural Formulas of Some Organic Disinfectant and Antiseptic Compounds

does not stain, sting, or harm tissue, but it is not as active as unbound iodine. Some common brand names include Betadine, Wescodyne, and Prepodyne. Iodophor formulations are widely used for skin degerming before insertion of needles, to prepare

TABLE 25.2
Characteristics of common disinfectants

CLASS	DISINFECTANT	ANTISEPTIC	OTHER PROPERTIES
Gas			
Ethylene oxide	+ 2 to + 4	0	Toxic; good penetration; requires relative humidity of 30% or more; bactericidal activity varies with apparatus used; absorbed by porous materials. Dry spores highly resistant; moisture must be present, and presoaking is desirable
Liquid			
Glutaraldehyde, aqueous	+ 3	0	Sporicidal; active solution unstable; toxic
Formaldehyde + alcohol	+ 2	0	Sporicidal; noxious fumes; toxic; volatile
Formaldehyde, aqueous	+ 1 to + 2	0	Sporicidal; noxious fumes; toxic
Phenolic compounds	+ 3	±	Stable; corrosive; little inactivation by organic matter; irritates skin
Chlorine compounds	+ 1	±	Flash action; much inactivation by organic matter; corrosive; irritates skin
Alcohol	+ 2	+ 3	Rapidly -cidal; volatile; flammable; dries and irritates skin
Iodine + alcohol	0	+ 4	Corrosive; very rapidly -cidal; causes staining; irritates skin; flammable
Iodophors	+ 1	+ 3	Somewhat unstable; relatively bland; staining temporary; corrosive
Iodine, aqueous	0	+ 2	Rapidly -cidal; corrosive; stains fabrics; stains and irritates skin
Quaternary ammonium compounds	+ 1	+ 2	Bland; inactivated by soap and anionics; absorbed by fabrics
Hexachlorophene	0	+ 2	Bland; insoluble in water, soluble in alcohol; not inactivated by soap; weakly -cidal.
Mercurial compounds	0	+ 1	Bland; much inactivated by organic matter; weakly -cidal.

Source: E.H. Spaulding, 1974, *Manual of Clinical Microbiology* (2nd ed.), Washington, D.C.: American Society of Microbiology.

The activity of each compound is rated from 0 to 4 on the basis of increasing effectiveness.

the site of surgical incisions, and for hand scrubbing. Iodine hypersensitivity should be watched for.

Phenolics

Phenol or carbolic acid was the first widely used disinfectant; it is still the standard against which others are measured. Five or 10 percent phenol kills all vegetative bacteria and most viruses. However, phenol may be used as a nutrient in lower concentrations (0.1 percent) by some bacteria such as the pseudomonads. Phenol's main use today is as the starting material from which substituted **phenolic** compounds are made. Both organic radicals and halogens may be added, and compounds with very strong -cidal activity can be produced. Three examples are orthophenylphenol, chlorophene, and hexachlorophene. Their good qualities include tuberculocidal ability, residual action, and great stability in the presence of organic matter. They are not active against spores. At least one phenolic (hexachlorophene) was shown to cause brain damage in rats. Its use in nurseries has been restricted to situations in which a staphylococcal infection or epidemic has been documented. Extensive skin absorption should be avoided. Lysol is a mixture of cresol and soap solution. Amphyl and O'syl are mixtures of phenolics. pHisohex is a mixture of hexachlorophene and surfactants.

Alcohols and Aldehydes

Alcohols   Ethyl alcohol and isopropyl alcohol are widely used disinfectants. In solutions of from 70–95 percent alcohol in water, they are effective against numerous agents including the tuberculosis organism and viruses with envelopes, but not against spores and hepatitis virus. Their mode of action is the coagulation of cellular protein. Water must be present for best results. Alcohols evaporate rapidly and have no residual action. Alcohols of higher molecular weight are more effective, but are seldom used because of their unpleasant odors and higher cost. Alcohols are often added to disinfectant mixtures with iodophors and cationic detergents for a combined bacteriocidal effect.

Aldehydes   The aldehydes are highly reactive compounds that alkylate proteins. Some are used as tanning and fixing agents; others are all too familiar as preservatives in dissection specimens. Formaldehyde in a 37 percent solution in water is called **formalin.** Formalin is often diluted in ethyl or isopropyl alcohol for use. When an adequate concentration is placed overnight in a closed container with contaminated inanimate objects, its vapor is a very effective sterilant. Bacteria or viruses for vaccine use are often killed by formalin treatment. Glutaraldehyde solutions (2%) are used for sterilizing endoscopes and respiratory equipment.

Gases

A sterilant gas may be utilized in a closed container under conditions similar to autoclaving. However, the times required are much longer. Gas sterilization is also called **fumigation.** Ethylene oxide (EtO) gas has found general acceptance. Many hospitals have EtO sterilizers (Fig. 25.5) as well as autoclaves in their central supply areas. EtO is an alkylating agent that readily penetrates paper or plastic wrappings. Therefore it can be used to sterilize prewrapped plastic disposables such as syringes and culture-collecting devices, respiratory therapy mouthpieces, and wrapped instruments. The exposure conditions must be carefully controlled. The gas, which by

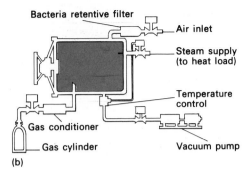

Bacteria retentive filter

Air inlet

Steam supply
(to heat load)

Temperature
control

Gas conditioner

Gas cylinder

Vacuum pump

(b)

25.5

Ethylene Oxide Sterilizer   (a) An EtO gas sterilization system.
(b) A diagram of a simplified gas sterilizer. When the chamber is
loaded and sealed, the air is replaced by the gas sterilant. The gas
is moistened and heated by steam. After sterilization is complete,
a vacuum pump removes the gas.

(a)

itself is explosive, is supplied in nonexplosive mixtures with CO_2 or Freon. The temperature must be elevated, the relative humidity must be maintained between 30–50
percent, and the gas concentration fixed. Depending on the unit, from 4–12 hours of
exposure are required. The gas blisters living tissues, so sterilized articles must be
aired from 12–24 hours afterward to allow the gas to dissipate. The gas cannot be
used to sterilize biochemicals because it chemically alters many of them.

β-propriolactone (BPL) is proposed as an alternate to EtO. It lacks the explosive
risk and acts more rapidly. Its human toxicity is greater, however, and it has not
been widely adopted.

Miscellaneous Chemicals

Metal ions   The ions of heavy metals such as mercury, silver, arsenic, and copper
have long been known to be toxic. They were used in insecticides, rat poisons, and
the earliest disinfectants. Since the advent of other substances with less toxicity and
more effectiveness against microorganisms, substances containing metal ions have
been displaced. Two exceptions are silver nitrate solution (infused into the eyes of
neonates to prevent ophthalmic gonorrhea) and silver sulfadiazine (used in the routine cleansing of severe burns). Copper sulfate, a fungicide and algicide, is used to
control microbial growth in swimming pools and lakes, and also in some agricultural applications.

The industrial and agricultural uses of heavy metals, once general, are now
restricted by law in order to reduce environmental hazards. Mercury compounds
were once used in vast quantities to control microbial growth in cooling water and
industrial processes such as pulp and paper mills. The mercury compounds became

incorporated into the bottom sediments of rivers or bays. There, certain environmental bacteria took in the mercury and methylated it, producing a more soluble form highly toxic to mammals. Fish from such waters, such as Minimata Bay in Japan, caused severe neurological disease due to mercury poisoning when consumed by human beings.

Peroxides   The decomposition of a peroxide molecule yields a chemically reactive free atom of oxygen. This nascent oxygen is toxic to anaerobes that lack enzymes (catalases or peroxidases) to safely bind it. Hydrogen peroxide solutions are used occasionally to clean superficial wounds. A medicinal zinc peroxide preparation may be used for deep wounds.

Acids   Strong acids are highly toxic to all cellular life. Nitric acid cautery was once used to cleanse rabid animal bites. A mixture of benzoic and salicylic acids (Whitfield's ointment) is used to control superficial mycosis. Undecylenic acid is the active ingredient of Desenex, a product used for the same purpose. Benzoic, propionic, sorbic, and acetic acids are static agents, widely used as food preservatives.

Athlete's foot; Chapter
17, pp. 507–508

CONTROL OF INFECTIOUS DISEASE

Disinfection and sterilization are intensively used in clinical facilities. There are three clearly defined goals. One is to reduce the population of microorganisms in general—to minimize microbial numbers throughout the facility. Another is to restrict the spread of pathogens from a patient with infectious disease to the surroundings, to other patients, or to the staff. A third goal is to prevent microorganisms from reaching and infecting individuals—especially compromised individuals.

Concurrent Disinfection

In an infectious disease, certain of the patient's body surfaces and secretions bear and transmit the agent, and some others do not. The immediate and ongoing disinfection of the contaminated secretions plus articles they have soiled is **concurrent disinfection**. Disinfection procedures are most often applied to discharges from the mouth and nose, soiled tissues, and eating utensils, or to feces, urine, urinals, or bedpans. The hands of the attendant must also be cleansed after giving patient care or handling soiled articles. Concurrent disinfection is the primary component in the various categories of precautions.

Terminal Disinfection

This occurs after the patient has vacated the unit. The routine, between-patient, cleaning procedures should be expanded when the last patient had a serious communicable disease. Terminal disinfection under these conditions should include thorough disinfection of walls, woodwork, floors, and furniture, and removal and disinfection of all containers or utensils used by the patient.

25.6

Warning Cards The card system utilizes a warning card placed on the room door of any patient for whom isolation or precautions are required. These warnings apply equally to all classes of hospital employees and to visitors. No one is exempt from following them. (a) Strict isolation is for the containment of a highly communicable disease agent such as diphtheria. Persons entering the room must protect themselves. Articles leaving the room contaminated must be specially handled. (b) Enteric precautions are used for communicable diseases, such as cholera, that are transmitted by fecal matter. The protective measures are designed to prevent contact with the contaminated wastes. (c) Protective isolation shields a patient who is unusually susceptible to infectious disease. In this case, gowns and masks are used to prevent the staff and visitors' microorganisms from reaching the patient.

(a) **Strict isolation**

Visitors – report to nurses' station before entering room

1. Private room *necessary*; door must be kept closed.
2. Gowns must be worn by all persons entering room.
3. Masks must be worn by all persons entering room.
4. Hands must be washed on entering and leaving room.
5. Gloves must be worn by all persons entering room.
6. Articles must be discarded, or wrapped before being sent to Central Supply for disinfection or sterilization.

(b) **Enteric precautions**

Visitors — report to nurses' station before entering room

1. Private room *necessary for children only*.
2. Gowns must be worn by all persons having direct contact with patient.
3. Masks not necessary.
4. Hands must be washed on entering and leaving room.
5. Gloves must be worn by all persons having direct contact with patient or with articles contaminated with fecal material.
6. Special precautions necessary for articles contaminated with urine and feces. Articles must be disinfected or discarded.

(c) **Protective isolation**

Visitors — report to nurses' station before entering room

1. Private room *necessary*; door must be kept closed.
2. Gowns must be worn by all persons entering room.
3. Masks must be worn by all persons entering room.
4. Hands must be washed on entering and leaving room.
5. Gloves must be worn by all persons having direct contact with patient.
6. Articles: see written instructions on unit.

Isolation

An isolated environment may be used for one of two reasons, either to prevent the dissemination of a communicable disease from a patient to other persons, or to protect a severely compromised patient from microorganisms from the surroundings. Different types of isolation are recommended for various conditions. Most hospitals presently use the classification system and techniques proposed by the United States Department of Health and Welfare. These include three categories of **isolation** (private room required) and four categories of **precautions** (private room preferred, but designed to block the spread of infection in multibed units as well).

Types of isolation   Practical isolation must be based on an exact knowledge of which body secretions are infective in the particular disease. The attending physician or other responsible party may order the adoption of precautions or isolation.

The **card system** is generally used. A card (Fig. 25.6) is placed on the door of the patient's unit. It presents a concise color-coded outline of the required isolation procedures. These may be checked by the staff member every time he or she prepares to enter.

Visitors are required to observe the same procedures as staff. They are requested to report to the nurses' station before entering, to receive instruction in precautionary technique.

TABLE 25.2
Precautionary measures recommended for specific conditions

TYPE OF PROCEDURE	STRICT ISOLATION	RESPIRATORY	PROTECTIVE	ENTERIC
Card Color	Yellow	Red	Blue	Brown
Diseases	Anthrax, inhalation	Measles	Agranulocytosis	Cholera
	Burn wound, major, infected with *Staphylococcus aureus* or Group A *Streptococcus*	Meningococcal meningitis	Dermatitis, noninfected vesicular, bullous, or eczematous disease when severe and extensive	Diarrhea, acute with suspected infectious etiology
		Meningococcemia		Enterocolitis, staphylococcal
	Congenital rubella syndrome	Mumps	Extensive noninfected burns (some)	Gastroenteritis caused by *Escherichia coli* *Salmonella* sp. *Shigella* sp. *Yersinia enterocolitica*
	Diphtheria	Pertussis	Lymphomas and leukemias (some)	
	Disseminated neonatal herpes simplex	Rubella		
		Tuberculosis		
	Disseminated Herpes Zoster			Hepatitis, viral, all
	Lassa fever			Typhoid fever
	Marburg disease			
	Plague, pneumonic			
	Pneumonia, *Staphylococcus aureus* or Group A *Streptococcus*			
	Rabies			
	Skin infection, major *Staphylococcus aureus*			
	Smallpox			
	Vaccinia			
	Varicella			

Source: Compiled from *Isolation Techniques for Use in Hospitals, 1975,* Washington, D.C.: United States Government Printing Office.

Techniques for isolation The exact techniques for each type of isolation vary in detail among hospitals. Each unit should have available complete and current procedures manuals. Newcomers to the unit should receive full orientation.

Problems with isolation The primary problem is noncompliance. Doctors and nurses are often hurried, and the time required to put on gloves and masks may seem excessive at the moment. It is important that all persons—including doctors, visitors, and housekeepers—entering an isolation room use the indicated precautions.

WOUNDED AND SKIN	BLOOD	LESIONS	ORAL	EXCRETION
Green	*None*	*None*	*None*	*None*
Infected burn wounds, other*	Arthropod-borne viral fever	Actinomycosis, draining lesions	Herpangina	Amebiasis
Gas gangrene	Malaria	Anthrax, cutaneous	Oral herpes	*Clostridium perfringens* food poisoning
Herpes-Zoster, localized		Brucellosis, draining lesions	Infectious mononucleosis	Enterobiasis
Melioidosis (draining sinuses)		Burn, skin, and wound infections, minor	Melioidosis, pulmonary	Giardiasis
Plague, bubonic			Pneumonia, *Mycoplasma*	Hand-foot-and-mouth disease
Puerperal sepsis			Pneumonia, bacterial, other	Herpangina
Wound or skin infections where discharge not adequately contained by dressings (other)			Psittacosis	Infectious lymphocytosis
			Q fever	Leptospirosis
			Respiratory infectious disease, acute, other	Meningitis, aseptic
			Scarlet fever	Pleurodynia
			Streptococcal pharyngitis	Poliomyelitis
				Staphylococcal food poisoning
				Tapeworm disease
				Viral diseases, other

*Exceptions are in columns to the left.

On the other side of the coin, because it requires much more time and effort to care for the isolated patient, there is a risk that he or she will not receive an adequate degree of care. An awareness of this tendency is usually sufficient to prevent this from occurring.

For the patient, isolation is a psychological stress. Human contact will be reduced. The people who enter the room are depersonalized behind masks, their voices are muffled, and they may perhaps be feeling and behaving acutely uncomfortably. It requires a special effort on the part of the nursing staff and visitors to project a feeling of warmth and involvement.

Surgical Asepsis

A modern surgical suite is very complex, not only in its architecture and equipment but also in the aseptic techniques practiced. The primary thrust is always to minimize the number of microorganisms at large in the environment prior to and during surgery.

Human flora   Both patient and surgical team bring their flora into the operating suite. Since these organisms cannot be eliminated, they must be contained so they cannot gain access to the incision. Entrance to the surgical suite is restricted to authorized persons only, and traffic flow is unidirectional. Beyond a central receiving area, everyone must wear special clothing. Gown or wash suit, mask, complete hair covering, and foot coverings replace street clothes. Their purpose is **occlusive**—they enclose the body and prevent surface microorganisms from becoming airborne.

The hands of the surgical team are prepared in a two-stage process. First, they are degermed by systematic scrub with a disinfectant soap. Then, sterile gloves are put on. It is estimated that at least 25 percent of surgical gloves are punctured during operations, so the hand degerming process is an important backup measure.

The patient's flora is also controlled by clean clothing and masks. The site of the incision is exactingly degermed and shaved, then sterile drapes are placed around the incision area. Skin should not be shaved many hours prior to surgery because the inevitable tiny nicks may in the interim develop small infections. Should the scalpel pass through one of these, the underlying tissues will be contaminated. Shaving should therefore be delayed until immediately prior to the procedure.

When an incision must pass through a mucous membrane (not practicably sterilizable) or into an existing focus of infection, contamination of the surgical site is inevitable. Infection may be prevented by the use of prophylactic antimicrobials given just prior to and during the operation, and for a short while thereafter.

Environmental controls   A constant flow of filtered conditioned air is supplied, at positive pressure, and used air is conducted to the exterior. Some surgical units are supplied with sterile air passed through an HEPA unit. In others, ultraviolet light is used to kill airborne microorganisms. The highest degree of environmental cleanliness is achieved by use of a "greenhouse," basically a large plastic dome placed over the operating table. These are used only for the highest-risk situations.

All materials brought into the operating suite must be sterile or be free of transmissible microorganisms. Drapes, dressing materials, sutures, instruments, rinsing solutions, and medications are all sterile. Many are purchased as sterilized disposables. Others are wrapped and sterilized by the surgical aides or by central supply. Reliable confirmation of sterility is required on each package. Each object is dated so that safe, sterile, shelf life will not be exceeded.

Following surgery, used materials are returned to be cleaned and resterilized. This function is usually carried out by the central supply service. All surfaces are

Box 25.2

Cleanliness or Compulsion?

In the grand old days of King Henry VIII or King Louis XIV, sanitary standards were rather relaxed. People (even kings) were bathed three times in their lives—when they were born, before their marriage, and after they died. Foods were heavily spiced in an effort to hide the taste of advanced microbial decomposition. The slop jars were emptied out the upstairs windows into the street or lugged to an open cesspool somewhere all too near at hand.

Our culture has made a 180° turnaround. We now pursue what may be completely unreasonable standards of hygiene. TV advertisements show us wild-eyed housewives stalking through their shining homes with cans of spray disinfectant. Should we find ourselves next to a person who clearly has not bathed too recently, we shy away with real fear as well as repugnance. We refuse to use milk that is a little sour even for cooking. The thought of using an outhouse disgusts us. We have developed a new fear—microbiophobia.

How much cleanliness is enough? How much grubbiness is too much?

To generalize, remember the grounds for susceptibility to infection. These are extreme youth, extreme age, and underlying ailments. If you are healthy, you can basically be infected only by pathogens. Those hairy little "germs" of the ad media—normal flora (yours or someone else's) and environmental organisms—are extremely unlikely to cause you any problems. They can cause problems for your newborn infant or your elderly grandmother, however.

Similarly, a healthy group of persons can get away with being very casual about passing their flora around. But when someone introduces a pathogen such as a wild-type polio virus or a *Shigella* bacillus to the community, everyone may pay the price for lax hygiene standards.

Clearly, there must be several standards of cleanliness. Certain things **are** essential. Clean drinking water, adequate sewage disposal, updated vaccinations for all ages, and properly handled food are the basic requirements. On the other hand, from a health standpoint, it literally doesn't matter whether you shower once a day or once a week. Those air freshener sprays can be promoted only for their esthetic values. If everyone is healthy and there are no apparent pathogens around, you may generally be as relaxed at home as your personal tastes permit. But when there is a high-risk person around, or pathogens are at hand, higher standards must be adopted for the duration. The very highest standards of cleanliness belong in the hospital or other clinical facility.

disinfected scrupulously between patients. Anesthesia and X-ray equipment receive special cleansing. Patients with infections are normally placed last in the operating schedule for the day so that there will be sufficient time for expanded terminal disinfection.

SUMMARY

1. Microbial control efforts can be directed toward one or more of several goals: to reduce the total number of microorganisms in the environment, as a general precautionary and esthetic measure; to establish protective barriers preventing infection in particularly susceptible body areas, persons, or populations; or to contain and inactivate pathogenic microorganisms from infective patients so they can spread no farther.

2. If the destruction of pathogens is required, a disinfectant treatment effective against the particular pathogen may be selected. If complete destruction is required, then sterilization is needed. Microbial death is logarithmic.

3. There is no absolute distinction among degerming, disinfection, and sterilization—there are simply differences in the degree of aggressiveness with which the microorganisms are assaulted.

4. A desired result can usually be obtained by many different physical or chemical treatments. Simple, inexpensive methods and equipment, scrupulously used, can be as effective as the largest and most expensive autoclaves, gas sterilizers, and gamma radiation sources.

5. Disinfection and sterilization nonspecifically destroy a cell's structure and its biochemical integrity. As a result, all harm human tissue to some extent, unless carefully used. Disinfectants cannot be taken internally as therapeutic agents. Their role is to interrupt the transmission of disease.

6. All disinfectant and sterilant treatments are effective only within certain defined limits of time, temperature, and concentration. If significant variation is introduced in their use, the treatments may become ineffective.

7. Unless specific checkup procedures are being done, the failure of a technique will not be revealed until some unfortunate person develops an avoidable infection. No technique or precaution is foolproof—all are exactly as good as the person using them is careful.

8. Many different chemical disinfectants are available. The most widely used proprietary preparations contain several different substances whose activities complement each other.

9. Isolation procedures combine a variety of disinfection and sterilization techniques with good aseptic technique. The type of isolation and the procedures used are selected according to the infection's mode of transmission.

10. In the surgical suite, procedures attempt to block all possible avenues of microbial access to the incision. Particular attention is paid to occluding human surface flora, providing sterile equipment and supplies, and working in a flow of clean or sterile air.

Study Topics

1. Compare the effects on microbial cells and endospores of moist heat, steam under pressure, and dry heat.

2. It is not theoretically possible to say that an object has been rendered absolutely sterile. Why?

3. A disinfectant preparation was added to a test tube containing a log-phase culture of a bacterium. Growth rapidly stopped. How can you tell if the disinfectant was bacteriocidal or bacteriostatic?

4. Design an in-use test to evaluate the effectiveness of a new degerming preparation for the skin.

5. Describe how a positive-pressure air supply could be incorporated into the protective isolation of a burn patient.

6. Referring to the epidemiology tables at the ends of Chapters 17 through 24, give examples of diseases that require concurrent disinfection of some sort; isolation; no precautions.

7. What sources of microorganisms remain in the scrupulously clean operating room?

Bibliography

Benenson, Abram S. (ed.), 1975, *Control of Communicable Diseases in Man* (12th ed.), Washington, D.C.: American Public Health Association.

Block, Seymour S. (ed.), 1977, *Disinfection, Sterilization, and Preservation* (2nd ed.), Philadelphia: Lea and Febiger.

Borick, Paul M. (ed.), 1973, *Chemical Sterilization*, Vol. 2, New York: Academic Press.

Center for Disease Control, 1975, *Isolation Techniques for Use in Hospitals* (2nd ed.), DHEW Publication 75–8043. Washington, D.C.: U.S. Government Printing Office.

Jay, James M., 1978, *Modern Food Microbiology* (2nd ed.) New York: Van Nostrand.

Perkins, J. J., 1973, *Principles and Methods of Sterilization in Health Sciences*, Springfield, Ill.: Charles C Thomas.

Physicians Desk Reference to Pharmaceutical Specialties and Biologicals. Published annually. Oradell, N.J.: Medical Economics, Inc.

Pike, Robert M., 1979, Laboratory-associated infections: Incidence, fatalities, causes, and prevention. *Ann. Rev. Microbiol.* **33:** 41–66.

26

Antimicrobial Chemotherapy

In the late nineteenth century, Paul Ehrlich defined the need for specific antimicrobial drugs—substances with a **differential toxicity.** This means they could be used therapeutically, with complete safety for the patient but with swift lethal effects on pathogens.

At first, a direct synthetic approach was tried, and chemists manufactured numerous compounds that were found, on evaluation, to be too toxic. The chemosynthetic attack was fruitless until the introduction of the sulfonamides (sulfa drugs) in the 1930s. Shortly thereafter, we discovered that nature had been millions of years ahead of us, and natural antibacterials were being produced by a variety of soil fungi and bacteria. In rapid succession, penicillin, streptomycin, tetracycline, and chloramphenicol were discovered. By the 1950s we were firmly launched into the Antibiotic Era. People who saw patients make prompt recoveries from previously hopeless diseases coined the term "miracle drug."

Of course, this period of unchecked enthusiasm was very short. Sobering evidence of the drugs' potential toxicity and the organisms' ability to develop resistance quickly pointed out that Ehrlich's ideal had certainly not been reached.

New antibiotics, mainly improved derivations of existing ones, are constantly synthesized in research labs around the world. A few become candidates for testing in human volunteers. Many fewer are released for clinical trial as experimental

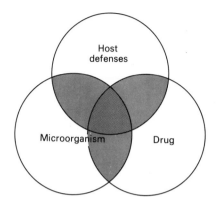

26.1
Factors in Antimicrobial Therapy   The effectiveness of antimicrobial therapy depends on three interacting factors. The **microorganism** must be susceptible to the drug used in the concentration that reaches its site of multiplication, the infection focus. The **drug** must be able to penetrate to the infection focus, enter the microbial cell, and block vital functions. Its effects on the host cells should be neutral. The **host's defenses** must be sufficiently potent to aid in removing not only the microbes killed or inactivated by the drug but also any small part of the microbial population the drug has not affected.

drugs. Very few (only three in 1978) win licensing by the Food and Drug Administration (FDA) for general use. In Europe and Japan, drugs are available that are not yet approved in the United States.

Overuse of antimicrobial drugs has always been an outstanding problem. Despite the ever-increasing complexities of therapy, drug prescriptions continue to increase. The global annual production of penicillin alone is over 100 million pounds. Methods of selecting the correct antimicrobial and monitoring its administration and action have increased in sophistication as our knowledge of the potential hazards has grown. Our armory of drugs can now be used far more effectively. To check unwarranted drug use, both professionals and the public need to modify their expectations for instant relief. Successful antimicrobial therapy is obtained when host, drug, and microorganism factors are all taken into consideration (Fig. 26.1).

CHARACTERISTICS OF AN ANTIMICROBIAL DRUG

Some Definitions

The term **antimicrobial** will be used as a general term for all substances that can be used systematically (introduced into the body) to inhibit or kill microbial pathogens, regardless of their origin. Some antimicrobials are of biological origin—they are synthesized by living microorganisms. These have traditionally been called **antibiotics;** a term that probably will continue to be used by the general public. Most widely prescribed antimicrobials are **semisynthetic;** that is, they are natural products that have been chemically modified by the addition of various atomic groupings to improve their characteristics. An increasing number, especially the antitubercular, antiviral, and antiprotozoan drugs, are fully **synthetic.** These different types of drugs are used interchangeably, depending solely on which is best suited for the individual case.

TABLE 26.1
Spectrum and effect of antimicrobial drugs

DRUG	LETHALITY IN THERAPEUTIC DOSAGE	SPECTRUM	USABLE AGAINST
Penicillin G	-cidal	Narrow	Gram-positive bacteria
Ampicillin	-cidal	Broad	Gram +, some Gram −
Carbenicillin	-cidal	Broad	Gram +, many Gram −
Methicillin	-cidal	Narrow	Gram +
Nafcillin	-cidal	Narrow	Gram +
Cephalosporins	-cidal	Narrow	Gram +, some Gram −
Erythromycin	-static	Narrow	Gram +
Clindamycin	-static	Narrow	Gram +, especially anaerobes
Streptomycin	-cidal	Broad	Gram +, Gram −, mycobacteria
Kanamycin	-cidal	Broad	Gram +, Gram −, mycobacteria
Gentamicin	-cidal	Narrow	Gram −
Tetracycline	-static	Broad	Most bacteria, rickettsia, chlamydia
Chloramphenicol	-static	Broad	Most bacteria, rickettsia, chlamydia
Polymyxin B	-cidal	Narrow	Gram −
Sulfonamides	-static	Broad	Bacteria, protozoans
Trimethoprim	-static	Broad	Bacteria, protozoans
Isoniazid	Dose-dependent	Narrow	Mycobacteria
Rifampin	-static	Broad	Gram +, mycobacteria
Ethambutol	-static	Narrow	Mycobacteria
Amphotericin B	-static	Narrow	Fungi

The distribution among antibacterial, antifungal, and antiviral agents is also becoming less absolute, as drugs appear that have diverse uses. The term **spectrum** refers to a drug's range of effectiveness (Table 26.1). A **broad-spectrum** drug would be one that can be used against many different species and types of organisms, such as bacteria, rickettsia and chlamydia, or fungi and protozoa. A **narrow-**

spectrum drug is useful against only one type of organism, such as either Gram-positive or Gram-negative bacteria, but not both. It must be pointed out that **no** drug safe enough for human use is effective against **all** true bacterial species, much less against all microbial pathogens.

Differential Toxicity

All drugs have different effects at different concentrations. A clinically useful antimicrobial must have differential toxicity. That is, it must damage the microbial cell at concentrations that do not significantly damage the human cell. At very low levels, it will not damage the pathogen; at levels that are too high, it will probably injure the patient.

Therapeutic range   For each drug, the level necessary for clinical control of a particular agent (**therapeutic dose**) and the level at which one gets unacceptable toxicity (**toxic dose**) can be determined. The ratio of the two is the **therapeutic index.** Between the two levels lies the safe margin for the clinician planning dosage—the **therapeutic range.** Any dosage that produces concentrations within that range is acceptable.

The penicillins have the widest therapeutic range of common antimicrobials. This allows considerable flexibility in dosage; it requires greatly excessive doses to provoke neurological toxicity. Amphotericin B has a minimal range; any dose large enough to be effective will produce serious toxicity.

The normal therapeutic range may be narrowed by patient factors that potentiate the drug's toxic effects. Neonates, for example, have little capacity to excrete chloramphenicol, so smaller doses must be used (with great caution).

Inhibitory in contrast to lethal effects   Like the disinfectants, some antimicrobials have irreversible effects that render the pathogen nonviable. Others inhibit growth, but the organisms recover once the drug is discontinued. The already familiar terms -cidal and -static are applied. A long-standing clinical rule, not universally accepted, is that a -cidal drug should be chosen in preference to a -static one whenever possible.

Bacteriocidal; Chapter 25, p. 723

Inhibitory or killing effects are concentration-dependent. A -cidal drug may prove only -static in inadequate amounts. Some standard terms are used to describe concentration-dependent effects. The **minimum inhibitory concentration** (MIC) is that which arrests growth. The **minimum lethal concentration** (MLC) is that which kills.

Host defenses   The question may arise: "Why does a bacteriostatic drug work if the organism can recover as soon as the drug is no longer taken?" The answer is that no drug works alone in eradicating infection. Infectious disease exists, as we have seen, because the host defenses have in some way been overwhelmed. However, in the normal individual, phagocytic and antibody-mediated immune systems do not quit when infectious disease is present. They continue to operate at some level, usually elevated. Drug–host infection control is cooperative. A -static antimicrobial keeps the microbial population from expanding, and the host defenses are then able to cope with the inhibited invaders.

In the immunosuppressed patient, this drug–host cooperation will be ineffective. Therefore, -cidal drugs must be given; even they may fail because of the difficulties of achieving the needed drug level in certain parts of the body, and the appearance of resistant strains during the course of therapy.

Classification

The antibacterial drugs are commonly classified by structural similarities. Other drugs are classed by type of pathogen they affect. The structural approach, although cumbersome for nonchemists, has the advantage that all drugs of a given class —**congeners**—tend to have very similar spectra, modes of action, and potential problems, which can be readily remembered.

MICROBIAL RESPONSE TO ANTIMICROBIAL DRUGS

Under defined exposure conditions, a given microbial strain may be either susceptible or resistant to a drug. In this section we will examine the mechanisms of differential toxicity. We will also discuss the strategems of microbial resistance. We will then compare various laboratory methods for evaluating susceptibility or resistance of clinical isolates.

Modes of Action of Antimicrobials

Procaryotic cell compared with eucaryotic cell; Chapter 3, pp. 91–97

Therapy directed against a bacterial pathogen is relatively easy because we can take advantage of the fact that the target is a procaryotic cell, while the host is eucaryotic. This advantage is lost when it comes to a fungal or protozoan parasite. Worst of all, when the target is a replicating virus, it is virtually impossible to separate the viral and host components within the host cell. Any drug that inactivates the virus is all too likely to stop the host as well. Thus the antibacterial drugs are far more numerous and satisfactory than any others (Table 26.2).

Peptidoglycan; Chapter 3, pp. 76–82

Inhibition of cell wall synthesis   The peptidoglycan layer of procaryotes is unique to them and vital to their survival. Neither the PG components nor the enzymes for PG synthesis are found in higher organisms. Thus anticell-wall drugs operate with complete specificity. They are the only group that does.

Penicillins, cephalosporins, and vancomycin block cell-wall synthesis at various steps. The first two block formation of the peptide bridges that link the carbohydrate chains together. Vancomycin blocks the incorporation of vital cell-wall proteins. An experimental class of drugs, the amino-acyl phosphonates, provide modified amino-acid substrates that confuse the cell-wall synthesizing enzymes.

TABLE 26.2
Modes of antimicrobial action

Cell-wall synthesis inhibitors	Penicillins Cephalosporins Vancomycin Amino-acyl phosphonates Cycloserine Bacitracin
Protein synthesis inhibitors	Aminoglycosides Tetracyclines Chloramphenicol Erythromycin Clindamycin Spectinomycin
Membrane-active	Amphotericin B Nystatin Polymyxin Some aminoglycosides
Nucleic acid synthesis inhibitors	Actinomycin D Rifampin Nalidixic acid
Antimetabolites	Sulfonamides Diaminopyrimidines Isoniazid Ethambutol Para-aminosalicylic acid

Inhibition of protein synthesis   The proteins of procaryotes are synthesized on smaller ribosomes than eucaryotic proteins. The bacterial 30S and 50S subunits are a target for many antimicrobials. The aminoglycosides bind to the 30S unit, whereas erythromycin and chloramphenical block the 50S subunit. The net effect is to arrest the translation process.

Protein synthesis inhibitors are mainly broad-spectrum drugs, inhibiting all procaryotes to varying extents. These drugs are mainly -static; translation resumes once they are removed. Their selectivity is incomplete and they are toxic to certain human organs in high doses or on prolonged exposure. This may be due to inhibitory effects within the human mitochondrion in which protein synthesis takes place on small ribosomes as it does in the procaryotic cell.

Membrane-active drugs   The membranes of various life forms differ in their sterol composition. The polyene class of antifungal agents binds preferentially to the ergosterol found in fungal cell membranes. Polyenes cause lethal leakage of cations and amino acids. They also bind to mammalian membrane sterols, causing marked toxicity.

Membranes; Chapter 3, pp. 82–84

26.2

Inhibitors of Folate Pathways   Bacteria can carry out Step 1, the conversion of PABA to FAH$_2$. The enzyme used is exclusively bacterial and has no mammalian counterpart. Thus sulfonanides acting as competitive inhibitors for Step 1 have a fully specific antibacterial mode of action. Step 2, the reduction of FAH$_2$ to FAH$_4$, occurs in both bacterial and mammalian cells. However, the bacterial enzyme is 10^5 times more sensitive to the inhibitory effects of trimethoprim than the mammalian enzyme. Since sulfonamide and trimethoprim block two steps in the same pathway, they have a synergistic effect and are usually used together in the combination called cotrimoxazole. Step 3 is the reduction of uridine to thymidine, a vital step toward DNA synthesis in all cells. The enzyme is blocked by methotrexate, an analog of FAH$_4$. All life is susceptible to this blocking effect. Methotrexate is a widely used antineoplastic drug. Its effect is to arrest cell division. Because the blockage is general, this drug has many severe side effects.

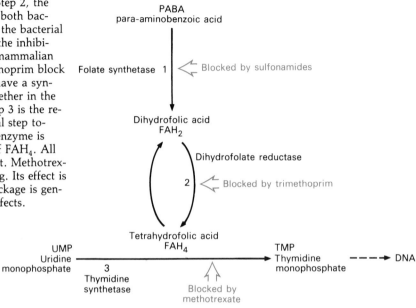

Inhibition of nucleic acid synthesis   Rifampin is an RNA synthesis inhibitor used for tuberculosis therapy. DNA synthesis inhibitors are few; they have so far seemed (with the exception of naladixic acid) to be undiscriminating.

Antimetabolites   Certain drugs act as competitive inhibitors, blocking microbial synthesis of an essential compound. The cell then stops growing; it will eventually die if the blockade continues. The best-understood set of antimetabolites interfere with the synthesis and utilization of folate (Fig. 26.2). This vitamin is essential for synthesis of purines and thymidine. When it is absent, nucleic acid synthesis is arrested.

Biosynthesis of purines and pyrimidines; Chapter 9, p. 250

The sulfonamides and diaminopyrimidines block folate synthesis. Most bacteria need to synthesize their own folate internally because they are impermeable to external folate. Thus their metabolism may be conveniently shut off. Because human cells do not synthesize folate (they receive their supply from dietary sources), they are not affected. Methotrexate, an antineoplastic drug prescribed in leukemia and Burkitt's lymphoma, is a folate analog. It blocks purine and pyrimidine synthesis, causing abnormal nucelic acid formation. Methotrexate is highly toxic, as are all drugs used in the chemotherapy of malignancy.

Microbial Resistance

Mechanisms of resistance   Certain antimicrobials have no effect whatever on some microbes. Penicillin G, for example, is rarely effective against Gram-negative organisms; few drugs touch the mycobacteria. These patterns reflect a major resistance mechanism—exclusion of the drug from the bacterial cell. The pathogen may also compensate for an inhibitory effect by shifting to alternate metabolic pathways or increasing output of the metabolite or enzyme under attack. The pathogen may also make an enzyme that can destroy the drug or an inhibitor that binds the drug and takes it out of commission.

Acquired resistance   An antimicrobial is presented to a large population of cells. All such populations are genetically heterogeneous. Whether or not the population has had previous drug exposure, some few random individuals will carry genes rendering them less susceptible. These genes are of no value in the absence of the drug. In fact, studies show that, in the normal flora, drug-resistant strains disappear when drug use is discontinued. In the presence of the drug, however, resistance-bearers rapidly emerge as dominant and the population acquires resistance.

Acquired resistance may be **multistep.** That is, over a period of months or years, successive mutant populations emerge tolerating increasing drug levels. The early history of gonorrhea therapy was marked by a gradual increase in the dose of penicillin needed for cure.

In some cases, **one-step** mutations confer absolute resistance. Recently the gonococcus acquired the gene for penicillinase production. These strains are now re-

TABLE 26.3
Clinically significant examples of acquired resistance

ORGANISM	SIGNIFICANT PERCENTAGE OF ISOLATES NOW RESISTANT TO
Hemophilus influenzae	Ampicillin
Neisseria meningitidis	Sulfonamides
Neisseria gonorrhoeae	Penicillin, spectinomycin
Streptococcus bovis	Penicillin
Streptococcus pneumoniae	Penicillin
Bacteroides fragilis	Tetracyclines
Pseudomonas aeruginosa	Gentamicin, carbenicillin
Plasmodium falciparum	Chloroquine
Serratia marcescens	All common antimicrobials

In these cases, the development of resistance to the drug of first choice often forces the use of less desirable second-choice drugs.

sistant to any dosage level. One-step resistance can arise when a drug has a unique site of action. A point mutation may change the conformation of the target site so that it no longer binds the drug. For example, ribosomal mutations confer resistance to the aminocyclotols. Alternatively, a modified enzyme may no longer be "fooled" by a competitive inhibitor.

Plasmids; Chapter 9, pp. 274–275

Resistance may be conferred by chromosomal genes, which are relatively stable, or by plasmid-borne genes. The plasmid **R-factors** are readily recombined among strains. Under control of transposons, individual resistance genes become bonded together to build multiple-drug resistance. Plasmids bearing four or more resistance genes are not uncommon. Some strains of *Serratia* have acquired combinations of chromosomal and plasmid information rendering them resistant to all commonly used antimicrobials.

The spread of R-factors among strains, species, and genera—the mechanism that explains the rapid spread of resistance that inevitably follows introduction of a new drug—can be effectively demonstrated in the bowel of normal hospital workers.

The most intensely studied resistance mechanism is the production of β-**lactamase** enzymes. These cleave and inactivate the central β-**lactam ring** of penicillins and cephalosporins. β-lactamase in some form is possessed by all peptidoglycan-bearing bacteria because it is essential for normal cell wall synthesis. The genetic information may be a stable chromosomal gene or a transmissible plasmid factor. β-lactamase confers resistance to β-lactam antibodies when it is excreted into the surrounding medium or accessible in the periplasmic space. Otherwise, the enzyme may not confer resistance. Clavulinic acid, oxacillin, and cloxacillin are competitive inhibitors of some β-lactamases. They are being studied as possible therapeutic aids.

Testing Susceptibility in the Laboratory

The techniques for routine testing of susceptibility in the clinical laboratory (the "sensitivity" part of the culture and sensitivity order) are in a state of flux. The ideal method should combine accuracy and reproducibility with speed and low cost. Several approaches are in general use.

Kirby–Bauer agar disc diffusion method　When an antimicrobial impregnated paper disc is placed upon the moist surface of an inoculated agar plate, the drug diffuses out. Drug concentration is highest nearest the disc and decreases with distance. The underlying microorganism is thus exposed to a drug concentration gradient. If the strain is susceptible, a **zone of inhibition** appears following incubation (Fig. 26.3). A small zone usually means the organism is inhibited only by the highest drug levels, while a large zone indicates susceptibility to lower levels. The actual zone diameters are affected by the water solubility of the drug. These principles are the basis of the standardized **Kirby–Bauer** method.

Mueller–Hinton agar of specified cation content is dispensed to yield plates of standard thickness and moisture content. The inoculum, a suspension of several

26.3
Disc Agar Diffusion Susceptibility Test Filter paper discs containing standardized concentrations of antimicrobial drugs are placed on the surface of a confluently inoculated agar plate. After incubation, susceptibility is shown by the presence of zones of inhibited growth around some discs.

colonies in broth, is adjusted to a specific visual turbidity and applied to the agar with a swab. Discs, which must be fresh, are aseptically applied. The plate is incubated overnight. The final zone size is then measured, and zones are compared with standard culture controls run daily. Using evaluation charts, the zone size of the isolate for a given drug is compared to the concentration of drug attainable in the patient serum. An isolate is considered **susceptible** only when it is inhibited by clinically attainable concentrations. If borderline, it is **intermediate.** Otherwise, it is reported as **resistant.** The Kirby–Bauer technique is semiquantitative. It is not useful for slow-growing or fastidious pathogens or for anaerobes. Minimal Inhibitory Concentration (MIC) methods (see below) are suggested for these.

The agar-overlay method Some labs prefer to add the standardized inoculum to melted, cooled liquid agar, then pour individual plates. Discs are added after the medium solidifies. Although more cumbersome, this method is said to give clearer zone margins.

Minimal inhibitory concentration (MIC) determinations There are a number of situations in which the clinician needs to know exactly how much drug will be needed to achieve control. These include neonatal sepsis and meningitis, where it is crucial to use no more drug than needed, and persistent infections such as osteomyelitis, endocarditis, and chronic bacteremia in which the organism is stubborn, and therapy will be completely useless unless a high enough dosage is given. The MIC can be determined by serial **tube dilution** or **microdilution** in microtiter wells.

Infective endocarditis; Chapter 24, pp. 693–696

TABLE 26.4
Interpretation of zone sizes

ANTIBIOTIC	DISC CONTENT	DIAMETER (MILLIMETERS) OF ZONE OF INHIBITION		
		RESISTANT	INTERMEDIATE	SUSCEPTIBLE
Ampicillin[1] when testing Gram-negative microorganisms and enterococci	10 mcg	11 or less	12–13	14 or more
Ampicillin[1] when testing staphylococci and penicillin G-susceptible microorganisms	10 mcg	20 or less	21–28	29 or more
Ampicillin[1] when testing *Hemophilus* species	10 mcg	19 or less	—	20 or more
Bacitracin	10 units	8 or less	9–12	13 or more
Carbenicillin when testing *Proteus* species and *Escherichia coli*	50 mcg	17 or less	18–22	23 or more
Carbenicillin when testing *Pseudomonas aeruginosa*	50 mcg	12 or less	13–14	15 or more
Cephalothin when reporting susceptibility to cephalothin, cephaloridine, and cephalexin	30 mcg	14 or less	15–17	18 or more[2]
Cephalothin when reporting susceptibility to cephaloglycin	30 mcg	14 or less	—	15 or more
Chloramphenicol	30 mcg	12 or less	13–17	18 or more
Clindamycin[3] when reporting susceptibility to clindamycin	2 mcg	14 or less	15–16	17 or more
Clindamycin when reporting susceptibility to lincomycin	2 mcg	16 or less	17–20	21 or more
Colistin	10 mcg	8 or less	9–10	11 or more
Erythromycin	15 mcg	13 or less	14–17	18 or more
Gentamicin	10 mcg	12 or less	—	13 or more
Kanamycin	30 mcg	13 or less	14–17	18 or more
Methicillin[5]	5 mcg	9 or less	10–13	14 or more
Neomycin	30 mcg	12 or less	13–16	17 or more
Novobiocin	30 mcg	17 or less	18–21	22 or more[6]
Oleandomycin[7]	15 mcg	11 or less	12–16	17 or more
Penicillin G when testing staphylococci[8]	10 units	20 or less	—	21 or more
Penicillin G when testing other microorganisms[8]	10 units	11 or less	12–21[9]	22 or more
Polymyxin B	300 units	8 or less	9–11	12 or more[4]
Rifampin when testing *Neisseria meningitidis* susceptibility only	5 mcg	24 or less	—	25 or more

ANTIBIOTIC	DISC CONTENT	DIAMETER (MILLIMETERS) OF ZONE OF INHIBITION		
		RESISTANT	INTERMEDIATE	SUSCEPTIBLE
Streptomycin	10 mcg	11 or less	12–14	15 or more
Tetracycline[10]	30 mcg	14 or less	15–18	19 or more
Vancomycin	30 mcg	9 or less	10–11	12 or more

[1]The ampicillin disc is used for testing susceptibility to both ampicillin and hetacillin.

[2]Staphylococci exhibiting resistance to the penicillinase-resistant penicillin class discs should be reported as resistant to cephalosporin class antibiotics. The 30 mcg cephalothin disc cannot be relied upon to detect resistance of methicillin-resistant staphylococci to cephalosporin class antibiotics.

[3]The clindamycin disc is used for testing susceptibility to both clindamycin and lincomycin.

[4]Colistin and polymyxin B diffuse poorly in agar and the accuracy of the diffusion method is thus less than with other antibiotics. Resistance is always significant, but when treatment of systemic infections due to susceptible strains is considered, it is wise to confirm the results of a diffusion test with a dilution method.

[5]The methicillin disc is used for testing susceptibility to all penicillinase-resistant penicillins; that is, methicillin, cloxacillin, dicloxacillin, oxacillin, and nafcillin.

[6]Not applicable to medium that contains blood.

[7]The oleandomycin disc is used for testing susceptibility to oleandomycin and troleandomycin.

[8]The penicillin G disc is used for testing susceptibility to all penicillinase-susceptible penicillins except ampicillin and carbenicillin; that is, penicillin G, phenoxymethyl penicillin, and phenethicillin.

[9]This category includes some organisms such as enterococci and Gram-negative bacilli that may cause systemic infections treatable with high doses of penicillin G. Such organisms should only be reported susceptible to penicillin G and not to phenoxymethyl penicillin or phenethicillin.

[10]The tetracycline disc is used for testing susceptibility to all tetracyclines; that is, chlortetracycline, demeclocycline, doxycycline, methacycline, oxytetracycline, rolitetracycline, minocycline, and tetracycline.

Source: *The Federal Register* **37**: 191, 20527–20529, 1972.

A series of dilutions of the drug is prepared, then a standardized inoculum is added. The MIC is the lowest concentration of drug in which no growth appears. If desired, these dilutions can be subcultured to drug-free media to determine the minimum lethal concentration (MLC). Several companies have developed automated dilution/inoculation/readout systems that provide results within a few hours rather than overnight. These will probably see increasing use in facilities whose work load is large enough to justify their cost.

Serum antimicrobial concentrations   Human beings are provokingly individual; if they receive equivalent drug doses, they may have quite different serum concentrations. This individuality has proved especially troublesome with the aminoglycosides, e.g., gentamicin, tobramicin, and amikacin. These drugs are toxic above certain serum levels while being ineffective below their MIC. Thus their therapeutic

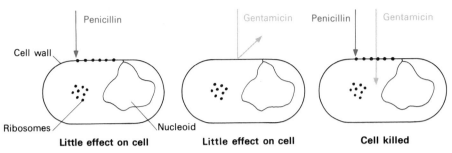

26.4
Synergism of Penicillin and Gentamicin on Gram-negative Organisms   By itself, penicillin does not kill this Gram-negative species. The peptidoglycan layer of its cell wall is somewhat weakened but there is little effect on growth. Gentamicin by itself cannot penetrate the cell and has no detectable effect on growth. When the two drugs are given together, the penicillin increases the bacterium's permeability. This allows entrance of gentamicin into the bacterial cytoplasm, where it blocks protein synthesis at the ribosomal level. The bacterium is killed.

range is quite narrow. Aminoglycosides are usually reserved for seriously ill, hospitalized patients. In the first days of drug therapy, the patient's serum drug level is monitored frequently. The dosage is thus correctly adjusted to that which gives an acceptable serum level. These techniques are being extended to other drugs that pose similar problems.

Combination Therapy

If one antimicrobial is good, will two be better? As you would expect, sometimes yes, sometimes no. Early in the antibiotic era, the pharmaceutical companies marketed a number of poorly designed, multidrug combinations. These did very little, and gave combination therapy a negative reputation. In the past few years, however, some very useful combinations have been developed.

Synergistic effects   There are several useful strategies of combination. The effects of both drugs combined should be greater than the sum of both of their effects separately. Their effect should be multiplied, or **synergistic.** Most accepted is the combination of a penicillin-class drug, which increases the cell wall permeability, with a protein-synthesis inhibitor (Fig. 26.4). These combinations are frequently successful against resistant organisms that neither drug could control alone.

In cases in which rapid one-step resistance often appears during therapy, combined antimicrobials are indicated for a different reason. The statistical probability of a single microbial cell's simultaneously acquiring two different favorable mutations and so surviving is virtually nil. Cotrimoxazole combines sulfisoxazole and trimethoprim, two different folate-synthesis inhibitors. Tuberculosis is treated with combinations of two or more drugs, such as isoniazid plus rifampin.

Tuberculosis therapy,
Chapter 19, pp.
560–561

Antagonism   In rare cases one drug will help an organism survive exposure to another drug, negating its effect. Drug **antagonism** is often noted when protein-synthesis inhibitors, such as erythromycin and gentamicin, are combined. In view of the possibility of antagonism, random antimicrobial combinations should not be used.

Therapy and Prophylaxis

Antimicrobial drugs are primarily used for therapy, that is, to treat obvious infectious disease. It is the consensus that their prophylactic use—to prevent infectious disease—should be very circumscribed. Prophylactic use is generally accepted in rheumatic fever and congenital heart disease, for protecting tuberculosis contacts, and in certain preoperative situations.

TABLE 26.5
Accepted synergistic combinations

COMBINATION	USED AGAINST
Penicillin G/streptomycin	Enterococci
Ampicillin/streptomycin	*Streptococcus* Groups B,D
Ampicillin/gentamicin	
Nafcillin/gentamicin	*Staphylococcus aureus*
Carbenicillin/gentamicin	*Pseudomonas aeruginosa*
Carbenicillin/amikacin	Indol-positive *Proteus*
Cotrimoxazole (sulfisoxazole/trimethoprim)	Enterobacteria, especially in prostate *Pneumocystis carinii*
Cotrimoxazole/polymyxin	*Pseudomonas maltophilia* *Pseudomonas cepacia* *Serratia marcescens*
Cotrimoxazole/gentamicin	*Escherichia coli* in neonate
Sulfonamide/pyrimethamine	*Plasmodium falciparum* *Toxoplasma gondii*
Rifampin/erythromycin	*Staphylococcus aureus* Enterobacteria
Rifampin/tetracycline	*Pseudomonas aeruginosa*
Isoniazid/rifampin	*Mycobacterium tuberculosis*
Isoniazid/streptomycin	
Isoniazid/ethambutol or many other combinations	

DYNAMICS OF DRUG ADMINISTRATION

Most of us have an overly simplistic view of what happens when we take a pill or receive a shot. We view our body as an open vessel and visualize the drug surging immediately to all parts for "instant relief." We are unaware that there are areas of the body in which the drug may be concentrated, or in which it may penetrate poorly or not at all. We also forget that the drug is a foreign substance. Our ungrateful body will immediately start to do what it does with all foreign chemicals—detoxify and/or excrete the drug. **Pharmacodynamics** is the study of how a drug distributes itself in the body, how long it remains active, and its effects on the body. The answers to these questions are ultimately as important as how the drug affects the pathogen.

Route of Administration

The choice of a suitable route of administration is influenced by many factors. Some drugs are supplied in various forms for a full range of uses, while others are restricted to a single route.

Topical administration   Whenever possible, superficial skin or mucous membrane infections should be treated topically, with antimicrobial ointments, salves, or soaking solutions. This procedure eliminates unnecessary systemic exposure to the drug. Many useful topical drugs (neomycin, bacitracin, nystatin) are too toxic or too poorly absorbed for other routes of administration.

Oral administration   This approach (*per os* or p.o.) is favored by both patient and practitioner because of its convenience and comfort.

Oral administration has limitations. Some drugs, such as penicillin G, are inactivated by contact with stomach acid. Others, such as the polymyxins, are not absorbed from the GI tract and appear unchanged in the feces. This means that they could not be used to treat a bladder infection since the drug would not get into the urine. However, since they do not leave the GI tract, such unabsorbed drugs may be very effective against intestinal infections such as enteropathogenic *Escherichia coli*. Gut irritation or ulceration may permit absorption of unusually high levels of a drug. The very ill, unconscious, or vomiting patient cannot take oral medication.

Intramuscular (IM) injection   This type of **parenteral** (into the tissues) administration creates a **depot** of drug, which is then gradually diffused into the circulation over a period of hours or days. An injection of a long-acting penicillin will be continuously effective for a month.

Some antimicrobial solutions are extremely painful on injection, and may cause tissue damage (myositis) on repeated use.

Intravenous (IV) infusion   For the hospitalized, seriously ill patient, intravenous infusion is usually best. The actual serum level of the drug remains constant, and the

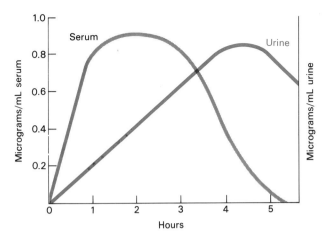

26.5
Serum and Urine Drug Concentrations   This graph shows a hypothetical time sequence for the transfer of a rapidly excreted antimicrobial from the circulation to the urine. After administration, the serum concentration rises rapidly and stays high for at least two hours. However, the kidneys immediately start to excrete the drug. This brings the serum level (black) down after the third hour, while the urine level (color) continues to climb. In order to maintain an effective level of the drug in serum, repeat doses must be given at least every four hours.

level can be readily adjusted to changing needs. A few drugs, such as erythromycin and certain cephalosporins, may cause thrombophlebitis at the needle site.

Intrathecal infusion   Because of the difficulties in achieving therapeutic drug levels in the cerebrospinal fluid, occasionally a drug is directly injected into the meningeal space.

Serum Levels of Antimicrobials in Therapy

Effective therapy requires that the MIC of drug be achieved right at the site of microbial attack. In practice this goal is usually pursued by maintaining the MIC or a selected higher concentration in the patient's serum. When a new drug is readied for licensing and marketing, dosage recommendations are prepared. These are based on studies of the drug's behavior in human volunteers. Manufacturers' recommendations take into account not only the initial dose needed (**loading dose**) but the frequency with which it will have to be replenished.

Following administration, there is an initial absorption period before the drug reaches the desired level (Fig. 26.5). After it reaches its peak concentration, the drug level begins to fall as the body eliminates it. The next dose is timed so that the falling drug level does not dip below the MIC, which would permit new microbial growth. Extreme oscillations are also undesirable because they place unnecessary stress on the organ systems. Strict adherence to medication schedules is not just a ritual—it fulfills biological needs.

Tissue and Fluid Distribution

Passage of drugs from the blood into tissues is promoted by small molecular size, neutral pH and charge, lipid solubility, and dissociation from serum proteins. Most antibiotics that are used parenterally enter tissues well and are also found in pleural fluid, the peritoneum, and the synovial fluid of the joints. Only certain drugs are effectively secreted in mucus and become available on oral and vaginal surfaces. Minocycline and the sulfonamides reach effective concentrations in all of the body fluids.

Blood–brain barrier; Chapter 22, p. 659

Not all body areas are equally accessible to all bloodborne substances. The central nervous system is one of the most restricted areas. Because of the permeability barrier, the CSF concentration of most medications will be only a small fraction of the serum concentration. Inflammation greatly increases CNS permeability.

The interior of the eyeball, the prostate gland, and certain other glands are also anatomically and/or physiologically isolated.

Devitalized tissue, vegetations, abscesses, and granulomas are poorly penetrated by most drugs. Some drugs are inactivated by pus. All impediments to drugs must be removed to obtain effective antimicrobial therapy.

Drug Clearance

Half-life   As soon as the chemotherapeutic agent is administered, the body starts clearing it. The **half-life** of a drug is the time needed for the serum level to be reduced by 50 percent. The half-life may differ from the norm in neonates, the very obese, or persons with renal or hepatic disease.

Cystitis; Chapter 21, pp. 621–622

Route of excretion   Most antimicrobials are excreted unchanged through the kidneys. Those with short half-lives, such as sulfisoxazole (Gantrisin), are so rapidly excreted that they go directly from the site of absorption (usually the intestine) via the blood to the urinary tract. Such drugs concentrate in the urine and are well suited for cystitis therapy. Therapeutic concentrations cannot be achieved elsewhere in the body, however.

Drugs excreted via the kidneys may be abnormally retained by patients with poor or absent kidney function. Drug retention will build unacceptable serum accumulation on the usual dosage schedule. Under these conditions the dosage must be much reduced or other drugs chosen. Excreting some drugs imposes heavy stress on both normal and damaged kidneys. This stress, called **nephrotoxicity**, may lead to temporary or permanent damage.

A few drugs are excreted into the bile by the liver, either unchanged, or in chemically modified form. Such drugs appear in the feces, as do all unabsorbed orally administered drugs. This group of drugs may be used in renal failure. Examples are erythromycin, some of the semisynthetic penicillins, and chloramphenicol.

Nonantimicrobial compounds can change the pharmacodynamics of an antimicrobial. For example, probenicid is included in certain penicillin formulations. Probenicid alters the function of the renal tubule and prolongs penicillin's residence time in the body. Procaine-penicillin, given intramuscularly, is absorbed much more slowly than penicillin alone and thus gives a prolonged effect.

ADVERSE REACTIONS

There is no medication without side effects. Antimicrobial therapy is continuously evolving. Investigation has been motivated by the search for drugs that can be given as safely as possible, and the need to fully define the safest conditions for their use. We have also seen growing awareness of the need to use these drugs only when they are truly needed, thus avoiding preventable toxic reactions. However, toxic antimicrobials must occasionally be used to save a life. It is important to be aware of toxic manifestations that might be encountered.

Liver and Kidney Toxicity

Moderate drug damage to the liver shows up as increased serum levels of liver enzymes or jaundice. This damage is usually reversible after short exposures. The risk of extensive or permanent damage increases as the dosage and duration of therapy increases. Isoniazid is occasionally associated with liver dysfunction.

Kidney damage may be either glomerular or tubular. A slight transient effect on creatinine clearance is observed during many drug regimens; damage is universal with amphotericin B and parenteral bacitracin, and common with the older aminoglyosides, neomycin and kanamycin. Newer substances cause trouble principally when serum levels become excessive.

Blood Dyscrasias

Blood cell populations may be depressed during antimicrobial therapy. In most cases, this effect reflects drug-induced damage to the hematopoietic (blood-forming) tissues of the bone marrow. In aplastic anemia, the production of all circulating blood cells ceases. This condition, which occasionally follows chloramphenicol use, is irreversible and fatal. Polyenes destabilize red blood cell membranes, leading to hemolysis. **Leukopenia** (inadequacy of circulating white blood cells) is common with systemic use of the antiviral agent, idoxuridine. Persons with the recessive genetic trait, glucose-6-phosphate dehydrogenase (G-6-PD) deficiency, are unusually susceptible to sulfonamides and may develop a hemolytic anemia when they are given these.

Neurological Disturbance

Dizziness, confusion, and muscular weakness may occur on administration of some drugs. Penicillin in high doses may cause seizures. Streptomycin has a direct effect on the auditory nerve and may lead to temporary or permanent loss of hearing. Quinine and its derivatives cause ringing ears and optic neuritis; these conditions stop when therapy is discontinued.

TABLE 26.6
Major side effects of some commonly used antimicrobials

ANTIMICROBIAL	COMMON	OCCASIONAL	UNCOMMON
Penicillin	Hypersensitivity		Hepatic injury Anemia Encephalopathy
Ampicillin	Hypersensitivity GI upset		Hepatic injury Anemia
Carbenicillin	Hypersensitivity		Hepatic injury Anemia Encephalopathy
Cephalothin		Hypersensitivity Thrombophlebitis	Renal injury Hepatic injury Leukopenia
Erythromycin		GI upset Hepatic injury	Hypersensitivity
Clindamycin		GI upset	Hypersenstivity Leukopenia
Gentamicin		Renal injury Vestibular dysfunction	Neuromuscular dysfunction Peripheral neuropathy Auditory dysfunction
Tetracycline	GI upset	Renal injury Hepatic injury	Hypersensitivity Intracranial hypertension Encephalopathy
Sulfonamides	Hypersensitivity	Renal injury Hepatic injury Anemia Thrombocytopenia Leukopenia	Encephalopathy
Isoniazid		Hypersensitivity GI upset Hepatic injury Peripheral neuropathy	Anemia Optic dysfunction Encephalopathy
Rifampin		Hepatic injury Auditory dysfunction	Hypersensitivity Thrombocytopenia
Chloroquine	GI upset	Dizziness and headache Anemia Optic dysfunction Auditory dysfunction	
Metronidazole		GI upset	Encephalopathy

Source: Compiled from Paul D. Hoeprich (ed.), 1977, *Infectious Disease* (2nd ed.), New York: Harper & Row.

Box 26.1

Antibiotic Misuse

Antimicrobial drugs are extensively and persistently overused. In 1960, three million pounds of medicinal antibiotic drugs were produced in this country. By 1972 production had jumped to 9.6 million pounds. This represents a growth of 360 percent over a period during which the population increased only 11 percent. Whereas the average person has an infectious disease requiring antimicrobial therapy on the average of only once every five to ten years, in 1972 enough doses of antibiotics were produced to treat two illnesses of average duration per year for every man, woman, and child in the United States. Some surveyed instances of questionable use:

In 1972, 49 percent of persons visiting a physician for the common cold, a viral infection which does not respond to antibiotics, received antibiotic prescriptions.

In a study of hospital discharges, there was no record of a bacteriological culture on 50.5 percent of patients who had received antimicrobial therapy.

In another hospital study, it was found that 62 percent of the hospitalized patients for whom antibiotics had been ordered had shown no clinical evidence of infection at the time.

Common errors include prescribing without a culture, without examining the patient, or over the telephone. Antibiotics are wrongly prescribed for viral infections or for prophylaxis (rarely justified). Broad-spectrum, toxic drugs are ordered in situations in which narrow-spectrum, less toxic drugs would do as well.

Physicians may overprescribe because of the pressure of patient expectations. Many patients are just the opposite of what they're called. They demand immediate relief from a disease that would be self-limiting within a few days without medication. Such individuals feel cheated if they leave a doctor's office without a prescription, and they will shop around until they find another physician who will fulfill their expectations. Practitioners also overprescribe to play it safe, or as a short-cut in the time-consuming business of dealing with "superficial" conditions. Sometimes drugs are started, and the order just never gets discontinued if the evidence shows the drug was not needed.

In the short term, antibiotic overuse is bad because 5 percent of antibiotic recipients experience side effects, and more than two percent develop superinfection. These outcomes may lengthen their illnesses. In the long term, overuse is slowly but steadily promoting drug-resistance, particularly in hospital strains of bacteria. Thus an antibiotic unnecessarily administered to one patient may increase the infection risk of the patient in the next bed.

How can the problem be solved? One important step would be to alert the public to the limitations of antibiotics. This could help to relieve the pressure on physicians to prescribe against their own good judgment. Detailed guidelines have been formulated for acceptable antibiotic use as part of a professional peer review process. Within the hospital, antibiotic use is monitored or audited through the pharmacy. Pharmacists can provide expert consultation services to clinicians.

Gastrointestinal Intolerance

Nausea, vomiting, or diarrhea follow oral intake of many antimicrobials. These conditions can often be avoided by taking the drug with food. However, food may seriously interfere with the absorption of drugs. Tetracycline should not be ingested with milk products because calcium ions prevent its absorption.

Hypersensitivity

Hypersensitivity mechanisms; Chapter 14, pp. 411–420

Some persons develop hypersensitivity to antimicrobials. Penicillins, cephalosporins, and sulfonamides are most notorious. On parenteral administration of an antimicrobial to a previously sensitized person, anaphylaxis is possible. Over 300

TABLE 26.7
Special-purpose or little-used antimicrobials

ANTIMICROBIAL	SOURCE	NOTES
Actinomycin D	Natural	Nonselective; used as antineoplastic drug
Bacitracin	Natural	Highly toxic; used in topical preparations against Gram-positive organisms
Furazolidone	Synthetic	Limited use against *Shigella, Giardia*
Fusidic acid	Natural	Used against *Staphylococcus aureus* in combination with other drugs
Lincomycin	Natural	Replaced by clindamycin for most uses
Methenamine	Synthetic	Limited use, urinary tract infection
Nalidixic acid	Synthetic	Limited use, urinary tract infection
Nifuroxime	Synthetic	Useful in vaginal creams against *Candida*
Nitrofurantoin	Synthetic	Useful in urinary tract infection if organism fails to respond to other drugs
Nitrofurazone	Synthetic	Topical application, bladder rinse, vaginal, urethral infections
Oleandomycin	Natural	With troleandomycin, seldom-used macrolides, effective against Gram-positive bacteria
Oxolinic acid	Synthetic	Occasional use in urinary tract infections
Paromomycin	Natural	Older aminoglycoside, useful against *Balantidium coli*
Spectinomycin	Natural	Used to treat penicillinase-positive *Neisseria gonorrhoeae*
Vancomycin	Natural	Used to treat penicillin-resistant bacterial endocarditis

persons a year in the United States suffer penicillin-induced shock. Anaphylaxis also has been reported with tetracyclines, streptomycin, and many other drugs. Oral or topical drug administration more often produces gastrointestinal upset or urticaria, respectively. A person sensitized to one penicillin-class drug is likely to react with any other, but there is no cross-reaction with cephalosporin-class drugs. However, a penicillin-sensitive person, who has a proven "atopic tendency," is all too likely to develop other allergies as other drugs are given.

Penicillin hypersensitivity may be assessed prior to prescription. Standardized antigen may be introduced via a scratch test or intradermal injection, in the absence of a medical history or to confirm a suspicion. These tests should never be carried out unless the tester is fully prepared to treat a possible anaphylactic response.

Superinfection

One of the great hazards of long-term antimicrobial therapy is emergence of a resistant replacement flora, usually derived from the normal microbial residents. In hospitals, a patient's GI tract rapidly becomes colonized with hospital "super-bugs" —enteric organisms bearing multiple-drug-resistant plasmids. These colonists will cause trouble only if the patient receives therapy selective for the plasmids. Otherwise, they will be eliminated by microbial competition shortly after the patient is discharged. Resistant enterobacteria can cause superinfection if they are selectively enhanced. The yeast *Candida albicans* also causes troublesome infestation of the mucous membranes.

Use of antimicrobials ineffective against anaerobes may lead to **antibiotic-associated colitis,** in which *Clostridium difficile* is the causative agent. This condition is most often encountered when the patient has had prolonged ampicillin, clindamicin, or cephalosporin therapy.

ANTIBACTERIAL DRUGS

The Penicillins

The penicillins are a diverse group of therapeutic agents, all of which share a central chemical feature, the β-lactam ring. This ring is essential to their antimicrobial activity. The main practical limitation to the use of penicillins is the instability of the crucial structure to the degradative effects of stomach acid and microbial enzymes. Organic chemists have been hard at work modifying the natural penicillins G and V. Over 10,000 semisynthetic and fully synthetic variants have been developed. Starting with the fungally synthesized 6-aminopenicillanic acid molecule, side-chains are added to confer desired attributes (Fig. 26.6). About a dozen penicillins are in general use, and there is a small number of promising new variants awaiting licensing.

General features   All penicillins interfere with cell-wall synthesis in actively growing cells. They do not affect metabolically inactive bacteria. In adequate dosage,

Cell wall structure;
Chapter 3, pp. 74–79

they are bactericidal. Penicillins are effective in organisms possessing an integral peptidoglycan layer, but only if they successfully penetrate the outer cell-wall layer to the site of enzymatic PG assembly, and if they are not inactivated by penicillinase. Penicillins penetrate tissues well and are readily absorbed from the site of administration. They do not enter the CNS too well, but since high dosage may be used safely, useful CNS levels can be achieved. Penicillins are secreted in all body fluids, but mainly in the urine.

True toxicity is extremely uncommon, and this is the main reason that the penicillins are the most widely used of all antimicrobials. Certain of them are safe to use in the patient with renal failure. Hypersensitivity is the major drawback; it develops quite commonly, and ranges from mild to fatal. A simple skin test may be used to determine a patient's hypersensitivity status. If he or she proves reactive to penicillin antigen, none of the penicillins may be safely used.

Classes   The natural penicillins G and V are synthesized directly from basic nutrients by the fungus *Penicillium*. Penicillin G (benzylpenicillin) is a highly potent, narrow-spectrum drug. It must be given parenterally, since it does not survive passage through the stomach. Penicillin G is widely used in community-acquired infections in which the agent is relatively unlikely to carry penicillinase resistance factors. It may be used for gonorrhea on a one-shot basis in the long acting procaine–penicillin or benzathene–penicillin forms.

Addition of side-chains to the natural penicillin structure may render it resistant to acid hydrolysis (Fig. 26.6). The fungus may add a phenoxymethyl group yielding penicillin V. Phenethicillin is another acid-resistant relative. These may be given orally.

Penicillinase resistance may be conferred if an added group hinders access of bacterial enzymes to the sensitive β-lactam ring. Methicillin, although penicillinase resistant, is acid-labile. Large parenteral doses are used for hospitalized patients with penicillinase-positive staphylococcal infections.

Combined acid and penicillinase resistance has been achieved with the addition of isoxazolyl groupings. Oxacillin, cloxacillin, and nafcillin are useful narrow-spectrum drugs with great flexibility of administration. However, they are of little use against Gram-negative bacteria.

Extended-spectrum penicillins are now available with anti-Gram-negative potential. Ampicillin is the most widely used of these, and it is usually effective against *Hemophilus*, *Salmonella*, *Shigella*, *Escherichia coli*, nonindol-producing *Proteus*, and many Group D streptococci. Ampicillin is given orally, although absorption is not efficient. Hetacillin is an oral penicillin that is converted to ampicillin on absorption. Amoxicillin has a similar spectrum, but it is better for oral administration because more of the drug is absorbed and higher serum levels are achieved.

Carbenicillin is one of very few drugs effective against *Pseudomonas aeruginosa*. About 75 percent of strains are susceptible on initial use. However, resistant strains arise freely, so the drug is frequently used in combination with gentamicin or

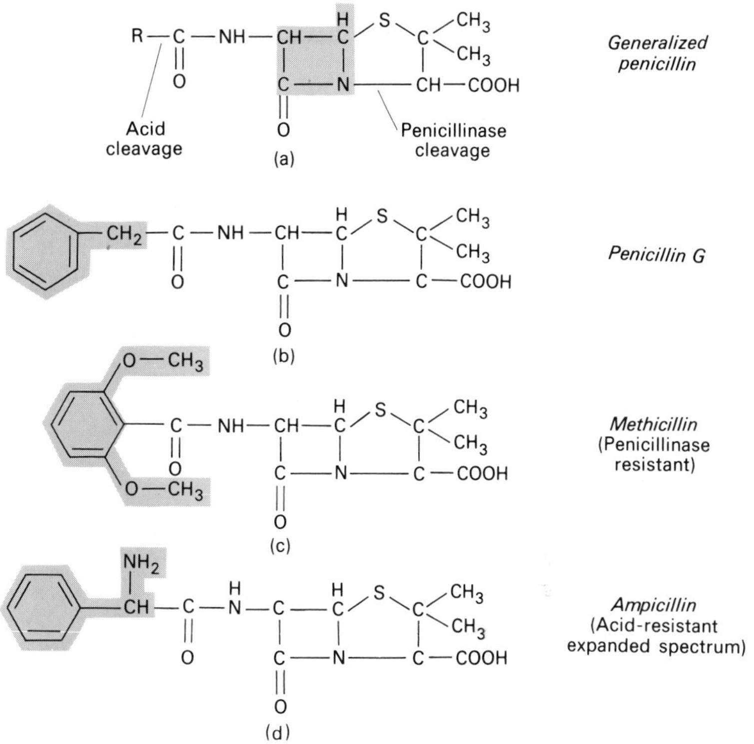

26.6
Penicillins (a) The generalized penicillin structure shows the 4-sided β-lactam ring (colored). This ring is essential for penicillin's activity. The ring is inactivated by penicillinase enzymes. An R group is attached to the left-hand carbonyl group. This bond is vulnerable to acid hydrolysis and prevents the pencillin from being effective if it is given by mouth. (b) In penicillin G the R group (colored) confers activity against Gram-positive bacteria. The molecule is sensitive both to acid and enzymatic attack. (c) The substitution of a bulky R group in methicillin renders it penicillin-resistant, perhaps because the enzyme's binding to the β-lactam ring is hindered. (d) Ampicillin is acid-resistant because of the substitution of an amino group on the vulnerable region where the R group is attached. It can be given by mouth. Ampicillin also has activity against Gram-negative bacteria.

Generalized penicillin

Penicillin G

Methicillin (Penicillinase resistant)

Ampicillin (Acid-resistant expanded spectrum)

other aminoglycosides. Ticarcillin is a very similar expanded-spectrum drug sometimes used as an alternative to carbenicillin.

The Cephalosporins

The cephalosporins are like the penicillins in many respects. They are of fungal origin, produced by the *Cephalosporium* genus, and active against the bacterial cell wall. Their active chemical structure is the 7-aminocephalosporanic acid nucleus, which contains a β-lactam ring. Like the penicillins, they differ in their stability to gastric juice and bacterial β-lactamases.

Chemical modification has led to a continuing succession of cephalosporins. Most of the first-generation cephalosporins are not used now. Where penicillin hypersensitivity is a problem, cephalosporins are most often used as first alternatives to the penicillins. They are extremely costly.

General features These drugs are bactericidal in adequate dosage. Their use (rarely) leads to toxicity, which may include GI upset, renal toxicity, or thrombosis on IV administration. These problems have been much minimized with the newer

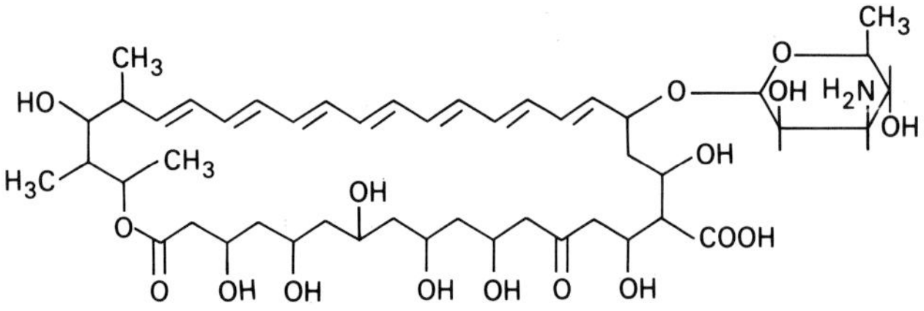

Amphotericin B

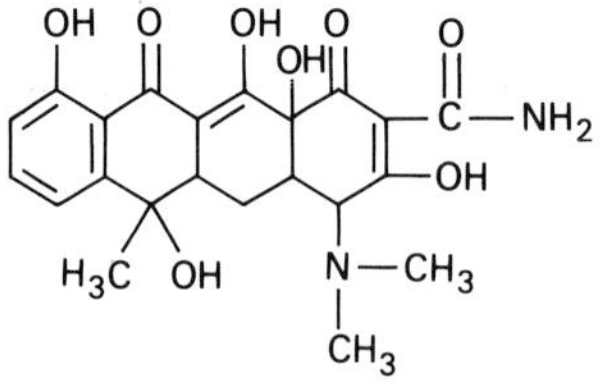

Tetracycline

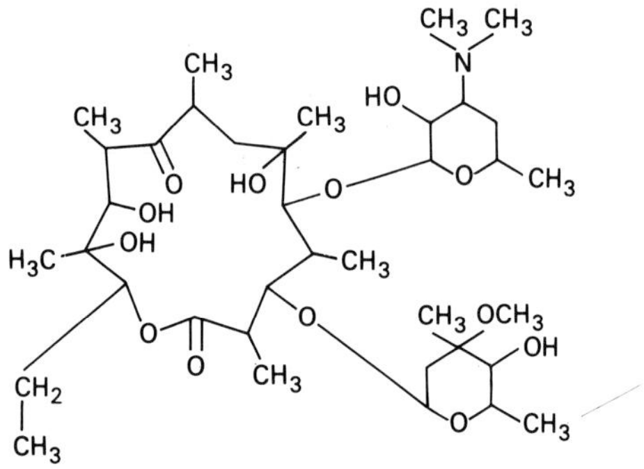

Erythromycin

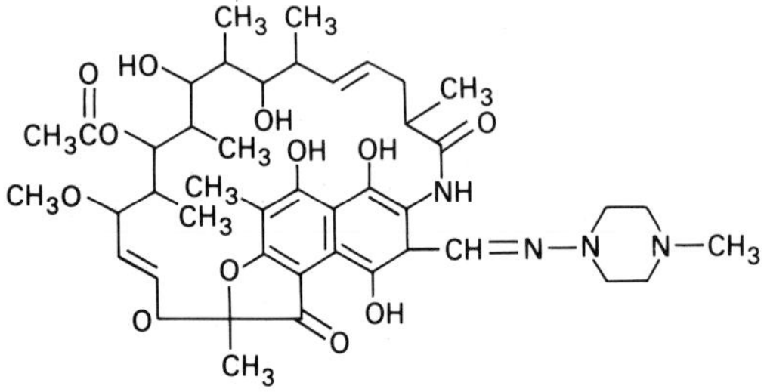

Rifampin

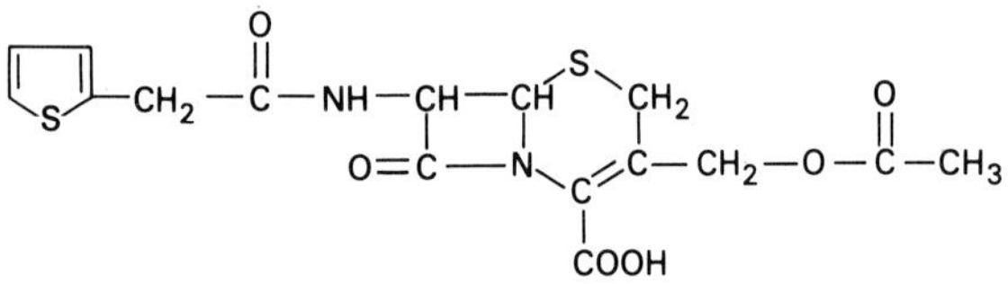

Cephalothin

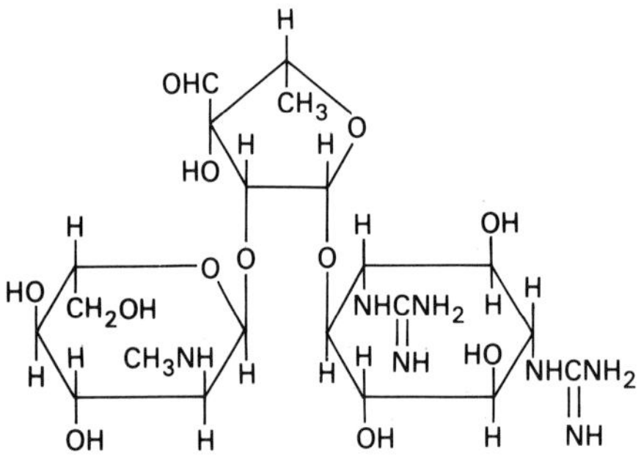

Streptomycin

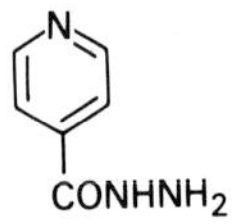

Isoniazid

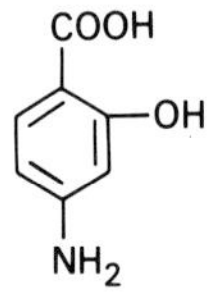

Para-aminosalicylic acid

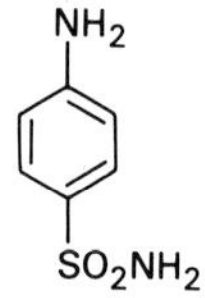

Sulfanilamide

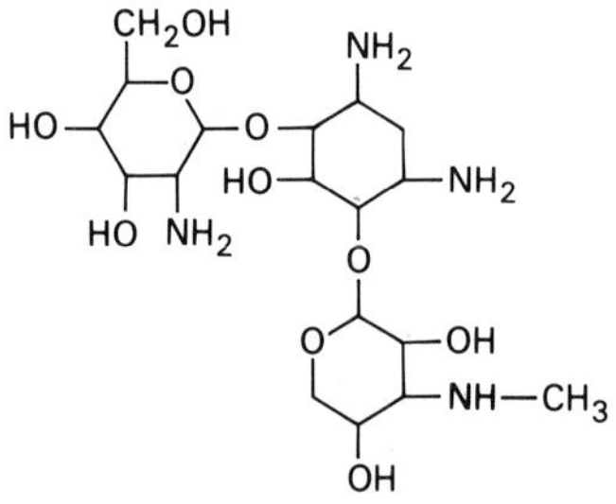

Gentamicin

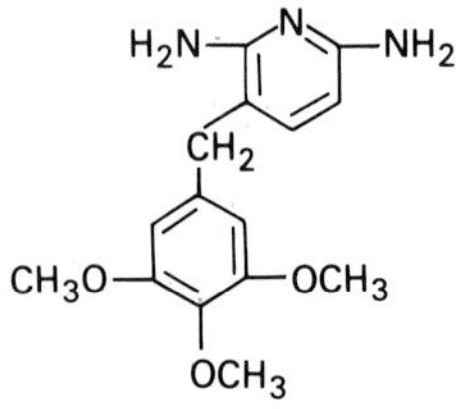

Trimethoprim

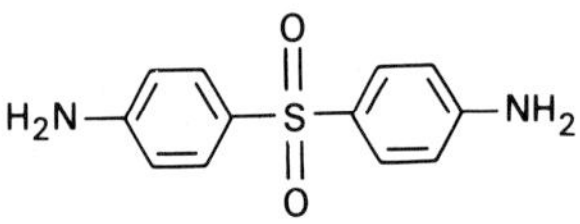

Dapsone

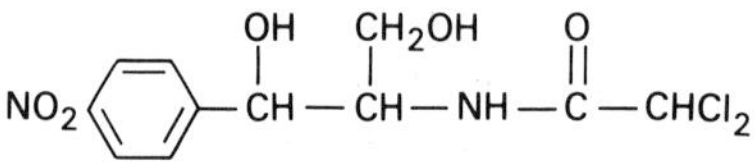

Chloramphenicol

26.7
Structural Formulas of Some Antimicrobials

additions to the family. Hypersensitivity occurs less commonly than with the penicillins. Urticaria is the usual manifestation and anaphylaxis is very uncommon.

Commonly used drugs There are currently nine members of the cephalosporin family. Most frequently cefazolin has been used for intramuscular administration and cephalexin or cephradine for oral use. The spectra of these three drugs are narrow. The drugs are primarily effective against Gram-positives, and they are β-lactamase sensitive. Cefamandole and cefoxitin have increased activity against Gram-negative anaerobes and are also β-lactamase resistant. Cefadroxil is a new oral drug, longer-acting and useful in urinary infections. SCE-129, an investigational drug, adds to these features antipseudomonad activity.

Erythromycin and Clindamycin

Erythromycin and clindamycin are used as penicillin alternates. Their spectra are similar but their modes of action are different. Pathogens rarely show combined resistance to all these drugs.

Erythromycin Erythromycin is a **macrolide**, containing a 14-membered ring structure. It is broad-spectrum, being effective not only against common Gram-positive pathogens but also against mycoplasma and chlamydia. Erythromycin is particularly good for treatment of diphtheria and *Listeria* infections, and is the drug of choice in treatment of *Legionella* pneumonias. It may be useful in combination with other chemicals against atypical mycobacteria.

Legionella; Chapter 19, pp. 557–558

Erythromycin is bactericidal and inhibitory to protein synthesis, acting at the ribosomal level. It can be administered by various routes, and achieves satisfactory tissue levels but poor CNS penetration. It may (rarely) cause liver toxicity, as it is excreted after breakdown by the liver.

Sixteen-member macrolides such as tylosin are widely used in animal feeds. There is a directed trend away from employing antimicrobials needed for human therapy in animal rations. This should reduce the spread of resistance factors from animal to animal and from animal to human being, and prevent inadvertent human hypersensitization.

Clindamycin Clindamycin, a nonmacrolide, is valuable in the hospitalized patient because it has unusual activity against anaerobes. It may also be used when there is poor kidney function. Oral absorption is rapid and complete, and clindamycin diffuses well into deep tissues and anaerobic foci. GI upset is common, and colitis resulting from alteration in the bowel flora may become severe.

TABLE 26.8
Bacterial pathogens: primary and secondary drugs of choice

PATHOGEN	PRIMARY	SECONDARY
Staphylococcus, penicillinase-negative	Penicillin G	Cephalosporins
Staphylococcus, penicillinase-positive	Penicillinase-resistant penicillins	β-lactamase-resistant cephalosporins
Streptococcus pneumoniae	Penicillin G	Penicillinase-resistant penicillin, cephalosporins
Streptococcus, Groups A–F	Penicillin G	Penicillinase-resistant penicillin, cephalosporins
Enterococci	Penicillin/gentamicin	Vancomycin
Clostridium	Penicillin	Erythromycin, tetracycline
Corynebacterium sp.	Erythromycin	Penicillin G
Actinomyces sp.	Penicillin	Erythromycin
Bacteroides fragilis	Clindamycin	Erythromycin, chloramphenicol
Other anaerobic bacteria	Penicillin	Erythromycin
Neisseria gonorrhoeae	Penicillin	Spectinomycin
Neisseria meningitidis	Penicillin	Chloramphenicol
Salmonella typhi	Chloramphenicol	Ampicillin
Other *Salmonella* sp.	Ampicillin	Chloramphenicol, only if very severe
Shigella sp.	Ampicillin	Cotrimoxazole, tetracycline
Escherichia coli	Ampicillin	Cephalosporins, tetracyclines
Proteus mirabilis (indol-negative)	Ampicillin	Cephalosporins
Other *Proteus* (indol-positive)	Carbenicillin	Gentamicin
Klebsiella pneumoniae	Cephalosporins	Tetracyclines
Vibrio cholerae	Tetracyclines	Antimicrobials not usually needed
Brucella sp.	Tetracyclines	Streptomycin
Yersinia pestis	Streptomycin	Tetracyclines
Hemophilus influenzae	Ampicillin	Chloramphenicol
Francisella tularensis	Streptomycin	Tetracycline, chloramphenicol
Pseudomonas aeruginosa	Carbenicillin/Gentamicin	Polymyxins
Serratia marcescens	Gentamicin	Kanamycin
Providencia sp.	Gentamicin	Kanamycin
Legionella pneumophila	Erythromycin	Tetracycline
Bordetella pertussis	Erythromycin	Ampicillin
Treponema pallidum	Penicillin (long-acting)	Erthromycin, tetracycline

The listing represents the usual susceptibilities of a majority of •clinical isolates. Laboratory evaluation may indicate a need for alternate selections.

The Aminoglycosides

The aminoglycoside family includes therapeutic agents primarily useful against Gram-negative bacteria. Each new addition to the family has found immediate acceptance, followed by a gradually increasing pattern of emerging bacterial resistance. The more recent drugs are less toxic than the earlier ones.

General features The aminoglycosides are protein synthesis inhibitors, interacting with the 30S ribosomal subunit. They also have a detergent action on bacterial membranes, and are considered bactericidal in normal use. One-step resistance develops rapidly, by means of altered ribosome structure, blocked permeation, or acquired inactivating enzymes. There is a cross-resistance; a bacterial strain resistant to one aminoglycoside may also be totally resistant to some or all of the others.

Classes The first-generation aminoglycosides—streptomycin, neomycin, kanamycin, and paromomycin—were all produced by the filamentous bacteria *Streptomyces.* They have only limited use at this time. Streptomycin, discovered in 1944, immediately became a mainstay of tuberculosis treatment. It is still included in some antitubercular regimens. It is recommended as primary therapy in plague and tularemia. Its main drawbacks are eighth cranial nerve damage resulting in deafness, and the rapid development of resistance.

Neomycin is poorly absorbed from the bowel. It was once used for presurgical bowel sterilization. However, because it results in kidney damage and bowel superinfection, its parenteral use has been discontinued. It is available for topical application only.

Kanamycin, highly active against susceptible Gram-negatives, can be given by any route, although oral absorption is poor. Its toxicity is considerable and may cause conditions including diarrhea, kidney damage, ototoxicity, and allergy symptoms. Kanamycin is now used principally when other drugs have been ruled out, as in retreatment of tuberculosis in which resistance to less toxic drugs has already developed.

The newer aminoglycosides are produced by a different bacterial genus, *Micromonospora,* and some have names ending in **-micin.** They are similar in mode of action and spectrum, but somewhat less toxic.

Gentamicin, tobramicin, and amikacin are quite similar in their characteristics. Absorption from the gut is poor and serum levels are highly variable. The therapeutic range of all aminoglycosides is narrow. To prevent toxic side effects (kidney, ear, GI tract), serum levels must be monitored. Preexisting kidney disease is a particular problem because these drugs are excreted largely unchanged by the kidney. A decrease in urinary output may produce a sudden spike in serum concentration exceeding toxic levels. This may in turn produce more kidney damage.

The use of aminoglycosides is indicated only in severe infections. Typical conditions would include bacteremia, necrotizing pneumonia, neonatal sepsis and meningitis, and burn sepsis.

TABLE 26.9
Promising antimicrobial drugs under development

DRUG	NOTES
Azlocillin Mezlocillin Piperacillin	Amino-substituted penicillins with antipseudomonad activity
Apalcillin	Extended-spectrum penicillin
Bacampicillin	Ampicillin derivative, better absorbed orally; hydrolyzed to yield ampicillin. Used in Europe
Alaphosphin	L-alanyl-L-1 aminoethyl-phosphoric acid; blocks cell-wall synthesis; half-life, 20 minutes
Rosamycin Josamycin	Macrolides, useful against gonococcus. Toxicity high
Netilmicin Sisomicin Episisomicin	Semisynthetic amino glycosides
Cinoxacin	Antibacterial; useful against urinary tract pathogens
Econozole Tioconozole	New imidazoles; spectrum includes pathogenic fungi and Gram-positive bacteria

Aminoglycosides are frequently combined with a penicillin or cephalosporin, and are effective against most Gram-negative pathogens. Amikacin has an expanded activity against *Pseudomonas aeruginosa*. Most anaerobes are not at all susceptible. In fact, treatment of anaerobic infections with aminoglycosides has been shown to **increase** the mortality rate.

The definite restrictions on the successful use of this family of drugs confines them almost completely to hospital-based medicine; they have limited use in community or out-patient applications.

The Tetracyclines

General features   Tetracyclines are broad-spectrum antimicrobials with a wide range of usefulness and flexibility. They inhibit protein synthesis, and are bacteriostatic. Therefore, tetracyclines are clinically successful only when there are active host defenses. Toxicities include nausea, photosensitivity, acute liver toxicity in severe infection in late pregnancy, renal toxicity with outdated preparations, and yellowing of teeth. Tetracycline accumulates in fast-growing tissues, such as areas of new bone growth. When tetracycline is taken by pregnant women or young children, first or second teeth may be visibly discolored. Tetracyclines are not recommended for children under eight years of age.

Common uses   The tetracyclines are very useful in the treatment of nonspecific urethritis or vaginitis because they are effective against mycoplasmas and chlamydia. They are the drug of choice for many rickettsial infections.

Nonspecific vaginitis; Chapter 12, p. 360

Oxytetracycline has very rapid renal clearance, and is useful in urinary tract infections because it concentrates in the urine. Doxycycline, which is not excreted by the kidney, can be used in the patient with renal failure. Minocycline enters body fluids and the CNS most effectively; it has a longer half-life and a prolonged effect. Other related drugs are chlortetracycline and methacycline.

Chloramphenicol

Chloramphenicol is a bacteriostatic, protein-synthesis inhibitor with a broad spectrum. It can be administered by any route, is readily absorbed, and distributes to all body compartments. It becomes concentrated in the CNS. Chloramphenicol is inactivated by the liver and excreted by the kidney. The useful qualities of the drug are offset by its unpredictable tendency to suppress the blood-forming tissues. Chloramphenicol may produce reversible dose-related anemia or irreversible aplastic anemia. In the neonate, immature liver and kidneys cannot clear the drug at adult rates. The **gray-baby syndrome,** a fatal anemia, may occur if the dose is excessive. GI upset, neurological disturbances, and hypersensitivity are also seen.

Chloramphenicol is used in life-threatening infections in which the risk of use is much smaller than the risk of death from withholding the drug. It is the drug of choice in typhoid fever, Rocky Mountain spotted fever, and typhus, some forms of bacterial meningitis, and melioidosis, a pseudomonad infection which occurs in the Far East. It may (rarely) be resorted to in other infections for which all other possible drugs have been ruled out. Chloramphenicol is widely used in veterinary medicine. In poorer countries its low cost makes it one of few affordable drugs. It can be used topically without risk in eye infections.

Typhoid fever; Chapter 20, pp. 586–590

The Polymyxins

Polymyxins are cyclical peptides, strongly charged, and destructive to bacterial membranes. Polymyxin B is used as alternate therapy for *Pseudomonas* infections unresponsive to other drugs. Topical preparations, sometimes in combination with neomycin and bacitracin, are used for external ear and eye infections. Colistin (polymyxin E methanesulfonate) is an injectable alternate. The toxicities include renal damage, peripheral nerve toxicity, and respiratory arrest. Bacitracin, a related peptide, is also reserved for topical use.

Inhibitors of Folate Synthesis

These are in two groups, the sulfonamides and the diaminopyrimidines. The two inhibit different steps in folate synthesis. Their effect is broad-spectrum but only bacteriostatic. Their effectiveness is reduced by pus or serum protein. Resistance

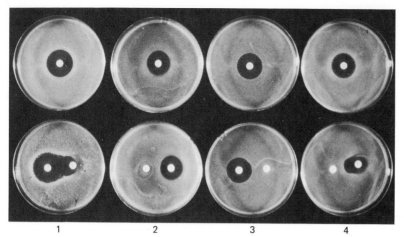

26.8
Sulfa-Trimethoprim Synergy (Top row) Four bacterial isolates show susceptibility to a disc combining sulfa and trimethoprim. (Bottom row) Each isolate was tested against individual discs of trimethoprim (left) and sulfamethoxazole (right). Isolate 1 shows individual susceptibility to both and synergy (fused zone). Isolates 2 and 3, susceptible to only one drug, show no added inhibition from the presence of the other drug. Isolate 4, while susceptible only to sulfa, does show some synergy, as evidenced by the skewed zone of inhibition.

develops rapidly, as does host hypersensitivity. However, these drugs have excellent diffusion into body fluids, including cerebrospinal fluid.

Sulfonamides   The primary differences among the sulfas are in half-life. The short-acting ones such as sulfisoxazole are widely used in uncomplicated bladder infection. Otherwise, these drugs are rarely used systemically by themselves. Sulfamylon (Mafenide) and silver sulfadiazine (Silvadene) are used for burn treatment.

Diaminopyrimidines   Trimethoprim may be combined with sulfamethoxazole, which is pharmacodynamically similar. This combination is marketed as cotrimoxazole (Bactrim or Septra) (Fig. 26.8). Cotrimoxazole is recommended in chronic prostatitis, some urinary tract and pulmonary infections, and salmonellosis. It is the preferred agent against *Pneumocystis carinii*.

Malaria treatment; Chapter 24, p. 712
A combination of pyrimethamine plus sulfadiazine (Fansidar) is effective against chloroquine-resistant malaria.

Trimethoprim may be toxic to the fetus and to small children. Sulfonamide toxicities are of all possible types. However, the actual frequency of serious toxicity is slight. Hypersensitivity to the sulfas is quite common.

OTHER ANTIMICROBIAL DRUGS

Antimycobacterial Drugs

The mycobacteria, in particular *M. tuberculosis* and *M. leprae*, respond to therapeutic agents quite different from those responded to by other bacteria. They pos-

TABLE 26.10
Antitubercular drugs

DRUG	ROUTE	SIDE EFFECTS	FREQUENCY OF SIDE EFFECTS	USED FOR
Isoniazid (INH)	Oral	Neuritis, hepatitis	Uncommon	Initial treatment
	Intramuscular	Eighth nerve, renal	Common	
Ethambutol (EMB)	Oral	Optic neuritis	Uncommon	
Para-aminosalicylic acid (PAS)	Oral	GI intolerance, hypersensitivity	Common	
Rifampin (RM)	Oral	Hepatic	Uncommon	
Pyrazinamide (PZA)	Oral	Hepatitis	Common	Retreatment
Ethionamide (ETA)	Oral	GI intolerance, hepatitis	Common	
Cycloserine (CS)	Oral	Psychosis, seizures	Common	
Viomycin (VM)	Intramuscular	Eighth nerve, renal	Common	
Kanamycin (KM)	Intramuscular	Eighth nerve, renal	Common	
Capreomycin (CM)	Intramuscular	Eighth nerve, renal	Occasional	

Source: Adapted from Paul D. Hoeprich, 1977, *Infectious Diseases* (2nd ed.), New York: Harper & Row.

sess unusual lipids that render their cell wall impermeable to the regular antimicrobials. Furthermore, mycobacteria are very slow-growing and not much affected by substances that kill rapidly growing bacteria. However, a number of effective antimycobacterial drugs have been developed. Because long periods of treatment are needed, the pathogen readily develops resistance to any drug offered singly. In active diseases, two or three drugs are combined. These should be selected on the basis of laboratory susceptibility tests on the causative organism.

Isoniazid is the single most used antitubercular drug. It can be administered orally, and toxicity, including neuritis and hepatitis, is uncommon. The mode of action is unclear, but perhaps involves interference with NAD synthesis. Pyridoxine (vitamin B_6) is often given concurrently to prevent the development of side effects.

Isoniazid is used alone for prophylactic treatment of tuberculin-positive contacts, and in combination with other drugs in therapy of active cases.

Rifampin is of microbial origin. Chemically it is a large, complex ring structure. Rifampin specifically inhibits bacterial RNA polymerase. Although it is currently used primarily for the treatment of mycobacterial infection, it is also effective against most Gram-positive bacteria and some Gram-negatives. Rifampin can be given orally, is well absorbed, and has a low incidence of toxicity. Bad effects may include jaundice, blood dyscrasias, and a flulike syndrome. Rifampin is available alone or in fixed-dose combination with INH.

Streptomycin revolutionized tuberculosis therapy because it was the first specific drug available. It is still widely used, although it has the disadvantage of needing intramuscular injection, impractical for the home-care patient.

Ethambutol, an RNA synthesis inhibitor, can be administered orally and is widely used for primary treatment.

Drugs used in treatment of leprosy The sulfone Dapsone is specific against *M. leprae*. Resistant strains have started to appear, however. Rifampin is also a promising therapeutic agent. The long-acting acedapsone and cofazimine are also available.

Antifungal Drugs

The eucaryotic fungal cell is most effectively controlled by polyene drugs that destroy its membrane. All antifungal drugs are incompletely specific and thus have appreciable human toxicity. Superficial fungal infections are treated with topically applied drugs. In this case, so long as the drug is not absorbed from the skin, its systemic toxicity is unimportant. Systemic mycoses are life-threatening if progressive, and the use of toxic drugs is justifiable.

Treatment of superficial mycosis The dermatomycoses (tinea capitis, pedis, etc.), candidiasis of the mucous membranes, and fungal infections of the eye and external ear are considered superficial; they can frequently be treated topically.

Imidazoles, such as clotrimazole and miconazole are supplied as creams, solutions, or vaginal applications. They are broad-spectrum antifungals, and their only adverse effect is local irritation in some patients. Clotrimazole is also useful against *Trichomonas.*

Candida; Chapter 17, pp. 510–511

Griseofulvin is effective against every dermatophyte except *Candida albicans.* It is given orally for extensive or persistent dermatomycosis. Although poorly absorbed, sufficient drug is deposited in the keratin-forming cells to arrest fungal spread on new growth of skin, hair, and nails. Headaches are a frequent side effect as are GI upset and hypersensitivity.

Nystatin, which is not absorbed at all from the gut, is widely used for candidiasis, alone or in combination with other drugs. Creams and vaginal tablets are available for topical use. A solution held in the mouth may control oral thrush; it is swallowed for gastrointestinal candidiasis. In combination with gentamicin, nystatin is used prophylactically in leukemia patients to control undesirable bowel flora leading to fatal rectal abscesses. Nausea and vomiting are the main side effects.

Natamycin is a recently licensed antimicrobial for fungal infections of the eye. Tolnaftate is useful against dermatophytes but not against *Candida.*

Systemic mycosis; Chapter 19, pp. 566–573

Treatment for systemic mycosis Amphotericin B, in the form of the less toxic methyl ester, is the standby for severe systemic mycosis. It causes irreversible leakage of fungal cell membranes. It is effective against *Blastomyces, Histoplasma* (Fig. 26.9), *Cryptococcus, Candida, Sporothrix,* and *Coccidioides,* but not against *Aspergillus* or the phycomycetes. Amphotericin is given intravenously, intrathecally, or by infusion, and most of the drug molecules bind to tissue lipid sites, giving

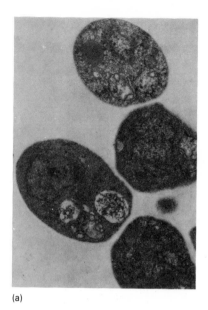
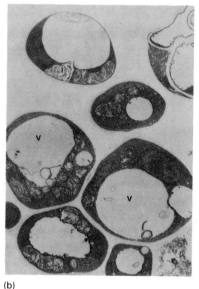

(a)   (b)

**26.9
Effect of Amphotericin B on Cells
of *Histoplasma capsulatum*** A suspension of yeast cells was exposed to 0.2 mg of drug/mL at 37°C. (a) At the beginning of the experiment, normal cells show dark, intact cytoplasm and distinct nuclei. (b) After 60 minutes the remaining cells, whose membrane permeability barrier has been destroyed by the drug, have developed large vacuoles (v). Nuclei are distorted or not visible.

slow release. Side effects are universal; they increase both with increasing dosage and prolonged time of administration. Fever, chills, weight loss, and GI upset are common. Some degree of renal damage and anemia are usually seen. Amphotericin B is incompatible with many other medications.

The substance 5-flucytosine, an inhibitor of nucleic acid synthesis, is effective against most systemic fungi, but unreliable with *Candida*; resistance develops readily. It may be given orally, and though side effects occur (bowel dysfunction, leukopenia, rare aplastic anemia), they are much less frequent than with amphotericin.

The sulfonamides are occasionally useful in systemic mycosis.

Antiviral Drugs

The viral replication cycle is almost completely dependent on host cell structures and enzymes. A few of the replicative enzymes, however, are coded by viral genes. These viral enzymes are essential for viral replication but not crucial to the host's survival. A drug to block virus growth must be an inhibitor sufficiently narrow in action to block only these viral enzymes. The search has met with limited success in combating DNA viruses, but with no success in combating RNA viruses.

Amantidine hydrochloride is approximately 70 percent effective in preventing influenza A, and it also moderates clinical illness. It may cause transitory nervousness, insomnia, or hallucinations in 2–5 percent of patients. A related drug, rimatidine, is preferred in the USSR.

Idoxuridine is licensed for topical use in herpetic keratitis, the leading cause of vision loss due to external eye infections. When tested experimentally for herpes

TABLE 26.11
Antiprotozoan drugs

DISEASE	AGENT	DRUGS
African trypanosomiasis	*Trypanosoma rhodesiense* *Trypanosoma gambiense*	Suramin Pentamidine isethionate Melarsoprol
Amebiasis	*Entamoeba histolytica*	Metronidazole Emetine Tetracycline Di-iodohydroxyquin
Balantidiasis	*Balantidium coli*	Tetracycline Metronidazole Paromomycin
Chagas' disease (American trypanosomiasis)	*Trypanosoma cruzi*	No proven therapy Nitrofurans ⎱ under trial Primaquine ⎰
Giardiasis	*Giardia lamblia*	Metronidazole Quinacrine
Leishmaniasis	*Leishmania donovani* *Leishmania tropica* *Leishmania brasiliensis*	Sulfadiazine/pyrimethamine Antimony sodium gluconate Quinacrine Cycloquanil pamoate Pentamidine isothionate (choice depends on geography, type of symptom)
Malaria	*Plasmodium vivax* *Plasmodium ovale* *Plasmodium malariae* *Plasmodium falciparum*, sensitive strains *Plasmodium falciparum*, resistant strains	Chloroquine Primaquine Sulfadiazine/pyrimethamine Quinine
Pneumocystosis	*Pneumocystis carinii*	Cotrimoxazole Sulfadiazine/pyrimethamine Pentamidine isothionate
Primary amoebic meningoencephalitis	*Naegleria fowleri* *Acanthamoeba* sp.	Amphotericin B Sulfonamides
Toxoplasmosis	*Toxoplasma gondii*	Sulfadiazine/pyrimethamine/folinic acid
Trichomoniasis	*Trichomonas vaginalis*	Metronidazole

encephalitis, idoxuridine was found not only ineffective but toxic. A sugar analog, 2-deoxy-D-glucose, is being tested for effectiveness in herpetic keratitis and has recently been approved for treatment of cervical herpes.

Cytosine arabinoside (ara-C) a cancer chemotherapeutic, was once used in experimental treatment of herpesviruses, including cytomegalovirus, varicella–zoster, and herpes simplex, but its use has been discontinued. Adenine arabinoside (ara-A) is less toxic than ara-C, and has shown promise in herpes encephalitis. In a carefully designed clinical study, ara-A reduced the mortality rate from 70 to 28 percent. It is currently being evaluated against progressive mucocutaneous and neonatal herpes.

Interferon, when it can be produced and delivered economically, holds promise as a highly specific, nontoxic inhibitor of viral replication. However, the problems remain great. Large-scale production of human leukocyte interferon is just beginning; drugs which may be taken by the patient to stimulate interferon production *in vivo* are of unacceptable toxicity.

Interferon; Chapter 11, pp. 324–326

However, recent developments in plasmid technology have paved the way for mass production of human interferon by bacteria. Exogenous interferon has been shown to have therapeutic effect in both rabies and serum hepatitis. Early clinical studies of its prophylactic value in the common cold appear promising. Human leukocyte interferon has also been shown experimentally to reduce the severity and extent of shingles and CMV infections in immunosuppressed patients. Full-scale national clinical trials of interferon as a viral prophylactic/therapeutic agent are currently in progress.

Antiprotozoan Drugs

Therapeutic options for the protozoan infections range from highly effective treatments for most forms of malaria to no available effective treatment for American trypanosomiasis. Many of the drugs used are highly toxic, and their administration requires special care. Antiprotozoan drugs are provided on an experimental basis with skilled consultation from the Center for Disease Control or other national public health agencies (Table 26.11).

Metronidazole (Flagyl) is a commonly encountered antiprotozoan drug in the United States. It is used in treatment of the three most common protozoan diseases in this country, *Entamoeba histolytica* dysentery, *Giardia lamblia* gastroenteritis, and *Trichomonas vaginalis* vulvovaginitis. The drug may cause nausea, drowsiness, a skin rash, or a bad taste in the mouth. Alcoholic beverages must be avoided, since there is a severe cross-reaction. The drug should not be used in early pregnancy or while the mother is breast-feeding.

SUMMARY

1. Specific antimicrobials have made it possible to significantly lower the case fatality rates of most bacterial, fungal, and protozoan diseases.

2. The value of antimicrobials is partially compromised by their occasional toxicity and hypersensitivity. Microorganisms also readily develop and transfer drug resistance.

3. Antimicrobials must be used so as to maximize good effects while minimizing bad ones. These drugs should never be used unnecessarily when the patient will experience as rapid and uncomplicated recovery without them, or when a susceptible microbial pathogen cannot be demonstrated.

4. Rational antimicrobial therapy requires that the causative agent(s) be isolated and its (their) antimicrobial susceptibilities determined. Both semiquantitative agar diffusion and quantitative microdilution techniques are in current use.

5. Antimicrobials that are -static inhibit microorganisms' growth while -cidal antimicrobials irreversibly inactivate them; -cidal drugs are considered preferable.

6. Each compound has a specific mode of action. Human toxicity is low when the mode of action is specific for the target pathogen, and spares human tissue. Toxicity is higher when the drug is relatively nonspecific.

7. Toxic effects are most commonly manifested in the gastrointestinal tract, liver and kidney, nervous system, and bone marrow. Hypersensitivity may also occur.

8. When a drug is administered, it follows specific kinetics of absorption, distribution, and excretion. These must be considered in planning drug delivery. The drug must reach the anatomical site of the infection. Effective drug levels must be continually maintained at that site.

9. Different groups of antimicrobials are recommended for different types of pathogens. Effective, relatively nontoxic therapy is available for almost all bacterial diseases. Choices are fewer and attended by significantly more risk for viruses, fungi, and protozoan pathogens. Antiviral therapy is just beginning.

10. Because microorganisms continually become resistant to newly introduced drugs and because many major diseases still lack specific therapeutic agents, we are far from the conquest of microbial pathogens by chemical means.

Study Topics

1. What is the meaning of **therapeutic range**? What restrictions does a narrow therapeutic range place on therapy?

2. What is the argument in favor of using bactericidal rather than bacteriostatic antimicrobial treatments? Are there any exceptions?

3. Give a number of reasons why antimicrobial therapy may fail.

4. Refer to Table 26.4. Why is it that even if a *Pseudomonas aeruginosa* isolate gives a zone of inhibition 12 mm in diameter against carbenicillin, it is still reported as resistant?

5. Why does antimicrobial therapy have to be chosen with careful consideration of the recipient's liver and kidney function?

6. In what areas of infectious disease management are more and better antimicrobials most urgently needed?

Bibliography

Balows, Albert (ed.), 1974, *Current Techniques for Antibiotic Susceptibility Testing.* Springfield, Ill.: Charles C Thomas.

Baron, S., and F. Dianzani (eds.), 1977. The interferon system; a current review to 1978, *Tex. Rep. Biol. Med.* **35:** 1–573.

Bauer, A. W.; W. M. M. Kirby; J. C. Sherris; and M. Turck, 1966, Antibiotic susceptibility testing by a standardized single disc method, *Am. J. Clin. Pathol.* **45:** 493–496.

Cumitech 6, 1978, *New Developments in Antimicrobial Agent Susceptibility Testing,* Washington, D.C.: American Society for Microbiology.

Galasso, G. J.; T. C. Merigan; and R. A. Buchanan (eds.), 1979, *Antiviral Agents and Viral Diseases of Man,* New York: Raven Press.

Goodman, L. S., and A. Gilman (eds.), 1979, *The Pharmacological Basis of Therapeutics,* New York: Macmillan.

Hoeprich, Paul D. (ed.), 1977, *Infectious Diseases* (2nd ed.), New York: Harper & Row.

Interqual, 1978, *Antibiotic Use Review and Infection Control—Evaluating Drug Use through Patient-care Audit,* Chicago: Interqual, Inc.

Kagan, Benjamin (ed.), 1974, *Antimicrobial Therapy* (2nd ed.), Philadelphia: W. B. Saunders.

Kolff, Cornelius A., and Ramon Sanchez, 1979, *Handbook for Infectious Disease Management,* Menlo Park, Calif.: Addison-Wesley, Chapters 3 and 4.

Kunin, C. M., et al., 1977, Audits of antimicrobial usage: Parts 1–12, *JAMA* **237:**

MICs: A Contemporary Perspective, 1978, Princeton, N.J.: Excerpta Medica.

Physician's Desk Reference. Oradell, N.J.: Medical Economics Co. Published annually: use current edition.

Simmons, H. E., and P. D. Stolley, 1974, This is medical progress? *JAMA* **227:** 1023–1028.

27

Immunization

The ultimate goal of all medical research is the prevention of disease. It is sometimes hard to keep this goal in view because the needs of the sick patient are so much more immediate and pressing than the needs of the well individual. Both research funds and practitioners' schedules are weighted heavily on the side of therapeutic, rather than prophylactic, activities. There are clear signs, however, that the balance is becoming more equal. With the revival of family-practice medicine and home health-care services, more attention will be directed toward providing healthy individuals with what they need to **stay** healthy. Immunization, a key to preventing infectious disease, is receiving more public attention and funds than it has in several decades.

IMMUNIZATION—ITS PRESENT STATUS

Some Terminology

Immunization is the conferring of a specific immunity by artificial means. When its goal is prevention, it is termed **immunoprophylaxis.** On occasions in which immunization is offered after symptoms are present, in the hopes of moderating the disease, it is called **immunotherapy.** Sadly, there are very few effective immunotherapies available—our knowledge of the human immune system is still too superficial to have suggested many useful ways of supplementing or redirecting it in an emergency.

You will recall that **active** immunity follows exposure to specific antigens, results in the body's acquiring the ability to produce humoral or cellular immune mediators, and is long-lasting. **Vaccination** is the process of inducing active immunity, by administering microorganisms or their products in a nonpathogenic form. A **vaccine** contains microorganisms; these may be killed or inactivated, or **attenuated.** An attenuated microorganism is one that has been cultivated under conditions that lead to an irreversible loss of pathogenicity. Usually this occurs during a long series of passages or transfers in media or tissues selective against its virulence factors.

Active immunity;
Chapter 13, p. 399

A **toxoid** is a microbial exotoxin, chemically treated so as to destroy its toxigenicity without reducing its antigenicity.

Combined vaccines contain two or more different immunizing substances. The most familiar is the DPT vaccine that contains two different toxoids (diphtheria and tetanus) and killed *Bordetella pertussis.* **Polyvalent** vaccines contain two or more strains of the same organism. The trivalent oral polio vaccine (TOPV) contains live, attenuated virus of all three pathogenic types.

An **antiserum,** or more accurately **immune serum,** is a product of animal or human origin. It contains a demonstrated titer of antibody against specified infectious agents or their toxins. Currently available human immune serum globulin (ISG) products are derived from normal human donor sera. The sera contain high concentrations of antibody against several diseases, as a result of either convalescence from disease or childhood vaccination. Other antisera are made by immunizing horses, cows, or rabbits with appropriate vaccines.

Role of Immunization in Public Health

Immunization of a whole population may be justifiable wherever a given infectious disease can be expected to strike a large number of people because the disease is endemic or predictably epidemic. In addition, the disease must be serious enough in its consequences for at least some individuals so that the risk from the disease is significantly greater than the risk of immunization. In the United States, these criteria are currently fulfilled by only seven infectious diseases—measles, mumps, rubella, diphtheria, whooping cough, tetanus, and poliomyelitis.

Routine immunization is carried out by systematically vaccinating all individuals, as they reach the optimum age, at a mutually convenient time. It is the most desirable way of fulfilling the need, since time can be provided to obtain an adequate physical examination and history. The American Academy of Pediatrics, working closely with the Center for Disease Control, has exerted admirable leadership in formulating and updating the recommendations for the immunization of preschool children.

Herd immunity;
Chapter 15, pp.
448–449

Box 27.1

What Is an Acceptable Risk?

Vaccination, like other medical actions, must conform to the Hippocratic oath's primary rule "First, do no harm." Yet, as has been amply documented, it is quite possible for vaccination to do far more harm than good. In 1978, several years after routine smallpox vaccination was discontinued in the United States, more than 4.4 million units of smallpox vaccine were still being distributed in this country. Probably only a few thousand of the recipients had valid need for the product. The Center for Disease Control documented the experiences of four unfortunate persons in whom smallpox vaccination gave rise to life-threatening complications. There was one fetal death. None of these people needed vaccination. For three of them, there were obvious contraindications, including malignancy, immunosuppressive drug therapy, and pregnancy.

It is very difficult to come to terms with the issue of "acceptable" risks. It is clear that the seven or eight annual deaths due to smallpox vaccination in the United States (prior to 1971) were "acceptable" because there was still a serious threat of a smallpox epidemic in which a thousand times that number might die. It is equally clear that, now that smallpox has been vanquished, the death of even one fetus is an unacceptable cost. More difficult ethical problems come in the gray areas in which neither benefits nor risks can be satisfactorily quantitated. If a vaccine for the common cold were developed, how great a risk (mortality per million doses administered) would you consider acceptable? Remember there are risks inherent in injecting or swallowing sterile distilled water, or even in the act of driving to the clinic!

Data from: *Morbidity and Mortality Weekly Reports* **28:** 265, 1979.

Individual immunization is used for the occasional individual who needs special protection against less common diseases. Such individuals may be veterinarians who seek rabies vaccination, or travelers who seek inoculation against foreign disease agents.

Mass immunization should occur only when an epidemic erupts in a community. When there is a polio outbreak, epidemiologists may be working 24 hours a day to carry out mass vaccination of all persons at risk. Because of the emergency nature of the operation, less attention can be paid to medical history and concurrent health problems, and there is more possibility of adverse reactions. The reason why routine vaccination is preferable becomes obvious.

History

In the 1890s, it was discovered that antibody-mediated immunity was passively transferrable. The serum of a person who had recovered from a dangerous disease could be administered to a patient, and it sometimes had an immediate therapeutic benefit. Until the advent of antimicrobial drugs, serum transfer was the only **specific** therapy available. It was widely used, with variable success. "Serum treatment" for streptococcal, staphylococcal, pneumococcal, and almost any other bacterial disease was attempted, often without attention to the standardization or even sterility of the product administered. Much of this therapy was useless. Some of it was lethal, as when serum grossly contaminated with *Staphylococcus aureus* was injected, or when the serum came from a chronic hepatitis carrier.

Sera for large-scale use were produced in horses. Since this was still the era of horse-drawn transportation, many people had preexisting hypersensitivity to horse antigens; serum sickness or fatal anaphylactic shock following first injections of equine serum were common, and became even more likely as these were repeated.

The era of widespread serum therapy ended with World War II and the discovery of penicillin. At present, serum therapy is used only in special instances and with significantly greater safety.

Current Production Methods

Modern immune sera are almost all of human origin. Human sera are preferable to equine sera for two reasons. They are effective for a longer period. The half-life of human serum proteins (in the human) is 26–30 days, while the half-life of foreign (horse) serum proteins is only seven days. Serum sickness following initial intramuscular injections of horse serum may occur in 40 percent of recipients. Side effects with human sera are much less common, restricted to occasional hypersensitivities or nonspecific distress.

Human immune sera are produced by pooling carefully selected donor sera. These are then alcohol-precipitated and fractionated to concentrate the globulins and remove some extraneous serum proteins. The preparation is then either redissolved to a final protein concentration of 16.5 percent or lyophilized. It is packaged aseptically. The lyophilized preparations are normally dispensed with a vial of sterile solvent of the correct volume to resuspend to the desired concentration. Careful attention to processing standards has largely eliminated the risk of transmitting infection via these products, although this may occasionally occur. Most antisera of human origin are expensive; many are in extremely short supply.

Available Immune Sera

Immune serum globulin (ISG) The name of this biological is nonspecific. However, selection of donor sera guarantees that the final product contains predetermined titers of antibody against measles, diphtheria, polio, and hepatitis A.

Antibody titer; Chapter 13, pp. 387–388

ISG may be given intramuscularly to persons who have been exposed to disease. It is recommended for household contacts of measles patients, particularly those under one year of age. When administered before or within two weeks after exposure to hepatitis A, it is 80–90 percent effective in preventing disease. ISG will

TABLE 27.1
Immune sera in general use

PRODUCT	SOURCE	INDICATIONS	RESTRICTIONS OR CONTRAINDICATIONS
Immune serum globulin	Human	Prophylaxis: hepatitis A Measles in exposed persons	Rubella Polio Diphtheria Mumps } Value unproven
Hepatitis B immune globulin	Human	Prophylaxis: patients in institutions or hospital units in which disease endemic; intimate or sexual contacts of patients Exposed neonates	
Rabies immune globulin	Human	Following documented exposure	Not to be used in cases in which probability of exposure is very slight
Varicella-zoster immune globulin	Human	Newborns; unimmunized, immunosuppressed persons	To be used only when there is documented exposure
Tetanus immune globulin	Human	May be used when wound cleansing has been delayed, wound is massive, and/or there is no immunization history within ten years	Not otherwise needed
Tetanus antitoxin	Equine	Not widely used	Do not use if human product is available
Mumps immune globulin	Human	For protection of unimmunized contacts; value not proven	
Pertussis immune globulin	Human	For protection of unimmunized contacts; value not proven	
Vaccinia immune globulin	Human	For prophylaxis of persons exposed to smallpox, therapy of persons with complications due to smallpox vaccination	
Botulinum antitoxin polyvalent ABE	Equine	Therapy if food containing toxin has been ingested	Useful only if given before symptoms are fully developed; use with care if hypersensitivity response is possible
Antivenins	Equine	After bite of respective venomous animal	Use with care if hypersensitivity response is possible

modify, but not prevent, viremia in rubella, and thus is of doubtful value in the exposed, pregnant woman.

Hepatitis B immune globulin (HBIG)   Donor sera especially high in antibody to hepatitis B are used in the manufacture of this product. It generally has an anti-HB titer greater than 1:100,000. It is of significant value in prophylactic treatment of patients in institutions and hemodialysis units (if hepatitis B is endemic), intimate contacts of hepatitis B patients, and infants born to mothers with acute hepatitis B in the third trimester.

Rabies immune globulin (RIG)   This is a product obtained by fractionation of the serum of hyperimmunized human donors. It is used, wherever available, to replace the older equine antirabies serum (ARS). RIG may be administered following the bite of any animal known or suspected to be rabid, escaped, or of unknown condition.

Tetanus immune globulin (TIG)   This human product is preferred to older equine products in wound management. If the patient's tetanus immunization history is not adequate and the wound is major, especially if wound cleansing has been delayed some hours, TIG will usually be given.

Varicella-zoster immune globulin (VZIG)   This product is prepared from the pooled sera of donors having high levels of anti-VZ antibody. It is still an investigational drug, and can be obtained only under very restricted conditions. It is for administration only to newborns or immunosuppressed persons without an immunization history, and with documented exposure. Like most of the IGs of human origin, the amounts available are critically small, limited by the number of human donors.

Other immune sera   Mumps immune globulin and pertussis immune globulin are available but their value is not proven and they are not widely used. Vaccinia immune globulin (human) is no longer much used in the United States because there is little need for it. Botulinum antitoxins A, B, and E (polyvalent, equine) is used when a person has ingested food containing toxin. It is provided by the CDC, and maintained in airport emergency depots around the country. It can be airlifted promptly to any community in which it is needed. A gas gangrene antitoxin (equine) used to be available but because it did not appear to be effective, it is no longer provided.

Antivenins   Antisera are also used to treat bites of poisonous snakes and spiders. The venoms of these animals contain toxins not unlike botulin in their action; if given promptly, antitoxin (equine) can be life-saving. There are three types, a coral snake antivenin, a trivalent crotaline (rattler, copperhead, moccasin) antivenin, and a black widow spider antivenin.

History

Active immunization started with the practice of **variolation,** a semicontrolled exposure to smallpox virus inhaled or scratched into the skin. Smallpox on the average kills 20–30 percent of sufferers, while variolation probably killed about 2–3 percent. That was considered a great enough improvement to induce untold thousands of persons to have the treatment.

Vaccination was introduced by Jenner, who adopted the use of virus-bearing lymph from cowpox-infected milkmaids. The related virus conferred sufficient cross-immunity to be protective, without being viruient enough to produce generalized infection in most persons. Rabies and anthrax vaccinations were next, both introduced by Pasteur.

The next immunizations promoted for general use were for diphtheria and tetanus. These were mixtures of toxin neutralized with equine antitoxin. Technical problems with the neutralization step led to many fatalities.

The discovery of reliable chemical means of converting toxins to toxoids at last opened the way for safe general immunization against these diseases. Methods of using killed bacteria for vaccination were also developed. Unfortunately, few killed-bacterial vaccines are completely satisfactory.

The next major breakthrough, viral vaccines, followed the development of cell-culture techniques. These allowed bulk production of viruses, and the development of both killed- and live-virus vaccines.

Criteria for an Acceptable Vaccine

There are several criteria that an immunizing product must fulfill in order to find general acceptance.

It should be easily produced by means that give reproducible results. The key antigen(s) should be conveniently assayed for potency. The antibody response achieved should in turn be readily quantitated.

The vaccine itself should not produce infectious disease nor have a significant level of side effects, immediate or delayed.

The preparation should be free of hazardous, contaminating substances, be they chemical or biological (live agents such as virus particles).

On administration the product should promote **seroconversion:** that is, the antibody titer should rise from a low to a specified higher level. This seropositivity should be long-lasting.

The immunity conferred should be protective: that is, when exposed to or challenged by the pathogen, vaccine recipients should resist infection.

If a live agent is used, it must be shown that the agent cannot be transmitted to the recipient's contacts resulting in disease. The agent should be incapable of causing fetal death or malformation when given to a pregnant woman.

No vaccine currently available fulfills all of these criteria. Thus various cautions are needed in their use.

27.1
Production of a Bacterial Vaccine
Pneumovax is a vaccine composed of a mixture of purified polysaccharide capsular materials from the most virulent strains of *Streptococcus pneumoniae.* Here a worker monitors the growth of the bacterial strains prior to their harvest.

When we have an infectious disease, the pathogen presents a mosaic of surface antigens to the body's lymphoid tissue. Some are strong antigens, and provoke a strong response, which may or may not be protective. Antibodies to surface antigens work at the pathogen's surface. They seem to either promote phagocytosis or prevent attachment of pathogen to host tissue. A vaccine is at its best when it most closely simulates the "natural" stimulus of disease, without causing any symptoms.

Production and Standardization

In the United States, vaccines are produced under control of the Bureau of Biologicals of the Food and Drug Administration. They are **stock** vaccines, produced from approved stock cultures.

Bacterial vaccines   These are, with the exception of the BCG vaccine for tuberculosis, all made with killed cells or purified bacterial products (cell-free) (Fig. 27.1). The agent is grown on standardized media, and the cells are collected, washed, and resuspended in sterile saline to a specified density. Then they are killed by exposure to heat; to chemicals, such as formalin, cresol, and Merthiolate; or to ultraviolet light. Careful sterility checks confirm the nonviability of the vaccine agent as well as the absence of viable contaminants. The material is then either dried or preferably lyophilized in sterile ampoules. Companion ampoules contain sterile diluent to rehydrate them. Normally each vial contains a small number of doses, each to contain between 10^8 to 10^9 bacteria per dose.

Toxoids   Stock toxigenic strains of bacteria are grown on high-yield media. After growth the bacterial cells are removed and discarded, and the culture medium is chemically treated to remove extraneous components and concentrate and purify

27.2
Production of a Viral Vaccine (a)
Embryonated eggs being placed on
a conveyor belt in preparation for
their inoculation with influenza
virus. (b) Harvesting the virus-laden
egg contents. Trays of eggs (background) pass to a manifold where
their contents are aspirated under
vacuum and collected in a sealed
vessel (foreground). This material
will be further purified and standardized before packaging.

(a)   (b)

the toxin. It is then treated with formalin and allowed to incubate until it has completely lost its toxicity. The toxoid is quantitated and adjusted to standard potency.

Rickettsial vaccines   The rickettsia must be cultivated in the yolk sacs of embryonated eggs. After growth, the microorganisms are separated from the yolk material and formalin-inactivated.

Virus culture; Chapter
11, pp. 313–318
Viral vaccines   Viral vaccines are produced in chick embryo, duck embryo, or various cell-culture lines (Fig. 27.2). Producing enough virus for vaccine to inoculate against a disease, such as mumps, is a difficult, costly business. Serial passage in an unnatural host cell attenuates the virus. Killed-virus vaccines are usually formalinized. In **split-virus** vaccines, the virion has been physically fractionated, and the major antigens, usually from capsid or envelope, are retained and concentrated.

All immunizing materials are subject to regulations requiring standardized dosage, lot testing, numbering, and labeling. Package inserts must be provided just as they must be with other licensed pharmaceuticals. All packages must carry an expiration date, after which the vaccine cannot be guaranteed.

Storage instructions must be carefully followed. Many vaccines lose potency rapidly if held at room temperature. These should be stored frozen or refrigerated. Allowing a liquid vaccine to stand at room temperature also carries the hazard that contaminating organisms introduced by poor technique will multiply and render the vaccine dangerous.

Administration of Vaccines

Routes of administration

The preferred route for most immunizations is intramuscular. Deposit of a small volume of vaccine in muscle leads to relatively slow, steady absorption. Larger volumes cause greater pain and muscle irritation.

Live-polio vaccine contains an enterovirus that is given orally, survives passage through the stomach, and infects gut-associated lymphoid tissue. Because it is attenuated, it does not invade nerve tissue.

Aerosolized vaccines contact and stimulate the lymphoid aggregates in the respiratory surfaces, resulting in local sIgA production. Experimentally, this route may be superior to the intramuscular route for vaccination against the respiratory viruses. However, it is not yet generally used.

Absorption of antigen

The extent of immune stimulation delivered by the vaccine depends on the duration of the stimulus, the intimacy of contact with the lymphoid tissue, and the antigenic mass. The duration of the stimulus can be prolonged by using a **depot** of slowly absorbed antigen. Vaccines may be combined with an inert, insoluble inorganic compound to delay absorption. **Alum-precipitated** or **aluminum hydroxide/aluminum phosphate-absorbed** formulations are common. The antigen may also be mixed with nonantigenic organic materials that delay absorption. These are called **adjuvants.** Some examples of adjuvants are mineral oils and peanut oils.

If the vaccine does not contain live organisms, the antigen will be eliminated fairly rapidly from the body and not replaced. Very large antigenic masses cannot be given at one time because the materials are too toxic. To maintain continuing stimulus, the initial immunization is a series of spaced injections. These, and the subsequent booster shots, take advantage of the memory function in immunity to stimulate higher antibody levels.

Multiplication in the host occurs if a live-agent vaccine is used. Particularly in viral vaccines, this gives an intimate stimulus that very closely simulates natural infection. Also, if a live organism is used, the intimate stimulus persists for a longer period. A large, antigenic mass may be developed by multiplication, far larger than could be tolerated in a single administration. The net result is that live-agent vaccines are much more effective than killed-agent vaccines.

Contraindications

There is a variety of reasons why a particular vaccine should not be administered to a certain individual at one time or another. These reasons are always detailed on the vaccine package insert, and should be reviewed before the responsibility of administering a vaccine is assumed. Some of the major contraindications are reviewed here.

Live viruses should not be given to pregnant women because the viruses cannot be completely absolved of fetal damage. Persons who are receiving steroids or who are known to have immunosuppression resulting from malignancy, drug therapy, or other causes should not be immunized. Persons with dermatitis or eczema should

TABLE 27.2
Contraindications for vaccination

CONDITION	VACCINES NOT GIVEN
Febrile illness*	Mumps, measles, rubella, DPT
Congenital immunodeficiency	Mumps, smallpox, yellow fever, polio (TOPV), measles, rubella
Leukemia, lymphoma, or other malignancy	Mumps, smallpox, yellow fever, polio (TOPV), measles, rubella, BCG
Immunosuppression	Mumps, smallpox, yellow fever, rabies, polio (TOPV), measles, rubella, BCG
Pregnancy	Mumps, smallpox, yellow fever, measles, rubella, BCG
Dermatitis in patient or family contact	Smallpox, yellow fever
Egg hypersensitivity	Yellow fever, rabies (DEV), influenza, typhus, Rocky Mountain spotted fever
Antibiotic hypersensitivity	Yellow fever, measles
Previous severe reaction	Cholera, diphtheria toxoid, tetanus toxoid, pertussis
Positive tuberculin test	BCG

*Normally, unless there is a severe, immediate challenge, vaccination of persons with acute illness of any sort should be deferred until they are recovered.

not receive smallpox vaccination. Table 27.2 presents other precautions and the reasons for them.

Currently Available Bacterial Vaccines

Only three bacterial vaccines are widely used, but there are a number of special-purpose immunizations that may also be of interest.

Diphtheria; Chapter 18, pp. 532–535

Diphtheria toxoid   Diphtheria toxoid is a formalin-treated toxin. It is supplied alone, or in combination with tetanus toxoid (Td), or combined with both tetanus toxoid and pertussis vaccine (DPT). For children up to six years of age, a relatively large dose of diphtheria toxoid is tolerated, but for older children and adults, the dosage is dropped tenfold because of the risk of severe reactions. Vaccines intended for early childhood use should never be administered to adults.

Pertussis; Chapter 19, p. 552

Pertussis vaccine   This may be either a killed suspension of whole bacteria or a bacterial fraction, depending on the manufacturer. The material contains endotoxin and may give rise to a number of nonspecific side effects such as fever and soreness. Severe side effects such as brain damage and death have also occurred rarely. It is about 62–65 percent effective in preventing pertussis when properly immunized children are exposed. Because whooping cough is of major concern only in small children, this material is not needed in children seven years of age or older.

Tetanus toxoid   This is a formalinized toxoid also. It provides long-lasting protection, and hypersensitivity reactions are uncommon. A few years ago it was common to give repeated boosters, often within six months, to persons with wounds. Under this sort of hyperimmunization some Arthus type hypersensitivity responses occurred. Now reimmunization is normally limited to every ten years.

Tetanus; Chapter 23, pp. 677–679

Other bacterial vaccines   BCG vaccine, used to prevent tuberculosis, is a freeze-dried, live-bacterial product derived from the avirulent Calmette–Guerin strain of *Mycobacterium bovis.* Although widely used in certain other countries, it has found limited acceptance in the United States, largely because of the effectiveness of prophylactic isoniazid in preventing new cases of tuberculosis. It may, however, be useful for health workers with unusually heavy occupational exposure, for populations in which regular and effective health care cannot be given, and for infants with household exposure. It should not be given to tuberculin-positive persons because they may have an overly emphatic reaction leading to ulceration at the intradermal injection site. Its efficacy has been variously assessed, but is never complete; the duration of immunity may be about 15 years.

Tuberculosis; Chapter 19, pp. 558–562

      Pneumococcal polysaccharide vaccine is a polyvalent product containing 50 µg each of the capsular polysaccharides from the 14 most prevalent pathogenic strains of *Streptococcus pneumoniae.* It is recommended for certain high-risk groups, including the elderly, those with abnormal or absent spleen function, chronic cardiopulmonary or other systemic disease, or any condition likely to increase the probability of secondary pneumonias. The duration of protection is not yet established. Side effects are frequent but mild.

Pneumococcal pneumonia; Chapter 19, pp. 554–555

      Other bacterial vaccines are summarized in Table 27.3.

Currently Available Viral Vaccines

Measles vaccine   This live-viral vaccine is prepared from the "further-attenuated strain" of the Edmonston B measles virus, cultivated in chick embryo cell culture. It is marketed alone or in combination with rubella virus (MR), or with rubella plus mumps virus (MMR). Measles immunization is recommended for all previously unimmunized children and adults, persons who received the earlier killed-virus vaccines (now known to be ineffective), children who were previously immunized before the age of 15 months (now shown to be incompletely protected), and all infants at 15 months of age or after. It should not be given to pregnant women, persons with febrile illness, tuberculosis, or immunodeficiency. About 90 percent of recipients develop protective immunity, and the effect seems to persist for more than 15 years. A moderate fever follows vaccination in 5–15 percent of recipients; neurological conditions appear in about one in a million doses.

Measles; Chapter 17, pp. 517–519

TABLE 27.3
Bacterial vaccines for special purposes

DISEASE	VACCINE	MATERIAL
Anthrax	Anthrax vaccine	Killed *Bacillus anthracis*
Tuberculosis	BCG vaccine	Attenuated live Bacille Calmette-Guerin
Cholera	Cholera vaccine	Killed Ogawa/Inaba strains *Vibrio cholerae*
Meningitis	Meningococcal polysaccharide vaccine, Monovalent A, Monovalent C, Bivalent A–C	Capsular polysaccharide *Neisseria meningitidis*
Plague	Plague vaccine	Killed *Yersinia pestis*
Pneumococcal pneumonia	Pneumococcal polysaccharide vaccine (Pneumovax)	Capsular polysaccharide *Streptococcus pneumoniae*
Rocky Mountain spotted fever	RMSF vaccine	Killed *Rickettsia rickettsii* from yolk sac
Tularemia	Tularemia vaccine	Killed *Francisella tularensis*
Typhoid fever	Typhoid vaccine	Killed *Salmonella typhi*
Typhus fever	Typhus vaccine	Killed *Rickettsia prowazekii* from yolk sac

Mumps; Chapter 18, pp. 541–542

Mumps vaccine This vaccine is prepared from live virus grown in chick embryo cell culture and is sold alone or in combination with measles and rubella vaccines. It is recommended by the American Academy of Pediatrics for universal childhood vaccination. However, it is not required by law in many states for school entry. Some physicians choose not to use it because of its relatively high cost. The vaccine gives over 90 percent protection; immunity is durable for over ten years.

RECOMMENDED FOR	ADVERSE EFFECTS
Veterinarians, tannery workers, herders, textile workers	Not evaluated
Exposed infants; at-risk professionals	Ulceration in tuberculin positive; rare generalized infection
Persons entering endemic areas, certain military units	Mild febrile immunity, short-lived
Military recruits; travelers to endemic areas; household contacts	Localized erythema
Travelers to Vietnam, Kampuchea (Cambodia), Laos; laboratory personnel; persons in contact with wild rodents in endemic areas	Mild pain, reddening; headache and fever on repeat doses; (rarely) sterile abscess
Asplenic or immunosuppressed persons; those with chronic metabolic disease; cardiopulmonary disease	Erythema, mild pain
Laboratory personnel; persons with occupational hazard in endemic areas; reimmunize yearly	Pain and tenderness; fever and malaise; egg hypersensitivity
Persons with occupational exposure; veterinarians; persons working with rabbits	Not evaluated
Household contacts of patients or carriers; travelers to endemic areas	Discomfort at site, fever, malaise, headache
Laboratory personnel; persons working in rural or remote areas who will have close contact with local population	Pain at site, fever and malaise; egg hypersensitivity

Side effects are minimal, but the vaccine should be avoided in the same persons for whom measles vaccine is contraindicated.

Rubella vaccine   This live, attenuated cell culture vaccine is universally recommended for early childhood administration. Although clinical rubella is always mild, the virus can pass the placenta and cause severe fetal malformation or abor-

Rubella; Chapter 17, pp. 519–520

Box 27.2

Measles Vaccination Campaign and Measles Mortality—Guatemala

From 1963 through 1971, Guatemala reported an average of 3,632 deaths due to measles each year, or 76.2 deaths per 100,000 population.

In 1972, the Ministry of Public Health and Social Assistance began an annual mass immunization campaign, using live, further-attenuated measles vaccine, directed during the first year at children one through four years of age and subsequently at children nine months to two years. The campaign's goal was to vaccinate 80 percent of children in the selected age range; the vaccine was given during one month each year, usually February.

From 1972 through 1974, reported measles deaths declined by 90 percent, with a low of 3.9 deaths per 100,000 population in 1973. Estimated coverage was high, increasing from 82 percent in 1972 to 94 percent in 1974. Coverage dropped from 1975 through 1978, and reported measles deaths rose disproportionately, reaching 102.1 per 100,000 in 1976, an incidence above the average for the period before the mass campaigns. The geographic distribution of reported deaths was not uniform: the highest rates occurred in the highland areas in which the majority of people are Mayan Indians.

Immunization levels and measles cases were evaluated in Santiago Sacatapequez, a town that had experienced measles outbreaks during the past two years in spite of presumably good vaccine coverage. Only 101 of 231 (44 percent) children one to four years old had a record of measles immunization; the records of 39 children had been lost, and 91 children had not received measles vaccine. Of 73 measles cases that occurred between January 1977

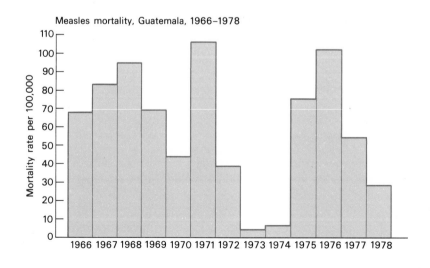

Measles mortality, Guatemala, 1966–1978

and May 1979, 48 were in children one to four years old; ten of these had a documented record of measles immunization.

Since all vaccinated children who developed measles had received vaccine either in 1976 or 1978, vaccine efficacy was calculated specifically for each of these years. Vaccine efficacy was low in both years, 54 percent for 1976 and 72 percent for 1978. By contrast, in the United States, vaccine efficacy is usually about 95 percent.

The vaccine cold chain (i.e., the process of refrigerated shipping and storing vaccine at various points from its manufacture to its ultimate destination) was evaluated in several areas in Guatemala. Storage of vaccine in the capital and in the regional health departments was generally satisfactory, as was shipment from the capital. However, vaccine was shipped from the health departments to rural health centers in poorly insulated containers that could not adequately protect the vaccine. Kerosene refrigerators in health centers often lacked replacement parts or fuel, and electricity for electric refrigerators was sometimes unreliable. Personnel were often inadequately trained in vaccine handling or refrigerator maintenance.

This study points out several potential difficulties in mass vaccination campaigns. One is the difficulty of maintaining public interest and momentum. The great gains made from 1972 to 1974 were partially lost in 1975–1977.

A second problem is actually achieving good vaccine coverage in rural areas. Clear evidence of measles vaccination was available for only 44 percent of children in the town studied.

Yet a third problem arises when the vaccine is not properly handled. Live-virus vaccines must be continuously refrigerated until administration. The low efficacy of the vaccination process in this town shows that there had been mishandling somewhere along the cold chain, although the exact failure could not in retrospect be pinpointed.

Source: *Morbidity and Mortality Weekly Reports* **28** (49), 1979.

---

tion. Since the vaccine has been introduced, congenital rubella syndrome has almost disappeared, with a huge saving in human suffering and health-care costs. The success rate with the vaccine is 95 percent, and the immunity is long-lasting. Rash, swollen glands, joint pain, and neuritis are possible side effects.

The contraindications are the same as for measles vaccine. Pregnant women customarily receive a diagnostic test for rubella antibody. If this is negative, they should not be vaccinated, but should carefully avoid possible sources of infection and seek immunization promptly after childbirth. More than 60 pregnant women have accidentally received rubella vaccine and, fortunately, there has been no indication that the vaccine virus is capable of fetal damage. Great care, however, is observed when rubella vaccine is given to postpubertal females. They must not be pregnant and must avoid pregnancy for three months after vaccination.

Polio vaccines   In the United States, the trivalent oral polio vaccine (TOPV) is universally recommended for all normal children. It should be given in three doses in the first year of life. It is easy to administer and confers a long-lasting immunity in over 90 percent of recipients. TOPV can be administered to the pregnant woman

Polio; Chapter 22, pp. 664–667

if she has been exposed, and will have a dual benefit of promoting active immunity for her and passive immunity for her child. However, when the vaccine is given to the breast-fed infant whose mother has active immunity, her colostral antibody may partially interfere with immunization. The course of administration should be prolonged so that it continues after breast-feeding has been concluded. There are no general side effects; however, in rare cases (approximately 12 a year in the United States) paralytic polio follows vaccination, apparently from reversion of the attenuated strain or unsuspected abnormalities of the recipient's defenses. The injectable inactivated polio vaccine (IPV) is produced in Canada and is available in the United States for immunization of immunodeficient individuals. The killed virus cannot cause active disease, but it does not promote as complete or as lasting an immunity. In some European countries it is universally used, however.

Influenza vaccines   Although there is continuing controversy about the usefulness of influenza vaccination programs, they can be readily justified for a certain group of persons. In any year where there is a moderate influenza A epidemic (and that

TABLE 27.4
Viral vaccines

DISEASE	VACCINE	MATERIAL
Rubeola	Measles	Further-attenuated live virus
Mumps	Mumps	Attenuated live virus
Rubella	Rubella	Attenuated live virus
Poliomyelitis	Polio (TOPV)	Attenuated live virus
Influenza	Influenza	Whole killed or split-virus preparations
Rabies	Rabies (DEV)	Killed virus
Yellow fever	Yellow fever	Live attenuated 17D strain virus
Smallpox	Smallpox	Live attenuated vaccine–strain virus

means almost every year in recent history) between 10,000 and 20,000 related deaths occur during the epidemic period. These are usually due to secondary bacterial pneumonia, and occur primarily in the elderly and chronically ill. Severe epidemics, even in the antibiotic era, have claimed 70,000 lives in one winter in the United States alone. Influenza vaccination can prevent or moderate the disease so that fatal complications become rare.

Influenza virus for vaccine is produced most often in embryonated eggs. It is available either as whole inactivated virus, recommended for persons over 13, and in a split-virus (chemically disrupted) form. The latter is less toxic, and produces fewer side effects. It is used in children, in whom the side effects would otherwise be more pronounced.

Influenza vaccines are the only type whose antigenic content is changed from year to year. Because of the constant drift in the antigenic nature of epidemic influenza A strains, an annual decision must be made as to the probable characteristics of the anticipated viral pathogen. This decision is made early in the spring, allowing sufficient time for the selection of antigenically appropriate stock viral strains, their

Influenza; Chapter 19, pp. 562–565

RECOMMENDED FOR	ADVERSE EFFECTS
All normal children at 15 months	Fever, transient rash, rare neurological disorders
All normal children	Rare; parotitis, allergic reactions, CNS effects
All normal children	Rash, swollen glands, joint pains, neuritis
All normal children, nonvaccinated adults	Rare; a few cases of vaccine-associated polio have been reported
High-risk children and adults, particularly those with underlying disease	Fever, malaise, myalgia; allergic reactions (rare); Guillain–Barré syndrome
Persons with occupational risk; persons exposed by bite or other means to rabid or suspect animal or human being	Common; pain, swelling, itching at site, fever, aching muscles, malaise Uncommon; anaphylaxis rare; neuroparalysis
Persons traveling or working in endemic areas, especially jungle areas; lab workers	Headache, fever, or other neurological symptoms
Lab workers; travelers to certain countries	Lesion at sites; severe to fatal complications in one in 110,000 recipients

mass production, and the testing, packaging, and distribution of the vaccine. These steps are completed in time to place the vaccine in physicians' hands by mid-fall, allowing time to identify, contact, and immunize the target population. Influenza epidemics normally occur between December and February.

The Guillain-Barré syndrome, usually a temporary ascending paralysis, is observed in about ten out of every million persons receiving influenza vaccine. Of this group, 5 percent retain permanent disability, and 5 percent die. The causal relationship between GB and influenza vaccination is unknown; GB also appears spontaneously, or following other vaccinations.

Rabies; Chapter 22, pp. 662–664

Rabies vaccines   There are two vaccines available for rabies prevention. Duck embryo vaccine (DEV) is a killed-virus vaccine. The virus is propagated in duck eggs, separated from the egg materials, and inactivated with beta-propriolactone. A highly effective human-tissue, cell-culture vaccine is now available, but it is currently in such short supply that its use is restricted to persons with documented exposure who are also hypersensitive to egg protein, or to persons who have not responded to DEV.

Preexposure rabies vaccine may be given to persons, such as veterinarians, dog officers, and game wardens, in high-risk occupations. It is also given postexposure to about 30,000 persons a year in the United States, most of whom have been bitten by a rabid or suspect animal. Its side effects are local reactions (common), fever and malaise in one-third, and neuroparalytic reactions (rare). Recipients should be evaluated after the vaccination sequence is complete to make sure that they have in fact developed antibody.

27.3
Smallpox Vaccination   Traveling by horse, mule, or camel, surveillance teams in Ethiopia can cover up to 50 kilometers a day. They search for cases and vaccinate people they meet en route.

Yellow fever vaccine   This is a live, attenuated product made from the 17D strain of virus grown in chick embryo. It is prescribed for those living in or traveling to endemic areas in Africa and South America.

Yellow fever vaccine; Chapter 24, p. 703

Smallpox vaccine   The live, attenuated vaccinia virus is available in the lyophilized form (preferred) or suspended in glycerin. It is recommended now only for persons working in laboratories in which smallpox virus is maintained, and for residents and persons traveling to an ever-decreasing number of countries in which smallpox is still considered a threat (Fig. 27.3). The multiple-pressure method of administration, during which the instrument does not puncture the tissues, is preferred. Successful vaccination has occurred only if there is a **major** reaction. This type of reaction has a region of induration and a central lesion in which the virus has actively multiplied.

Smallpox; Chapter 17, p. 520

Serious side effects from smallpox vaccination are common enough to cause concern. These include postvaccinal encephalitis, generalized vaccinia, and eczema vaccinatum. Cumulatively, these produced one severe reaction per 110,000 recipients in the United States, and seven or eight deaths annually when smallpox vaccination was universally required.

Problems in Vaccination

Vaccination failures   A certain percentage of persons receiving vaccine fail to become completely immunized. This may be made evident either by a failure to seroconvert or by susceptibility to the disease on challenge. The failure rate varies widely—from as much as 40 percent with pertussis vaccine to less than 5 percent with TOPV. Failures may be due to transient factors such as an ineffective lot of vaccine (outdated or incorrectly stored), improper injection procedure, or sometimes solely to sloppy record keeping that states an individual has received an immunization when he or she has not. Equally, failure may be caused by the recipient's temporarily inactive immune system. The recipient may eliminate the antigen without having a sufficient immune response.

Contamination   Periodically, lots of vaccine have been released that contain bacterial or viral contaminants. A highly publicized case was the early IPV. This vaccine contained not only poliovirus but also passenger viruses from the monkey cells in which they were grown. The monkey viruses, now called SV40, were more resistant to chemical inactivation than the polio virus. They remained fully viable in many millions of doses of injected IPV. Fortunately, they do not seem to have caused any untoward effects. Gross bacterial contamination in commercially produced vaccines is now essentially unknown. However, faulty aseptic techniques in offices and clinics occasionally lead to contamination of vaccine vials.

SV40 virus; Chapter 11, p. 330

Toxicity   Vaccines prepared from Gram-negative bacterial cells contain much endotoxin. This can cause nonspecific discomfort that may be of several days duration.

Hypersensitivity   Substances present in vaccines may provoke allergic responses. Viral vaccines are commonly produced in chicken or duck eggs. Many persons are allergic to proteins of avian origin, and will react strongly to the traces of egg protein remaining in the vaccine.

Other viral vaccines are produced in tissue culture. Cell culture media always contain antibiotics to keep down chance bacterial contaminants; traces of these find their way into the vaccines too. Allergic responses to neomycin are a relatively common problem occurring when vaccines prepared in cell culture are given.

Various neurological disturbances, rare sequels to vaccination, are believed by some to represent autoimmune conditions. They are in some way provoked either by the process of vaccination or by antigenic substances present in the vaccine itself.

CURRENT RECOMMENDATIONS

In the United States, several organizations cooperate to formulate and update immunization recommendations. These include professional organizations such as the American Academies of Pediatrics and Internal Medicine, the American Public Health Association, plus governmental agencies such as the United States Public Health Service Advisory Committee on Immunization Practices. Internationally, the recommendations differ widely from country to country.

Recommendations for Infants and Children

At present, it is recommended that all normal children be immunized against seven potentially common and serious diseases. A suggested schedule for administration is shown in Table 27.5. The timing is dictated by the need to offer initial protection at the ages when most needed, most effective, and least toxic. Later, boosters are provided at regular intervals. When immunization has not been started at the usual time, or when there is uncertainty that the child will be brought in for the scheduled series of immunizations, the sequence may be altered.

Measles vaccination campaign; Box 17.2, p. 518

The greatest problem with childhood vaccination programs is compliance. There is widespread parent apathy about the threat of renewed polio or measles activity. Sadly, the percentage of children protected against these diseases has actually been declining. However, universal vaccination has been reemphasized as a key public health policy. In most states, proof of vaccination is required for school admission. When students are actually excluded from school for nonvaccination, there is a sudden motivation to complete the required immunization series.

Adult Immunization

Vaccination is for kids—so goes the prevailing public opinion. But this is clearly wrong. Few childhood vaccinations are truly permanent. Since protective antibody levels are eventually lost, many adults are inadequately protected. Most cases of

TABLE 27.5
Recommended childhood immunizations

AGE	IMMUNIZATIONS
2 months	DPT, TOPV
4 months	DPT, (TOPV optional)
6 months	DPT, TOPV
15 months	MMR, tuberculin test
1½ years	DPT, TOPV
4–6 years	DPT, TOPV
14–16 years	Td

Legend:
DPT = Diphtheria and tetanus toxoids combined with pertussis vaccine childhood dosage
TOPV = Trivalent oral polio vaccine
MMR = Combined measles, mumps, and rubella live-virus vaccine
Td = Combined tetanus and diphtheria toxoids, adult type

diphtheria and tetanus in this country occur in older adults, many of them with childhood immunizations far in their past. Adults **do** need certain booster vaccinations. They also require specialized occupational and postexposure immunizations. Table 27.6 shows the immunizations that may be needed for certain groups of adults. In particular, note that all adults should receive tetanus–diphtheria toxoid (adult dosage) every ten years.

Immunization for Travelers

In certain areas of the globe, diseases unknown in the United States are endemic. Travelers to these areas may or may not need special vaccination. The need depends largely on the traveler's style. If you stick to the more urban areas and standard tourist accommodations, reasonable precautions will normally be sufficient. However, if you are planning campouts in rural areas and will have close contact with local populations, immunization may be essential. The traveler must recall that he or she has less resistance than the local residents. Updated lists of suggested immunizations for each foreign country or geographical area are prepared by the Center for Disease Control and can be obtained by contacting local or state public health agencies. This should be done well in advance of the date of departure, since series of immunizations may sometimes be needed and there may be some delay in obtaining the less commonly used vaccines. The requirements of the countries to be visited should be carefully followed and all necessary certificates obtained. Otherwise, the tourist may experience serious difficulties at border crossings.

TABLE 27.6
Immunization for adults

IMMUNIZATION BEFORE DISEASE EXPOSURE

All adults
Tetanus-diphtheria toxoid	Must be given every ten years

Certain adults
Influenza vaccine	Different types available each year

Women of child-bearing age
Rubella vaccine	Only if shown susceptible by antibody testing and if pregnancy can be prevented for at least three months postvaccination

Some postpubertal men
Mumps vaccine	Most adults have natural immunity

Travelers to foreign countries
Typhoid vaccine	Two injections plus booster every three years
Cholera vaccine	Needed within two months of departure to endemic area
Yellow fever vaccine	Endemic areas in Africa, South America
Typhus vaccine	Only for extended stay in endemic area
Polio vaccine	Tropical areas or developing countries
Plague vaccine	Endemic in Vietnam, Kampuchea, Laos
Immune serum globulin for viral hepatitis	Extended stay in tropical areas and developing countries
Smallpox vaccine	Only for travel to countries requiring valid International Certificate of Vaccination

Persons with unusual occupational exposure
Smallpox vaccine	Certain laboratory personnel
Plague vaccine	Certain laboratory personnel, field-workers
Rabies vaccine	Certain laboratory personnel, veterinarians
Anthrax vaccine	Certain laboratory personnel, textile workers, veterinarians
Rocky Mountain spotted fever vaccine	Certain laboratory personnel, persons repeatedly exposed to ticks in endemic areas
Tularemia vaccine	Certain laboratory personnel, repeated contacts with enzootic areas
Venezuelan equine encephalitis vaccine	Certain laboratory personnel, field-workers
Eastern equine encephalitis vaccine	Certain laboratory personnel, field-workers

Special epidemiological situations
Bacille Calmette–Guerin vaccine	About 80% efficacy in certain situations
Meningococcal vaccine	Only for serogroups A and C
Adenovirus vaccine	Some military personnel

IMMUNIZATION AFTER EXPOSURE TO DISEASE BUT BEFORE ONSET OF SYMPTOMS	
Rabies vaccine and rabies immune globulin	Most animal cases now in wildlife
Immune serum globulin for viral hepatitis	Probably indicated for type B as well as type A
Immune serum globulin for measles	Only for high-risk susceptibles with exposure
IMMUNIZATION AFTER ONSET OF CLINICAL ILLNESS	
Tetanus immune globulin (and tetanus–diphtheria toxoids)	Human preferable over equine antitoxin
Diphtheria antitoxin	Give immediately after clinical diagnosis
Botulism antitoxin	Three preparations available

Source: Adapted from David Rimland; John E. McGowan, Jr.; and James A. Schulman, 1976, Immunization for the internist, *Ann. Int. Med.* **85**:622–629.

FUTURE DEVELOPMENTS IN IMMUNIZATION

There remains a number of serious and common diseases for which no vaccines exist, and several for which the present vaccines are inadequate. In the future, two advances are needed—better immunizing products, and better methods of delivery.

Potential Improvements in Vaccines

Subunit vaccines   By careful disassembly of the bacterial cell or viral particle, it is possible to prepare vaccines containing only a few of the pathogen's antigens. One can then test these to find out what antigen combination is most stimulatory, gives greatest protection, and has the fewest toxic side effects on administration. By concentrating large amounts of the key substances in one dose, one can achieve the desired large antigenic mass. Because there are no live organisms, any chance of developing the actual disease is taken away.

Strongly antigenic capsular polysaccharides are the basis of the pneumococcal and meningococcal vaccines. Intact Gram-negative bacteria have an outer membrane rich in lipopolysaccharide (LPS). This substance in its native state is unfortunately not only a weak antigen but also toxic. However, extracted and/or fractionated LPS preparations are receiving a great deal of attention and show some promise. For example, all members of the enterobacteriaceae share a portion of the LPS molecule, making a single, nonspecific vaccine theoretically possible. LPS vaccines are also under study for treatment of gonorrhea.

Bacterial capsule; Chapter 3, pp. 74–76

Another promising subunit approach is the use of ribosomal preparations. These are being tested for *Pseudomonas aeruginosa, Hemophilus influenzae*, the mycobacteria, streptococci, and neisseriae. It is unclear how antibodies to ribosomes actually penetrate the bacterial cell to inactivate it.

Improved adjuvants   The duration and intensity of antigenic stimulus can be increased by the addition of adjuvants. This practice is customary with vaccines tested in animals, but most adjuvants have so many side effects that they have not been used for human beings. The search continues for adjuvants that may be used with acceptable safety.

Urgently Needed Vaccines

There remain several diseases for which immunization does not exist. Gonorrhea is a key example. Worldwide, it is one of the most common infectious diseases. However, the causative agent is so nonimmunogenic that recovery from infection confers no immunity whatever. This same property carries over to hamper vaccine production. The causative organism will have to be altered, either in its surface characteristics or its presentation to the body, to get a protective response. Despite years of research, there is no vaccine ready for large-scale testing.

Hepatitis B is a serious threat to certain populations. It has a high incidence in persons occupationally exposed to human blood and blood products, such as dentists, doctors, and lab technicians. It is also prevalent in male homosexuals. Efforts to produce a vaccine have been blocked by the absence of a reliable source of antigen. The virus has never been cultivated *in vitro*. The only antigen source is the Dane particle, found in the serum of patients and chronic carriers. Studies are underway using partially purified particles from volunteers to immunize high-risk persons.

Common cold; Chapter 18, pp. 538–539   The common cold is, from an economic point of view, a very costly infectious disease; it causes millions of lost workdays per year. Although each cold virus is actively antigenic, there are so many (over 100 antigenically distinct variants) that to concoct a complete vaccine would be highly impractical. Investigators propose that cold vaccines be administered intranasally, to induce formation of local sIgA antibodies.

There are no vaccines for fungal or protozoan infections. There is no single reason for this lack. However, we can note that natural fungal infections seem to be contolled primarily by cell-mediated immunity, while concurrent antibody-mediated immunity is of uncertain value. Vaccinations mainly stimulate the appearance of AMI, and only to a lesser extent, CMI.

Some pathogenic protozoa, in particular the agents of malaria and Chagas' disease, "hide" from the immune system. Plasmodia and trypanosomes can coat themselves with a layer of host blood proteins—camouflage that contributes notably to their long-term survival in the host. At present, intensive research is underway to identify suitable vaccine materials for malaria prophylaxis.

The Ethics of Vaccine Development and Testing

The World Health Organization has led the way in preparing international chemical and biological standards for vaccines, toxoids, and antisera for human use. They have also proposed ethical codes to guide physicians conducting research, including vaccine trials on human beings. The Declaration of Helsinki (1964, revised 1975) is the standard to which it is hoped all researchers will conform. Additional criteria for the testing of biologicals require careful design of field trials. They should provide definitive, not ambiguous, results. Trials should be continuously evaluated for early detection of any adverse effects, and should be terminated as soon as the results are conclusive. The human research subjects must give informed and free consent. Special attention must be paid to the rights of children, of emotionally or physically handicapped persons, and of those such as prisoners who may be subjected to coercion. The test process should be open to outside surveillance. Once a product is shown to be valuable, it should not continue to be withheld from control groups.

SUMMARY

1. We can systematically utilize the body's natural defenses for its protection. Immunization against many infectious diseases has become practical. The general use of vaccination has contributed, along with sanitary engineering, antibiotics, and other factors, to the great twentieth-century increase in human life expectancy.

2. Immunization is passive when immune sera bearing desirable antibodies are administered. Such protection is immediate but transitory.

3. The best antisera are produced from the blood of human donors, minimizing the chance of hypersensitivity reactions. Such sera are expensive and in short supply. Passive immunization is normally reserved for those faced with an immediate threat of serious infection not readily controlled by antimicrobials.

4. Vaccination is active immunization. An altered, less-hazardous antigen is administered. Immunizing products may include toxoids, live attenuated organisms, killed organisms, subunits or chemical derivatives of organisms.

5. After administration, there is a delay before measurable antibody appears. It may be several weeks before full protection is established. The immunity lasts for a period that varies with both nature of the antigen and the individual peculiarities of the recipient.

6. Extensive research, field testing, and clinical testing go into the development of a new vaccine before it can be licensed. Careful controls over the production process are designed to guarantee the safety and effectiveness of each vaccine lot. However, conscientious attention to the health status and history of each recipient, as well as scrupulous aseptic technique during administration, are also required in order to make the vaccination process maximally safe.

7. In the United States, routine vaccination of children for three bacterial and four viral diseases is recommended and, to varying degrees, required by law.

8. In adulthood, periodic receipt of vaccinations is wise. Travelers, persons with certain occupational hazards, and contacts of persons with certain communicable diseases may require immunization. We must be as careful not to give unneeded immunizations as to protect where protection is needed.

9. Immunization is still needed for some of the remaining unconquered infectious diseases. Some targets include gonorrhea, hepatitis, malaria, and Chagas' disease.

Study Topics

1. Under what clinical epidemiological conditions are immune sera, in contrast to vaccines, particularly useful?

2. What problems remain with human-origin immune sera?

3. By reviewing previous chapters, search for other bacterial diseases in which the development of toxoid vaccines could be useful.

4. The pneumococcal polysaccharide vaccine and the influenza vaccine are recommended for about the same group of recipients. What types of persons and why?

5. The live viral vaccines are generally preferred to killed-virus vaccines. Why? Why also do they have a large group of contraindications?

6. Prepare a rationale to explain the reason why each of the seven childhood vaccinations is recommended, based on the characteristics of the disease it prevents and the expected benefits/risks.

Bibliography

Beale, A. John, and Robert J. C. Harris, 1979, Microbial technology: production of vaccines. In *Microbial Technology: Current State, Future Prospects*, Symposium 29, Cambridge: Society for General Microbiology, pp. 151–162.

Benenson, Abram S., 1975. *Control of Communicable Diseases in Man* (12th ed.), Washington, D.C.: American Public Health Association.

Eickhoff, Theodore C., 1977, The current status of BCG immunization against tuberculosis, *Ann. Rev. Medicine* **28:** 411–423.

Physicians Desk Reference, current edition, Oradell, N.J.: Medical Economics Co.

Recommendation of the Committee on Infectious Diseases, 1974, (The "Red Book"), Evanston, Ill.: American Academy of Pediatrics.

Recommendations of the Public Health Service Advisory Committee on Immunization Practice (ACIP), published periodically in *Morbidity and Mortality Weekly Report*, Atlanta: Center for Disease Control.

Rimland, David; John E. McGowan, Jr.; and James A. Schulman, 1976, Immunization for the internist, *Ann. Int. Med.* **85:** 622–629.

Voller, A., and H. Freedman, 1978, *New Trends and Developments in Vaccines*, University Park Press.

WHO, *The Role of the Individual and the Community in the Research, Development and Use of Biologicals with Criteria for Guidelines: A Memorandum*, 1976, Geneva: World Health Organization.

28

Infection Control
in the Clinical Environment

I nfectious diseases in the community tend to occur when a virulent pathogen penetrates the inherently normal defenses of a healthy individual. In an acute-care hospital or in a long-term care facility such as a nursing home, the basic pattern is different. A relatively nonvirulent opportunist penetrates (or is assisted to penetrate by catheterization or surgery) the reduced or absent defenses of persons already seriously ill. These hospital-acquired infections—**nosocomial** infections—are more difficult to prevent, more unpredictable to treat, and more resistant to cure than are community-acquired infections.

THE SCOPE OF THE PROBLEM

Although the emphasis on nosocomial infections as a problem to be solved is new, the problem itself is not. In medieval hospitals, called "pesthouses," several patients—sick, dying, and dead—often shared the same bed. However, medical practice was so primitive that hospitalized patients either died or got well and fled so fast that disease transmission **within** the hospital probably escaped notice. Semmelweis, who in 1850 documented the fact that physician-attended childbirth in a hospital actually reduced the mother's chance for survival, proved the highly unpopular point that hospitals could be hazardous to your health. In the past, the hazard was caused either by a lack of awareness of aseptic techniques or by gross negligence. Nowadays, with few exceptions, hospitals follow reasonable standards of care and their staffs are conscientious in carrying them out. Why, then, do nosocomial infections continue to occur, little diminished by an "epidemic" of scrubbing, disinfecting, and surveillance? The reason is, paradoxically, that modern medicine has improved so much. Persons who might once have died rapidly from cancers, burns, heart attacks, diabetes, kidney failure, and the like are now kept alive. In many cases, they eventually return to their homes and enjoy additional years of useful and happy life. But first they must pass through a prolonged period of complex, highly technical therapy during which infection is a constant threat. In an imperfect world, no benefit is without cost. Hospital infections are one price paid for medical advances.

Some Definitions

A nosocomial infection is one acquired in the hospital. A good deal of attention is paid to establishing exactly where and how it was acquired. If the infection was present in an active, latent, or incubating state on admission, it is not nosocomial. Infections that become apparent after discharge are classed as nosocomial if, on counting back, their incubation period began during the hospital stay.

The clear separation of community-acquired from hospital-acquired infections has numerous consequences. If an infection is hospital-acquired, the hospital has medical responsibility for its treatment, ethical responsibility for tracing and eliminating the source and means of transmission, and legal responsibility if there is a question of negligence or malpractice.

An **iatrogenic** disease is, literally translated, physician-caused. That is, it is the direct result of a diagnostic or therapeutic procedure. Many nosocomial infections, such as bacteremia resulting from implantation of a replacement heart valve, are iatrogenic. Others, such as a pneumonia developing in the immobile elderly patient, are not. Iatrogenic infections need not always be hospital-acquired. An immunosuppressed kidney transplant recipient may be sent home and there acquire an opportunistic *Candida* infection.

A variable proportion of patients are more or less **compromised** hosts. Persons become compromised when any or all of their three layers of defense—anatomical integrity, phagocytic network, or immune system—are altered.

Opportunists; Chapter 12, p. 347

Opportunist microorganisms do not cause infection in normal humans. Instead, their usual habitats are external to the tissues, on body surfaces or in air, liquids, or food. Such organisms tend to be very hardy. Because they are adapted for life outside the body, their environmental persistence under heat, cold, or drying conditions tends to be excellent. On the other hand, without the usual virulence factors, they require lowered bodily defenses to enter and survive in human tissues. Opportunistic bacterial species in a clinical facility are noteworthy in their ability to resist multiple drugs.

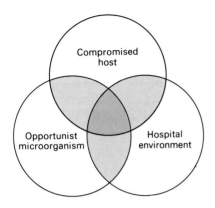

28.1
Interactive Factors in Nosocomial Infection   The compromised host has minimal resistance to microbial colonization or invasion. Opportunistic microorganisms, many of which possess multiple drug resistances, are selectively retained within the hospital. The environment of a hospital is a simplified ecosystem. It provides many of the usual opportunities for disease transmission plus some, such as surgery, that are unique to the location.

Interactive Factors in Hospital Infections

Nosocomial infections occur because three factors interact (Fig. 28.1). First, there must be a significant number of opportunistic or pathogenic microorganisms. Second, these must have access to compromised hosts. Third, there must be an environment that brings and keeps the first two components together. This environment is provided by the unique features of the modern hospital—its architecture, its population density, its staffing patterns, and the types of specialized intimate contact that occur within its walls.

Incidence of Nosocomial Infection

Surveys have established the United States nosocomial infection rate. About 5 percent of persons admitted to a hospital will develop one or more nosocomial infections as a result. The rate is about 8 percent in long-term care facilities. Oddly enough, large university-associated research and teaching hospitals tend to have high rates too. This is not because of inferior care—quite the opposite. These are referral units admitting the most severely ill patients requiring the highest-risk procedures.

The personal and economic costs of nosocomial infections are staggering. In 1977, about 1.84 million patients in the United States developed infections that added an average of three days to their hospital stays, at an average cost of $174 a day. The total bill borne by insurers, private persons, and the government was a shattering $966 million.

To attack the problem, the Joint Commission on Accreditation of Hospitals (JCAH) and the American Hospital Association (AHA) have worked with the American College of Surgeons, other professional organizations, the Infection Control Division of the Center for Disease Control, universities, and teaching hospitals. A large volume of productive research has been, and continues to be, done leading to constant improvements in procedures. Certain basic requirements for infection con-

Box 28.1

The Progressively Compromised Patient

For the failing patient, there may be a progressive sequence of compromising events. An initial organ failure or a malignancy may lead to hospitalization that in turn may lead to a primary nosocomial infection. These problems may in turn progress to secondary infections or organ failures. The odds on dealing successfully with each additive trial get poorer.

This pattern is revealed by a recent study of hospital-acquired fungemia. In one series of cases, the predisposing factors overlapped so markedly that it is quite clear how the cards were stacked against the patients. At the time fungemia developed, all patients had prior antibiotic therapy and were receiving fluids by intravenous catheter. (Two-thirds were getting total parenteral nutrition.) All except one had

urinary catheters. Thus this was a group of people that was more or less immobilized in bed. Almost 90 percent had concurrent bacterial infections despite the antibiotic therapy, suggesting that the therapy was ineffective in the face of poor natural defenses. Almost 70 percent had had recent surgery. The exact sequence of these added factors of course differed from patient to patient, but seemingly each factor, by solving one immediate and pressing problem, set the stage for another problem.

The infective agent in most of these cases was *Candida albicans.* Almost all the patients died. As you can see, it might be very difficult to assign an exact cause of death in cases such as these.

trol must be met before a hospital can obtain JCAH accreditation; these are quite stringent. However, licensing requirements for other types of facilities are much less definitive and vary greatly from place to place.

OPPORTUNISTIC MICROORGANISMS

Organisms with Limited Virulence

It has already been emphasized that even the species viewed as "true pathogens" differ widely in virulence. This is expressed in the **attack rate**, or percentage of those at risk that get the disease. If the attack rate for an infectious agent is as low as 5 percent, one may suspect that the organism is truly an opportunist, victimizing only that subgroup in the population that is especially susceptible.

Another way of quantitating the pathogen versus opportunist issue is by the inoculum size needed to initiate infection. This may be ten organisms for one microbial species and several million for another. These numbers are an expression of the likelihood of each individual organism's successfully penetrating host defenses. When the defenses are poor, a smaller inoculum may cause disease. Opportunistic infection can occur when the microbial burden that the host can tolerate without infection has been exceeded.

Common opportunists   We will consider any organism to be an opportunist if it either causes disease only in compromised hosts, attacks them more often than normal hosts, or causes markedly more severe and generalized infection in compromised individuals. A very large number of microbial agents fall in this category. These include bacteria, fungi, protozoan parasites, and viruses. See Tables 28.1–28.3 for summary presentations.

TABLE 28.1
Opportunistic bacteria

ORGANISM	MICROSCOPIC APPEARANCES	SOURCE	INFECTIONS IN COMPROMISED HOSTS
Acinetobacter calcoaceticus	Gram-negative diplococcus	Soil, water, skin	Meningitis, septicemia
Aeromonas hydrophila		Water	Wound infection, diarrhea, septicemia
Alcaligenes faecalis		GI tract	Benign septicemias
Anaerobic bacteria: *Bacteroides*	Gram-positive rod, some sporulating Gram-variable filamentous rod	UR, GI, GU tracts	Abdominal
Clostridium			Respiratory, gynecological infections, brain abscess
Fusobacterium			Septicemia
Actinomyces spp.			Actinomycosis
Bacillus subtilis, cereus, licheniformis	Gram-positive sporulating rod	Soil, water, skin	Eye, wound, systemic infection
Brucella spp.	Gram-negative rod	Infected animals	Brucellosis especially likely in Hodgkin's disease
Campylobacter fetus	Gram-negative curved rod	GI tract	Septicemia
Cardiobacterium hominis	Gram-negative rod	UR and GI tracts	Endocarditis
Corynebacterium spp.	Gram-negative rod	Skin, mucous membranes	Endocarditis, meningitis, bacteremia, osteomyelitis
		Soil	Pneumonitis, erythasma

continued

TABLE 28.1 continued

ORGANISM	MICROSCOPIC APPEARANCES	SOURCE	INFECTIONS IN COMPROMISED HOSTS
Eikenella corrodens	Gram-negative rod	UR and GI tracts	Abscesses
Flavobacterium meningosepticum	Gram-negative rod	Soil, water, disinfectant solutions Cooling ice	Meningitis and septicemia (newborns) Urinary tract infections,
Hemophilus spp.	Gram-negative rod	URT	Septicemia, pneumonia, meningitis, endocarditis
Lactobacillus spp.	Gram-negative rod	Mucous membranes	Endocarditis
Listeria monocytogenes	Gram-positive rod	Soil, vegetation, feces, vagina	Septicemia, meningitis
Moraxella spp.	Gram-negative rod	Normal mucosal flora	Endocarditis, chronic bronchitis, pneumonia, septicemia, meningitis
Mycobacterium spp.	Acid-fast rod	Human, soil	Chronic or acute lesions; lung, disseminated
Mycoplasma pneumoniae	Tiny, plemorphic	URT	Pneumonia (severe)
Neisseria spp.	Gram-negative diplococcus	URT	Meningitis, endo-carditis, septicemia
Nocardia spp.	Filamentous or mycelial		
Pasteurella multocida	Gram-negative rod	Soil	Pulmonary, CNS, skin
Propionibacterium acnes	Gram-positive rod	Dogs and cats	Septicemia and meningitis
Pseudomonas aeruginosa	Gram-negative rod	Skin	Endocarditis, septicemia, CNS, osteomyelitis
Pseudomonas cepacia	Gram-negative rod	Colonized human, moist items in hospital environment	Pneumonia, burn infection, septicemia, any other organ
Staphylococcus aureus	Gram-positive coccus	Same	Urinary and respiratory infec-tions, endocarditis
Staphylococcus epidermidis	Gram-positive coccus	Ubiquitous	Wound infection, septicemia, any other organ
Streptococcus spp.	Gram-positive coccus	Ubiquitous Human or animal UR or GI tract	Wound infection, septicemia, endocarditis See Table 24.2

TABLE 28.2
Opportunistic enterobacteriaceae

ORGANISM	INFECTIONS
Escherichia coli	Urinary tract infection, neonatal meningitis, peritonitis, bacteremia
Enterobacter cloacae and *aerogenes*	Lower respiratory and urinary tract infections
Enterobacter agglomerans	Septicemia associated with contaminated IV fluids
Klebsiella spp.	Urinary tract infection, pneumonia, septicemia
Serratia marcescens	Urinary tract infection, bacteremia, endocarditis
Proteus spp.	Urinary tract infection, rare pneumonia
Providencia spp.	UTI, pneumonia and septicemia
Citrobacter freundii	UTI, septicemia
Citrobacter diversus	Neonatal meningitis and septicemia, UTI
Shigella spp.	GI, more common and severe in compromised host, especially if gastric acidity is reduced
Salmonella spp.	GI, septicemia developing in compromised individual, meningitis in infants
Arizona hinshawii	GI, systemic disease in immunosuppression
Edwardsiella tarda	GI, systemic disease in compromised hosts
Yersinia enterocolitica	Enterocolitis progressing to septicemia in patients with underlying disease

Dominant forms The most significant microbial hazards vary among hospitals and from service to service within the hospital. At present, nationwide, the enterobacteriaciae are the most numerous and bothersome group. *Escherichia coli* is most frequent in a nursery, where it causes neonatal meningitis, and in other services as the leading cause of bladder infection. *Klebsiella pneumoniae* is frequently encountered in urinary tract infections; it is quickly transmitted among catheterized patients within a unit. It also causes a pneumonia leading to septicemia and death. *Serratia marcescens* is found in similar circumstances. The hospital strains of this organism, most of which lack the characteristic red pigment, readily acquire and exchange multiple-drug resistance plasmids.

Pseudomonas aeruginosa is the primary invader of burn wounds. However, this organism can also invade any wound or orifice and, in the absence of active phagocytosis, can infect and destroy any organ. Recurrent *P. aeruginosa* infections are common in cystic fibrosis and in any immunosuppressive condition.

TABLE 28.3
Opportunistic fungi, protozoans, and viruses

GROUP	ORGANISM	PREDISPOSING FACTOR	FORM OF DISEASE
Fungi	*Candida albicans, tropicalis, parapsilosis*	Prosthetic heart valve, immunosuppression allograft	Mucocutaneous candidiasis, dissemination, candidemia
	Aspergillus oryzae, fumigatus	Same conditions	Pulmonary infection, infection of prosthetic heart valves
	Histoplasma capsulatum	Immunosuppression may promote reactivated infection	Severe disseminated histoplasmosis
	Cryptococcus neoformans	Hodgkin's disease and other lymphomas	Severe disseminated cryptococcosis; CNS involvement
Protozoans	*Toxoplasma gondii*	Pregnancy, immuno-suppression, allograft	Asymptomatic in pregnant female; congenital malformation of fetus; mononucleosis-like infection of adult
	Pneumocystis carinii	Immunosuppression, allograft	Rapidly progressive pulmonary infiltration
Viruses	Herpes simplex	Immunosuppression, allograft	Generalized cutaneous eruption; pneumonia; encephalitis
	Varicella–zoster	Hodgkin's disease, other malignancies, immuno-suppression, allograft	Shingles, pneumonia, disseminated visceral disease, hemorrhage
	Cytomegalovirus	Immunosuppression, allograft, blood transfusion	Fever, pneumonitis, hepatitis
	Vaccine viruses— smallpox, measles, polio, others	Vaccination during immuno-suppression; malignancy	Severe generalized disease; symptoms like most severe form of infection with virulent virus in normal host

Surgical infections; Chapter 23, p. 684 and Box 23.2

Staphylococcus aureus, long viewed as the big nosocomial threat, has now been partly controlled and is therefore less common than those above. It remains important as a postsurgical complication. The general decline in prevalence is a result of specific infection control techniques, such as surveillance and disinfection, aimed at its elimination. Similar approaches have not as yet been successful with the Gram-negative rods.

Candida albicans and other *Candida* species (Fig. 28.2) cause generalized infection when cell-mediated immunity is subnormal, in patients with immunosuppressive malignancies or who are receiving immunosuppressive therapies. Fungal infections in general are responsible for about half of the deaths following renal transplant.

The DNA viruses, especially the herpes group, establish a latent state following the initial infection. In the immunologically incompetent person, the initial infection itself may be overwhelming. As a normal person becomes immunodeficient, reactivation may occur leading to severe, generalized infection. Vaccination with any live-virus vaccine, totally harmless to a normal person, may produce fatal illness in the compromised host.

Multiple-drug resistance   Plasmid resistance factors were discussed in Chapter 9. Plasmids from various sources come into close proximity *in vivo*, in the intestines of hospitalized patients. Genetic recombination continually produces new gene complexes. Under the selective influence of systemic antimicrobial drugs, plasmids containing an ever-increasing number of resistance factors provide a survival advantage for their bacterial hosts. In the patient's gut, plasmids may be exchanged among the various closely related enterobacterial species. These "superbugs" become permanent members of the floras of patients and staff.

Hospital strains   In nosocomial infections it is sometimes not the name of the organism, but its origin that matters. A strain of *Serratia marcescens*, isolated in the hospital and endemic among patients and staff, will have a totally different antibiogram from one isolated in the community. A hospital-acquired *Staphylococcus aureus* infection is automatically more serious than a community-acquired one because difficulties in treatment are predictable. An extended range of antibiotic resistance is expected in any hospital-acquired bacterial infection.

It must also be pointed out that the hospital patient is at risk not **only** from opportunists. His or her susceptibility to the true pathogens is equal to or greater than that of a normal person.

Colonization

When a patient enters the hospital, his or her life changes. At the least, there is a diet different from home and a shift in the timing and extent of daily activity. These

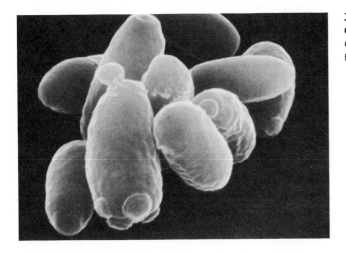

28.2
Candida   This is one of several opportunistic *Candida* species found in some nosocomial infections.

changes cause alterations in the normal flora. At the same time, the patient contacts the microorganisms indigenous to the hospital. The effect resembles what happens to the newborn baby. Within a brief period the skin, mucous membranes and gut become colonized with the hospital microorganisms. Antimicrobial drugs reduce the normal microbial competition and facilitate colonization by resistant organisms. Resistant strains enter into a close relationship with the host that can easily progress one step farther, if the defenses sag, to true disease.

A colonizing agent is one that does not belong in the normal flora, but does not inflict local damage **at the time** of colonization. The colonized state, not in itself harmful, usually precedes overt nosocomial infection.

THE COMPROMISED HOST

The list of compromising factors (Tables 28.4 and 28.5) is long. It grows longer with each new medical advance requiring extensive intervention.

TABLE 28.4
Factors predisposing to infection in the compromised host

GROUP	FACTOR
Circumvents external anatomical barriers	Burns, extensive dermatitis
	Catheters, intravenous and urinary
	Decubiti
	Impairment of the tracheobronchial mucociliary escalator by: cold smoke drugs (alcohol, anesthetics)
	Influenza (necrotizing tracheitis)
	Injections, diagnostic procedures
	Neurological disorders altering consciousness
	Surgery
	Trauma
Impairs granulocyte behavior and cellular immunity	Adrenal steroid therapy
	Alcoholism
	Antineoplastic drugs
	Anergic states, as in sarcoidosis
	Burns
	Collagen-vascular diseases
	Complement (C3 and C5) deficiencies (chemotaxis deficiencies)
	Congenital defects of cellular immunity

GROUP	FACTOR
	Congenital disorders of leukocyte or lysosomal function
	Congestive heart failure
	Diabetes mellitus, especially with acidosis
	Foreign bodies (valves, grafts, shunts, catheters)
	Hemoglobinopathies
	Immunologic deficiency syndromes
	Immunosuppressive drugs
	Ionizing radiation
	Leukemia
	Lymphoma, Hodgkin's disease, myeloma
	Malignancies, especially gastrointestinal or advanced
	Malnutrition
	Neutropenia (granulocytopenia <1,000/mm)
	Obstruction of bronchi, biliary tract, ureter
	Old age
	Organ transplants with use of immunosuppressive techniques
	Shock
	Splenectomy
	Surgical creation of dead spaces, hematomas, tissue necrosis
	Uremia, especially with acidosis
	Vascular insufficiency, gangrene
Impairs immunoglobulin defenses	Cirrhosis
	Complement defects
	Congenital and acquired immunoglobulin deficiencies
	Dysproteinemias
	Splenectomy
Affects microbial load or dose	Antimicrobial drug-induced alterations in flora
	Changes in normal flora with serious disease per se
	Disseminating carriers
	Environmental reservoirs contaminated equipment contaminated medications
	Food and water
	Improper isolation procedures
	Nursing and housekeeping practices

TABLE 28.5
Mechanisms by which therapeutic measures predispose to infection

THERAPY	EFFECT
Antimicrobial drugs	Alter normal flora and select resistant types
	May encourage growth of some fungi
	May alter immunoglobulins in sera
Corticosteroids	Depress antibody formation
	Depress reticuloendothelial cellular function
	Promote the diabetic state
	Suppress granulocyte responses
	Suppress interferon production
	Suppress lymphocytes and mononuclear cells needed for cellular immunity reactions
Ionizing radiation and antimetabolites and antineoplastic drugs	Depress antibody formation
	Depress bone marrow production of leukocytes
	Depress granulocyte function
	Injure tissues, producing portals of entry and other structural derangements hindering drainage and defense mechanisms
	Suppress the reticuloendothelial system
	Produce changes similar to ionizing radiation above
Surgical procedures and insertion of foreign material	Provide portal of entry to highly susceptible tissues
	Provide a mechanism for persistence of organisms

Some Indicators of Susceptibility

Natural defenses; Chapter 12, pp. 356–369

It is the responsibility of all persons having contact with the ill individual to be aware of the patient's increased susceptibility and to take steps to protect him or her. This is as true of family and visitors as of staff.

The health-care professional approaching a new patient can form a fairly accurate estimate of the patient's degree of susceptibility by observing certain key indicators. The first four can be readily noted at the bedside.

Age of patient   Persons who are very young or very old are always most susceptible to infection because their immunological defenses are not fully functional. Neonates, especially if premature or suffering from birth trauma, and the very aged, are automatically to be considered at risk.

Anatomical damage   Whenever anatomical integrity is disrupted, an abnormal portal of entry is available to microbial invaders. If the patient has wounds, burns, skin ulceration, surgical incisions, intravenous cut-downs, or a tracheostomy, special infection control consideration is needed.

Intubation   At any point at which a tube enters the body, there will be a disturbance of the normal surface defenses. If the patient has urinary, intravenous, arterial, or umbilical catheters or a tracheostomy tube, these add to the risk of infection. The risk increases the longer the tube remains in place.

Malnutrition   An extremely thin patient with severe wasting of tissue has been inadequately nourished for a prolonged period. This state invariably indicates that there will be inadequate antimicrobial defenses.

Two other prime indicators of susceptibility, leukopenia and metabolic imbalance, can be readily spotted on examination of the patient's laboratory reports.

Leukopenia   Reduction in the number of circulating white blood cells indicates inadequate blood cell formation in the bone marrow. In its extreme form, **agranulocytosis,** there is a severe reduction in polymorphonuclear cells (PMNs). When these fall below 500/cc, the nosocomial infection risk becomes very great.

Box 28.2

Hospitals Aren't So Dangerous If You're Really Healthy

Beginning in 1944, an Englishman named McIlroy began a long career as an imaginary invalid, one in whom a remarkable ability to magnify or fake symptoms was linked to the pathological compulsion to be hospitalized. On various occasions, he was treated for acute respiratory failure, a collapsed lung, a variety of neurological disorders, severe abdominal pain, and urine retention. He had 207 documented hospital admissions, under many fictitious names, and the British National Health Service spent several million pounds on his "ailments."

McIlroy underwent surgical exploration of the skull, abdominal and orthopedic operations, hundreds of X-rays and blood tests, at least 48 spinal taps, three encephalograms, numerous myelograms, and innumerable intravenous and urinary catheterizations. He would have seemed to have been the world's fittest candidate for death from nosocomial infection. However, he walked out of his last hospital in 1978. At last report, he was residing peacefully in an old-age home.

Source: Excerpted from *Scientific American* **241**: 80, 1979, p. 80.

Metabolic imbalance Uncontrolled metabolic imbalances show up in the blood chemistry. Elevated blood sugar, high blood urinary nitrogen, and acidosis indicate physiological malfunction that is usually associated with deficient phagocytosis, complement activity, or CMI. When the underlying disease state is successfully brought into control, the threat of infection diminishes.

The physician's admitting diagnosis, the patient's history, and the orders for therapy will provide a complete picture of the degree of compromise of the patient. This of course changes as the patient's condition improves or deteriorates. Good aseptic techniques should be followed as a matter of course with all patients. Special attention should be paid when caring for a patient identified as highly compromised—a **high-risk patient.** These persons also require careful observation in order that developing infection can be detected at the earliest possible time.

Special Units and Their Risks

Because of the nature of hospital services, patients with high risks are often grouped. This has the advantage that specialized care can be conveniently rendered, along with the appropriate infection control measures. Its disadvantage is that it may set the stage for epidemic spread of nosocomial infections. Clusters of compromised patients occur in the newborn nursery, intensive care units, hemodialysis unit, surgical suite and recovery areas, burn unit, cancer chemotherapy units, and other special-care areas.

Hazardous Procedures

Certain diagnostic and therapeutic procedures add significantly to the patient's infection risk. Anesthesia, by altering the respiratory reflexes, may contribute to pneumonia. Inhalation therapy poses a hazard because it is very difficult to keep the equipment from becoming colonized with bacteria that may then be blown forcefully into the lungs during treatment. Tracheostomy poses a double threat—an abnormal airway for microbial entrance and a surgical wound at the skin surface.

The urinary catheter or Foley is the most clearly documented of risks. About 60 percent of all hospital-acquired infections involve the urinary tract; most follow urinary catheterization. The tips of intravenous catheters, which enter the vein through a skin puncture, readily become colonized with skin flora. If left in place for more than 48 hours, the risk of overt local or systemic infection increases sharply and progressively.

The infection risk may be amplified if a nutrient fluid is being infused. Dextrose-saline solutions, or complete fluid diets (total parenteral nutrition or TPN fluids) are perfectly adequate microbial media. They may serve as an infection source if they were contaminated during preparation. Also, microorganisms can migrate from the catheter tip up the tubing to the reservoir and there multiply to a high population density. The addition of special drugs to infusion fluids poses another risk. This should be done under strict aseptic precautions in a centralized location such as the pharmacy.

Surgical dressings pose an infection hazard to both patient and others. Dressing management protocols are designed to give mutual protection. These are much more stringent when the wound is known to be infected.

Immunosuppressive therapy is widely employed. It can relieve inflammation, reduce pulmonary edema, permit the acceptance of allografts, or treat malignancy. However, a side effect is always an increased infection risk. Recent or current use of steroids (such as prednisone) cytotoxic cancer drugs, antilymphocyte serum, or radiation therapy is a warning flag of patient compromise.

HOSPITAL-ACQUIRED INFECTIONS

Certain types of infection, such as URT or GI infection, develop with roughly equivalent frequency inside and outside the hospital. Others, such as urinary tract infections, are more prevalent in the hospital. Postsurgical and intravenous site infections are almost always hospital-acquired.

Types of Infection

In 1975–1976, **urinary tract infections** (UTI) were the most common nosocomial infections. They constituted 53 percent of infections on medical services, 38 percent on surgical services, and 17 percent on pediatric services. Acquisition of a UTI is strongly correlated with catheterization. Drug-resistant bacteria produce a stubborn infection that may eventually lead to ascending kidney infection or septicemia.

Upper respiratory infections (URI) are common on pediatric services, paralleling the frequent occurrence of these conditions in well children.

Lower respiratory infections (LRI) are common in both medical and surgical patients, especially if they are immobile. Anesthesia or prolonged unconsciousness create special risks. In the debilitated, LRIs are more likely to progress to septicemia.

Gastroenteritis is most common in the newborn nursery or pediatric ward. Enteropathogenic *Escherichia coli* is a prominent causative agent. Unsatisfactory food sanitation practices in the diet kitchen can produce large institutionwide outbreaks of *Salmonella* and other common gastrointestinal pathogens as well.

Skin infections are the outstanding problem in the nursery, where *Staphylococcus aureus* may be endemic. Generalized herpetic infections appear most often in units that specialize in tissue transplantation.

Surgical wound infections are significant because they are the second most numerous nosocomial infection affecting about 125 persons per 10,000 hospital discharges, and because they are potentially preventable.

All severe **burns** become colonized. The risk of infection depends on the size and depth of the burn and the difficulties experienced in covering the wound.

TABLE 28.6
Causative agents most commonly identified in various types of nosocomial infection

SITE OF INFECTION	AGENT	ISOLATES/1,000 PATIENTS DISCHARGED
Urinary tract	*Escherichia coli*	55.0
	Group D streptococcus	24.2
	Proteus-Providencia spp.	19.3
Surgical wound	*Staphylococcus aureus*	20.4
	Escherichia coli	19.6
	Group D streptococcus	13.6
Lower respiratory tract*	*Klebsiella* spp.	7.9
	Staphylococcus aureus	7.4
	Escherichia coli	5.5
Cutaneous	*Staphylococcus aureus*	8.8
	Escherichia coli	2.2
	Group D streptococcus	1.8

*The number of cases in which there was no culture was 117, illustrating the point that LRI pathogens may be very difficult to recover.
Source: Data from the National Nosocomial Infections Study Report, 1975–1976.

Uncommon but potentially life-threatening developments may be caused by **bacteremia,** which may be **primary** (microorganisms not previously infecting the patient) or **secondary** (infection derived from a concurrent infection at another site). The secondary form often results from a primary nosocomial infection, such as a URI or LRI.

Epidemiology of Hospital-acquired Infections

In studying the transmission of infectious disease, we have learned that we must first identify the source or reservoir of the agent and then the means by which it is transmitted to the new victim. The same rules apply to nosocomial infection.

Reservoirs; Chapter 15, pp. 441–450 **Reservoirs and sources**   In a modern hospital the number of potential sources and reservoirs of infection is somewhat restricted as the patient usually has no contact with soil, animals, or insects. Three basic categories remain. **People** are, as always, the most significant reservoir. Carriers among the staff exchange microorganisms with colonized patients; personnel carry the organisms from one bed to the next. There is no doubt that direct person-to-person contact is the main route of spread. Of course, patient care is impossible without continuous close contact. Therefore, interrupting the cycle of infection is entirely dependent on personal cleanliness and aseptic technique. All authorities agree that frequent and conscientious hand-washing is the most important single step in breaking the network of infection.

Fomites—inanimate and nonbiological objects in the patient's environment —may be equally important as reservoirs and sources in causing infection. Non-sterile dressing materials, contaminated linens, and unclean equipment may all cause problems. It is usually the occasional slip-up in an otherwise satisfactory service that causes trouble. This underlines the substantial role of human error in hospital infection.

Foods and **medications** also occasionally become contaminated and cause outbreaks among patients. Furthermore, foods that have been contaminated by the infective patient (for example, in hepatitis where saliva is infective) may be a hazard to the kitchen workers. Contaminated utensils may remain a hazard if they are not adequately disinfected during dishwashing.

A few of the many nosocomial factors that can spread infection are presented in Table 28.7. This table is far from complete; it is intended solely to alert you to the almost unending range of possibilities.

Patterns of transmission   Three basic patterns can be identified. **Patient–patient** transmission might be by droplets, as when an undiagnosed, active tuberculosis case infects a roommate. More usually there is an intermediate, such as a piece of shared equipment, and the route is indirect. In nursery staphylococcal outbreaks, the agent spreads rapidly among the infants, usually via nurses' hands. Direct **patient–staff** transmission has been observed. Recently, an attending physician in an obstetrical clinic contracted rubella as a result of exposure to a patient. Indirect patient–staff transmission occurred when a nurse contracted hepatitis from patient serum.

Staff–patient transmission may be direct. The physician mentioned above inadvertently exposed about 200 pregnant patients to rubella during the period when his disease was incubating and unsuspected. Indirect staff–patient transfer occurred when an intestinal carrier of *Salmonella* infected food supplies that were later served to a number of patients.

PROGRAMS FOR CONTROL OF HOSPITAL INFECTION

Conceptual Approach

No method of control is likely to completely eradicate nosocomial infections in the immediate future. However, by improving the standard of care from that of "general practice" to that of "best practice," a significant reduction may occur. There are two conceptual approaches that should be implemented simultaneously.

The first approach is to reduce the number of organisms to which the patients are exposed. The means to this end include greater overall cleanliness, removal or education of staff carriers, and more sophisticated use of asepsis and isolation.

The other basic approach is to maintain and improve host resistance, or at least to refrain from actions (unless the actions are clearly justified) that lower resistance.

TABLE 28.7
Some examples of sources of infection in the hospital

TYPE OF SOURCE	EXAMPLES	AGENT	CASE(S) DOCUMENTED
Personnel	Surgeon's hair	*Staphylococcus aureus*	Yes
	Nurse's hands	*Citrobacter diversus*	Yes
	Obstetrics clinic staff	Rubella virus	No
	Surgical assistant	*Streptococcus pyogenes*	Yes
Medications	Steroid cream	*Pseudomonas aeruginosa*	No
	TPN solution	*Candida albicans*	Yes
	Pancreatin	*Salmonella agona*	Yes
	Platelet preparation	*S. cholerae-suis*	Yes
	Corneal transplant	Rabies virus	Yes
	Cryoprecipitate	*Pseudomonas cepacia*	Yes
Equipment	Stethoscope	*Staphylococcus aureus*	No
	Oxygen humidifying bottles	*Pseudomonas aeruginosa*	No
	Endoscopes	*Salmonella typhimurium*	Yes
	Breast-milk pump	*P. aeruginosa*	Yes
	Shaving brush	*P. aeruginosa*	Yes
	Resuscitator	*P. aeruginosa*	Yes
	Anesthesia equipment	*P. aeruginosa*	Yes
	Elastoplast wound dressing	*Rhizopus oryzae*	Yes
Environment	Floor scrubbing machine	*P. aeruginosa*	No
	Carmine dye in food	*Salmonella cubana*	Yes
	Deionized distilled water	*P. aeruginosa*	Yes
	Hand lotion	*Klebsiella pneumoniae*	Yes
	Sink traps, nursery and surgery	*P. aeruginosa*	Yes
	Tomatoes	*P. aeruginosa*	No
	Chrysanthemum leaves	*P. aeruginosa*	No
	Faucet aerator	*P. aeruginosa*	No

Some actions that might be avoided are (1) prescribing immunosuppressive medication when an alternate approach could be tried; (2) using antimicrobial drugs indiscriminately; (3) catheterizing when it is not medically necessary but done simply for the convenience of the attendants; and (4) using invasive procedures when noninvasive ones are at hand.

These principles have been expanded into infection-control programs with structured administrative bases and networks of routine activity. Such programs reach into all phases of hospital operation.

Infection-control Programs

Rationale Correctable nosocomial problems often stem from lack of essential knowledge, lack of up-to-date techniques, or lack of consistent motivation among personnel. Collecting information, providing in-service education, increasing and maintaining employee motivation, and correcting the noncompliant staff person (often a delicate task) are called for. These processes work much more smoothly within an administrative framework. The specific responsibilities of each employee can be clearly defined and assigned; the individual employee's performance can be consistently monitored. Ongoing data collection provides an early warning of developing situations and objective information as to how well each service is doing.

Standards The Joint Committee on Accreditation of Hospitals requires at a minimum that each institution provide

- [ ] definitions of nosocomial infections, for uniform surveillance, reporting, and identification;

- [ ] a practical system for collection and review of infection records for patients and personnel;

- [ ] ongoing review of written aseptic, sanitation, and isolation procedures;

- [ ] written isolation policies, and a commitment to equal standards of care for the isolated patient;

- [ ] appropriate procedures for safe operation of service areas not directly involved with patient care, such as laundry and waste disposal;

- [ ] all necessary laboratory support, especially microbiological and serological;

- [ ] an employee health program;

- [ ] orientation and in-service education relative to infection control; and

- [ ] coordination with medical staff on antibiotic usage.

Structure Most of these responsibilities are assumed by the **Infection Control Committee,** which sees that the jobs get done. Committee representation includes hospital administration, internal medicine, microbiology (pathology), nursing, obstetrics, pediatrics, and surgery. Other departments that may be included are blood bank, central supply, dentistry, dietetics, employee health, housekeeping, house staff, outpatient services, and pharmacy.

Two persons are given specific duties. A **hospital epidemiologist,** usually either a doctor or the microbiology director, provides overall supervision and acts as a liaison with the medical staff. An **infection control nurse** or infection control clinician does the day-to-day work of surveillance and in-service education.

Activities   **Case-finding** surveillance ensures that all nosocomial infections are discovered, treated, and analyzed. The infection control nurse may make daily visits to each nursing unit and to the microbiology laboratory. By reviewing patient charts and laboratory reports, all infection incidents are discovered. Fever is regarded as the primary clinical indicator of an infectious process. Positive microbiological cultures confirm this. Daily and monthly summaries of nosocomial infections are prepared. Surveillance of employee health and follow-up of discharged patients may also be necessary.

Routine microbiological screening of personnel to detect carriers and routine enumeration of microorganisms from environmental samples are no longer recommended. However, when a problem appears, microbiological sampling is often required to establish the source. The RODAC plate, in which the agar projects above the rim, is widely used for environmental samples. Air and water filtration methods are also generally adopted. Periodic sampling may also be highly educational if the results are used to remind personnel of the easily forgotten microbial flora.

The employee health service contributes to infection control in several ways. It provides preemployment physical examination to detect, among other things, communicable disease such as tuberculosis. It provides vaccination, if needed, for diphtheria, tetanus, measles, mumps, polio, and rubella. Vaccinations against influenza, smallpox, or tuberculosis are given only in high-risk situations. Pregnant employees should have serological proof of rubella immunity; they should not be assigned to hemodialysis units or other blood-handling positions, to avoid exposure to cytomegalovirus and hepatitis virus. When an employee comes to work ill, he or she may need to be transferred to tasks without patient contact or sent home, following treatment if indicated. These decisions should not result in loss of pay if employee cooperation is desired.

Hospital visitors may be a special problem, as they frequently have little understanding of infectious disease or its prevention. If the patient has an infectious condition, visitors should be required to check in at the nursing station to receive instructions before entering the room. In certain types of isolation, visiting is restricted or prohibited. Visiting may also be prohibited during a community epidemic. Everyone should be aware that it is a dubious favor to visit a friend or family member if they bring an infectious disease with them along with the flowers!

Educational programs take place at several levels. The new employee receives orientation that should include a discussion of his or her role in the control of infection. Procedure manuals should include details of infection control procedures. Review material or new information is presented in in-service programs. Supervisors educate by informal discussion of failures in technique, and teach colleagues by good example.

Antimicrobial drugs are still widely overprescribed, and inappropriate use of broad-spectrum drugs instead of narrow-spectrum drugs is common. One estimate is that between 40–50 percent of all antibiotic prescriptions are of questionable value. The main problem is the unjustified use of "prophylactic" antibiotics. These abuses can be discovered with systematic auditing of antibiotics dispensed. If alterations in use patterns seem indicated, these can be referred to the individual doctor or medical staff at large.

Outcome   The combined effect of the various components of these programs has been rewarding. In hospitals in which the infection control staff enjoys the full cooperation of all services, the attack rate for nosocomial infection has been rapidly lowered from the 5 percent average to below 2 percent. If such achievements could become general, this would mean a reduction of several millions of cases per year.

SUMMARY

1. Hospital-acquired infections occur because the hospital houses and groups together patients whose antimicrobial defenses are less than adequate. Ill health, in whatever form, increases the susceptibility to infection.

2. The hospital environment has a high population density, extensive use of common facilities, and a large amount of human traffic—visitors, shift changes of staff, and support workers.

3. The widespread use of antimicrobial drugs in the clinical settings has led to the rapid selection of multiply-resistant hospital microflora. These hospital strains colonize the hospital staff, and transfer to patients within a few hours of their admission.

4. Opportunistic microorganisms cause disease in the compromised host. Many of these agents are not exclusively restricted to a life within host tissues. They survive indefinitely in or on equipment and supplies in the hospital.

5. A helpful approach to the issue of host susceptibility is to evaluate the degree of compromise. Such analysis may be based on the number of compromising factors present and their severity. This allows the identification of patients for whom extra precautions are warranted. Compromising factors are additive in effect.

6. All accredited hospitals in the United States follow infection-control guidelines established by the Joint Committee on Accreditation of Hospitals. The guidelines require the formation of an interdisciplinary infection-control committee and the appointment of two or more infection-control officers.

7. A successful infection-control program will include active case-finding, preparation of daily and monthly reports on observed nosocomial infections, surveillance of employee health, orientation and in-service education in the control of infectious disease, the preparation and up-dating of procedures manuals for all areas of the hospital, and antimicrobial drug audits.

8. When all constituencies within the hospital work together, the incidence of nosocomial infections can be markedly reduced.

Study Topics

1. Why are nosocomial infections particularly difficult to prevent?

2. List aspects of clinical therapy that are directly linked to increased infection risk. Pair them with suggestions to minimize their effect.

3. What might be the role of each of the following staff persons in spreading nosocomial infection—dietician, laundry worker, laboratory technician, physical therapist?

4. In case-finding, what might an infection-control nurse look for as clues to infection?

5. Discuss the contributions of antimicrobial drug therapy to **increasing** nosocomial infections; to **controlling** them.

Bibliography

American Hospital Association, 1979, *Infection Control in the Hospital* (4th ed.), Chicago: American Hospital Association.

Bennett, John V., and Philip S. Brachman, 1979, *Hospital Infections*, Boston: Little, Brown.

Burke, John F., and Gavin Y. Heldrich-Smith, 1978, *The Infection-prone Hospital Patient*, Boston: Little, Brown.

Center for Disease Control, 1978, Infection surveillance and control programs in United States Hospitals: an assessment, 1976, *Morbidity and Mortality Weekly Report* **27:** 139–145.

Center for Disease Control, 1975, *Isolation Techniques for Use in Hospitals* (2nd ed.), Washington, D.C.: United States Government Printing Office.

Center for Disease Control, 1977, National Nosocomial Infections Study—United States, 1975–1976. *Morbidity and Mortality Weekly Report* **26:** 377–383.

Dixon, Richard E., 1978, Effect of infections on hospital care, *Ann. Int. Medicine* **89:** 749–753.

Joint Commission on Accreditation of Hospitals, 1975, *Infection-control Standards Adopted by the Board of Commissioners*, JCAH, Washington, D.C.

Veterans Administration Ad Hoc Interdisciplinary Advisory Committee on Antimicrobial Drug Abuse, 1977, Audits of Antimicrobial Usage 1–18, *JAMA* **237,** serialized in issues of March 7 through May 2 inclusive.

Von Graevenitz, Alexander, 1977, The role of opportunistic bacteria in human disease, *Ann. Rev. Microbiol.* **31:** 447–472.

Glossary

In every scientific textbook, there are terms that readers will find unfamiliar and difficult to pronounce. This glossary supplies a pronunciation guide (in parentheses) after difficult terms. The syllables printed in CAPITAL LETTERS should be given greatest emphasis as the words are pronounced. The key below is divided into four columns. Column one lists letters and diacritical marks along with examples of simple words showing how each is pronounced. Column two shows how the sounds are indicated in this glossary. Columns three and four list examples of words that contain the sound and their pronunciation guides.

Letter and diacritical mark		Appears in glossary as	Example	Guide
ă	(mat)	a	mastoid	MAST oyd
ā	(may)	ay	brain	BRAYN
âr	(fare)	ai	nares	NAIR eez
ä	(father)	ah	common	CAHM uhn
ch	(churn)	ch	charge	CHAHRJ
ĕ	(met)	eh	petal	PEHT uhl
ē	(meet)	ee	amebic	uh MEE bihk
g	(gang)	g	gamete	GAM eet

Letter and diacritical mark		Appears in glossary as	Example	Guide
ĭ	(sit)	ih	gingira	JIHN juh vuh
ī	(mite)	y	microbe	MY krohb
j	(jug)	j	germ	JURM
k	(kit, chorus)	k	chitin	(KYT ihn)
ŏ	(hot)	ah	operon	(AHP uh rohn)
ō	(flow)	oh	oocyst	(OH uh syt)
ô	(cough)	aw	autoclave	(AW toh klayv)
oi	(boil)	oy	oil	(OYL)
ŏŏ	(book)	uh	medulla	(muh DUHL uh)
ōō	(root)	oo	croup	(KROOP)
ou	(ouch)	ow	Cowper's gland	(KOW purz GLAND)
s	(seem)	s	vaccine	(vak SEEN)
sh	(shin, fish)	sh	addition	(uh DIHSH uhn)
ŭ	(but)	uh	pertussis	(pur TUHS uhs)
ûr	(urge)	er	allergy	(AL ur jee)
z	(zebra)	z	zygote	(ZY goht)
zh	(measure)	zh	seroconversion	(sihr oh kuhn VUR zhuhn)
a	(around)	uh	undulant	(UHN juh luhnt)

ABO blood group   The classical blood group system defined by the agglutination reactions of erythrocytes to the natural isoantibodies anti-A and anti-B and related antisera (Landsteiner, 1900).

A site   The site upon the ribosome where tRNAs bearing single amino acids attach.

acetylcoenzyme A (uh SEET uhl coh EHN zym AY)   Acetyl-CoA; condensation product of coenzyme A and acetic acid; symbolized as $CoAS\text{-}COCH_3$; intermediate in transfer of two-carbon fragment, notably in its entrance into the tricarboxylic acid cycle.

acid   Sour; having properties opposed to those of the alkalis. A substance that dissociates in water giving rise to hydrogen ions or protons.

acid decalcification (AS ihd dee kal sih fih KAY shuhn)   The process whereby calcium salts are removed from the tooth structure in the presence of the acids.

acid hydrolase   A hydrolytic enzyme that functions most rapidly at acidic pH values.

acidic amino acid   An organic acid in which one of the CH hydrogen atoms has been replaced by COOH.

acidogenic flora (uh sihd oh JEHN ihk FLAWR uh)   Microbial bacteria that process large amounts of organic acids from metabolic activity.

acquired pellicle (PEHL ihk uhl)   A thin skin or film, such as a thin film on the surface of a liquid, produced by influences originating outside the organism, or the thin film forming on newly cleaned teeth.

acquired resistance   Resistance to infection acquired as a result of natural disease or immunization.

actinomycosis (ak tuh noh my KOH suhs)   An infectious disease caused by *Actinomyces israelii* in human beings and by *A. bovis* in cattle.

activated macrophage (AK tuh vay tuhd MAK roh fayj)   A macrophage that, as a result of contact with a T-lymphocyte, has developed increased phagocytic and digestive capacity.

activated sludge   Sludge from well-aerated sewage, that, being well supplied with oxidizing bacteria, ensures the presence of sufficient oxidizing organisms to activate the next tankful of sewage.

activation energy   The minimum amount of energy that must be supplied from an outside source before a chemical reaction will proceed.

active immunity   Acquired immunity attributable to the presence of antibody or of immune lymphoid cells formed in response to antigenic stimulus.

active transport   The movement of materials across cell membranes and epithelial layers resulting directly from the expenditure of metabolic energy.

acute bacterial endocarditis (uh KYOOT bak TEER ee uhl ehn doh cahr DYT uhs)   A febrile systemic disease marked by focal bacterial or fungal (mycotic) infection of the heart valves, with formation of bacteria- or fungi-laden vegetations. The **acute** form, characterized by abrupt onset and a rapidly progressive course, is due to virulent organisms— such as *Staphylococcus aureus*, Gram-negative bacteria, and fungi—capable of invading other tissues.

acute bacterial pharyngitis (uh KYOOT bak TEER ee uhl fahr uhn JYT uhs)   Inflammation with pain in the throat, especially on swallowing, dryness, followed by moisture of the pharynx, congestion of the mucous membrane, and fever.

acute necrotizing ulcerative gingivitis (ANUG) (uh KYOOT NEHK ruh tyz ihng UHL sur uh tihv jihn juh VYT uhs)   Trench mouth; an acute or chronic gingival infection characterized by redness and swelling, by necrosis extending from the interdental papillae along the gingival margins, and by pain, hemorrhage, a necrotic odor, and often a pseudomembrane.

acute rejection   An immune reaction against grafted tissue. In acute rejection, the response occurs after the sixth day and then proceeds rapidly. It is characterized by loss of function of the transplanted organ and by pain and swelling, with leukocytosis and thrombocytopenia.

acute respiratory disease (ARD)   An acute, self-limiting infectious disease frequently caused by adenoviruses.

acute rhinitis (rih NIT uhs)   Coryza, or cold in the head; an acute congestion of the mucous membrane of the nose, marked by dryness, followed by increased mucous secretion from the membrane, impeded respiration through the nose, and some pain.

addition mutation   A point mutation in which the change is the insertion of one additional base pair into the normal DNA sequence.

adenoid   Hypertrophy of the adenoid tissue that normally exists in the nasopharynx of children and is known as the pharyngeal tonsil.

adenosine triphosphate (ATP) (uh DEHN uh seen try FAWS fayt)   A nucleotide compound occurring in all cells, where it represents energy storage.

adenovirus (ad uhn oh VY rahs)   One of a group of viruses found in all parts of the world, causing disease of the upper respiratory tract and conjunctiva, and also present in latent infections in normal persons.

adenyl cyclase (AD uhn ihl SYK layz)   An enzyme found in the liver and muscle cell membranes that catalyzes the conversion of adenosine triphosphate (ATP) to cyclic adenosine monophosphate (AMP) plus pyrophosphate, and is activated by many hormones.

A disc   A filter paper disc impregnated with the antimicrobial drug bacitracin. Used for the identification of strains of Group A streptococci.

adjuvant (AJ uh vuhnt)   A substance that can be added to a vaccine to slow down its absorption and increase its effectiveness.

aerobic (Ai(uh)r ROH bihk)   Having molecular oxygen present.

aerobic respiration   The oxidative transformation of certain substrates into secretory products, the released energy being used in the process of assimilation.

aerotolerant   Of an anaerobe, not being inhibited by O_2.

afferent signal (AF ur uhnt SIHG nuhl)   A signal that is moving toward the cell body from the point of origin.

aflatoxin (af luh TAHK suhn)   A toxic factor, $C_{17}H_{12}O_6$, produced by *Aspergillus flavus* and *A. parasiticus*, molds contaminating groundnut seedlings. It causes aflatoxicosis, which is responsible for the deaths of fowl and other farm animals fed with infected groundnut meal. Experimentally, it is regularly able to produce hepatomas in ducklings and rats.

African Burkitt's Lymphoma (ABL)   Any form of undifferentiated malignant lymphoma, usually found in Central Africa, but also reported from other areas, and manifested most often as a large osteolytic lesion in the jaw or as an abdominal mass.

agammaglobulinemia (ay gam uh glahb yuh luh NEE mee uh)   Hypogamma-globulinemia; antibody deficiency disease or syndrome; a condition characterized by extremely low levels of γ-globulin in the blood and frequent occurrence of sup-purative bacterial infections.

agar-overlay method   A method for determining antimicrobial susceptibilities by incorporating the organism to be tested into agar and pouring a layer of this inoculated agar over the discs.

agglutination (uh gloot uhn AY shuhn)   The combination of antigen and antibody to form relatively large clumps or flakes visible to the naked eye.

agranulocyte (ay GRAN yuh loh syt)   A nongranular leukocyte.

agranulocytosis   A symptom complex characterized by marked decrease in the number of granulocytes and by lesions of the throat and other mucous membranes, of the gastrointestinal tract, and of the skin.

alcohol   Chemical name: ethanol. A transparent, colorless, mobile, volatile liquid, C_2H_5OH, miscible with water, ether, and chloroform, obtained by fermentation of carbohydrates with yeast. An organic compound characterized by one or more OH groups.

aldehyde (AL duh hyd)   Any one of a large class of substances, derived from the primary alcohols by oxidation, and containing the group CHO.

algal blooms   Overabundant growth of algae, often as a result of enrichment with a nutrient that would otherwise be scarce enough to limit growth.

alimentary tract (or canal)   The passage in the body through which the food passes and in which it is digested, extending from the mouth through the esophagus, stomach, and intestines to the anus.

alkali   Any of a class of compounds which act as proton acceptors; bases.

alkalosis (al kuh LOH sihs)   A pathogenic condition resulting from ac-cumulation of base in, or loss of acid from, the body.

alkylation (al kuh LAY shuhn)   The substitution of an alkyl group for an active hydrogen atom in an organic com-pound.

allele (uh LEE(UH)L)   One of two or more alternative forms of a gene occupy-ing corresponding sites (loci) on homolo-gous chromosomes, any two of which may be carried by a given individual and that determine alternative characters in inheritance.

allergy (AL ur jee)   A hypersensitive state acquired through exposure to a par-ticular allergen, reexposure eliciting an altered capacity to react.

allograft (AL oh graft)   A graft of tissue between individuals of the same species but of disparate genotype.

allosteric change (al oh STEHR ihk)   Denoting a macromolecule (an enzyme) whose reactivity with another molecule is altered by combination with a third mole-cule.

allosteric interaction   Refers to an enzyme whose activity is inhibited by a substance, usually a normal biochemical, when this substance combines at a site on the enzyme other than the active site.

alpha feto-protein (AL fuh FEE toh-PROH teen)   A fetal antigen that also occurs in adults in certain diseases; a_1-fetoprotein appears in the serum of patients with hepatoma and embryonal adeno-car-cinoma.

alpha helix (AL fuh HEE lihks)   The helical secondary structure of many pro-teins in which each NH group is con-nected to a CO group by a hydrogen bond at a distance equivalent to 3 amino acid residues; the helix makes a complete turn for each 3.6 residues.

alpha-ketoglutaric acid (AL fuh-kee toh gloo TAIR ihk)   A five-carbon inter-mediate in the Krebs cycle; may be con-verted to glutamic acid.

alphatoxin (AL fuh tawks ihn)   A membrane-active toxin produced by some strains of *Staphylococcus aureus*.

alphavirus (al fuh VY ruhs)   An RNA virus transmitted by insect vectors; some varieties cause encephalitis in human beings.

alternate pathway   A pathway for the activation of the complement cascade that does not require initial antigen–antibody combination.

alum precipitation   A procedure used in the preparation of toxoids for vaccine use.

alveolar bone (al VEE uh lur)   The thin layer of bone making up the bony pro-cesses of the maxilla and mandible, and

surrounding and containing the teeth; it is pierced by many small openings through which blood vessels, lymphatics, and nerve fibers pass.

alveolar macrophage   Alveolar phagocyte, found in the alveoli of the lungs.

Amantidine (uh MAN tah deen)   A drug used in the treatment and prevention of influenza.

amastigote (uh MAS tuh gaht)   The nonflagellate, intracellular, morphologic stage in the development of certain hemo-flagellates, resembling the type adult form of *Leishmania*.

amebic dysentery (uh MEE bihk DIHS uhn tehr ee)   A form of dysentery caused by *Entamoeba histolytica* and known also as intestinal amebiasis and amebic colitis.

American trypanosomiasis (try puh noh soh MY uh sihs)   Chagas' disease.

American Type Culture Collection (ATCC)   An association that maintains, collects, and sells guaranteed pure cul-tures of microorganisms.

Ames test   A test used to determine the carcinogenic potential of chemical sub-stances. This test measures the degree to which the test chemical promotes muta-tion in a particular strain of *Salmonella*.

amino acid   Any one of a class of organic compounds containing the amino (NH_2) group and the carboxyl (COOH) group.

amino acid activation   Reaction of amino acids with ATP, converting them to the activated compound aminoacyl AMP.

aminoglycoside (uh mee noh GLY kuh syd)   Any of a group of bacterial antibi-otics (e.g., streptomycin and genta-mycin), derived from various species of *Streptomyces*, that interfere with the function of bacterial ribosomes.

amino group   The NH_2 grouping, char-acteristic of amino acids but also found in many other classes of organic compound. It functions as a proton acceptor, thus acts as a weak base.

amoeba (uh MEE buh)   Also spelled ameba. A minute protozoon of the sub-phylum *Sarcodina*. It is a single-celled nucleated mass of protoplasm that changes shape by extending cytoplasmic processes, called pseudopodia, by means of which it moves about and absorbs nourishment.

ammonification (uh moh nuh fih KAY shuhn)   The formation of ammonia by the action of bacteria on proteins.

ammonium ion   The grouping NH_4^+, a cation.

amphoteric (am foh TEHR ihk)   Having opposite characters; capable of acting either as an acid or as a base; combining with both acids and bases; affecting both red and blue litmus.

amphotericin B (am foh TEHR uh sihn BEE)   An antibiotic substance derived from strains of *Streptomyces nodosus*, occurring as a yellow to orange odorless powder, and used in treatment of meningitis caused by *Cryptococcus* and of systemic fungus infections.

anaerobiosis (an uh roh by OH suhs)   Life only in the absence of molecular oxygen.

anal canal   The terminal portion of the alimentary canal, extending from the rectum to its distal opening.

anamnestic response (an am NEHS tihk rih SPAHNS)   In immunology, the rapid reappearance of antibody in the blood following the administration of an antigen to which the subject had previously developed a primary immune response.

anaphylaxis (an uh fah LAK suhs)   An unusual or exaggerated allergic reaction of an organism to foreign protein or other substances.

animal reservoir   A wild or domestic animal that serves as a host for an infective agent, and from which the agent may periodically be transmitted to human beings.

anionic detergents (an y AHN ihk dih TUR juhnts)   Detergents, such as soaps (alkali metal salts of long-chain fatty acids) that carry a negative electrical charge on a lipid-like molecule and exert a limited antibacterial effect.

anoxygenic photosynthesis (an ahks ih JEHN ihk foh toh SIHN thuh suhs)   Photosynthesis that takes place in the absence of oxygen and during which oxygen is not produced.

antenna (an TEHN uh)   A cluster of light-absorbing pigment molecules in the photosystem of a photosynthetic organism.

anterior horn   The anterior portion of each gray column of the spinal cord; contains many cell bodies of motor neurons.

antibiogram (an tih BY oh gram)   A record of the resistance of microbes to various antibiotics.

antibiotic   A chemical substance produced by a microorganism that has the capacity, in dilute solutions, to inhibit the growth of or to kill other microorganisms.

antibody   An immunoglobulin molecule that has a specific amino acid sequence by virtue of which it interacts only with the antigen that incited its synthesis in lymphoid tissue, or with antigen closely related to it.

antibody-mediated immunity (AMI)   Immunity produced by activation of the B-lymphocyte population, leading to the production of immunoglobulins of several classes.

anticellular response   A type of cellular immunity in which the T-lymphocytes specifically attack and kill antigen-bearing cells.

anticodon   A triplet of nucleotides in transfer RNA that is complementary to the codon in messenger RNA that specifies the amino acid.

antigen   Any substance capable, under appropriate conditions, of inducing the formation of antibodies and of reacting specifically in some detectable manner with the antibodies so induced.

antigen depot   Tissue site in which antigen is injected, and from which it is slowly absorbed.

antigen determinant   The structural component of an antigen molecule that is responsible for specific interaction with antibody (immunoglobulin) molecules elicited by the same or a related antigen.

antigenic mass   The total amount of antigen encountered by the organism; the larger the antigenic mass, the stronger the response in most cases.

antilymphocyte globulin (an tee LIHM foh syt GLAH byoo lihn)   The gamma-globulin fraction of antilymphocyte serum, containing antibodies against lymphocyte surface antigens.

antimetabolite (an tee muh TAB oh lyt)   A substance bearing a close structural resemblance to one required for normal physiological functioning, and exerting its effect by interfering with the utilization of the essential metabolite.

antimicrobial audit   A study that may be carried out by pharmacists or other interested parties to determine extent and patterns of antimicrobial drug use and/or abuse.

antiseptic   (1) An approach to surgical procedures that seeks to destroy all microorganisms within the area. (2) A chemical substance that can be applied to skin or tissue to kill microorganisms.

antiserum (ANT ih sihr uhm)   A serum that contains antibody or antibodies; it may be obtained from an animal that has been immunized either by injection of antigen into the body or by infection with microorganisms containing the antigen.

antistreptolysin O (ASO) (an tee strehp toh LY suhn OH)   A circulating antibody often produced by rheumatic fever patients, specific against one of the streptococcal streptolyses.

antitoxin   Antibody to the toxin of a microorganism (usually the bacterial exotoxins), to a zootoxin (e.g., spider or bee venom), or to a phytotoxin (e.g., ricin of the castor bean), that combines specifically with the toxin, *in vivo* and *in vitro*, with neutralization of toxicity.

antivenin (AN tee vehn uhn)   A proteinaceous material used in the treatment of poisoning by animal venom.

anus (AY nuhs)   The distal or terminal orifice of the alimentary canal.

apical pore   The pore at the apex of the tooth root, through which the tooth receives its vascular and nervous connections.

aplastic anemia (ay PLAS tihk uh NEE mee uh)   A form of anemia generally unresponsive to specific antianemia therapy, often accompanied by granulocytopenia and thrombocytopenia, in which the bone marrow may not necessarily be acellular or hypoplastic but fails to produce adequate numbers of peripheral blood elements.

apoenzyme (ap oh EHN zym)   The protein component of an enzyme that is separable from the prosthetic group (or coenzyme) but that requires the presence of the prosthetic group to form the functioning compound (holoenzyme).

appendix   A general term used in anatomical nomenclature to designate a supplementary accessory, or dependent part attached to a main structure. A blind sac extending from the cecum.

arachnoid (uh RAHK noyd)   The serous membrane between the dura mater and the pia mater.

arbovirus (AHR boh vy ruhs)   Any group of viruses, including the causative agents of yellow fever, viral encephalitis, and certain febrile infections, which are transmitted to human beings by various mosquitoes and ticks.

arenavirus (uh REE nah vy ruhs)   Any of a group of morphologically similar, ether-sensitive viruses that seem to contain RNA.

armed macrophage   A macrophage that has undergone contact with a specifically stimulated T-cell, and has acquired reactivity toward a particular antigen.

artery   A vessel through which the blood passes away from the heart to the various parts of the body.

arthropod-borne disease   An infectious disease transmitted either in the saliva or the feces of an arthropod such as a mosquito or tick.

arthrospore (AHR throh spawr)   A type of spore formed in a close sequence in the hyphae of certain fungi, and leading to hyphae fragmentation.

artificial passive immunity   A form of immunity acquired as the result of transfer of **preformed protective substances,** artificially as in the administration of convalescent serum from a person who has recovered from the disease, or hyperimmune serum prepared by means of active immunization of an animal.

Arthus phenomenon   The development of an inflammatory lesion, classically an ulcer, marked by edema, hemorrhage, and injection of an antigen to which the animal already has precipitating antibody. It is generally considered an immediate hypersensitivity, or is classed as a type III reaction. Antigen–antibody complexes formed in the presence of complement adhere to the vascular epithelium and are encircled by fibrin, blood platelets, and polymorphonuclear neutrophils. Plugging of the vessels is followed by exudation into surrounding tissues of fluid laden with the neutrophils.

ascending infection   An infection that originates at one point in the body, then ascends to other tissues or organs, e.g. a bladder infection may ascend to the kidneys.

ascospore (AS koh spawr)   A sexual spore formed within a special sac, or ascus, as in ascomycetous fungi.

ascus (AS kuhs)   The sporangium or spore case of lichens and fungi, consisting of a single terminal cell.

asepsis (ay SEHP suhs)   Freedom from infection.

aseptic meningitis (ay SEPH tihk mehn ihn JY tihs)   Meningitis due to various viruses, such as the coxsackieviruses and the virus of lymphocytic choriomeningitis, characterized by malaise, fever, headache, nausea, cerebrospinal fluid pleocytosis (principally lymphocytic), abdominal pain, stiffness of the neck and back, and a short uncomplicated course. Called also acute aseptic m., benign lymphocytic m., and lymphocytic m.

assimilation (uh sihm uh LAY shuhn)   The transformation of food into living tissue.

asthma   A condition marked by recurrent attacks of paroxysmal dyspnea, with wheezing due to spasmodic contraction of the bronchi.

atomic mass unit (amu)   A unit mass equal to 1/12 the mass of the nuclide of carbon-12.

atomic number   The whole number representing the number of units of positive charge on the nucleus of an atom.

atomic weight   The weight of an atom of a substance as compared with the weight of an atom of carbon-12 isotope, which is taken as 12.00000.

atopy (AT uh pee)   A clinical hypersensitivity state, or allergy, with a hereditary predisposition; that is, the tendency to develop some form of allergy is inherited, but the specific form, e.g., hay fever, asthma, eczema, is not.

attenuated virus   One whose pathogenicity has been reduced by serial animal passage or by other means.

attenuation (uh tehn yoo AY shuhn)   The alteration of the virulence of a pathogenic microorganism by passage through another host species, decreasing the virulence of the organism for the native host and decreasing it for the new host.

atypical mycobacteria (ay TIHP ih kuhl my coh bak TEER ee uh)   Acid-fast bacteria resembling the tubercle bacilli, found in pulmonary infections, usually of a chronic nature, in human beings, for which species names have not been established. They are divided into the chromogens and nonchromogens, the former including photochromogens (Group I), which produce yellow pigment in the presence of light, and scotochromogens (Group II), which produce orange pigment independent of light. The nonchromogens are subdivided into filamen-

tous forms (Group III) and rapid growers (Group IV).

Australia antigen   An antigen found in the sera of patients with acute serum hepatitis and rarely in patients with infectious hepatitis.

autoantibody   An antibody (immunoglobulin) formed in response to, and reacting against, one of the individual's own normal antigenic endogenous body constituents.

autoclave (AW toh klayv)   An apparatus for effecting sterilization by steam under pressure.

autograft (AW toh graft)   A graft of tissue derived from another site in or on the body of the organism receiving it.

autoimmune disease   Any of a group of disorders in which tissue injury is associated with humoral or cell-mediated responses to body constituents.

autoimmune hemolytic anemia (aw toh ih MYOON hee moh LIH tihk uh NEE mee uh)   Acquired hemolytic anemia in which serum antibodies, usually of the IgG class, react with erythrocytes, and the Coombs test is positive. It occurs in several autoimmune diseases, including systemic lupus erythematosus. The antibodies may be of the cold or warm type. In the former case, the course is usually chronic and mild, with cyanosis, Raynaud's phenomenon, and hemoglobinuria. The warm type may be more severe, with icterus, splenomegaly, and macrocytic anemia.

autotrophic (aw toh TROH fihk)   Self-nourishing; said of organisms that can build their organic constituents from carbon dioxide and inorganic salts.

auxotroph (AWK suh trohf)   An organism lacking one or more genetic functions, and requiring in consequence that additional specific nutrients be supplied in the growth medium.

avidity   A term denoting the firmness with which an antigen is bound by an antibody sample.

axenic culture (ay ZEEN ihk)   A culture that is not contaminated by or associated with any foreign organisms.

axilla (ag ZIHL uh)   A small pyramidal space between the upper lateral part of the chest and the medial side of the arm, and including, in addition to the armpit, axillary vessels, the brachial plexus of nerves, a larger number of lymph nodes, and fat and loose alveolar tissue.

axon The portion of a neuron leading away from the cell body toward the point of connection with other neurons or effectors.

B-lymphocyte A lymphocyte that can proliferate into a clone of plasma cells, all of which will produce one specific antibody.

B memory cell A specifically stimulated B-lymphocyte, not actively multiplying, but able to commence multiplication and production of plasma cells on subsequent antigenic stimuli.

B particles Intracellular viral particles having a particular electron microscopic appearance; murine mammary tumor viruses are of this type.

bacillary dysentery A form of dysentery caused by bacteria of the bacillus form, usually *Shigella* species.

bacillus (buh SIHL uhs) In general, any rod-shaped bacterium.

bactericidal (bak TEER ih syd uhl) Destructive to bacteria, leading to their irreversible loss of viability.

bacteriocin (bak TEER ee oh sihn) Any of a group of substances, e.g., colicin, released by certain bacteria, that kill but do not lyse other strains of bacteria.

bacteriophage A virus with specific affinity for bacteria.

bacteriuria (bak teer ee (Y)OOR ee uh) The passage of bacteria in the urine.

baeocysts Reproductive cells formed by certain groups of cyanobacteria.

basal body Blepharoplast, structure at base of flagella and cilia in eucaryotic organisms.

base A substance that associates with protons or hydrogen ions, removing them from solution. When added to an aqueous solution, a base raises the pH value of the solution.

base analog A chemical whose structure is similar (analogous) to one of the purine or pyrimidine bases found in DNA or RNA.

base pair The complex of two heterocyclic bases, one a pyrimidine and the other a purine, brought about by hydrogen bonding between positions 1 and 6 of the purine and positions 3 and 4 of the pyrimidine.

basement membrane Membrane propria; basilemma; a thin layer, from 6μ thick in the adult trachea to less than 1μ in other regions, that intervenes between epithelium and connective tissue.

basic amino acid An amino acid bearing two or more amino groups; these act as weak basic groups.

basidium (buh SIHD ee uhm) The clublike organ of the fungal class *Basidiomycetes* which, following karyogamy and meiosis, bears the basidiospore.

basidiospore A spore formed on a basidium.

basophile (BAY zuh fyl) A structure, cell, or other histologic element staining readily with basic dyes.

BCG vaccine A preparation used as an active immunizing agent against tuberculosis, consisting of a dried, living culture of the Calmette-Guerin strain of *Mycobacterium tuberculosis* var. *bovis*. It is usually administered intracutaneously, but may also be given by multiple punctures with a special instrument, or by scarification through a suspension applied to the skin.

Bence-Jones protein (Behns Johnz PROH teen) A low-molecular weight, thermosensitive urinary protein, which is found almost exclusively in multiple myeloma and constitutes the light-chain component of myeloma globulin. Its outstanding characteristic is its unique property of coagulating on heating at 45°–55°C. and of redissolving partially or wholly on boiling.

benign tumor A swelling, one of the cardinal signs of inflammation; morbid enlargement that is not malignant; not recurrent; favorable for recovery.

benthos (BEHN thohs) The flora and fauna at the bottom of oceans.

β-lactamase The general name for a group of enzymes, such as penicillinase, that attack the β-lactam ring in the penicillin-class antimicrobials.

β-lactam ring The portion of the penicillin molecule that conveys its antimicrobial activity.

β-oxidation Oxidation of the β carbon (carbon 3) of fatty acid, forming the β-keto acid analogue; of importance in fatty acid catabolism.

β structure One of the higher levels of protein structure; the pleated-sheet arrangement of certain structural proteins such as collagen.

bile A fluid secreted by the liver and poured into the small intestine via the bile ducts.

bile solubility A characteristic of the pneumococci; when suspended in a solution of bile salts, they undergo lysis.

binomial system A system of naming organisms in which the "scientific" name consists of two parts (the first designates genus and the second, species); originated by Linnaeus and sometimes called the Linnaean System.

biological vector A vector, such as the mosquito for the malarial agents or the tsetse fly for the agents of African sleeping sickness, that is essential in the life cycle of the pathogenic organism.

biomass The entire assemblage of living organisms, both animal and vegetable, of a particular region, considered collectively.

biosynthesis The building up of a chemical compound in the physiologic processes of a living organism.

biphasic Having two phases or forms that may interconvert; certain fungi may exist in either the yeast or the mycelial growth tube.

black water In sanitation, a term applied to waste water heavily contaminated with human wastes.

blackwater fever A severe form of falciparium malaria in which hemoglobin excreted in the urine gives it a black appearance.

black plague Black death. Term applied to the worldwide epidemic of the 14th century, of which some 60 million persons are said to have died; the descriptions indicate that it was pneumonic plague.

blastogenesis (blas toh JEHN uh sihs) The development of many lymphocytes (a clone) from a single progenitor that has become activated to divide by antigenic stimulus.

blocking antibodies Antibodies that inhibit certain specific cell-mediated immune functions.

blood–brain barrier A selective mechanism opposing the passage of substances, such as bacteria and their toxins, from the blood to the cerebrospinal fluid and brain.

blood culture A fluid culture in which whole blood is used as the inoculum, used to detect bacteremia and septicemia.

blood dyscrasia (dihs KRAY zee uh) A diseased state of the blood; usually refers to abnormal cellular elements of more or less permanent character.

bone resorption (BOHN rih ZAWRP shuhn)   The removal of calcium salts from the bone.

botulin (BAHCH uh lihn)   A highly active neurotoxin sometimes found in imperfectly preserved or canned meats and vegetables.

botulism   A type of food poisoning caused by a neurotoxin (botulin) produced by the growth of *Clostridium botulinum* in improperly canned or preserved foods.

Bowman's capsule (BOH muhnz KAP suhl)   The expanded, cup-shaped beginning of a renal tubule.

branched pathway   A metabolic pathway in which a single intermediate may be used as a substrate by two or more different enzymes, leading to the appearance of several different products.

bright-field   A type of light microscopy in which the object is illuminated directly by a focused beam of light.

Brill-Zinsser disease (BRIHL ZIHN sur)   A recrudescence of typhus occurring as long as 70 years after the initial acute episode of epidemic typhus. In the United States, it is seen primarily in middle-aged or elderly immigrants from Russia, Poland, and neighboring countries. It is milder than the primary infection, owing to the persistence of specific antibody of the IgG class. Called also Brill's disease, benign typhus, and recrudescent typhus.

broad spectrum   A term indicating a broad range of activity of an antibiotic against a wide variety of microorganisms.

bronchial constriction   A reduction in the cross-sectional area of the bronchial vessels, caused by contraction of smooth muscle bands in their walls.

bronchial dilation   An increase in the cross-sectional area of the bronchial vessels, caused by a relaxation of smooth muscle bands in their walls.

bronchiole   One of the finer (1 mm or less) subdivisions of the branched bronchial tree, differing from the bronchi in having no cartilage plates and having cuboidal epithelial cells.

bronchitis (brawng KYT uhs)   Inflammation of one or more bronchi.

bronchus   Any of the larger air passages of the lungs, having an outer fibrous coat with irregularly placed plates of hyaline cartilage, an interlacing network of smooth muscle, and a mucous membrane of columnar ciliated epithelial cells.

brucellosis (broo suh LOH suhs)   A generalized infection of human beings involving primarily the reticuloendothelial system, caused by species of *Brucella*. Undulant fever.

bubo (B(Y)OO boh)   An enlarged and inflamed lymph node, particularly in the axilla or groin, due to such infections as plague, syphilis, gonorrhea, lymphogranuloma venereum, and tuberculosis.

bubonic plague   The usual form of plague marked by inflammatory enlargement of the lymphatic glands in the groins, axillae, or other parts.

budding   A form of asexual reproduction in which the body divides into two unequal parts, the larger part being considered the parent and the smaller one the bud.

buffer   A chemical system that prevents change in the concentration of another chemical substance, e.g., proton donor and acceptor systems serve as buffers, preventing marked changes in hydrogen ion concentration.

Bursa of Fabricius (BUR suh uhv fab REE see uhs)   An epithelial outgrowth of the cloaca in chick embryos, which develops in a manner similar to that of the thymus in mammals, atrophying after five or six months and persisting as a fibrous remnant in sexually mature birds. It contains lymphoid follicles, and before involution is a site of formation of lymphocytes associated with antibody-mediated immunity.

C particle   A noninfectious RNA virus that is postulated to be a normal inhabitant of all living cells and has been postulated as the single cause of all forms of cancer.

calculus   An abnormal concretion occurring within the animal body and usually composed of mineral salts. Commonly applied to calcified plaque on the teeth.

Calvin cycle   The set of reactions through which carbon dioxide is reduced to carbohydrates during the dark phase of photosynthesis.

canaliculus (kan uh LIHK yoo luhs)   An extremely narrow tubular passage or channel; used as a general term in anatomical nomenclature for various small channels.

candidiasis (kan dih DY uh suhs)   Infection with a fungus of the genus *Candida*.

capillary (KAP uh lair ee)   Pertaining to or resembling a hair; one of a network of tiny blood vessels connecting the arteries to the veins.

capillary action   The movement of water or any liquid along a surface; results from the combined effect of cohesion and adhesion.

capillary permeability   The permeability of the capillary to the passage of fluids and blood cells into the surrounding tissues.

capping   The movement of adherent antigenic material to a polar region of a lymphocyte before it is internalized.

capsid (KAP sihd)   The shell of protein that protects the nucleic acid of a virus; it may have helical or cubic symmetry and is composed of structural units, or capsomers.

capsule   (1) An exterior polymeric layer enclosing the cells of certain bacteria. (2) An anatomical structure enclosing an organ or body part.

carbohydrate   An aldehyde or ketone derivative of a polyhydric alcohold, particularly of the pentahydric and hexahydric alcohols.

carbon cycle   The steps by which carbon (in the form of carbon dioxide) is extracted from the atmosphere by living organisms and ultimately returned to the atmosphere.

carbon dioxide fixation   The process whereby carbon dioxide is metabolically converted into organic compounds, primarily via the Calvin cycle in autotrophic organisms.

carboxysome   An intracellular granule in many autotrophic procaryotes associated with carbon dioxide fixation.

carcinoembryonic antigen (CEA) (kahr sih noh ehm bree AHN ihk)   A cancer-specific glycoprotein antigen of colon carcinoma, also present in many adenocarcinomas of endodermal origin and in normal gastrointestinal tissues of human embryos.

cardiac sphincter (KAHR dee ak SFIHNGK tur)   Muscle fibers about the opening of the esophagus into the stomach.

carditis   Inflammation of the heart.

cariogenic diet (kair ee oh JEHN ihk)   A diet that measurably increases the incidence of dental caries in the consumer.

carious lesion   A focus of dental caries within the tooth substance.

caseation (kay see AY shuhn)   A form of necrosis in which the tissue is changed into a dry, amorphous mass resembling cheese.

case-finding   In epidemiology, the process of investigating the contacts of a known case of infectious disease in order to find any other undiscovered or incubating cases.

catabolite repression (kuh TAB uh lyt)   The inhibition of genes governing a pathway of degradation by substances that can be efficiently utilized as sources of energy.

catalase (KAT uh layz)   Hydrogen-peroxide:hydrogen-peroxide oxido-reductase. A crystalline enzyme that specifically catalyzes the decomposition of hydrogen peroxide, which is found in practically all cells except certain anaerobic bacteria.

cavitation   The formation of cavities, as in pulmonary tuberculosis.

cecum (SEE kuhm)   The first part of the large intestine, a dilated pouch into which open the ileum, the colon, and the appendix vermiformis.

cell culture   A growth of cells *in vitro;* although the cells proliferate, they do not organize into tissue.

cell envelope   The outer layers of the Gram-negative bacterial cell, composed of the outer membrane and a peptidoglycan layer.

cell-mediated immunity (CMI)   Specific acquired immunity in which the role of small lymphocytes of thymic origin is predominant; it is responsible for resistance to infectious diseases caused by certain bacteria and by viruses, certain aspects of resistance to cancer, delayed hypersensitivity reactions; certain autoimmune diseases, and allograft rejection, and plays a role in certain allergies.

cell-mediated lymphocytolysis test   An in vitro test to assay the potential for lymphocyte populations from two different individuals to develop a cytolytic immune response against each other.

cell membrane   The outermost membrane of the cell; also called the plasma membrane.

Cell Theory   The doctrine that all living matter is composed of cells and that activity is the essential process of life.

cell wall   A structure outside of and protecting the cell membrane, present in all plant cells (composed chiefly of cellulose) and in many bacteria and other types of cells.

cementum (suh MEHN tuhm)   The bonelike connective tissue covering the root of a tooth and assisting in tooth support.

Center for Disease Control   A branch of the United States Public Health Service, devoted to the collection and dissemination of epidemiological information.

central nervous system (CNS)   The portion of the nervous system consisting of the brain and spinal cord.

centrifugation   The process of separating the lighter portions of a solution, mixture, or suspension from the heavier portions by centrifugal force.

centriole (SEHN tree ohl)   Either of the two minute organelles that migrate to opposite poles of a cell during cell division and serve to organize the alignment of the spindles.

cephalosporin (sehf ah loh SPAWR ihn)   Any of a group of broad-spectrum, penicillinase-resistant antibiotics from *Cephalosporium,* including cephalexin, cephaloridine, cephaloglycin, and cephalothin, which share the nucleus 7-amino-cephalosporanic acid.

cerebellum   The part of the metencephalon that occupies the posterior cranial fossa behind the brainstem, being a fissured mass consisting of a median lobe (vermis) and two lateral lobes (the hemispheres) connected with the brain stem by three pairs of peduncles. It is concerned in the coordination of movements.

cerebrospinal fluid (CSF) (suh ree broh SPYN uhl FLOO uhd)   The fluid contained within the four ventricles of the brain, the subarachnoid space, and the central canal of the spinal cord.

cerebrum   The main portion of the brain, occupying the upper part of the cranial cavity.

cervix   The opening of the uterus.

Chagas' disease (SHAHG uhs dihz EEZ)   A form of trypanosomiasis that runs an acute course in children and a chronic course in adults.

charging enzymes   Cytoplasmic enzymes that specifically catalyze the attachment of individual amino acid molecules to individual transfer RNA molecules.

chemical bond   An interaction between two atoms, based on specific properties of each.

chemiosmotic hypothesis   The hypothesis that the living cell establishes a proton and electrical gradient across its membrane, and that by controlled reentry of protons into the cell, the energy to carry out several different types of endergonic processes may be obtained.

chemoautotroph (kee moh AW toh trohf)   A microorganism deriving its energy source from the oxidation of inorganic chemicals.

chemoheterotroph (kee moh HEHT uh roh trohf)   A microorganism, parasitic or saprophytic, deriving its energy and most of its carbon from the oxidation of preformed organic compounds.

chemotactic factor (kee moh TAK tihk)   The substance that brings about chemotaxis, or oriented movement either toward or away from the chemical.

chemotherapy   The treatment of disease by chemical agents; first applied to use of chemicals that affect the causative organism unfavorably but do not harm the patient.

chitin (KY tuhn)   A white, insoluble, horny polysaccharide, $C_{30}H_{50}O_{19}N_4$, the principal constituent of the shells of arthropods and the shards of beetles and found in certain fungi.

chloramphenicol (klawr am FEHN ih kahl)   An antibiotic substance, $C_{11}H_{12}Cl_2N_2O_5$, originally derived from cultures of *Streptomyces venezuelae,* and later produced synthetically; it occurs as fine white to grayish or yellowish-white, needle-like crystals or elongated plaques, and is used as an antibacterial and anti-rickettsial.

chlorhexidine (klawr HEHKS ih deen)   A substance which has been investigated as a mouthwash component with the goal of reducing microbial growth in the mouth and arresting dental caries.

chlorophyll   The green coloring matter of plants by which photosynthesis is accomplished.

chloroplast   Any one of the chlorophyll-bearing bodies of plant cells.

chloroquine (KLAWR uh kween) Chemical name: 7-chloro-4-[[diethyl-amino-1-methylbutyl]amino]quinoline. A compound, $C_{18}H_{26}ClN_3$, occurring as a white or slightly yellow, crystalline powder with a bitter taste and freely soluble in water; used as an antimalarial and lupus erythematosus suppressant, and in the treatment of lupus erythematosus.

cholera   An acute infectious disease caused by *Vibrio cholerae* and characterized by severe diarrhea with extreme fluid and electrolyte depletion, and by vomiting, muscle cramps, and prostration.

choleragen (KAL ur uh jehn)   The exotoxin produced by the cholera vibrio, which is thought to stimulate electrolyte and water secretion into the small intestine in Asiatic cholera.

chondroitin sulfatase (kahn DROYT uhn SUHL fuh tays)   An acid mucopolysaccharide that contains acetylgalactosamine, glucuronic acid, and sulfuric acid.

chorioamnionitis (kawr ee oh am nee uh NY tuhs)   Inflammation of fetal membranes caused by bacterial infection.

chorion   In human embryology, the cellular, outermost extraembryonic membrane, composed of trophoblast lined with mesoderm; it develops villi about two weeks after fertilization, is vascularized by allantoic vessels a week later, gives rise to the placenta, and persists until birth.

chromatography   A method of chemical analysis in which the solution to be analyzed is poured into a vertical glass tube containing an adsorbent (or stationary phase), the different solutes moving through the stationary phase at different velocities according to the degree of attraction to it, and producing bands of separated substance at different levels of the adsorption column.

chromatophore (kroh MAT oh fawr) Any pigmentary cell or color-producing plastid, such as those of the cutis or deep layers of the epidermis.

chronic infection   Invasion and multiplication of microorganisms in body tissues, resulting in local cellular injury due to competitive metabolism, toxins, intracel-

lular replication, or antigen–antibody response persisting over a long period of time.

chronic rejection   Gradual progressive loss of function of the transplanted organ.

chytrids (KY trihdz)   A type of primitive fungi known as the water molds, found often in fresh water on decaying leaves.

-cidal   Of a killer or killing.

-cidal treatment   A treatment which renders the treated organism permanently incapable of reproduction or growth.

cilium (SIHL ee uhm)   (1) An eyelid or its outer edge. (2) Any of the hairs growing on the edges of the eyelids; called also eyelash. (3) A minute vibratile, hairlike process attached to the free surface of a cell.

classic pathway   Activation of the complement cascade by initial reaction of complement with an antigen–antibody combination.

clean catch   The procedure most commonly used for obtaining a urine specimen for culture.

clitoris (KLIHT uh ruhs)   A small, elongated, erectile body, situated at the anterior angle of the rima pudendi; homologous with the penis in the male.

clonal selection theory   A selective theory of antibody formation proposed by Burnet, according to which a complement of clones of lymphoid cells capable of reacting with all possible antigenic determinants is present in the normal individual. During fetal life, those clones that react against self-antigens are suppressed on contact with the antigen.

clone   The asexual progeny of a single cell.

coagulase (koh AG yoo lays)   An antigenic substance of bacterial origin, produced chiefly by the staphylococci.

coccidioidin (kahk sihd ee OY dihn)   A sterile solution containing the by-products of growth products of *Coccidioides immitis*, injected intracutaneously as a rest for coccidioidomycosis.

coccidiodomycosis (kahk sihd ee oy doh my KOH suhs)   A fungal disease caused by infection with *Coccidioides immitis*, occurring in a primary and secondary form.

coccus (KAHK uhs)   A spherical bacterial cell, usually slightly less than 1μ in diameter.

codon   A series of three adjacent bases in one polynucleotide chain of a DNA or RNA molecule, which codes for a specific amino acid.

coenocyte (SEE nuh syt)   A mass of protoplasm containing several nuclei formed from an original cell with one nucleus.

coenzyme (koh EHN zym)   An organic, dialyzable, thermostable molecule, usually containing phosphorus and some vitamins and sometimes separable from the enzyme protein.

coenzyme A   Pantothenic acid. Acts as carrier for acetyl groups entering the Krebs cycle.

cold agglutinin (KOHLD uh GLOOT uh nihn)   An agglutinin that acts only at relatively low temperatures (ranging from 0° to 20°C).

coliform (KOH luh fawrm)   Resembling or being *Escherichia coli*.

coliform bacteria   Gram-negative rods, including *Escherichia coli* and similar species, that normally inhabit the colon. Commonly included in the coliform group are *Enterobacter aerogene*, *Klebsiella* sp., and other related bacteria.

colinear   Two related linear information sequences having their units arranged so that one may be directly taken from the other without rearrangement.

collagenase (kuh LAJ uh nays)   An enzyme that hydrolyzes peptides containing proline, including collagen and gelatin.

collagen fibers (KAHL uh juhn FY burz) The soft, flexible white fibers that are the most characteristic constituent of all types of connective tissue.

colon (KOH luhn)   The part of the large intestine that extends from the cecum to the rectum; sometimes used inaccurately as a synonym for the entire large intestine.

colonization   The process by which strains of bacteria become part of a host's normal flora.

colony   A collection or group of bacteria in a culture derived from the increase of an isolated single organism or group of organisms.

colony-forming unit (CFU)   The unit of viable cells, usually one but sometimes a small clump, that will initiate colonial growth and develop into a visible colony.

colorimeter (kuh uh RIHM uh tur)   An instrument for measuring color differences; especially one for measuring the color of the blood in order to determine the proportion of hemoglobin. Called also chromometer.

colostrum (kuh LAHS trahm)   The thin, yellow, milky fluid secreted by the mammary glands a few days before or after parturition.

commensalism (kuh MEHN suhl ihz uhm)   Symbiosis (q.v.) in which one population (or individual) gains from the association and the other is neither harmed nor benefited.

community-acquired infection   An infection that has been acquired by a person in the course of normal life, apart from contact with a hospital or other clinical facility.

competition   The phenomenon in which two structurally similar molecules "compete" for a single binding site on a third molecule. The process in which two or more similar groups of organisms compete for limited supplies of nutrients and space.

competitive inhibition   The inhibition of enzyme activity caused by the competition of the inhibitor with the substrate for the active (catalytic) site on the enzyme. Impairment of function of an enzyme is due to its reaction with a substance chemically related to its normal substrate.

complement (KAHM pluh mehnt)   A complex series of enzymatic proteins occurring in normal serum that interact to combine with antigen–antibody complex, producing lysis when the antigen is an intact cell.

complement fixation   When antigen unites with its specific antibody, complement, if present, is taken into the complex and becomes inactive or fixed.

compound microscope   A viewing instrument that consists of two lens systems, one above the other, in which the image formed by the system nearer the object (objective) is further magnified by the system nearer the eye (eyepiece).

compromised host   An individual who, by reason of injury, age, underlying disease, or a number of other factors, has increased susceptibility to infectious disease.

concentration   (1) An increase in strength by evaporation. (2) A measure of the amount of a dissolved substance contained per unit of volume.

concurrent disinfection   Immediate disinfection and disposal of discharges and infective matter all through the course of a disease.

condensation   The act of rendering, or process of becoming, more compact; the process of passing from a gaseous to a liquid or solid phase; the formation of a covalent bond by the elimination of a molecule of water.

condenser   A set of lenses within the microscope used for focusing or directing light from the light source onto the object.

confluent growth   Microbial growth on solid media so densely inhabited that there is no space between colonies and one merges into another.

congener (KAHN juh nur)   Something closely related to another thing, as a muscle having the same function as another, or a chemical compound closely related to another in composition, and exerting similar or antagonistic effects, or something derived from the same source or stock.

congenital rubella syndrome   A congenital syndrome due to intrauterine rubella infection (German measles), characterized most commonly by cataracts, cardiac anomalies (especially patent ductus arteriosus), deafness, microcephaly, and mental retardation.

congenital syphilis   Syphilis acquired *in utero*, and manifested variously by any of several characteristic malformations of the teeth or bones known as stigmata and by active mucocutaneous syphilis at the time of birth or shortly afterward, ocular changes, such as interstitial keratitis, or by neurological changes, such as deafness.

conidium (kuh NIHD ee uhm)   An asexual fungal spore shed at maturity (deciduous), and formed by splitting off from the summit of a conidiophore; also conidiospore.

conjugation (kahn joo GAY shuhn)   The act of joining together. In bacterial genetics, a form of sexual reproduction in which a donar bacterium (male) contributes some, or all, or its DNA (in the form of a replicated set) to a recipient (female) which then incorporates differing genetic information into its own chromosome by recombination, and passes the recombined set on to its progeny by replication.

conjunctiva (kahn JUHNK tih vuh)   The delicate membrane that lines the eyelids and covers the exposed surface of the sclera.

conjunctival sac   The sac that includes the lining of the eyelids and is continuous with the cornea.

consolidation   The process of becoming or the condition of being solid, as when the lung becomes firm as air spaces are filled with exudate in pneumonia.

constant region   In the immunoglobulin molecule, that portion of the polypeptide for which the amino acid sequence does not vary among immunoglobulins of the same type but of different antigenic specificities.

constitutive (kuhn STIHCH uht ihv)   Basic; an essential function that is continuously present.

constitutive enzyme   An enzyme produced by a microorganism regardless of the presence or absence of a specific stimulus.

consumers   In ecology, those organisms that derive energy primarily by ingesting the flesh of other organisms.

contact dermatitis   An acute allergic inflammation of the skin caused by contact with various substances of a chemical, animal, or vegetable nature to which delayed hypersensitivity has been acquired.

contractile vacuole (kuhn TRAK tuhl VAK yoo ohl)   A small cavity containing water fluid, seen in the protoplasm of certain unicellular organisms; it gradually increases in size and then collapses. Its function is thought to be respiratory or excretory.

convalescent carrier   An excreter of pathogenic organisms who has had a clinically recognizable attack of the disease.

Coombs test (KOOMZ)   A test using various antisera, usually employed to detect the presence of proteins (usually, but not always, antibodies) on the surface of red cells, as in the test for erythroblastosis fetalis.

coproantibody (kahp roh ANT ih bahd ee)   An antibody (chiefly IgA) present in the intestinal tract, associated with immunity to enteric infection.

cord factor A substance produced by pathogenic mycobacteria in liquid media leading to ropy growth; it has been suggested to be a virulence factor but its exact role is not clear.

core antigen An antigen associated with the core of the hepatitis B virus.

coronavirus (kawr oh nuh VY ruhs) Any of a group of morphologically similar, ether-sensitive viruses, probably RNA viruses, causing infectious bronchitis in birds, hepatitis in mice, gastroenteritis in swine, and respiratory infections in human beings.

corticosteroid (kawr tih koh STEHR oyd) Any of the steroids produced by the adrenal cortex, including cortisol, corticosterone, aldosterone.

coupled reaction Two or more chemical reactions that occur simultaneously, joined by the presence of common chemical reactants. One reaction usually provides the energy needed to drive the other.

covalent bond A chemical bond formed between atoms as a result of the sharing of one or more pairs of electrons.

Cowper's gland (KOW purz) Bulbourethral gland; either of two glands embedded in the substance of the sphincter of the male urethra, just posterior to the membranous part of the urethra.

coxsackievirus (kahk SAK ee VY ruhs) (From Coxsackie, N.Y., where it was first identified.) One of a heterogeneous group of enteroviruses producing, in human beings, a disease resembling poliomyelitis, but without paralysis; separable into two groups: A, producing degenerative lesions of striated muscle and B, producing leptomeningitis in infant mice.

C-reactive protein (CRP) A globulin that forms a precipitate with the somatic C-polysaccharide of the pneumococcus *in vitro*; its demonstration in the serum is a sensitive indicator of inflammation of infectious or noninfectious origin.

crista (KRIHS tuh) (Plural is cristae.) The ridge or ridges of the inner mitochondrial membrane bearing the coenzymes of the electron transport system.

critical pH The pH value, usually around 5.6, below which calcium salts in the tooth start to dissociate into soluble ions.

cross-matching Determination of the compatibility of the blood of a donor and that of a recipient before transfusion by placing red cells of the donor in the recipient's serum and red cells of the recipient in the donor's serum. Absence of agglutination indicates that the two blood specimens are compatible.

croup (KROOP) A condition resulting from acute obstruction of the larynx caused by allergy, foreign body, infection, or new growth, occurring chiefly in infants and children, and characterized by resonant barking cough, hoarseness, and persistent stridor.

cryptococcosis (krihp tuh kah KOH suhs) An infection by *Cryptococcus neoformans* that may involve the skin, lungs, or other parts, but has a predilection for the brain and meninges.

cryptomonad (krihp tah MOH nad) Any of the algae of the order Cryptophyceae.

crypts of Lieberkuhn (LEE bur kun) Tubular glands in the lining of the small and large intestines.

crystalloid (1) Resembling a crystal. (2) A noncolloid substance that, in solution, passes readily through animal membranes, lowers the freezing point of the solvent containing it, and is generally capable of being crystallized.

cuvette (kyoo VEHT) A glass container, generally possessing well-defined characteristics with regard to dimensions (particularly thickness) and optical properties, and generally used to examine colored and colorless solutions free of turbidity, but also used to examine the light scattering of turbid suspensions, such as bacterial suspensions.

cyanobacteria (sy an oh bak TEER ee uh) A major subdivision of the procaryotes, all of which carry out oxygenic photosynthesis.

cyanophycin An intracellular reserve material formed by cyanobacteria, made of aspartic acid and arginine.

cyclic adenosine monophosphate (cAMP) A regulatory substance found in all cells. It is formed from ATP, and its concentration is raised or lowered by environmental stimuli, leading in turn to alterations in the rate of key intracellular processes.

cyclical pathway A metabolic pathway, such as the Krebs cycle, that regenerates some of its own starting material in addition to other products.

cyclic nucleotide A nucleotide in which the phosphate residues are not linear, but bonded in a cyclical fashion.

cyst Any closed cavity or sac, normal or abnormal, lined by epithelium, and especially one that contains a liquid or semisolid material.

cytochrome (SY tuh krohm) Any of a class of hemoproteins whose principal biologic function is electron transport by virtue of a reversible valency change of its heme iron.

cytochrome oxidase Ferrocytochrome c:oxygen reductase. A hemoprotein enzyme found in cells, usually attached to mitochondria, associated with copper.

cytolytic cycle A cycle of viral replication in which the destruction of the host cycle regularly occurs.

cytolytic infection A viral infection in which the replication of the virus leads to the destruction of tissue.

cytomegalovirus (CMV) A group of herpes-viruses infecting human beings and other animals. Many of these viruses have special affinity for salivary glands and cause enlargement of cells of various organs and development of characteristic inclusions in the cytoplasm or nucleus.

cytopathic effect (CPE) (sy toh PATH ihk) Observable changes in cells *in vitro* produced by viral action; for example, -lysis of cells or fusion of cells.

cytoplasm (SY toh plaz uhm) The protoplasm of a cell exclusive of that of the nucleus; it consists of a continuous aqueous solution (cytosol) and the organelles and inclusions suspended in it and is the site of the chemical activities of the cell.

cytoplasmic membrane The flexible structure immediately surrounding the cytoplasm in all cells.

dalton A unit of mass, being one sixteenth of the mass of the oxygen atom, or approximately 1.65×10^{-24} gm.

Dane particle A viral particle found in the serum of hepatitis B patients, believed to be either the causative virus or a subunit of it.

dark-field microscopy A microscopic technique in which no light is visible except that which is reflected from an object.

dark reactions In photosynthesis, the fixation of CO_2, into carbohydrate, that is independent in place and time on the absorption of light.

deamination (dee am uh NAY shuhn) Removal of the amino group, NH_2, from a compound.

death phase The stage in which the number of viable bacteria in a population decreases at an exponential rate.

debridement (dih BREED muhnt) The removal of foreign material and devitalized or contaminated tissue from or adjacent to a traumatic or infected lesion until surrounding tissue is exposed.

deciduous teeth (dih SIHJ uh wahs) Falling off or shed at maturity; the term is used to designate the teeth of the first dentition in animals and human beings.

decomposers Organisms (bacteria, fungi) in an ecosystem or community that convert dead organic material into plant nutrients.

decubitus ulcer (dee KOO bih tuhs) An ulceration caused by prolonged pressure in a patient allowed to lie too still in bed for a long period of time; called also decubital sore, bed sore, and pressure sore.

defective virus A virus that cannot be completely replicated or cannot form a protein coat; in some cases replication can proceed if missing gene functions are supplied by other (helper) viruses.

degranulation (de gran yuh LAY shuhn) The process of losing granules; said of certain granular cells such as mast cells.

deintegration The process by which a viral genome (prophage or provirus) is removed from the host genome.

delayed hypersensitivity (DHS) A slowly developing increase in cell-mediated immune response to a specific antigen; it is involved in the graft-rejection phenomenon, autoimmune disease, and contact dermatitis, as well as in antimicrobial immunity.

demyelination (de my uh luh NAY shuhn) Destroying or removing the myelin sheath of a nerve or nerves.

denaturation The loss of the native configuration of a macromolecule resulting, for instance, from heat treatment, extreme pH changes, chemical treatment, or other denaturing agents. It is usually accompanied by loss of biological activity.

dendrite One of the threadlike extensions of the cytoplasm of a neuron.

dendritic macrophage Any of the large, highly phagocytic cells with a small, oval, sometimes indented nucleus and inconspicuous nucleoli, occurring in the walls of blood vessels (adventitial cells) and in loose connective tissue (histiocytes, phagocytic reticular cells), branched like a tree.

dengue (DEHNG gee) An infectious, eruptive, febrile disease, marked by severe pains in the head, eyes, muscles, and joints, sore throat, catarrhal symptoms, and sometimes a cutaneous eruption and painful swellings of the parts.

denitrification Conversion of nitrate into nitrogen gases under anaerobic conditions, resulting in loss of nitrogen from ecosystems.

dental caries (DEHNT uhl KA(UH)R eez) The defects (lesions) produced by destruction of enamel and dentin in the teeth.

dentifrice (DEHNT uh frahs) Any preparation for cleaning teeth.

dentin The chief substance or tissue of the teeth, which surrounds the tooth pulp and is covered by enamel on the crown and by cementum on the roots of the teeth.

deoxyribonucleic acid (DNA) A nucleic acid originally isolated from sperm and thymus gland, but later found in all living cells; on hydrolysis it yields adenine, quanine, cytosine, thymine, deoxyribose, and phosphoric acid. It is the carrier of genetic information for all organisms except the RNA viruses.

depth filter A type of filter that removes particles from solution by trapping them in the matrix of a thick layer of porous material.

dermatomycosis (dur muh to my KOH suhs) A superficial infection of the skin or its appendages by fungi.

dermatophyte (dur MAT uh fyt) A fungus parasitic upon the skin; the term embraces the imperfect fungi of the genera *Microsporum*, *Epidermophyton*, and *Trichophyton*.

dermis The skin layer beneath the epidermis; contains nerve endings and blood vessels.

desensitization A condition in which the organism does not react immunologically to a specific antigen; also the process by which this is brought about.

desquamation (dehs kwuh MAY shuhn) The shedding of epithelial elements, chiefly of the skin, in scales or small sheets; exfoliation.

dextran (DEHK stran) A water-soluble polysaccharide of glucose (dextrose) produced by the action of bacteria on sucrose; used as a plasma volume extender; contributes to the formation of dental plaque.

diaminopimelic acid (DAP) A naturally occurring amino acid discovered in 1950, and a component of peptidoglycan.

diaminopyrimidines A group of chemotherapeutic agents that includes trimethoprim.

diatom Any unicellular microscopical form of alga having a wall of silica and belonging to the family *Diatomaceae*.

dicaryon A fungal cell or structure containing two complete nuclei from different mating types, formed by fusion.

differential media Bacteriological media on which growth of specific types of organism leads to readily visible changes in the appearance of the media.

differential stain Contrast stain; a dye used to color one portion of a tissue or cell that remained unaffected when the other part was stained by a dye of a different color.

differentiation The distinguishing of one thing or disease from another. The process by which embryonic tissues assume specialized structures and functions.

diffuse lymphoid tissue Widely distributed latticework of reticular tissue the interspaces of which contain lymphocytes; lymphoid tissue may be diffuse, or densely aggregated as in lymph nodules and nodes.

DiGeorge's syndrome A congenital syndrome in which absence of the thymus and parathyroids due to defective development of the third and fourth embryonic pharyngeal pouches is associated with impairment of cell-mediated immunity and normal levels of immunoglobulins; it is characterized clinically by neonatal tetany, hypocalcemia, frequent viral and fungal infections, and, often, deformities of the ears, nose, mouth, and great vessels.

dinoflagellate (dy noh FLAJ uh layt)   Of the order *Dinoflagellata*; a unicellular algal cell bearing two flagella.

diphtheritic membrane (dihf thuh RIHT ihk)   A false membrane characteristic of diphtheria and resulting from coagulation necrosis.

diphtheroid (DIHF thuh royd)   Coryne-bacterial species resembling the diphtheria bacillus.

diploid cell   An individual or cell having two full sets of homologous chromosomes.

direct count   The microscopic analysis of the sample that leads to an enumeration of the number of microorganisms present per specified volume of sample.

direct immunofluorescence   Immuno-fluorescence in which the antibody specific for the antigen being tested for is fluorescein-labeled.

direct transmission   Infectious disease transmission from host to recipient by a single step or a short sequence of events.

disaccharide   Any of a class of sugars that yield two monosaccharides on hydrolysis and have the general formula $C_n(H_2O)_{n-1}$ or $C_{12}H_{22}O_{11}$.

disclosing stain   A transitory stain incorporated in a mouth rinse; when held briefly in the mouth, it dyes areas of plaque, rendering them clearly visible and making it easier to direct removal efforts.

dissimilation   The act of decomposing a substance into simpler compounds, for the production of energy or of materials that can be eliminated.

disulfide bond   The —S—S link binding two peptide chains (or different parts of one peptide chain). It occurs as part of the molecule of amino acid, cystine, and is important as a structural determinant in many protein molecules, notably keratin, insulin, and oxytocin.

diverticulum (dy vur TIHK yuh luhm)   A circumscribed pouch or sac of variable size occurring normally or created by her-niation of the lining mucous membrane through a defect in the muscular coat of a tubular organ.

DNAase   Deoxyribonuclease; applies to a group of diverse enzymes all of which hydrolyze DNA in various ways.

DNA base ratio   The ratio in mole per-cent of adenine plus thymine to guanine plus cytosine.

DNA-dependent RNA polymerase   The enzyme that makes RNA transcripts of the informational DNA.

DNA ligase   An enzyme whose catalytic function is DNA repair. It links up single-stranded ends of free DNA.

DNA modification   After the DNA has been synthesized, certain of the bases are chemically modified, as by the addition of methyl groups to the cytosines. This masks areas of potential attack by the cell's DNAases.

DNA replication   The process of copying the DNA double helix exactly.

dorsal root   The posterior, dorsal, or sensory root of a spinal nerve.

double bond   A type of covalent bond in which two pairs of electrons are shared.

doubling time   The time required for the number of cells in a population to double in number, essentially the same as the generation time.

DPT vaccine   Diphtheria-pertussis-tetanus (vaccine); a formulation contain-ing diphtheria and tetanus toxoids plus killed cells of the pertussis organism. Used for routine immunization of small children.

droplet nuclei   Dried residue of a droplet; droplet nuclei from sneezing and coughing that carry airborne infection.

drug clearance   The removal of a drug from the body by liver detoxification, excretion, or both.

dry heat   Heat used as a sterilizing method in the absence of water vapor or steam.

dry socket   A condition sometimes oc-curring after tooth extraction, resulting in exposure of bone with localized osteo-myelitis of an alveolar crypt, and symp-toms of severe pain.

dry weight   The net weight of a biologi-cal or chemical sample after drying, that is, without water.

duck embryo vaccine (DEV)   Vaccine prepared from embryonated duck eggs infected with modified rabies virus.

duodenum (d(y)oo uh DEE nuhm)   The first or proximal portion of the small intestine, extending from the pylorus to the jejunum; so called because it is about 12 fingerbreadths in length.

dura mater (D(Y)UR uh MAYT ur)   The outermost, toughest, and most fibrous of the three membranes (meninges) covering the brain, composed of two mostly fused layers: an endosteal outer layer (endocra-nium) adherent to the inner aspect of the cranial bones, and an inner, meningeal layer.

D-value   The decimal reduction time, or time required for 90 percent of the organisms to be killed at a certain temperature.

early proteins   The proteins of viral origin that appear early in the viral repli-cation cycle.

eclipse phase   In virology, that period of the infective cycle during which infected bacterial cells contain no detectable infec-tive bacteriophage.

ecosystem (EE coh sihs tuhm)   The fundamental unit in ecology, comprising the living organisms and the nonliving elements interacting in a certain defined area.

ecosystem homeostasis (hoh mee oh STAY suhs)   Maintenance of a relatively stable internal physiological environment or internal equilibrium in an organism, population, or ecosystem.

ectosymbiosis   A type of symbiosis in which one organism lives attached to the surface of another.

elastase (ih LAS tays)   An enzyme capable of catalyzing the digestion of elas-tic tissue.

elastin (ih LAS tihn)   A yellow sclero-protein, the essential constituent of yel-low elastic connective tissue.

electrode potential value   A measure-ment in millivolts of the redox potential of a solution.

electrolyte   A substance that dissociates into ions when fused or in solution, and thus becomes capable of conducting elec-tricity; an ionic solute.

electrolyte balance   The bodily content of sodium and potassium and the concen-trations of these ions in extracellular and intracellular fluids.

electron   The unit or "atom" of negative electricity.

electron acceptor   Substance that accepts electrons during an oxidation-reduction reaction.

electron donor   Substance that gives up electrons during an oxidation-reduction reaction; an energy source.

electron microscope (EM) One in which an electron beam, instead of light, forms an image for viewing on a fluorescent screen, or for photography.

electron transport system A sequence of electron transport coenzymes, arranged linearly in a membrane, which transport electrons from a high energy level to a final combination with a lower-energy electron acceptor.

electrophoresis (ih lehk troh fuh REE suhs) The movement of charged particles suspended in a liquid under the influence of an applied electric field.

element An aggregation of atoms of one kind only.

elementary body One of the stages in the intracellular replication of the chlamydia.

elongation In protein synthesis, the addition of amino acids to a growing peptide chain.

empirical Based on experience or direct observation.

empirical formula A chemical formula that expresses the proportions of the elements present in a substance. For substances composed of discrete molecules, it expresses the relative numbers of atoms present in a molecule of the substance in the smallest whole numbers.

empyema (ehm py EE muh) Accumulation of pus in a cavity of the body.

enamel The white, compact, and very hard substance that covers and protects the dentin of the crown of a tooth.

encapsulated node A lymph node that is surrounded by a membranous capsule.

endemic Present in a community at all times.

endemic typhus A milder form of typhus, caused by *Rickettsia prowazekii,* which is transmitted from person to person by rat or mouse fleas.

endergonic (ehn dur GAHN ihk) Characterized by or accompanied by the absorption of energy.

endocardium The endothelial lining membrane of the heart and the connective tissue bed on which it lies.

endoenzyme (ehn doh EHN zym) An intracellular enzyme; an enzyme that is retained in a cell and does not normally diffuse out of the cell into the surrounding medium.

endogenous C particle A noninfectious RNA virus that is postulated to be a normal inhabitant of all living cells and has been postulated as the single cause of all forms of cancer (growths from within).

endogenous infection Infection caused by reactivation of organisms present in a dormant focus, as occurs in tuberculosis, histoplasmosis, coccidiodomycosis, etc.

endogenous pyrogen (PY ruh jehn) A fever-producing substance liberated by the host's own tissues.

endometrium (ehn doh MEE tree uhm) The mucous membrane of the uterus, the thickness and structure of which vary with the phase of the menstrual cycle.

endosymbiosis The state achieved between a virus and its host cell in which cellular division is inhibited but the cell is not immediately destroyed. The state achieved by many if not all eucaryotic cells in which procaryotic cells or cell remnants live within.

endotoxin A heat-stable toxin present in the Gram-negative bacterial cell but not in cell-free filtrates of cultures of intact bacteria.

endotoxin shock Sudden circulation failure associated with overwhelming bacterial infection, usually due to Gram-negative bacteria.

enrichment culture The addition of nutrients, as to culture media; the medium resulting from such addition.

Enterobacteriaceae (ehn tur oh bak teer ee AH see ay) A family of Gram-negative rod-shaped organisms, occurring as plant or animal parasites, or as saprophytes.

enteromammary immunization Transfer of immunogenic substances from the mammary gland of the mother to the intestinal tract of the infant.

enterotoxin A toxin specific for the cells of the intestinal mucosa.

enterovirus One of a subgroup of the picornaviruses infecting the gastrointestinal tract and discharged in the excreta, including poliovirus, the coxsackieviruses, and the echoviruses.

entropy (**EHN truh pee**) The measure of the unavailable energy in a thermodynamic system.

envelope An encompassing structure or membrane.

enzyme A protein, capable of accelerating or producing by catalytic action some change in a substrate for which it is often specific.

enzyme-linked immunosorbent assay (ELISA) A versatile immunological test, in which the extent of antigen–antibody combination is revealed by the amount of a readily assayed enzymatic change.

eosinophil (ee oh SIHN uh fihl) A structure, cell, or histologic element readily stained by eosin, especially a granular leukocyte with a nucleus that usually has two lobes connected by a slender thread of chromatin, and cytoplasm containing coarse, round granules that are uniform in size.

epidemic Attacking many people in any region at the same time; widely diffused and rapidly spreading.

epidemic hemorrhagic fever An acute infectious disease characterized by fever, purpura, peripheral vascular collapse, and acute renal failure, caused by a filtrable agent thought to be transmitted to human beings by mites or chiggers.

epidemiology The study of the relationships of the various factors determining the frequency and distribution of diseases in a human community.

epidermis The outermost and nonvascular layer of the skin, derived from the embryonic ectoderm, varying in thickness from 1/200 to 1/20 inch, and made up from within outward, of five layers.

epididymis (ehp uh DIHD uh muhs) The elongated cordlike structure along the posterior border of the testis, in the ducts of which the spermatozoa are stored. It consists of a head, a body, and a tail.

epididymitis (ehp uh dihd uh MY tuhs) Inflammation of the epididymis.

epiglottis The lidlike cartilaginous structure overhanging the entrance to the larynx and serving to prevent food from entering the larynx and trachea while swallowing.

epiglottitis (ehp uh glah TY tuhs) Inflammation of the epiglottis.

episome (EHP uh sohm) In bacterial genetics, any accessory extrachromosal replicating genetic element that can exist either autonomously or integrated with the chromosome.

Epstein–Barr virus (DBV) A herpesvirus originally isolated from Burkitt lymphomas and believed to be the etiological agent of infectious mononucleosis or closely related to it.

ergot (UR guht)   The dried sclerotium of *Claviceps purpurea*, which is developed on rye plants. Contains a mixture of highly toxic substances that can cause ergotism when contaminated grains are ingested.

ergotism   Chronic poisoning from excessive or misdirected use of ergot as a medicine, or from eating ergotized grain; it is marked by cerebrospinal symptoms, spasms, cramps, or by a kind of dry gangrene.

erysipelas (ehr uh SIHP uh luhs)   A contagious disease of skin and subcutaneous tissue due to infection with *Streptococcus pyogenes* and marked by redness and swelling of affected areas, with constitutional symptoms; sometimes accompanied by vesicular and bulbous lesions.

erythroblastosis fetalis (ih rihth roh blas TOH suhs fee TAL uhs)   Hemolytic anemia of the fetus or newborn infant, caused by the transplacental transmission of maternally formed antibody, usually secondary to an incompatibility between the blood group of the mother and that of her offspring, characterized by increased numbers of nucleated red cells in the peripheral blood, hyperbilirubinemia, and extramedullary hematopoiesis.

erythrocyte   One of the elements found in peripheral blood, called also red blood or corpuscle. Normally, in the human being, the mature form is a nonnucleated, biconcave disk, adapted, by virtue of its configuration and its hemglobulin content, to transport oxygen.

erythrogenic toxin (ih rihth ruh JEHN ihk)   An exotoxin produced by many, but not all, strains of *Streptococcus pyogenes* and that produces an erythematous reaction on intradermal inoculation in human beings and to a lesser extent in the rabbit, and is responsible for the scarlatiniform rash of scarlet fever.

eschar (EHS kahr)   A slough produced by a thermal burn, by a corrosive application, or by gangrene.

esophagus   The musculomembranous passage extending from the pharynx to the stomach.

ethambutol (ehth AM byoo tahl)   Ethambutol hydrochloride, $C_{10}-H_{24}N_2O_2 \cdot HCl$, is used as a tuberculostatic agent.

ethylene oxide (ETO)   A highly toxic gas used in sterilization.

etiological agent   The microbial or viral causative agent of an infectious disease.

eucaryotic (yoo kar ee AHT ihk)   Pertaining to a eucaryon or to a eucaryote. A complex cell type containing a highly structured, membrane-bounded nucleus.

euphotic zone   In the ocean, the top layer of water through which sufficient light penetrates to support the growth of photosynthetic organisms.

eustachian tube (yoo STAY sh(ee)uhn)   A tube leading from the middle ear to the nasopharynx.

exergonic (ehk sur GAHN ihk)   Characterized or accompanied by the release of energy.

exfoliation   A falling off in scales or layers.

exoenzyme   An extracellular enzyme; an enzyme that acts outside of the cells in which it originates.

exogenous pyrogen   A fever-inducing substance that originates from without the organism.

exosporium (ehk soh SPAWR ee uhm)   The external layer of the envelope of a spore.

exotoxin   A toxic substance formed by bacteria that is found outside the bacterial cell, or free in the culture medium.

exotoxin A   A substance produced by *Pseudomonas aeruginosa* believed to be a major factor in its virulence.

exponential phase   Period during the growth cycle of a population in which growth increases at an exponential rate.

extended spectrum   A term used to describe an antimicrobial drug that has a range of effectiveness somewhat greater than its congeners.

extracellular enzyme   An enzyme that exists outside of the cell secreting it.

exudate, purulent (EHK su dayt, PYOO roo luhnt)   Consisting of or containing pus that has escaped from blood vessels and has been deposited in tissues or on tissue surfaces, usually as a result of inflammation.

exudate, serous (EHK su dayt, SEE ruhs)   Pertaining to or resembling serum; associated with the formation of or caused by pus.

exudative phase   In tuberculosis, the early phase of the disease in which the bacilli are widely disseminated in lymphoid tissues.

F^+ cell   A bacterial cell that can transfer the extrachromosomal F (fertility) particle to a recipient (F^-) cell.

F^- cell   A bacterial cell that does not contain an F particle, but can act as a recipient and receive one from an F^+ cell.

F' cell   A bacterial cell that carries an extrachromosomal F particle to which is attached a fragment of chromosomal DNA.

Fab [fragment, antigen-binding]   Either of two segments of the IgG molecule, obtained by treatment of the antibody molecule with papain, that retains the ability to combine with antigen.

facultative anaerobe (FAK uhl tayt ihv AN uh rohb)   Microorganisms that are able to grow under either anaerobic or aerobic conditions.

fallopian tube (fuh LOH pee uhn)   Either of two slender tubes that carry ova from the ovaries to the uterus.

Fansidar   Trade name for a chemotherapeutic drug combination useful against chloroquine-resistant malaria.

farmer's lung disease   A hypersensitivity in which the individual becomes overresponsive to the antigens carried on fungal spores such as found in hay dust.

fastidious organism   Used to refer to an organism that requires many growth factors.

fat   Adipose tissue; a white or yellowish tissue that forms soft pads between various organs of the body, serves to smooth and round out bodily contours, and furnishes a reserve supply of energy.

fatty acid, unsaturated   Fatty acid that cannot be formed in the body and therefore must be provided by the diet; the most important are linoleic acid, linolenic acid, and arachidonic acid.

Fc [fragment, crystallizable]   One of the two segments of the IgG molecule obtained by treatment of the antibody molecule with papain; it is crystallizable and contains most of the antigenic determinants.

feedback inhibition   End product inhibition. Inhibition of enzyme activity resulting from the effect of a critical concentration of the end product (allosteric effector) of a biosynthetic pathway on an enzyme whose activity is crucial to the formation of an intermediate metabolite; it is a self-regulatory mechanism by which the rate of synthesis of a metabolite is controlled.

fermentation   Enzymatic decomposition, especially of carbohydrates as used in the production of alcohol, bread, vinegar, and other food or industrial materials.

fermentative metabolism   Metabolism in which the attack on carbohydrates is anaerobic and leads to the production of organic acids and sometimes gases.

ferredoxin (fehr uh DAHK suhn)   A nonheme iron-containing protein, having a high sulfide content, that serves as an acceptor molecule in electron transport from the chlorophyll during the formation of NADPH in photosynthesis.

ferritin (FEHR uh tihn)   The iron-apoferritin complex, which is one of the chief forms in which iron is stored in the body; it occurs in the gastrointestinal mucosa, liver, spleen, bone marrow, and reticuloendothelial cells generally.

fever blister   Vesicular lesion caused by Herpes simplex, usually on the face.

fibril (FYB ruhl)   A minute fiber or filament; often a component of a compound fiber.

fibrin (FY bruhn)   The insoluble protein formed from fibrinogen by the proteolytic action of thrombin during normal clotting of the blood. Fibrin forms the essential portion of the blood clot.

fibrinogen (fy BRIHN uh juhn)   [USP] a sterile compound derived from normal human plasma and dried from the frozen state; used to increase the coagulability of the blood.

fibrinolysin (fy bruhn uhl YS uhn)   A commercial preparation of proteolytic enzyme formed from profibrinolysin by the action of physical agents or by specific bacterial kinases; used to promote dissolutions of thrombi. An extracellular virulence factor produced by several pathogenic bacteria.

fibroblast   A connective tissue cell; a flat elongated cell with cytoplasmic processes at each end, having a flat, oval, vesicular nucleus.

filament   A delicate fiber or thread; the long undulant portion of the flagellum.

filamentous phage   A type of long, slender bacteriophage particle, replicating by a nonlytic pathway.

fimbria (FIHM bree uh)   A fringe, border, or edge. [NA] A general term for such a structure. In microbiology, one of the minute filamentous appendages of certain bacteria; they are considerably smaller and less rigid than flagella and are associated with antigenic properties of the cell surface.

first law of thermodynamics (thur moh dy NAM ihks)   Energy is conserved in any

process; i.e., the energy gained (or lost) by a system is exactly equal to the energy lost (or gained) by the surroundings.

five kingdoms   The classification scheme devised by Whittaker, in which all living things are placed in five groups.

fixation   The act or operation of holding, suturing, or fastening in a fixed position. In microscopy, the chemical or physical process used to fix structures in their original shape and location during staining and observation.

fixed cells   Phagocytic cells of the reticuloendothelial system that are more or less permanently incorporated into localized sites.

flagella (fluh JEHL uh)   Plural of flagellum; organs of motility in some unicellular microorganisms.

flagellin   A protein occurring in the flagella of bacteria; it is similar to keratin, myosin, and fibrinogen.

flatus (FLAYT uhs)   Gas or air in the gastrointestinal track.

flavivirus (FLAY vuh vy ruhs)   A subcategory of togaviruses; the type species is the yellow fever virus.

flavoprotein (FP)   An electron transport coenzyme that contains riboflavin.

fluid balance   The state of the body in relation to ingestion and excretion of water and electrolytes.

fluid mosaic   A conceptual model of membrane structure, based on a bilayer of phospholipid freely intermixed with proteins at all sites, with all components free to migrate and form transitory associations for specific functions.

fluorescein (floo(u)r EHS ee uhn)   Chemical name: resorcinolphtalein. The simplest of the fluorane dyes and the parent compound of eosin; used intravenously in tests to assess by its fluorescence the adequacy of the circulation, and combined with radioactive iodine in localization of brain tumors, etc.

fluorescence   The property of emitting light while exposed to light, the wavelength of the emitted light being longer than that of the light absorbed.

fluorescent antibody (FA)   Immunoglobulin molecule that has been coupled with a fluorescent dye so that it exhibits the property of fluorescence.

fomite (FOH myt)   Fomes—an object, such as a book, wooden object, or an article of clothing, that is not in itself harmful, but is able to harbor pathogenic

microorganisms and thus may serve as an agent of transmission of an infection.

food chain   A sequence (as grass, rabbit, fox) of organisms in a community in which each member of the chain feeds on the member below it.

forespore   One of the more readily identified stages in the process of sporulation, identified as a refractile body that is not yet resistant to heat.

formalin (FAWR muh lihn)   A solution of formaldehyde in water, containing not less than 37 percent of formaldehyde; used as a disinfectant. Called formaldehyde solution and formalin.

fossil fuel   The remains of once living organisms that are burned to release energy. Examples are coal, oil, and natural gas.

frameshift mutation   A point mutation, such as an addition or subtraction, which alters the reading frame of the codons to follow.

Frascatoro (fra skuh TAWR oh)   An Italian scientist of the seventeenth century, first to provide experimental evidence against the spontaneous generation theory.

free energy   The energy equal to the maximum amount of work that can be obtained from a process occurring under conditions of fixed temperature and pressure.

freeze-etching   A method used to study unfixed cells by electron microscopy, in which the object to be studied is placed in 20 percent glycerol, frozen at $-100°C$, and then mounted on a chilled holder.

fruiting body   A specialized fungal structure, as an apothecium, that produces spores.

fundus   The bottom or base of anything; a general term for the bottom or base of an organ, or the part of a hollow organ farthest from its mouth.

fungi   Plural of fungus. A general term used to denote a group of eucaryotic protists, including mushrooms, yeasts, rusts, molds, smuts, etc. that are characterized by the absence of chlorophyll and by the presence of a rigid cell wall composed of chitin, mannans, and sometimes cellulose.

fungus balls   Aggregates of fungal mycelium microscopically visible in the tissues of individuals with certain systemic mycoses.

fusiform (FYOO zuh fawrm)   Spindle-shaped, descriptive of certain bacteria.

gallbladder   The pear-shaped reservoir for the bile on the posterioinferior surface of the liver, between the right and the quadrate lobe.

gamete (GAM eet)   A reproductive element; one of two cells, male (spermatozoon) and female (ovum), whose union is necessary, in sexual reproduction, to initiate the development of a new individual.

ganglion (GANG glee uhn)   A knot, or knotlike mass.

gastric juice   A strongly acidic secretion of the glandular stomach lining.

gastrointestinal tract (GI)   The stomach and intestines in continuity.

gas vesicle   A gas-filled structure in certain procaryotes that confers ability to float. Sometimes called gas vacuole.

general-purpose medium   A microbiological growth medium formulated to support the growth of the maximum number of different microorganisms.

generation time   The time elapsing from one generation to the next, or in bacteria the time from one fission to the next. Also the reciprocal of the growth rate, which is less than the doubling time since growth is exponential rather than linear.

genetic code   The arrangement of nucleotides in the polynucleotide chain of a chromosome that governs the transmission of genetic information to proteins, i.e., determines the sequence of amino acids in the polypeptide chain making up each protein synthesized by the cell.

genetic cross   A mating between two different sexually reproducing individuals leading to exchange or recombination of genetic materials.

genital herpes (JEHN ih tuhl HUR peez)   Herpes genitalis. An infection by herpes simplex virus marked by the eruption of one or more groups of vesicles on the vermillion border of the lips, at the external nares, or on the glans, prepuce, or vulva.

genome (JEE nohm)   The complete set of hereditary factors, as contained in the haploid assortment of chromosomes.

genotype   The entire genetic constitution of an individual; also the alleles present at one or more specific loci.

genus   A taxonomic category subordinate to a tribe (or subtribe) and superior to a species (or subgenus).

geochemical activities   Chemical processes that contribute to significant changes in the chemistry of the minerals or organic constituents of the earth's crust.

geotrichosis (jee oh trih KOH suhs)   Infection by *Geotrichum candidum*, which may attack the bronchi, lungs, mouth, or intestinal tract; its manifestations resemble those of candidiasis.

germ-free animal   An animal which has no normal flora; all its tissues and surfaces are sterile, and it is maintained in that condition by being housed and fed in a sterile environment.

germinal follicles   In lymph nodes, regions in which lymphocytes are rapidly dividing.

germ tubes   Characteristic structures formed by yeast cells of *Candida* under certain cultural conditions.

giardiasis (jee ar DY uh sihs)   Infection with *Giardia*.

gingiva (JIHN juh vuh)   The gums: the mucous membrane, with the supporting fibrous tissue, that overlies the crowns of unerupted teeth and encircles the necks of those that have erupted.

gingival sulcus (SUHL kuhs)   A furrow surrounding a tooth, bounded internally by the surface of the tooth and externally by the epithelium lining the free gingiva.

gingivitis   Inflammation involving the gingival tissue only.

globular protein   Protein soluble in water, usually with added acid, alkali, salt, or ethanol, and roughly so classified (albumins, globulins, histones, protamines).

glomerulus (gloh MEHR yoo luhs)   A tuft or cluster; used in anatomical nomenclature as a general term to designate such a structure, as one composed of blood vessels or nerve fibers.

glutaraldehyde (gloo tuh RAL duh hyd)   A compound, $CHO \cdot (CH_2)_3 \cdot CHO$, used as a tissue fixative for light and electron microscopy because ot its preservation of fine structural detail and localization of enzyme activity.

glyceraldehyde-3-phosphate   A triosephosphate that results from the decomposition of hexosephosphate in the chemistry of muscle contraction; an intermediate in glycolysis.

glycerol   A trihydric sugar alcohol, $CH_2OH \cdot CHOH \cdot CH_2OH$, being the alcoholic component of the fats; it is soluble in water and alcohol.

glycolysis (gly KAHL uh sihs)   The breaking down of sugars into simpler compounds, chiefly pyruvate or lactate.

gnotobiotic (noh toh by AHT ihk)   Pertaining to an animal delivered and reared aseptically, and possessing no normal flora.

goblet cell   A form of epithelial cell containing a large globule of mucin and bulged out like a goblet.

Golgi complex (GAHL jee)   A complex cuplike structure within cells, made up of several elements, each consisting of a number of saccules. They are membrane sites of the formation of the carbohydrate side-chains of glycoproteins and mucopolysaccharides, and for other substances.

gonococcal conjunctivitis (gahn uh KAHK uhl kuhn juhnk tuh VY tuhs)   A severe form of inflammation of the conjunctiva, generally consisting of conjunctiva hyperemia associated with a discharge, caused by infection with gonococci.

gonorrhea   Infection due to *Neisseria gonorrhoeae* transmitted venereally in most cases, but also by contact with infected exudates in neonatal children at birth, or by infants in households with infected inhabitants.

graft-vs.-host disease   An immunological disease in which transplanted bone marrow cells develop an immune response towards the host's histocompatibility antigens.

Gram-negative   Losing the stain or decolorized by alcohol in Gram's method of staining, a primary characteristic of certain microorganisms.

Gram-positive   Retaining the stain or resisting decolorization by alcohol in Gram's method of staining, a primary characteristic of certain microorganisms.

grana (GRAY nuh)   Bodies within the chloroplasts of plant cells that contain layers composed of chlorophyll and phosphatides.

granulocyte   A mature granular leukocyte, including neutrophilic, acidophilic, and basophilic types of polymorphonuclear leukocytes, i.e., respectively, neutrophils, eosinophils, and basophils.

granuloma   A tumor-like mass or nodule of granulation tissue, with actively growing fibroblasts and capillary buds, due to a chronic inflammatory process associated with an infectious disease, such as tuberculosis, syphilis, and lymphogranuloma linguinale, or invasion by nonliving foreign bodies.

gray matter The ganglionic or cellular portion of the brain and spinal cord.

gray water In sanitary engineering, a term used to describe water that is contaminated with human flora from washing activities, as from showers, kitchen sinks, and laundries.

Group A streptococci Streptococci classified as Group A by the Lancefield scheme; particularly *S. pyogenes.*

growth curve The curve obtained by plotting increase in size or numbers against the elapsed time, as a measure of the growth of a child, or the multiplication of microorganisms.

growth limitation A characteristic of normal cells in culture; when closely surrounded by other similar cells they stop growing.

growth range A range of particular environmental condition over which growth can occur, such as a temperature range.

guanosine 5′-triphosphate GTP; similar to ATP; immediate precursor of quanine nucleotides in RNA.

Guillain-Barré syndrome Acute febrile polyneuritis.

gumma An infectious granuloma that is characteristic of tertiary syphilis, but does not develop with regularity and is observed only infrequently.

hair follicle (FAHL uh kuhl) One of the tubular invaginations of the epidermis that enclose the hairs, and from which the hairs grow.

half-life The time in which the radioactivity originally associated with a sample of isotopes will be reduced by one half through radioactive decay.

halogen (HAL uh juhn) An element of a closely related chemical family, all of which form similar (saltlike) compounds in combination with sodium and most other metals. Includes fluorine, chlorine, bromine, and iodine.

halophile (HAL uh fyl) A microorganism that requires a high concentration of salt for optimal growth.

H antigen The antigen of the flagella of motile bacteria.

hand-foot-and-mouth disease A mild, highly infectious viral disease of children, characterized by vesicular lesions in the mouth and on the hands and feet.

haploid cell An individual or cell having only one member of each pair of homologous chromosomes.

hapten (HAP tuhn) A specific protein-free substance whose chemical configuration is such that it can interact with specific combining groups on an antibody, but that, unlike antigenic determinants, does not itself elicit the formation of a detectable amount of antibody.

hard chancre (SHANG kur) The primary sore of syphilis; a painless, indurated, eroded papule occurring at the site of entry of the infection.

HBVs antigen The surface antigen of the hepatis B particle.

heat-labile enterotoxin (LT) A toxic substance produced by *Escherichia coli,* essential to its enteropathogenic activity.

heat of vaporization Heat of evaporation. The heat absorbed in the evaporation of water, sweat, or other liquid.

heavy chain Any of the large polypeptide chains of five classes that, paired with the light chains, make up the antibody molecule.

helper virus A virus (e.g., the Rous-associated virus) that aids the development of a defective virus by supplying or restoring the activity of a viral gene or enabling a defective virus (e.g., the Rous virus) to form a protein coat.

hemagglutination (hee muh gloot uhn AY shuhn) Agglutination of erythrocytes, which may be caused by antibodies, by certain virus particles, or by substances such as high-molecular-weight dextrans.

hemagglutination inhibition (HI) Inhibition of nonimmune hemagglutination by antibody specific for the nonspecific hemagglutination.

hemagglutinin (hee muh GLOOT ihn ihn) An antibody that agglutinates erythrocytes, classified according to the cells which it agglutinates.

hematogenous spread Dispersion of an infectious agent through the tissues via the bloodstream.

hematopoietic tissue (hih mat uh poy EHT ihk) Tissue that takes part in the production of the formed elements of the blood.

hematopoietic stem cell A primitive cell in the bone marrow that gives rise through division and differentiation to mature blood cells.

hemolysin (hee muh LYS uhn) A substance that liberates hemoglobulin from red blood corpuscles by interrupting their structural integrity.

hemorrhage The escape of blood from the vessels; bleeding.

hepatitis Inflammation of the liver.

hepatitis A Infectious hepatitis; a viral disease in which there is some liver damage, usually transmitted by the oral–fecal route.

hepatitis A virus (HAV) An RNA virus that causes infectious hepatitis. It is most frequently transmitted by the oral–fecal route.

hepatitis B Serum hepatitis; a more serious liver infection transmitted predominantly by blood.

hepatitis B immune globulin (HBIG) A prophylactic human serum product containing antibodies against Hepatitis B virus, used to treat exposed contacts.

hepatitis B virus (HBV) A DNA virus, the agent causing serum hepatitis, transmitted by inadequately sterilized syringes and needles, or through infectious blood plasma, or certain blood products.

hepatitis virus non-A non-B One or more viruses discrete from the HAV or HBV viruses, frequent causes of serum hepatitis following transfusions.

hepatitis, infectious An acute viral illness of worldwide distribution, occurring most commonly in children and young adults. It is usually transmitted by oral ingestion of infected material, but may also be transmitted by blood transfusion. It is caused by Hepatitis A virus.

hepatitis, serum An acute viral illness, formerly considered to be transmitted only by parenteral exposure, but now known to be transmitted by oral ingestion of contaminated material.

herd immunity The resistance of a group to attack by a disease because of the immunity of a large proportion of the members and the consequent lessening of the likelihood of an affected individual coming into contact with a susceptible individual.

herpangina (hur pan JY nuh) A specific infectious disease characterized by sudden onset of fever of short duration and appearance of typical vesicular or ulcerated lesions in the facial area or on the soft palate.

herpes virus (HUR peez) Any of a large group of DNA viruses found in many animal species, with a nucleocapsid about 100 mμ in diameter, composed of 162 capsomers, and sometimes enclosed in a loose membrane.

heterocysts Specialized cells in some cyanobacteria, carrying out the fixation of atmospheric nitrogen gas into amino acids.

heterograft Xenograft; grafted tissue from a species different than the recipient.

heterophile antibody (HEHT uh roh fyl) An accessory antibody produced by the injection of a heterogenetic or heterophilic antigen. Its appearance during infectious mononucleosis may be diagnostic.

heterotrophic Not self-sustaining; said of organisms that require a reduced form of carbon for energy and synthesis.

hexose A monosaccharide containing six carbon atoms in a molecule.

hierarchical classification A taxonomic scheme in which the categories are ordered in a sequence of increasing size and inclusiveness.

high-efficiency particulate air (HEPA) **filter** An air filter that removes very small particles and can deliver sterile air.

high-energy bond A chemical bond the hydrolysis of which yields high levels of free energy; such bonds involve phosphate or sulfur or other mixed anhydride types of chemical structure.

high frequency of recombination (Hfr) A strain of bacterium that has the ability to transfer its chromosome to an F$^-$ cell. The F particle is integrated into the chromosome of the Hfr cell.

high-risk patient A compromised host; an individual with a much higher-than-average risk of contracting infectious disease.

high-yield mutant A mutant that has acquired the potential for producing unusually high concentrations of a desired product, such as penicillin.

histamine A depressor amine derived from histidine by decarboxylation. Present in ergot and in animal tissues, also formed from histidine by putrefaction; powerful stimulant of gastric secretion and constrictor of bronchial smooth muscle.

histocompatibility antigens (hihs toh kuhm pat uh BIHL uht ee) Genetically determined isoantigens, present on the lipoprotein membranes of nucleated cells of most tissues, that incite an immune response when grafted onto a genetically disparate individual and thus determine the compatibility of tissues in transplantation.

histone (HIHS tohn) A simple protein containing many basic groups, soluble in water and insoluble in dilute ammonia.

histoplasmin (hihs toh PLAZ mihn) A sterile broth filtrate of a culture of *Histoplasma capsulatum*, injected intracutaneously as a test for histoplasmosis.

histoplasmosis (hihs toh plaz MOH suhs) Infection resulting from inhalation or, infrequently, the ingestion of spores of *Histoplasma capsulatum*.

histoplasmosis, chronic cavitary A form of the disease in which cellular immune responses are too weak to control the spread of the disease; a cavity forms and gradually enlarges.

histoplasmosis, primary acute The most common form of the disease, in which a period of rapid fungal growth is followed by development of effective CMI and the disease process is arrested at an early, usually inapparent stage.

histoplasmosis, severe disseminated A rapidly spreading, unchecked form of the disease in which immune responses are essentially nonfunctional; death will occur unless effective chemotherapy is supplied.

holdfast An adhesive material at a localized position on a cell, enabling the cell to attach to a surface. An attachment structure in the macroscopic algae.

holoenzyme (hoh loh EHN zym) The functional compound formed by the combination of an apoenzyme and its appropriate coenzyme.

hook A portion of the structure of the bacterial flagellum, bent at a right angle to the cell.

horizontal transmission The spread of an infectious agent from one individual to another, usually through contact with excreta, e.g., sputum, containing the agent.

hospital-acquired infection An infection acquired while the individual was within a clinical facility.

hospital strain A bacterial strain endemic to hospital staff and patients; often it possesses resistance to multiple drugs.

human diploid cell vaccine (HDCV) A rabies vaccine produced by culturing the virus within human epithelian cells.

humoral immunity (HYOOM (uh) ruhl) Acquired immunity in which the role of circulating antibodies (immunoglobulins) is predominant.

humus (HYOO muhs) Decayed vegetable material. A residue containing many complex polymers that are difficult to degrade. Humus forms the nutritive and water-retaining component of topsoil.

hyaluronic acid (hyl yoo RAHN ihk) A mucopolysaccharide that is a polymer of acetylglucosamine and glucuronic acid.

hyaluronidase (hyl yoo RAHN uh days) Hyaluronate glycanohydrolase: An enzyme that catalyzes the hydrolysis of hyaluronic acid, the cement substance of the tissues.

hydrolase One of the six main groups of enzymes, comprising those that catalyze the hydrolytic cleavage of a compound such as esters, peptides, glycosides, and amides.

hydrolysis The splitting of a compound into fragments by the addition of water, the hydroxyl group being incorporated in one fragment, and the hydrogen atom in the other.

hydrolyzed casein (kay SEEN) A substance used in many microbiological media, derived by enzymatic digestion of casein, the protein of milk.

hydrophilic Readily absorbing moisture, or mixing with water, or dissolving in water.

hydrophobia Rabies.

hydrophobic Not readily mixing with water, not soluble in water.

5-hydroxymethyl cytosine (FYV-hy drahk see MEHTH uhl SYT uh seen) A pyrimidine that replaces cytosine in the DNA of certain coliphages.

hyperacute rejection An unusual form of graft rejection in which the patient has been previously sensitized to antigens of the grafted tissue, and rejection begins within minutes or hours.

hypersensitivity A state of altered reactivity in which the body reacts with an exaggerated response to a foreign agent.

hypertonic A biological term denoting a solution that when bathing body cells causes a net flow of water across the semipermeable cell membrane out of the cell.

hyphae (HY fee)   Plural of hypha. The filaments or threads composing the mycelium of a fungus.

hypotonic   Denoting a solution that, when bathing body cells, causes a net flow of water across the semipermeable cell membrane into the cell.

hypovolemia (hy poh voh LEE mee uh)   Abnormal decrease in the volume of circulating fluid (plasma) in the body.

iatrogenic disease (y a truh JEHN ihk)   A disease resulting from the activity of physicians. Originally applied to disorders induced in the patient by autosuggestion based on the physician's examination, manner, or discussion, the term is now applied to any adverse condition in a patient occurring as the result of treatment by a physician or surgeon.

icosahedron   A geometrical shape occurring in many virus particles, with 20 triangular faces and 12 corners.

icteric phase   In hepatitis or other liver disease, the period where retained pigmented waste products cause the skin to become jaundiced.

ID$_{50}$ Median infective dose, being that amount of pathogenic microorganisms that will produce infection in 50 percent of the test subjects.

idiopathic thrombocytopenia purpura (ihd ee uh PATH ihk thrahm buh syt uh PEE nee uh PUR p(y)uh rah)   Thrombocytopenic purpura unassociated with any definable systematic disease but often accompanied by the presence of a serum antiplatelet factor, now characterized as an IgG immunoglobulin.

IgA, IgD, IgE, IgG, IgM   Five classes of immunoglobulins.

ileum (IHL ee uhm)   The distal portion of the small intestine, extending from the jejunum to the cecum.

immediate hypersensitivity (IHS)   A hypersensitivity mediated by immunoglobulins.

immobilization (TPI) **test**   *Treponema pallidum* immobilization test. A specific serological test for syphilis, utilizing an antigen obtained by differential centrifugation of *Treponema pallidum* from infected rabbit tissue, cryolysis, and ammonium sulfate precipation.

immune adherence   A complement-dependent phenomenon in which antigen–antibody complexes or particulate antigens coated with antibody (e.g., antibody-coated bacteria) adhere to red blood cells when complement component C3 is bound.

immune-complex disease   A state in which circulating antigen–antibody complexes, formed by coexisting immune reactions, induce vascular injury.

immune effectors   May be either immunoglobulins or immune cells (lymphocytes, macrophages) that attack specific targets and carry out the destructive effects of an immune response.

immune electron microscopy (IEM)   A procedure in which some subcellular structures are immunologically labeled with an electron-opaque marker.

immune serum   A serum containing one or more antibodies, especially one in which the antibody content has been increased by recovery from its specific infection or by injection with its specific antigen.

immune serum globulin (ISG)   A sterile solution of globulins, derived from the blood plasma of normal, adult human donors, that is prepared from immune serum globulin that complies, after dilution if necessary, with the measles antibody requirements of the United States Public Health Service.

immune surveillance   A hypothesis for the immunological control of neoplasms that states that the immune system continually searches for, detects, and destroys tumor cells bearing new antigens.

immune tolerance   A process whereby the immune system is rendered permanently nonreactive to a specific antigen.

immunization   The process of rendering a subject immune, or of becoming immune.

immunodeficiency   A deficiency in immune response, either in that mediated by humoral antibody or in that mediated by immune lymphoid cells.

immunoelectrophoresis (ihm yuh noh uh lehk trah fuh REE suhs)   A method combining electrophoresis and double diffusion for distinguishing between proteins and other materials by means of differences in their electrophoretic mobility and antigenic specificities.

immunoglobulin   A protein of animal origin endowed with known antibody activity; produced by plasma cells.

immunological enhancement   Greatly increased survival of incompatible grafts (tumors or normal tissue) caused by specific humoral antibodies.

immunology   That branch of biomedical science concerned with the response of the organism to antigenic challenge, the recognition of self from not self, and all the biological (*in vivo*), serological (*in vitro*), and physical chemical aspects of immune phenomena.

immunopathology   That branch of biomedical science concerned with immune reactions associated with disease, whether the reactions be beneficial, without effect, or harmful.

immunoprophylaxis (ihm yuh noh proh fuh LAK suhs)   The prevention of disease by the use of vaccines or therapeutic antisera.

immunosuppression   The artificial prevention or diminution of the immune response, as by irradiation or by administration of antimetabolites, antilymphocyte serum, or specific antibody.

immunotherapy   Passive immunity of an individual conferred by administration of preformed antibodies (serum or gamma globulin) actively produced in another individual. Treatment of disease by specific or nonspecific elevation of immune response.

inactivated polio vaccine (IPV)   A vaccine in which the polio virus has been rendered completely incapable of replication in the recipient.

inbred line   In mice or other laboratory animals, a line or strain derived from many generations of brother–sister matings. All individuals are genetically identical.

inclusion body   Round-, oval-, or irregular-shaped bodies occurring the cytoplasm and nuclei of cells of the body, as in disease caused by virus infection such as rabies, smallpox, herpes, etc.

inclusion conjunctivitis   Conjunctivitis caused by *Chlamydia trachomatis*; primarily it affects newborn infants, beginning as an acute purulent conjunctivitis that leads to papillary hypertrophy of the palpebral conjunctiva.

incubation period   The interval of time required for development; the period of time between the moment of entrance of the infecting organism into the body and the first symptoms of the consequent disease, or between the movement of entrance into a vector and the time at which the vector is capable of transmitting the disease.

incubatory carrier   Carrier of an infectious agent who, without exhibiting symptoms, is already excreting the agent into the surroundings.

index case   The case of the original patient (propositus, or proband) that provides the stimulus for study of other members of the family, to acertain possible factors in causation of the presenting condition.

indicator culture   A differential culture medium designed to reflect or accentuate physiological characteristics of a bacterium by characteristic colonial morphology.

indicator organism   An organism whose presence or behavior in its environment is a sensitive indicator of some environmental change.

indicator system   In the complement fixation test, the sheep RBC and antisheep hemolysis added to indicate the presence of unfixed complement.

indicator tape   A form of adhesive tape impregnated with a temperature-sensitive dye. When placed on objects to be autoclaved, a color change in this dye indicates successful completion of the autoclave run.

indirect immunofluorescence (ihm yuh noh flu(uh)r EHS uhn(t)s)   Immunofluorescence in which neither the antigen nor its specific antibody is fluorescein-labeled, but the combination of antibody with antigen is determined by the use of fluorescein-labeled antiglobulin.

indirect transmission   Transmission of an infectious disease involving two or more steps or intermediates, often delayed.

individual immunization   Immunization of an individual for protection against some less common occupational or other microbial hazard.

induced mutation   A genetic mutation caused by external factors that are experimentally or accidentally produced.

inducible enzyme   One whose production requires or is markedly stimulated by a specific small molecule, the

inducer, that is the substrate of the enzyme or a compound structurally related to it.

induction   The act of process of inducing or causing to occur; in metabolism, the induction of synthesis of certain enzymes by contact with their substrates.

induration (ihn d(y)uh RAY shuhn)   The quality of being hard; the process of hardening.

infant botulism   A form of botulism in which spores of the agent, ingested by an infant, have germinated and *C. botulinum* has become part of the bowel flora, producing toxin.

infection   Invasion and multiplication of microorganisms in body tissues, resulting in local cellular injury due to competitive metabolism, toxins, intracellular replication, or antigen–antibody response.

infection control   A systematic approach to the control of sources of infection and means of transmission such as within the hospital.

infection control nurse   A nurse practitioner with special training in epidemiology, who carries out a variety of specialized infection control activities.

infection control team   A group of individuals representing various constituencies within the hospital staff, working together to carry out infection control measures.

infectious disease   A disease caused by organisms ranging in size from viruses to parasitic worms; it may be contagious in origin, result from nosocomial (acquired in hospitals) organisms, or be due to endogenous microflora from the nose and throat, skin, or bowel.

infectious (infective) endocarditis   A febrile systemic disease marked by focal bacterial or fungal (mycotic) infection of the heart valves, with formation of bacteria- or fungi-laden vegetations.

infectious mononucleosis (IM) (mahn oh n(y)oo klee OH suhs)   An acute infectious disease associated with the Epstein–Barr virus and characterized by fever, malaise, sore throat, hepatic dysfunction, lymphadenopathy, hepatosplenomegaly, atypical lymphocytes (resembling monocytes) in the peripheral blood, and high titers of agglutinins against sheet cells (heterophile titer).

inflammatory response   A nonspecific response to injury characterized by red-

ness, heat, swelling, and pain in the affected area.

influenza   An acute viral infection involving the respiratory tract, occurring in isolated cases, in epidemics, or in pandemics striking many continents simultaneously or in sequence.

informational macromolecule   A macromolecule made up of a variable sequence of subunits, where that sequence is essential to its function, and where the sequence is established by synthesis of the macromolecule on a template.

infrared (IR)   Denoting thermal radiation of wavelengths greater than that of the red end of the spectrum, between the red waves and the radio waves, having wavelengths between 7700 and 120,000 angstroms.

initial body   The first intracellular form in chlamydial replication.

initiation   In protein synthesis, the stage at which the translating complex of message, ribosome, and first transfer RNA assemble.

initiation complex   The combination of messenger RNA, ribosomal subunits and transfer RNA that initiates protein synthesis.

initiation factors   A group of proteins, cytoplasmic and ribosomal, necessary to initiate translation.

innate resistance   A form of resistance to a particular infective agent based on genetic factors of the race of species.

inoculum (ihn AHK yuh luhm)   The substance used in inoculation.

inoculum size   A quantitative measurement of the number of viable particles of an agent needed to initiate infection on the average.

inorganic   Describing chemical compounds lacking carbon.

insect vector   An insect or other arthropod that transmits an infectious disease, in saliva, feces, or mechanically on its body parts.

intercalation (ihn tur kuh LAY shuhn)   Insertion of extraneous material; the insertion of dye molecules into the DNA double helix, leading to spatial distortion.

interferon (ihnt uhr FIH(UH)R ahn)   A class of soluble small proteins that inhibit virus multiplication; they are produced by cells infected by almost any animal virus containing either DNA or RNA, in tissue culture or in the animal.

intermittent sterilization (ihnt ur MIHT uhnt stehr uh luh ZAY shuhn) Destruction of microbial viability by successive application of the procedure at intervals, to allow spores to develop into vegetative forms, which are more easily destroyed.

intertrigo (ihn tur TRY goh) Superficial dermatitis occurring on apposed surfaces of the skin, as the creases of the neck, folds of the groin, and armpit, and beneath pendulous breasts.

intervascular coagulation The coagulation of blood within the vessels, usually in the capillary bed surrounding an infection focus or a wound.

intoxication Poisoning; the state of being poisoned.

intrauterine infection Infection of the fetus within the uterus.

in-use tests Protocols that test the effectiveness of a disinfectant under conditions similar or identical to actual applications.

in utero Within the uterus.

invasiveness The ability of a microorganism to enter the body and to spread more or less widely throughout the tissues.

in vitro Within a glass; observable in a test tube; in an artificial environment.

in vivo Occurring within a living organism.

iodophore (y OHD uh faw(uh)r) A chemical substance that binds iodine.

ionic bond (y AHN ihk) A weak chemical bond between molecules having ionic groups that contain one or more units of net positive or negative charge.

ionization The disocciation of a substance in solution into ions.

ionizing radiation High-energy radiation (X-rays and gamma rays) which interacts to produce ion pairs in matter.

iridovirus (ihr uh doh VY rahs) Any of a group of large, morphologically similar DNA viruses that infect the larvae of various insects, giving an iridescent appearance to the infected insect.

isograft A graft between genetically identical individuals. Typically, isografts are grafts between identical twins, between animals of a single highly inbred strain, or between the F_1 hybrid produced by crossing inbred strains.

isolation The process of isolating, or the state of being isolated, such as (1) the physiologic separation of a part, as by tissue culture or by interposition of inert material; (2) the chemical extraction of an unknown substance in pure form from a tissue; (3) the separation from contact with others of patients having a communicable disease; or (4) the successive propagation of a growth of microorganisms until a pure culture is obtained.

isomerase (y SAHM uh rays) A major class of enzymes comprising those that catalyze the process of isomerization, such as the interconversion of aldoses and ketoses, or the shift of a double bond.

isoniazid (y suh NY uh zuhd) Chemical name: isonicotinic acid hydrazide. Colorless or white crystals or powder, $C_6H_7N_3O$, of tuberculosis.

isoosmotic (y soh ahz MAHT ihk) Having the same osmotic pressure.

isotope A chemical element having the same atomic number as another (i.e., the same number of nuclear protons) but possessing a different atomic mass (i.e., a different number of nuclear neutrons).

jaundice A condition characterized by abnormally large amounts of certain hemoglobulin breakdown products in the blood, skin, and mucous membranes, giving them a yellow color.

jejunum (jih JOO nuhm) The middle part of the small intestine, between the duodenum and the ileum.

Jenner, Edward English physician who introduced vaccination.

kappa chain One of the type of polypeptide found in the immunoglobulin complex.

kappa particle An intracellular symbiont of *Paramecium*.

keratin One of the group of tough, fibrous proteins formed by certain epidermal tissues and especially abundant in skin, claws, hair, feathers, and hooves.

keratitis Infection of the cornea of the eye.

kidney In vertebrates, the organ that regulates the balance of water and solutes in the blood and the excretion of nitrogen wastes in the form of urine.

kinetic theory Used to describe the dynamics of ongoing processes, such as growth of bacteria on enzyme reactions.

kinins (KY nihnz) A group of peptide substances found in blood and tissue in inactive form, which upon activation play a role in causing inflammation, blood clotting, and other effects.

Kirby–Bauer method The standardized semiquantitative disc-diffusion method of determining antimicrobial susceptibility.

Koch, Robert A German scientist of the Golden Age of bacteriology. He and his co-workers made primary contributions to the germ theory, isolated many important pathogens, developed vaccines, and contributed many of the basic laboratory methods we use today.

Koch's postulates A sequence of four logical experimental steps necessary to establish the causative role of an infective agent in a particular disease.

Koplik's spots Small, red lesions on the oral membranes, appearing as the earliest sign of measles in most cases.

Krebs cycle State of cellular respiration in which pyruvate fragments are completely broken down into carbon dioxide; molecules reduced in the process can be used in ATP formation.

Kupffer cells Fixed macrophages embedded along the sinusoids of the liver.

labia (LAY bee uh) Plural of labium. A fleshy border or edge; used in anatomical nomenclature as a general term to designate such a structure. In the plural, often used alone to designate the labia majora and minora pudendi.

lacrimal glands Glands that produce lacrimal secretions or tears.

lactic-acid bacteria Bacteria of several groups, especially the streptococci and lactobacilli, with fermentative metabolism producing large amounts of lactic acid as a by-product.

lactoferrin An iron-binding protein found in milk and other body fluids. By holding iron in a form unavailable to microorganisms, this protein inhibits growth of microbial invaders.

lagging strand In DNA synthesis the strand that is synthesized somewhat after the other.

lag phase A brief period in the course of a bacterial culture, especially at the beginning, during which the growth is very slow or scarcely appreciable.

lambda chain One of the polypeptide chains found in immunoglobulins.

laminar flow hood A laboratory hood in which a layer of air is directed across the work area.

Lancefield typing scheme A method for antigenic typing of the surface antigens of the streptococci, by which they may be placed in one of many lettered groups.

large intestine The intestine is the digestive tube passing from the stomach to the anus. It is divided primarily into the small and large intestine. The large intestine is divided into cecum and appendix, ascending, transverse, descending, and sigmoid colons, and rectum.

laryngitis Laryngeal catarrh; inflammation of the mucous membrane of the larynx.

larynx The organ of voice production.

late proteins In viral infection, those proteins coded by viral genes that appear late in the infective process; usually becoming part of the assembled virion.

lawn A confluent growth of bacteria on an agar plate; used to test the effects of antimicrobial drugs or bacteriophages.

LD$_{50}$ Abbreviation for the median lethal dose, one that is fatal to 50 percent of the test animals.

lecithinase (LEHS uh thuh nays) Phospholipase.

Leeuwenhoek, (LAY wuhn hohk) **Anton van** Dutch microscopist, 1632–1723.

Legionnaire's disease A form of pneumonia caused by *Legionella pneumophila.*

lepromatous leprosy (lehp ROH muh tuhs LEHP rah see) The malignant form with modular cutaneous lesions that are infiltrated and have ill-defined borders. The lesions are bacteriologically positive. Lepromin test is negative.

lepromin (LEHP roh mihn) An extract of infected tissue used in skin tests of resistance to leprosy.

L–E test A test done on blood freshly drawn from a patient suspected of having lupus erythematosis.

leukemia Progressive proliferation of abnormal leukocytes found in hemopoietic tissues, other organs, and usually in the blood in increased numbers.

leukencephalapathy (PML) A rare degenerative disease of the brain, believed to be caused by slow viruses of the polyoma group.

leukocidin A substance able to destroy phagocytes.

leukocyte Any colorless, ameboid cell mass.

leukocytosis (loo kuh sy TOH suhs) A transient increase in the number of leukocytes in the blood, resulting from various causes, as hemorrhage, fever, infection, inflammation, etc.

leukopenia (loo kuh PEE nee uh) Reduction in the number of leukocytes in the blood, the count being 5000 or less.

levan (LEE van) A hexosan from various grasses that on hydrolysis yields levulose.

L-form L-phase variant. A bacterial variant that has a defective cell wall but can multiply on hypertonic medium.

ligament A band of fibrous tissue that connects bones or cartilages, serving to support and strengthen joints.

ligase Any of a class of enzymes that catalyze the joining together of two molecules coupled with the breakdown of a pyrophosphate bond in ATP or a similar triphosphate.

light chain Either of two small polypeptide chains (molecular weight 22,000) that, when linked to heavy chains by disulfide bonds, make up the antibody molecule; they are of two types, kappa and lambda, that are unrelated to immunoglobulin class differences.

light-harvesting pigments Pigments, such as the carotenoids, found in association with chlorophyll in photosynthetic reaction centers.

light microscope One in which the specimen is viewed under considerable light.

light reaction That portion of photosynthesis in which protons are captured and ATP and reducing power are generated.

lipid Any of a group of organic substances that are insoluble in water, but soluble in alcohol, ether, chloroform, and other fat solvents and which have a greasy feel.

lipid A The lipid component of the lipopolysaccharide of the Gram-negative bacterial outer membrane. Has toxic effects.

lipid bilayer The most visible structural component of a cytoplasmic membrane; two layers of oriented phospholipids, in a mosaic arrangement with molecules of other lipids, proteins, and coenzymes.

lipopolysaccharide (ly poh pahl ih SAK uh ryd) A molecule or compound in which lipids and polysaccharides are linked, as in cell membranes.

lithotroph (LIHTH uh trohf) A microorganism that lives in and obtains energy from the oxidation of inorganic materials.

liver A large gland of a dark-red color situated in the upper part of the abdomen on the right side.

logarithmic death curve A graphic representation and mathematical analysis of the death of a bacterial population under controlled conditions.

log phase The stage of growth of a bacterial culture when the cells are multiplying exponentially.

lower respiratory tract (LRT) That portion of the respiratory tract from the epiglottis down to and including the lungs.

luciferase (loo SIHF uh rays) An enzyme, of which there are many forms, that catalyzes the bioluminescent reaction in certain animals capable of luminescence.

luciferin A heterocyclic phenol that can be reduced and oxidized. A component of the light-emitting photochemical system of luminescent organisms.

Ludwig's angina (LUD wihgz an JY nuh) Diffuse purulent inflammation of the floor of the mouth, its facial spaces, muscles, and glands, usually due to streptococcal infection.

lumen (LOO muhn) The cavity or channel within a tube or tubular organ.

lungs The organs of respiration.

lyase Any of a class of enzymes that remove groups from their substrates (other than by hydrolysis), leaving double bonds, or that conversely add groups to double bonds.

lymphatic Pertaining to lymph or a lymph vessel.

lymph capillary One of the most minute vessels of the lymphatic system, having a caliber slightly greater than that of a capillary of the blood circulatory system.

lymph node Any of the accumulations of lymphoid tissue organized as definite lymphoid organs, varying from 1 to 25 mm in diameter, situated along the course of lymphatic vessels, and consisting of an outer cortical and an inner medullary part.

lymphocyte A mononuclear leukocyte 7μ to 20μ in diameter, with a deeply staining nucleus containing dense chromatin, and a pale-blue-staining cytoplasm.

lymphocyte processing   The developmental process by which fetal lymphocytes become conditioned and differentiated as either B-cells or T-cells.

lymphoid tissue   A latticework of reticular tissue the interspaces of which contain lymphocytes.

lymphokines (LIHM fuh kynz)   A general term for soluble protein mediators postulated to be released by sensitized lymphocytes on contact with antigen and believed to play a role in macrophage activation, lymphocyte transformation, and cell-mediated immunity.

lyophilization (ly ahf uh luh ZAY shuhn)   Freeze drying.

lysogenic conversion (ly suh JEHN ihk)   The change in the properties of bacteria as a result of their carrying a prophage.

lysogenic cycle   Bacteria carry a bacteriophage genome integrated into the bacterial chromosome. The viral genome may subsequently begin to replicate independently and set up an active cycle of infection, causing lysis of the bacterial cells.

lysosome   One of the minute bodies seen with the electron microscope in many types of cells, containing various hydrolytic enzymes and normally involved in the process of localized intracellular digestion.

lysozyme   A crystalline, basic enzyme present in saliva, tears, egg white, and many animal fluids and which functions as an antibacterial agent, especially effective in lysing *Micrococcus lysodeikticus.*

M protein   The streptococcus M antigen.

macroconidium (mak roh kuh NIHD ee uhm)   A large, frequently multicelled conidium or exospore.

macrogametocyte (mak roh guh MEET uh syt)   The infected red blood cell containing the female form of the malarial parasite that, transferred from human beings to the mosquito, becomes a macrogamete.

macromolecule   A very large molecule having a polymeric chain structure, as in proteins, polysaccharides, and other natural and synthetic polymers.

macronucleus   In ciliate protozoa, the larger of the two types of nucleus in each cell that is required for vegetative but not for sexual reproduction.

macrophage   Any of the large, highly phagocytic cells with a small, oval, some-times idented nucleus and inconspicuous nuclei, occurring in the walls of blood vessels (adventitial cells) and in loose connective tissue (histiocytes, phagocytic reticular cells).

macrophage inhibitory factor (MIF)   A lymphokine that immobilizes macrophages at the site of a cellular immune response.

magnification   Apparent increase in size as under the microscope.

major histocompatibility complex (MHC)   The genetic region in human beings that controls not only histocompatibility but apparently also all aspects of the development and activation of the immune resonses.

major reaction   When a smallpox vaccination has been received, the strong local skin reaction that indicates that the vaccine virus has established itself and is replicating; thus, that immunization will occur.

malaria   An infectious febrile disease caused by protozoa of the genus **Plasmodium** that are transmitted by the bites of infected mosquitoes of the genus **Anopheles.**

malignant tumor   A tumor that has the properties of invasion and metastasis and that shows a greater degree of anaplasia than do benign tumors.

mammary tumor virus (MTV)   A retrovirus of the B type, found to cause mammary tumors in mice of certain strains.

manganese nodules (MANG guh neez NAHJ oo(uh)lz)   Irregularly shaped nodules of crude mixed metallic ore in which manganese predominates, found in large numbers on the ocean floor in several areas of the world.

Mantoux test (for tuberculosis)   Give intracutaneous injection of 0.1 mL of desired dilution of tuberculin and successive injections of gradually increasing concentration until a reaction occurs.

Marek's disease   A lymphoproliferative disease of chickens caused by a herpesvirus.

mass immunization   General immunization of all unprotected members of a population threatened with epidemic disease, usually a crash program.

mast cell   A connective tissue cell that releases histamine in Type I hypersensitivity.

mastitis (ma STYT uhs)   Inflammation of the mammary gland, or breast.

mastoid cells   Numerous small intercommunicating cavities in the mastoid process of the temporal bone.

mastoiditis (mas toyd YT uhs)   Inflammation of the mastoid antrum and cells.

matrix   The groundwork on which anything is based, or the basic material from which a thing develops.

matter   (1) Substance; anything that occupies space. (2) Pus.

mechanical vector   An arthropod vector that transmits an infective organism from one host to another but that is not essential to the life cycle of the parasite.

medulla (muh DUHL uh)   The inmost part; a general term for the inmost portion of an organ or structure.

megakaryocyte (mehg uh KAR ee oh syt)   The giant cell of bone marrow.

meiosis (my OH suhs)   A special method of cell division, occurring in maturation of the sex cells, by means of which each daughter nucleus receives half the number of chromosomes characteristic of the somatic cells of the species.

membrane filter   A type of filter formed from cellulose acetate or other polymer, precipitated from solution under controlled conditions to form a meshwork of predetermined pore size.

membrane lysis   Lysis of a cell due to irreversible weakening of its cytoplasmic membrane.

meninges (muh NIHN jeez)   The three membranes that envelop the brain and spinal cord: the dura mater, pia mater, and arachnoid.

meningococcemia (muh nihng guh kahk SEE mee uh)   Invasion of the blood stream by meningococci.

merozoite (mehr uh ZOH yt)   One of the forms derived from the splitting up of the schizont in the human cycle of the malarial plasmodium. It is released into the circulating blood and invades new erythrocytes.

mesentery (MEHZ uhn tehr ee)   A membranous fold attaching various organs to the body wall. Commonly used with specific reference to the peritoneal fold attaching the small intestine to the dorsal body wall.

mesophile (MEHZ uh fyl)   An organism that grows best at temperatures between 20° and 45° C.

mesosome (MEHZ uh sohm)　An invagination of the cell membrane occurring in certain bacteria. Various memosomes are associated with DNA replication and with protein secretion.

messenger RNA (mRNA)　Ribonucleic acid. Messenger RNA is an RNA fraction of intermediate molecular weight, with a base ration corresponding to the DNA of the same organism, that transmits information from DNA to the protein-forming system of the cell.

metabolism　The sum of all the physical and chemical processes by which living organized substance is produced and maintained (anabolism), and also the transformation by which energy is made available for the uses of the organism (catabolism).

metastasis (muh TAS tuh suhs)　The transfer of disease from one organ or part to another not directly connected with it.

metronidazole (meh trah NIHD uh zohl)　White to pale yellow, odorless crystals or crystalline powder, $C_6H_9N_3O_3$, used against a trichomonas.

Michaelis–Menton equation　The fundamental equation describing the relationship of enzymes, their substrates, and their products.

microaerophile (my kroh AR oh fyl)　A microaerophilic microorganism; one which preferentially grows in environments with less than atmospheric oxygen levels and elevated carbon dioxide.

microbioassay　Determination of the active power of a nutrient or other factor by noting its effect upon the growth of a microorganism, as compared with the effect of a standard preparation.

microconidium (my kroh kuh NIHD ee uhm)　A small, usually single-celled conidium or exospore; sometimes used interchangeably with microaleurospore.

microdilution　A technique for carrying out serial dilution of substances using very small volumes of solutions, used in many serological tests and in MIC determinations.

microenvironments　Specialized areas of small volume in which conditions may be significantly different from those immediately surrounding.

microfilament　Any of the filaments about 60 Å in diameter found in the cytoplasmic ground substance.

microgametocyte (my kroh guh MEET uh syt)　The male gametocyte.

micronucleus　In ciliate protozoa, the smaller of two types of nuclei in each cell that is required for sexual but not for vegetative reproduction.

microorganism　A minute living organism, usually microscopic.

microtubule　A cylindrical hollow-appearing structure in the cytoplasmic ground substance of many motile cells, especially erythrocytes; microtubules, which increase in number during mitosis, are found in the mitotic spindle.

middle ear　The space immediately medial to the tympanic membrane.

midstream sample　A urine sample taken after the patient first voids a little to clear out much of the normal flora.

MIF test　A test done to test for the presence of macrophage migration inhibitory factor.

mineralization　Return of the elemental components of decomposing organic matter to an inorganic or mineral form.

minimum inhibitory concentration　The concentration of a particular antimicrobial drug necessary to inhibit growth of a particular strain of microorganism.

minimum lethal concentration　The concentration of an antimicrobial needed to irreversibly inactivate a microorganism, as tested by subculture to antimicrobial-free medium after exposure to the drug.

mitochondrion (myt uh KAHN dree uhn)　Singular of mitochondria. Small spherical to rod-shaped components found in the cytoplasm of cells, enclosed in a double membrane, the inner one having infoldings called cristae.

mitosis (my TOH suhs)　A method of indirect division of a cell, consisting of a complex of various processes, by means of which the two daughter nuclei normally receive identical complements of the number of chromosomes characteristic of the somatic cells of the species.

mixed lymphocyte reaction　An *in vitro* test of tissue compatibility that evaluates the reactivity of lymphocytes from donor and recipient when cultured together.

mixture　A combination of different drugs or ingredients, as a fluid resulting from mixing a fluid with other fluids, or with solids, or a suspension of a solid in a mixture.

mode of transmission　The means or method by which infective agents move

or are moved from reservoir to susceptible host.

moderate anaerobe　An anaerobe that may tolerate only moderately reducing conditions and/or traces of oxygen.

moist heat　Heat used in sterilization during which the atmosphere is saturated with water vapor or steam.

mold　Any of a large group of parasites and saprophytic fungi that cause mold or moldiness, also the deposit or growth produced by such fungi.

molecule　A very small mass of matter; the smallest amount of a substance that can exist alone and still be the substance; an aggregation of atoms; specifically, a chemical combination of two or more atoms that forms a specific chemical substance.

monochromatic filter　A light filter tinted with various colored materials that allows passage of light of only one color and a limited range of wavelengths.

monoclonal immunoglobulin　An immunoglobulin in which all molecules are identical in polypeptide structure because all originated from lymphocytes of the same clone.

monoculture　A term used in agriculture to denote the planting of large areas of land exclusively in a single species.

monocyte (MAHN uh syt)　A mononuclear phagocyte leukocyte, 13μ to 25μ in diameter, with an ovoid or kidney-shaped nucleus, containing lacy, linear chromatin, and abundant gray-blue cytoplasm filled with fine, reddish azurophilic granules.

monolayer　Pertaining to or consisting of a single layer, such as a monolayer sheet of cells in culture in studies of viruses.

monomer　A simple molecule of a compound of relatively low molecular weight.

monosaccharide　A simple sugar; a carbohydrate that cannot be decomposed by hydrolysis.

monotypic virus　A virus that exists in nature in only one form; there is only a single strain.

Montagu, Lady Mary Wortley　An Englishwoman who traveled to China and brought back the practice of variolation.

morbidity rate　The number of cases of a given disease occurring during a specified period per 1,000, 10,000 or 100,000 of population.

mortality rate　Death rate.

motile Having spontaneous but not conscious or volitional movement.

motor neuron (N(Y)OO rahn) Any neutron possessing a motor function; an efferent neuron conveying motor impulses.

mouse leukemia virus (MLV) Several RNA rodent viruses that produce leukemia and sometimes lymphosarcomas in mice, and include the Gross, Moloney, Friend, and Raucher strains.

mucociliary elevator The movement of mucous secretions, carried by ciliary action, up from the bronchial tree to the pharynx.

mucocutaneous candidiasis A severe form of candidiasis in which lesions are found widely distributed on the skin and mucous membranes; usually found only in immunosuppressed persons.

mucopolysaccharide (myoo koh pahl ih SAK uh ryd) A group of polysaccharides that contain hexosamine, which may or may not be combined with protein and which, dispersed in water, form many of the mucins.

multicellular (1) Composed of many cells. (2) Containing many hollow spaces.

multiple drug resistance Simultaneous resistance to several antimicrobials, usually conferred by the presence of a plasmid bearing an assembled sequence of resistance genes.

multistep resistance Antimicrobial resistance acquired over time in many steps, each dictating the need to use a larger dosage of the drug to control the agent.

mumps A contagious myxovirus disease occurring mainly in children and conferring a resultant persistent immunity.

mutagen (MYOOT uh juhn) A chemical or physical agent that induces genetic mutations.

mutation A change in form, quality, or some other characteristic.

mutualism Symbiosis in which both populations (or individuals) gain from the association and are unable to survive without it.

mycelium (my SEE lee uhm) The mass of threadlike processes (hyphae) constituting the fungal thallus.

mycetomes Specialized organs within some insects, housing microbial endosymbionts.

mycoplasma Any member of the genus

Mycoplasma. A taxonomic name given a genus including pleuropneumonia-like organisms (PPLO) and separated into 15 species on the basis of source, glucose fermentation, and growth on agar media.

mycorrhiza (my kuh RY zuh) A growth occurring as a result of the symbiotic relationship between certain fungi and the roots of plants and trees.

mycosis (my KOH suhs) Any disease caused by a fungus.

myelin (MY uh luhn) The lipid substance forming a sheath around certain nerve fibers.

myelinated fiber Grayish, white nerve fibers whose axons are encased in a myelin sheath, which in turn is enclosed by a neurilemma.

myeloma (my uh LOH muh) A tumor composed of cells of the type normally found in the bone marrow.

myeloperoxidase system A catalytic reaction mixture with bacteriocidal effects, contained and released by the intracellular granules of phagocytic cells and having its effect within the phagolysosome.

myocardium The middle and thickest layer of the heart wall, composed of cardiac muscle.

myxospore (MIHK soh spawr) One of a number of spores occurring embedded in a gelatinous mass, noted in certain fungi and protozoan organisms.

myxoviruses A general name for a large group of viruses, including the viruses of influenza, parainfluenza, mumps, and Newcastle disease.

N-acetyl glucosamine (NAG) An amino-substituted hexose used in the synthesis of peptidoglycan.

N-acetyl muramic acid (NAM) An amino-substituted hexose that, with NAG, is incorporated into peptidoglycan.

nares The external orifices of the nose.

narrow spectrum An antimicrobial drug that is effective against only a small number of microorganisms or one or more specific groups.

nasal turbinates The convolutions of the anterior nasal passages formed by protrusions of the cranial bones.

nasolacrimal ducts (nay zoh LAK ruh muhl) The passages leading downward from the lacrimal sac on each side to the anterior portion of the nasopharynx.

nasopharyngeal carcinoma (NPC) A form of carcinoma in which the Epstein-Barr virus has been implicated as the causative agent.

nasopharynx (nay zoh FAR ihng(k)s) The part of the pharynx that lies above the level of the soft palate.

natural active immunity The capacity of the normal animal to respond immunologically by producing an active immunity following a natural exposure to antigen, such as disease.

natural taxonomy A taxonomy that is based insofar as possible on direct estimation of genetic relatedness.

negative chemotaxis (kee moh TAK suhs) Movement of an organism from a region of high to a region of low concentration of a specific chemical compound or element; tactile irritability.

negative feedback The condition of maintaining a constant output of a system by exertion of an inhibitory control on a key step in the system by a product of that system.

negative pressure area A room or unit which the air pressure is maintained below that of the surrounding area by means of exhaust fans.

negative selection Selection by removal of unfit or inappropriate members of a population; as, the removal of self-responding lymphocytes during the fetal period.

negative staining Staining of the background and not the organism, to facilitate the microscopic study of bacteria.

neonatal gonorrhea An infection of the eyes of the newborn, contracted usually during the birth process.

neonate (NEE uh nayt) A new-born infant; in the first four weeks of life.

neoplasm Any new and abnormal growth.

nephritogenic antigen An antigen found on the cells of certain strains of *Streptococcus pyogenes* that contributes to the development of a hypersensitivity leading to glomerulonephritis.

nephron The anatomical and functional unit of the kidney, consisting of the renal corpuscle, the proximal convoluted tubule, the descending and ascending limbs of Henle's loop, the distal convoluted tubule, and the collecting tubule.

nerve A cordlike structure, visible to the naked eye, comprising a collection of nerve fibers that convey impulses between a part of the central nervous system and some other region of the body.

nerve sheath   The thick layer of connective tissue that surrounds the nerve processes, insulating the bundle from its surroundings.

neuraminidase (n(y)ur uh MIHN uh days)   An enzyme contained in the surface coat of myxoviruses that destroys the neuraminic acid of the cell surface during attachment, thereby preventing hemagglutination.

neuroglia (n(y)u RAHG lee uh)   The supporting structure of nervous tissue.

neuron   Any of the conducting cells of the nervous system.

neurotransmitter   A chemical substance released by the axons of neurons that facilitates or inhibits the activity of adjacent and connecting neurons.

neutralism   The absence of interaction between coexisting organisms of different species.

neutralization   Rendering biologically inactive, as in neutralizing a virus or toxin; altering the pH of a solution toward neutrality.

neutron   An electrically neutral or uncharged particle of matter existing along with protons in the atoms of all emenets except the mass 1 isotope of hydrogen.

neutrophil (N(Y)OO truh fihl)   A granular leukocyte having a nucleus with three to five lobes connected by slender threads of chromatin, and cytoplasm containing five inconspicuous granules.

nicotinamide adenine dinucleotide (NAD) (nihk uh TEE nuh myd AD uhn een dy N(Y)OO klee uh tyd)   The dinucleotide of nicotinamide and of adenine, a coenzyme widely found in nature, and involved in numerous enzymatic reactions.

nicotinamide adenine dinucleotide phosphate (NADP)   A coenzyme required for a limited number of reactions, and similar to nicotinamide-adenine dinucleotide, except for the three phosphate units.

9 + 2 arrangement   A descriptive term for the pattern of microtubules that compose eucaryotic flagella or cilia.

nitrate ion   A highly oxidized form of inorganic nitrogen, NO_3.

nitrification (ny truh fuh KAY shuhn)   The bacterial oxidation of ammonia to nitrate and nitrite in the soil.

nitrogen cycle   The steps by which nitrogen is extracted from the nitrates of soil and water, incorporated as amino acids and proteins in living organisms, and ultimately reconverted to nitrates.

nitrogen fixation   The union of the free atmospheric nitrogen with other elements to form chemical compounds, such as ammonia and nitrates or amino groups.

nitrogenous base (ny TRAHJ uh nuhs)   An aromatic, nitrogen-containing molecule that serves as a proton acceptor, e.g., purine or pyrimidine.

nodule   A small boss or node that is solid and can be detected by touch.

nonbacterial pharyngitis   Pharyngitis in which the causative agent is a virus.

nonbiodegradable   A term that indicates that a substance is degraded slowly, if at all, by microorganisms in the environment; persistent.

noncompetitive inhibitor   A chemical substance that combines irreversibly with an enzyme, permanently blocking its function.

nongonococcal urethritis (NGU)   Urethritis in which, although the symptoms may be similar to gonorrhea, the causative agent is not the gonococcus; mycoplasmas and chlamydias are the most common causes.

non-Group A streptococci   Streptococcal species other than the highly pathogenic *S. pyogenes.*

noninfectious disease   A disease that is not transmitted from one individual to another.

nonoxidizing atmosphere   An atmosphere from which oxygen is absent.

nonpermissive cell   A cell type that does not support productive viral replication.

nonpolar amino acid   An amino acid, such as isoleucine or valine, in which the side-chain is of a nonpolar nature.

non-self   In immunology, any antigen that is not part of the normal antigenic makeup of the individual; foreign.

nonsense condons   A sequence by three purines and pyrimidines in messenger RNA that does not have a corresponding complementary sequence of three purines and pyrimidines in tRNA. Therefore, the presence of a nonsense codon terminates the synthesis of a polypeptide chain.

nonspecific resistance   Resistance to infection that is not based on past experi-

ence and that is effective against many different agents.

nontreponemal tests   Serological tests for syphilis in which the test antigen is not the syphilis organism or substances derived from it.

normal flora   Those microorganisms normally resident on the skin and mucous membranes that do not cause adverse changes by their activities.

Northern American blastomycosis   Gilchrist's disease or mycosis; a suppurative granulomatus chronic disease caused by *Blastomyces dermatitidis*; it starts as a respiratory infection and disseminates usually with cutaneous, osseous, and pulmonary manifestations.

Norwalk agent   A DNA virus that causes much self-limited gastroenteritis among adults.

nosocomial infection (nahz uh KOH mee uhl)   An infection originating in a hospital.

notifiable disease   Denoting that disease that should be made known. Said of diseases that are required to be made known to the Board of Health, Center for Disease Control, or other agency.

nuclear membrane   The condensed double layer of lipids and proteins that encloses the nucleoplasm, separating it from the cytoplasm and comprising an inner and outer membrane closely apposed.

nucleocapsid (n(y)oo klee oh KAP suhd)   A unit of viral structure, consisting of a capsid with the enclosed nucleic acid.

nucleoid (N(Y)OO klee oyd)   Resembling a nucleolus.

nucleolus (n(y)oo KLEE uh luhs)   A vacuole-like achromatin body, rich in ribonucleic acid, within the nucleus of a cell.

nucleosomes   Structural arrangements of the encaryotic DNA, consisting of a length of DNA of about 140 base pairs, in association with eight histone molecules.

nucleotide   One of the compounds into which nucleic acid is split by the action of nuclease.

nucleus   A spheroid body within a cell, consisting of a member of characteristic organelles visible with the optical microscope, a thin nuclear membrane, nucleoli, irregular granules of chromatin and lining, and a diffuse nucleoplasm.

nutrient cycles Transfers of key elements, such as carbon and nitrogen, from one group of organisms to another, eventually returning the nutrients to the environment in mineral form.

O-antigen The antigen that occurs in the lipopolysaccharide layer of the wall of Gram-negative bacteria.

obligate intracellular parasite An infective agent that can complete its replication cycle only within the host cell.

occlusive covering Serving to close; denoting a bandage or dressing that closes a wound and protects it from the air.

oil An unctuous, combustible substance that is liquid, or easily liquefiable on warming, and is soluble in ether, but insoluble in water.

oil immersion The covering of the microscopical objective and the object with oil.

omentum (oh MEHNT uhm) A fold of peritoneum extending from the stomach to adjacent organs in the abdominal cavity.

oncofetal antigen An antigen that appears in the serum and the body fluids in some forms of neoplasm, similar or identical to antigens found only during the fetal period in nondifferentiated tissues.

oncogene hypothesis The hypothesis that integrated viral material may serve as a gene or genes to induce transformation.

oncogenic virus A cancer-inducing virus.

one-step resistance Antimicrobial drug resistance of a type in which complete resistance to all doses of the drug is obtained in a single step, as by acquisition of the genetic capacity to produce an enzyme to degrade the drug.

oocyst (OH uh sihs) The encysted or encapsulated egg cell in the wall of a mosquito's stomach; also, the analogous stage in the development of any sporozoan.

operator region The gene that combines with an active repressor molecule.

operon (AHP uh rahn) In genetic theory, a segment of a chromosome comprising an operator gene and closely linked structural genes having related functions, the activity of the latter being controlled by the operator gene through its interaction with a regular gene.

opportunistic pathogen An organism, normally nonpathogenic, capable of causing disease only when the host's resistance is lowered, e.g., by other diseases or drugs.

opsonin (AHP suh nuhn) An antibody that renders bacteria and other cells susceptible to phagocytosis.

opsonization (ahp suh nuh ZAY shuhn) The rendering of bacteria and other cells subject to phagocytosis.

oral–fecal route A mode of transmission in which infective fecal material is ingested, either by contact with dirty hands or in fecally contaminated food or water.

orbital A region in an atom that may contain either one or two opposite spin electrons.

organ culture A laboratory technique for maintaining sections of animal organs in culture for periods of days or weeks.

organelle (awr guh NEHL) A specific particle of membrane-bound organized living substance present in eucaryotic cells, including mitochondria, the Golgi complex, endoplasmic reticulum, lysosomes, ribosomes, centrioles, and the cell center.

organic Pertaining to an organ or the organs. Pertaining to chemical compounds that contain carbon atoms bonded to hydrogen.

organic electron acceptor An organic substance that can accept electrons in metabolism, becoming reduced.

oropharynx (awr oh FAR ihng(k)s) That division of the pharynx that lies between the soft palate and the upper edge of the epiglottis.

orthomyxovirus (awr thuh MIHK suh vy rahs) A subgroup of the myxoviruses that includes the viruses of human and animal influenza.

osmosis (ahz MOH suhs) The passage of pure solvent from a solution of lesser to one of greater solute concentration when the two solutions are separated by a membrane that selectively prevents the passage of solute molecules, but is permeable to the solvent.

osmotic pressure The potential pressure of a solution directly related to its solute osmolar concentration; it is the maximum pressure developed by osmosis in a solution separated from another by a semipermeable membrane, i.e., the pressure

that will just prevent osmosis between two such solutions.

osteoclast-activating factor A lymphokine, the effect of which is to activate osteoclasts, cells active in bone resorption.

otitis media (oh TYT uhs MEED ee uh) Inflammation of the middle ear; tympanitis.

outgrowth The stage in the development of endospores into vegetative cells that follows endospore germination.

ovum The female reproductive cell which, after fertilization, develops into a new member of the same species.

ovary The female gonad: one of the two sexual glands in which the ova are formed.

oxalacetic acid Ketosuccinic acid; an important intermediate in the tricarboxylic acid cycle; the product formed when aspartic acid acts as amine donor in transamination reactions.

oxidase test A test for the presence of the cytochrome oxidase complex, widely used in the presumptive identification of the *Neisserias* and useful in distinguishing enterobacteria from the aerobic Gram-negative rods.

oxidation The act of oxidizing or state of being oxidized. Chemically it consists in the increase of positive charges on an atom or the loss of negative charges.

oxidation–reduction The chemical reaction whereby electrons are removed (oxidation) from atoms of the substance being oxidized and transferred to atoms being reduced (reduction).

oxidative decarboxylation The process of carbohydrate metabolism in which pyruvic acid is converted to acetyl CoA.

oxidative phosphorylation The formation of high-energy phosphate bonds by phosphorylation of ADP to ATP, in which electrons are transferred from the substrate to oxygen; it occurs in mitochondria.

oxidizing atmosphere An atmosphere containing free molecular oxygen.

oxidoreductase (ahk suhd oh rih DUHK tays) A class of enzymes that catalyze the reversible transfer of electrons from one substance to another.

oxygenic photosynthesis Photosynthesis in which both Photosystems I and II are operational, and in which water molecules are split to yield molecular oxygen as a by-product.

paired sera Sera taken from the patient at an interval of ten days to two weeks. When the titer of antibody in the two samples is compared, it will show if there has been recent or concurrent infection.

palatine tonsil A small, almond-shaped mass between the palatoglossal and palatopharyngeal arches on either side, composed mainly of lymphoid tissue, covered with mucous membrane and containing various crypts and many lymph follicles.

palisades Bacilli arranged in parallel rows or groupings like a fence.

pancreas A large, elongated, racemose gland situated transversely behind the stomach, between the spleen and the duodenum.

pandemic Widely epidemic; distributed thoroughout a region or continent, or globally.

papain fragments Polypeptide fragments derived from immunoglobulin molecules by treatment with papain.

Papanicolaou smear An exfoliative cytological staining procedure for the detection and diagnosis of various conditions, particularly malignant and premalignant conditions of the female genital tract (cancer of the vagina, cervix, and endometrium), in which cells desquamated from the genital epithelium are obtained by smears, fixed and stained, and examined under the microscope for evidence of pathologic changes.

papillomavirus (pap uh loh muh VY ruhs) Any of a subgroup of the papovaviruses causing papillomata in human beings, and in rabbits, cows, dogs, pigs, and various other animals.

paralytic shellfish poison (PSP) *Gonyaulax* poision. The neurotoxic principle produced by members of the genus *Gonyaulax* and related dinoflagellates; ingestion of shellfish that feed on these organisms causes a severe neurologic reaction which may end in paralysis and death.

paramyxovirus (pa(uh)r uh MIHK soh vy rahs) A subgroup of the myxoviruses, including the viruses of human and animal parainfluenza, mumps, and Newcastle disease.

parasite A plant or animal that lives upon or within another living organism at whose expense it obtains some advantage.

paratyphoid A *Salmonella* infection in which the symptoms are similar to those of true typhoid fever, but the causative agent is a different species.

parenteral (puh REHNT uh rahl) Not through the alimentary canal but rather by injection through some other route such as subcutaneous, intramuscular, intraorbital, intracapsular, intraspinal, intrasternal, and intravenous.

parietal peritoneum (puh RY uh tuhl pehr uht uhn EE uhm) The peritoneum that lines the abdominal and pelvic walls and the undersurface of the diaphragm.

paroxysmal phase (par ahk SISH muhl) The phase in whooping cough in which episodes of severe uncontrollable coughing, with the whooping sound of inhaled air, are present.

partial diploid An organism, normally haploid, in which a small number of duplicate genes from a sexual conjugation have become integrated.

parvovirus A small DNA virus.

passive transfer Transfer of an immunity (temporary) by transfer of serum from an immunized person.

passive transport Transport of a substance into a cell without the expenditure of energy.

Pasteur, Louis (pas CHOOR, LOO ee) French chemist and bacteriologist, 1822–1895, who founded the science of microbiology and developed the technique of vaccination by attenuated virus, and whose discoveries embrace the entire field of microbial activity.

pasteurization The process of heating milk or other liquids, e.g., urine, to a moderate temperature for a definite time. This exposure kills most species of pathogenic bacteria and considerably delays other bacterial development.

pathogen Any microorganism capable of initiating disease in a normal host.

pathogenicity The quality of producing or the ability to produce pathogenic changes or disease.

patient–patient transmission A means for the spread of hospital-acquired infections from one patient to another.

patient–staff transmission Transmission of an infection, in the hospital, to a healthy staff member.

pellicle (PEHL ih kuhl) A thin skin or film, such as a thin film on the surface of a liquid.

pemphigus neonatorum (PEHM(P) fih guhs nee oh nuht AWR uhm) Impetigo in the neonate caused usually by *Staphylococcus aureus*. Often rapidly spreading and more severe than in older children or adults.

penicillin An antibiotic substance extracted from cultures of certain molds of the genera *Penicillium* and *Aspergillus* that have been grown on special media.

penicillinase (pehn uh SIHL uh nays) Penicillin Amido-β-lactamhydrolase: an enzyme produced by certain bacteria that converts penicillin to an inactive product and thus increases resistance to the antibiotic.

penis The male organ of copulation and of urinary excretion, comprising a root, body, and an extremity, or glans penis.

pentose A monosaccharide containing five carbon atoms in a molecule.

peplomer (PEHP luh mur) A subunit of the peplos, or envelope, of a virion.

peptide bond The common link (-CO-NH-) between amino acids in proteins, actually a form of amide linkage.

peptidoglycan (PG) The rigid backbone of the bacterial cell wall, consisting of two major subunits, N-acetyl muramic acid and N-acetyl glucosamine, and a number of amino acids.

peptone A derived protein, or a mixture of cleavage products, produced by the partial hydrolysis of a native protein either by an acid or by an enzyme.

periapical abscess Acute apical periodontal abscess.

perinephric abscess Abscess in the tissue immediately around the kidney.

periodontal disease Any disease or disorder of the periodontium, or tissues surrounding the tooth.

periodontal pocket A gingival sulcus pathologically deepended by periodontal disease.

periodontitis (pehr ee oh dahnt YT uhs) Inflammatory reaction of the tissues surrounding a tooth, usually resulting from the extension of gingival inflammation into the periodontium.

periosteum (pehr ee AHS tee uhm) A specialized connective tissue covering all the bones of the body, and possessing bone-forming potentialities.

peripheral nervous system (PNS) That portion of the nervous system consisting of the nerves and ganglia outside the brain and spinal cord.

periplasmic space The area between the plasma membrane and the cell wall, containing certain enzymes involved in nutrition.

peristalsis The wormlike movement by which the alimentary canal or other tubular organs provided with both longitudinal and circular muscle fibers propel their contents.

peritoneal cavity (pehr uht uhn EE uhl) Cavum peritonei. The abdominal cavity, in which intestinal mass, urinary and reproductive organs are found.

peritoneal macrophages Wandering macrophages found in the peritoneal cavity.

peritrichous flagella Bacteria with flagella around the entire surface are peritrichous.

perlèche (pehr LEHSH) Cracking of the corners of the mouth, often with secondary infection, accompanied by sensations of dryness and burning at the corners of the mouth.

permissive cell A cell type that supports productive replication of a virus.

person-to-person transmission The conveyance of infectious disease from one person to another.

pertussis Whooping cough.

petechia (puh TEE kee uh) A pinpoint, nonraised, perfectly round, purplish-red spot on the skin or mucous membranes, caused by intradermal or submucous hemorrhage.

Peyer's patches Aggregates of lymphoid tissue in the wall of the small intestinal tract.

pH The symbol relating the hydrogen ion (H⁻) concentration or activity of a solution to that of a given standard solution.

phage fd A filamentous bacteriophage that infects *Escherichia coli.*

phage lambda A temperate bacteriophage that infects and becomes integrated in cells of *Escherichia coli*, and that mediates transduction.

phage typing Characterization of bacteria, extending to strain differences, by demonstration of susceptibility to one or more (a spectrum) of types of bacteriophage.

phagocytosis (fag uh suh TOH suhs) The engulfing of microorganisms, other cells, and foreign particles by phagocytes.

phagolysosome (fag uh LY suh sohm) The digestive vacuole formed when the membranes of preexistent lysosomes within the cytoplasm merge with the phagosome; the lysosomes then discharge their hydrolytic enzymes, which digest the phagocytized material.

phagosome The membrane-bounded vesicle in a phagocyte formed by invagination of the cell membrane and the phagocytized material.

pharmacodynamics The study of the actions of drugs on living systems.

pharyngeal tonsil (far uhn JEE uhl) A mass of lymphoid tissue located in the rear of the pharynx.

pharynx (FAR ihng(k)s) The musculomembranous sac between the mouth and nares and the esophagus. It is continuous below with the esophagus, and above it communicates with the larynx, mouth, nasal passages, and auditory tubes.

phase contrast A form of high-resolution light microscopy, in which the phase variation of the light source is used to heighten the contrast of the image.

phenol coefficient (PC) (FEE nohl) A measure of the bactericidal activity of a chemical compound in relation to phenol.

phenolic (fih NOH lihk) Pertaining to or derived from phenol.

phenotype The entire, physical, biochemical, and physiological makeup of an individual as determined both genetically and environmentally. Also, any one or any group of such traits.

phospholipid (fahs foh LIHP uhd) A lipid that contains phosphorus. On hydrolysis yields fatty acids, glycerin, and a nitrogenous compound.

phosphorylation The process of introducing the trivalent PO₄ group into an organic molecule.

photoautotroph (foht oh AHT uh trohf) An organism that depends on light for its energy and principally on carbon dioxide for its carbon.

photochemical reaction center In a photosynthetic organism, the structural unit containing the photosynthetic pigments and coenzymes.

photoheterotroph (foht oh HEHT uh rah trohf) An organism that depends on light for its energy and principally on organic compounds for its carbon.

photon A particle (quantum) of radiant energy.

photophosphorylation The formation of ATP occurring in chloroplasts during photosynthesis; it is analogous to oxidative phosphorylation.

photoreduction Reductive reaction driven by light energy.

photosynthesis A chemical combination caused by the action of light; specifically the formation of carbohydrates from carbon dioxide and water in the chlorophyll tissue of plants under the influence of light.

Photosystem I A photosynthetic complex that generates ATP, but does not split water or reduce organic substances directly.

Photosystem II A photosynthetic complex which, in connection with Photosystem I, produces ATP, reduces NADP, and splits water.

phycobilin (fy coh BIHL uhn) Any of a group of protein-linked pigments including phycoerythrin (red pigment) and phycocyanin (blue pigment), which are found in the red algae and the cyanobacteria.

phytoplankton The minute plant (vegetable) organisms that, with those of the animal kingdom, make up the plankton of natural waters.

pia mater (PY uh MAYT ur) The innermost of the three membranes (meninges) covering the brain and the spinal cord, investing them closely and extending into the depths of the fissures and sulci; it consists of reticular, elastic, and collagenous fibers.

pili (PY ly) Plural of pilus. Short tubular structures on the surface of many types of bacteria. Serve specialized roles in attachment of the bacterial cell to surfaces.

pilin The protein composing the subunits of bacterial pili.

pilosebaceous unit (py luh sih BAY shuhs) The hair follicle and the associated apocrine gland.

pinkeye A popular term for an epidemic, contagious conjunctivitis.

placenta An organ characteristic of true mammals during pregnancy, joining mother and offspring, providing endocrine secretion and selective exchange of soluble, but not particular, bloodborne substances through an apposition of uterine and trophoblastic vascularized parts.

plague An acute, febrile, infectious disease with a high fatality rate, caused by *Yersinia pestis.*

plankton A collective name for the minute free-floating organisms, vegetable and animal, that live in practically all natural waters.

plaque count A method for the estimation of the number of lytic virus particles present by counting the number of plaques of destroyed cells formed on a lawn of susceptible bacteria.

plaque-forming units Number of viable phage particles capable of forming plaques.

plasma The fluid portion of blood, in which particulate components are suspended.

plasma cell A spherical or ellipsoidal cell, with a single eccentrically placed nucleus containing clumped chromatin, an area of perinuclear clearing, and generally abundant, sometimes vacuolated cytoplasm. Produces immunoglobulins.

plasmid A segment of extrachromosomal DNA found in the cytoplasm of certain bacteria. Also, a generic term for all types of intracellular inclusions that can be considered as having genetic functions.

plasmodesmata (plaz muh DEHZ muht uh) Plural of plasmodesm. Bridges of cytoplasm passing between the pores of plasma membranes of adjacent cells in some plant cell complexes.

plasmodium (plaz MOHD ee uhm) An acellular mass of protoplasm of indefinite size and shape, formed by the acellular slime molds.

platelet A disc-shaped structure, 2 to 4μ in diameter, found in the blood of all mammals and chiefly known for its role in blood coagulation.

pleomorphic Occurring in various distinct forms; exhibiting plemorphism.

pleura The serous membrane investing the lungs and lining the thoracic cavity, completely enclosing a potential space known as the pleural cavity.

pleurisy Inflammation of the pleura, with exudation into its cavity and upon its surface.

pneumonia, aspiration Pneumonia due to the entrance of foreign matter, such as food particles, into the respiratory passages (bronchi).

pneumonia, giant cell A rare, usually fatal, form of interstitial pneumonia, caused by the measles virus, affecting

children with disease of the reticulo-endothelial system (such as leukemia), and marked by the presence of multinucleate giant-cell inclusion bodies.

pneumonia–influenza (P–I) deaths A disease classification used by the USPHS to describe pneumonia deaths in which influenza preceded and predisposed to the secondary bacterial infection.

pneumonia, interstitial A chronic form of pneumonia with increase of the interstitial tissue and decrease of the proper lung tissue, with induration.

pneumonia, lobar An acute febrile disease produced by *Streptococcus pneumoniae*, and marked by inflammation of one or more lobes of the lung, together with consolidation.

pneumonia, primary atypical An acute infectious primary disease caused by *Mycoplasma pneumoniae* and various viruses, including adenoviruses and parainfluenza virus.

pneumonia, secondary bacterial A pneumonia of bacterial origin that is secondary to a previous respiratory tract infection, most frequently of viral origin, or to focal infections of all types.

polar amino acid An amino acid, such as cysteine or threonine, in which the side-chain contains polar covalent bonds.

polar covalent bond A covalent bond in which unequal electronegativities contribute to unequal distribution of the pair of electrons.

polar flagellation Condition of having flagella attached at one end or both ends of the cell.

polarity The fact or condition of having poles. A characteristic of certain covalent bonds in which slight electrical charges can be detected at the opposite ends of the bond.

poly-β-hydroxybutyrate Storate reserve of carbon and energy that occurs as long chains of individual subunits forming readily visible granules.

polymer A compound, usually of high molecular weight, formed by the linear combination of simpler repeating molecules, or monomers.

polymerase Any enzyme that catalyzes polymerization.

polymerization The act or process of forming a compound (polymer), usually of high molecular weight, by the combination of simpler molecules.

polymicrobic infection A mixed infection, one in which the pathogenic effects are the result of the activities of several microbial species together.

polymorphonuclear cell (PMN) (polymorphonuclear leukocyte) (pahl ih mawr fah N(Y)OO klee ur LOO kuh syt) A phagocyte with a lobed nucleus; also called neutrophilic granulocyte, poly or neutrophil.

polymyxin (pahl in MIHK suhn) A generic term used to designate a number of antibiotic substances derived from strains of the soil bacterium *Bacillus polymyxa*.

polymyxovirus A DNA virus group, some of which have been implicated in tumors of nonhuman animals.

polypeptide A peptide that on hydrolysis yields more than two amino acids; called tripeptides, tetrapeptides, etc., according to the number of amino acids contained.

polysaccharide A carbohydrate that on hydrolysis yields more than ten monosaccarides.

polysome (PAHL ih sohm) Polyribosome; the functional unit of translation consisting of messenger RNA plus attached ribosomal subunits.

polyunsaturated Having many unsaturated bonds and tending to be liquid at room temperature; said of fats and oils.

polyvalent vaccine A bacterial vaccine prepared from cultures of more than one strain or species of bacteria.

Pontiac fever A relatively mild, short-term illness caused by *Legionella* different from legionellosis both in symptoms and severity.

positive chemotaxis (kee moh TAK suhs) Movement of an organism from a region of low to a region of high concentration of a specific chemical compound or element.

positive pressure area A work area in which the air pressure is maintained slightly higher than the surroundings by means of pumped air supply.

positive selection Selection by enhancement of the growth of desired cells or species.

poststreptococcal complications Conditions such as rheumatic fever or glomerulonephritis, arising after and as a result of primary streptoccal infection.

potassium tellurite An inhibitory substance added to selective media for the isolation of *Corynebacterium diphtheriae.*

pour plate A bacterial culture poured into a petri dish from a test tube in which the medium has been inoculated.

poxvirus Any of a group of relatively large, morphologically similar, and immunologically related DNA viruses, including the viruses of vaccinia (cowpox), variola (smallpox), and those producing pox diseases in lower animals.

precipitation The act or process of precipitating or separating out from solution.

predator An organism that derives elements essential for its existence from organisms of other species, which it consumes and destroys.

prednisone (PREHD nuh sohn) A synthetic glucocorticoid occurring as a white odorless crystalline powder, $C_{21}H_{26}O_5$, slightly soluble in methanol; used like prednisolone.

preerythrocytic cycle A stage of replication of the malarial parasite prior to its entrance to the red blood cells.

preicteric phase (pree ihk TEHR ihk) Denoting the phase of hepatic disease before jaundice (icterus) appears.

primary amebic meningoencephalitis (PAM) (muh nihng goh uhn sehf uh LYT uhs) A disease caused by free-living ameobas of the genus *Naegleria*, characterized by a rapidly deteriorating course and death in three to five days.

primary cell culture A cell culture derived from a mixed biopsy of normal tissue; may be subcultured only a few times.

primary immunodeficiency An immunodeficiency of genetic origin, present at birth.

primary lymphoid tissue A latticework of reticular tissue the interspaces of which contain lymphocytes; lymphoid tissue may be diffuse, or densely aggregated as in lymph nodules and nodes.

primary stage In syphilis, the first stage for the infection, characterized by the presence of a hard, painless chancre at the site of infection.

primary structure The sequence of amino acids in a protein molecule.

primary treatment The initial stage of sewage treatment, in which the bulk of the solids are removed mechanically.

primoquine A chemotherapeutic agent used in treatment of some forms of malaria.

process A prominence or projection, as of bone.

producer An organism that fixes atomospheric carbon dioxide into organic matter.

procaryote (proh KAR ee oht) An organism possessed of a simple cellular structure in which the genetic material is not enclosed in a nuclear membrane.

proliferative disease A disease characterized by uncontrolled proliferation of certain populations of cells.

promoter The region within the operon to which the transcribing enzyme attaches prior to transcription.

properdin (proh PURD uhn) A relatively heat-labile, normal serum protein (a euglobulin) that, in the presence of complement component C3 and magnesium ions, acts nonspecifically against Gram-negative bacteria and viruses and plays a role in lysis of erthrocytes.

prophage The latent stage of a phage in a lysogenic bacterium, in which the viral genome becomes inserted into a specific portion of the host chromosome and is duplicated each cell generation.

prophylaxis (proh fuh LAK suhs) The prevention of disease; preventive treatment.

prostaglandins (prahs tuh GLAN duhnz) A group of naturally occurring, chemically related, long-chain hydroxy fatty acids that stimulate contractility of the uterine and other smooth muscle and have the ability to lower blood pressure and to affect the action of certain hormones.

prostate gland (PRAHS tayt) A gland in the male that surrounds the neck of the bladder and the urethra.

protective isolation An isolation procedure instituted to protect a compromised patient from infection; reverse isolation.

protein I A protein component of the outer membrane of Gram-negative bacteria, believed to make up the pores.

prothrombin (proh THRAHM buhn) Factor II:—coagulation factor—a protein present in the plasma, that is theoretical hematology is converted to thrombin by extrinsic prothrombin-converting principle.

Protista (proh TIHS tuh) A kingdom comprising bacteria, algae, slime molds, fungi, and protozoa.

proton A positive charged subatomic particle bearing one atomic mass unit.

proton motive force (PMF) Energized state of a membrane induced as a result of electron transport and extrusion from the cell of hydrogen ions.

protoplast A bacterial cell from which the cell wall has been entirely removed.

provirus The genome of an animal virus integrated (by crossing over) into the chromosome of the host cell, and thus replicated in all of its daughter cells.

provocative contact Contact with an antigen by a previously sensitized host leading to development of hypersensitivity such as anaphylaxis.

pseudomycelia (sood oh my SEE lee uh) Plural of pseudomycelium. A mass of fungal filaments resembling hyphae but composed of loosely united, elongated cells formed by budding.

pseudopodia (sood uh POHD ee uh) Plural of pseudopodium. A temporary protrusion of the cytoplasm of an ameboid cell, serving for purposes of locomotion or to engulf food.

P site The second site on the ribosome, which holds a transfer RNA bearing the growing peptide.

psychrophile (SYK roh fyl) An organism that grows best at low temperatures.

puerperal fever (pyoo UR p(uh) ruhl) Septicemia accompanied by fever, in which the focus of infection is a lesion of the mucous membrane of the birth canal due to trauma during childbirth.

pulp Any soft, juicy animal or vegetable tissue, such as that contained within the spleen or the pulp chamber of a tooth.

pulpitis Inflammation of the dental pulp.

Purified Protein Derivative (PPD) A purified antigenic material derived from *Mycobacterium tuberculosis* culture medium, used in the tuberculin test.

purine A colorless, crystalline heterocyclic compound, $C_5H_4N_4$, that is not found free in nature, but is variously substituted to produce a group of compounds known as *purines* or *purine bases* (purine bodies), of which uric acid is a metabolic end product; examples are adenine and guanine.

pyelonephritis, acute (py (uh) loh nih FRYT uhs)   Pyelonephritis of sudden onset characterized by fever, shaking chills, pain in the costovertebral region or flanks, and symptoms of bladder inflammation; it is a self-limited bacterial disease caused most often by Gram-negative enteric bacilli.

pyelonephritis, chronic   Pyelonephritis attributed to effects of a previous infection or to recurring or progressive infection. Typically, it is of insidious onset, manifested by symptoms of chronic renal insufficiency, with fatigue, headache, loss of appetite, weight loss, excessive thirst, and polyuria.

pyloric sphincter   The muscular band at the junction of the stomach and small intestine.

pyocyanine $C_{13}H_{10}ON_2$; antibiotic crystalline substances isolated from peptone broth cultures of *Pseudomonas aeruginosa*: active against Gram-negative and Gram-positive bacteria, but highly toxic to animal tissues.

pyorrhea (py uh REE uh)   A discharge of pus, especially around the teeth.

pyrimidine (py RIHM uh deen)   An organic compound which is the fundamental form of the pyrimidine bases; examples are thymine, cytosine, and uracil.

pyrogen   A fever-producing substance.

pyrophosphate (PPi)   Any salt of pyrophosphoric acid. Liberated from ATP by cleavage of the molecule between the first and second phosphate units.

pyruvic acid   A colorless liquid $CH_3CO \cdot CO_2H$, with an odor like acetic acid, formed during glycolysis, from glucose.

Q fever (Q for Query)   A febrile rickettsial infection, usually respiratory, described in Australia but worldwide in distribution and including Balkan grippe in the Mediterranean area; caused by *Coxiella burnetii*.

quaternary structure   A feature of protein structure; the formation of an active unit by aggregation of two or more polypeptides.

Quellung reaction (Neufeld's reaction)   When pheumococci are mixed with specific immune serum there occurs in addition to agglutination a swelling (quellung) of the peripheral zones of the organisms. This may be observed under the microscope.

R-factor   The bacterial plasmid (R-plasmid) responsible for resistance to antibiotics; it is transmitted to other bacterial cells by conjugation, as well as to the progeny of any cell containing it.

rabies immune globulin (RIG)   A passive immunization given to persons who have been directly exposed to rabies virus.

radiation   Emission from a radioactive source; may be in the form of electromagnetic energy or streams of subatomic particles.

radioactivity   The quality of emitting radiations consequent to nuclear disintegration, a natural property of all chemical elements of atomic number above 83, and possible of induction in all other known elements.

radioimmunoassay (RIA) (rayd ee oh ihm yuh noh AS ay)   Determination of antigen or antibody concentration by means of a radioactive-labeled substance that reacts with the substance under test.

rapid plasma reagin test   A nontreponemal test for syphilis antibodies in the sera of human beings.

rat-bite fever   An infectious disease following the bite of a rat or other rodent, occurring in two forms: Haverhill fever and sodoku.

reactant   An original substance entering into a chemical reaction.

reaction   The interaction of two or more chemical substances resulting in their chemical alteration.

reagin (ree AY juhn)   Antibody of a specialized immunoglobulin class (IgE) that attaches to tissue cells of the same species from which it is derived, and that interacts with its antigen to induce the release of histamine and other vasoactive amines.

recombinant   The new cell that results from genetic recombination.

recombinant DNA   DNA that results from the combining of DNA molecules from two or more sources. More recently it connotes DNA formed by combining DNA molecules *in vitro*.

recombination   In genetics, the formation of a genetically new individual by combining portions of the original genetic material from two or more individuals.

rectum   The distal portion of the large intestine, beginning anterior to the third sacral vetebra as a continuation of the sigmoid and ending at the anal canal.

recurring infection   An infection which keeps reactivating, due to a focus of the infectious agent remaining within the body through periods of apparent cure.

redox (REE dahks)   In chemistry, mutual reduction and oxidation.

redox couple   A paired electron acceptor and electron donor that participates in redox reactions.

red tide   The proliferation in sea water of the genus of protozoa of the order *Dinoflagellata*, found in salt, fresh, or brackish waters, and having yellow to brown chromatophores.

reduction   In chemistry, the addition of hydrogen to a substance, or more generally, the gain of electrons.

reference culture   A stock culture of carefully described nature, maintained as a permanent reference for comparison and study.

reflection   In physics, the turning back of a ray of light, sound, or heat when it strikes against a surface that it does not penetrate.

reflux (REE fluhks)   A backward or return flow.

refraction   The act or process of refracting. The change of direction of a ray of light in passing obliquely from one medium into another in which its speed is different.

refractive index   The refractive power of a medium compared with that of air.

regulatory gene   In genetic theory, a gene that synthesizes repressor, a substance that, through interaction with the operator gene, switches off the activity of the structural genes associated with it in the operon.

rejection   An immune reaction against grafted tissue.

renal pelvis   The expansion from the upper end of the ureter into which the calices of the kidney open.

reovirus (ree oh VY ruhs)   A group of ether-resistant RNA viruses, formerly classified as a subgroup of the echoviruses.

replacement flora   The microbial population that arises during the course of antimicrobial therapy; its component organisms are resistant to the drug in use.

replica plating A technique for the simultaneous transfer of organisms of a large number of separated colonies from one medium to others. Used in the direct selection of bacterial mutants.

replication Formation of new copies of a structure; used in relationship to the copying of new double helices of DNA.

replication fork The Y-shaped structure in the replicating DNA molecule; the site of DNA replication.

replicative form A double-strand form of nucleic acid, produced during multiplication of single-stranded viruses and consisting of the single strand of nucleic acid and complementary strand. The replicative form serves as the intermediate in the synthesis of new single strands of viral nucleic acid.

repressible A term applied to those enzymes the synthesis of which is arrested when their end products are present in adequate supply.

repressible enzyme One whose rate of production is decreased as the concentration of certain metabolites is increased.

reservoir The source or sources of an infectious agent in nature.

resistance factor (R) A gene that confers antimicrobial drug resistance, usually incorporated into a plasmid.

resistance transfer factor (RTF) A set of genes associated with R factors in transmissible plasmids. These genes carry the information required for the transmission process.

resolution The subsidence of a pathologic state, as the subsidence of an inflammation, or the softening and disappearance of a swelling.

resolving power The ability of the eye or of a lens to make small objects that are close together, separately visible, thus revealing the structure of an object.

respiration The exchange of oxygen and carbon dioxide between the atmosphere and the cells of the body. The complete oxidation of the glucose molecule to carbon dioxide under conditions in which the resulting electrons are disposed of via electron transport.

respiratory metabolism The exchange of respiratory gases in the lungs and the oxidation of foodstuffs in the tissues with the production of carbon dioxide and water.

respiratory syncytial virus (RSV) A virus isolated from children with bronchopneumonia and bronchitis, characteristically causing syncytium formation in tissue culture; first isolated from chimpanzees with symptoms of respiratory disease.

reticuloendothelial system (RE) A functional rather than anatomical system that serves as an important body defense mechanism; composed of highly phagocytic cells having both endothelial and reticular attributes and the ability to take up particles of collodial dyes; these cells include macrophages, Kupffer's cells, reticular cells, cells lining the blood sinuses of the pituitary and suprarenal glands, monocytes, and probably the microglia.

reticulum A network, especially a protoplasmic network in cells, as the flattened double membrane sheets of the endoplasmic reticulum. Also the cellular structural network that composes much of lymphoid tissues, in which lymphocytes are retained.

retrovirus A group or RNA viruses, all of which cause their host cells to express the reverse transcriptase enzyme.

reverse transcriptase An enzyme that synthesizes DNA complementary to an RNA template. RNA-dependent DNA polymerase.

Reye syndrome An acute and often fatal childhood syndrome of encephalopathy and fatty degeneration of the liver, marked by rapid development of brain swelling and hepatomegaly and by distributed consciousness and seizures. Has been observed to follow upon cases of influenza types A and B, as well as other viral infections.

rhabdovirus (RAB duh vy ruhs) Any of a group of morphologically similar, bullet-shaped or bacilliform RNA viruses, including the viruses of vesicular stomatitis and rabies.

rheumatic fever A febrile disease occurring as a delayed sequel of infections with Group A hemolytic streptococci and characterized by multiple focal inflammatory lesions of connective tissue structures, especially of the heart, blood vessels, and joints, and by the presence of Aschoff bodies in the myocardium and skin.

rheumatogenic antigen An antigen, possessed by some strains of *Streptococcus pyogenes*, that causes sensitization and may lead to the development or rheumatic fever.

rhinovirus (ry noh VY ruhs) Any of a subgroup of the picornaviruses considered to be etiologically associated with the common cold and certain other upper respiratory ailments.

rhizosphere Region around the plant root immediately adjacent to the root surface where microbial activity is usually high.

Rh system A group of red blood cell surface antigens including the several forms of the rhesus antigen.

rho A subunit of RNA polymerase, involved in the termination of transcription.

ribonucleic acid (RNA) A nucleic acid originally isolated from yeast, but later found in all living cells. On hydrolysis it yields adenine, quanine, cytosine, uracil, ribose, and phosphoric acid.

ribosomal RNA (rRNA) A type of RNA found in ribosomes.

ribosome Intracellular ribonucleoprotein particles concerned with protein synthesis; they consist of reversibly dissociable units and are found either bound to membranes or free in the cytoplasm.

ribulose diphosphate A key intermediate in the dark reactions of photosynthesis.

rickettsia (rihk EHT see uh) An individual organism of the family *Rickettsiaceae*.

rifampicin (ry FAM puh suhn) A semisynthetic antibacterial, $C_{42}H_{58}N_4O_{12}$, derived from rifamycin SV, used in the treatment of pulmonary tuberculosis and carriers of *Neisseria meningitidis*.

rising titer An increasing level of serum antibody, indicating a recent or concurrent infection.

RNA-dependent DNA polymerase An enzyme found in the white cells of leukemia patients, but not in normal patients. Also called the reverse transcriptase.

RNA-dependent RNA polymerase Enzyme found in cells infected with RNA viruses, needed to replicate the viral genome.

Rocky Mountain spotted fever Infection with *Rickettsia rickettsii*, transmitted by the ticks *Dermacentor andersoni; D. variabilis, Amblyomma americanum*, and the rabbit tick *Haemaphysalis leporispalustris*.

root caries Decay of the root of a tooth, occurring when the root becomes exposed due to advanced periodontal disease or some other cause.

root nodules Aggregates of cellular material containing nitrogen-fixing endosymbiotic bacteria, found on the roots of leguminous plants.

rotavirus The viruses most commonly found in epidemic infant diarrhea.

rough endoplasmic reticulum (ER) An ultramicroscopic organelle of nearly all cells of higher plants and animals, consisting of a more or less continuous system of membrane-bound cavities that ramify throughout the cytoplasm of a cell.

rough strain The strain that results from bacterial dissociation; R colonies have a dull, uneven surface and irregular border, the growth in fluid media tends to flake out, no capsules are seen, and the culture tends to be less virulent.

Rous sarcoma virus (RSV) A leukovirus producing fibrosarcoma in fowl, especially chickens; some strains have been shown to produce tumors in other animals.

routine immunization Immunization of each member of a population at a certain age regardless of immediate need as a general protective measure.

rubella German measles: A mild viral infection characterized by a pink discrete and confluent macular rash.

rubeola (roo bee OH luh) A synonym for measles in English and of German measles in French and Spanish. A moderate to severe viral infection characterized by fever, rash, photophobia, and sometimes followed by severe complications.

rumen The first stomach of a ruminant, or cud-chewing animal; also called paunch.

saliva The clear, alkaline, somewhat viscid secretion from the parotid, submaxillary, sublingual, and smaller mucous glands of the mouth.

salmonellosis (sal muh nehl OH suhs) Infection with certain species of the genus *Salmonella*, usually caused by the ingestion of food containing the organisms or their products and marked by violent diarrhea attended by cramps and tenesmus and/or paratyphoid fever.

sanitize To clean and sterilize, as eating or drinking utensils.

saprophytic (sap ruh FIHT ihk) Of the nature of or pertaining to a saprophyte; growing on dead or decomposing organic matter.

saturation The act of saturating or condition of being saturated. May refer to a solution in which the maximum amount of solute has been dissolved. Also indicates an organic molecule in which there are the maximum number of hydrogen atoms, such as certain fatty acids.

scanning electron microscope (SEM) A microscope in which the object is examined point by point directly by an electron beam and an image formed on a television screen.

scarlet fever Infection due to Group A β-hemolytic streptococci, and rarely to other serological types of β-hemolytic streptococci that elaborate erythrogenic toxin.

Schick test A skin test to detect whether or not an individual has immunity to diphtheria toxin.

schizogony (skihz AHG uh nee) The asexual cycle of sporozoa; particularly the life cycle of the malarial parasite *(Plasmodium)* in the blood corpuscle of human beings.

Schwann cell One of the large nucleated masses of protoplasm lining the inner surface of the sheath of Schwann or neurilemma.

scrub typhus A self-limited, febrile disease of two weeks' duration, caused by *Richettsia tsutsugamushi*, transmitted by larval mites of the genus *Trombicula*, especially *T. akamushi* and *T. deliensis*; it is characterized by sudden onset of fever with a primary skin lesion and development of a rash about the fifth day.

sebaceous gland Any of thousands of glands located in the dermis of the skin that secrete sebum.

sebum The secretion of the sebaceous glands; a thick, semifluid substance composed of fat and epithelial debris from the cells of the malpighian layer.

second law of thermodynamics There is always an increase in entropy in any naturally occurring (spontaneous) process.

second signal The system through which lymphocytes become activated by contact with their antigen; the antigen is bound by receptors not only on the primary responder cell but also by less specific recep-

tors on a second lymphocyte. Both must bind before an immune response may begin.

secondary immunodeficiency An immunodeficiency that is not genetic, but arises secondary to some disease or treatment suppressive to hematopoietic or lymphoid tissues.

secondary stage In syphilis, a state in the disease progression commencing several weeks after the primary stage has ended. Characterized by rash, multiple lesions over the mucous membranes, and high infectivity.

secondary structure In protein structure, the patterns of hydrogen bonding into which the primary amino acid sequence may enter, including the alpha helix and the pleated sheet arrangements.

secondary treatment A stage of sewage treatment characterized by continuing aerobic degradation of the wastes in both effluent and sludge and elimination of human-origin microorganisms. The end product is largely free of pathogens after chlorination, and its bulk and total organic content has been significantly reduced.

secretory IgA IgA immunoglobulin molecules bearing the secretory polypeptide, secreted across mucous membranes and effective in protecting the exterior surfaces of mucous membranes from the attachment of potential infective agents.

selection technique Any technique for selecting desired members of a mixed population, such as penicillin selection for bacterial mutants or HAT selection for somatic cell hybrids.

selective environment A natural environment that supports extensive growth of only certain well-adapted microorganisms.

selective medium A laboratory medium that contains selective agents, so that only certain types of microorganisms will be able to grow.

self An expression used by Burnet and Fenner to denote an animal's own antigenic constituents, in contrast to "not self," denoting foreign antigenic constituents.

self-assembly A characteristic of many subcellular structures, especially those made up of a repeating molecular structure. The molecular components are so shaped as to readily form weak bounds with those molecules synthesized as their neighbors. Thus the structure assembles spontaneously. Examples are bacterial flagella, eucaryotic microtubules, and the bacteriophage tails.

self-limited   Limited by its own peculiarities, and not by outside influence; said of a disease that runs a definite limited course.

semen   The thick, whitish secretion of the reproduction organs in the male; composed of spermatozoa in their nutrient plasma, secretions from the prostate, seminal vesicles, and various other glands, epithelial cells, and minor constituents.

semiconservative model   The model for DNA replication that states that each old strand of the helix serves as a template for the synthesis of a new strand.

seminal vesicles   Sacs associated with the vas deferens; their secretions contribute to the formation of semen.

semipermeable   Permitting the passage of certain molecules and hindering that of others.

semipermissive cell   A host cell type in which viral replication only occasionally results in productive infection.

semisynthetic   Produced by chemical manipulation of naturally occurring substances.

sensitizing contact   The primary contact with an antigen, which stimulates the development of a hypersensitivity. Symptoms do not develop, however, until further antigenic contacts occur.

sensory neuron   Any neuron possessing a sensory function; an afferent neuron conveying sensory impulses.

separation   A procedure that allows the purification of a number of closely similar chemical entities.

sepsis   The presence in the blood or other tissues of pathogenic microorganisms or their toxins.

septum (SEHP tuhm)   A dividing wall or partition.

septic shock   Shock developing in the presence of severe infections.

septic tank   A tank for the receipt of sewage, there to remain for a time in order that the solid matter may settle out and a certain amount of putrefaction occurs from the action of the anaerobic bacteria present in the sewage.

serial cell culture   A form of cell culture in which the cell line can be propagated indefinitely without dying out or changing its character.

serial dilution   (1) The progessive dilution of a substance in a series of tubes in predetermined ratios. (2) A method of obtaining a pure bacterial culture by rapid transfer of an exceedingly small amount of material from one nutrient medium to a succeeding one of the same volume.

seroconversion (sihr oh kuhn VUR zhuhn)   The development of antibodies in response to the administration of a vaccine.

serology   The study of antigen-antibody reactions *in vitro*.

seropositive   Serologically positive; showing positive results on serological examination.

serum   The clear portion of any animal liquid separated from its more solid elements; especially the clear liquid that separates in the clotting of blood from the clot and the corpuscles.

serum sickness   A hypersensitivity reaction occurring 8 to 12 days following a single, relatively large injection of foreign serum and marked by urticarial rashes, edema, adenitis, joint pains, high fever, and prostration.

severe combined immunodeficiency (SCID)   A primary immunodeficiency of genetic origin in which lymphoid tissues entirely fail to develop.

sewage   The matter found in sewers; it consists of the excreta of human beings and animals, and other waste material from homes and other inhabitated structures.

sexduction (sehks DUHK shuhn)   In bacterial genetics, the process whereby part of the bacterial chromosome is attached to the autonomous f factor (sex factor) and thus is transferred with high frequency from the donor (male) bacterium to the recipient (female).

sex pili   Bacterial surface structures used to bind donor and recipient cell together during genetic exchanges.

sexually transmitted disease (STD)   Any infectious disease transmitted primarily by sexual contact, e.g., gonorrhea.

shadowing   A process for increasing the contrast and adding a three-dimensional character to preparations for electron

microscopy; metal ions are vaporized across the specimen at an angle.

sieve filter   A filter that acts like a sieve; objects larger than the pores are retained on its surface.

sigma (SIHG muh)   A subunit of the enzyme RNA polymerase, necessary for transcription to begin.

simian virus 40 (SV40)   Vacuolating virus. A DNA virus found as a passenger virus in cultured monkey kidney tissue. May be oncogenic to neonatal animals.

simple stain   A microbiological staining procedure employing a single dye application.

sinuses   A general term for such spaces as the dilated channels for venous blood in the cranium, or the air cavities in the cranial bones.

sinusitis   Inflammation of a sinus.

sinusoid   (1) Resembling a sinus. (2) A form of terminal blood channel consisting of a large, irregular anastomosing vessel, having a lining of reticuloendothelium.

skin test   Cutaneous reaction. A reaction produced by applying to an abrasion or by injecting into the skin a solution of a protein or a pollen to which the patient is sensitive.

slide culture   Cultivation of fungi on a block of solid medium mounted on a slide. The growth of microscopic fungal structures can thus be observed directly.

slow infections   Viral infections such as SSPE that take months or years to develop to the symptomatic stage.

slow virus   Any virus causing a disease characterized by a very long preclinical course and very gradual progression once the symptoms appear; slow viruses include etiologic agents of kuru, scrapie, and Aleutian disease of mink.

small intestine   The proximal portion of the intestine.

small lymphocyte   Circulating lymphocyte; may be of either the B or T type.

smooth ER   Endoplastic reticulum that does not bear ribosomes, involved in the collection and distribution of proteins in the eucaryotic cell.

smooth strain   The smooth strain that results from bacterial dissociation. The S colonies have a smooth surface and an unbroken border; growth in fluid media tends to be diffuse; capsules, if present at all, are found in this strain; and the culture tends to be more virulent.

sneaking through A hypothesis for the means by which a tumor establishes itself as a small cluster of cells, without triggering an immune response until the tumor mass is too large to be controlled.

soap Any compound of one or more fatty acids, or their equivalents, with an alkali.

soft plaque The early form of dental plaque, in which the aggregate of dextran and microorganisms is still soft and readily removed by brushing.

solute A substance dissolved in a solvent.

solution A homogeneous mixture of one or more substances (solutes) dispersed molecularly in a sufficient quantity of dissolving medium (solvent).

solvent Dissolving; effecting a solution.

species A taxonomic category subordinate to a genus (or subgenus) and superior to a subspecies or variety, composed of individuals possessing common characters distinguishing them from other categories of individuals of the same taxonomic level.

species diversity A characteristic of all natural ecosystems, wherein many different species of microorganisms are present and in competition for resources.

specific immunity Immunity against a particular disease; e.g., scarlet fever, or against a particular antigen.

specificity The quality or state of being specific. When applied to an enzyme, or an antibody, the ability to bind to only one unique molecular structure.

specific macrophage activating factor (SMAF) A factor released by sensitized lymphocytes that induces in macrophages an increased content of lysosomal enzymes, more aggressive phagocytosis, and increased mitotic activity.

spectrophotometer (spehk troh fuh TAHM uht ur) An apparatus for measuring the light sensed by means of a spectrum.

spectrum A charted band of wavelengths of electromagnetic vibrations obtained by refraction and diffraction.

spheroplast (SFIHR uh plast) A spherical bacterial or plant cell, produced in hypertonic media under conditions that result in partial absence of the cell wall that no longer serves as a supporting structure.

spherule (1) A small sphere. (2) A spherical multinucleate cell of the parasitic stage of *Coccidioides immitis,* in which endospores are developed.

spinal column The vertebrae, making up a supporting column for attachment of the muscles of the trunk and enclosing and protecting the spinal cord and spinal nerve origins.

spinal cord That part of the central nervous system lodged in the vertebral canal. The spinal cord conducts impulses to and from the brain, and controls many automatic muscular activities (reflexes).

spinal nerves The 31 pairs of nerves arising from the spinal cord.

spirillum (spih RIHL uhm) A genus of microorganisms of the family *Spirillaceae.* Long spirals, or portions of a turn.

spleen A large glandlike but ductless organ situated in the upper part of the abdominal cavity on the left side and lateral to the cardiac end of the stomach.

split-virus vaccine A vaccine, such as certain influenza vaccines, composed of subunits of the viral particles.

spontaneous generation The discredited concept of continuous generation of living organisms from nonliving matter; abiogenesis.

spontaneous mutants Mutants that arise without the addition of recognized mutagenic agents.

sporadic Not widely diffused or epidemic; occurring only occasionally.

sporangium (spuh RAN jee uhm) Any encystment containing spores or sporelike bodies, as in certain of the fungi.

sporangiospore (spuh RAN jee uh spawr) A spore contained in a sporangium.

spore The reproductive element of one of the lower organisms, such as protozoa, fungi, algae, etc. A resisting body produced by certain species of bacteria, noted for its resistance to heat inactivation.

spore coat A layer of highly resistant material enclosing a bacterial endospore.

spore septum A structure, developing during sporulation in bacteria, that separates the developing spore from the rest of the cell contents.

sporogony (spuh RAHG uh nee) The sexual cycle of sporozoa, especially in the life cycle of the malarial parasite *Plasmodium* in the stomach and body of the mosquito.

sporozoite (spawr uh ZOH yt) A spore formed after fertilization; any one of the sickle-shaped nucleated germs formed by division of the protoplasm of a spore of a sporozoan organism.

sputum Matter ejected from the lungs, bronchi, and trachea, through the mouth. Formed as a result of inflammation of the lower respiratory mucosa.

staff–patient transmission Transmission of an infection from a staff member to the patient under care.

stagnation The retardation or cessation of flow of blood in the vessels; passive congestion; accumulation in any part of a normally circulating fluid.

staphylococcal scalded-skin syndrome (SSSS) A condition of neonates, in which infection by staphylococcal strains producing exfoliatin leads of a condition in which the exterior epidermal layers separate from the skin surface.

-static treatment Growth-arresting effect.

stationary phase The stage of the growth of a bacterial culture at which multiplication of organisms gradually decreases, the number of bacterial remaining practically constant.

steam A vapor, fume, or exhalation. Water in the gaseous phase.

stem cell A generalized mother cell whose descendants specialize, often in different directions, such as an undifferentiated mesenchymal cell that may be considered to be a progenitor of the blood and fixed-tissue cells of the bone marrow.

sterilization The complete elimination of microbial viability.

steroid A group name for compounds that contain a hydrogenated cyclopentophenanthrene-ring system. Some of the substances included in this group are progesterone, adrenocortical hormones, the gonadal hormones, cardiac aglycones, bile acids, sterols (such as cholesterol), toad poisons, saponins, and some of the carcinogenic hydrocarbons.

sterol Steroids with long (8–10 carbons) aliphatic side-chains at position 17, and at least one hydroxyl group, usually at position 3. They have lipid-like solubility. Examples are cholesterol and ergosterol.

stock culture A permanent culture from which transfers may be made.

stock vaccine A vaccine of which all lots are derived from a carefully maintained stock culture.

stomach The musculomembraneous expansion of the alimentary canal between the esophagus and the duodenum.

straight pathway In metabolism, a pathway progressing without branches from a starting material to an end product.

streak plate A plate of solid culture medium in which the infectious material is inoculated in streaks across the surface.

streptolysin O (strep tohl LYS ihn OH) A hemolysin that is inactive in the oxidized state but is readily activated by treatment with mild reducing agents, such as sulfite.

streptolysin S A hemolysin sensitive to treatment with heat or acid, but not activated by oxygen.

streptomycin (strehp tuh MYS uhn) A bacterial antibiotic of the aminoglycoside class, produced by the soil antinomycete, *Streptomyces griseus*, and effective against most Gram-negative and acid-fast bacteria and some Gram-positive forms, but used chiefly in the treatment of tuberculosis.

strict aerobe An organism requiring oxygen for growth.

strict anaerobe An organism that grows only in the absence of oxygen.

stroma The supporting tissue or matrix of an organ, as distinguished from its functional element, or parenchyma.

strong acid An inorganic acid that dissociates completely to yield potentially high contractions of hydrogen ion.

strong base An inorganic base such as sodium hydroxide that dissociates completely to yield high concentrations of OH^- ion.

structural formula A chemical formula telling how many atoms of each element are present in a molecule of a substance, which atom is linked to which, and the type of linkages involved.

structural gene A gene that specifies the amino acid sequence of a polypeptide chain.

structural protein A protein that functions as part of a cellular structure.

sty A suppurative infection of a marginal gland of the eyelid.

subacute bacterial endocarditis (SBE) Bacterial endocarditis. A febrile systemic disease marked by focal bacterial or fungal (mycotic) infection of the heart valves, with formation of bacteria- or fungi-laden vegetations. The subacute form, marked by insidious onset and protracted course, is caused by various bacteria, most often α-hemolytic streptococci.

subacute sclerosing panencephalitis (SSPE) A rare and devastating form of leukoencephalitis usually affecting children and adolescents. Insidious in onset, it characteristically produces progressive cerebral dysfunction over a course of several weeks or months and death within a year. Follows measles after a lapse of several years.

subarachnoid space The space between the arachnoid and the pia mater.

subatomic particle Particle found within the atomic structure, e.g., a proton.

subculture A culture of bacteria derived from another culture.

subcutaneous Beneath the skin.

subsoil Layer of primarily inorganic material found below the topsoil.

substitution mutation A type of point mutation in which one base is substituted for another as a DNA strand is being synthesized.

substrate A substance upon which an enzyme acts.

substrate-level phosphorylation Synthesis of high-energy phosphate bonds through reaction of inorganic phosphate with an activated (usually) organic substrate.

30S subunit The smaller subunit of the procaryotic ribosome.

40S subunit The smaller subunit of the eucaryotic ribosome.

50S subunit The larger subunit of the procaryotic ribosome.

60S subunit The larger subunit of the eucaryotic ribosome.

subunit vaccine A vaccine prepared from a cellular subunit, such as a ribosomal vaccine.

Sudden Infant Death Syndrome (SIDS) The sudden and unexpected death of an apparently healthy infant, typically occurring between the ages of three weeks and five months and not explained by careful postmortem studies.

sugar A sweet carbohydrate of various kinds, and of both animal and vegetable origin.

sulfhydryl group The grouping –SH, found in the amino acid cystaine, and forming disulfied bonds.

sulfonamides (suhl FAHN uh mydz) The chemical group SO_2NH_2; the sulfonamide compounds are groups of compounds with one or more benzene rings, amino groups, and a sulfonamide group, and include a group of antibacterial drugs closely related to sulfanilamide.

sulfur granules Peculiar granular bodies of a yellow color found in actinomycotic lesions and discharges.

sulfur oxidation Conversion of elemental sulfur to sulfur oxides.

superficial mycosis Fungal infection of the skin and related keratinous structures.

superinfection A new infection complicating the course of antimicrobial therapy of an existing infectious process and resulting from invasion by bacteria or fungi resistant to the drug(s) in use. It may occur at the site of the original infection or at a remote site.

surface-active agent Surfactant. Includes substances commonly referred to as wetting agents, surface tension depressants, detergents, dispersing agents, emulsifiers, quaternary ammonium antiseptics, etc.

surface antigen An antigen associated with the surface of a cell or structure, binding antibody to the surface.

surface tension The tension or resistance that acts to preserve the integrity of a surface, such as the tension or resistance to rupture possessed by the surface film of a liquid or the tension or strain upon the surface of a liquid in contact with another substance with which it does not mix.

surface/volume ratio Ratio of the square centimeters of surface area of a cell to the cubic centimeters of its volume.

surface water Water found in surface features of land, such as streams, ponds, lakes.

surgical asepsis The techniques of preventing infection during surgery by excluding microorganisms from the surgical field.

surveillance In immunological theory, the constant monitoring of the body tissues by the immune (T-lymphocyte) system for abnormal cells. In epidemiology, the constant watching for the appearance of new cases of a specific disease.

susceptibility The state of being readily affected by infectious disease. In immunology, the condition may be an acquired, familial, individual, inherited, racial, specific, etc., the same as is immunity.

susceptibility testing Testing a pure culture of a microorganism to evaluate its response to an antimicrobial drug.

swarmer cells   The cells of *Rhizobium* that penetrate the root hairs to set up the symbiotic association with legumes. They are more nearly spherical than the rod-shaped cells of *Rhizobium*.

sweat glands   Glands in the skin secreting sweat.

sylvatic reservoir   Reservoir of an infective agent in a wild animal species.

symbiosis (sihm by OH suhs)   The living together or close association of two dissimilar organisms, each of the organisms being know as a *symbiont*.

symbiotic   Associated in symbiosis; living together.

synapse (SIHN aps)   The anatomical relation of one nerve cell to another; the region of junction between processes of two adjacent neurons, forming the place at which a nervous impulse is transmitted from one neuron to another.

synchronous growth   A laboratory growth pattern in which most or all of the cells in a bacterial culture divide at the same time.

syncytial (sihn SIHSH (ee) uhl)   Of, pertaining to, or producing a syncytium or giant cell containing many nuclei and produced by fusion of many cells.

synergism (SIHN ur jihz uhm)   The joint action of agents so that their combined effect is greater than the algebraic sum of their individual effects.

synthetase (SIHN thuh tays)   Ligase.

synthetic   Pertaining to, or the nature of, or participating in synthesis. Also, referring to a chemical substance entirely manufactured by human efforts from simple starting materials.

syphilis   A contagious veneral disease leading to many structural and contagious lesions, caused by the spirochete *Treponema pallidum* and transmitted by direct intimate contact or *in utero*.

systemic   Pertaining to or affecting the body as a whole.

systemic lupus erythematosis (SLE)   An autoimmune disease characterized by the appearance of LE cells, antinuclear antibodies, and generalized severe symptoms.

systemic mycosis   A fungal infection of deep tissues such as the lung, intestinal tract or brain. Very serious if unchecked.

tail   Any slender appendage. A structure found in some groups of bacteriophage, used in attachment to the host cell.

tail fibers   Protein fibers forming the attachment mechanism to the t-even group of bacteriophage.

taxonomy   The orderly classification of organisms into appropriate categories (taxa) on the basis of relationships among them, with the application of suitable and correct names.

T-cytotoxic cell   A type of T-lymphocyte that kills antigen-bearing target cells on contact.

T-delayed hypersensitivity cells   A group of T-lymphocytes that release lymphokines on contact with antigen-bearing targets.

teichoic acid   Acidic polysaccharide containing glycerol or ribitol, connected by phosphate diester bonds. Found in the walls of Gram-positive bacteria.

temperate phage   Bacteriophage that enter the prophage state in the host cell.

template   A pattern or mold. In dentistry, a curved or flat plate used as an aid in setting teeth for a denture. In theoretical immunology, an antigen that determines the configuration of combining (antigen-binding) sites of antibody molecules. In genetics, a strand of DNA that specifies the synthesis of a complementary strand of RNA (mRNA), that in turn serves as a template over the synthesis of nucleic acids or proteins.

teratogen (tuh RAT uh juhn)   A drug or other agent that causes abnormal development.

terminal disinfection   Disinfection of the room occupied by a patient and the materials used by the patient after discharge.

termination   The completion or interruption of protein synthesis.

terminators   Codons that do not code for an amino acid, and that cause the polypeptide chain to detach from the polysome.

terpene (TUR peen)   Any hydrocarbon of the formula $C_{10}H_{16}$, derivable chiefly from essential oils, resins, and other vegetable aromatic products.

tertiary treatment   The final stage in sewage treatment, in which almost all organic and inorganic contaminants are removed from the water and the water is recovered in potable condition.

testis   The male gonad; the paired egg-shaped glands normally situated in the scrotum.

test system   In the complement fixation test, the first antigen–antibody pair to which a measured amount of complement is added.

tetanospasmin   The neurotoxin of *Clostridium tetani*; causes the characteristic signs and symptoms of tetanus. The chief action is on the anterior horn cells, and the spasms seem to be due to action at inhibitory synapses.

tetanus   An infectious disease in which tonic muscle spasm and hyperreflexia result in trismus ("lockjaw"), generalized muscle spasm, and seizures.

tetracycline   An antibiotic substance, $C_{22}H_{24}N_2O_8$, isolated from the elaboration products of certain species of *Streptomyces* on suitable media; used as an antiamebic, antibacterial, and antirickettsial.

tetrapyrrol nucleus   A molecule of four pyrrol nuclei, as in porphyrin.

thallus (THAL uhs)   (1) A simple plant body not differentiated into root, stem, and leaf, that is characteristic of mycelial fungi and some algae. (2) The actively growing vegetative organism as distinguished from reproductive or resting portions, as in fungi.

T-even phage   Lytic bacteriophage parasitic on *Escherichia coli*.

T-helpers   A class of T-lymphocytes that contribute to or augment reactions of other lymphocytes.

therapeutic range   The range of concentration that lies between the effective dose and the toxic dose of a drug.

thermal death point (TDP)   The degree of heat required to kill a given microorganism in a stated length of time.

thermal death time (TDT)   The duration of exposure required to kill a bacterium at a stated temperature.

thermodynamics   The branch of science that deals with heat, energy, and the interconversion of these, and with related problems.

thermophile (THUR muh fyl)   A microorganism that grows at temperatures above 40°C.

thioglycollate   A salt or ester of thioglycolic acid, incorporated into bacteria media to provide anaerobic conditions.

thoracic cavity   The portion of the ventral body cavity situated between the neck and the respiratory diaphragm.

thrombi   Fibrin clots forming within blood vessels.

thrombocyte (THRAHM buh syt) A synonym for platelet.

thromboplastin (thrahm boh PLAS tuhn) A substance having procoagulant properties or activity.

thrush Candidiasis of the mucous membranes of the mouth of infants (sometimes of adults), characterized by the formation of aphthae, or whitish spots in the mouth.

thylakoid Membranous structures containing the photosynthetic pigments.

thymocyte A lymphocyte arising in the thymus.

thymus A ductless glandlike body situated in the anterior mediastinal cavity that reaches its maximum development during the early years of childhood and then undergoes involution.

tincture An alcoholic or hydroalcoholic solution prepared from animal or vegetable drugs or from chemical substances.

tinea pedis (TIHN ee uh PEE duhs) Athlete's foot. A chronic superficial fungal infection of the skin of the foot, especially of that between the toes and on the soles, caused by species of *Trichophyton* or by *Epidermophyton floccosum*.

tine test, tine tuberculin test (Rosenthal) Four tines or prongs 2 mm long, attached to a plastic handle and coated with dip-fried Old Tuberculin (O.T.) (or PPD) are pressed into the skin of the volar surface of the forearm, where they deposit a dose of the tuberculin on the outer layer. The skin is checked 48 to 72 hours later for the presence of palpable induration; if the induration around one or more of the puncture wounds is 2 mm or more in diameter, the test is considered positive.

tissue An aggregation of specialized cells united in the performance of a particular function.

tissue cysts Inactive forms of protozoan parasites localized in the tissues of the host.

tissue matching Determining the histocompatibility of donor and recipient tissues prior to grafting.

titer The quantity of a substance required to produce a reaction with a given volume of another substance, or the amount of another substance. Also the quantity of antibody present in a serum sample as measured by the degree of reactivity with antigen.

T-lymphocyte. A lymphocyte functional in cell-mediated immune responsiveness.

tobacco mosaic virus (TMV) A plant virus containing RNA that causes mosaic disease of tobacco.

tonsil A small rounded mass of tissue, especially of lymphoid tissue.

tonsilitis Inflammation of the tonsils, especially the palatine tonsils.

tooth decay Dental caries; a degenerative condition caused by acid erosion of tooth structure.

topical Pertaining to a particular surface area, as a topical anti-infective applied to a certain area of the skin and affecting only the area to which it is applied.

topsoil The uppermost layer of soil, containing humus and living organisms, and being rich in plant nutrients.

TORCH series A series of serological tests to detect antibodies to four of the major intrauterine infective agents, *Toxoplasma*, rubella virus, cytomegalovirus, and herpes virus.

toxin neutralization Combination of antibody with toxin that results in a loss of toxigenicity.

toxoid A modified bacterial exotoxin that has lost toxicity but retains the properties of combining with, or stimulating the formation of, antitoxin.

toxoplasmosis (tahk suh plaz MOH suhs) A protozoan disease of human beings caused by *Toxoplasma gondii*.

trachea The cartilaginous and membranous tube descending from the larynx and branching into the right and left main bronchi.

tracheitis (tray kee YT uhs) Inflammation of the trachea.

tracheobronchitis (tray kee oh brahng KYT uhs) Inflammation of the trachea and bronchi.

trachoma (truh KOH muh) A chronic infectious disease of the conjunctiva and cornea, producing photophobia, pain, and lacrimation, caused by an organism once thought to be a virus but now classified as a strain of the bacteria *Chlamydia trachomatis*.

transamination The reversible transfer of an amino group from an amino acid to what was originally an α-keto acid, forming a new keto acid and a new amino acid, without the appearance of ammonia in the free state.

transcription The process by which genetic information contained in DNA produces a complementary sequence of bases in an RNA chain.

transduction A method of genetic recombination in bacteria, in which DNA from a lysed bacterium is transferred to another bacterium by bacteriophage, thereby changing the genetic constitution of the second organism.

transferase Any of a class of enzymes that catalyze the transfer, from one molecule to another, of a chemical group that does not exist in the free state during the transfer.

transfer factor A factor occurring in sensitized lymphocytes that has the capacity to transfer delayed hypersensitivity to a normal (nonreactive) individual.

transferrin (tran(t)s FEHR uhn) Serum β-globulin that binds and transports iron.

transfer RNA (tRNA) Soluble short chain RNA molecules present in cells in at least 20 varieties, each variety capable of combining with a specific amino acid.

transformation Change of form or structure; conversion from one form to another. In bacteria, a genetic change caused by the transfer of DNA fragments from donor to recipient cell. In oncology, the acquisition by normal cells of characteristics that mark the cells as abnormal or neoplastic.

transitory bacteremia Bacteremia lasting for only a few minutes, following minor trauma.

translation A removal or change of place; in genetics, the formation of a polypeptide chain in the specific amino acid sequence directed by the genetic information carried by messenger RNA.

translocation Removal to another place. In genetics, the shifting of a segment of fragment of one chromosome into another part of a homologous chromosome, or into a nonhomologous chromosome.

transmission A transfer, as of a disease or neural impulse; the communication of inheritable qualities to offspring.

transmission cycle The pattern of transmission of infectious disease from one individual to another, resulting in the maintenance of the disease organism in the endemic area.

transmission electron microscope (TEM) A type of electron microscope in which the image is formed by a stream of electrons passing through the object being studied.

transovarial transmission (tran(t)s oh VAR ee uhl)   Referring to transmission of pathogens from the maternal organism, by invasion of the ovary and infection of eggs, to individuals of the next generation, as may occur in infections of arthropods, especially mites and ticks.

transplantation   The grafting of tissues taken from the same body or from another.

transport   The movement of materials in biological systems, particularly into and out of cells and across epithelial layers.

transport culture   A culture inoculated on a special transport medium, designed to maintain its viability during the time required to transport it to a suitable laboratory.

transposon   A genetic element that can move from place to place and that contains an insertion element at each end.

trauma   A wound or injury, whether physical or psychic.

trenchmouth   Acute necrotizing ulcerative gingivitis or Vincent's angina; an acute inflammation of the gums.

treponemal immobilization test (TPI)   A test based upon the fact that an antibody present in the serum of a syphilitic patient in the presence of complement causes the immobilization of actively motile *Treponema pallidum* obtained from the testes of a rabbit infected with syphilis.

treponemal tests   Treponemal hemagglutination test. A passive serological test for syphilis using as antigen disrupted *Treponema pallidum* coated onto tannin-treated sheep erthrocytes that are agglutinated in the presence of specific antibody.

trichome (TRIHK ohm)   A filamentous or hairlike structure.

triglyceride   A compound consisting of three molecules of fatty acid esterified to glycerol; it is a neutral fat synthesized from carbohydrates for storage in animal adipose cells.

trivalent oral polio vaccine (TOPV)   A vaccine containing live polio virus of the three strains found in nature, attenuated so that they are rendered nonvirulent.

trophic level   The position of a species in the food web or chain; a step in the movement of biomass or energy through a system.

trophoblast   A layer of extraembryonic ectodermal tissue on the outside of the blastocyst.

trophozoite (troh fuh ZOH yt)   The active, motile, feeding stage of a protozoan organism, as contrasted with the nonmotile encysted stage.

trypomastigote (try poh MAS tih goht)   The morphologic stage in the development of certain hemoflagellates of vertebrate hosts, resembling the typical adult form of *Trypanosoma* and characterized by having a flagellum arising from posteriorly situated blepharoplast and running forward to form the outer edge of the undulating membrane; it may be free at the anterior end of the body.

trypticase   Hydrolyzed protein, a trysin digest of casein.

tryptone (TRIHP tohn)   A peptone produced by proteolytic digestion with trypsin.

T-suppressors   T-lymphocytes specialized to interact with other lymphocytes in ways that moderate their responsiveness.

tube dilution   A testing technique in which a sample is transfered serially from one tube to others, leading to progressive dilution.

tubercle   Any small, rounded nodule produced by infection with *Mycobacterium tuberculosis.*

tuberculin test   A test for the existence of tuberculosis, consisting in the subcutaneous injection of 5 mg of tuberculin.

tuberculoid leprosy   That type of leprosy, in which, as a result of high cell-mediated resistance to the infection, *Mycobacterium leprae* are few or lacking by ordinary methods of examination, and nerve damage occurs very early, so that all skin lesions are denervated from the start, often with dissociation of sensation. This form of the disease progresses very slowly.

tuberculosis   Any of the infectious diseases of human beings and animals caused by species of *Mycobacterium* and characterized by the formation of tubercles and caseous necrosis in the tissues.

tuberculosis, miliary (acute miliary)   An acute form of tuberculosis in which minute tubercles are formed in a number of organs throughout the body, due to dissemination of the bacilli throughout the body by the blood stream. So named because tubercles resemble millet seeds.

tuberculosis, primary   Pulmonary tuberculosis, formerly known as the childhood type, occurring when the individual is first infected with the disease.

tuberculosis; reactivation   A form of the disease in which a previously healed lesion reopens, and the mycobacteria start to multiply again.

tubular gland   Any gland made up of or containing a tubule or a number of tubules.

tubulin   The protein of the microtubules or eucaryotes.

tularemia (t(y)u luh REE mee uh)   A disease of rodents, resembling plague, that is transmitted by the bites of flies, fleas, ticks, and lice, and may be acquired by human beings through handling of contaminated animal products (leather) or infected animals, or the bites of fleas, ticks, and the deerfly.

tumor   A new growth of tissue in which the multiplication of cells is uncontrolled and progressive. A swelling, one of the cardinal signs of inflammation; morbid enlargement.

tumor antigen (T antigen)   Any novel antigen that appears on the surface of the tumor cell, but is not found on normal cells of a similar type.

tumor-specific antigens   Cell-surface antigens of tumors that elicit a specific immune response in the host.

tumor-specific transplantation antigen   Any of the cell-surface histocompatibility antigens of any given tumor, that evoke a specific immune response on transplantation to a syngeneic host.

turbidimentry (tur buh DIHM uh tree)   The measurement of the turbidity of a fluid.

turbulent flow   A fluid flow over an uneven surface, developing eddies and swirls.

tympanum (TIHM puh nuhm)   The cavity of the middle ear, located just medial to the tympanic membrane and containing the auditory ossicles and connecting with the mastoid cells and auditory tube.

Tyndallization (tihn dahl ih ZAY shuhn)   The technique of eliminating endospores by treating a material with heat sufficient to kill vegetative organisms, incubating to allow any spores to germinate and develop into vegetative cells, and then reheating to kill the new vegetative cells.

typhoid fever   Infection by *Salmonella typhosa,* involving primarily the lymphoid follicles of the ileum.

typhus fever Any of a group of related arthropod-borne infectious diseases caused by species of *Rickettsia* and marked by malaise, severe headache, sustained high fever, and macular or maculopapular eruption that appears from the third to seventh day.

typing In transplantation immunology, a method of measuring the degree of organ, solid tissue, or blood compatibility between individuals, in which specific histocompatibility antigens are detected by means of suitable isoimmune antisera.

ubiquinone (yoo buh kwihn OHN) Coenzyme A, one of the coenzymes of electron transport.

ultrafiltration Filtration through filters with minute pores, thus allowing the separation of extremely minute particles.

ultrasound Mechanical radiant energy, with a frequency greater than 20,000 cycles per second.

ultrathin section A cell or tissue section cut with extreme care to a thickness of less than a micrometer.

ultraviolet Beyond the violet end of the spectrum; said of rays or radiation between the violet rays and the roentgen rays, that is, with wavelengths between 1800 and 3900 angstroms.

ultraviolet microscope A microscope that utilizes reflecting optics of quartz and other ultraviolet-transmitting lenses, with radiation of less than 400 mμ wavelength as the image-forming energy.

umbilical vessels The artery and vein serving the placenta and carrying the blood flow from it to the fetus within the umbilical cord.

uncoating The process by which the envelope and/or capsid is removed from a virus particle after it has entered the host cell.

undulant fever (UHN juh luhnt) Brucellosis.

unicellular Made up of but a single cell, as the bacteria.

United States Public Health Service A branch of the federal department of Health and Human Services specifically charged with maintaining the public health.

unsaturated Denoting a solution in which the solvent is capable of dissolving more of the solute. In organic chemistry, denoting presence of double or triple covalent bonds in a molecule.

upper respiratory tract (URT) The portion of the respiratory tract from the nostrils to the epiglottis.

upwelling The process whereby warm, mineral-laden waters rise from the ocean floor to the surface along the coasts of certain continents.

ureter The fibromuscular tube that conveys the urine from the kidney to the bladder.

urethra The membranous canal conveying urine from the bladder to the exterior of the body.

uterus The hollow muscular organ in female mammals in which the fertilized ovum normally becomes embedded and in which the developing embryo and fetus is nourished.

vaccination Administration of an antigenic material derived from a microorganism or virus to induce the development of active immunity.

vaccine A suspension of attenuated or killed microorganisms, administered for the prevention, amelioration, or treatment of infectious diseases.

vagina (1) A sheath, or sheathlike structure; used as a general term in anatomical nomenclature. (2) The canal in the female, extending from the vulva to the cervix uteri, that receives the penis in copulation.

valve A membranous fold in a canal or passage that prevents the reflux of the contents passing through it.

varicella (var uh SEHL uh) Chickenpox.

variolation (vehr ee uh LAY shuhn) Deliberate inoculation with the virus of unmodified smallpox to produce immunity to the naturally occurring disease.

vasculitis (vas kyuh LYT uhs) Inflammation of a vessel.

vector A carrier, especially an animal, that transfers an infective agent from one host to another.

vegetation Any plantlike fungoid neoplasm or growth; a luxuriant growth of pathologic tissue such as forms on interior surfaces of the heart in endocarditis.

vein A vessel through which blood passes from various organs or parts back to the heart.

venereal disease A contagious disease, most commonly acquired in sexual intercourse or other genital contact.

vertical transmission Transmission from one generation to another. The term is restricted by some to genetic transmission and extended by others to include also transmission of infection from one generation to the next, e.g., through milk or through the placenta.

V factor A factor required for the growth of *Hemophilus influenzae*.

villus A small vascular process or protrusion, especially such a protrusion from the free surface of a membrane.

Vincent's angina Acute necrotizing ulcerative gingivitis.

viroid An infective agent smaller than a virus consisting of nucleic acid alone.

virion The complete viral particle, found extracellularly and capable of surviving in crystalline form and infecting a living cell.

virulence The degree of pathogenicity of a microorganism as indicated by case fatality rates and/or its ability to invade the tissues of the host.

visceral peritoneum The layer of peritoneum investing the abdominal organs.

volutin (VAHL yuh tihn) Bacterial polymetaphosphate occurring as cytoplasmic granules having a marked affinity for basic dyes.

vulva The region of the external genital organs of the female, including the labia majora, labia minora, mons pubis, bulb of the vestibule, vestibule of the vagina, greater and lesser vestibular glands, and vaginal orifice.

wandering cells Cells capable of ameboid movement, such as free macrophages, lymphocytes, mast cells, and plasma cells.

wavelength The distance between the top of one wave and the identical phase of the succeeding one.

wax A lipid substance deposited by insects or obtained from plants and bacteria.

Weil–Felix reaction The diagnostic agglutination of *Proteus* bacteria by the blood sera of typhus fever cases due, apparently, to the presence of a common antigen.

white matter of nervous system Substantia alba. The conducting portion of the brain and spinal cord composed of nerve fibers.

whooping cough   An infectious disease caused by *Bordetella pertussis* and characterized by catarrh of the respiratory tract and peculiar paroxysms of cough, ending in a prolonged crowing or whooping respiration.

Widal reaction   Widal's syndrome, test (reaction, serum test). For typhoid fever. Agglutination of *Salmonella typhosa* by dilutions of the patient's serum.

wound   A bodily injury caused by physical means, with disruption of the normal continuity of structures.

X factor (Factor X)   A cofactor required for the growth of *Hemophilus influenzae*.

X ray   Roentgen ray. Electromagnetic radiation capable of penetrating soft tissue.

xylitol (ZY luh tahl)   An alcohol $CH_2OH(CHOH):CH_2OH$, from xylose.

yeast   A general term including single-celled, usually rounded, fungi that reproduce by budding (blastospore formation).

yeast extract   A powder prepared from a water-soluble peptone-like derivative of yeast cells.

yellow fever   An acute infectious disease due to a virus, transmitted to human beings that acquire the infection either from human beings or from animals.

zone of inhibition   The area of no growth surrounding a source of growth inhibitor, such as an antibiotic impregnated disc.

zoonosis (zoh AHN uh suhs)   A disease of animals that may be transmitted to human beings.

zooplankton (zoh uh PLANG(K) tuhn)   The minute animal organisms that, with those of the vegetable kingdom (phytoplankton), make up the plankton of natural waters.

zoospore   A motile, reproductive spore, as in aquatic phycomycetous fungi and other fungi.

zoster   A girdle, or encircling structure or pattern. The pattern of eruptions or shingles seen in herpes zoster.

zygospore (ZY guh spawr)   A spore formed by the conjugation of two cells (isogametes) that are morphologically identical, or in the zygomycetes, from the fusion of like gametangia.

zygote   The cell resulting from the union of a male and a female gamete, until it divides; the fertilized ovum.

Photograph Acknowledgments

1, Courtesy of Barbara Cooksy and Don Manszalek.

3, From J.M. Sieburth, *Microbial Seascapes*. Baltimore: University Park Press (1975); Plate 2.10.

5, From M. Haberey, *Microskopie* (1971) **27**: 226–234, published by Georg Fromme & Co., Vienna.

7, Courtesy of A.H. Knoll and E.S. Berghoorn, Archean microfossils showing cell division from the Swaziland system of South Africa, *Science* (1977) **198**: 396–398, Fig. 1.

14, Courtesy of Institute Pasteur, Paris, France.

15, Photo courtesy of Merck, Sharp, Dohme

40, Courtesy of D.L. Balkwill, D. Maratea, and R. Blakemore.

59, Courtesy of Rainin Instrument Company.

60, Courtesy of Bio-Rad Laboratories.

69, From T.D. Brock, *Biology of Microorganisms* (3rd ed.). Englewood Cliffs, N.J.: Prentice-Hall, p. 15. (Fig. 2.1).

72, From H.C. Tsien, G.D. Shockman, and M.L. Higgins, *J. Bacteriol.* (1978) **133**: 372–386.

73, From K. Yamada and M. Matsuhashi, Interconversion of *Micrococcus rubens*, *J. Bacteriol.* (1977) **129**: 1513–1517, Fig. 2A.

75, From B.E Brooker, *J. Bacteriol.* (1977) **131**: 288–292. Figs. 1, 2, and 3.

76, From P.A. Meacock, R.H. Pritchard, and E.M. Roberts, Effect of thymine on cell shape in *E. coli. J. Bacteriol.* (1978) **133**: 320–328 Fig. 1.

79, From A.T. Hastie and C.C. Brinton, Tetragonal layers of *Bacillus sphaericus. J. Bacteriol.* (1979) **138**: 999–1009, Fig. 4A.

87, Courtesy of *Journal of Bacteriology* **132** (3): 950 (December 1977), Roger M. Cole.

88, From M.A. Listgarten and S.S. Socronsky, *Oral spirochetes, J. Bacteriol.* (1964) **88**: 1087.

104, Courtesy of Drs. R.Y. Stanier and G. Cohen-Bazire, *Ann Rev. Microbiol.* **33**: 236. Reproduced, with permission, from The Annual Review of Microbiology © 1977 by Annual Reviews Inc.

111, (a)/(b) Courtesy of Dr. Germaine Cohen-Bazire, Directeur de Recherche CNRS—Paris. (c) Permission granted by the American Society for Microbiology. Fine Structure of *Ectothiorhodospira mobilis pelsh*, C.C. Remsen, S.W. Watson, J.B. Waterbury, and H.G. Truper, *J. Bacteriol.* **95**: 2376 (d) Courtesy of S.W. Watson, Woods Hole Oceanographic Institution, Woods Hole, Mass. (e) Courtesy of Dr. Jessup M. Shively.

114, (a) Courtesy of Drs. James T. Staley and J.P. Dalmasso. (b) Permission granted by the American Society for Microbiology. Van Vien et al. The *Sphaerotilus–Leptothrix* group. *Microbiol. Rev.* (1978) **43**, Fig. 8. (c) Courtesy of J.S. Poindexter, The Public Health Research Institute of New York City.

115, From P.L. Grillone and J. Pangborn, *J. Bacteriol.* (1975) **124**: 1558–1565.

116, Courtesy of Drs. J.T. Staley and J.P. Dalmasso.

117, Courtesy of F. Scanga, *Atlas of Electron Microscopy; Biological Applications,* New York: Elsevier Publishing Co. (1964) Fig. 111.

119, Courtesy of Z. Yoshii et al., *Atlas of Scanning Electron Microscopy in Microbiology,* Igaku Shoin, Ltd., Tokyo, 1976.

121, From Dr. J.G. Zeikus, Dept. of Biology, University of Wisconsin.

122, Courtesy of Z. Yoshii et al., *Atlas of Scanning Electron Microscopy in Microbiology,* Igaku Shoin, Ltd., Tokyo, 1976. (b) From D.L. Shungu, J.B. Cornett, and G.D. Shockman, Autolytic mutants of *S. faecium, J. Bacteriol.* (1979) **138**: 601, Fig. 1a.

123, Courtesy of Dr. Carl Franz Robinow with permission from Harper & Row.

126, From W.W. Gregory et al., *J. Bacteriol.* (1979) **138**: 242, Fig. 1a.

127, (a) From Joseph E. Gallagher and Keith Rhodes, *J. Bacteriol.* (1979) **137**: 974, Fig. 3b. (b) Shmuel Razin, The mycoplasmas, Fig. 3. *Microbiol Revs.* (1978) **42**: 414–470.

131, (a) Courtesy of Prof. Jeremy Pickett-Heaps, University of Colorado; (b) Andrew Staehelin, University of Colorado; (c) Courtesy of Patrick Echline.

134, Carolina Biological Supply Company.

135, (Top), From J.M. Sieburth, *Microbial Seascapes,* Baltimore: University Park Press (1975) Plate 2.6.

(Bottom), From J.M. Sieburth, *Microbial Seascapes,* Baltimore: University Park Press (1975) Plate 2.8.

136, From J.M. Sieburth, *Microbial Seascapes,* Baltimore: University Park Press (1975).

138, From Kenneth Watson and Helen Arthur, *J. Bacteriol.* (1977) **130**: 312–317, Fig. 5.

142, From James S. Lovett. Growth and development of *Blastocladiella emersonii, Bacteriol. Revs.* (1975) **39**: 347, Fig. 3. Reprinted with permission of W.E. Barstow.

152, From J.M. Sieburth, *Microbial Seascapes,* Baltimore: University Park Press (1975).

160, Courtesy of (a) John W. Kimball, *Biology,* (2d ed.) and (b) © Carolina Biological Supply Company.

174, Courtesy of D.H. Marx, USDA.

175, (a) From F.B. Dazzo and W.J. Brill, *J. Bacteriol.* (1979) **137**: 1362–1373, (b) Courtesy of The Nitragen Co., Milwaukee, Wisc.

179, Courtesy of *GERMFREE* Laboratories, Inc., Miami, Florida.

189, Photo courtesy of Millipore Corporation, Bedford, Massachusetts

190, From S.M. Finegold, W.J. Martin, and E.G. Scott, *Bailey and Scott's Diagnostic Microbiology* (5th ed.) C.V. Mosby, 1978, Fig. 4–2A,B, p. 19.

191, From R.R. Gillies and T.C. Dodds, *Bacteriology Illustrated* (4th ed.) (1976). Churchill Livingstone: Edinburgh, London and New York.

193, Courtesy of the Virtis Co., Inc., Gardiner, New York.

196, Pickles courtesy of Candace L. Smith.

198, Courtesy of BBL Microbiology Systems, Division of Becton, Dickinson and Company.

203, Courtesy of Millipore Corporation, Bedford, Massachusetts.

204, Courtesy of Vitek Systems, Inc.

206, From I.D.J. Burdett, *J. Bacteriol.* (1980) **137:** 1399, with permission from the author and ASM.

237, Courtesy of the University of Alabama.

240, Courtesy of the Wine Institute.

249, Drawing modified from R.Y. Stanier and H.J. Rogers (eds.), *Relations between Structure and Function in the Prokaryotic Cell,* Society for General Microbiology Symposium: No. 28, p. 225.

261, From O.L. Miller, Jr., B.A. Hamkalo, and C.A. Thomas, Jr., Visualization of bacterial genes in action, *Science* (1970) **169:** 392–395, Fig. 24.

274, Courtesy of Dr. R. Laufs, Universität Hamburg from the *Journal of Bacteriology.*

286, Diagram with permission from D.E. Bradley, *Bacteriol. Revs.* (1967) **31:** 230–314, Fig. 1.

288, From Dale Fay and Bernard U. Bowman, Mycobacteriophage R1., *J. Virol.* (1978) **27:** 432–435, Fig. 2A.

290, From F.W. Studier, *Science* (1972) **176:** 367–376.

292, 294, Courtesy of Prof. Jonathan King, Department of Biology, M.I.T., Massachusetts.

295, From I.M. Smith, *Staphylococcal infection.* Year Book, 1958.

300, From M.K. Corbett, *Virology* (1964) **22:** 539–543. By permission.

301, From Katherine Esau, *Viruses in Plant Hosts.* Madison: The University of Wisconsin Press; © 1968 by the University of Wisconsin, p. 92.

306, From Samuel Dales, *J. Cell. Biol.* (1963) **18:** 51.

312, From Lelio Orci and Alain Parrelet, *Freeze-Etch Histology,* New York: Springer-Verlag (1975) Plate 25.

315, Courtesy of Bio-Rad Laboratories.

316, Courtesy of Forma Scientific.

316, From Charles Grose et al., Cell-free *Varicella-zoster* virus, *J. Gen. Virol.* (1979) **43:** 15–27, Fig. la.

317, Courtesy of Merck, Sharp & Dohme.

318, Courtesy of the Jackson Laboratory.

321, From Fumio Uno, *J. Electron Microscopy* (1979) **28:** 83–92 Figs. 7a and 8a. Reproduced by permission of author and the Japanese Society of Electron Microscopy.

343, From W.H. Fahrenbach in W. Montagna and P.F. Parakkal, 1974, *The Structure and Functions of Skin.* New York: Academic Press.

346, From Z. Yoshii et al., *Atlas of Scanning Electron Microscopy in Microbiology.* Tokyo, Igaku Shoin, Ltd., 1976. Reproduced by permission.

351, From G.P. Youmans and P.Y. Paterson, *Biological and Clinical Basis of Infectious Diseases.* Philadelphia: Saunders (1975) Fig. 24-4, p. 338.

361, From Gerard J. Tortora and Nicholas P. Anagnostakos. *Principles of Anatomy and Physiology,* 2nd ed. (1978). Copyright © 1975, 1978 by Gerard J. Tortora and Nicholas P. Anagnostakos. By permission of Harper & Row, Publishers, Inc.

366, (a) From W.M. Copenhaver, D.E. Kelly, R.L. Wood, *Bailey's Textbook of Histology,* 17th ed., Williams & Wilkins (1978) Fig. 16-83 (p. 540).

375, From R.O. Greep and L. Weiss, *Histology* New York: McGraw-Hill, (1973) (3rd ed.) Figs. 18–57, 19–17.

384, Ivan Roitt, *Essential Immunology.* Blackwell, 1977 Figs. 13–16a, p. 73.

387, Photo courtesy of Wellcome Reagents Division, Burroughs Wellcome Co.

388, Courtesy of Michael Katz, Medical Technology Corp., Hackensack, N.J.

392, Coagglutination as seen in Phadebact ® Gonococcus Test. Courtesy of Pharmacia Diagnosis, Piscataway, N.J.

398, Courtesy of Dr. Alwin H. Warfel from Sloan-Kettering Institute for Cancer Research and of Experimental and Molecular Pathology (in press, 1980).

415, From B.F. Feingold, *Introduction to Clinical Allergy,* Springfield: Charles C Thomas, (1973) Figs. 17-2, 17-3.

418, Dr. Leon J. LeBeau, University of Illinois Medical Center.

433, From Ivan Roitt, *Essential Immunology,* Blackwell Scientific Publications (1977) Fig. 9-4.

439, 461, Photos courtesy of the World Health Organization.

461, Photo courtesy of the World Health Organization.

465, Courtesy of Dr. C. Mouton

469, From S.N. Bhaskar (ed.), *Orban's Oral Histology and Embryology,* St. Louis: C.V. Mosby (1972), Fig. 41.

471, Courtesy of Myron Nevins, D.D.S., Swampscott.

476 and 478 (Top), Photographs courtesy of S.J. Jones, 1972, *The Dental Practitioner,* Vol. 22, No. 12, August 1972 "The Tooth Surface in Periodontal Disease," pp. 462–473.

478 (Bottom), Photo courtesy of Dr. Tryggve Lie, Dept. of Periodontology, University of Bergen, Bergen, Norway.

481, Reprinted by courtesy of New Zealand Dental Journal.

483, Malcolm A. Lynch (ed.), *Burket's Oral Medicine; Diagnosis and Treatment* (7th ed.), Philadelphia: J.B. Lippincott (1977) Fig. 10.2(a), p. 178.

497, From Lee Langley and John B. Christensen, *Structure and Function of the Human Body,* Burgess (1978). Reprinted by permission.

489, 499, From W.H. Fahrenbach in W. Montagna and P.F. Parakkal, *The Structure and Function of Skin.* New York: Academic Press, 1974.

509, From C.W. Emmons et al., *Medical Mycology* (3rd ed.), Philadelphia: Lea and Febiger, 1977.

510, From W.H. Fahrenbach in Montagna and P.F. Parakkal, *The Structure and Function of Skin.* New York: Academic Press, 1974.

512, From K.-P. Chang, *Leishmania donovani*—promastigate–macrophage interaction *in vitro, Exp. Paristol.* (1979) **48:** 175–189, Figs. 17, 18.

517, (a) From M.A. Chernesky, The role of electron microscopy in diagnostic virology. In D.A. Lennette et al., *Diagnosis of Viral Infections: The Role of the Clinical Laboratory,* University Park Press. Copyright © 1979 by University Park Press. (b) From P. Gold and S. Dales, *PNAS*(US) (1968) **60:** 845. (c) Courtesy of Dr. June D. Almeida, The Wellcome Research Laboratories, Langley Court, Beckenham, Kent, England. (d) Reproduced from C.R. Madeley, *Virus Morphology* by permission of the author and Churchill Livingstone, Edinburgh.

527, From Greisheimer and Weiderman, *Physiology and Anatomy* (9th ed.) Lippincott, 1972, p. 409, Fig. 12-3.

539, (a, b, d) From C.R. Madeley, *Virus Morphology,* Baltimore: The Williams & Wilkins Co., 1972, p. 100, plate 41; p. 98, plate 40; p. 60, plate 21. (c) Courtesy of Dr. H.D. Mayor, Department of Microbiology, Bayler College of Medicine, Houston, Texas.

548, From R.G. Kessel and C.Y. Shih, *Scanning Electron Microscopy in Biology: A Student's Atlas on Biological Organization,* New York: Springer-Verlag. Copyright 1974.

556, Courtesy of Michael G. Gabridge, Ph.D.; W. Jones Cell Science Center, Lake Placid, N.Y.

557, From F.W. Chandler et al., *Legionnaire's Disease, A.J.C.P.* (1979) **71** (1): 47, Fig. 4. Courtesy of Dr. F.W. Chandler, Center for Disease Control, Atlanta.

563, Photo courtesy of Dr. W.G. Laver, The John Curtin School of Medical Research, Canberra City, Australia.

564, Photo courtesy of J.L. Carson, Ph.D., Infectious Disease Division, Department of Pediatrics, University of North Carolina.

567, 568, From C.W. Emmons et al., *Medical Mycology*, Philadelphia: Lea and Febiger, 1977.

577, From Lelio Orci and Alain Perrelet, 1975, *Freeze-Etch Histology.* New York: Springer-Verlag, Plate 142 (p. 85) and Plate 77 (p. 156).

584, (a) From H.E. Jones and R.W.A. Park, The influence of medium composition on the growth and swarming of *Proteus, J. Gen. Microbiol.* (1967) **47**: 369. (b) Courtesy of Drs. T. Eda and Y. Kanda, from *J. Electr. Microsc.* **27** (2), 119–126, 1978.

587, From R.R. Gillies and T.C. Dodds, 1976, *Bacteriology Illustrated*, Edinburgh, London and New York: Churchill Livingstone.

592, Courtesy of M. Neal Guentzel, D. Guerrero, and T.V. Gay, Allied Health and Life Sciences, University of Texas, San Antonio.

593, From ASM slide collection.

594, From P.J. Pead, *J. Med. Microbiol.* (1979) **12**: 383, Fig. 1.

595, (a) From G. Chabanon, C.C. Hartley, and M.H. Richmond, Adhesion of *E. coli* strains, *J. Clin. Microbiol.* (1979) **10**: p. 564, Fig. 1. (b) Center of Disease Control, Atlanta.

597, From Jose Esparza and Francisco Gil, A study on the ultrastructure of human rotavirus, *Virology* (1978) **91**: 141–150. By permission.

598, Reprinted from the Center for Disease Control's June 1978 *Morbidity and Mortality Report* **27** (24): Fig. 1.

600, Reproduced from C.R. Madeley, *Virus Morphology*, by permission of the author and Churchill Livingstone, Edinburgh.

604, From Erich Scholtyseck, *Fine Structure of Parasitic Protozoa*, New York: Springer-Verlag, (1979) Fig. 6, pp. 50–51.

605, Courtesy of Carey S. Callaway, Pathology Division, Center for Disease Control, Atlanta.

606, Reprinted from the Center for Disease Control's May 1978 *Morbidity and Mortality Report* **27** (19): p. 155, Fig. 1.

611, Photographs by Carrie M. Lewis, Bigelow Laboratory for Ocean Sciences.

624, Dr. Leon J. LeBeau, Dept. of Pathology, University of Illinois Med. Center, Chicago.

628, Courtesy of Dr. Nicholas J. Fiumara, Director, Division of Communicable and Venereal Diseases, State Department, Mass. D.P.H.

633, From G.H. Cassell et al., *Pathobiology of Mycoplasmas*, in *Microbiology*—1978. American Society for Microbiology, Figs. 1, 2 (p. 400).

634, From J. Robertson and E. Smook, *J. Bacteriology* (1976) **128**: 658–660.

640, From E. Scholtyseck, *Fine Structure of Parasitic Protozoa*, New York: Springer-Verlag, (1979) Fig. 41, pp. 120–121.

668, Courtesy of Dr. Thomas Brown and the editors of the *Journal of Medical Microbiology.*

674, Courtesy of Becton, Dickinson and Company VACUTAINER Systems, Rutherford, N.J.

675, Designed and manufactured by LAB-LINE® INSTRUMENTS, INC.

693, Photo courtesy of Becton, Dickinson and Company VACUTAINER Systems, Rutherford, N.J.

700, Courtesy of David J. Silverman and Charles L. Wissman, Jr., *R. rickettsia* cytopathology, *Infection and Immunity* (1979) **26**: 714. By permission of ASM Publications.

701, Courtesy of Carolina Biological Supply Co., Burlington, N.C.

712, Courtesy of the World Health Organization.

713, From Erich Scholtyseck, *Fine Structure of Parasitic Protozoa*, New York: Springer-Verlag, (1979) Fig. 2, p. 42.

716, Courtesy of the World Health Organization.

719, From W.N. Arnold, A.T. Pringle, and R.G. Gerrison, Amphotericin B and *H. capsulatum, J. Bacteriol.* (1980) **141**: 354, Figs. 6A & D.

733, Courtesy of the Baker Company, Inc., Sanford, Maine.

739, Courtesy of AMSCOL/American Sterilizer Co., Erie, Pennsylvania.

778, Reproduced with permission from *Cultura;* September 1978, p. 1. Published by Oxoid Ltd. Basingstoke, Hampshire, RG 24 OPW, United Kingdom.

781, From W.N. Arnold, A.T. Pringle, and R.G. Garrison, Amphotericin B and *H. capsulatum, J. Bacteriol.* (1980) **141**: 354, Figs. 6A & D.

793, 794, Courtesy of Merck, Sharp & Dohme.

804, Courtesy of the World Health Organization.

Index

883